T0213107

Lecture Notes in Computer Science　　9869

Commenced Publication in 1973
Founding and Former Series Editors:
Gerhard Goos, Juris Hartmanis, and Jan van Leeuwen

Yury Kochetov · Michael Khachay
Vladimir Beresnev · Evgeni Nurminski
Panos Pardalos (Eds.)

Discrete Optimization
and Operations Research

9th International Conference, DOOR 2016
Vladivostok, Russia, September 19–23, 2016
Proceedings

 Springer

Editors
Yury Kochetov
Sobolev Institute of Mathematics
Novosibirsk
Russia

Michael Khachay
Krasovsky Institute of Mathematics and
    Mechanics
Ekaterinburg
Russia

Vladimir Beresnev
Sobolev Institute of Mathematics
Novosibirsk
Russia

Evgeni Nurminski
Far Eastern Federal University
Vladivostik
Russia

Panos Pardalos
University of Florida
Gainesville, FL
USA

ISSN 0302-9743   ISSN 1611-3349   (electronic)
Lecture Notes in Computer Science
ISBN 978-3-319-44913-5   ISBN 978-3-319-44914-2   (eBook)
DOI 10.1007/978-3-319-44914-2

Library of Congress Control Number: 2016948223

LNCS Sublibrary: SL1 – Theoretical Computer Science and General Issues

Printed on acid-free paper

This Springer imprint is published by Springer Nature
The registered company is Springer International Publishing AG Switzerland

Preface

This volume contains the proceedings of the 9th International Conference on Discrete Optimization and Operations Research (DOOR 2016), held in Vladivostok, Russia, during September 19–23, 2016. It was organized by the Far Eastern Federal University, Sobolev Institute of Mathematics, Krasovsky Institute of Mathematics and Mechanics, Novosibirsk State University, and the Higher School of Economics in Nizhny Novgorod.

Previous conferences took place at the Sobolev Institute of Mathematics, Novosibirsk, in 1996, 1998, 2000, 2002, and 2004. The 6th conference was held in the Russian Far East in a picturesque setting on the shore of the Japanese Sea near Vladivostok in 2007. The 7th one, in 2010, was held in the Altay Mountains. The 8th event took place in Novosibirsk again. DOOR is part of a series of annual international conferences on optimization and operations research that covers a wide range of topics in mathematical programming and its applications, integer programming and polyhedral combinatorics, bi-level programming and multi-criteria optimization, optimization problems in machine learning and data mining, discrete optimization in scheduling, routing, bin packing, locations, and optimization problems on graphs, computational complexity, and polynomial time approximation. The main purpose of the conference is to provide a forum where researchers can exchange ideas, identify promising directions for research and application domains, and foster new collaborations.

In response to the call for papers, DOOR 2016 received 181 submissions. Papers included in this volume were carefully selected by the Program Committee on the basis of reports from two or more reviewers. Only 39 submissions were selected for inclusion in this volume. Nine invited talks by eminent speakers are also included here.

We thank all the Program Committee members and external reviewers for their cooperation. We also thank the Organizing Committee members for their efforts. Finally, we thank our sponsors, the Russian Foundation for Basic Research, the Far Eastern Federal University, Novosibirsk State University, the Laboratory of Algorithms and Technologies for Networks Analysis (LATNA), the Higher School of Economics in Nizhny Novgorod, and Alfred Hofmann from Springer for supporting our project.

September 2016

Yury Kochetov
Michael Khachay
Vladimir Beresnev
Evgeni Nurminski
Panos Pardalos

Organization

Program Chairs

Evgeni Nurminski   Far Eastern Federal University, Russia
Vladimir Beresnev   Sobolev Institute of Mathematics, Russia
Panos Pardalos   University of Florida, USA

Program Committee

Ekaterina Alekseeva   Sobolev Institute of Mathematics, Russia
Edilkhan Amirgaliev   Suleyman Demirel University, Kazakhstan
Oleg Burdakov   Linköping University, Sweden
Igor Bykadorov   Sobolev Institute of Mathematics, Russia
Emilio Carrizosa   Universidad de Sevilla, Spain
Yair Censor   University of Haifa, Israel
Ivan Davydov   Sobolev Institute of Mathematics, Russia
Vladimir Deineko   The University of Warwick, UK
Stefan Dempe   TU Bergakademie Freiberg, Germany
Anton Eremeev   Sobolev Institute of Mathematics, Russia
Adil Erzin   Sobolev Institute of Mathematics, Russia
Yury Evtushenko   Dorodnicyn Computing Centre, Russia
Edward Gimadi   Sobolev Institute of Mathematics, Russia
Alexandr Grigoriev   Maastricht University, The Netherlands
Florian Jaehn   Universität Augsburg, Germany
Josef Kallrath   TU Darmstadt, Germany
Valery Kalyagin   Higher School of Economics, Russia
Alexander Kelmanov   Sobolev Institute of Mathematics, Russia
Michael Khachay   Krasovsky Institute of Mathematics and Mechanics, Russia
Oleg Khamisov   Melentiev Energy Systems Institute, Russia
Andrey Kibzun   Moscow Aviation Institute, Russia
Yury Kochetov   Sobolev Institute of Mathematics, Russia
Alexander Kolokolov   Sobolev Institute of Mathematics, Russia
Alexander Kononov   Sobolev Institute of Mathematics, Russia
Mikhail Kovalev   Belarusian State University, Belarus
Nikolay Kuzyurin   Institute for System Programming, Russia
Bertrand Lin   National Chiao Tung University, Taiwan
Bertrand Mareschal   Université Libre de Bruxelles, Belgium
Athanasios Migdalas   Luleå University of Technology, Sweden
Nenad Mladenović   University of Valenciennes, France
Urfat Nuriyev   Ege University, Turkey

Alexandr Plyasunov   Sobolev Institute of Mathematics, Russia
Artem Pyatkin   Sobolev Institute of Mathematics, Russia
Soumyendu Raha   Indian Institute of Science, India
Konstantin Rudakov   Dorodnicyn Computing Centre, Russia
Yaroslav Sergeev   Università della Calabria, Italy
Sergey Sevastianov   Sobolev Institute of Mathematics, Russia
Vadim Shmyrev   Sobolev Institute of Mathematics, Russia
Petro Stetsuk   Institute of Cybernetics, Ukraine
Alexander Strekalovsky Matrosov Institute for System Dynamics and Control
                       Theory, Russia
Maxim Sviridenko   Yahoo, USA
El-Ghazali Talbi   University of Lille, CNRS, Inria, France
Yury Zhuravlev   Dorodnicyn Computing Centre, Russia

Organizing Committee

Natalia Shamry   Far Eastern Federal University, Russia
Yury Kochetov   Sobolev Institute of Mathematics, Russia
Mikhail Khachay   Krasovsky Institute of Mathematics and Mechanics, Russia
Timur Medvedev   Higher School of Economics, Russia
Evgeniya Vorontsova   Far Eastern Federal University, Russia
Nina Kochetova   Sobolev Institute of Mathematics, Russia
Polina Kononova   Sobolev Institute of Mathematics, Russia
Andrey Velichko   Far Eastern Federal University, Russia

Sponsors

Russian Foundation for Basic Research
Far Eastern Federal University, Vladivostok
Novosibirsk State University
Higher School of Economics, Nizhny Novgorod

Additional Reviewers

Adamczyk, Marek   Basturk, Nalan   Dessevre, Guillaume
Ageev, Alexander   Berg, Kimmo   Di Pillo, Gianni
Ago, Takanori   Bevern, René Van   Dudko, Olga
Aizenberg, Natalia   Broersma, Hajo   Duplinskiy, Artem
Alaev, Pavel   Buzdalov, Maxim   Dzhafarov, Vakif
Alekseeva, Ekaterina   Chernykh, Ilya   Ellero, Andrea
Aleskerov, Fuad   Chubanov, Sergei   Filatov, Alexander
Antipin, Anatoly   Custic, Ante   Fomin, Fedor
Antsyz, Sergey   Davidović, Tatjana   Funari, Stefania
Bagirov, Adil   Davydov, Ivan   Gaudioso, Manlio
Baklanov, Artem   Dempe, Stephan   Glebov, Aleksey

Golikov, Alexander
Gorelik, Victor
Gornov, Aleksander
Goubko, Mikhail
Griewank, Andreas
Grigoriev, Alexander
Grishagin, Vladimir
Gruzdeva, Tatiana
Hung, Hui-Chih
Ignatov, Dmitry
Ivanko, Evgeny
Ivanov, Sergey
Jung, Verena
Kalashnykova, Nataliya
Karakitsiou, Athanasia
Kateshov, Andrey
Katueva, Yaroslava
Kazakov, Alexander
Kazakovtsev, Lev
Kazansky, Alexander
Khachay, Michael
Khamidullin, Sergey
Khamisov, Oleg
Khandeev, Vladimir
Khromova, Olga
Kobylkin, Konstantin
Kokovin, Sergey
Konnov, Igor
Kononov, Alexander
Kononova, Polina
Konstantinova, Elena
Koshel, Konstantin
Kovalenko, Julia
Kreuzen, Vincent
Kutucu, Hakan
Kuzjurin, Nikolay
Kvasov, Dmitri

Lagutaeva, Daria
Letsios, Dimitrios
Levanova, Tatiana
Lin, Bertrand M.T.
Lucarelli, Giorgio
Malyshev, Dmitriy
Mareschal, Bertrand
Mazalov, Vladimir
Melnikov, Andrey
Migdalas, Athanasios
Mondrus, Olga
Morozov, Andrei
Namm, Robert
Naumov, Andrey
Nishihara, Ko
Norkin, Vladimir
Nurminski, Evgeni
Oosterwijk, Tim
Orlovich, Yury
Oron, Daniel
Panin, Artem
Pekarskii, Sergey
Philpott, Andy
Plyasunov, Alexander
Popov, Leonid
Popova, Ewgeniya
Predtetchinski, Arkadii
Raha, Soumyendu
Rapoport, Ernst
Romanovsky, Joseph
Sandomirskaia, Marina
Savvateev, Alexei
Semenikhin, Konstantin
Semenov, Vladimir
Servakh, Vladimir
Shafransky, Yakov
Shary, Sergey

Shenmaier, Vladimir
Shikhman, Vladimir
Sidorov, Alexander
Sidorov, Denis
Sidorov, Sergei
Simanchev, Ruslan
Skarin, Vladimir
Sloev, Igor
Sobol, Vitaly
Stetsyuk, Petro
Strusevich, Vitaly
Sudhölter, Peter
Sukhorukova, Nadia
Suslov, Nikita
Suvorov, Anton
Takhonov, Ivan
Timmermans, Veerle
Tiskin, Alexander
Trionfetti, Federico
Tsai, Yen-Shing
Tsidulko, Oxana
Tsoy, Yury
Ugurlu, Onur
Valeeva, Aida
Vasilyev, Igor
Vasin, Alexander
Vegh, Laszlo
Wijler, Etiënne
Winokurow, Andrej
Yanovskaya, Elena
Zabotin, Igor
Zalyubovskiy, Slava
Zambalaeva, Dolgor
Zhurbenko, Nickolay
Zolotykh, Nikolai

Abstracts of Invited Talks

Algorithmic Issues in Energy-Efficient Computation

Evripidis Bampis

Sorbonne Universités, UPMC Univ Paris 06, UMR 7606, LIP6, Paris, France
evripidis.bampis@lip6.fr

Abstract. Energy efficiency has become a crucial issue in Computer Science. New hardware and system-based approaches are explored for saving energy in portable battery-operated devices, personal computers, or large server farms. The main mechanisms that have been developed for saving energy are the ability of transitioning the device among multiple power states, and the use of dynamic voltage scaling (speed scaling). These last years, there is also an increasing interest in the development of algorithmic techniques for finding tradeoff-solutions between energy consumption and performance. In this talk, we will focus on algorithmic techniques with provably good performances for fundamental power management problems. Among the different models that have been developed in the literature, we will focus on the speed scaling model, the power-down model and the combination of these two models that we will call the power-down with speed scaling model.

Linear Superiorization for Infeasible Linear Programming

Yair Censor and Yehuda Zur

Department of Mathematics, University of Haifa, Mt. Carmel, 3498838,
Haifa, Israel
yair@math.haifa.ac.il

Abstract. Linear superiorization (abbreviated: LinSup) considers linear programming (LP) problems wherein the constraints as well as the objective function are linear. It allows to steer the iterates of a feasibility-seeking iterative process toward feasible points that have lower (not necessarily minimal) values of the objective function than points that would have been reached by the same feasiblity-seeking iterative process without superiorization. Using a feasibility-seeking iterative process that converges even if the linear feasible set is empty, LinSup generates an iterative sequence that converges to a point that minimizes a proximity function which measures the linear constraints violation. In addition, due to LinSup's repeated objective function reduction steps such a point will most probably have a reduced objective function value. We present an exploratory experimental result that illustrates the behavior of LinSup on an infeasible LP problem.

Modern Trends in Parameterized Algorithms

Fedor Fomin

University of Bergen, Bergen, Norway
fomin@ii.uib.no

We overview the recent progress in solving intractable optimization problems on planar graphs as well as other classes of sparse graphs. In particular, we discuss how tools from Graph Minors theory can be used to obtain

- subexponential parameterized algorithms
- approximation algorithms, and
- preprocessing and kernelization algorithms

on these classes of graphs.

Short Survey on Graph Correlation Clustering with Minimization Criteria

Victor Il'ev[1,2], Svetlana Il'eva[2], and Alexander Kononov[1]

[1] Sobolev Institute of Mathematics, Novosibirsk, Russia
[2] Omsk State University, Omsk, Russia
iljev@mail.ru, alvenko@math.nsc.ru

Abstract. In clustering problems one has to partition a given set of objects into some subsets (called clusters) taking into consideration only similarity of the objects. One of the most visual formalizations of clustering is the graph clustering, that is, grouping the vertices of a graph into clusters taking into consideration the edge structure of the graph whose vertices are objects and edges represent similarities between the objects.

In this short survey, we consider the graph correlation clustering problems where the goal is to minimize the number of edges between clusters and the number of missing edges inside clusters. We present a number of results on graph correlation clustering including results on computational complexity and approximability of different variants of the problems, and performance guarantees of approximation algorithms for graph correlation clustering. Some results on approximability of weighted versions of graph correlation clustering are also presented.

Wardrop Equilibrium for Networks with the BPR Latency Function

Jaimie W. Lien[1], Vladimir V. Mazalov[2], Anna V. Melnik[3],
and Jie Zheng[4]

[1] Department of Decision Sciences and Managerial Economics,
The Chinese University of Hong Kong, Shatin, Hong Kong, China
jaimie.academic@gmail.com
[2] Institute of Applied Mathematical Research,
Karelian Research Center, Russian Academy of Sciences,
11, Pushkinskaya Street, Petrozavodsk, 185910, Russia
vmazalov@krc.karelia.ru
[3] Saint-Petersburg State University, Universitetskii Prospekt 35,
Saint-petrsburg, 198504, Russia
a.melnik@spbu.ru
[4] Department of Economics, School of Economics and Management,
Tsinghua University, Beijing, 100084, China
zhengjie@sem.tsinghua.edu.cn

Abstract. This paper considers a network comprised of parallel routes with the Bureau of Public Road (BPR) latency function and suggests an optimal distribution method for incoming traffic flow. The authors analytically derive a system of equations defining the optimal distribution of the incoming flow with minimum social costs, as well as a corresponding system of equations for the Wardrop equilibrium in this network. In particular, the Wardrop equilibrium is applied to the competition model with rational consumers who use the carriers with minimal cost, where cost is equal to the price for service plus the waiting time for the service. Finally, the social costs under the equilibrium and under the optimal distribution are compared. It is shown that the price of anarchy can be infinitely large in the model with strategic pricing.

Location Modeling in the Presence of Firm and Customer Competition

Athanasios Migdalas

ETS Institute, Lulea University of Technology, 971 87 Lulea, Sweden
athmig@ltu.se

Location problems form a wide class of mathematical programming models, of great interest of both in practice and from the point of view of optimization theory. Facility location problem aims at determining the optimal sites to locate facilities such as plants, warehouses, and/or distribution centers. Competitive location models (CFL) additionally incorporate the fact that location decisions have been or will be made by independent decision-makers who will subsequently compete with each other for market share, profit maximization etc. In addition decisions such as customers' allocation and pricing policies may also be incorporated to the basic model.

The first paper dealing with the effect of competition in the location decisions is due Hotelling. Since then, a vast number of publications have been devoted to the subject. Sequential CFL problems are usually modeled as hierarchical or multi-level programming models. Such models are concerned with decision making problems that involve multiple decision makers ordered within a hierarchical structure. The most well-known case is the so-called Stackelberg game in which decision makers of two different levels with different, often conflicting, objectives are involved.

The research work dealing with the bi-level formulation of location problems is limited only to the competition among the locators. Customers are passively assigned to the facilities according to some criteria. A first attempt to study the influence of market competition on location decisions is due to Tobin and Friesz.

In this talk we formulate and study a class of location problems where the autonomous decisions of the customers regarding the facilities from which they will be served influence the locations decisions. The conditions under which customers make their choice of facilities to be served are in general complicated. We assume here that every customer will choose the facilities that minimize their own total transportation and waiting for service cost. Thus, concerning mathematical modeling we investigate facility location problems not only in the presence of firm competition but also in the presence of customer competition with respect to the quality level of the provided services. We derive bi-level programming models which are interpreted and analyzed in game theoretic terms. The issues of optimality conditions, computational complexity and solution algorithms are also discussed.

References

1. Karakitsiou, A., Migdalas, A.: Locating facilities in a competitive environment, Optim. Lett. (2016). doi:10.1007/s11590-015-0963-7
2. Karakitsiou, A., Migdalas, A.: Nash type games in competitive facilities location. Int. J. Decision Support Syst. (2016, in print)
3. Karakitsiou, A.: Modeling Discrete Competitive Facilities Location. Springer Briefs in Optimization (2015)

A Review on Network Robustness
from an Information Theory Perspective

Tiago Schieber[1,2], Martín Ravetti[1], and Panos M. Pardalos[3,4]

[1] Departamento de Engenharia de Produção, Universidade Federal de Minas
Gerais, Belo Horizonte, MG, Brazil
tischieber@gmail.com
[2] Departamento de Engenharia de Produção,
Pontifícia Universidade Católica de Minas Gerais, Belo Horizonte, MG, Brazil
[3] Center for Applied Optimization, Industrial and Systems Engineering,
University of Florida, Gainesville, FL, USA
[4] Laboratory of Algorithms and Technologies for Network Analysis,
National Research University Higher School of Economics,
Nizhny Novgorod, Russia

Abstract. The understanding of how a networked system behaves and keeps its topological features when facing element failures is essential in several applications ranging from biological to social networks. In this context, one of the most discussed and important topics is the ability to distinguish similarities between networks. A probabilistic approach already showed useful in graph comparisons when representing the network structure as a set of probability distributions, and, together with the Jensen-Shannon divergence, allows to quantify dissimilarities between graphs. The goal of this article is to compare these methodologies for the analysis of network comparisons and robustness.

An Iterative Approach for Searching an Equilibrium in Piecewise Linear Exchange Model

Vadim I. Shmyrev

Sobolev Institute of Mathematics, Novosibirsk, Russia
shmyrev.vadim@mail.ru

Abstract. The exchange model with piecewise linear separable concave utility functions is considered. This consideration extends the author's original approach to the equilibrium problem in a linear exchange model and its variations. The conceptual base of this approach is the scheme of polyhedral complementarity. It has no analogs and made it possible to obtain the finite algorithms for some variations of the exchange model. Especially simple algorithms arise for linear exchange model with fixed budgets (Fisher's model). This is due to monotonicity property inherent in the models and potentiality of arising mappings. The algorithms can be considered as a procedure similar to the simplex-method of linear programming. It is natural to study applicability of the approach for more general models. The considered piecewise linear version of the model reduces to a special exchange model with upper bounds on variables and the modified conditions of the goods' balances. For such a model the monotonicity property is violated. But it remains, if upper bounds are substituted by financial limits on purchases. This is the idea of proposed iterative algorithm for initial problem. It is a generalization of an analogue for linear exchange model.

Handling Scheduling Problems
with Controllable Parameters by Methods
of Submodular Optimization

Akiyoshi Shioura[1], Natalia V. Shakhlevich[2],
and Vitaly A. Strusevich[3]

[1] Tokyo Institute of Technology, Tokyo, Japan
[2] University of Leeds, Leeds, UK
[3] Univeristy of Greenwich, London, UK
V.Strusevich@greenwich.ac.uk

Abstract. In this paper, we demonstrate how scheduling problems with controllable processing times can be reformulated as maximization linear programming problems over a submodular polyhedron intersected with a box. We explain a decomposition algorithm for solving the latter problem and discuss its implications for the relevant problems of preemptive scheduling on a single machine and parallel machines.

Contents

Invited Talks

Algorithmic Issues in Energy-Efficient Computation 3
   Evripidis Bampis

Linear Superiorization for Infeasible Linear Programming 15
   Yair Censor and Yehuda Zur

Short Survey on Graph Correlation Clustering with Minimization Criteria 25
   Victor Il'ev, Svetlana Il'eva, and Alexander Kononov

Wardrop Equilibrium for Networks with the BPR Latency Function 37
   Jaimie W. Lien, Vladimir V. Mazalov, Anna V. Melnik, and Jie Zheng

A Review on Network Robustness from an Information Theory Perspective . . . 50
   Tiago Schieber, Martín Ravetti, and Panos M. Pardalos

An Iterative Approach for Searching an Equilibrium in Piecewise Linear
Exchange Model. 61
   Vadim I. Shmyrev

Handling Scheduling Problems with Controllable Parameters by Methods
of Submodular Optimization. 74
   Akiyoshi Shioura, Natalia V. Shakhlevich, and Vitaly A. Strusevich

Discrete Optimization

Constant-Factor Approximations for Cycle Cover Problems 93
   Alexander Ageev

Precedence-Constrained Scheduling Problems Parameterized
by Partial Order Width 105
   René van Bevern, Robert Bredereck, Laurent Bulteau,
   Christian Komusiewicz, Nimrod Talmon, and Gerhard J. Woeginger

A Scheme of Independent Calculations in a Precedence Constrained
Routing Problem. 121
   Alexander G. Chentsov and Alexey M. Grigoryev

On Asymptotically Optimal Approach to the m-Peripatetic Salesman
Problem on Random Inputs 136
   Edward Kh. Gimadi, Alexey M. Istomin, and Oxana Yu. Tsidulko

Efficient Randomized Algorithm for a Vector Subset Problem 148
   Edward Gimadi and Ivan Rykov

An Algorithm with Approximation Ratio 5/6 for the Metric Maximum
m-PSP. 159
   Aleksey N. Glebov and Anastasiya V. Gordeeva

An Approximation Algorithm for a Problem of Partitioning a Sequence
into Clusters with Restrictions on Their Cardinalities. 171
   Alexander Kel'manov, Ludmila Mikhailova, Sergey Khamidullin,
   and Vladimir Khandeev

A Fully Polynomial-Time Approximation Scheme for a Special Case
of a Balanced 2-Clustering Problem 182
   Alexander Kel'manov and Anna Motkova

PTAS for the Euclidean Capacitated Vehicle Routing Problem in R^d 193
   Michael Khachay and Roman Dubinin

On Integer Recognition over Some Boolean Quadric Polytope Extension 206
   Andrei Nikolaev

Variable Neighborhood Search-Based Heuristics for Min-Power Symmetric
Connectivity Problem in Wireless Networks 220
   Roman Plotnikov, Adil Erzin, and Nenad Mladenovic

On the Facets of Combinatorial Polytopes 233
   Ruslan Simanchev and Inna Urazova

A Branch and Bound Algorithm for a Fractional 0-1 Programming Problem. . . 244
   Irina Utkina, Mikhail Batsyn, and Ekaterina Batsyna

Scheduling Problems

Approximating Coupled-Task Scheduling Problems with Equal
Exact Delays 259
   Alexander Ageev and Mikhail Ivanov

Routing Open Shop with Unrelated Travel Times 272
   Ilya Chernykh

The 2-Machine Routing Open Shop on a Triangular Transportation
Network. 284
   Ilya Chernykh and Ekaterina Lgotina

Mixed Integer Programming Approach to Multiprocessor Job Scheduling
with Setup Times 298
   Anton V. Eremeev and Yulia V. Kovalenko

On Speed Scaling Scheduling of Parallel Jobs with Preemption   309
Alexander Kononov and Yulia Kovalenko

Facility Location

Facility Location in Unfair Competition   325
Vladimir Beresnev and Andrey Melnikov

Variable Neighborhood Descent for the Capacitated Clustering Problem   336
*Jack Brimberg, Nenad Mladenović, Raca Todosijević,
and Dragan Urošević*

A Leader-Follower Hub Location Problem Under Fixed Markups   350
Dimitrije D. Čvokić, Yury A. Kochetov, and Aleksandr V. Plyasunov

Tabu Search Approach for the Bi-Level Competitive Base Station
Location Problem   364
Ivan Davydov, Marceau Coupechoux, and Stefano Iellamo

Upper Bound for the Competitive Facility Location Problem
with Quantile Criterion   373
Andrey Melnikov and Vladimir Beresnev

Mathematical Programming

Fast Primal-Dual Gradient Method for Strongly Convex Minimization
Problems with Linear Constraints   391
Alexey Chernov, Pavel Dvurechensky, and Alexander Gasnikov

An Approach to Fractional Programming via D.C. Constraints Problem:
Local Search   404
Tatiana Gruzdeva and Alexander Strekalovsky

Partial Linearization Method for Network Equilibrium Problems
with Elastic Demands   418
Igor Konnov and Olga Pinyagina

Multiple Cuts in Separating Plane Algorithms.   430
Evgeni Nurminski

On the Parameter Control of the Residual Method for the Correction
of Improper Problems of Convex Programming   441
Vladimir D. Skarin

On the Merit and Penalty Functions for the D.C. Optimization   452
Alexander S. Strekalovsky

Mathematical Economics and Games

Application of Supply Function Equilibrium Model to Describe the
Interaction of Generation Companies in the Electricity Market   469
  Natalia Aizenberg

Chain Store Against Manufacturers: Regulation Can Mitigate Market
Distortion.   480
  Igor Bykadorov, Andrea Ellero, Stefania Funari, Sergey Kokovin,
  and Marina Pudova

On the Existence of Immigration Proof Partition into Countries
in Multidimensional Space   494
  Valeriy M. Marakulin

Search of Nash Equilibrium in Quadratic n-person Game.   509
  Ilya Minarchenko

Applications of Operational Research

Convergence of Discrete Approximations of Stochastic Programming
Problems with Probabilistic Criteria.   525
  Andrey I. Kibzun and Sergey V. Ivanov

A Robust Leaky-LMS Algorithm for Sparse System Identification   538
  Cemil Turan and Yedilkhan Amirgaliev

Extended Separating Plane Algorithm and NSO-Solutions
of PageRank Problem   547
  Evgeniya Vorontsova

Short Communications

Location, Pricing and the Problem of Apollonius   563
  André Berger, Alexander Grigoriev, Artem Panin,
  and Andrej Winokurow

Variable Neighborhood Search Approach for the Location
and Design Problem   570
  Tatyana Levanova and Alexander Gnusarev

On a Network Equilibrium Problem with Mixed Demand   578
  Olga Pinyagina

Author Index   585

Invited Talks

Algorithmic Issues in Energy-Efficient Computation

Evripidis Bampis[✉]

Sorbonne Universités, UPMC Univ Paris 06, UMR 7606, LIP6, Paris, France
evripidis.bampis@lip6.fr

1 Introduction

Energy efficiency has become a crucial issue in Computer Science. New hardware and system-based approaches are explored for saving energy in portable battery-operated devices, personal computers, or large server farms. The main mechanisms that have been developed for saving energy are the ability of transitioning the device among multiple power states, and the use of dynamic voltage scaling (speed scaling). These last years, there is also an increasing interest in the development of algorithmic techniques for finding tradeoff-solutions between energy consumption and performance. In this talk, we will focus on algorithmic techniques with provably good performances for fundamental power management problems. Among the different models that have been developed in the literature, we will focus on the *speed scaling* model, the *power-down* model and the combination of these two models that we will call the *power-down with speed scaling* model. In the *speed scaling* model [54], the speed of the processor (machine) may be dynamically changed over time. When a processor runs at speed s, then the rate with which the energy is consumed (i.e., the power) is $f(s)$ with f a non-decreasing function of the speed. The energy is the integral of the power over time. According to the well-known cube-root rule for CMOS devices, the speed of a device is proportional to the cube-root of the power and hence $f(s) = s^3$, but in the literature, many works consider that the power is $f(s) = s^\alpha$ where $\alpha > 1$ is a constant, or an arbitrary convex function. In the *power-down* model [26], the processors run at a *fixed* speed but are equipped with a *sleep state*. This means that the processor has two states ON and OFF and may be suspended during its idle time. However, during its wake-up from state OFF to state ON, there is a *start-up* energy consumption, denoted by L. Hence, suspending the processor is only beneficial when the idle periods are long enough to compensate the consumed start-up energy. The *power-down with speed scaling* model [43] combines the previous two models by considering speed scalable processors with a sleep state. Here, the power function is $g(s) = f(s) + c$ where $f(s)$ is defined as in the speed-scaling model and $c > 0$ is a constant that specifies the power consumed when the processor is in the ON state.

We will be interested in some recent developments for scheduling a set of jobs on a (set of) processor(s) focusing in the *offline* context. Most of the problems considered are *deadline-based* problems: we are given a set of n jobs, where each

© Springer International Publishing Switzerland 2016
Y. Kochetov et al. (Eds.): DOOR 2016, LNCS 9869, pp. 3–14, 2016.
DOI: 10.1007/978-3-319-44914-2_1

job j is characterized by its release date r_j, its deadline d_j and its processing volume (work) p_j. Two important families of instances that have been studied in the literature are the *agreeable* and the *laminar* instances. We call an instance *agreeable* if earlier released jobs have earlier deadlines, i.e., for each j and j' with $r_j \leq r_{j'}$ then $d_j \leq d_{j'}$. In a *laminar instance*, for any two jobs j and j' with $r_j \leq r_{j'}$ it holds that either $d_j \geq d_{j'}$ or $d_j \leq r_{j'}$. The processing volume of a job is the number of CPU-cycles required by the job. If job j is executed with speed s then its processing time is $\frac{p_j}{s}$. In the case of processors with fixed speed the processing volume equals the processing time of the job. A *feasible schedule* in this context is a schedule in which each job is executed in the interval between its release date and its deadline. The problems in this setting are bi-objective by nature. For instance, we wish to minimize the energy consumption while at the same time we aim to determine a feasible schedule. A lot of other objectives have been studied when we consider a given budget of energy: the *throughput*, i.e. the number of jobs that complete before their deadlines, the *makespan*, i.e. the time at which the last job completes its execution, the *sum of (weighted) completion times*, the *sum of flow times*, ... Finally, in some works, the objective is the minimization of a linear combination of the energy and of some scheduling criterion (e.g. sum of completion times).

2   Speed Scaling

2.1   Single Machine

Energy minimization. Yao et al. [54] considered the problem of scheduling a set of n jobs on a single machine, where the *preemption*, i.e. the possibility to interrupt the execution of a job and resume it later, was allowed. They proposed an optimal $O(n^3)$-time algorithm. Later, Li et al. [47] proposed a faster algorithm with time complexity $O(n^2 \log n)$. Other algorithms with better time complexities than the one of [54] have been proposed in [37] for agreeable instances, and in [36] for general instances. These algorithms exploit the relation of the energy minimization problem with the computation of shortest paths. When the instances are restricted to be laminar, Li et al. [46] showed that the problem can be solved in $O(n)$ time.

Antoniadis and Huang [16] were the first to consider the non-preemptive energy minimization problem. They proved that it is strongly $\mathcal{NP}$-hard even for laminar instances. They also presented a $2^{4\alpha-3}$-approximation algorithm for laminar instances and a $2^{5\alpha-4}$-approximation algorithm for general instances. Furthermore, the authors noticed that the problem can be solved optimally in polynomial time when the instances are agreeable by observing that the optimal preemptive schedule produced by the algorithm in [54] executes the jobs non-preemptively. A series of papers improved the approximation ratio of the non-preemptive case. In [21], an approximation algorithm of ratio $2^{\alpha-1}(1+\epsilon)^\alpha \tilde{B}_\alpha$ has been proposed where $\tilde{B}_\alpha = \sum_{k=0}^{\infty} \frac{k^\alpha e^{-1}}{k!}$ is the generalized Bell number which is defined for any $\alpha \in \mathbb{R}^+$ and corresponds to the α-th (fractional) moment of

Poisson's distribution. This algorithm improved the ratio given in [16] for any $\alpha < 114$. Then, an approximation algorithm of ratio $(12(1 + \epsilon))^{\alpha-1}$ was given in [34], improving the approximation ratio for any $\alpha > 25$. In [23], an approximation algorithm of ratio $(1 + \varepsilon)^{\alpha-1}\tilde{B}_\alpha$ has been presented which became the best algorithm, at that moment, for any $\alpha \leq 77$. Recently, a $(1 + \epsilon)$-approximation algorithm which runs in $n^{O(polylog(n))}$ time has been proposed in [42].

Moreover, the relation between preemptive and non-preemptive schedules in the energy-minimization setting has been studied in [20]. The authors showed that starting from the optimal preemptive solution obtained using the algorithm of [54], it is possible to obtain a non-preemptive solution which guarantees an approximation ratio of $(1 + \frac{p_{max}}{p_{min}})^\alpha$, where p_{max} and p_{min} are the maximum and the minimum processing volumes of the jobs. In the special case where all jobs have equal processing volumes this leads to a constant factor approximation of 2^α. For this special case, Angel et al. [9] and Huang and Ott [41], independently, proposed an optimal polynomial-time algorithm based on dynamic programming.

Throughput. Angel et al. studied the throughput maximization problem in the offline setting in [12]. They provided a polynomial time algorithm to solve optimally the single-machine problem for agreeable instances. More recently in [11], they proved that there is a pseudo-polynomial time algorithm for solving optimally the preemptive single-machine problem with arbitrary release dates, deadlines and processing volumes. For the weighted version, the problem is $\mathcal{NP}$-hard even for instances in which all the jobs have common release dates and deadlines. Angel et al. [12] showed that the problem admits a pseudo-polynomial time algorithm for agreeable instances. Furthermore, Antoniadis et al. [18] considered a related problem. More precisely, they studied a generalization of the classical knapsack problem where the objective is to maximize the total profit of the chosen items minus the cost incurred by their total weight. The case where the cost functions are convex can be translated in terms of a weighted throughput problem where the objective is to select the most profitable set of jobs taking into account the energy costs. They presented a fully polynomial time approximation scheme (FPTAS) and a fast 2-approximation algorithm for the non-preemptive problem where the jobs have no release dates or deadlines.

Sum of Completion Times. Pruhs et al. [49] considered the problem of minimizing the average completion time under a budget of energy. They proposed an $O(n^2 \log \frac{E}{\varepsilon})$ polynomial time algorithm for jobs with equal processing volumes, where E is the energy budget and ε the desired accuracy. Albers et al. [6] proposed a simplified algorithm for the problem of minimizing the average completion time plus energy for jobs with equal processing volumes which is based on dynamic programming. Megow et al. [48] considered the weighted version of the average completion time objective. When all the jobs have equal release dates, they established a polynomial time approximation scheme (PTAS). They also showed that the non-preemptive version of the problem is equivalent to the fixed-speed single-machine problem where the objective function is: $\sum w_j(C_j)^{\frac{\alpha-1}{\alpha}}$, where w_j is the weight of job j and C_j its completion time. This result has also been obtained independently by Vásquez [53]. For the preemptive problem where the jobs have arbitrary release dates, Megow et al. [48] proposed a $(2 + \varepsilon)$-approximation algorithm.

Makespan. Bunde [29] proposed an optimal polynomial-time algorithm for the problem of scheduling a set of jobs with arbitrary release dates and deadlines, under a given budget of energy, so that the makespan to be minimized.

Maximum Lateness. In [25], the non-preemptive problem of minimizing the maximum lateness, under a given budget of energy, has been studied. An optimal combinatorial polynomial-time algorithm has been proposed for the case in which the jobs have common release dates. For arbitrary release dates, the problem is shown to be strongly $\mathcal{NP}$−hard. The authors study also the problem where the objective is the minimization of a linear combination of maximum lateness and energy. The results for the budget variant can be adapted to this case. More interestingly, a 2-approximation algorithm is presented when the jobs are subject to release dates.

2.2 Multiple Machines

When more than one machines are available, we distinguish again between two cases: the *preemptive* and the *non-preemptive* cases. In the preemptive case, the execution of the jobs may allow the *migration* of the jobs, i.e. the possibility to execute a job on more than one machines, without allowing its parallel execution. This case is known as the *migratory* case. In the preemptive *non-migratory case*, the execution of a job must be done on the same machine.

We have also to distinguish between *homogeneous* and *heterogeneous* environments. In the homogeneous case, the characteristics of each job (release date, deadline and processing volume) are independent of the machine on which it is executed and the speed-to-power function is the same for all the machines. In the heterogeneous case, we consider the following subcases: In the *fully heterogeneous environment* both, the jobs' characteristics are machine-dependent and every machine has its own power function. Formally, the problem is as follows: we are given a set $\mathcal{J}$ of n jobs and a set $\mathcal{P}$ of m parallel machines. Every machine $i \in \mathcal{P}$ obeys to a different speed-to-power function, i.e., it is associated with a different $\alpha_i \geq 1$ and hence if a job runs at speed s on machine i, then the power is $f(s) = s^{\alpha_i}$. Each job $j \in \mathcal{J}$ has a different release date $r_{i,j}$, deadline $d_{i,j}$ and processing volume $p_{i,j}$ in each machine $i \in \mathcal{P}$. In the *power-heterogeneous environment*, the characteristics of each job are independent of the machine on which the job is executed, while every machine has its own speed-to-power function. Finally, in the *unrelated-heterogeneous* environment the processing volumes of the jobs are machine-dependent while all the other characteristics are independent of the machine on which each job is executed.

Energy minimization. Chen et al. [30] were the first to study a multiprocessor energy-efficient scheduling problem involving speed scaling. More specifically, they proposed an $O(n \log n)$-time algorithm for solving optimally the homogeneous-migratory problem when the release dates and deadlines are identical for all the jobs. Later, Bingham et al. [28] constructed an optimal algorithm for the homogeneous-migratory problem when the jobs have arbitrary release dates and deadlines. The algorithm in [28] makes repetitive calls of a black-box algorithm for solving linear programs. Then, independently, Albers et al. [4] and Angel et al. [15] presented combinatorial algorithms based on a series of maximum flow computations that allow the partition of the set of jobs into subsets in which all the jobs are executed at the same speed. The optimality of these algorithms is based on a series of technical lemmas showing that this partition and the corresponding speeds lead to the minimization of the energy consumption. In [24], it has been shown that both the algorithms and their analysis can be greatly simplified. In order to do this, the problem has been formulated as a convex cost flow problem in an appropriate flow network. Furthermore, it has been shown that this approach is useful to solve other problems in the dynamic speed-scaling setting. As an example, the authors consider the *preemptive open-shop* speed-scaling problem and they propose a polynomial-time algorithm for finding an optimal solution based on the computation of convex cost flows. In [52], Shioura et al. consider the same formulation as convex cost flow for the homogeneous-migratory problem and they propose a method for reducing the running time of the algorithm. For the migratory problem in a *fully heterogeneous* environment, an algorithm using a configuration linear programming (LP) formulation, has been proposed in [21]. This algorithm returns a solution which is within an additive factor of ε far from the optimal solution and runs in time polynomial to the size of the instance and to $1/\varepsilon$. However, the algorithm proposed in [21] is based on the solution of a configuration linear program using the Ellipsoid method. Given that this method may not be very efficient in practice, an alternative polynomial-time algorithm based on a compact linear programming formulation which solves the problem within any desired accuracy was proposed in [5]. This algorithm does not need the use of the Ellipsoid method and it applies for more general than convex power functions; it is valid for a large family of continuous non-decreasing power functions. Furthermore, in the same work, a max-flow based algorithm has been proposed for the migratory problem in a *power-heterogeneous* environment, in which jobs' densities are lower bounded by a small constant, producing a solution arbitrarily close to the optimal.

For the *homogeneous non-migratory* problem, Albers et al. [7] considered the case of a set of jobs with unit processing volumes. They showed that the problem can be solved optimally in polynomial time if the instance is agreeable. Moreover, they established an $\mathcal{NP}$-hardness proof for the unit-work case when the release dates and the deadlines of the jobs are arbitrary. They proposed an $\alpha^\alpha 2^{4\alpha}$-approximation algorithm for this special case. They have also presented an algorithm of the same approximation ratio for arbitrary-work instances when the jobs have either equal release dates or equal deadlines. Next, Greiner et al. [39] presented a $B_{\lceil\alpha\rceil}$-approximation algorithm for the problem with jobs having arbitrary processing volumes, release dates and deadlines, where $B_{\lceil\alpha\rceil}$ is the $\lceil\alpha\rceil$-th Bell number. Cohen-Addad et al. [34] proved that the non-migratory problem is APX-hard for the *unrelated-heterogeneous* model even if all the jobs have the same release dates and deadlines. For the non-migratory problem in a *fully heterogeneous* environment, an approximation algorithm of ratio $(1 + \varepsilon)\tilde{B}_\alpha$ based on a randomized rounding of a configuration LP relaxation has been presented in [21].

For the non-preemptive problem in a *homogeneous* environment, Albers et al. [7] observed that the problem is $\mathcal{NP}$-hard even in the special case where the jobs have the same release dates and deadlines. Moreover, they showed that, for this special case, there exists a polynomial time approximation scheme (PTAS). For arbitrary release dates, deadlines and processing volumes; an approximation algorithm with ratio $m^\alpha(\sqrt[m]{n})^{\alpha-1}$ has been presented in [20]. Cohen-Addad et al. [34] presented an algorithm of ratio $(\frac{5}{2})^{\alpha-1}\tilde{B}_\alpha((1+\varepsilon)(1+\frac{p_{\max}}{p_{\min}}))^\alpha$. This algorithm leads to an approximation ratio of $2(1 + \varepsilon)^\alpha 5^{\alpha-1}\tilde{B}_\alpha$ when all jobs have equal processing volumes. It has to be noticed that he authors in [34] observed that their algorithm can be used for the non-preemptive problem in the unrelated-heterogeneous model by loosing an additional factor of $(\frac{p_{\max}}{p_{\min}})^\alpha$. Finally, a $(1 + \epsilon)$-approximation algorithm which runs in $n^{O(polylog(n))}$ time, for the non-preemptive problem in a *homogeneous* environment, has been presented in [42].

Throughput. The throughput maximization problem has been studied in the case of a fully heterogeneous environment in [14]. For the *fully heterogeneous non-migratory* problem, Angel et al. presented a greedy algorithm which is based on the primal-dual scheme that approximates the optimum solution within a factor depending on the speed-to-power functions (the factor is constant for functions of the form $f(s) = s^\alpha$). Then, they focused on the *homogeneous non-preemptive* problem for which they considered a *fixed* number of machines and two important families of instances: (1) instances with equal processing volume jobs; and (2) agreeable instances. For both cases they presented optimal pseudo-polynomial-time algorithms.

Sum of Completion Times. A polynomial-time algorithm for minimizing a linear combination of the sum of the completion times of the jobs and the total energy consumption, for the non-preemptive multiprocessor speed-scaling problem has been proposed in [24]. Instead of using convex cost flows, the proposed algorithm is based on the computation of a minimum weighted maximum matching in an appropriate bipartite graph.

Makespan. Shabtay and Kaspi [51] proved that the problem is NP-hard even if all the jobs have the same release dates. Pruhs et al. [50] observed that when all the jobs have the same release dates then a PTAS can be obtained using the load balancing algorithm of Alon et al. [8] for the minimization of the L_α norm of loads. Pruhs et al. considered in [50] the problem of scheduling a set of jobs on a set of speed scalable machines subject to precedence constraints among the jobs. The goal is to minimize the makespan of the schedule without exceeding a given energy budget. The approach in [50] is based on *constant power schedules*, which are schedules that keep the total power of all processors constant over time. Based on this property and by performing a binary search to determine the value of the power, they transformed the problem to the classical problem of minimizing the makespan for scheduling a set of jobs with precedence constraints on related parallel processors, in which each processor runs at a single predefined speed. The proposed algorithm has an approximation ratio of $O(\log^{1+2/\alpha} m)$, where m is the number of the machines. This ratio has been improved in [22] where a simple $(2 - \frac{1}{m})$-approximation algorithm has been presented. The idea of this algorithm is the following: first, a convex programming relaxation for the speed scaling problem is given. The solution of this convex program defines a speed and hence a processing time for each job. Given that the obtained processing times respect the energy budget, it is then sufficient to use the classical list scheduling algorithm. This approach may be used for a more general problem where in addition to the precedence constraints the jobs are subject to release dates and/or precedence delays. For these generalizations, the approximation ratio of the algorithm remains asymptotically smaller than 2.

3 Power down

3.1 Single Machine

Chrétienne [33] proved that it is possible to decide in polynomial time whether there is a schedule with no idle time. Baptiste [26] proposed an $O(n^7)$-time dynamic programming algorithm for unit-time jobs and general L. For that, he proved a dominance property showing that there are only a few relevant starting points for the jobs in some optimal schedule and he proposed a clever decomposition of the problem. Then, Baptiste et al. [27] proposed an $O(n^5)$-time dynamic programming algorithm for the preemptive case with jobs of arbitrary processing times. They also proposed an $O(n^4)$ algorithm for unit-time jobs. A simpler dynamic programming with the same time-complexity for unit-time jobs has been proposed in [32]. Given the high time complexity of the algorithms in the general case, Gururaj et al. [40] improved the time-complexity by restricting their attention to agreeable instances. They proposed an $O(n \log n)$ algorithm for jobs with arbitrary lengths and with unit start-up energy consumption, i.e. $L = 1$. For arbitrary L and unit-time jobs, they proposed an $O(n^3)$ algorithm. In [10], this result has been improved by providing an $O(n^2)$ algorithm for arbitrary L and arbitrary processing times. In [31], a simple greedy algorithm has been presented that approximates the optimum solution within a factor of 2 and it has been shown that its analysis is tight. The algorithm runs in time $O(n^2 \log n)$ and needs only $O(n)$ memory. More recently in [32], different variants of the minimum-gap scheduling problem have been studied. These variants include the maximization of the throughput given a budget for gaps or the minimization of the number of gaps given a throughput requirement. Other objective functions are also studied. For instance, maximizing the number of gaps. For the model without deadlines, the authors focus on the tradeoff between the number of gaps and flow time.

3.2   Multiple Machines

The algorithm of [26] has been generalized for the multiple machines case in [35]. The time complexity becomes $O(n^7 m^5)$, where n is the number of jobs and m is the number of machines. For agreeable instances, Gururaj et al. [40] proposed an $O(n^3 m^2)$ algorithm for unit-time jobs and unit start-up energy consumption, $L = 1$. This result has been improved in [10], where an $O(n^2 m)$ algorithm has been proposed.

4   Power-Down with Speed Scaling

While in the speed scaling model, it is always beneficial for the energy consumption to lower the speed of a job as far as the schedule remains feasible, this is not the case for the power-down with speed scaling model. Indeed, by increasing the speed of a job we may increase the length of some idle period and in that way be able to gain in energy consumption by turning off the machine. A central notion in this model is the notion of *critical speed* which, roughly speaking, is the speed minimizing the energy consumption while jobs are processed.

Irani et al. [43] proposed a 2-approximation algorithm for general convex power functions. The rough idea of the algorithm is the following: first, a schedule is produced using the algorithm of Yao et al. [54] for the speed scaling model. Given this schedule, the set of jobs is partitioned into two subsets: the first subset contains all the jobs that are executed with a speed higher than the critical speed, while the second subset contains the jobs that are executed with a speed smaller than the critical one. The schedule returned by the algorithm of Irani et al. [43] executes all the jobs of the first subset using the algorithm of Yao et al. [54], while all the jobs of the second subset are executed with the critical speed. Only recently, Albers and Antoniadis [3] and Kumar and Shannigrahi [45] proved that the problem is $\mathcal{NP}$-hard. For agreeable instances, an $O(n^3)$-time algorithm has been provided in [19]. This algorithm is based in a combination of the algorithm of Yao et al. and the use of dynamic programming for the jobs that are executed with a speed smaller than the critical speed. For general convex power functions, Albers and Antoniadis [3] derived a $\frac{4}{3}$-approximation algorithm. Their algorithm is also a combination of the algorithm of [54] and the use of dynamic programming. Here the partition of the jobs is not based on the critical speed, but on some appropriate value s_0. All the jobs executed in the schedule produced by the algorithm of [54] with a speed lower than s_0 are scheduled with speed s_0. The schedule of these jobs is derived by the dynamic program for the power-down model of Baptiste et al. [27]. All the other jobs are scheduled using the algorithm of [54]. Albers and Antoniadis have also obtained an approximation factor of $\frac{137}{117} < 1.171$ for power functions of the form $g(s) = \beta s^\alpha + c$, where s

is the speed and $\beta, c > 0$ as well as α are constants. More recently, in [17] a fully polynomial-time approximation scheme (FPTAS) for the problem has been proposed.

Finally, the single-machine non-preemptive throughput maximization problem has been studied in [13]. More precisely, optimal polynomial-time algorithms have been presented for two types of instances: (1) agreeable instances and (2) instances with arbitrary release dates and deadlines, but equal processing volumes. Both algorithms are based on dynamic programming.

To the best of our knowledge, no results are known for multiple machines.

5   Concluding Remarks

We gave a quick overview of some recent developments in the context of energy-efficient scheduling focusing on the offline setting. A huge literature exists for the online setting. For more results in this area, the interested reader is invited to consult the recent surveys in [1, 2, 38, 44].

Acknowledgments. This work has been partially supported by the COFECUB project CHOOSING (n. 828/15).

References

1. Albers, S.: Energy efficient algorithms. Commun. ACM **53**(5), 86–96 (2010)
2. Albers, S.: Algorithms for dynamic speed scaling. In: International Symposium of Theoretical Aspects of Computer Science (STACS 2011), LIPIcs, vol. 9, pp. 1–11. Schloss Dagstuhl (2011)
3. Albers, S., Antoniadis, A.: Race to idle: new algorithms for speed scaling with a sleep state. In: Symposium on Discrete Algorithms (SODA), pp. 1266–1285. ACM-SIAM (2012)
4. Albers, S., Antoniadis, A., Greiner, G.: On multi-processor speed scaling with migration. In: Symposium on Parallelism in Algorithms and Architectures (SPAA), pp. 279–288. ACM (2011)
5. Albers, S., Bampis, E., Letsios, D., Lucarelli, G., Stotz, R.: Scheduling on power-heterogeneous processors. In: Kranakis, E., Navarro, G., Chávez, E. (eds.) LATIN 2016. LNCS, vol. 9644, pp. 41–54. Springer, Heidelberg (2016)
6. Albers, S., Fujiwara, H.: Energy efficient algorithms for flow time minimization. ACM Trans. Algorithms **3**(4), 49 (2007)
7. Albers, S., Muller, F., Schmelzer, S.: Speed scaling on parallel processors. In: Symposium on Parallelism in Algorithms and Architectures (SPAA), pp. 289–298. ACM (2007)
8. Alon, N., Azar, Y., Woeginger, G.J., Yadid, T.: Approximation schemes for scheduling. In: Proceedings of 8th Annual ACM-SIAM Symposium on Discrete Algorithms, New Orleans, Louisiana, 5–7 January 1997, pp. 493–500 (1997)
9. Angel, E., Bampis, E., Chau, V.: Throughput maximization in the speed-scaling setting. CoRR, abs/1309.1732 (2013)
10. Angel, E., Bampis, E., Chau, V.: Low complexity scheduling algorithms minimizing the energy for tasks with agreeable deadlines. Discret. Appl. Math. **175**, 1–10 (2014)

11. Angel, E., Bampis, E., Chau, V.: Throughput maximization in the speed-scaling setting. In: 31st International Symposium on Theoretical Aspects of Computer Science (STACS 2014), Lyon, France, 5–8 March 2014, pp. 53–62 (2014)

12. Angel, E., Bampis, E., Chau, V., Letsios, D.: Throughput maximization for speed-scaling with agreeable deadlines. In: Chan, T.H.H., Lau, L.C., Trevisan, L. (eds.) TAMC 2013. LNCS, vol. 7876, pp. 10–19. Springer, Heidelberg (2013)

13. Angel, E., Bampis, E., Chau, V., Thang, N.K.: Nonpreemptive throughput maximization for speed-scaling with power-down. In: Proceedings of Euro-Par: Parallel Processing 21st International Conference on Parallel and Dis-tributed Computing, Vienna, Austria, 24–28 August 2015, pp. 171–182 (2015)

14. Angel, E., Bampis, E., Chau, V., Thang, N.K.: Throughput maximization in multiprocessor speed-scaling. Theor. Comput. Sci. **630**, 1–12 (2016)

15. Angel, E., Bampis, E., Kacem, F., Letsios, D.: Speed scaling on parallel processors with migration. In: Kaklamanis, C., Papatheodorou, T., Spirakis, P.G. (eds.) Euro-Par 2012. LNCS, vol. 7484, pp. 128–140. Springer, Heidelberg (2012)

16. Antoniadis, A., Huang, C.C.: Non-preemptive speed scaling. J. Sched. **16**(4), 385–394 (2013)

17. Antoniadis, A., Huang, C.C., Ott, S.: A fully polynomialtime approximation scheme for speed scaling with sleep state. In: Proceedings of 26th Annual ACM-SIAM Symposium on Discrete Algorithms, SODA, San Diego, CA, USA, 4–6 January 2015, pp. 1102–1113 (2015)

18. Antoniadis, A., Huang, C.-C., Ott, S., Verschae, J.: How to pack your items when you have to buy your knapsack. In: Chatterjee, K., Sgall, J. (eds.) MFCS 2013. LNCS, vol. 8087, pp. 62–73. Springer, Heidelberg (2013)

19. Bampis, E., Dürr, C., Kacem, F., Milis, I.: Speed scaling with power down scheduling for agreeable deadlines. Sustain. Comput.: Inform. Syst. **2**(4), 184–189 (2012)

20. Bampis, E., Kononov, A.V., Letsios, D., Lucarelli, G., Nemparis, I.: From preemptive to non-preemptive speed-scaling scheduling. Discret. Appl. Math. **181**, 11–20 (2015)

21. Bampis, E., Kononov, A.V., Letsios, D., Lucarelli, G., Sviridenko, M.: Energy efficient scheduling and routing via randomized rounding. In: IARCS Annual Conference on Foundations of Software Technology and Theoretical Computer Science, FSTTCS, Guwahati, India, 12–14 December 2013, pp. 449–460 (2013)

22. Bampis, E., Letsios, D., Lucarelli, G.: A note on multiprocessor speed scaling with precedence constraints. In: 26th ACM Symposium on Parallelism in Algorithms and Architectures, SPAA 2014, Prague, Czech Republic, 23–25 June 2014, pp. 138–142 (2014)

23. Bampis, E., Letsios, D., Lucarelli, G.: Speed-scaling with no preemptions. In: Ahn, H.-K., Shin, C.-S. (eds.) ISAAC 2014. LNCS, vol. 8889, pp. 259–269. Springer, Heidelberg (2014)

24. Bampis, E., Letsios, D., Lucarelli, G.: Green scheduling, flows and matchings. Theoret. Comput. Sci. **579**, 126–136 (2015)

25. Bampis, E., Letsios, D., Milis, I., Zois, G.: Speed scaling for maximum lateness. Theor. Comput. Syst. **58**(2), 304–321 (2016)

26. Baptiste, P.: Scheduling unit tasks to minimize the number of idle periods: a polynomial time algorithm for offline dynamic power management. In: Symposium on Discrete Algorithms (SODA), pp. 364–367. ACM-SIAM (2006)

27. Baptiste, P., Chrobak, M., Dürr, C.: Polynomial-time algorithms for minimum energy scheduling. ACM Trans. Algorithms **8**(3), 26 (2012)

28. Bingham, B.D., Greenstreet, M.R.: Energy optimal scheduling on multiprocessors with migration. In: International Symposium on Parallel and Distributed Processing with Applications (ISPA), pp. 153–161. IEEE (2008)
29. Bunde, D.P.: Power-aware scheduling for makespan and flow. J. Sched. **12**(5), 489–500 (2009)
30. Chen, J.J., Hsu, H.R., Chuang, K.H., Yang, C.L., Pang, A.C., Kuo, T.W.: Multiprocessor energy efficient scheduling with task migration considerations. In: Euromicro Conference on Real-Time Systems (ECRTS), pp. 101–108. IEEE (2004)
31. Chrobak, M., Feige, U., Taghi Hajiaghayi, M., Khanna, S., Li, F., Naor, S.: A Greedy approximation algorithm for minimum-gap scheduling. In: Spirakis, P.G., Serna, M. (eds.) CIAC 2013. LNCS, vol. 7878, pp. 97–109. Springer, Heidelberg (2013)
32. Chrobak, M., Golin, M., Lam, T.-W., Nogneng, D.: Scheduling with gaps: new models and algorithms. In: Paschos, V.T., Widmayer, P. (eds.) CIAC 2015. LNCS, vol. 9079, pp. 114–126. Springer, Heidelberg (2015)
33. Chrétienne, P.: On single-machine scheduling without intermediate delays. Discret. Appl. Math. **156**(13), 2543–2550 (2008). In: 5th Conference, Honour of Peter Hammer's and Jakob Krarup's 70th Birthday, Graphs and Optimization, Fifth International Conference on Graphs and Optimization (GOV 2006)
34. Cohen-Addad, V., Li, Z., Mathieu, C., Milis, I.: Energy-efficient algorithms for non-preemptive speed-scaling. In: Bampis, E., Svensson, O. (eds.) WAOA 2014. LNCS, vol. 8952, pp. 107–118. Springer, Heidelberg (2015)
35. Demaine, E.D., Ghodsi, M., Hajiaghayi, M.T., Sayedi-Roshkhar, A.S., Zadimoghaddam, M.: Scheduling to minimize gaps and power consumption. In: Symposium on Parallelism in Algorithms and Architectures (SPAA), pp. 46–54. ACM (2007)
36. Gaujal, B., Navet, N.: Dynamic voltage scaling under EDF revisited. Real-Time Syst. **37**(1), 77–97 (2007)
37. Gaujal, B., Navet, N., Walsh, C.: Shortest-path algorithms for real-time scheduling of FIFO tasks with minimal energy use. ACM Trans. Embed. Comput. Syst. **4**(4), 907–933 (2005)
38. Gerards, M.E.T., Hurink, J.L., Hölzenspies, P.K.F.: A survey of offline algorithms for energy minimization under deadline constraints. J. Sched. **19**(1), 3–19 (2016)
39. Greiner, G., Nonner, T., Souza, A.: The bell is ringing in speed scaled multiprocessor scheduling. In: Symposium on Parallelism in Algorithms and Architectures (SPAA), pp. 11–18. ACM (2009)
40. Gururaj, Jalan, and Stein. Unpublished work, see survey of m. chrobak
41. Huang, C.-C., Ott, S.: New results for non-preemptive speed scaling. In: Csuhaj-Varjú, E., Dietzfelbinger, M., Ésik, Z. (eds.) MFCS 2014, Part II. LNCS, vol. 8635, pp. 360–371. Springer, Heidelberg (2014)
42. Im, S., Shadloo, M.: Brief announcement: a QPTAS for non-preemptive speed-scaling. In: Proceedings of ACM SPAA (2016)
43. Irani, S., Gupta, R.K., Shukla, S.K.: Competitive analysis of dynamic power management strategies for systems with multiple power savings states. In: Conference on Design, Automation and Test in Europe (DATE), pp. 117–123. IEEE (2002)
44. Irani, S., Pruhs, K.: Algorithmic problems in power management. ACM SIGACT News **36**(2), 63–76 (2005)
45. Kumar, G., Shannigrahi, S.: On the NP-hardness of speed scaling with sleep state. Theor. Comput. Sci. **600**, 1–10 (2015)
46. Li, M., Liu, B.J., Yao, F.F.: Min-energy voltage allocation for tree-structured tasks. J. Comb. Optim. **11**(3), 305–319 (2006)

47. Li, M., Yao, A.C., Yao, F.F.: Discrete and continuous min-energy schedules for variable voltage processors. Proc. Nat. Acad. Sci. U.S.A. **103**(11), 3983–3987 (2006)
48. Megow, N., Verschae, J.: Dual techniques for scheduling on a machine with varying speed. In: Fomin, F.V., Freivalds, R., Kwiatkowska, M., Peleg, D. (eds.) ICALP 2013, Part I. LNCS, vol. 7965, pp. 745–756. Springer, Heidelberg (2013)
49. Pruhs, K., Uthaisombut, P., Woeginger, G.J.: Getting the best response for your erg. ACM Trans. Algorithms **4**(3), 38 (2008)
50. Pruhs, K., van Stee, R., Uthaisombut, P.: Speed scaling of tasks with precedence constraints. Theor. Comput. Syst. **43**(1), 67–80 (2008)
51. Shabtay, D., Kaspi, M.: Parallel machine scheduling with a convex resource consumption function. Eur. J. Oper. Res. **173**(1), 92–107 (2006)
52. Shioura, A., Shakhlevich, N., Strusevich, V.: Energy optimization in speed scaling models via submodular optimization. In: 12th Workshop on Models and Algorithms for Planning and Scheduling Problems (MAPSP) (2015)
53. Vásquez, O.C.: Energy in computing systems with speed scaling: optimization and mechanisms design (2012). arXiv:1212.6375
54. Yao, F.F., Demers, A.J., Shenker, S.: A scheduling model for reduced cpu energy. In: Symposium on Foundations of Computer Science (FOCS), pp. 374–382. IEEE (1995)

Linear Superiorization for Infeasible Linear Programming

Yair Censor$^{(\boxtimes)}$ and Yehuda Zur

Department of Mathematics, University of Haifa, Mt. Carmel, 3498838 Haifa, Israel
`yair@math.haifa.ac.il`

Abstract. Linear superiorization (abbreviated: LinSup) considers linear programming (LP) problems wherein the constraints as well as the objective function are linear. It allows to steer the iterates of a feasibility-seeking iterative process toward feasible points that have lower (not necessarily minimal) values of the objective function than points that would have been reached by the same feasiblity-seeking iterative process without superiorization. Using a feasibility-seeking iterative process that converges even if the linear feasible set is empty, LinSup generates an iterative sequence that converges to a point that minimizes a proximity function which measures the linear constraints violation. In addition, due to LinSup's repeated objective function reduction steps such a point will most probably have a reduced objective function value. We present an exploratory experimental result that illustrates the behavior of LinSup on an infeasible LP problem.

Keywords: Superiorization · Perturbation resilience · Infeasible linear programming · Feasibility-seeking · Simultaneous projection algorithm · Cimmino method · Proximity function

1 Introduction: The General Concept of Superiorization

Given an algorithmic operator $\mathcal{A} : X \to X$ on a Hilbert space X, consider the iterative process

$$x^0 \in X, \; x^{k+1} = \mathcal{A}\left(x^k\right), \text{ for all } k \geqslant 0, \tag{1}$$

and let $SOL\,(P)$ denote the solution set of some problem P of any kind. The iterative process is said to solve P if, under some reasonable conditions, any sequence $\left\{x^k\right\}_{k=0}^{\infty}$ generated by the process converges to some $x^* \in SOL\,(P)$. An iterative process (1) that solves P is called perturbation resilient if the process

$$y^0 \in X, \; y^{k+1} = \mathcal{A}\left(y^k + v^k\right), \text{ for all } k \geqslant 0, \tag{2}$$

also solves P, under some reasonable conditions on the sequence of perturbation vectors $\left\{v^k\right\}_{k=0}^{\infty} \subseteq X$. The iterative processes of (1) and (2) are called "the basic algorithm" and "the superiorized version of the basic algorithm", respectively.

Y. Kochetov et al. (Eds.): DOOR 2016, LNCS 9869, pp. 15–24, 2016.
DOI: 10.1007/978-3-319-44914-2_2

Superiorization aims at identifying perturbation resilient iterative processes that will allow to use the perturbations in order to steer the iterates of the superiorized algorithm so that, while retaining the original property of converging to a point in $SOL\,(P)$, they will also do something additional useful for the original problem P, such as converging to a point with reduced values of some given objective function. These concepts are rigorously defined in several recent works in the field, we refer the reader to the recent reviews [5,13] and references therein. More material about the current state of superiorization can be found also in [6,14,19].

A special case of prime importance and significance of the above is when P is a convex feasibility problem (CFP) of the form: Find a vector $x^* \in \cap_{i=1}^{I} C_i$ where $C_i \subseteq R^J$, the J-dimensional Euclidean space, are closed convex subsets, and the perturbations in the superiorized version of the basic algorithm are designed to reduce the value of a given objective function ϕ.

In this case the basic algorithm (1) can be any of the wide variety of feasibility-seeking algorithms, see, e.g., [2,7,8], and the perturbations employ nonascent directions of ϕ. Much work has been done on this as can be seen in the Internet bibliography at [4].

The usefulness of this approach is twofold: First, feasibility-seeking is, on a logical basis, a less-demanding task than seeking a constrained minimization point in a feasible set. Therefore, letting efficient feasibility-seeking algorithms "lead" the algorithmic effort and modifying them with inexpensive add-ons works well in practice.

Second, in some real-world applications the choice of an objective function is exogenous to the modeling and data acquisition which give rise to the constraints. Thus, sometimes the limited confidence in the usefulness of a chosen objective function leads to the recognition that, from the application-at-hand point of view, there is no need, neither a justification, to search for an exact constrained minimum. For obtaining "good results", evaluated by how well they serve the task of the application at hand, it is often enough to find a feasible point that has reduced (not necessarily minimal) objective function value[1].

2   Linear Superiorization

2.1   The Problem and the Algorithm

Let the feasible set M be

$$M := \{x \in R^J \mid Ax \le b, \ x \ge 0\} \tag{3}$$

---

[1] Some support for this reasoning may be borrowed from the American scientist and Noble-laureate Herbert Simon who was in favor of "satisficing" rather then "maximizing". Satisficing is a decision-making strategy that aims for a satisfactory or adequate result, rather than the optimal solution. This is because aiming for the optimal solution may necessitate needless expenditure of time, energy and resources. The term "satisfice" was coined by Herbert Simon in 1956 [20], see: https://en.wikipedia.org/wiki/Satisficing.

where the $I \times J$ real matrix $A = (a_j^i)_{i=1,j=1}^{I,J}$ and the vector $b = (b_i)_{i=1}^I \in R^I$ are given.

For a basic algorithm we pick a feasibility-seeking projection method. Here projection methods refer to iterative algorithms that use projections onto sets while relying on the general principle that when a family of, usually closed and convex, sets is present, then projections onto the individual sets are easier to perform than projections onto other sets (intersections, image sets under some transformation, etc.) that are derived from the individual sets.

Projection methods may have different algorithmic structures, such as block-iterative projections (BIP) or string-averaging projections (SAP) (see, e.g., the review paper [9] and references therein) of which some are particularly suitable for parallel computing, and they demonstrate nice convergence properties and/or good initial behavior patterns.

This class of algorithms has witnessed great progress in recent years and its member algorithms have been applied with success to many scientific, technological and mathematical problems. See, e.g., the 1996 review [2], the recent annotated bibliography of books and reviews [7] and its references, the excellent book [3], or [8].

An important comment is in place here. A CFP can be translated into an unconstrained minimization of some proximity function that measures the feasibility violation of points. For example, using a weighted sum of squares of the Euclidean distances to the sets of the CFP as a proximity function and applying steepest descent to it results in a simultaneous projections method for the CFP of the Cimmino type. However, there is no proximity function that would yield the sequential projections method of the Kaczmarz type, for CFPs, see [1].

Therefore, the study of feasibility-seeking algorithms for the CFP has developed independently of minimization methods and it still vigorously does, see the references mentioned above. Over the years researchers have tried to harness projection methods for the convex feasibility problem to LP in more than one way, see, e.g., Chinneck's book [11].

The mini-review of relations between linear programming and feasibility-seeking algorithms in [17, Sect. 1] sheds more light on this. Our work in [6] and here leads us to study whether LinSup can be useful for either feasible or infeasible LP problems.

The objective function for linear superiorization will be

$$\phi(x) := \langle c, x \rangle \tag{4}$$

where $\langle c, x \rangle$ is the inner product of x and a given $c \in R^J$.

In the footsteps of the general principles of the superiorization methodology, as presented for general objective functions ϕ in previous publications, we use the following linear superiorization (LinSup) algorithm. The algorithm and its implementation details follow closely those of [6] wherein only feasible constraints were discussed.

The input to the algorithm consists of the problem data A, b, and c of (3) and (4), respectively, a user-chosen initialization point $\bar{y}$ and a user-chosen parameter (called here kernel) $0 < \alpha < 1$ with which the algorithm generates the step-sizes $\beta_{k,n}$ by the powers of the kernel $\eta_\ell = \alpha^\ell$, as well as an integer N that determines the quantity of objective function reduction perturbation steps done per each feasibility-seeking iterative sweep through all linear constraints. The perturbation direction $-\frac{c}{\|c\|_2}$ used in step 10 of Algorithm 1 is a nonascend direction of the linear objective function, as required by the general principles of the superiorization methodology, see, e.g., [14, Subsect. II.D].

Algorithm 1. The Linear Superiorization (LinSup) Algorithm

---

1. **set** $k = 0$
2. **set** $y^k = \bar{y}$
3. **set** $\ell_{-1} = 0$
4. **while** stopping rule not met **do**
5.   **set** $n = 0$
6.   **set** $\ell = rand(k, \ell_{k-1})$
7.   **set** $y^{k,n} = y^k$
8.   **while** $n < N$ **do**
9.   **set** $\beta_{k,n} = \eta_\ell$
10.   **set** $z = y^{k,n} - \beta_{k,n}\frac{c}{\|c\|_2}$
11.   **set** $n \leftarrow n + 1$
12.   **set** $y^{k,n} = z$
13.   **set** $\ell \leftarrow \ell + 1$
14.   **end while**
15.   **set** $\ell_k = \ell$
16.   **set** $y^{k+1} = \mathcal{A}\left(y^{k,N}\right)$
17.   **set** $k \leftarrow k + 1$
18. **end while**

All quantities in this algorithm are detailed and explained below, except for the choice of the basic algorithm for the feasibility-seeking operator represented by $\mathcal{A}$ in step 16 of Algorithm 1 which appear in the next subsection.

Step-sizes of the Perturbations. The step sizes $\beta_{k,n}$ in Algorithm 1 must be such that $0 < \beta_{k,n} \leq 1$ in a way that guarantees that they form a summable sequence $\sum_{k=0}^\infty \sum_{n=0}^{N-1} \beta_{k,n} < \infty$, see, e.g., [10]. To this end Algorithm 1 assumes that we have available a summable sequence $\{\eta_\ell\}_{\ell=0}^\infty$ of positive real numbers generated by $\eta_\ell = \alpha^\ell$, where $0 < \alpha < 1$. Simultaneously with generating the iterative sequence $\{y^k\}_{k=0}^\infty$, a subsequence of $\{\eta_\ell\}_{\ell=0}^\infty$ is used to generate the step sizes $\beta_{k,n}$ in step 9 of Algorithm 1. The number α is called the kernel of the sequence $\{\eta_\ell\}_{\ell=0}^\infty$.

Controlling the Decrease of the Step-sizes of Objective Function Reduction. If during the application of Algorithm 1 the step sizes $\beta_{k,n}$ decrease too fast then too little leverage is allocated to the objective function reduction activity that is interlaced into the feasibility-seeking activity of the basic algorithm. This delicate balance can be controlled by the choice of the index ℓ updates and separately by the value of α whose powers α^ℓ determine the step sizes $\beta_{k,n}$ in step 9. In our work we adopt a strategy for updating the index ℓ that was proposed and implemented for total variation (TV) image reconstruction from projections by Prommegger and by Langthaler in [18, p. 38 and Table 7.1 on p. 49] and in [15], respectively. This strategy advocates to set ℓ at the beginning of every new iteration sweep (steps 5 and 6) to a random number between the current iteration index k and the value of ℓ from the last iteration sweep, i.e., $\ell_k = rand(k, \ell_{k-1})$.

The Proximity Function. To measure the feasibility-violation (or level of disagreement) of a point with respect to the target set M we used the following proximity function

$$\Pr(x) := \frac{1}{2I} \sum_{i=1}^{I} \frac{\left(\left(\langle a^i, x \rangle - b_i \right)_+ \right)^2}{\sum_{j=1}^{J} \left(a_j^i \right)^2} + \frac{1}{2J} \sum_{j=1}^{J} \left((-x_j)_+ \right)^2 \tag{5}$$

where the plus notation means, for any real number d, that $d_+ := \max(d, 0)$.

The Number N of Perturbation Steps. This number N of perturbation steps that are performed prior to each application of the feasibility-seeking operator $\mathcal{A}$ (in step 16) affects the performance of the LinSup algorithm. It influences the balance between the amounts of computations allocated to feasibility-seeking and those allocated to objective function reduction steps. A too large N will make Algorithm 1 spend too much resources on the perturbations that yield objective function reduction.

Handling the Nonnegativity Constraints. The nonnegativity constraints in (3) are handled by projections onto the nonnegative orthant, i.e., by taking the iteration vector in hand after each iteration of Cimmino's feasibility-seeking algorithm applied to all I row-inequalities of (3) and setting its negative components to zero while keeping the others unchanged.

2.2 Cimmino's Feasibility-Seeking Algorithm as the Basic Algorithm

We use the simultaneous projections method of Cimmino for linear inequalities, see, e.g. [12], as the basic algorithm for the feasibility-seeking operator represented by $\mathcal{A}$ in step 16 of Algorithm 1. Denoting the half-spaces represented by individual rows of (3) by H_i,

$$H_i := \{ x \in R^J \mid \langle a^i, x \rangle \leq b_i \}, \tag{6}$$

where $a^i \in R^J$ is the i-th row of A and $b_i \in R$ is the i-th component of b in (3), he orthogonal projection of an arbitrary point $z \in R^J$ onto H_i, has the closed-form

$$P_{H_i}(z) = \begin{cases} z - \dfrac{\langle a^i, z \rangle - b_i}{\|a^i\|^2} a^i, & \text{if } \langle a^i, z \rangle > b_i, \\ z, & \text{if } \langle a^i, z \rangle \leq b_i. \end{cases} \tag{7}$$

Algorithm 2. The Simultaneous Feasibility-Seeking Projection Method of Cimmino

Initialization: $x^0 \in R^J$ is arbitrary.

Iterative step: Given the current iteration vector x^k the next iterate is calculated by

$$x^{k+1} = x^k + \lambda_k \left(\sum_{i=1}^{I} w_i \left(P_{H_i}(x^k) - x^k \right) \right) \tag{8}$$

with weights $w_i \geq 0$ for all $i \in I$, and $\sum_{i=1}^{I} w_i = 1$.

Relaxation parameters: The parameters λ_k are such that $\epsilon_1 \leq \lambda_k \leq 2 - \epsilon_2$, for all $k \geq 0$, with some, arbitrarily small, fixed, $\epsilon_1, \epsilon_2 > 0$.

This Cimmino simultaneous feasibility-seeking projection algorithm is known to generate convergent iterative sequences even if the intersection $\cap_{i=1}^{I} H_i$ is empty, as the following, slightly paraphrased, theorem tells.

Theorem 1. *[12, Theorem 3] For any starting point $x^0 \in R^J$, any sequence $\{x^k\}_{k=0}^{\infty}$, generated by the simultaneous feasibility-seeking projection method of Cimmino (Algorithm 2) converges. If the underlying system of linear inequalities is consistent, the limit point is a feasible point for it. Otherwise, the limit point minimizes $f(x) := \sum_{i=1}^{I} w_i \| P(x) - x \|^2$, i.e., it is a weighted (with the weights w_i) least squares solution of the system.*

3   An Empirical Result

Employing MATLAB 2014b [16], we created five test problems each with 2500 linear inequalities in R^J, $J = 2000$. The entries in 1250 rows of the matrix A in (3) were uniformly distributed random numbers from the interval $(-1, 1)$. The remaining 1250 rows were defined as the negatives of the first 1250 rows, i.e., $a_j^{1250+t} = -a_j^t$ for all $t = 1, 2, \ldots, 1250$ and all $j = 1, 2, \ldots, 2000$. This guarantees that the two sets of rows represent parallel half-spaces with opposing normals. For the right-hand side vectors, the components of b associated with the first set of 1250 rows in (3) were uniformly distributed random numbers from the interval $(0, 100)$. The remaining 1250 components of each b were chosen as follows: $b_{1250+t} = -b_t - rand(100, 200)$ for all $t = 1, 2, \ldots, 1250$. This guarantees that the distance between opposing parallel half-spaces is large making them inconsistent, i.e., having no point in common, and that the whole system is infeasible.

For the linear objective function, the components of c were uniformly distributed random numbers from the interval $(-2, 1)$. All runs of Algorithms 1 and 2 were initialized at $\bar{y} = 10 \cdot \mathbf{1}$ and $x^0 = 10 \cdot \mathbf{1}$, respectively, where $\mathbf{1}$ is the vector of all 1's.

We ran Algorithm 1 on each problem until it ceased to make progress, by using the stopping rule

$$\frac{\left\| y^k - y^{k-1} \right\|}{\left\| y^k \right\|} \le 10^{-4}. \tag{9}$$

The same stopping rule was used for runs of Algorithm 2. The relaxation parameters in Cimmino's feasibility-seeking basic algorithm in step 16 of Algorithm 1 were fixed with $\lambda_k = 1.99$ for all $k \geqslant 0$. Based on our work in [6] we used $N = 20$ and $\alpha = 0.99$ in steps 8 and 9 of Algorithm 1, respectively, where $\eta_\ell = \alpha^\ell$.

The three figures, presented below, show results for the five different (but similarly generated) families of inconsistent linear inequalities along with nonnegativity constraints. Figures 1 and 2, in particular, show that the perturbation steps 5–15 of the LinSup Algorithm 1 initially work and reduce the objective function value powerfully during the first ca. 500 iterative sweeps (an iterative sweep consists of one pass through steps 5–17 in Algorithm 1 or one pass through all linear inequalities and the nonnegativity constraints in Algorithm 2). As iterative sweeps proceed the perturbations in Algorithm 1 loose steam because of the decreasing values of the $\beta_{k,n}$s and later the algorithm proceeds toward feasibility at the expense of some increase of objective function values. However, even at those later sweeps the objective function values of LinSup remain well

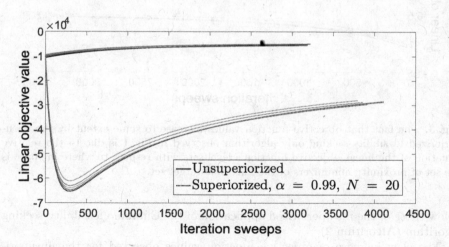

Fig. 1. Linear objective function values plotted against iteration sweeps. LinSup has reduced objective function values although the effect of objective function reducing perturbations diminishes as iterations proceed.

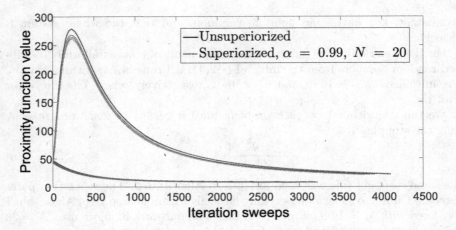

Fig. 2. Proximity function values plotted against iteration sweeps. The unsuperior-ized feasibility-seeking only algorithm does a better job than LinSup here which is understandable. LinSup's strive for feasibility comes at the expense of some increase in objective function values, as seen in Fig. 1.

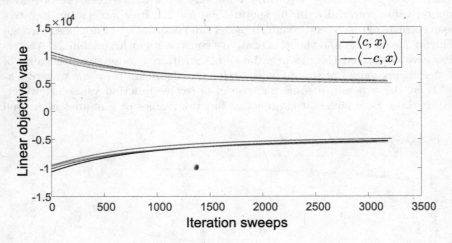

Fig. 3. The fact that objective function values increase to some extent by the unsu-periorized feasibility-seeking only algorithm observed in Fig. 1 is due to the relative situation of the linear objective function's level sets with respect to where in space is the set of proximity minimizers of the infeasible target set.

below those of the unsuperiorized application of the Cimmino feasibility-seeking algorithm (Algorithm 2).

The slow increase of objective function values observed for the unsuperi-orized application of the Cimmino feasibility-seeking algorithm seems intrigu-ing because the feasibility-seeking algorithm is completely unaware of the given objective function $\phi(x) := \langle c, x \rangle$. But this is understood from the fact that the unsuperiorized algorithm has an orbit of iterates in R^J which, by proceeding in space toward proximity minimizers, crosses the linear objective function's level sets in a direction that either increases or decreases objective function values. It would keep them constant only if the orbit was confined to a single level set of ϕ which is not a probable thing to happen. To clarify this we recorded in Fig. 3 the values of $\langle c, x \rangle$ and $\langle -c, x \rangle$ at the iterates x^k produced by the Cimmino feasibility-seeking algorithm (Algorithm 2).

Concluding Comments

We proposed a new approach to handle infeasible linear programs (LPs) via the linear superiorization (LinSup) method. To this end we applied the feasibility-seeking projection method of Cimmino to the original linear infeasible constraints (without using additional variables). This Cimmino method is guaranteed to converge to one of the points that minimize a proximity function that measures the violation of all constraints. We used the given linear objective function to superiorize Cimmino's method to steer its iterates to proximity minimizers with reduced objective function values. Further computational research is needed to evaluate and compare the results of this new approach to existing solution approaches to infeasible LPs.

Acknowledgments. We thank Gabor Herman, Ming Jiang and Evgeni Nurminski for reading a previous version of the paper and sending us comments that helped improve it. This work was supported by Research Grant No. 2013003 of the United States-Israel Binational Science Foundation (BSF).

References

1. Baillon, J.-B., Combettes, P.L., Cominetti, R.: There is no variational characterization of the cycles in the method of periodic projections. J. Funct. Anal. **262**, 400–408 (2012)
2. Bauschke, H.H., Borwein, J.M.: On projection algorithms for solving convex feasibility problems. SIAM Rev. **38**, 367–426 (1996)
3. Cegielski, A.: Iterative Methods for Fixed Point Problems in Hilbert Spaces, vol. 2057. Springer, Heidelberg (2012)
4. Censor, Y.: Superiorization and Perturbation Resilience of Algorithms: A Bibliography compiled and continuously updated by Y. Censor. http://math.haifa.ac.il/yair/bib-superiorization-censor.html
5. Censor, Y.: Weak and strong superiorization: Between feasibility-seeking and minimization. Analele Stiintifice ale Universitatii Ovidius Constanta-Seria Matematica **23**, 41–54 (2015)
6. Censor, Y.: Can linear superiorization be useful for linear optimization problems? Inverse Probl. (2016). under review
7. Censor, Y., Cegielski, A.: Projection methods: an annotated bibliography of books and reviews. Optimization **64**, 2343–2358 (2015)

8. Censor, Y., Chen, W., Combettes, P.L., Davidi, R., Herman, G.T.: On the effective-ness of projection methods for convex feasibility problems with linear in equality constraints. Comput. Optim. Appl. **51**, 1065–1088 (2012)

9. Censor, Y., Segal, A.: Iterative projection methods in biomedical inverse problems. In: Censor, Y., Jiang, M., Louis, A.K. (eds.) Mathematical Methods in Biomedical Imaging and Intensity-Modulated Radiation Therapy (IMRT), pp. 65–96. Edizioni della Normale, Pisa (2008)

10. Censor, Y., Zaslavski, A.J.: Strict Fejér monotonicity by superiorization of feasibility-seeking projection methods. J. Optim. Theory Appl. **165**, 172–187 (2015)

11. Chinneck, J.W.: Feasibility and Infeasibility in Optimization: Algorithms and Com-putational Methods. Springer Science+Business Media, LLC, New York, NY, USA (2008)

12. De Pierro, A.R., Iusem, A.N.: A simultaneous projections method for linear inequalities. Linear Algebra Appl. **64**, 243–253 (1985)

13. Herman, G.T.: Superiorization for image analysis. In: Barneva, R.P., Brimkov, V.E., Šlapal, J. (eds.) IWCIA 2014. LNCS, vol. 8466, pp. 1–7. Springer, Heidelberg (2014)

14. Herman, G.T., Garduño, E., Davidi, R., Censor, Y.: Superiorization: an optimiza-tion heuristic for medical physics. Med. Phys. **39**, 5532–5546 (2012)

15. Langthaler, O.: Incorporation of the superiorization methodology into biomed-ical imaging software. Marshall Plan Scholarship Report, Salzburg University of Applied Sciences, Salzburg, Austria, and The Graduate Center of the City Univer-sity of New York, NY, USA, 76 p. (2014)

16. MATLAB. A high-level language and interactive environment system by The Mathworks Inc., Natick, MA, USA. http://www.mathworks.com/products/matlab/

17. Nurminski, E.A.: Single-projection procedure for linear optimization. J. Glob. Optim., 1–16 (2015). doi:10.1007/s10898-015-0337-9

18. Prommegger, B.: Verification, evaluation of superiorized algorithms used in bio-medical imaging: Comparison of iterative algorithms with and without superioriza-tion for image reconstruction from projections. Marshall Plan Scholarship Report, Salzburg University of Applied Sciences, Salzburg, Austria, and The Graduate Center of the City University of New York, NY, USA, 84 p. (2014)

19. Reem, D., De Pierro, A.: A new convergence analysis, perturbation resilience of some accelerated proximal forward-backward algorithms with errors. arXiv preprint arXiv:1508.05631 (2015)

20. Simon, H.A.: Rational choice and the structure of the environment. Psychol. Rev. **63**, 129–138 (1956)

Short Survey on Graph Correlation Clustering with Minimization Criteria

Victor Il'ev[1,2](✉), Svetlana Il'eva[2], and Alexander Kononov[1]

[1] Sobolev Institute of Mathematics, 4 Acad. Koptyug Avenue,
630090 Novosibirsk, Russia
iljev@mail.ru, alvenko@math.nsc.ru
[2] Omsk State University, 55a Mira Avenue, 644077 Omsk, Russia
iljeva@mail.ru

Abstract. In clustering problems one has to partition a given set of objects into some subsets (called clusters) taking into consideration only similarity of the objects. One of the most visual formalizations of clustering is the graph clustering, that is, grouping the vertices of a graph into clusters taking into consideration the edge structure of the graph whose vertices are objects and edges represent similarities between the objects.

In this short survey, we consider the graph correlation clustering problems where the goal is to minimize the number of edges between clusters and the number of missing edges inside clusters. We present a number of results on graph correlation clustering including results on computational complexity and approximability of different variants of the problems, and performance guarantees of approximation algorithms for graph correlation clustering. Some results on approximability of weighted versions of graph correlation clustering are also presented.

Keywords: Graph clustering · Computational complexity · Approximation algorithm · Performance guarantee

1 Introduction

The objective of *clustering problems* is to partition a given set of objects (data elements) into a family of subsets (called *clusters*) such that objects within a cluster are more like to one another than objects in different clusters. The similarity measure may be defined in different ways for different settings of the problem [19,32].

One of the most visual formalizations of clustering is *graph clustering* [27,32], that is, grouping the vertices of a graph into clusters taking into consideration the edge structure of the graph whose vertices are objects and edges represent similarities between the objects. We consider a minimization version of graph clustering. In this version the goal is to minimize *disagreements* (the number of edges between clusters plus the number of missing edges inside clusters). In other words, given an undirected graph G, the minimization version of GRAPH

© Springer International Publishing Switzerland 2016
Y. Kochetov et al. (Eds.): DOOR 2016, LNCS 9869, pp. 25–36, 2016.
DOI: 10.1007/978-3-319-44914-2_3

CORRELATION CLUSTERING problem asks to transform G into a vertex-disjoint union of cliques by a minimum number of edge modifications.

Note that the maximization version of GRAPH CORRELATION CLUSTERING was also studied in the literature (see [4]), where the goal is to maximize *agreements* (the number of edges inside clusters plus the number of missing edges between clusters). This version is equivalent to the problem of minimizing disagreements with respect to optimal solution but differs from the point of view of approximation.

In this survey, we focus on the minimization version of GRAPH CORRELATION CLUSTERING only.

GRAPH CORRELATION CLUSTERING seems to have been first defined by Harary [20] in 1955. Harary introduced *the signed graph*, i. e., an undirected graph with $+1$ or -1 labels on the edges, and considered a psychological interpretation of the problem: positive edges correspond to pairs of people who like one another, and negative edges to pairs who dislike one another. Harary's aim was to find two nearest to complete groups. Apart from social psychology, the study of signed graphs has many other applications, notably in statistical mechanics and biological networks.

First theoretical results on GRAPH CORRELATION CLUSTERING were obtained in sixties-seventies of the last century.

In 1964, Zahn [40] considered the problem of finding an equivalence relation E, which "best approximates" a given symmetric relation R in the sense of minimizing the number of elements of $(E - R) \cup (R - E)$. Zahn called it the *approximating symmetric relations by equivalence relation*. This problem might be regarded as a simplified model of a situation in which an interconnected structure or organization must be partitioned (perhaps for cataloging or for formal administrative purposes) in a way which reflects the actual interconnections as well as possible. Zahn solved this problem for a special class of relations R representing two-level and three-level "hierarchical" structures.

In seventies, GRAPH CORRELATION CLUSTERING was studied under the name of GRAPH APPROXIMATION PROBLEM. Lyapunov [28] came to this problem in connection with the problem of hierarchical classification of control systems. In the same years, Fridman [15] showed that GRAPH APPROXIMATION PROBLEM for the triangle-free graphs can be reduced to the maximum matching problem. In [17], he considered GRAPH APPROXIMATION PROBLEM as a version of the minimum cut problem on the complete signed graph. In subsequent years, this problem was considered as a special case of the hierarchical-tree clustering problem [26].

In the past two decades, GRAPH CORRELATION CLUSTERING have been repeatedly rediscovered under the different names by many authors independently (CORRELATION CLUSTERING [4], CLUSTER EDITING [6,9,33], TRANSITIVE GRAPH PROJECTION [31], etc.) In these and other works the variants of the problem in which the number of clusters is bounded were studied and also more general settings of the problem were considered.

Interest to these problems was revived due to its numerous applications in different areas of life and science. In 1999, Ben-Dor, Shamir and Yakhimi [6] considered the problem equivalent to GRAPH CORRELATION CLUSTERING when studying some computational biology questions. In 2002–2004, Bansal, Blum and Chawla [4] independently defined CORRELATION CLUSTERING problem one version of which is equivalent to GRAPH CORRELATION CLUS-TERING. They were motivated by some machine learning problems concerning document clustering problem. Chen, Jiang and Lin [11] considered GRAPH CORRELATION CLUSTERING as a special case of the closest phylogenetic k-th root problem. Shamir, Sharan and Tsur [33] mentioned numerous applications of the equivalent CLUSTER EDITING problem in computational biology, image processing, VLSI design and other fields.

As the graph clustering problem is a popular topic of research it is not surprising that there are surveys dedicated to this topic. For example, Schaeffer [32] discussed the different definitions of clusters and measures of cluster quality and presented some general approaches to solving the clustering problems. Recently, Böcker and Baumbach [9] reviewed exact methods for GRAPH CORRELATION CLUSTERING. However, approximation algorithms were outside the focus of their review. The main topics of [9] are exact algorithms based on Integer Linear Programming, parametrized algorithms and data reduction. Because of space limitation and to avoid repetitions we omit these topics in our servey. See [5,7] for the recent results on parametrized algorithms.

In this short survey we present a number of results on minimizaton version of GRAPH CORRELATION CLUSTERING including results on computational complexity and approximability of different variants of the problem, and performance guarantees of approximation algorithms for GRAPH CORRELATION CLUSTERING. Some results on approximability of weighted versions of GRAPH CORRELATION CLUSTERING are also presented.

2   Different Settings of GRAPH CORRELATION CLUSTERING

We consider only *simple* graphs, i. e., the graphs without loops and multiple edges. A graph is called a *matroidal graph* (*M-graph*) [36] or a *cluster graph* [33] if each of its connected components is a complete graph.

Let V be a finite set. Denote by $\mathcal{M}(V)$ the set of all cluster graphs on the vertex set V; let $\mathcal{M}_k(V)$ stand for the set of all cluster graphs on V consisting of exactly k nonempty connected components, and let $\mathcal{M}_{1,k}(V)$ be the set of all cluster graphs on V consisting of at most k connected components, $2 \leq k \leq |V|$.

If $G_1 = (V, E_1)$ and $G_2 = (V, E_2)$ are graphs on the same vertex set V, then the *distance* $d(G_1, G_2)$ between them is defined as follows:

$$d(G_1, G_2) = |E_1 \setminus E_2| + |E_2 \setminus E_1|,$$

i.e., $d(G_1, G_2)$ is the number of noncoinciding edges in G_1 and G_2.

In our survey we focus on the following variants of GRAPH CORRELATION CLUSTERING.

GCC. Given a graph $G = (V, E)$, find a graph $M^* \in \mathcal{M}(V)$ such that

$$d(G, M^*) = \min_{M \in \mathcal{M}(V)} d(G, M).$$

GCC$_k$. Given a graph $G = (V, E)$ and an integer k, $2 \leq k \leq |V|$, find a graph $M^* \in \mathcal{M}_k(V)$ such that

$$d(G, M^*) = \min_{M \in \mathcal{M}_k(V)} d(G, M).$$

GCC$_{1,k}$. Given a graph $G = (V, E)$ and an integer k, $2 \leq k \leq |V|$, find a graph $M^* \in \mathcal{M}_{1,k}(V)$ such that

$$d(G, M^*) = \min_{M \in \mathcal{M}_{1,k}(V)} d(G, M).$$

In these settings a clustering can be understood as a cluster graph M whose connected components correspond to clusters.

Bansal, Blum and Chawla [4] proposed the following problem equivalent to **GCC**.

CORRELATION CLUSTERING (**CC**). Given a complete graph $G = (V, E)$ with edges labelled $+1$ (similar) or -1 (different), find an *optimal clustering*, i. e., a partition of the vertex set of G into clusters minimizing *disagreements* (the number of -1 edges inside clusters plus the number of $+1$ edges between clusters).

The variants of **CC** in which the number of clusters is bounded were also studied [4, 33].

It is easy to see that **CC** is equivalent to **GCC**.

3 Bounds on Distances to Optimal Solutions

Let $\tau(G)$, $\tau_k(G)$, and $\tau_{1,k}(G)$ be the distances from a graph G to optimal solutions in **GCC**, **GCC$_k$**, and **GCC$_{1,k}$**, respectively, i. e.,

$$\tau(G) = \min_{M \in \mathcal{M}(V)} d(G, M), \quad \tau_k(G) = \min_{M \in \mathcal{M}_k(V)} d(G, M), \quad \tau_{1,k}(G) = \min_{M \in \mathcal{M}_{1,k}(V)} d(G, M).$$

It is obvious that for any n-vertex graph G and $k \geq 2$

$$\tau(G) \leq \tau_{1,k}(G) \leq \tau_k(G) \leq \frac{n(n-1)}{2},$$

where the latter inequality is a trivial bound on the number of edges in a graph.

An n-vertex graph G is called τ-critical, if it has a maximum value of $\tau(G)$ among all n-vertex graphs. The τ_k-critical and $\tau_{1,k}$-critical graphs are defined similarly.

The following upper bounds on quantities $\tau(G)$, $\tau_k(G)$, $\tau_{1,k}(G)$, and the corresponding critical graphs are known. In 1974, Fridman [16,17] proved that for any n-vertex graph G the following holds:

$$\tau(G) \leq \left\lfloor \frac{(n-1)^2}{4} \right\rfloor,$$

and up to isomorphism the only τ-critical n-vertex graphs are the complete bipartite graphs $K_n^1 = (X_1, Y_1; U_1)$ and $K_n^2 = (X_2, Y_2; U_2)$, where $0 \leq |X_1| - |Y_1| \leq 1$ and $|X_2| - |Y_2| = 2$.

Tomescu [34,35] independently obtained similar result for problem $\mathbf{GCC}_{1,k}$: for any n-vertex graph G

$$\tau_{1,k}(G) \leq \left\lfloor \frac{(n-1)^2}{4} \right\rfloor.$$

Besides that, for $k = 2$ the $\tau_{1,k}$-critical graphs are all the complete bipartite graphs and only they. For $k \geq 3$ the only $\tau_{1,k}$-critical graphs (up to isomorphism) are the graphs K_n^1 and K_n^2.

Later, Il'ev and Fridman [21] proved that for each $k \geq 2$ and any n-vertex graph G with $n \geq 5(k-1)$

$$\tau_k(G) \leq \left\lfloor \frac{(n-1)^2}{4} \right\rfloor.$$

For $n \geq 7$ the τ_2-critical graphs are the complete bipartite graphs and only they. For $k \geq 3$ and $n \geq 5k - 1$ the only τ_k-critical graphs are the graphs K_n^1 and K_n^2.

4 Computational Complexity of GRAPH CORRELATION CLUSTERING

To the best of our knowledge, NP-hardness of problem $\mathbf{GCC}$ was first proved by Křivánek and Morávek [26] in 1986. They considered the binary hierarchical-tree clustering problem. Given a set of elements $\Omega = \{\omega_1, \omega_2, \ldots, \omega_n\}$ and a symmetric integer $n \times n$ matrix $\Delta = (\delta_{ij})$ such that

$$\delta_{ij} \in \{1, 2\} \text{ if } i \neq j, \text{ and } \delta_{ij} = 0 \text{ if } i = j.$$

A *hierarchical tree* T over Ω is defined as a finite sequence of pairs $T = ((P_1, l_1), (P_2, l_2), \ldots, (P_q, l_q))$, where $P_1, P_2, \ldots, P_q$ are partitions of Ω; $l_1, l_2, \ldots, l_q$ are integers, $0 = l_1 < l_2 < \cdots < l_q$; P_k is a proper refinement of P_{k+1}; $P_1 = \{\{\omega_1\}, \{\omega_2\}, \ldots, \{\omega_n\}\}$ and $P_q = \{\Omega\}$. The integer q is called the *height* of T. The function $u\langle T \rangle : \Omega \times \Omega \to \mathbb{N}_0$ is defined as follows:

$$u\langle T \rangle(\omega_i, \omega_j) = \min\{l_k \mid \exists M \in P_k \text{ such that } \{\omega_i, \omega_j\} \subseteq M\}.$$

The problem is to find a hierarchical tree T such that

$$F(T) = \sum_{i,j} |\delta_{ij} - u\langle T \rangle(\omega_i, \omega_j)|$$

is minimal.

Křivánek and Morávek [26] proved that the binary hierarchical tree clustering problem is NP-hard even for hierarchical trees of height 3. Moreover, if T is a solution to the problem, then $l_2 = 1$ and $l_3 = 2$. Let us consider a graph $G = (V, E)$ and $V = \{v_1, \ldots, v_n\}$. We associate V and Ω, and set $\delta_{ij} = 1$ if $(v_i, v_j) \in E$ and $\delta_{ij} = 2$ otherwise. Then the partition P_2 uniquely determines a cluster graph M on the vertex set V and $F(T) \equiv d(G, M)$. Thus, **GCC** is strongly NP-hard.

Seventeen years later, Chen, Jiang and Lin [11] considered the closest phylogenetic root problems and also proved NP-hardness of **GCC**. Given a graph $G = (V, E)$, a *k-th root phylogeny* is a tree T with no internal degree 2 vertices such that the leaves of T are in one to one correspondence with V, and two vertices of G are adjacent if and only if the corresponding leaves in T are at distance at most k in T. In the closest phylogenetic k-th root problem it is required to edit a graph in a minimum way, so that it becomes a k-th root phylogeny. Chen, Jiang and Lin [11] proved that the problem is NP-complete for any $k \geq 2$. It is not too difficult to see that a graph has a phylogenetic 2-th root if and only if it is a cluster graph.

In the mid 2000s, several groups of authors were independently dealing with different versions of **GCC**. Bansal, Blum and Chawla [4] using a reduction from partition into triangles problem showed that **GCC** is NP-hard even if all clusters are of size at most 3.

Shamir, Sharan and Tsur [33] independently showed NP-hardness of problem **GCC** by a reduction from the 3-exact 3-cover problem. They also reduced the known NP-complete problem of 2-coloring of a 3-uniform hypergraph to problem **GCC$_2$** and as a result they showed that problem **GCC$_k$** is NP-hard for any fixed $k \geq 2$. In both cases Shamir, Sharan, and Tsur used rather complicated reduction. Later, Giotis and Guruswami [18] published a more simple proof of the same result by a polynomial reduction from the graph bisection problem.

At the same time, Ageev, Il'ev, Kononov and Talevnin [1] independently proved that problems **GCC$_2$** and **GCC$_{1,2}$** are NP-hard on cubic (i. e., 3-regular) graphs and deduced from this that all the above-mentioned variants of GRAPH CORRELATION CLUSTERING (including **GCC$_{1,k}$**) are NP-hard on general graphs.

In 2012, Komusiewicz and Uhlmann showed that **GCC** remains NP-hard on graphs with maximum degree six [25]. It is an open question whether **GCC** on graphs with maximum degree three, four or five is NP-hard or solvable in polynomial time. Bastos et al. [5] showed that **GCC** is NP-hard even restricted to graphs of diameter two.

GRAPH CORRELATION CLUSTERING can be solved in polynomial time for particular graph types. In 1971, Fridman [15] proved that if a graph G is triangle-free, i. e., G contains no complete 3-vertex graph as a subgraph, then problem **GCC** can be reduced to the maximum matching problem. Therefore, **GCC** on triangle-free graphs is solvable in polynomial time. This directly implies a polynomial-time solvability of **GCC** for bipartite graphs and graphs of maximum degree two, namely paths and circles. Mannaa [29] presented a polynomial-time dynamic programming algorithm to solve **GCC** for proper interval graphs. Deciding whether **GCC** is NP-hard in the case of interval graphs is still an open question. Xin [39] provided a linear time algorithm for **GCC** with bounded treewidth.

5 Approximation Algorithms for GRAPH CORRELATION CLUSTERING

In 2004, Bansal, Blum and Chawla [4] presented a simple polynomial time 3-approximation algorithm for $\mathbf{GCC}_{1,2}$. For each $v \in V$ their algorithm considers the following pair of clusters. The first cluster contains v and all neighbors of v in $G = (V, E)$. The second cluster contains all other vertices. The algorithm outputs the pair that minimizes the number of mismatched edges.

In 2006, Ageev, Il'ev, Kononov and Talevnin [1] proved the existence of a randomized $PTAS$ for problem $\mathbf{GCC}_{1,2}$ by reducing this problem to the graph bisection problem on dense instances, and Giotis and Guruswami [18] presented a randomized $PTAS$ for problem $\mathbf{GCC}_k$ (for any fixed $k \geq 2$). Navrotskaya, Il'ev and Talevnin [22, 30] considered a local search algorithm for problem $\mathbf{GCC}_{1,2}$. They showed that if for a given graph $G = (V, E)$ we have $|E| = o(|V|^2)$, then the worst-case ratio of the local search algorithm tends to 1 as $|V| \to \infty$, i. e., if the number of edges in a graph is subquadratic of the number of vertices, then the local search algorithm is asymptotically exact.

In 2008, Coleman, Saunderson and Wirth [12] pointed out that complexity of $PTAS$ from [18] makes it practically useless. They presented a 2-approximation algorithm for problem $\mathbf{GCC}_{1,2}$ applying local search to the feasible solution obtained by the 3-approximation algorithm from [4].

For problem $\mathbf{GCC}_2$ Il'ev, Il'eva and Navrotskaya [23] presented a polynomial time approximation algorithm with the worst-case approximation ratio of $3 - 6/|V|$.

As to problem **GCC**, in 2005, Charicar, Guruswami and Wirth [10] proved that the problem is APX-hard. They also constructed a 4-approximation algorithm for problem **GCC** by rounding a natural LP relaxation using the region growing technique. In 2008, Ailon, Charicar and Newman [2] improved the latter result. First, they presented an elegant iterative 3-approximation algorithm using an idea similar to that presented in [4]. At each iteration the algorithm randomly picks an arbitrary vertex $v \in V$ and forms a cluster of v and its neighbors. The above procedure is repeated for all vertices still ungrouped. Second, Ailon, Charicar and Newman proposed a randomized 2.5-approximation algorithm. At each iteration the algorithm randomly picks an arbitrary vertex and randomly decides for all other vertices whether to include it to the cluster or not based on the solution of the LP problem. Williamson and van Zuylen [41] derandomized both algorithms from [2] with matching approximation guarantees.

6   Weighted GRAPH CORRELATION CLUSTERING

Weighted versions of GRAPH CORRELATION CLUSTERING are the natural generalizations of these problems. Bansal, Blum and Chawla [4] considered two variants of weighted GRAPH CORRELATION CLUSTERING.

In the first variant the weight function $w : V \times V \to \mathbb{R}_+$ is defined and the *weighted distance* $wd(G_1, G_2)$ between graphs $G_1 = (V, E_1)$ and $G_2 = (V, E_2)$ equals to the total weight of noncoinciding edges in graphs G_1 and G_2:

$$wd(G_1, G_2) = \sum_{(u,v) \in E_1 \setminus E_2} w(u, v) + \sum_{(u,v) \in E_2 \setminus E_1} w(u, v).$$

WEIGHTED GRAPH CORRELATION CLUSTERING (**WGCC**). Given a graph $G = (V, E)$, find a graph $M^* \in \mathcal{M}(V)$ such that

$$wd(G, M^*) = \min_{M \in \mathcal{M}(V)} wd(G, M).$$

NP-hardness of this problem follows from NP-hardness of **GCC**. For the case of unbounded weights Bansal, Blum and Chawla [4] proved APX-hardness of **WGCC** by a simple reduction from the multiway cut problem.

Demaine, Emanuel, Fiat and Immorlica [13] and independently Charikar, Guruswamy and Wirth [10] used a linear-programming formulation of **WGCC** to design an $O(\log |V|)$-approximation algorithm. In both papers it was shown that this problem is equivalent to the minimum multicut problem, yielding another implicit $O(\log |V|)$-approximation as well as an APX-hardness result even if all weights belong to $\{-1, 0, 1\}$. Demaine, Emanuel, Fiat and Immorlica [13] presented a modification of this approach that yields an $O(r^3)$-approximation for $K_{r,r}$-minor-free graphs.

Voice, Polukarov and Jennings [37] proved that **WGCC** is NP-hard even for planar graphs. Actually, they presented the NP-hardness proof for a more general edge sum graph coalition structure generation problem. However, as noted in [3,24] their proof does carry over to **WGCC** on planar graphs. Recently for planar graphs, Klein, Mathieu and Zhou showed that **WGCC** reduces to the two-edge-connected augmentation problem and presented a $PTAS$ for both problems [24].

Bansal, Blum and Chawla [4] also proposed another generalization of **GCC**. This version of weighted **GCC** is a direct generalization of CORRELATION CLUSTERING (**CC**). An instance of this problem can be represented by a complete graph $G = (V, E)$ each edge $(i, j) \in E$ of which is assigned two fractional weights $w_{ij}^+ \geq 0$ and $w_{ij}^- \geq 0$. The cost $c(G, M)$ of a clustering M will now be the sum of w_{ij}^+ over all i, j in the different clusters plus the sum of w_{ij}^- over all i, j in the same cluster. In other words, $c(G, M)$ equals to the sum of w_{ij}^+ over the edges $(i, j) \in E$ between connected components of cluster graph M plus the sum of w_{ij}^- over the edges $(i, j) \in E$ inside components.

WEIGHTED CORRELATION CLUSTERING (**WCC**). Given a complete graph $G = (V, E)$, find a graph $M^* \in \mathcal{M}(V)$ such that

$$c(G, M^*) = \min_{M \in \mathcal{M}(V)} c(G, M).$$

Obviously, the unweighted case can be encoded as a 0/1 weighted case, therefore NP-hardness of **CC** implies NP-hardness of **WCC**.

Bansal, Blum and Chawla [4] and later, Ailon, Charikar and Newmann [2] analyzed problem **WCC** with *probability constraints*:

$$w_{ij}^+ + w_{ij}^- = 1 \text{ for all } i, j \in V.$$

Bansal, Blum and Chawla [4] showed, that any algorithm finding a good clustering in unweighted case also works well on the weighted problem. They proved the following result. Let A be an algorithm that produces a clustering for **CC** with approximation ratio ρ. Then we can construct an algorithm A' for **WCC** that achieves a $(2\rho + 1)$-approximation.

The best of known results for **WCC** was obtained by Ailon, Charikar and Newmann [2]. They proposed a 5/2-approximation algorithm for **WCC** with probability constraints and a 2-approximation algorithm for **WCC** with probability constraints and *triangle inequality*:

$$w_{ik}^- \le w_{ij}^- + w_{jk}^- \text{ for all } i, j, k \in V.$$

A special case of **WCC** with probability constraints is known as the CONSENSUS CLUSTERING problem. Recently, more attention has been given to this problem because of its application in bioinformatics, in particular, microarray data analysis [14]. In CONSENSUS CLUSTERING, we are given a list of k clusterings $M_1, ..., M_k \in \mathcal{M}(V)$ on the same ground set V. The goal is to find a clustering $M \in \mathcal{M}(V)$ that minimizes the number of pairwise disagreements with the given k clusterings, i. e., M minimizes $\sum_{i=1}^{k} d(M, M_i)$.

Consider an instance of the CONSENSUS CLUSTERING. For each pair of vertices $i, j \in V$ define an edge (i, j) with weight w_{ij}^+ equal to the average number of input partitions containing i and j in the same set and weight w_{ij}^- equal to the average number of input partitions containing i and j in different sets. Now we obtain an instance of **WCC** with probability constraints.

CONSENSUS CLUSTERING was studied extensively in the literature, its NP-hardness is well known [26,38]. In [14] it was observed that the problem is polynomially solvable for instances of at most 2 clusters. Bonizzoni, Vedova, Dondi and Jiang [8] showed that CONSENSUS CLUSTERING is APX-hard even on instances with 3 input clusterings. CONSENSUS CLUSTERING admits a 11/7-approximation algorithm for the general case [2].

7   Conclusion

By now, graph correlation clustering problems with minimization criteria are known for more than 50 years. In a series of rediscoveries and extensive studies, many independent research groups from a variety of fields obtained a lot of impressive and intriguing results.

This short survey lists some of the results related to graph correlation clustering problems with minimization criteria. However, this research area is very reach and there are, of course, many other closely related problems and papers.

Acknowledgements. The research of the first and the third authors (Sects. 1–4, 6, 7) was supported by the RSF grant 15-11-10009.

References

1. Ageev, A.A., Il'ev, V.P., Kononov, A.V., Talevnin, A.S.: Computational complexity of the graph approximation problem. Diskretnyi Analiz i Issledovanie Operatsii. Ser. 1 **13**(1), 3–11 (2006). (in Russian). English transl. in: J. Appl. Ind. Math. **1**(1), 1–8 (2007)
2. Ailon, N., Charikar, M., Newman, A.: Aggregating inconsistent information: ranking and clustering. J. ACM **55**(5), 1–27 (2008)
3. Bachrach, Y., Kohli, P., Kolmogorov, V., Zadimoghaddam, M.: Optimal coalition structure generation in cooperative graph games. In: 27th AAAI Conference on Artificial Intelligence, pp. 81–87. AAAI Press (2013)
4. Bansal, N., Blum, A., Chawla, S.: Correlation clustering. Mach. Learn. **56**, 89–113 (2004)
5. Bastos, L., Ochi, L.S., Protti, F., Subramanian, A., Martins, I.C., Pinheiro, R.G.S.: Efficient algorithms for cluster editing. J. Comb. Optim. **31**, 347–371 (2016)
6. Ben-Dor, A., Shamir, R., Yakhimi, Z.: Clustering gene expression patterns. J. Comput. Biol. **6**(3–4), 281–297 (1999)
7. van Bevern, R., Froese, V., Komusiewicz, C.: Parameterizing edge modification problems above lower bounds. CSR 2016. LNCS, vol. 9691, pp. 57–72. Springer, Heidelberg (2016). doi:10.1007/978-3-319-34171-2_5
8. Bonizzoni, P., Vedova, G.D., Dondi, R., Jiang, T.: On the approximation of correlation clustering and consensus clustering. J. Comput. Syst. Sci. **74**, 671–696 (2008)
9. Böcker, S., Baumbach, J.: Cluster editing. In: Bonizzoni, P., Brattka, V., Löwe, B. (eds.) CiE 2013. LNCS, vol. 7921, pp. 33–44. Springer, Heidelberg (2013)
10. Charikar, M., Guruswami, V., Wirth, A.: Clustering with qualitative information. J. Comput. Syst. Sci. **71**(3), 360–383 (2005)
11. Chen, Z.-Z., Jiang, T., Lin, G.: Computing phylogenetic roots with bounded degrees and errors. SIAM J. Comput. **32**(4), 864–879 (2003)
12. Coleman, T., Saunderson, J., Wirth, A.: A local-search 2-approximation for 2-correlation-clustering. In: Halperin, D., Mehlhorn, K. (eds.) ESA 2008. LNCS, vol. 5193, pp. 308–319. Springer, Heidelberg (2008)
13. Demaine, E., Emanuel, D., Fiat, A., Immorlica, V.: Correlation clustering in general weighted graphs. Theoret. Comput. Sci. **361**, 172–187 (2006)
14. Filkov, V., Skiena, S.: Integrating microarray data by Consensus clustering. In: 15th IEEE International Conference on Tools with Artificial Intelligence (ICTAI), pp. 418–425 (2003)
15. Fridman, G.Š.: A graph approximation problem. Upravlyaemye Sistemy **8**, 73–75 (1971). (in Russian)

16. Fridman, G.Š.: On an inequality in the graph approximation problem. Kibernetika **3**, 151 (1974). (in Russian). English transl. in: Cybernetics **10**, 554 (1974)
17. Fridman, G.Š.: Analysis of a classification problem on graphs. Metody Modelirovaniya i Obrabotka Informatsii, pp. 147–177. Nauka, Novosibirsk (1976). (in Russian)
18. Giotis, I., Guruswami, V.: Correlation clustering with a fixed number of clusters. Theory Comput. **2**(1), 249–266 (2006)
19. Hastie, T., Tibshirani, R., Friedman, J.: The Elements of Statistical Learning: Data Mining, Inference, and Prediction. Springer Series in Statistics. Springer, New York (2009)
20. Harary, F.: On the notion of balance of a signed graph. Michigan Math. J. **2**, 143–146 (1955)
21. Il'ev, V.P., Fridman G.Š,: On the problem of approximation by graphs with fixed number of components. Doklady AN SSSR **264**(3), 533–538 (1982). (in Russian). English transl. in: Soviet Math. Dokl. **25**(3), 666–670 (1982)
22. Il'ev, V.P., Navrotskaya, A.A., Talevnin, A.S.: Polynomial time approximation scheme for the graph approximation problem. Vestnik Omskogo Universiteta **4**, 24–27 (2007). (in Russian)
23. Il'ev, V.P., Il'eva, S.D., Navrotskaya, A.A.: Approximation algorithms for graph approximation problems. Diskretnyi Analiz i Issledovanie Operatsii **18**(1), 41–60 (2011). (in Russian). English transl. in: J. Appl. Ind. Math. **5**(4), 1–15 (2011)
24. Klein, P.N., Mathieu, C., Zhou, H.: Correlation clustering and two-edge-connected augmentation for planar graphs. In: 32nd Symposium on Theoretical Aspects of Computer Science (STACS 2015. Leibniz International Proceedings in Informatics (LIPIcs), vol. 30, pp. 554–567. Schloss Dagstuhl - Leibniz-Zentrum für Informatik GmbH, Dagstuhl Publishing, Saarbrücken/Wadern (2015)
25. Komusiewicz, C., Uhlmann, J.: Cluster editing with locally bounded modifications. Discrete Appl. Math. **160**(15), 2259–2270 (2012)
26. Křivánek, M., Morávek, J.: NP-hard problems in hierarchical-tree clustering. Acta informatica **23**, 311–323 (1986)
27. Kulis, B., Basu, S., Dhillon, I., Mooney, R.: Semi-supervised graph clustering: a kernel approach. Mach. Learn. **74**(1), 1–22 (2009)
28. Lyapunov, A.A.: The structure and evolution of the control systems in connection with the theory of classification. Problemy Kibernetiki **27**, 7–18 (1973). Nauka, Moscow (in Russian)
29. Mannaa, B.: Cluster editing problem for points on the real line: a polynomial time algorithm. Inform. Process. Lett. **110**, 961–965 (2010)
30. Navrotskaya, A.A., Il'ev, V.P., Talevnin, A.S.: Asymptotically exact algorithm for the problem of approximation of nondense graphs. In: III All-Russian Conference "Problemy Optimizatsii i Ekonomicheskiye Prilozheniya", p. 115. Izd. OmGTU, Omsk (2006). (in Russian)
31. Rahmann, S., Wittkop, T., Baumbach, J., Martin, M., Truß, A., Böcker, S.: Exact and heuristic algorithms for weighted cluster editing. In: 6th Annual International Conference on Computational Systems Bioinformatics (CSB 2007), vol. 6, pp. 391–401. Imperial College Press, London (2007)
32. Schaeffer, S.E.: Graph clustering. Comput. Sci. Rev. **1**(1), 27–64 (2007)
33. Shamir, R., Sharan, R., Tsur, D.: Cluster graph modification problems. Discrete Appl. Math. **144**(1–2), 173–182 (2004)
34. Tomescu, I.: Note sur une caractérisation des graphes done le degreé de deséquilibre est maximal. Mathematiques et Sciences Humaines **42**, 37–40 (1973)

35. Tomescu, I.: La reduction minimale d'un graphe à une reunion de cliques. Discrete Math. **10**(1–2), 173–179 (1974)
36. Tyshkevich, R.I.: Matroidal decompositions of a graph. Diskretnaya Matematika **1**(3), 129–139 (1989). (in Russian)
37. Voice, T., Polukarov, M., Jennings, N.R.: Coalition structure generation over graphs. J. Artif. Intell. Res. **45**, 165–196 (2012)
38. Wakabayashi, Y.: The complexity of computing defians of relations. Resenhas **3**(3), 323–349 (1998)
39. Xin, X.: An FPT algorithm for the correlation clustering problem. Key Eng. Mater. Adv. Mater. Comput. Sci. **474–476**, 924–927 (2011)
40. Zahn, C.T.: Approximating symmetric relations by equivalence relations. J. Soc. Ind. Appl. Math. **12**(4), 840–847 (1964)
41. van Zuylen, A., Williamson, D.P.: Deterministic pivoting algorithms for constrained ranking and clustering problems. Math. Oper. Res. **34**(3), 594–620 (2009)

Wardrop Equilibrium for Networks with the BPR Latency Function

Jaimie W. Lien[1], Vladimir V. Mazalov[2(✉)], Anna V. Melnik[3], and Jie Zheng[4]

[1] Department of Decision Sciences and Managerial Economics,
The Chinese University of Hong Kong, Shatin, Hong Kong, China
jaimie.academic@gmail.com
[2] Institute of Applied Mathematical Research, Karelian Research Center,
Russian Academy of Sciences, 11, Pushkinskaya Street, Petrozavodsk, Russia 185910
vmazalov@krc.karelia.ru
[3] Saint-Petersburg State University,
Universitetskii Prospekt 35, Saint-petrsburg, Russia 198504
a.melnik@spbu.ru
[4] Department of Economics, School of Economics and Management,
Tsinghua University, Beijing 100084, China
zhengjie@sem.tsinghua.edu.cn

Abstract. This paper considers a network comprised of parallel routes with the Bureau of Public Road (BPR) latency function and suggests an optimal distribution method for incoming traffic flow. The authors analytically derive a system of equations defining the optimal distribution of the incoming flow with minimum social costs, as well as a corresponding system of equations for the Wardrop equilibrium in this network. In particular, the Wardrop equilibrium is applied to the competition model with rational consumers who use the carriers with minimal cost, where cost is equal to the price for service plus the waiting time for the service. Finally, the social costs under the equilibrium and under the optimal distribution are compared. It is shown that the price of anarchy can be infinitely large in the model with strategic pricing.

Keywords: Traffic flow · BPR latency function · Wardrop equilibrium · Price of anarchy

1 Introduction

The road traffic distribution problem possesses a rich history. Starting from the 1950s, this field of research has employed models with different optimality principles and corresponding numerical methods. For instance, in 1952, Wardrop hypothesized that any transport system reaches an equilibrium state after some period of time, as well as formulated two principles of equilibrium traffic flow distribution [15]. According to the Wardrop principle, the trip time along all existing routes is the same for all road users and is smaller than the trip time of any road user in the case of route diversion. Moreover, the average trip time

© Springer International Publishing Switzerland 2016
Y. Kochetov et al. (Eds.): DOOR 2016, LNCS 9869, pp. 37–49, 2016.
DOI: 10.1007/978-3-319-44914-2_4

is minimized. Currently, the concept of Wardrop equilibrium represents a major tool in the theory of traffic flows [2,3,14,16].

Wardrop's ideas can be further developed by assuming that not only trip time, but also that the total costs of road users on all routes are the same and minimal. This coincides with recent trends in operations research: investigators incorporate the behavioral features guiding agents into the mathematical models [6]. The cost function may include service price, the average trip time, risks and other relevant factors. A series of publications adhered to this approach within the framework of queueing theory in the following way [5,7–9,11–13]. For a transport flow of intensity λ, the latency was defined as the average service time $1/(\mu - \lambda)$, i.e., the cited works expressed the expected sojourn time of a user in a queueing system $M/M/n$. In paper [10] the Wardrop principle was applied to networks of general topology and the BPR (Bureau of Public Road) latency functions [1].

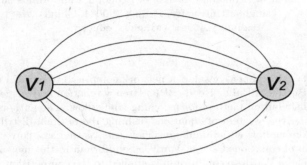

Fig. 1. Parallel routs

Analysis of the social costs in an equilibrium and in the case of centralized control forms an extremely relevant issue for modern transport and communication systems. The ratio of these costs is called the price of anarchy, also known as the coordination ratio. This ratio has been defined in the seminal work by Koutsoupias and Papadimitriou [8]. The price of anarchy was evaluated for different classes of latency functions. Rourghgarden and Tardos showed that the price of anarchy is exactly 4/3 in case of linear latency functions [13]. In the paper [7] the price of anarchy in the linear case was found for oblivious and selfish users. The oblivious users route their flow through the shortest path connecting their origin to their destination while selfish users minimize the personal costs. A network model with linear player-specific latency functions was investigated via the potential theory in [5]. For polynomial latency functions of maximum degree d, Roughgarden [12] showed that the price of anarchy is $\frac{(d+1)^{1+1/d}}{(d+1)^{1+1/d}-d}$. Dumrauf and Gairing [4] proved upper and lower bounds on the price of anarchy for polynomial latency functions with an upper bound d and a lower bound s on the degree of all monomials that appear in the polynomials. The price of anarchy was also studied for latency functions that involve a delay function of $M/M/1$ queues [9,12].

Our study focuses on a network of parallel routes with a certain incoming traffic flow. First, we utilize the Karush–Kuhn–Tucker theorem to demonstrate how the optimal distribution of the traffic flow by network routes can be found; here optimality is defined in the sense of social cost minimization. Based on this theorem, we introduce an evaluation method for the number of optimal distribution routes and derive a system of equations for calculating these distribution flows. Second, we obtain the Wardrop equilibrium traffic flow using the connection of this game to potential games and the Karush–Kuhn–Tucker theorem. Third, we use the Wardrop principle to find the equilibrium for the model with rational consumers who select the service carrier with minimal costs. Costs here are calculated as the price for the ticket plus waiting time for the service.

In the concluding section of the paper, we illustrate how the proposed optimal distribution approach of incoming traffic flows can be adopted to evaluate the price of anarchy in a network with the BPR latency function. We also consider the case where the routes of the network have different capacities. The evaluation of the price of anarchy includes the parameters of the BPR latency functions and can be useful in practice.

2 The Transportation Network Model

For the transportation network model, let us consider a network composed of n parallel routes (Fig. 1). Transportation networks analysis often utilizes the following empirical relationship known as the BPR (Bureau of Public Road) latency function:

$$f(x) = t\left(1 + \alpha \left(\frac{x}{c}\right)^{\beta}\right),$$

where t indicates the trip time on an unoccupied route, c specifies the capacity of a route, and the constants α, β capture route-specific features which may affect the impact of flow to capacity ratio on travel time. α, β are evaluated based on the statistical data of a route. Generally, β takes values from 1 to 4, (see [1]).

Suppose that the latency function in route i has the form

$$f_i(x) = t_i\left(1 + \alpha_i \left(\frac{x}{c_i}\right)^{\beta_i}\right), \quad i = 1, ..., n.$$

Without loss of generality, we may assume that all routes are renumbered so that

$$t_1 \leq t_2 \leq ... \leq t_n.$$

Assume that the network of n parallel routes receives an incoming flow of volume X. The incoming flow is decomposed into n subflows running through each corresponding route. Denote by $x_i, i = 1, ..., n$ the values of the subflows; actually, $x_i \geq 0$ and

$$\sum_{i=1}^{n} x_i = X. \tag{1}$$

If some route has no flow, then $x_i = 0$. Denote the profile of subflows as $x = (x_1, ..., x_n)$.

In the analysis, we are concerned with two problems. The first problem lies in optimal flow distribution within the network in order to minimize the social costs

$$SC(x) = \sum_{i=1}^{n} x_i f_i(x) = \sum_{i=1}^{n} x_i t_i \left(1 + \alpha_i \left(\frac{x_i}{c_i}\right)^{\beta_i}\right)$$

under the condition (1). Social costs minimization calls for applying external control actions, which often incurs appreciable additional costs. Therefore, it is a useful starting point to find the social costs when each road user acts independently by minimizing his or her individual costs. In this case, we naturally arrive at the game-theoretic statement of the problem and it is necessary to evaluate an appropriate equilibrium. Throughout the paper, the concept of a Wardrop equilibrium is used. The above transportation game with the described latency functions represents a potential game [11]. In such games, equilibrium evaluation is reduced to potential minimization. In the current notation, the potential acquires the following form:

$$P(x) = \sum_{i=1}^{n} \int_{0}^{x_i} t_i \left(1 + \alpha_i \left(\frac{u}{c_i}\right)^{\beta_i}\right) du.$$

At the outset, we show a solution procedure for the first problem.

3   Cooperative Solution

Consider the optimization problem

$$SC(x) = \sum_{i=1}^{n} x_i t_i \left(1 + \alpha_i \left(\frac{x_i}{c_i}\right)^{\beta_i}\right) \rightarrow \min, \tag{2}$$

subject to the constraints

$$\sum_{i=1}^{n} x_i = X,$$

$$x_i \geq 0, \forall i = 1, ..., n.$$

Its solution corresponds to the cooperative behavior of the carriers.

Lemma 1. The subflows profile x^* is the solution of the problem (2) if there exists a nonnegative number λ (the Lagrange multiplier) such that

$$t_i \left(1 + \alpha_i(\beta_i + 1)\left(\frac{x_i}{c_i}\right)^{\beta_i}\right) \begin{cases} = \lambda, & if \quad x_i > 0, \\ \geq \lambda, & if \quad x_i = 0. \end{cases}$$

$$for \ \forall \, i \in \{1, n\}.$$

Proof. The idea is to involve the conditions of the Karush–Kuhn–Tucker theorem. Due to the convexity of the objective function (2) and the admissible solution domain, the Karush–Kuhn–Tucker conditions play the role of necessary and sufficient conditions simultaneously. We construct the Lagrange function for the problem (2):

$$L(x, \lambda) = \sum_{i=1}^{n} t_i \left(1 + \alpha_i \left(\frac{x_i}{c_i} \right)^{\beta_i} \right) x_i$$

$$+ \lambda (X - \sum_{i=1}^{n} x_i) + \sum_{i=1}^{n} \lambda_i (-x_i).$$

By applying the first-order necessary optimality conditions with respect to x_i, we obtain the equations

$$t_i \left(1 + \alpha_i \left(\frac{x_i}{c_i} \right)^{\beta_i} \right) + \alpha_i \beta_i t_i \left(\frac{x_i}{c_i} \right)^{\beta_i} - \lambda_i = \lambda,$$

$$i = 1, ..., n.$$

The complementary slackness condition yields the equalities

$$\lambda_i x_i = 0, \quad i = 1, ..., n.$$

The last equality takes place if at least, one of the multipliers is zero. Therefore, if for some i we have $x_i > 0$, then $\lambda_i = 0$ and subsequently,

$$t_i \left(1 + \alpha_i \left(\frac{x_i}{c_i} \right)^{\beta_i} \right) + \alpha_i \beta_i t_i \left(\frac{x_i}{c_i} \right)^{\beta_i} = \lambda.$$

In the case of $x_i = 0$, the inequality $\lambda_i \geq 0$ is immediate and

$$\lambda = t_i - \lambda_i.$$

This concludes the proof of Lemma 1.

Recall that the routes are renumbered so that

$$t_1 \leq t_2 \leq ... \leq t_n. \tag{3}$$

We introduce the notation

$$g_i(x) = t_i \left(1 + \alpha_i (\beta_i + 1) \left(\frac{x}{c_i} \right)^{\beta_i} \right), i = 1, ..., n.$$

The next result follows directly from Lemma 1. The optimal flow is distributed among the first k routes if for some value λ such that

$$t_1 \leq t_2 \leq ... \leq t_k < \lambda \leq t_{k+1} \leq ... \leq t_n \tag{4}$$

we have

$$g_1(x_1) = g_2(x_2) = \ldots = g_k(x_k)$$
$$= \lambda \le g_{k+1}(0) = t_{k+1}. \tag{5}$$

In fact, the conditions (5) determine the optimal distribution flows $x_i, i = 1, \ldots, k$ $(k = 1, \ldots, n-1)$.

The parameter λ represents a function of the flow volume X. The quantity $\lambda(X)$ is a continuous nondecreasing function of X. The number of optimal distribution routes increases from k to $k+1$ as the function $\lambda(X)$ crosses the point t_{k+1}.

Interestingly, the functions $g_i(x)$, $i = 1, \ldots, n$ increase monotonically on the interval $[0; +\infty)$. Hence, there exist the inverse functions $g^{-1}(y)$ with the growth property on $[0; +\infty)$. Denote x_{ij} as the solution to the equation

$$g_i(x) = t_j, \quad j = i+1, \ldots, n.$$

By virtue of the monotonicity of the functions $g_i(x)$ and the condition (2), we arrive at the inequality

$$x_{i,i+1} \le x_{i,i+2} \le \ldots \le x_{in}, \forall i. \tag{6}$$

Set

$$V_k = \sum_{i=1}^{k} x_{i,k+1}, \quad k = 1, \ldots, n-1, V_0 = 0.$$

It appears from (6) that

$$V_1 \le V_2 \le \ldots \le V_{n-1}.$$

Note that $V_1 = x_{12}$ meets the equation

$$g_1(x) = t_2.$$

If the incoming flow is such that

$$X \le V_1,$$

then $g_1(X) \le t_2$ and the whole flow runs through route 1. In the case of $g_1(X) > t_2$, the flow gets decomposed into two subflows and some part of the flow corresponds to route 2.

Let us demonstrate that, as the flow volume X crosses the value V_k, the number of optimal distribution routes varies from k to $k+1$.

Suppose that the optimal flow has been distributed among k routes. The optimal flow satisfies the conditions (4)–(5). The expression (5) and the monotonicity of the functions $g_i^{-1}(y)$ lead to

$$x_i = g_i^{-1}(\lambda) \le g_i^{-1}(t_{k+1}) = x_{i,k+1}, i = 1, \ldots, k.$$

Consequently,

$$X = \sum_{i=1}^{k} x_i = \sum_{i=1}^{k} g_i^{-1}(\lambda) \le \sum_{i=1}^{k} x_{i,k+1},$$

which leads to the inequality $X \leq V_k$. On the other hand, by assuming that the optimal flow runs through route $k + 1$, we obtain the following result. The conditions (4)–(5)

$$g_1(x_1) = g_2(x_2) = ... = g_k(x_{k+1}) = \lambda > t_{k+1}$$

dictate that

$$X = \sum_{i=1}^{k+1} x_i = \sum_{i=1}^{k+1} g_i^{-1}(\lambda) > \sum_{i=1}^{k+1} x_{i,k+1} > V_k.$$

Therefore, an important result is stated in the following Theorem.

Theorem 1. For the optimal flow to be distributed among the first k routes, it is necessary and sufficient to have

$$V_{k-1} < X \leq V_k = \sum_{i=1}^{k} x_{i,k+1}, \quad k = 1, 2, ..., n.$$

Moreover, the optimal distribution x_{opt} represents the solution to the system of equations

$$\begin{aligned} x_1 + x_2 + ... + x_k &= X, \\ g_1(x_1) = g_2(x_2) = ... &= g_k(x_k). \end{aligned} \tag{7}$$

4   Wardrop Equilibrium

Now, we direct our attention to considering the competitive equilibrium. As mentioned earlier, the equilibrium evaluation problem is reduced to the optimization problem of the function

$$P(x) = \sum_{i=1}^{n} \int_0^{x_i} t_i \left(1 + \alpha_i \left(\frac{u}{c_i}\right)^{\beta_i}\right) du,$$

subject to the constraints

$$\sum_{i=1}^{n} x_i = X,$$

$$x_i \geq 0, \forall i = 1, ..., n.$$

By analogy, we employ the Karush–Kuhn–Tucker theorem. Construct the Lagrange function

$$L(x, \lambda) = P(x) + \lambda \left(X - \sum_{i=1}^{n} x_i\right) + \sum_{i=1}^{n} \lambda_i(-x_i)$$

and apply the first-order necessary optimality conditions with respect to x_i to obtain

$$t_i \left(1 + \alpha_i \left(\frac{x_i}{c_i}\right)^{\beta_i}\right) - \lambda_i = \lambda, \quad i = 1, ..., n.$$

By repeating the same line of reasoning as above, except that the functions $g_i(x), i = 1, ..., n$ are replaced by the latency functions

$$f_i(x) = t_i \left(1 + \alpha_i \left(\frac{x}{c_i} \right)^{\beta_i} \right), i = 1, ..., n,$$

we naturally establish the following result.

Theorem 2. For the optimal flow to be distributed among the first k routes, it is necessary and sufficient to have

$$V'_{k-1} < X \le V'_k, \quad k = 1, 2, ..., n.$$

Here $V'_k = \sum_{i=1}^{k} x'_{i,k+1}$, $k = 1, 2, ..., n$ and x'_{ij} satisfies the system of equations

$$f_i(x) = t_j, \quad j = i + 1, ..., n.$$

Moreover, the optimal distribution x_{eq} represents the solution to the system of equations

$$\begin{cases} x_1 + x_2 + ... + x_k = X, \\ f_1(x_1) = f_2(x_2) = ... = f_k(x_k). \end{cases} \tag{8}$$

5   Competition Model with Rational Consumers

Let us consider the competition model with rational consumers who use the carriers with minimal cost, where cost is equal to the price for service plus the waiting time for the service.

Consider a network composed of two parallel routes and imagine two carriers serving the two parallel routes with prices p_1 and p_2, respectively. Depending on the incoming flow intensity X, the flow runs through (faster) route 1 or is distributed between the both routes. If $t_1 < t_2$ (route 2 has a higher latency than route 1), passengers do not choose carrier 2 even under zero price in the case of low incoming flow; therefore, player 2 is eliminated from competition.

That is, under the condition

$$p_1 + t_1 \left(1 + \alpha_1 \left(\frac{X}{c_1} \right)^{\beta_1} \right) \le t_2,$$

the incoming flow runs through route 1 only. The optimal price of player 1 is given by

$$p_1^* = t_2 - t_1 \left(1 + \alpha_1 \left(\frac{X}{c_1} \right)^{\beta_1} \right). \tag{9}$$

For a sufficiently large flow, X runs through both routes and player 2 also assigns some price for its service. A two-player game thus arises on the described transportation network [10]. Two players establish prices p_1 and p_2 for their services. The incoming passenger flow of intensity X is decomposed into two subflows of intensities x_1 and x_2 so that $x_1 + x_2 = X$ and

$$p_1 + t_1 \left(1 + \alpha_1 \left(\frac{x_1}{c_1} \right)^{\beta_1} \right) = p_2 + t_2 \left(1 + \alpha_2 \left(\frac{x_2}{c_2} \right)^{\beta_2} \right). \tag{10}$$

The payoffs of the players have the form

$$H_1 = p_1 x_1, \quad H_2 = p_2 x_2.$$

Solving the pricing game, we obtain the equilibrium prices

$$p_1^* = x_1^* \left(\frac{\alpha_1 t_1 \beta_1}{c_1} \left(\frac{x_1^*}{c_1} \right)^{\beta_1 - 1} + \frac{\alpha_2 t_2 \beta_2}{c_2} \left(\frac{x_2^*}{c_2} \right)^{\beta_2 - 1} \right), \tag{11}$$

$$p_2^* = x_2^* \left(\frac{\alpha_1 t_1 \beta_1}{c_1} \left(\frac{x_1^*}{c_1} \right)^{\beta_1 - 1} + \frac{\alpha_2 t_2 \beta_2}{c_2} \left(\frac{x_2^*}{c_2} \right)^{\beta_2 - 1} \right), \tag{12}$$

where x_1^*, x_2^* satisfy to (10).

Theorem 3. Under the condition

$$X \le c_1 \left(\frac{t_2 - t_1}{\alpha_1 t_1 (1 + \beta_1)} \right)^{\frac{1}{\beta_1}}$$

all traffic runs through route 1 and the optimal price of player 1 equals (9). Otherwise, the incoming flow is distributed between the both routes and the equilibrium prices have the form (11)–(12).

Remark. In the linear case, for sufficiently large flow, the pricing game has the equilibrium

$$p_1^* = x_1^* \left(\frac{\alpha_1 t_1}{c_1} + \frac{\alpha_2 t_2}{c_2} \right), \quad p_2^* = x_2^* \left(\frac{\alpha_1 t_1}{c_1} + \frac{\alpha_2 t_2}{c_2} \right). \tag{13}$$

It yields

$$p_1^* = \frac{1}{3} \left(t_1 \left(\frac{\alpha_1}{c_1} X - 1 \right) + t_2 \left(1 + 2 \frac{\alpha_2}{c_2} X \right) \right), \tag{14}$$

$$p_2^* = \frac{1}{3} \left(t_2 \left(\frac{\alpha_2}{c_2} X - 1 \right) + t_1 \left(1 + 2 \frac{\alpha_1}{c_1} X \right) \right). \tag{15}$$

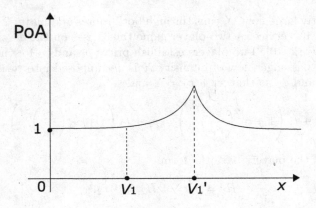

·**Fig. 2.** The price of anarchy

6   The Price of Anarchy

The price of anarchy (PoA) is the ratio of social costs under equilibrium to social costs under cooperation. We have considered here two types of equilibria. Let us compare the prices of anarchy for both considered cases: for the Wardrop equilibrium without pricing x_{eq} and for the Wardrop equilibrium with strategic pricing x'_{eq}. Denote the cooperative distribution of routing as x_{opt}.

The BPR latency function is a particular case of polynomial latency function. For polynomial latency functions of maximum degree β, Roughgarden [12] showed that the price of anarchy is $\frac{(\beta+1)^{1+1/\beta}}{(\beta+1)^{1+1/\beta}-\beta}$. In particular, for the linear latency function the PoA is equal $4/3$.

Remark. In fact, the price of anarchy depends on the parameters of the network. For simplicity, consider here the case of two routes and a linear latency function. The price of anarchy has the form shown in Fig. 2. The maximum of this function is achieved at $X = V'_1 = (t_2 - t_1)c_1/(\alpha_1 t_1)$. For this value the optimal solution x_{opt} prescribes to use both routes but under the equilibrium x_{eq} the traffic flow runs only through route 1. According to Theorem 2

$$SC(x_{eq}) = V'_1 t_2,$$

and by Theorem 1 $x_{opt} = (x_1^{opt}, x_2^{opt})$ satisfies to conditions

$$x_1^{opt} + x_2^{opt} = V'_1, \quad t_1(1 + 2\frac{\alpha_1}{c_1}x_1^{opt}) = t_2(1 + 2\frac{\alpha_2}{c_2}x_2^{opt}).$$

This yields

$$x_1^{opt} = \frac{t_2 - t_1 + 2k_2 V'_1}{2(k_1 + k_2)},$$

where $k_1 = \alpha_1 t_1 / c_1$, $k_2 = \alpha_2 t_2 / c_2$. The social costs are

$$SC(x_{opt}) = x_1^{opt} t_1 \left(1 + \alpha_1 \frac{x_1^{opt}}{c_1}\right) + x_2^{opt} t_2 \left(1 + \alpha_2 \frac{x_2^{opt}}{c_2}\right) =$$

$$(t_2 - t_1) \frac{4k_2 t_2 + k_1 t_1 + 3k_1 t_2}{4k_1(k_1 + k_2)}.$$

Finally, we obtain

$$PoA = \frac{SC(x_{eq})}{SC(x_{opt})} = 1 + \frac{k_1(t_2 - t_1)}{4k_2 t_2 + k_1 t_1 + 3k_1 t_2}.$$

For any parameters PoA is bounded by the value 4/3.

A key question is whether the price of anarchy in the model with strategic pricing is bounded. We show below that in fact it can be infinitely large.

Assume that the incoming traffic flow is sufficiently large (according to Theorem 1 it is larger than $c_1(t_2 - t_1)/(2\alpha_1 t_1)$) and the flow is distributed between both routes.

First, find $SC(x_{eq'})$. According (14), (15) for large X

$$p_1^* \approx \frac{1}{3}(k_1 + 2k_2)X, \quad p_2^* \approx \frac{1}{3}(k_2 + 2k_1)X.$$

Consequently,

$$x_1^{eq} \approx \frac{1}{3} \frac{k_1 + 2k_2}{k_1 + k_2} X, \quad x_2^{eq} \approx \frac{1}{3} \frac{k_2 + 2k_1}{k_1 + k_2} X,$$

and social costs in the equilibrium x'_{eq} are

$$SC(x'_{eq}) = x_1^{eq} t_1 \left(1 + \alpha_1 \frac{x_1^{eq}}{c_1}\right) + x_2^{eq} t_2 \left(1 + \alpha_2 \frac{x_2^{eq}}{c_2}\right) \approx$$

$$\left(k_1(k_1 + 2k_2)^2 + k_2(k_2 + 2k_1)^2\right) \frac{X^2}{9(k_1 + k_2)^2}.$$

Now we find $SC(x_{opt})$. From Theorem 1, the optimal distribution (x_1^{opt}, x_2^{opt}) satisfies the condition $x_1^{opt} + x_2^{opt} = X$ and

$$t_1 + 2k_1 x_1^{opt} = t_2 + 2k_2 x_2^{opt}.$$

For large X

$$x_1^{opt} \approx \frac{k_2}{k_1 + k_2} X, \quad x_2^{opt} \approx \frac{k_1}{k_1 + k_2} X,$$

Consequently,

$$SC(x_{opt}) \approx x_1^{opt} t_1 \left(1 + \alpha_1 \frac{x_1^{opt}}{c_1}\right) + x_2^{opt} t_2 \left(1 + \alpha_2 \frac{x_2^{opt}}{c_2}\right) \approx$$

$$\left(k_1 k_2^2 + k_2 k_1^2\right) \frac{X^2}{(k_1 + k_2)^2}.$$

Finally, for large X we obtain

$$PoA = \frac{SC(x_{eq'})}{SC(x_{opt})} \approx \frac{1}{9} \frac{k_1(k_1 + 2k_2)^2 + k_2(k_2 + 2k_1)^2}{k_1 k_2^2 + k_2 k_1^2}.$$

It yields

$$PoA \approx \frac{7}{9} + \frac{1}{9}\left(\frac{k_1}{k_2} + \frac{k_2}{k_1}\right).$$

For large k_1/k_2 or k_2/k_1 this ratio can be infinitely large.

7　Conclusion

The present paper has proposed an optimal distribution method for the incoming traffic flow of a network composed of parallel routes with the BPR latency function. We have analytically derived a system of equations defining the optimal flows. This allows comparison of the subflows in the cases of social costs minimization and independent decision-making of route suppliers based on their individual costs. Costs of consumers are calculated as a price for service plus waiting time for the service. It is shown that price of anarchy in this case can be infinitely large. The high price of anarchy suggests a potential role for transportation policy in this framework.

Acknowledgements. This research is supported by the Russian Fund for Basic Research (projects 16-51-55006, 16-01-00183), the National Natural Science Foundation of China (project 61661136002) and Tsinghua University Initiative Scientific Research Grant (project 20151080397).

References

1. U.S. Bureau of Public Roads. Traffic Assignment Manual. U.S. Department of Commerce, Washington, D.C (1964)
2. Altman, E., Boulogne, T., El-Azouzi, R., Jimànez, T., Wynter, L.: A survey onnetworking games in telecommunications. Comput. Oper. Res. **33**, 286–311 (2006)
3. Correa, J.R., Stier-Moses, N.E.: Wardrop Equilibria. In: Wiley Encyclopedia of Operations Research and Management Science, pp. 1–12 (2011)
4. Dumrauf, D., Gairing, M.: Price of anarchy for polynomial wardrop games. In: Spirakis, P.G., Mavronicolas, M., Kontogiannis, S.C. (eds.) WINE 2006. LNCS, vol. 4286, pp. 319–330. Springer, Heidelberg (2006)

5. Gairing, M., Monien, B., Tiemann, K.: Routing (Un-) splittable flow in games with player-specific linear latency functions. In: Bugliesi, M., Preneel, B., Sassone, V., Wegener, I. (eds.) ICALP 2006. LNCS, vol. 4051, pp. 501–512. Springer, Heidelberg (2006)

6. Hammalainen, R., Luoma, J., Saarinen, E.: On the importance of behavioral operationalresearch: the case of understanding and communicating about dynamicsystems. Eur. J. Oper. Res. **228**(3), 623–634 (2013)

7. Karakostas, G., Kim, T., Viglas, A., Xia, H.: On the degradation of performance for traffic networks with oblivious users. Transp. Res. Part B Methodol. **45**, 364–371 (2011)

8. Koutsoupias, E., Papadimitriou, C.: Worst-case equilibria. In: Proceedings of the 16th Annual Symposium on Theoretical Aspects of Computer Science, pp. 404–413 (1999)

9. Mazalov, V., Monien, B., Schoppmann, F., Tiemann, K.: Wardrop equilibria and price of stability for bottleneck games with splittable traffic. In: Spirakis, P.G., Mavronicolas, M., Kontogiannis, S.C. (eds.) WINE 2006. LNCS, vol. 4286, pp. 331–342. Springer, Heidelberg (2006)

10. Mazalov, V., Melnik, A.: Equilibrium prices and flows in the passenger traffic problem. Int. Game Theory Rev. **18**(1), 1–19 (2016)

11. Monderer, D., Shapley, L.S.: Potential Games. Games Econ. Behav. **14**, 124–143 (1996)

12. Roughgarden, T.: The price of anarchy is independent of the network topology. J. Comput. Syst. Sci. **67**, 341–364 (2003)

13. Roughgarden, T., Tardos, E.: How bad is selfish routing? J. ACM **49**(2), 236–259 (2002)

14. Sheffy, Y.: Urban Transportation Networks: Equilibrium Analysis with Mathematical Programming Methods. Prentice-Hall, Englewood Cliffs (1985)

15. Wardrop, J.G.: Some theoretical aspects of road traffic research. ICE Proc. Eng. Divisions **1**, 325–362 (1952)

16. Zakharov, V.V., Krylatov, A.Y.: Competitive routing of traffic flows by navigation providers. Autom. Remote Control **7**, 179–189 (2016)

A Review on Network Robustness from an Information Theory Perspective

Tiago Schieber[1,2(✉)], Martín Ravetti[1], and Panos M. Pardalos[3,4]

[1] Departamento de Engenharia de Produção,
Universidade Federal de Minas Gerais, Belo Horizonte, MG, Brazil
tischieber@gmail.com
[2] Departamento de Engenharia de Produção,
Pontifícia Universidade Católica de Minas Gerais, Belo Horizonte, MG, Brazil
[3] Center for Applied Optimization, Industrial and Systems Engineering,
University of Florida, Gainesville, Florida, USA
[4] Laboratory of Algorithms and Technologies for Network Analysis,
National Research University Higher School of Economics, Nizhny Novgorod, Russia

Abstract. The understanding of how a networked system behaves and keeps its topological features when facing element failures is essential in several applications ranging from biological to social networks. In this context, one of the most discussed and important topics is the ability to distinguish similarities between networks. A probabilistic approach already showed useful in graph comparisons when representing the network structure as a set of probability distributions, and, together with the Jensen-Shannon divergence, allows to quantify dissimilarities between graphs. The goal of this article is to compare these methodologies for the analysis of network comparisons and robustness.

1 Introduction

Quantification of dissimilarities between graphs has been a central subject in graph theory for many decades. With the complex networks field, we witness a burst of applications on real systems where the measure of graph or subgraph similarities have played a major role. Several methods for this quantification have become increasingly addressed, where most approaches are based on invariant measurements under graph isomorphism [1–6]. Although there exists in the literature a quasi-polynomial time algorithm to solve graph isomorphism [7], still, an efficient way to decide if two structures are isomorphic continues an open problem, as the search for efficient pseudo-distances between networks.

Representing a network as a set of stochastic measures (probability distributions associated with a given set of measurements) showed useful to characterize network evolution, robustness and efficiently treat the graph isomorphism problem [6, 8–10].

These characteristics are useful to define a pseudo-metric between networks via the Jensen-Shannon divergence, an Information Theory quantifier that

© Springer International Publishing Switzerland 2016
Y. Kochetov et al. (Eds.): DOOR 2016, LNCS 9869, pp. 50–60, 2016.
DOI: 10.1007/978-3-319-44914-2_5

already showed very effective in measuring small network topology changes [6,9–11]. When comparing n probability distributions, it is given by the Shannon entropy of the average minus the average of the Shannon entropies and, it was proven to be a bounded *square of a metric* between probability distributions [28], here defined for the discrete case:

$$JS(\mathbf{P}_1, \mathbf{P}_2, \ldots, \mathbf{P}_n) = H\left(\frac{\sum_{i=1}^{n} \mathbf{P}_i}{n}\right) - \frac{\sum_{i=1}^{n} H(\mathbf{P}_i)}{n} \tag{1}$$

being $H(\mathbf{P}) = -\sum_i p_i \log p_i$ the Shannon entropy of $\mathbf{P}$.

The JS divergence (Eq. 1) possesses a lower bound equals zero and an upper bound equals $\log n$. The zero value means that all probabilities are equal to the same distribution $\mathbf{P}_1 = \mathbf{P}_2 = \cdots = \mathbf{P}_n = \mathbf{P}$. A $\log n$ value gives the biggest uncertainty when comparing $\mathbf{P}_1, \mathbf{P}_2, \ldots, \mathbf{P}_n$ since $\log n$ is the biggest entropy value achieved only by the uniform distribution.

The metric property of the square root of the JS divergence, together with stochastic measures on networks, allows to define two pseudo-metrics between networks: one given only by global properties (D^g) representing the network as a single probability distribution and, the other, more precise but more computationally expensive (D), considering local network characteristics by representing the network as a set of probability distributions.

The analysis of properties of complex networks, therefore, relies on using stochastic measurements capable of expressing the most relevant topological features. Depending on the network and application, a specific set of stochastic measures could be chosen. This article presents a survey of such measurements. It includes classical complex network measurements, applications on network evolution, comparisons and robustness.

2 Methodology

A network G is a pair $(V, \mathbb{E})$, where V is a set of nodes (or vertices), and $\mathbb{E}$ is a set of ordered pairs of distinct nodes, which we call edges. A *weighted network* associates a weight (ω_e) to every edge $e \in \mathbb{E}$, characterizing not only the connections among vertices but also the strength of these connections.

Exists, in the literature, several measurements representing network connectivity. In particular, most real networks present small average distance between elements and high-density communities.

The in-degree (out-degree) of a node, k^{in} (k^{out}), is the number of incoming (outgoing) edges. The in-weight (out-weight) of a node, ω^{in} (ω^{out}), is the sum of all incoming (outgoing) edge weights. Following [29] it is possible to define a degree centrality measure considering both degree and weight by relating them to a tuning parameter $\alpha \in [0, 1]$ as:

$$\kappa_\alpha^{in}(v) = (k_v^{in})^{1-\alpha}(\omega_v^{in})^\alpha \quad \text{and} \quad \kappa_\alpha^{out}(v) = (k_v^{out})^{1-\alpha}(\omega_v^{out})^\alpha. \tag{2}$$

If $\alpha = 0$, the weights are forgotten to obtain the node degree. As α increases the number of connections loses in importance and, when α reaches 1, the centrality is given by the total vertex weight.

For any two vertices i, $j \in V(G)$, the distance $d(i,j)$ is the length of the shortest path between i and j, if there is no path between them, $d(i,j) = \infty$. In a weighted network, there are several distances measures in literature because the strength of these connections sometimes implies in small distances between the nodes. In an e-mail network, a bigger edge weight value may represent a frequent communication and, therefore, a small distance between them. Here, we consider the same approach used in [12] transforming weights into costs by inverting them and computing shortest paths between pairs of nodes. Readers should refer to [13] for a deeper discussion on the topic.

The network diameter (average path length) is the maximum (average) distance between all pairs of connected nodes.

The clustering coefficient (C), also known as transitivity, characterizes triangles in the network. It is the fraction of the number of triangles and the number of connected triples. Thus, a complete graph possesses $C = 1$ and, a tree graph, $C = 0$. Analogously, the vertex clustering coefficient, C_v, is given by:

$$C_v = \frac{3n_\Delta(v)}{n_e(v)},$$

being, $n_\Delta(v)$ the number of triangles involving node i and $n_3(v)$ the number of connected triples having v as a central vertex. A node clustering coefficient value equals 1 means that there is a connection between all pairs of its first neighbors, and a zero value represents the lack edges between them.

The closeness centrality measure of a node is the sum of the inverse of all pairs of distances from it:

$$c_v = \sum_{j,\, j \neq v} \frac{1}{d(v,j)}.$$

A high closeness centrality value means that the node possesses a lower total distance from all other nodes.

Betweenness centrality quantifies node importance in terms of interactions via the shortest paths among all other nodes:

$$B_v = \sum_{i \neq j \in V(G)} \frac{n(i,j,v)}{2n(i,j)},$$

being, $n(i,j)$ the number of shortest paths connecting i and j and $n(i,j,x)$ the number of shortest paths connecting i and j passing through x.

See Table 1 for space and time computational complexity of the above mentioned measures.

Given two networks G_1 and G_2 and two stochastic measurements $\mathbf{P}_{G_1}$ and $\mathbf{P}_{G_2}$, the global pseudo-metric

$$D_{\mathbf{P}}^g(G_1, G_2) = \sqrt{\frac{JS(\mathbf{P}_{G_1}, \mathbf{P}_{G_2})}{\log 2}}. \tag{3}$$

measures how far away two networks are via probability distributions.

Table 1. Space/time computational complexity in a network with N nodes and E edges.

	Space	Time
Degree	$O(N)$	$O(N^2)$
All pairs of distances (unweighted)	$O(N^2)$	$O(N^2 + NE)$
Local clustering coefficient	$O(N)$	$O(N^3)$
Closeness	$O(N)$	$O(NE)$
Betweenness	$O(N)$	$O(NE)$

The degree distribution $\mathbf{P}_{deg}(k)$ is the fraction of nodes with degree k. The network distance distribution, $\mathbf{P}_{\delta}(d)$, gives the fraction of pairs of nodes at distance d. Analogously, $\mathbf{P}_{B_v}$, $\mathbf{P}_c$ and $\mathbf{P}_C$ are given, respectively, by distributions of the betweennesss, closeness and local clustering coefficient.

Here, we consider five variations of the D^g function (Eq. (3)) associated with the stochastic measures given by the degree ($D^g_{\mathbf{P}_{deg}}$), distance ($D^g_{\mathbf{P}_\delta}$), closeness ($D^g_{\mathbf{P}_c}$), betweenness ($D^g_{\mathbf{P}_{bet}}$) and clustering coefficient ($D^g_{\mathbf{P}_C}$) distributions.

We can also obtain local information from the stochastic measure. We focus our attention on the node distance distribution ($\mathbf{P}_{\delta,v}(d)$) given by fraction of nodes at distance d from each node v. The network node dispersion (NND), a network quantifier related to the heterogeneity of nodes, introduced in [10] to a network G of size n:

$$NND(G) = JS(\mathbf{P}_{\delta,1}, \mathbf{P}_{\delta,2}, \ldots, \mathbf{P}_{\delta,n})$$

allows, together with the global pseudo-metric associated with the distance distribution ($D^g_{\mathbf{P}_\delta}$), to have an efficient size independent pseudo-metric between networks:

$$D(G_1, G_2) = \frac{1}{2}D^g_{\mathbf{P}_\delta}(G_1, G_2) + \frac{1}{2}\left|\sqrt{\frac{NND(G_1)}{\log n}} - \sqrt{\frac{NND(G_2)}{\log m}}\right|, \quad (4)$$

being, n and m, the sizes of networks G_1 and G_2, respectively.

Each global dissimilarity measure captures different characteristics. Most real networks present a degree distribution following a power-law $\mathbf{P}_{deg}(k) \sim k^{-\gamma}$ [16] but, there exist several networks with different topologies sharing the same degree distribution. The clustering based dissimilarity measures how far away two networks are comparing connected communities densities but, it fails to characterize properly tree-like structures. Distance based measures capture important features on networks: from the distance distribution, it is possible to obtain the network diameter, average path length, and average degree. From the node distance distribution perspective, as more information are available, we also get the node degree, closeness centrality, among others.

3 Applications

3.1 Distance Between Null Models

Here we compare how well-known networks null models are away from each other using the D^g and D functions. We consider four of the most commonly used models: K-regular [14], Erdös-Renyi (ER) [15], Barabási-Albert (BA) [16], Exponential (EXP) [17] and Watts-Strogatz rewiring model (WS) [18].

The K-regular consists in generating random networks with a constant degree K. ER is the random graph generation given by a connection probability $p \in [0, 1]$. Both BA and EXP are models of evolving networks: at each time step a new node is added and connected to m other existing nodes but, in the Exponential model, the new node is connected at random and the BA uses a preferential attachment mechanism[1]. WS model generates random networks by rewiring, with a given probability, links from a regular lattice.

The experiment consists in generate 10000 independent samples of each model with a fixed size $N = 1000$ computing averaged stochastic measures for each null model and then get comparisons via D^g and D. We set the parameters aiming to preserve the average degree of all generated networks: 10-Regular, BA and EXP with parameter $m = 5$, ER with $p = 10/999$ and WS with $k = 5$ and different rewiring probabilities $p = 0.2, 0.4, 0.6, 0.8$. Figure 1 shows the multidimensional scaling map [19] performed over the outcomes.

All of the analyzed measures were able to capture the scale-free behavior of the BA model ($P(k) \sim k^{-3}$) identifying significant structural differences even when compared with a similar growing model like the EXP, highlighting how different is the preferential attachment procedure in growing networks. It is also possible to see that bigger rewiring probability values imply higher proximity between WS and ER models [11]. As p increases, the randomness of the WS networks also increases. Figures 1B and C show the dissimilarity function importance: B shows that the average of the distance distributions of the ER network approaches the distance distribution of the regular graph meaning that, on average, a random graph behaves like a regular one but, the NND value is zero in most regular networks (Fig. 1C).

3.2 Critical Element Detection Problem and Network Robustness

The knowledge about how the network behaves after failures is of paramount importance and, therefore, the detection critical elements are important to plan efficient strategies to protect or even to destroy networks.

Given a network and an integer k, the critical element detection problem is to find a set of at most k elements (nodes or edges), whose deletion generates the biggest topological difference when comparing the residual and the original networks [20–22].

Here, we consider finding the critical 3 nodes in the Infectious Sociopatterns network whose deletion generates the biggest $D^g_{\mathbf{P}_{deg}}$, $D^g_{\mathbf{P}_\delta}$, $D^g_{\mathbf{P}_{bet}}$, $D^g_{\mathbf{P}_c}$, $D^g_{\mathbf{P}_C}$ and

---

[1] Higher degree nodes have a bigger probability of getting new connections.

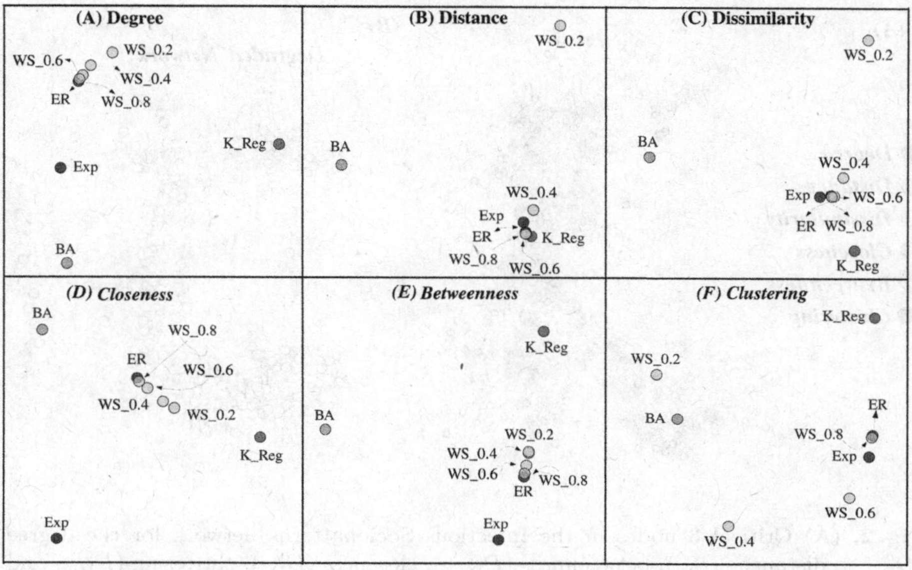

Fig. 1. Multidimensional scaling map performed over the outcomes of (A) $D^g_{\mathbf{P}_{deg}}$, (B) $D^g_{\mathbf{P}_\delta}$, (C) D, (D) $D^g_{\mathbf{P}_C}$, (E) $D^g_{\mathbf{P}_{bet}}$ and (F) $D^g_{\mathbf{P}_C}$ between all pairs of network null models: BA, EXP, K-regular and WS for different rewiring probability values ($WS\_0.2$, $WS\_0.4$, $WS\_0.6$ and $WS\_0.8$ consider the rewiring probability given by 0.2, 0.4, 0.6 and 0.8, respectively).

D values. The Infectious Sociopatterns network consists the face-to-face behavior of people during the exhibition INFECTIOUS: STAY AWAY in 2009 at the Science Gallery in Dublin. Nodes represent exhibition visitors; edges represent face-to-face contacts that were active for at least 20 seconds. The network has the data from the day with the highest number of interactions and is consider undirected and unweighted [23,24]. Figure 2A shows the outcomes. It is interesting to see that the betweenness and distance distributions share the same 3 critical elements. The dissimilarity function, on the other way, shares only two elements with the betweenness distribution sharing the third element with the clustering coefficient distribution. Figure 2B shows the degraded network after the removal of the critical elements found in A. When comparing the original and the degraded network, the last possesses a larger diameter (11), average path length (4.213771) and a small global clustering coefficient (0.436811).

The critical element detection problem is proven to be NP-hard in the general case for nodes and/or edges and, thus, the real case problems usually need heuristic approaches. The most common in the literature [25] is the strategy given by attacking the most central nodes (targeted attack[2]). Table 2 compares the values obtained by using 4 strategies of targeted attacks: higher degree,

---

[2] The nodes fail in decreasing order of centrality.

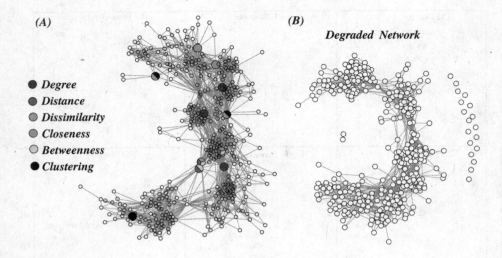

Fig. 2. (A) Critical 3 nodes in the Infectious Sociopatterns network for the degree $(D^g_{\mathbf{P}_{deg}})$, distance $(D^g_{\mathbf{P}_\delta})$, betweenness $(D^g_{\mathbf{P}_{bet}})$, closeness $(D^g_{\mathbf{P}_c})$, clustering $(D^g_{\mathbf{P}_C})$ and dissimilarity (D). (B) the degraded network obtained by the disconnecting the critical nodes.

closeness, betweenness, and clustering coefficient and the strategy of selecting the best combination of elements, we call it *Best* and it is computed by a brute force algorithm.

None of the above-mentioned targeted attack strategies achieved the network degradation given by the *Best* strategy, indicating that only one centrality measure is not enough as strategy to efficiently destroy the network.

Table 2. Comparing D^g and D values between targeted attacks (degree, closeness, clustering and betweeness) and the best strategy given by the critical node detection problem solution.

	$D^g_{\mathbf{P}_{deg}}$	$D^g_{\mathbf{P}_\delta}$	D	$D^g_{\mathbf{P}_c}$	$D^g_{\mathbf{P}_{bet}}$	$D^g_{\mathbf{P}_C}$
Degree	0.1468	0.1290	0.0745	0.2413	0.0860	0.0293
Closeness	0.1228	0.1790	0.0968	0.2471	0.0952	0.0333
Betweenness	0.1204	0.1666	0.0932	0.2462	0.1040	0.0333
Clustering	0.0638	0.0858	0.0509	0.2295	0.0115	0.0285
Best	0.1867	0.2288	0.12563	0.3811	0.1291	0.0456

Network failures may not occur all at once, but, at different time instances. Two sequences of failures may result in the same degraded network, even though, one may have caused a bigger topological destruction at the beginning of the attack. Therefore, the critical element detection problem fails in capturing this time-dependence of the failures.

In order to capture this time dependence of the failure process, following [9], a sequence of failures is defined as a sequence of time-indexed networks (G_t) where $G_0 = 0$ and G'_t is a subgraph of G_t for all $t' > t$ (as time increases, the network became more degraded).

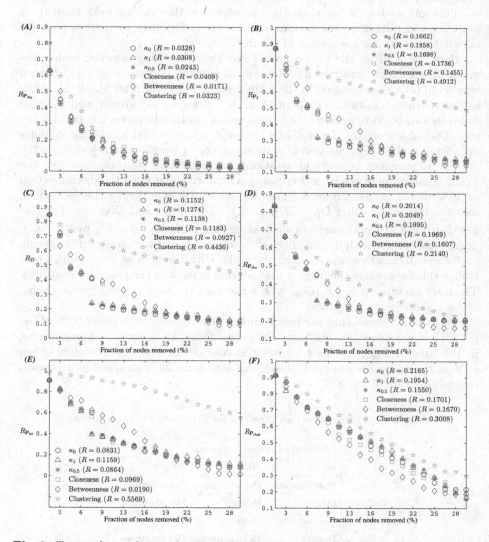

Fig. 3. Targeted attacks on the Train Bombing network. (A) $R_{\mathbf{P}_{deg}}$, (B) $R_{\mathbf{P}_{\delta}}$, (C) $R_{\mathbf{P}_D}$, (D) $R_{\mathbf{P}_c}$, (E) $R_{\mathbf{P}_{bet}}$ and (F) $R_{\mathbf{P}_C}$.

It is possible then, to define the *robustness* of G, for any given sequence of n failures $(G_t)_{t \in \{1, 2, ..., n\}}$ as:

$$R_P(G|(G_t)_{t \in \{1, 2, ..., n\}}) = \prod_{t=1}^{n} R(G_{t-1}|G_t), \qquad (5)$$

being $R(G_t|(G_{t-1})_{t \in \{1, 2, ..., n\}}) = 1 - D(G_t, G_{t-1})$.

This formulation is based on the consideration that the network robustness is a measure related to the distance that a given topology is apart from itself cumulatively during a sequence of failures.

Here, we analyze the robustness of the Train bombing network under targeted attacks. This undirected and weighted network contains contacts between suspected terrorists involved in the train bombing in Madrid on March 11, 2004, as reconstructed from newspapers. A node represents a terrorist and an edge between two terrorists shows that there was a contact between the two terrorists. The edge's weight denotes how "strong" a connection was. This includes friendship and co-participation in training camps or previous attacks [23, 26, 27].

The experiment consists in attacking, at each time step, one node of the Train bombing network by a decreasing centrality value until the disconnection of 30 % of the nodes. Figure 3 shows the outcomes considering the robustness measure computed using $D^g_{\mathbf{P}_{deg}}$, $D^g_{\mathbf{P}_\delta}$, $D^g_{\mathbf{P}_{bet}}$, $D^g_{\mathbf{P}_c}$, $D^g_{\mathbf{P}_C}$ and D values. The targeted attacks are performed in decreasing order of: degree (κ^0), weight (κ^1), degree and weight with importance ($\kappa^{0.5}$), closeness, betweenness and clustering coefficient. In most cases, targeting the nodes with the highest betweenness centrality value generates the highest degradation in most of the analyzed measures. The only exemption is for $R_{\mathbf{P}_C}$, where the best strategy is given by attacking nodes considering $\kappa^{0.5}$ values.

Table 3 also shows that the best strategy after the degradation of 30 % of the network is not necessary the best when considering 20 % or 10 %. For example, in the case of $\mathbf{P}_{bet}$, the best strategy is considering the nodes' weight when 10 % of the nodes are removed, the degree attack for 20 % and, the betweenness centrality strategy for 30 %.

Table 3. Best targeted attack strategy for the Train Bombing network.

	Fraction of nodes removed		
	10 %	20 %	30 %
$R_{\mathbf{P}_{deg}}$	Betweenness	Betweenness	Betweenness
$R_{\mathbf{P}_\delta}$	Degree	Degree	Betweenness
R_D	Degree	Degree	Betweenness
$R_{\mathbf{P}_c}$	Degree	Betweenness	Betweenness
$R_{\mathbf{P}_{bet}}$	Weight	Degree	Betweenness
$R_{\mathbf{P}_C}$	Betweenness	Betweenness	$\kappa^{0.5}$
	Best attack strategy		

4   Concluding Remarks

In this work, we review a methodology to quantify graph dissimilarities based on Information Theory quantifiers that possess important properties. One of them is the flexibility of choosing the network measurement depending on the purpose of the analysis or application.

Acknowledgments. Research is partially by supported by the Laboratory of Algorithms and Technologies for Network Analysis, National Research University Higher School of Economics, CNPq and FAPEMIG, Brazil.

References

1. Bunke, H.: Recent developments in graph matching. In: Proceedings of the 15th International Conference on Pattern Recognition, vol. 2 (2000). http://dx.doi.org/10.1109/ICPR.2000.906030
2. Dehmer, M., Emmert-Streib, F., Kilian, J.: A similarity measure for graphs with low computational complexity. Appl. Math. Comput. **182**(1), 447–459 (2006)
3. Rodrigues, L., Travieso, G., Boas, P.R.V.: Characterization of complex networks: a survey of measurements. Adv. Phys. **56**(1), 167–242 (2006)
4. Schaeffer, S.E.: Survey: graph clustering. Comput. Sci. Rev. **1**(1), 27–64 (2007)
5. Bai, L., Hancock, E.R.: Graph kernels from the Jensen-Shannon divergence. J. Math. Imaging Vis. **47**(1–2), 60–69 (2013)
6. Schieber, T.A., Ravetti, M.G.: Simulating the dynamics of scale-free networks via optimization. PLoS ONE **8**(12), e80783 (2013)
7. Babai, L.: Graph isomorphism in quasipolynomial time. Arxiv, January 2016. http://arxiv.org/abs/1512.03547
8. Carpi, L.C., Saco, P.M., Rosso, O.A., Ravetti, M.G.: Structural evolution of the tropical pacific climate network. Eur. Phys. J. B **85**(11), 1–7 (2012). http://dx.doi.org/10.1140/epjb/e2012-30413-7
9. Schieber, T.A., Carpi, L., Frery, A.C., Rosso, O.A., Pardalos, P.M., Ravetti, M.: Information theory perspective on network robustness. Phys. Lett. A **380**(3), 359–364 (2016)
10. Schieber, T.A., Carpi, L., Ravetti, M., Pardalos, P.M., Massoler, C., Diaz Guilera, A.: A size independent network difference measure based on information theory quantifiers (2016, Unpublished)
11. Carpi, L.C., Rosso, O.A., Saco, P.M., Ravetti, M.: Analyzing complex networks evolution through information theory quantifiers. Phys. Lett. A **375**(4), 801–804 (2011). http://www.sciencedirect.com/science/article/pii/S037596011001577X
12. Newman, M.E.J., Strogatz, S.H., Watts, D.J.: Random graphs with arbitrary degree distributions and their applications. Phys. Rev. E **64**, 026118 (2001)
13. Deza, M.M., Deza, E.: Encyclopedia of Distances, p. 590. Springer, Heidelberg (2009)
14. Lewis, T.G.: Network Science: Theory and Applications. Wiley Publishing, Hoboken (2009)
15. Erdös, P., Rényi, A.: On random graphs. Publ. Math. **6**(290), 290–297 (1959)
16. Albert, R., Barabási, A.: Statistical mechanics of complex networks. Rev. Mod. Phys. **74**, 47–97 (2002). http://arxiv.org/abs/cond-mat/0106096

17. Frank, O., Strauss, D.: Markov graphs. J. Am. Stat. Assoc. **81**(395), 832–842 (1986)
18. Watts, D.J., Strogatz, S.H.: Collective dynamics of small-world networks. Nature **393**(1), 440–442 (1998)
19. Cox, T.F., Cox, T.F.: Multidimensional Scaling, 2nd edn. Chapman and Hall/CRC, Boca Raton (2000). http://www.amazon.com/Multidimensional-Scaling-Second-Trevor-Cox/dp/1584880945
20. Arulselvan, A., Commander, C.W., Elefteriadou, L., Pardalos, P.M.: Detecting critical nodes in sparse graphs. Comput. Oper. Res. **36**(7), 2193–2200 (2009). http://dx.doi.org/10.1016/j.cor.2008.08.016
21. Dinh, T.N., Xuan, Y., Thai, M.T., Pardalos, P.M., Znati, T.: On new approaches of assessing network vulnerability: hardness and approximation. IEEE/ACM Trans. Netw. **20**(2), 609–619 (2012)
22. Walteros, J.L., Pardalos, P.M.: A decomposition approach for solving critical clique detection problems. In: Klasing, R. (ed.) SEA 2012. LNCS, vol. 7276, pp. 393–404. Springer, Heidelberg (2012)
23. Kunegis, J.: KONECT - the Koblenz network collection. In: Proceedings of International Web Observatory Workshop (2013)
24. Isella, L., Stehlé, J., Barrat, A., Cattuto, C., Pinton, J.F., den Broeck, W.V.: What's in a crowd? analysis of face-to-face behavioral networks. J. Theor. Biol. **271**(1), 166–180 (2011)
25. Iyer, S., Killingback, T., Sundaram, B., Wang, Z.: Attack robustness and centrality of complex networks. PLoS ONE **8**(4), e59613 (2013)
26. Train bombing network dataset - KONECT, January 2016
27. Hayes, B.: Connecting the dots. Can the tools of graph theory and social-network studies unravel the next big plot? Am. Sci. **94**(5), 400–404 (2006)
28. Lin, J.: Divergence measures based on the Shannon entropy. IEEE Trans. Inf. Theory **37**(1), 145–151 (1991)
29. Opsahl, T., Agneessens, F., Skvoretz, J.: Node centrality in weighted networks: generalizing degree and shortest paths. Soc. Netw. **32**(3), 245–251 (2010)

An Iterative Approach for Searching an Equilibrium in Piecewise Linear Exchange Model

Vadim I. Shmyrev[✉]

Sobolev Institute of Mathematics, Novosibirsk, Russia
shmyrev.vadim@mail.ru

Abstract. The exchange model with piecewise linear separable concave utility functions is considered. This consideration extends the author's original approach to the equilibrium problem in a linear exchange model and its variations. The conceptual base of this approach is the scheme of polyhedral complementarity. It has no analogs and made it possible to obtain the finite algorithms for some variations of the exchange model. Especially simple algorithms arise for linear exchange model with fixed budgets (Fisher's model). This is due to monotonicity property inherent in the models and potentiality of arising mappings. The algorithms can be considered as a procedure similar to the simplex-method of linear programming. It is natural to study applicability of the approach for more general models. The considered piecewise linear version of the model reduces to a special exchange model with upper bounds on variables and the modified conditions of the goods' balances. For such a model the monotonicity property is violated. But it remains, if upper bounds are substituted by financial limits on purchases. This is the idea of proposed iterative algorithm for initial problem. It is a generalization of an analogue for linear exchange model.

Keywords: Exchange model · Economic equilibrium · Fixed point · Linear programming · Polyhedral complementarity · Monotonicity · Iterative algorithm

1 Introduction

It is known that the problem of finding an equilibrium in the linear exchange model can be reduced to the linear complementarity problem [5]. But the dimension of this complementarity problem is relatively large. The polyhedral complementarity approach [1,3] is based on a fundamentally different idea, that reflects more the character of economic equilibrium as a concordance the consumers' preferences with financial balances. In algorithmic aspect it may be treated as a realization of the main idea of the simplex-method of linear programming. It has no analogs and made it possible to obtain the finite algorithms not only for the linear exchange model, but also for some it's variations [13–15]. The most

© Springer International Publishing Switzerland 2016
Y. Kochetov et al. (Eds.): DOOR 2016, LNCS 9869, pp. 61–73, 2016.
DOI: 10.1007/978-3-319-44914-2_6

simple are the algorithms for models with fixed budgets [2,22], more known as Fisher's problem. For this case the convex programming formulation, given by Eisenberg and Gale [6,7], is well known. This result was used by many authors for study computational aspects of the problem. Some review of that can be found in [11]. In this article a polynomial time algorithm is proposed, that uses the idea of the primal-dual scheme and max-flow techniques. The polyhedral complementarity approach gave an alternative convex program for the Fisher's problem [1,2]. (Russian version of the last article (1983) was later translated into English (2006) [8].) The obtained by this way procedures [2,22] use only well known elements of algorithms for transportation problem. A version with primal-dual scheme for transportation problem (the Hungarian method) was also considered [9,10]. These simple algorithms may be used for getting iterative methods for more complicate models. The first one was proposed for the general linear exchange model in [4]. In presenting consideration we extend this approach on the model with separable piecewise linear concave utility functions. The obtained algorithm uses that fact, that the algorithms for linear Fisher's model with additional restrictions on purchases are as simple as without them [17]. At each step, we study a model with initial utility functions and with the appropriate (changing) restrictions on purchases. This model is known as spending constraint model and was introduced in [18]. The polyhedral complementarity approach is applicable in this case as well. It should be noted, that a generalization of our reduction of Fisher's model [1,2] for the spending constraint model was given in [12,19]. A strongly polynomial algorithm for this problem, as well as for the Fisher's model, was proposed in [24]. The simplex-like algorithm for spending constraint model was presented in [19]. The algorithm is based on reduction of the initial problem to a linear complementarity one and use a special algorithmic technique, which is similar to that of well known Lemke's algorithm. A detailed analysis of the polynomial solvability of the problem is given. The approach of linear complementarity for equilibrium searching, begun in [5], was extended on the general linear exchange model with piecewise linear separable concave utility functions in [20].

2   Model

The classical linear exchange model has the following description.

Consider a model with n commodities and m consumers. Let $J = \{1, \ldots, n\}$ and $I = \{1, \ldots, m\}$ be the index sets of commodities and consumers respectively. Each consumer $i \in I$ possesses a vector of initial endowments $w^i \in R^n_+$ and must choose a consumption vector $x^i \in R^n_+$ maximizing his linear utility function (c^i, x^i). The exchange is realized with respect to some nonnegative prices p_j, forming a price vector $p \in R^n_+$.

Thus we have the following problem of consumer i:

$$(c^i, x^i) \to \max,$$
$$(p, x^i) \leq (p, w^i),$$
$$x^i \geq 0.$$

Let $\tilde{x}^i$ be a vector x^i that solves this program. A price vector $\tilde{p} \neq 0$ is an *equilibrium price vector* if there exist solutions $\tilde{x}^i$, $i = 1, \ldots, m$, for the individual optimization problems such that

$$\sum_{i \in I} \tilde{x}^i = \sum_{i \in I} w^i.$$

In the mentioned description the consumers' utility functions are linear $(c^i, x^i) = \sum_{j \in J} c_j^i x_j^i$. Here c_j^i, x_j^i are the components of the vectors c^i, x^i. In the piecewise linear version of the model these functions replace by separable piecewise linear concave functions $\sum_{j \in J} c_j^i(x_j^i)$ (Fig. 1).

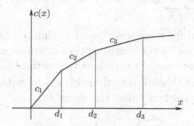

Fig. 1. Piecewise linear utility function

In what follows we normalize the initial endowment of each commodity to 1, i.e., $\sum_i w^i = (1, \ldots, 1) \in R^n$. The sum of p_j is also normalized to 1, restricting the price vector p to lie in the unit simplex

$$\sigma = \left\{ p \in R_+^n \,\middle|\, \sum_{j \in J} p_j = 1 \right\}.$$

In the case $w^i = \lambda_i(1, \ldots, 1)$ we have $(p, w^i) = \lambda_i$. Thus the budgets are fixed. Such a model is named Fisher's model.

For the linear model the author has proposed an original approach to obtaining finite algorithms for searching equilibrium. This approach is based on reduction the initial problem to the fixed-point one for piecewise constant point-to-set mapping, which leads to the polyhedral complementarity problem. The piecewise linear version of the model can be reduced to the linear one but with additional restrictions to the consumption volumes and modified goods' balances in equilibrium.

More in details. It is well known, that the maximization problem with linear restrictions and piecewise linear separable concave goal function can be reduced to the linear problem with upper bounds on the variables.

Let the problem be as follows:

$$\sum_j c_j(x_j) \to \max$$

$$\sum_j a^j x_j = b, \quad x_j \geq 0.$$

Let the linearity intervals of the function $c_j(x_j)$ are $[0, d_{j1}], [d_{j1}, d_{j2}], \ldots, [d_{jn_j}, \infty)$, and on $[d_{j(s-1)}, d_{js}]$ we have $c_j(x_j) = const + c_{js}x_j$. For the simplicity can be assumed n_j be the same for all j. Introduce the lengths of intervals $\Delta_{js} = d_{js} - d_{j(s-1)}$ and new variables x_{js}. We obtain an equivalent problem by the change of variables $x_j = \sum_s x_{js}$, $\quad 0 \leq x_{js} \leq \Delta_{js}$:

$$\sum_j \sum_s c_{js}x_{js} \to \max$$

$$\sum_j \sum_s a^j x_{js} = b, \quad 0 \leq x_{js} \leq \Delta_{js}.$$

In this way we obtain for piecewise linear exchange model the linear one but with additional restrictions to the consumption volumes and modified conditions of goods' balances. In consumer's problem the variable x_j^i replaces by $\sum_s x_{js}^i$. For simplicity, we assume that all functions $c_j^i(x_j^i)$ have the same quantity of the linearity intervals. The problem takes the form:

$$\sum_j \sum_s c_{js}^i x_{js}^i \to \max,$$

$$\sum_j p_j \sum_s x_{js}^i \leq (p, w^i),$$

$$0 \leq x_{js}^i \leq \Delta_{js}.$$

Respectively change the conditions of goods' balances in equilibrium:

$$\sum_i \sum_s \tilde{x}_{js}^i = 1, \quad j \in J.$$

If all functions $c_j^i(x_j^i)$ have only one linearity interval we obtain the linear exchange model with upper bounds for variables. The exchange model with upper bounds for volumes of purchases turns out to be qualitatively more complicated then the linear one without additional restrictions. The monotonicity property of the arising point-to-set piecewise constant mapping fails. However the polyhedral complementarity approach can be applied, but the obtained algorithm needs a spacial start from the price simplex boundary [16].

The searching equilibrium procedure under consideration for piecewise linear model is based on it's approximation by a simple model with fixed budgets and additional restrictions on many for buying each commodity. For such a model the monotonicity property remains and the simple algorithms can be proposed [17]. These algorithms are very closed to the simplex procedures of linear programming.

3   The Main Idea of the Approach for the Linear Model

Here we follow [23].

1°. <u>The parametric transportation problem of the model.</u>
Let be $c^i > 0$ for all $i \in I$. Given a price vector p consider the following *transportation problem of the model:*

$$\sum_{i \in I} \sum_{j \in J} z_{ij} \ln c_j^i \to \max$$

under conditions

$$\{z_{ij}\} \in Z(p) \left| \begin{array}{ll} \sum_{j \in J} z_{ij} = (p, w^i), & i \in I, \\ \sum_{i \in I} z_{ij} = p_j, & j \in J, \\ z_{ij} \geq 0, & (i,j) \in I \times J. \end{array} \right.$$

The equations of this problem represent the financial balances for the consumers and commodities. The variables z_{ij} are introduced by $z_{ij} = p_j x_j^i$.

This is the classical transportation problem. The price vector p is a parameter of the problem. Under the mentioned assumption about $\{w^i\}$ this problem is solvable for each $p \in \sigma$.

2°. <u>Reduction to a fixed point problem.</u>
Consider the restrictions of the corresponding dual problem:

$$u_i + v_j \geq \ln c_j^i, \quad i \in I, j \in J.$$

Let $V(p)$ be the set of optimal vectors $v = (v_1, ..., v_n)$. For $v \in V(p)$ introduce a vector $g(v) = (\exp(v_1), ..., \exp(v_n)) / \sum_{j=1}^n \exp(v_j)$. We have $g(v) \in \sigma^\circ$ (the relative interior of the price simplex σ). Introduce the set

$$G(p) = \{g(v) | v \in V(p)\}$$

Thus we obtain the point-to-set mapping $G : \sigma \to 2^{\sigma^\circ}$. For this mapping all conditions of Kakutani's theorem are fulfilled and so the fixed points of the mapping exist.

Theorem 1. *The fixed points of the mapping G, and only they, give the equilibrium price vectors of the model.*

The polyhedral complementarity approach make it possible to proposer a finite method for searching an equilibrium price vector in the general linear exchange model [1,3]. Later it was named as *method of meeting paths.*

4   The Model with Fixed Budgets

1°. Reduction to an optimization problem.

In the set of linear exchange models the subset of the models with fixed budgets is selected. In such a model it is assumed, that the participants have not initial endowments but some money. The right sides (p, w^i) in the budget conditions are replaced by the constants λ_i. This model is known as Fisher's model. In this case the equilibrium problem was reduced to an optimization one [6,7]. But any special algorithms were proposed for it.

For the model with fixed budgets the equilibrium problem is simplified because in this case the mapping G is in some sense potential. The $G(p)$ is induced by the set $V(p)$, and this set is the subdifferential of the concave function f, which indicate for each $p \in \sigma$ the optimal value of the goal function in the transportation problem of the model: $V(p) = \partial f(p)$. Here the subdifferetial for a concave function f is defined by such one of convex function $(-f)$:

$$\partial f(p) = -\partial(-f)(p).$$

Introduce the convex function $h(p)$ on σ as $h(p) = (p, \ln p) = \sum_{j \in J} p_j \ln p_j$ for $p > 0$ and $h(p) = 0$ for $p \not> 0$.

Theorem 2. *The equilibrium problem for the linear exchange model with fixed budgets is equivalent to this one of minimization on σ for the convex function $\varphi(p) = h(p) - f(p)$.*

The function φ is very simple and for the minimization can be used the suboptimization approach [21]. In this way we obtain the finite algorithm [22] for searching the equilibrium in the model.

Another algorithm for searching an equilibrium price vector can be obtained if we take into account that the mapping G and the inverse mapping G^{-1} have the same fixed points. For the introduced concave function f consider the conjugate function f^*:

$$f^*(y) = \inf_z \{(y, z) - f(z)\}$$

Theorem 3. *The equilibrium price vector in the model with fixed budgets is the minimum point of the function $\psi(q) = -f^*(\ln q)$ on $\sigma°$.*

The obtained in this way algorithm [8] is in some sense duel to the algorithm [22], obtained by minimizing φ.

It is to note, that the mapping G for the model with fixed budgets has a special monotonicity property: the inequality

$$(p^1 - p^2, \ln q^1 - \ln q^2) \leq 0$$

is fulfilled for each $p^1, p^2 \in \sigma$ and $q^1 \in G(p^1), q^2 \in G(p^2)$. We can say, that G is the logarithmic monotone decreasing mapping [2,23].

2°. Polyhedral complementarity problem.

For algorithmic realization we have to consider the mapping G more in detail. The consideration is based on the new notion of consumption *structure*.

Definition 1. A set $\mathcal{B} \subset I \times J$ is named *a structure*, if for each $i \in I$ there exists $(i,j) \in \mathcal{B}$.

This notion is analogous to the basic index set in linear programming. Two sets of the price vectors can be considered for each structure $\mathcal{B}$ of the model: *preference zone* $\Xi(\mathcal{B})$ and *balance zone* $\Omega(\mathcal{B})$.

$\Xi(\mathcal{B})$ is the set of prices by which the participants prefer the connections of the structure, ignoring the budget conditions and balances of goods.

$\Omega(\mathcal{B})$ is the set of prices by which the budget conditions and balances of goods are possible when the connections of the structure are respected, but the participants preferences are ignored.

Let $\mathfrak{B}$ be *the collection of all dual feasible basic index sets of the transportation problem and of all their subsets being structures.*

For $\mathcal{B} \in \mathfrak{B}$ we obtain the description of $\Omega(\mathcal{B})$ and $\Xi(\mathcal{B})$ in the following way:

$$\mathcal{B} \in \mathfrak{B} \Longrightarrow \begin{vmatrix} a) & \Omega(\mathcal{B}) \subset \sigma \text{ is the balance zone of the structure:} \\ & \Omega(\mathcal{B}) = \{p \in \sigma \mid \exists z \in Z(p), z_{ij} = 0, (i,j) \notin \mathcal{B}\}; \\ \\ b) & \Xi(\mathcal{B}) \subset \sigma^\circ \text{ is the preference zone of the structure:} \\ & \Xi(\mathcal{B}) = \left\{ q \in \sigma^\circ \mid \max_k \dfrac{c_k^i}{q_k} = \dfrac{c_j^i}{q_j}, \ \forall (i,j) \in \mathcal{B} \right\}. \end{vmatrix}$$

It is clear that

$$p^* \text{ is an equilibrium price vector} \iff (\exists \mathcal{B}) p^* \in \Omega(\mathcal{B}) \cap \Xi(\mathcal{B}).$$

It is easy to give the adduced descriptions of $\Xi(\mathcal{B})$ and $\Omega(\mathcal{B})$ in more detail. For $q \in \Xi(\mathcal{B})$ we have:

$$\frac{q_k}{c_k^i} = \frac{q_j}{c_j^i} \qquad (i,k) \in \mathcal{B}, \quad (i,j) \in \mathcal{B},$$

$$\frac{q_l}{c_l^i} \geq \frac{q_j}{c_j^i} \qquad (i,l) \notin \mathcal{B}, \quad (i,j) \in \mathcal{B}.$$

Thus $\Xi(\mathcal{B})$ is the intersection of a polyhedron with σ°.

To obtain the description of $\Omega(\mathcal{B})$ we should use the well known tools of transportation problems theory. Given $\mathcal{B} \in \mathfrak{B}$, introduce a graph $\Gamma(\mathcal{B})$ with the set of vertices $G = \{1, 2, \ldots, m+n\}$ and the set of edges $\{(i, m+j) | (i,j) \in \mathcal{B}\}$. Let τ be the number of components of this graph, let G_ν be the set of vertices of ν-th component, $I_\nu = I \cap G_\nu$ and $J_\nu = \{j \in J | (m+j) \in G_\nu\}$. It is not difficult to show that the following system of linear equations must hold for $p \in \Omega(\mathcal{B})$:

$$\sum_{j \in J_\nu} p_j = \sum_{i \in I_\nu} \lambda_i, \qquad \nu = 1, \ldots, \tau.$$

Under these conditions the values z_{ij} can be obtained from

$$z \in Z(p), z_{ij} = 0, (i,j) \notin \mathcal{B}$$

presenting linear functions of p: $z_{ij} = z_{ij}(p)$. Now for $p \in \Omega(\mathcal{B})$ we have in addition the system of linear inequalities

$$z_{ij}(p) \geq 0, \qquad (i,j) \in \mathcal{B}.$$

Thus $\Omega(\mathcal{B})$ is described by a linear system of equalities and inequalities. So it is also a polyhedron.

It is clear that $\omega = \{\Omega(\mathcal{B})\}$ and $\xi = \{\Xi(\mathcal{B})\}$ are polyhedral complexes and the pairs $(\Omega(\mathcal{B}), \Xi(\mathcal{B}))$, $\mathcal{B} \in \mathfrak{B}$, form a one-to-one correspondence by which these complexes are in duality. So we have the polyhedral complementarity problem [23]. Figure 2 illustrate the polyhedral complexes in a model with 3 commodities and 2 consumers. Each of both complexes has 17 elements. Figure 3 illustrate the arising complementarity problem with it's solution: $c^{12} \in G(c^{12})$.

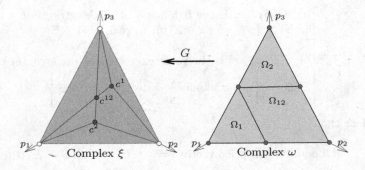

Fig. 2. Polyhedral complexes in exchange model

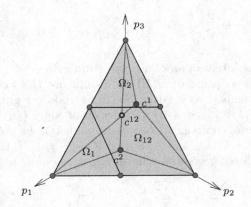

Fig. 3. Complementarity problem: c^{12} is the solution.

3°. Algorithm.

We describe here one step of the algorithm, which is based on Theorem 3 [8]. The alternative version, based on Theorem 2, can be seen in [22].

At the beginning of the recurrent step $(k+1)$ we have a structure $\mathcal{B}_k \in \mathfrak{B}$ and a point $q^k \in \Xi(\mathcal{B}_k)$.

Let $L(\mathcal{B}) \supset \Xi(\mathcal{B})$ and $M(\mathcal{B}) \supset \Omega(\mathcal{B})$ be the affine hulls of these zones. We obtain the "ideal" point $r = r(\mathcal{B})$, solving the system:

$$\frac{r_k}{c_k^i} = \frac{r_j}{c_j^i} \quad (i,k),(i,j) \in \mathcal{B},$$
$$\sum_{j \in J_\nu} r_j = \sum_{i \in I_\nu} \lambda_i, \quad \nu = 1,\ldots,\tau.$$

It is easy to solve this system. It decomposes in the subsystems, corresponding to the components of the graph $\Gamma(\mathcal{B}_k)$. For the ν-th component we have

$$\frac{r_k}{c_k^i} = \frac{r_j}{c_j^i} \quad (i,k),(i,j) \in \mathcal{B}, i \in I_\nu, j \in J_\nu,$$
$$\sum_{j \in J_\nu} r_j = \sum_{i \in I_\nu} \lambda_i,$$

The first equations give proportions between r_j on the component and determined r_j up to a multiplier. The multiplier can be obtained from the last equation.

The point r is the point of the intersection $L(\mathcal{B}) \cap \mathcal{M}(\mathcal{B})$. It is the minimum point of the function $\psi(q)$ on $L(\mathcal{B})$. If $r \in \Xi(\mathcal{B}) \cap \Omega(\mathcal{B})$, the point r defines the equilibrium prise vector. Otherwise two possibilities may occur.

(a) $q^k = r$. It means, that q^k being the minimum point on $L(\mathcal{B})$, lies in $M(\mathcal{B})$ In this case $z_{ij}(r) = z_{ij}^k$ can be obtained. If all they are nonnegative, $r \in \Omega(\mathcal{B})$ and we have the equilibrium price vector $p^* = r$. Otherwise we choose certain $z_{i_o,j_o}^k < 0$ and go to the next step with

$$\mathcal{B}_{k+1} = \mathcal{B}_k \backslash \{(i_o, j_o)\}, \quad q^{k+1} = r.$$

(b) $q^k \neq r$. We consider the moving point $q(t) = q^k + t(r - q^k)$ and search

$$t^* = \{t \in [0,1] \mid q(t) \in \Xi(\mathcal{B}_k)\}.$$

If $t^* = 1$, we take $\mathcal{B}_{k+1} = \mathcal{B}_k, q^{k+1} = r$ and on the next step we will have case (a). In the case $t^* < 1$ certain of the inequalities from description $\Xi(\mathcal{B}_k)$ limits t^*. Let (i',j') be the pair corresponding to that inequality. We take

$$\mathcal{B}_{k+1} = \mathcal{B}_k \cup \{(i',j')\}, \quad q^{k+1} = q(t^*).$$

Theorem 4. *Under the dual nondegeneracy condition for the transportation problem of the model the algorithm gives the equilibrium price vector.*

5   The Model with the Finance Restrictions on the Consumption

As it was mentioned, the equilibrium problem for the model with separable piecewise linear utility functions of participants reduces to the linear model, but with additional restrictions on the consumption volumes for each commodity: $x_j^i \leq d_j^i$. These additional restrictions cause significantly quality changes. The mapping G looses the monotonicity property even in models with fixed budgets. The polyhedral complementarity approach gives the algorithm [16], which requires a special start (as in the triangulation algorithms for searching fixed points). We have more simple problem in the model with additional restrictions on the finance for buying each commodity: $p_j x_j^i \leq \beta_{ij}$. In the transportation problem of the model appear the upper bounds for variables: $z_{ij} \leq \beta_{ij}$. The problem turn out to be solvable not for all $p \in \sigma$ and thus the mapping G is defined not on the hole simplex σ. But it has the monotonicity property as before, and Theorems 2 and 3 remain valid. This make it possible to propose the simple algorithms as in model without the additional restrictions [17]. We only have to use the well known modification of the simplex method for the problem with upper bounds. For these problems basic set supplements by a set, that indicates the variables fixed on upper bounds. In our consideration a structure $\mathcal{B}$ supplements by a set W, and we have *substructure* $U = \{\mathcal{B}, W\}$. For $(i, j) \in W$ the inequalities of the transportation dual problem turn to inverse. The preference zone and algorithm descriptions change respectively.

6   The Iterative Approach for the Equilibrium Searching

The algorithmic realization of proposed methods for general case of the linear exchange model needs to solve on each step a linear system of equations. The procedure becomes significantly simple for the model with fixed budgets. In this case we need to solve only triangular systems, as in the transportation problem algorithms. For general case of the model (with variable budgets) the iterative method of successive approximations was proposed [4]. Given a price vector p^k on the step k we consider an approximation to the initial model by a model with fixed budgets $\lambda_i^k = (p^k, w^i)$. The equilibrium price vector of this model is taken as p^{k+1}. It was proofed that the process converges to the equilibrium as a geometric progression.

This approach occur not applicable for the model with upper bounds $x_j^i \leq d_j^i$. The approximating model with fixed budgets is to complicate in this case for the equilibrium searching. It is more simple if we take for approximation the model with additional restrictions on the finance for the commodities buying. In this version of the method we calculate on current step besides the budgets λ_i^k also the financial upper bounds $\beta_{ij}^k = p_j^k d_j^i$. We take the equilibrium price vector of the obtained model as next approximation p^{k+1}.

The numerical tests show the convergence of this process to the equilibrium of the initial model.

7 Application to the Piecewise Linear Model

The above described idea of approximating model can be applied for searching an equilibrium in the initial piecewise linear model. Given on the step k a price vector p^k, we calculate the budgets $\lambda_i^k = (p^k, w^i)$ and the financial upper bounds $\beta_{js}^{ki} = p_j^k \Delta_{js}^i$. In this way we obtain the approximating model with initial piecewise linear utility functions, fixed budgets and with additional financial constraints on purchases which correspond to the linearity intervals of utility functions. The parametric transportation problem of this approximating model is as follows:

$$\sum_i \sum_j \sum_s z_{js}^i \ln c_{js}^i \to \max$$

$$\sum_j \sum_s z_{js}^i = \lambda_i^k, \quad i \in I,$$

$$\sum_i \sum_s z_{js}^i = p_j, \quad j \in J,$$

$$0 \le z_{js}^i \le \beta_{js}^{ki}.$$

We can use the scheme of polyhedral complementarity approach. This yield us the finite algorithms for equilibrium searching. The obtained equilibrium price vector we take as p^{k+1}.

Thus we obtain the iterative method for searching an equilibrium in the exchange model with piecewise linear separable concave utility functions of participants.

8 Conclusion

The exchange model with piecewise linear separable concave utility functions of participants is considered. The consideration extends an original approach to the equilibrium problem in a linear exchange model and its variations. The conceptual base of this approach is the scheme of polyhedral complementarity. It has no analogs and made it possible to obtain the finite algorithms for some variations of the linear exchange model. One of them we use for getting an approximation for the piecewise linear model. To be exact, this is the linear exchange model with fixed budgets and additional restrictions on the finance for buying each commodity. For this model a finite algorithm was developed. In this way we obtain an iterative method for the model under consideration. On each step of the process the approximating model is formed. The equilibrium price vector of this model is used to get the next approximation. This approach was justified for general case of linear exchange model [4]. The numerical tests confirm it's validity for the piecewise linear model.

Acknowledgments. The author is very grateful to unknown referee for substantive comments and useful information. This work was supported by the Russian Foundation for Basic Research, project 16-01-00108 A.

References

1. Shmyrev, V.I.: On an approach to the determination of equilibrium in elementary exchange models. Sov. Math. Dokl. **27**(1), 230–233 (1983)
2. Shmyrev, V.I.: Algorithms for searching an equilibrium in fixed budget exchange models. Optimization **31**(48), 137–155 (1983). (in Russian)
3. Shmyrev, V.I.: An algorithm for the search of equilibrium in the linear exchange model. Sibirian Math. J. **26**, 288–300 (1985)
4. Shmyrev, V.I., Shmyreva, N.V.: An iterative algorithm for searching an equilibrium in the linear exchange model. Siberian Adv. Math. **6**(1), 87–104 (1996)
5. Eaves, B.C.: A finite algorithm for linear exchange model. J. Math. Econ. **3**(2), 197–204 (1976)
6. Eisenberg, E., Gale, D.: Consensus of subjective probabilities: the pari-mutuel method. Ann. Math. Stat. **30**(1), 165–168 (1959)
7. Gale, D.: The Theory of Linear Economic Models. McGraw-Hill, New York (1960)
8. Shmyrëv, V.I.: An algorithmic approach for searching an equilibrium in fixed budget exchange models. In: Driessen, T.S.H., van der Laan, G., Vasil'ev, V.A., Yanovskaya, E.B. (eds.) Russian Contributions to Game Theory and Equilibrium Theory. Theory and Decision Library C: vol. 39, pp. 217–235. Springer, Heidelberg (2006)
9. Shmyrev, V.I., Volenko, Y.: The Hungarian method for finding equilibrium in a linear exchange model with fixed budgets. Diskretn. Anal. Issled. Oper. Ser. 2 **6**(1), 61–77 (1999). (in Russian)
10. Shmyrev, V.I.: A new version of the Hungarian method for finding equilibrium in a linear exchange model. Diskretn. Anal. Issled. Oper. Ser. 2 **6**(1), 43–60 (1999). (in Russian)
11. Devanur, N.R., Papadimitriou, C.H., Saberi, A., Vazirani, V.V.: Market equilibrium via a primal-dual algorithm for a convex program. J. ACM (JACM) **55**(5), 22 (2008)
12. Birnbaum, B., Devanur, N.R., Xiao, L.: Distributed algorithms via gradient descent for fisher markets. In: 12th ACM conference on Electronic commerce, pp. 127–136. ACM New York, USA (2011)
13. Shmyrev, V.I.: A generalized linear exchange model. J. Appl. Ind. Math. **2**(1), 125–142 (2008)
14. Shmyrev, V.I.: A method of meeting paths for the linear production-exchange model. J. Appl. Ind. Math. **6**(4), 490–500 (2012)
15. Shmyrev, V.I.: A linear production-exchange model, polyhedral complexes and a criterion for an equilibrium. J. Appl. Ind. Math. **6**(2), 240–247 (2012)
16. Shmyrev, V.I.: About the equilibrium searching in the linear exchange model with upper bounds for variables. Optimization **42**(59), 86–111 (1988). (in Russian)
17. Shmyrev, V.I.: About the searching in the linear exchange model with fixed butgets and additional restrictions on purchases. Optimization **45**(62), 66–86 (1989). (in Russian)
18. Vazirani, V.V.: Spending constraint utilities, with applications to the adwords market. Math. Oper. Res. **35**(2), 458–478 (2010)
19. Garg, J., Mehta, R., Sohoni, M., Vishnoi, N.K.: Towards polynomial simplex-like algorithms for market equilibria. In: 24th Annual ACM-SIAM Symposium on Discrete Algorithms, pp. 1226–1242. SIAM, Philadelphia (2013)
20. Garg, J., Mehta, R., Sohoni, M., Vazirani, V.V.: A complementarity pivot algorithm for market equilibrium under separable, piecewise-linear concave utilities. SIAM J. Comput **44**(6), 1820–1847 (2015)

21. Rubinstein, G.S., Shmyrev, V.I.: Methods for minimization quasiconvex function on polyhadron. Optimization **1**(18), 82–117 (1971). (in Russian)
22. Shmyrev, V.I.: An algorithm for finding equilibrium in the linear exchange model with fixed budgets. J. Appl. Ind. Math. **3**(4), 505–518 (2009)
23. Shmyrev, V.I.: Polyhedral complementarity and equilibrium problem in linear exchange models. Far East J. Appl. Math. **82**(2), 67–85 (2013)
24. Vegh, L.A.: Strongly polynomial algorithm for a class of minimum-cost flow problems with separable convex objectives. In: 44th ACM Symposium on Theory of Computing, New York, USA, pp. 27–40 (2012)

Handling Scheduling Problems with Controllable Parameters by Methods of Submodular Optimization

Akiyoshi Shioura[1], Natalia V. Shakhlevich[2], and Vitaly A. Strusevich[3(✉)]

[1] Tokyo Institute of Technology, Tokyo, Japan
[2] University of Leeds, Leeds, UK
[3] Univeristy of Greenwich, London, UK
V.Strusevich@greenwich.ac.uk

Abstract. In this paper, we demonstrate how scheduling problems with controllable processing times can be reformulated as maximization linear programming problems over a submodular polyhedron intersected with a box. We explain a decomposition algorithm for solving the latter problem and discuss its implications for the relevant problems of preemptive scheduling on a single machine and parallel machines.

1 Introduction

In *Scheduling with Controllable Processing Times* (*SCPT*), the actual durations of the jobs are not fixed in advance, but have to be chosen from a given interval. For an SCPT model, two types of decisions are required: (i) each job has to be assigned its actual processing time, and (ii) a schedule has to be found that provides a required level of quality. A penalty is applied for assigning shorter actual processing times. The quality of the resulting schedule is measured with respect to the cost of assigning the actual processing times that guarantee a certain scheduling performance. This area of scheduling has been active since the 1980s, see surveys [20] and [26].

Nemhauser and Wolsey were among the first who noticed that the SCPT models could be handled by methods of *Submodular Optimization* (*SO*); see, e.g., Example 6.1 (Sect. 6 of Chap. III.3) of their book [19]. A systematic development of a general framework for solving the SCPT problems via submodular methods has been initiated by Shakhlevich and Strusevich [27,28] and further advanced in [29]. As a result, a powerful toolkit of the SO techniques [4,25] can be used for design and justification of algorithms for solving a wide range of the SCPT problems. In this paper, we present convincing examples of a positive mutual influence of scheduling and submodular optimization, mainly based on our recent work [31–34].

2 Scheduling with Controllable Processing Times

The jobs of set $N = \{1, 2, \ldots, n\}$ have to be processed either on a single machine M_1 or on parallel machines $M_1, M_2, \ldots, M_m$, where $m \geq 2$. For each job $j \in N$,

© Springer International Publishing Switzerland 2016
Y. Kochetov et al. (Eds.): DOOR 2016, LNCS 9869, pp. 74–90, 2016.
DOI: 10.1007/978-3-319-44914-2_7

its processing time $p(j)$ is not given in advance but has to be chosen from a given interval $[\underline{p}(j), \overline{p}(j)]$. That selection process is often seen as *compressing* the longest processing time $\overline{p}(j)$ down to $p(j)$. The value $x(j) = \overline{p}(j) - p(j)$ is called the *compression amount* of job j. Compression may decrease the completion time of each job j but incurs additional cost $w(j)x(j)$, where $w(j)$ is a given non-negative unit compression cost. The total cost is represented by the linear function $W = \sum_{j \in N} w(j)x(j)$.

Each job $j \in N$ is given a *release date* $r(j)$, before which it is not available, and a *deadline* $d(j)$, by which its processing must be completed. In the processing of any job, *preemption* is allowed, so that the processing can be interrupted on any machine at any time and resumed later, possibly on another machine. It is not allowed to process a job on more than one machine at a time, and a machine processes at most one job at a time. Given a schedule, let $C(j)$ denote the completion time of job j. A schedule is called *feasible* if the processing of a job $j \in N$ takes place in the time interval $[r(j), d(j)]$.

We distinguish between the *identical* parallel machines and the *uniform* parallel machines. In the former case, the machines have the same speed. If the machines are uniform, then it is assumed that machine M_h has speed s_h, $1 \leq h \leq m$. Throughout this paper we assume that the machines are numbered in non-increasing order of their speeds, i.e.,

$$s_1 \geq s_2 \geq \cdots \geq s_m. \tag{1}$$

Adapting standard notation for scheduling problems [14], we denote a generic problem of our primary concern by $\alpha | r(j), p(j) = \overline{p}(j) - x(j), C(j) \leq d(j), pmtn | W$. Here, in the first field α we write "1" for a single machine, "P" in the case of $m \geq 2$ identical machines and "Q" in the case of $m \geq 2$ uniform machines. In the middle field, the item "$r(j)$" implies that the jobs have individual release dates; this parameter is omitted if the release dates are equal. We write "$p(j) = \overline{p}(j) - x(j)$" to indicate that the processing times are controllable and $x(j)$ is the compression amount to be found. The condition "$C(j) \leq d(j)$" reflects the fact that in a feasible schedule the deadlines should be respected; we write "$C(j) \leq d$", if all jobs have a common deadline d. The abbreviation "$pmtn$" is used to point out that preemption is allowed. Finally, in the third field we write the objective function to be minimized, which is the total compression cost $W = \sum w(j)x(j)$.

If the processing times $p(j)$, $j \in N$, are fixed then the corresponding counterpart of problem $\alpha | r(j), p(j) = \overline{p}(j) - x(j), C(j) \leq d(j), pmtn | W$ is denoted by $\alpha | r(j), C(j) \leq d(j), pmtn | \circ$. In the latter problem, it is required to verify whether a feasible preemptive schedule exists. If the deadlines are equal, then the counterpart of problem $\alpha | r(j), p(j) = \overline{p}(j) - x(j), C(j) \leq d, pmtn | W$ with fixed processing times can be denoted by $\alpha | r(j), pmtn | C_{\max}$, so that it is required to find a preemptive schedule that for the corresponding settings minimizes the makespan $C_{\max} = \max \{C(j) | j \in N\}$: if the optimal makespan is larger than d the required feasible schedule does not exist; otherwise, it exists.

Below we give examples of two most popular interpretations of the SCPT models. Alternative interpretations can be found, e.g., in [11] and [17].

In computing systems that support imprecise computation [15], some computations can be run partially, producing less precise results. In our notation, to produce a result of reasonable quality, the mandatory part of each task j must be completed, and this takes $p(j)$ time units. If instead of an ideal computation time $\overline{p}(j)$ a task is executed for $p(j) = \overline{p}(j) - x(j)$ time, then computation is imprecise, and $x(j)$ corresponds to the error of computation. The objective function $W = \sum w(j)x(j)$ is interpreted as the total weighted error.

The SCPT problems serve as mathematical models in make-or-buy decision-making [29], where it is required to determine which part of each order j is manufactured internally and which is subcontracted. For this model, $p(j) = \overline{p}(j) - x(j)$ is understood as the chosen actual time for internal manufacturing, where $x(j)$ shows how much of the order is subcontracted and $w(j)x(j)$ is the cost of this subcontracting. Thus, we need to minimize the total subcontracting cost and to find a deadline-feasible schedule for internally manufactured orders.

These and other versions of the SCPT problems can be formulated as SO models and handled by SO methods.

3 Submodular Polyhedra and Decomposition Algorithm

In this section, we present some basic facts related to submodular optimization. Unless stated otherwise, we follow a comprehensive monograph on this topic by Fujishige [4], see also [13,25]. We also describe a decomposition algorithm for solving a linear programming problem subject to submodular constraints.

For a positive integer n, let $N = \{1, 2, \ldots, n\}$ be a ground set, and let 2^N denote the family of all subsets of N. For a subset $X \subseteq N$, let $\mathbb{R}^X$ denote the set of all vectors $\mathbf{p}$ with real components $p(j)$, where $j \in X$. For two vectors $\mathbf{p} = (p(1), p(2), \ldots, p(n)) \in \mathbb{R}^N$ and $\mathbf{q} = (q(1), q(2), \ldots, q(n)) \in \mathbb{R}^N$, we write $\mathbf{p} \leq \mathbf{q}$ if $p(j) \leq q(j)$ for each $j \in N$. For a vector $\mathbf{p} \in \mathbb{R}^N$, define $p(X) = \sum_{j \in X} p(j)$ for every set $X \in 2^N$.

A set-function $\varphi : 2^N \rightarrow \mathbb{R}$ is called *submodular* if the inequality $\varphi(X) + \varphi(Y) \geq \varphi(X \cup Y) + \varphi(X \cap Y)$ holds for all sets $X, Y \in 2^N$. For a submodular function φ defined on 2^N such that $\varphi(\emptyset) = 0$, the pair $(2^N, \varphi)$ is called a *submodular system* on N, while φ is referred to as its *rank function*.

For a submodular system $(2^N, \varphi)$, define two polyhedra $P(\varphi) = \{\mathbf{p} \in \mathbb{R}^N \mid p(X) \leq \varphi(X), \quad X \in 2^N\}$ and $B(\varphi) = \{\mathbf{p} \in \mathbb{R}^N \mid \mathbf{p} \in P(\varphi), \quad p(N) = \varphi(N)\}$, called the *submodular polyhedron* and the *base polyhedron*, respectively, associated with the submodular system. The main problem of our interest is as follows:

$$(\text{LP}) : \max \sum_{j \in N} w(j)p(j) \tag{2}$$
$$\text{s.t.} \quad p(X) \leq \varphi(X), \qquad X \in 2^N,$$
$$\underline{p}(j) \leq p(j) \leq \overline{p}(j), j \in N,$$

where $\varphi : 2^N \rightarrow \mathbb{R}$ is a submodular function with $\varphi(\emptyset) = 0$, $\mathbf{w} \in \mathbb{R}^N_+$ is a nonnegative weight vector, and $\overline{\mathbf{p}}, \underline{\mathbf{p}} \in \mathbb{R}^N$ are upper and lower bound vectors, respectively. We refer to (2) as *Problem (LP)*. This problem serves as a mathematical model for many SCPT problems, as demonstrated in Sects. 4 and 5.

Problem (LP) can be classified as a problem of maximizing a linear function over a submodular polyhedron intersected with a box. As shown in [29], Problem (LP) can be reduced to optimization over a base polyhedron.

Theorem 1 (cf. [29]). *If Problem (LP) has a feasible solution, then the set of its maximal feasible solutions is a base polyhedron $B(\tilde{\varphi})$ associated with the submodular system $(2^N, \tilde{\varphi})$, where the rank function $\tilde{\varphi} : 2^N \to \mathbb{R}$ is given by*

$$\tilde{\varphi}(X) = \min_{Y \in 2^N} \{\varphi(Y) + \overline{p}(X \backslash Y) - \underline{p}(Y \backslash X)\}. \tag{3}$$

Notice that the computation of the value $\tilde{\varphi}(X)$ for a given $X \in 2^N$ reduces to minimization of a submodular function, which can be done in polynomial time [10, 24]. However, the running time of known general algorithms is fairly large. In many special cases of Problem (LP), including its applications to the SCPT problems, the value $\tilde{\varphi}(X)$ can be computed more efficiently, as shown later.

Throughout this paper, we assume that Problem (LP) has a feasible solution, which is equivalent to the conditions $\underline{p} \in P(\varphi)$ and $\underline{p} \leq \overline{p}$. Theorem 1 implies that Problem (LP) reduces to the following problem:

$$\max \sum_{j \in N} w(j)p(j) \tag{4}$$

$$\text{s.t. } \mathbf{p} \in B(\tilde{\varphi}),$$

where the rank function $\tilde{\varphi} : 2^N \to \mathbb{R}$ is given by (3).

An advantage of the reduction of Problem (LP) to a problem of the form (4) is that the solution vector can be obtained essentially in a closed form by a greedy algorithm. To determine an optimal vector $\mathbf{p}^*$, the algorithm starts with $\mathbf{p}^* = \underline{p}$, considers the components of the current $\mathbf{p}^*$ in non-increasing order of their weights and gives the current component the largest possible increment that keeps the vector feasible.

Introduce the sequence $\sigma = (\sigma(1), \sigma(2), \ldots, \sigma(n))$ such that $w(\sigma(1)) \geq w(\sigma(2)) \geq \cdots \geq w(\sigma(n))$ and define $N_t(\sigma) = \{\sigma(1), \ldots, \sigma(t)\}$, $1 \leq t \leq n$, where, for completeness, $N_0(\sigma) = \emptyset$.

Theorem 2 (cf. [4]). *Vector $\mathbf{p}^* \in \mathbb{R}^N$ given by*

$$p^*(\sigma(t)) = \tilde{\varphi}(N_t(\sigma)) - \tilde{\varphi}(N_{t-1}(\sigma)), \ t = 1, 2, \ldots, n,$$

is an optimal solution to problem (4) (and also to the problem (2)).

Now we describe a decomposition algorithm for solving Problem (LP), which for certain classes of the problem may overperform the greedy algorithm.

We say that a subset $\hat{N} \subseteq N$ is a *heavy-element subset* of N with respect to the weight vector $\mathbf{w}$ if it satisfies the condition $\min_{j \in \hat{N}} w(j) \geq \max_{j \in N \backslash \hat{N}} w(j)$. For completeness, we also regard the empty set as a heavy-element subset of N.

For a given set $X \subseteq N$, in accordance with (3) define a set $Y_* \subseteq N$ such that the equality

$$\tilde{\varphi}(X) = \varphi(Y_*) + \overline{p}(X \backslash Y_*) - \underline{p}(Y_* \backslash X) \tag{5}$$

holds. In the remainder of this paper, we call Y_* an *instrumental* set for set X.

Lemma 1 (cf. [32,34]). *Let $\hat{N} \subseteq N$ be a heavy-element subset of N with respect to* $\mathbf{w}$, *and $Y_* \subseteq N$ be an instrumental set for set $\hat{N}$. Then, there exists an optimal solution $\mathbf{p}^*$ of Problem (LP) such that*

(a) $p^*(Y_*) = \varphi(Y_*)$,   (b) $p^*(j) = \overline{p}(j),\ j \in \hat{N} \backslash Y_*$,   (c) $p^*(j) = \underline{p}(j),\ j \in Y_* \backslash \hat{N}$.

In what follows, we use two fundamental operations on a submodular system $(2^N, \varphi)$, as defined in [4, Sect. 3.1]. For a set $A \in 2^N$, define a set-function $\varphi^A : 2^A \to \mathbb{R}$ by $\varphi^A(X) = \varphi(X)$, $X \in 2^A$. Then, $(2^A, \varphi^A)$ is a submodular system on A and it is called a *restriction* of $(2^N, \varphi)$ to A. On the other hand, for a set $A \in 2^N$ define a set-function $\varphi_A : 2^{N \backslash A} \to \mathbb{R}$ by $\varphi_A(X) = \varphi(X \cup A) - \varphi(A)$, $X \in 2^{N \backslash A}$. Then, $(2^{N \backslash A}, \varphi_A)$ is a submodular system on $N \backslash A$ and it is called a *contraction* of $(2^N, \varphi)$ by A.

Theorem 3 (cf. [32,34]). *Let $\hat{N} \subseteq N$ be a heavy-element subset of N with respect to* $\mathbf{w}$, *and Y_* be an instrumental set for set $\hat{N}$. Let $\mathbf{p_1} \in \mathbb{R}^{Y_*}$ and $\mathbf{p_2} \in \mathbb{R}^{N \backslash Y_*}$ be optimal solutions of the linear programs (LPR) and (LPC), respectively, given by*

$$(\text{LPR}) : \max \sum_{j \in Y_*} w(j)p(j)$$
$$\text{s.t.}\ \ p(X) \leq \varphi(X), \qquad X \in 2^{Y_*},$$
$$\underline{p}(j) \leq p(j) \leq \overline{p}(j),\ j \in Y_* \cap \hat{N},$$
$$p(j) = \underline{p}(j), \qquad j \in Y_* \backslash \hat{N}.$$

$$(\text{LPC}) : \max \sum_{j \in N \backslash Y_*} w(j)p(j)$$
$$\text{s.t.}\ \ p(X) \leq \varphi(X \cup Y_*) - \varphi(Y_*),\ X \in 2^{N \backslash Y_*},$$
$$\underline{p}(j) \leq p(j) \leq \overline{p}(j), \qquad j \in (N \backslash Y_*) \backslash \left(\hat{N} \backslash Y_* \right),$$
$$p(j) = \overline{p}(j), \qquad j \in \hat{N} \backslash Y_*.$$

Then, the vector $\mathbf{p}^* \in \mathbb{R}^N$ given by the direct sum $\mathbf{p}^* = \mathbf{p_1} \oplus \mathbf{p_2}$, where

$$(p_1 \oplus p_2)(j) = \begin{cases} p_1(j), \text{ if } j \in Y_*, \\ p_2(j), \text{ if } j \in N \backslash Y_*. \end{cases}$$

is an optimal solution of Problem (LP).

Notice that Problem (LPR) is obtained from Problem (LP) as a result of restriction to Y_* and the values of components $p(j), j \in Y_* \backslash \hat{N}$, are fixed to their lower bounds in accordance with Property (c) of Lemma 1. Similarly, Problem (LPC) is obtained from Problem (LP) as a result of contraction by Y_* and the values of components $p(j)$, $j \in \hat{N} \backslash Y_*$, are fixed to their upper bounds in accordance with Property (b) of Lemma 1.

Now we explain how the original Problem (LP) can be decomposed recursively based on Theorem 3, until we obtain a collection of trivially solvable problems with no non-fixed variables. As described in [32,34], in each stage of the recursive procedure, we need to solve a subproblem that can be written in the following generic form:

$$\text{LP}(H, F, K, \mathbf{l}, \mathbf{u}) : \max \sum_{j \in H} w(j)p(j) \tag{6}$$
$$\text{s.t.} \quad p(X) \leq \varphi_K^H(X) = \varphi(X \cup K) - \varphi(K), \; X \in 2^H,$$
$$l(j) \leq p(j) \leq u(j), \qquad\qquad j \in H \backslash F,$$
$$p(j) = u(j) = l(j), \qquad\qquad j \in F,$$

where $H \subseteq N$ is the index set of components of vector $\mathbf{p}$; $\mathbf{l} = (l(j) \mid j \in H)$ and $\mathbf{u} = (u(j) \mid j \in H)$ are, respectively, the current vectors of the lower and upper bounds on variables $p(j), j \in H$; $F \subseteq H$ is the index set of fixed components, i.e., $l(j) = u(j)$ holds for each $j \in F$; $K \subseteq N \backslash H$ is the set that defines the rank function $\varphi_K^H : 2^H \to \mathbb{R}$ such that $\varphi_K^H(X) = \varphi(X \cup K) - \varphi(K)$, $X \in 2^H$.

Suppose that Problem $\text{LP}(H, F, K, \mathbf{l}, \mathbf{u})$ of the form (6) contains at least one non-fixed variable, i.e., $|H \backslash F| > 0$. We define a function $\widetilde{\varphi}_K^H : 2^H \to \mathbb{R}$ by

$$\widetilde{\varphi}_K^H(X) = \min_{Y \in 2^H} \{\varphi_K^H(Y) + u(X \backslash Y) - l(Y \backslash X)\}. \tag{7}$$

By Theorem 1, the set of maximal feasible solutions of Problem $\text{LP}(H, F, K, \mathbf{l}, \mathbf{u})$ is given as a base polyhedron $B(\widetilde{\varphi}_K^H)$ associated with the function $\widetilde{\varphi}_K^H$. Therefore, if $|H \backslash F| = 1$ and $H \backslash F = \{j'\}$, then an optimal solution $\mathbf{p}^* \in \mathbb{R}^H$ is given by

$$p^*(j) = \begin{cases} \widetilde{\varphi}_K^H(\{j'\}), j = j', \\ u(j), \qquad j \in F. \end{cases} \tag{8}$$

Suppose that $|H \backslash F| \geq 2$. Then, we call a recursive Procedure $\text{DECOMP}(H, F, K, \mathbf{l}, \mathbf{u})$ explained below. Let $\hat{H} \subseteq H$ be a heavy-element subset of H with respect to the vector $(w(j) \mid j \in H)$, and $Y_* \subseteq H$ be an instrumental set for set $\hat{H}$, i.e.,

$$\widetilde{\varphi}_K^H(\hat{H}) = \varphi_K^H(Y_*) + u(\hat{H} \backslash Y_*) - l(Y_* \backslash \hat{H}). \tag{9}$$

Without going into implementation details, we follow [32,34] and give a formal description of the recursive procedure. For the current Problem $\text{LP}(H, F, K, \mathbf{l}, \mathbf{u})$, we compute optimal solutions $\mathbf{p_1} \in \mathbb{R}^{Y_*}$ and $\mathbf{p_2} \in \mathbb{R}^{H \backslash Y_*}$ of the two subproblems by calling Procedures $\text{DECOMP}(Y_*, F_1, K, \mathbf{l_1}, \mathbf{u_1})$ and $\text{DECOMP}(H \backslash Y_*, F_2, K \cup Y_*, \mathbf{l_2}, \mathbf{u_2})$. By Theorem 3, the direct sum $\mathbf{p}^* = \mathbf{p_1} \oplus \mathbf{p_2}$ is an optimal solution of Problem $\text{LP}(H, F, K, \mathbf{l}, \mathbf{u})$, which is the output of Procedure $\text{DECOMP}(H, F, K, \mathbf{l}, \mathbf{u})$.

Procedure Decomp$(H, F, K, \mathbf{l}, \mathbf{u})$

Step 1. If $|H \backslash F| = 0$, then output the vector $\mathbf{p}^* = \mathbf{u} \in \mathbb{R}^H$ and return. If $|H \backslash F| = 1$ and $H \backslash F = \{j'\}$, then compute the value $\tilde{\varphi}_K^H(\{j'\})$, output the vector $\mathbf{p}^*$ given by (8) and return.

Step 2. Select a heavy-element subset $\hat{H}$ of $H \backslash F$ with respect to $\mathbf{w}$, and determine an instrumental set $Y_* \subseteq H$ for set $\hat{H}$ satisfying (9).

Step 3. Define the vectors $\mathbf{l}_1, \mathbf{u}_1 \in \mathbf{R}^{Y_*}$ and set F_1 by

$$l_1(j) = l(j), j \in Y_*, \quad u_1(j) = \begin{cases} l(j), \ j \in Y_* \backslash \hat{H}, \\ u(j), \ j \in Y_* \cap \hat{H}, \end{cases} ; \quad F_1 = Y_* \backslash \hat{H},$$

Call Procedure DECOMP$(Y_*, F_1, K, \mathbf{l}_1, \mathbf{u}_1)$ to obtain an optimal solution $\mathbf{p}_1 \in \mathbb{R}^{Y_*}$ of Problem LP$(Y_*, F_1, K, \mathbf{l}_1, \mathbf{u}_1)$.

Step 4. Define the vectors $\mathbf{l}_2, \mathbf{u}_2 \in \mathbf{R}^{H \backslash Y_*}$ and set F_2 by

$$l_2(j) = \begin{cases} u(j), j \in \hat{H} \backslash Y_*, \\ l(j), j \in H \backslash (Y_* \cup \hat{H}), \end{cases} \quad u_2(j) = u(j), j \in H \backslash Y_*;$$
$$F_2 = (\hat{H} \cup (H \cap F)) \backslash Y_*.$$

Call Procedure DECOMP$(H \backslash Y_*, F_2, K \cup Y_*, \mathbf{l}_2, \mathbf{u}_2)$ to obtain an optimal solution $\mathbf{p}_2 \in \mathbb{R}^{H \backslash Y_*}$ of Problem LP$(H \backslash Y_*, F_2, K \cup Y_*, \mathbf{l}_2, \mathbf{u}_2)$.

Step 5. Output the direct sum $\mathbf{p}^* = \mathbf{p}_1 \oplus \mathbf{p}_2 \in \mathbb{R}^H$ and return.

The original Problem (LP) is solved by calling Procedure DECOMP$(N, \emptyset, \emptyset, \mathbf{p}, \overline{\mathbf{p}})$. Its actual running time depends on the choice of a heavy-element subset $\hat{H}$ in Step 2 and on the time complexity of finding an instrumental set Y_*. As proved in [32], if at each level of recursion a heavy-element set is chosen to contain roughly a half of the non-fixed variables, then the overall depth of recursion of Procedure DECOMP applied to Problem LP$(N, \emptyset, \emptyset, \mathbf{p}, \overline{\mathbf{p}})$ is $O(\log n)$.

For a typical iteration of Procedure DECOMP applied to Problem LP$(H, F, K, \mathbf{l}, \mathbf{u})$ with $|H| = h$ and $|H \backslash F| = g$, let $T_{Y_*}(h)$ denote the running time for computing the value $\tilde{\varphi}_K^H(\hat{H})$ for a given set $\hat{H} \subseteq H$ and finding an instrumental set Y_* in Step 2. In Steps 3 and 4, Procedure DECOMP splits Problem LP$(H, F, K, \mathbf{l}, \mathbf{u})$ into two subproblems: one with h_1 variables among which $g_1 \leq \min\{h_1, \lceil g/2 \rceil\}$ variables are not fixed, and the other one with $h_2 = h - h_1$ variables, among which $g_2 \leq \min\{h_2, \lfloor g/2 \rfloor\}$ variables are not fixed. Let $T_{\text{Split}}(h)$ denote the time complexity for setting up the instances of these two subproblems. It is shown in [32,34] that Problem (LP) can be solved by Procedure DECOMP in $O((T_{Y_*}(n) + T_{\text{Split}}(n)) \log n)$ time.

4   SCPT Problems with a Common Deadline

In this section, we review the results on the SCPT problems $\alpha | r(j), p(j) = \overline{p}(j) - x(j), pmtn, C(j) \leq d | W$, where $\alpha \in \{1, P, Q\}$, provided that the jobs have a common deadline d. We also report the results on the makespan min-imization versions $\alpha|r(j), pmtn|C_{\max}$ and the bicriteria versions $\alpha|r(j), p(j) = \overline{p}(j) - x(j), pmtn| (C_{\max}, W)$. For the latter type of problems, it is required to minimize both objective functions $C_{\max}$ and W simultaneously, in the Pareto sense, so that the solution is delivered in the form of an efficiency frontier.

First, assume that the release dates are equal to zero, so that the prob-lems with a single machine are trivial. Solving problem $P|pmtn|C_{\max}$ with fixed processing times can be done by a linear-time algorithm [18]. As shown in [12], problem $P|p(j) = \overline{p}(j) - x(j), pmtn, C(j) \le d|W$ reduces to a continu-ous knapsack problem and can be solved in $O(n)$ time. The bicriteria problem $P|p(j) = \overline{p}(j) - x(j), pmtn| (C_{\max}, W)$ is solved in [27] by an $O(n \log n)$-time algorithm, which is the best possible.

In the case of uniform machines, the best known algorithm for solving problem $Q|pmtn|C_{\max}$ with fixed processing times is due to [6]. For problem $Q|p(j) = \overline{p}(j) - x(j), pmtn, C(j) \le d|W$, it is shown in [21] how to find the actual processing times in $O(nm + n \log n)$ time. For the latter problem, Shakhlevich and Strusevich [28] use the SO reasoning to design an algorithm of the same running time and extend it to solving a bicriteria problem $Q|p(j) = \overline{p}(j) - x(j), pmtn| (C_{\max}, W)$. The fastest algorithms for solving problems $Q|p(j) = \overline{p}(j) - x(j), pmtn| (C_{\max}, W)$ and $Q|p(j) = \overline{p}(j) - x(j), pmtn, C(j) \le d|W$ are discussed in Sect. 4.2 and in Sect. 4.3, respectively; their respective running times are $O(nm \log m)$ and $O(n \log n)$.

For the models with different release dates, problem $1|r(j), p(j) = \overline{p}(j) - x(j), pmtn, C(j) \le d(j)|W$ is one of the most studied SCPT problems. The first algorithm that requires $O(n \log n)$ time and provides all implementation details is developed in [27].

Problem $P|r(j), pmtn|C_{\max}$ with fixed processing times on m identical par-allel machines can be solved in $O(n \log n)$ time, as proved in [22]. For problem $Q|r(j), pmtn|C_{\max}$, an algorithm for that requires $O(mn + n \log n)$ time is due to [23]. Prior to work of our team on the links between the SCPT problems and SO [31], no purpose-built algorithms had been known for problems $Q|p(j) = \overline{p}(j) - x(j), pmtn, C(j) \le d|W$ and $\alpha|r(j), p(j) = \overline{p}(j) - x(j), pmtn, C(j) \le d|W$ with $\alpha \in \{P, Q\}$, as well as for their bicriteria counterparts. We consider these problems in Sects. 4.3 and 4.2, respectively.

4.1   Production Capacity Rank Functions

In this subsection, we present reformulations of the SCPT problems with a com-mon deadline in terms of Problem (LP) with appropriately defined rank func-tions.

We assume that if the jobs have different release dates, they are numbered to satisfy

$$r(1) \le r(2) \le \ldots \le r(n). \tag{10}$$

If the machines are uniform, they are numbered in accordance with (1). We denote

$$S_0 = 0, \qquad S_k = s_1 + s_2 + \cdots + s_k, \quad 1 \le k \le m. \tag{11}$$

S_k represents the total speed of k fastest machines; if the machines are identical, $S_k = k$ holds.

For each problem $Q|p(j) = \overline{p}(j) - x(j), C(j) \leq d, pmtn|W$, $P|r(j), p(j) = \overline{p}(j) - x(j), C(j) \leq d, pmtn|W$ and $Q|r(j), p(j) = \overline{p}(j) - x(j), C(j) \leq d, pmtn|W$, we need to find the actual processing times $p(j) = \overline{p}(j) - x(j)$, $j \in N$, such that all jobs can be completed by a common deadline d and the total compression cost $W = \sum_{j \in N} w(j)x(j)$ is minimized. Each of these problems can be formulated as Problem (LP) with $p(j)$, $j \in N$, being decision variables, and the objective function to be maximized being $\sum_{j \in N} w(j)p(j) = \sum_{j \in N} w(j)(\overline{p}(j) - x(j))$. Since each decision variable $p(j)$ has a lower bound $\underline{p}(j)$ and an upper bound $\overline{p}(j)$, the set of constraints of Problem (LP) includes the box constraints of the form $\underline{p}(j) \leq p(j) \leq \overline{p}(j)$, $j \in N$. Besides, the inequality

$$p(X) \leq \varphi(X) \tag{12}$$

should hold for each subset $X \subseteq N$ of jobs, where a meaningful interpretation of a rank function $\varphi(X)$ is the largest capacity available for processing the jobs of set X.

To determine $\varphi(X)$ for each of these SCPT problems, an important generic condition is available, which, according to [2] can be stated as follows: for a given deadline d a feasible schedule exists if and only if: (i) for each $k, 1 \leq k \leq m - 1$, k longest jobs can be processed on k fastest machines by time d, and (ii) all n jobs can be completed on all m machines by time d.

For example, problem $Q|p(j) = \overline{p}(j) - x(j), C(j) \leq d, pmtn|W$, in which all jobs are simultaneously available from time zero, reduces to Problem (LP) of the form (2) with the rank function

$$\varphi(X) = dS_{\min\{|X|, m\}} = \begin{cases} dS_{|X|}, & \text{if } |X| \leq m - 1, \\ dS_m, & \text{if } |X| \geq m. \end{cases} \tag{13}$$

It is clear that the conditions $p(X) \leq \varphi(X)$, $X \in 2^N$, for the function $\varphi(X)$ defined by (13) correspond to the conditions (i) and (ii) above, provided that $|X| \leq m - 1$ and $|X| \geq m$, respectively. As proved in [28], function φ is submodular.

Given problem $Q|r(j), p(j) = \overline{p}(j) - x(j), C(j) \leq d, pmtn|W$, for a set of jobs $X \subseteq N$, we define $r_i(X)$ to be the i-th smallest release date in set $X \in 2^N$, $1 \leq i \leq |X|$. Then, for a non-empty set X of jobs, the largest processing capacity available on the fastest machine M_1 is $s_1(d - r_1(X))$, the total largest processing capacity on two fastest machines M_1 and M_2 is $s_1(d - r_1(X)) + s_2(d - r_2(X))$, etc. We deduce that

$$\varphi(X) = \begin{cases} dS_{|X|} - \sum_{i=1}^{|X|} s_i r_i(X), & \text{if } |X| \leq m - 1, \\ dS_m - \sum_{i=1}^{m} s_i r_i(X), & \text{if } |X| \geq m, \end{cases} \tag{14}$$

which in the case of problem $P|r(j), p(j) = \overline{p}(j) - x(j), C(j) \leq d, pmtn|W$ simplifies to

$$\varphi(X) = \begin{cases} d|X| - \sum_{i=1}^{|X|} r_i(X), & \text{if } |X| \leq m - 1, \\ dm - \sum_{i=1}^{m} r_i(X), & \text{if } |X| \geq m. \end{cases} \tag{15}$$

It can be verified that functions (14) and (15) are submodular.

Thus, each of the three SCPT problems above reduces to Problem (LP) and in principle can be solved by the greedy algorithm discussed in Theorem 2. Similar reductions can be provided for other SCPT problems [29]. In what follows, we show that using the decomposition algorithm from Sect. 3, these problems can be solved faster.

Notice that the greedy reasoning has always been the main tool for solving the SCPT problems. However, in early papers on this topic, each individual problem was considered separately and a justification of the greedy approach was often lengthy and developed from the first principles. In fact, as seen from above, the greedy nature of the solution approaches is due to the fact that many SCPT problems can be reformulated in terms of linear programming problems with submodular constraints.

4.2   Solving Bicriteria Problems by Submodular Methods

Theorem 2 provides the foundation to an approach that finds the efficiency frontier of the bicriteria scheduling problems $Q|p(j) = \overline{p}(j) - x(j), pmtn| (C_{\max}, W)$ and $\alpha m|r(j), p(j) = \overline{p}(j) - x(j), pmtn| (C_{\max}, W)$ with $\alpha \in \{P, Q\}$ in a closed form [31].

Given an instance of the problem, let $S^*(d)$ denote a schedule with a makespan $C_{\max} = d$ that minimizes the total compression cost. The solution to the bicriteria problem will be delivered as a collection of break points of the efficiency frontier $(d, W(d))$, where d is a value of the makespan of schedule $S^*(d)$ and $W(d)$ is a (piecewise-linear in d) function that represents the total optimal compression cost. Let also $p^*(j, d)$ denote the optimal value of the actual processing time of job j in schedule $S^*(d)$. It follows that

$$W(d) = \sum_{t=1}^{n} w\left(\sigma(t)\right) p^*\left(\sigma(t), d\right). \tag{16}$$

For the problems under consideration, due to (13), (14) and (15), the rank function $\varphi(X)$ as well as the function $\tilde{\varphi}(X)$ of the form (3) are functions of d; therefore in this paper we may write $\varphi(X, d)$ and $\tilde{\varphi}(X, d)$ whenever we want to stress that dependence.

Given a value of d such that all jobs can be completed by time d, define a function

$$\psi_t(d) = \tilde{\varphi}(N_t(\sigma), d), \quad 1 \le t \le n, \tag{17}$$

computed for this value of d. By (5),

$$\psi_t(d) = \overline{p}(N_t(\sigma)) + \min_{Y \in 2^N} \left\{\varphi(Y) - \overline{p}(N_t(\sigma) \cap Y) - \underline{p}(Y \backslash N_t(\sigma))\right\}.$$

For all scheduling problems under consideration, due to (13), (14) and (15), there are m expressions for $\varphi(X)$, depending on whether $|X| \le m-1$ or $|X| \ge m$, and $\psi_t(d)$ can be represented as a piecewise-linear function of the form of an envelope with $m + 1$ pieces.

Setting for completeness $w\left(\sigma(n+1)\right)=0$, we deduce from Theorem 2 that

$$W(d)=\sum_{t=1}^{n} w\left(\sigma(t)\right)\left(\psi_t(d)-\psi_{t-1}(d)\right)=\sum_{t=1}^{n}\left(w\left(\sigma(t)\right)-w\left(\sigma(t+1)\right)\right)\psi_t(d). \tag{18}$$

Thus, in order to be able to compute the (piecewise-linear) function $W(d)$, we first have to compute the functions $\psi_t(d)$, $1 \le t \le n$, for all relevant values of d. It is shown in [31], that after the functions $\psi_t(d)$, $1 \le t \le n$, for all relevant values of d are found, their weighted sum (18) can be computed in $O\left(nm\log n\right)$ time. Thi s (piecewise-linear) function $W(d)$ fully defines the efficiency frontier for the corresponding bicriteria scheduling problem.

Adapting this general framework to problem $Q|p(j) = \overline{p}(j) - x(j)$, $pmtn|(C_{\max}, W)$, it can be proved that all required functions $\psi_t(d)$, $1 \le t \le n$, can be found in $O\left(n\log n + nm\right)$ time, and the overall problem is solvable in $O\left(nm\log m\right)$ time, while problems $\alpha m|r(j), p(j) = \overline{p}(j) - x(j), pmtn|\left(C_{\max}, W\right)$ can be solved in $O\left(n^2\log m\right)$ time and in $O(n^2 m)$ time for $\alpha = P$ and $\alpha = Q$, respectively.

4.3   Solving Single Criterion Problems by Decomposition

We now show that problems $Q|p(j) = \overline{p}(j) - x(j), pmtn, C\left(j\right) \le d|W$ and $\alpha m|r(j), p(j) = \overline{p}(j) - x(j), pmtn, C\left(j\right) \le d|W$ with $\alpha \in \{P, Q\}$ can be solved faster than is guaranteed by the algorithms for the respective bicriteria problems considered in Sect. 4.2. This is achieved by adapting the decomposition algorithm based on Procedure DECOMP presented in Sect. 3. The crucial issue here is the computation of the instrumental set Y_* in each iteration.

For an initial Problem $LP(N, \emptyset, \emptyset, \mathbf{l}, \mathbf{u})$, associated with one of the three scheduling problems above, assume that the following preprocessing is done in $O(n\log n)$ time before calling Procedure DECOMP$(N, \emptyset, \emptyset, \mathbf{l}, \mathbf{u})$: the jobs are numbered in non-decreasing order of their release dates in accordance with (10); the machines are numbered in non-increasing order of their speeds in accordance with (1), and the partial sums S_v are computed for all v, $0 \le v \le m$, by (11); the lists $(l(j) \mid j \in N)$ and $(u(j) \mid j \in N)$ are formed and their elements are sorted in non-decreasing order.

In a typical iteration of Procedure DECOMP applied to Problem $LP(H, F, K, \mathbf{l}, \mathbf{u})$ of the form (6) related to the rank function $\varphi_K^H(Y) = \varphi(Y \cup K) - \varphi(K)$, it is shown in [32] that for a given set $X \subseteq H$ the function $\widetilde{\varphi}_K^H : 2^H \to \mathbb{R}$ can be computed as

$$\widetilde{\varphi}_K^H(X) = u(X) - \varphi(K) + \min_{Y \in 2^H}\left\{\varphi(Y \cup K) - b(Y)\right\}, \tag{19}$$

where φ is the initial rank function associated with the scheduling problem under consideration, and

$$b(j) = \begin{cases} u(j), & \text{if } j \in X, \\ l(j), & \text{if } j \in H \backslash X. \end{cases} \tag{20}$$

Notice that if the minimum in the right-hand side of (19) is achieved for $Y = Y_*$, then Y_* is an instrumental set for set X.

For Problem $LP(H, F, K, \mathbf{l}, \mathbf{u})$ associated with problem $Q|p(j) = \bar{p}(j) - x(j), pmtn, C(j) \le d|W$ due to (13) and (19) we deduce that

$$\tilde{\varphi}_K^H(X) = u(X) - dS_{\min\{m,k\}} + \min\{\Phi', \Phi''\}. \tag{21}$$

Here, $\Phi' = +\infty$ if $h > m - k - 1$; otherwise

$$\Phi' = \min_{0 \le v \le m-k-1} \{dS_{v+k} - \sum_{i=1}^{v} b_i\}, \tag{22}$$

where b_i is the i-th largest value in the list $(b(j) \mid j \in H)$, while $\Phi'' = +\infty$ if $h \le m - k - 1$; otherwise $\Phi'' = dS_m - b(H)$. In any case, in terms of the notions introduced in Sect. 3 we deduce that $T_{Y_*}(h) = T_{\text{Split}}(h) = O(h)$, so that the overall running time needed to solve problem $Q|p(j) = \bar{p}(j) - x(j), pmtn, C(j) \le d|W$ by the decomposition algorithm based on recursive applications of Procedure DECOMP is $O(n \log n)$. An alternative implementation of the same approach, also presented in [32], does not involve a full preprocessing and requires $O(n + m \log m \log n)$ time.

Problem $P|r(j), p(j) = \bar{p}(j) - x(j), C(j) \le d, pmtn|W$ and Problem $Q|r(j), p(j) = \bar{p}(j) - x(j), C(j) \le d, pmtn|W$ can be solved by the decomposition algorithm in $O(n \log m \log n)$ time and in $O(nm \log n)$ time, respectively.

5   SCPT Problems with Distinct Deadlines

We start with a brief review of the feasibility problems $\alpha|r(j), pmtn, C(j) \le d(j)|\circ$, where $\alpha \in \{1, P, Q\}$, in which for each job $j \in N$ the processing time $p(j)$ is fixed and the task is to verify the existence of a deadline-feasible preemptive schedule.

Divide the interval $[\min_{j \in N} r(j), \max_{j \in N} d(j)]$ into subintervals by using the release dates $r(j)$ and the deadlines $d(j)$ for $j \in N$. Let $T = (\tau_0, \tau_1, \ldots, \tau_\gamma)$, where $1 \le \gamma \le 2n - 1$, be the increasing sequence of distinct numbers in the list $(r(j), d(j) \mid j \in N)$. Introduce the intervals $I_k = [\tau_{k-1}, \tau_k]$, $1 \le k \le \gamma$. Denote the length of interval I_k by $\Delta_k = \tau_k - \tau_{k-1}$.

For a set of jobs $X \subseteq N$, let $\varphi(X)$ be a set-function that represents the total production capacity available for the feasible processing of the jobs of set X.

For a particular problem, the function $\varphi(X)$ can be suitably defined. Interval I_k is *available* for processing job j if $r(j) \le \tau_{k-1}$ and $d(j) \ge \tau_k$. For a job j, denote the set of the available intervals by $\Gamma(j)$. For a set of jobs $X \subseteq N$, introduce the set-function

$$\varphi_1(X) = \sum_{I_k \in \cup_{j \in X} \Gamma(j)} \Delta_k. \tag{23}$$

Then for problem $1|r(j), pmtn, C(j) \le d(j)|\circ$ a feasible schedule exists if and only if inequality (12) holds for all sets $X \subseteq N$ for $\varphi(X) = \varphi_1(X)$.

Such a statement (in different terms) was first formulated in [7] and [9]. For the problems on parallel machines, the corresponding representation of the total processing capacity in the form of a set-function is defined in [28]. For all versions of the problem, with a single or parallel machines, the set-function φ is submodular.

The single machine feasibility problem $1|r(j), pmtn, C(j) \leq d(j)|\circ$ in principle cannot be solved faster than finding the ordered sequence $T = (\tau_0, \tau_1, \ldots, \tau_\gamma)$ of the release dates and deadlines. The best possible running time $O(n \log n)$ for solving problem $1|r(j), pmtn, C(j) \leq d(j)|\circ$ is achieved by the EDF (Earliest Deadline First) algorithm designed in [9].

For parallel machine problems, it is efficient to reformulate the problem of checking the inequalities (12) in terms of finding the maximum flow in a special bipartite network; see, e.g., [3]. Using an algorithm from [1], such a network problem can be solved in $O(n^3)$ time and in $O(mn^3)$ time, for problem $P|r(j), pmtn, C(j) \leq d(j)|\circ$ and problem $Q|r(j), pmtn, C(j) \leq d(j)|\circ$, respectively.

Most studies on the SCPT problems $\alpha|r(j), p(j) = \overline{p}(j) - x(j), pmtn, C(j) \leq d(j)|W$, where $\alpha \in \{1, P, Q\}$, have been conducted within the body of research on imprecise computation scheduling [15]; however, the best known algorithms have been produced by alternative methods.

For the SCPT problems on parallel machines, the most efficient algorithms are based on reductions to the parametric max-flow problems in bipartite networks. McCormick [17] develops an extension of the parametric flow algorithm in [5] and this approach gives the running times of $O(n^3)$ for problem $P|r(j), p(j) = \overline{p}(j) - x(j), pmtn, C(j) \leq d(j)|W$ and of $O(mn^3)$ for problem $Q|r(j), p(j) = \overline{p}(j) - x(j), pmtn, C(j) \leq d(j)|W$, matching the best known times for the corresponding problems with fixed processing times.

Notice that in most papers on imprecise computation scheduling it is claimed that $P|r(j), p(j) = \overline{p}(j) - x(j), pmtn, C(j) \leq d(j)|W$ can be solved faster, in $O(n^2 \log^2 n)$ time, by reducing it to the min-cost flow problem in a special network; see [15]. However, as demonstrated in [33], such a representation, although possible for a single machine problem, cannot be extended to the parallel machines models, so that the best known running time for solving problem $P|r(j), p(j) = \overline{p}(j) - x(j), pmtn, C(j) \leq d(j)|W$, as well as its counterpart with fixed processing times, is $O(n^3)$.

Problem $1|r(j), p(j) = \overline{p}(j) - x(j), pmtn, C(j) \leq d(j)|W$, for many years has been an object of intensive study, mainly within the body of research on imprecise computation. The history of studies on this problem is a race for developing an $O(n \log n)$-time algorithm, matching the best possible estimate achieved for a simpler feasibility problem $1|r(j), pmtn, C(j) \leq d(j)|\circ$.

Hochbaum and Shamir [8] present two algorithms, one solves problem $1|r(j), p(j) = \overline{p}(j) - x(j), pmtn, C(j) \leq d(j)|W$ in $O(n^2)$ time and the other solves its counterpart with the unweighted objective function in $O(n \log n)$ time. An algorithm for problem $1|r(j), p(j) = \overline{p}(j) - x(j), pmtn, C(j) \leq d(j)|W$ developed in [16] requires $O(n \log n + \kappa n)$ time, where κ is the number of distinct weights $w(j)$, while an algorithm in [30] takes $O\left(n \log^2 n\right)$ time, provided that the numbers $\overline{p}(j), \underline{p}(j), r(j), d(j)$ are integers.

5.1   Solving Single Machine Problem by Decomposition

The time complexity of problem $1|r(j), p(j) = \overline{p}(j) - x(j), pmtn, C(j) \leq d(j)|W$ is finally settled in [34], where an $O\left(n \log n\right)$-time algorithm is given. The algorithm is based on a decomposition algorithm for Problem (LP) and uses an algorithm from [8] as a subroutine.

The efficient implementation of the decomposition algorithm developed in [34] is based on the following statement.

Theorem 4 (cf. [4, Corollary 3.4]). *For a submodular system* $(2^H, \varphi)$ *and a vector* $\mathbf{b} \in \mathbb{R}^H$, *the equality*

$$\min_{Y \in 2^H} \{\varphi(Y) + b(H \backslash Y)\} = \max\{p(H) \mid \mathbf{p} \in P(\varphi), \ \mathbf{p} \leq \mathbf{b}\}$$

holds. In particular, if $\mathbf{b} \geq \mathbf{0}$ *and* $\varphi(X) \geq 0$ *for all* $X \subseteq N$ *then the right-hand side is equal to* $\max\{p(H) \mid \mathbf{p} \in P(\varphi), \ \mathbf{0} \leq \mathbf{p} \leq \mathbf{b}\}$.

Given Problem $LP(H, F, K, \mathbf{l}, \mathbf{u})$ of the form (6), for a set $X \subseteq H$ define the vector $\mathbf{b} \in \mathbb{R}^H$ by (20), and for a set $X \subseteq H$ represent $\tilde{\varphi}_K^H(X)$ in the form

$$\tilde{\varphi}_K^H(X) = \min_{Y \in 2^H} \{\varphi_K^H(Y) + u(X \backslash Y) - l(Y \backslash X)\}$$

$$= -l(H \backslash X) + \min_{Y \in 2^H} \{\varphi_K^H(Y) + b(H \backslash Y)\}.$$

Since $-l(H \backslash X)$ is a constant, in order to find an instrumental set Y_* that defines $\tilde{\varphi}_K^H(X)$ it suffices to find the set-minimizer for $\min_{Y \in 2^H} \{\varphi_K^H(Y) + b(H \backslash Y)\}$. By Theorem 4, the latter minimization problem is equivalent to the following auxiliary problem:

$$(\text{AuxLP}) : \max \sum_{j \in H} q(j) \tag{24}$$
$$\text{s.t.} \quad q(Y) \leq \varphi_K^H(Y), \ Y \in 2^H;$$
$$0 \leq q(j) \leq b(j), \ j \in H.$$

Let $\mathbf{q}_* \in \mathbb{R}^H$ be an optimal solution to Problem (AuxLP) with the values $b(j)$ defined with respect to a set $X \subseteq H$. It is proved in [34] that a set Y_* is the required instrumental set for Problem $LP(H, F, K, \mathbf{l}, \mathbf{u})$ of the form (6) if and only if

$$q_*(Y_*) = \varphi_K^H(Y_*); \qquad q(j) = b(j), \ j \in H \backslash Y_*.$$

Problem $1|r(j), p(j) = \overline{p}(j) - x(j), pmtn, C(j) \leq d(j)| \sum w(j)x(j)$ reduces to Problem (LP) with the rank function $\varphi = \varphi_1$ defined by (23). Consider a typical iteration of Procedure DECOMP applied to Problem $LP(H, F, K, \mathbf{l}, \mathbf{u})$ of the form (6) related to the rank function $\varphi_K^H(Y) = \varphi(Y \cup K) - \varphi(K)$.

For a set $X \subseteq H$ of jobs, a meaningful interpretation of $\varphi_K^H(X)$ is the total length of the time intervals originally available for processing the jobs of set $X \cup K$ after the intervals for processing the jobs of set K have been completely used up.

Select a heavy-element set $\hat{H}$ and define the values $b(j)$ by (20) applied to $X = \hat{H}$. Our goal is to find an instrumental set Y_* for set $\hat{H}$. As described above, for this purpose we may solve the auxiliary Problem (ULP)

$$(\text{ULP}) : \max \sum_{j \in H} q(j) \tag{25}$$
$$\text{s.t.} \quad q(X) \leq \psi(X), \ X \in 2^H,$$
$$0 \leq q(j) \leq b(j), \ j \in H.$$

Problem (ULP) can be seen as a version of a scheduling problem $1|r(j), q(j) = b(j) - x(j), pmtn, C(j) \leq d(j)| \sum x(j)$, in which it is required to determine the actual processing times $q(j)$ of jobs of set H to maximize the total (unweighted) actual processing time, provided that $0 \leq q(j) \leq b(j)$ for each $j \in H$. It can be solved by an algorithm developed by Hochbaum and Shamir [8], which uses the UNION-FIND technique and guarantees that the actual processing times of all jobs and the corresponding optimal schedule are found in $O(h)$ time, provided that the jobs are renumbered in non-increasing order of their release dates. The algorithm is based on the latest-release-date-first rule. Informally, the jobs are taken one by one in the order of their numbering and each job $j \in H$ is placed into the current partial schedule to fill the available time intervals consecutively, from right to left, starting from the right-most available interval. The assignment of a job j is complete either if its actual processing time $q(j)$ reaches its upper bound $b(j)$ or if no available interval is left. Only a slight modification of the Hochbaum-Shamir algorithm is required to find not only the optimal values $q_*(j)$ of the processing times, but also an associated instrumental set. The running time of modified algorithm is still $O(h)$.

In terms of the notions introduced in Sect. 3 we deduce that $T_{Y_*}(h) = T_{\text{Split}}(h) = O(h)$, so that the overall running time needed to solve problem $1|r(j), p(j) = \overline{p}(j) - x(j), pmtn, C(j) \leq d(j)|W$ by the decomposition algorithm based on recursive applications of Procedure DECOMPis $O(n \log n)$.

6   Conclusions

In this paper, we demonstrate how the SCPT problems on parallel machines can be solved efficiently by applying methods of submodular optimization. For single criterion SCPT problems to minimize the total compression costs a developed decomposition recursive algorithm for maximizing a linear function over a submodular polyhedron intersected with a box is especially useful, since it leads to fast algorithms, some of which are the best possible. Another area of applications of submodular reformulations of the SCPT problems includes bicriteria problems, for which either faster than previously known algorithms are obtained or first polynomial algorithms are designed.

We intend to extend this approach to other scheduling models with controllable processing parameters, in particular to speed scaling problems. It will be interesting to identify problems, including those outside the area of scheduling, for which an adaptation of our approach is beneficial.

Acknowledgement. This research was supported by the EPSRC funded project EP/J019755/1 "Submodular Optimisation Techniques for Scheduling with Controllable Parameters". The first author was partially supported by the Humboldt Research Fellowship of the Alexander von Humboldt Foundation and by Grant-in-Aid of the Ministry of Education, Culture, Sports, Science and Technology of Japan.

References

1. Ahuja, R.K., Orlin, J.B., Stein, C., Tarjan, R.E.: Improved algorithms for bipartite network flow. SIAM J. Comput. **23**, 906–933 (1994)
2. Brucker, P.: Scheduling Algorithms, 5th edn, p. 371. Springer, Heidelberg (2007)
3. Federgruen, A., Groenevelt, H.: Preemptive scheduling of uniform machines by ordinary network flow techniques. Manag. Sci. **32**, 341–349 (1986)
4. Fujishige, S.: Submodular Functions and Optimization. Annals of Discrete Mathematics, vol. 58, 2nd edn. Elsevier, Amsterdam (2005)
5. Gallo, G., Grigoriadis, M.D., Tarjan, R.E.: A fast parametric maximum flow algorithm and applications. SIAM J. Comput. **18**, 30–55 (1989)
6. Gonzales, T.F., Sahni, S.: Preemptive scheduling of uniform processor systems. J. ACM **25**, 92–101 (1978)
7. Gordon, V.S., Tanaev, V.S.: Deadlines in single-stage deterministic scheduling. Optimization of Systems for Collecting, Transfer and Processing of Analogous and Discrete Data in Local Information Computing Systems. Materials of the 1st Joint Soviet-Bulgarian seminar (Institute of Engineering Cybernetics of Academy of Sciences of BSSR - Institute of Engineering Cybernetics of Bulgarian Academy of Sciences, Minsk), pp. 53–58 (1973) (in Russian)
8. Hochbaum, D.S., Shamir, R.: Minimizing the number of tardy job units under release time constraints. Discr. Appl. Math. **28**, 45–57 (1990)
9. Horn, W.: Some simple scheduling algorithms. Naval Res. Logist. Q. **21**, 177–185 (1974)
10. Iwata, S., Fleischer, L., Fujishige, S.: A combinatorial, strongly polynomial-time algorithm for minimizing submodular functions. J. ACM **48**, 761–777 (2001)
11. Janiak, A., Kovalyov, M.Y.: Single machine scheduling with deadlines and resource dependent processing times. Eur. J. Oper. Res. **94**, 284–291 (1996)
12. Jansen, K., Mastrolilli, M.: Approximation schemes for parallel machine scheduling problems with controllable processing times. Comput. Oper. Res. **31**, 1565–1581 (2004)
13. Katoh, N., Ibaraki, T.: Resource allocation problems. In: Du, D.-Z., Pardalos, P.M. (eds.) Handbook of Combinatorial Optimization, vol. 2, pp. 159–260. Kluwer, Dordrecht (1998)
14. Lawler, E.L., Lenstra, J.K., Rinnooy Kan, A.H.G., Shmoys, D.B.: Sequencing and scheduling: algorithms and complexity. In: Graves, S.C., Rinnooy Kan, A.H.G., Zipkin, P.H. (eds.) Handbooks in Operations Research and Management Science. Logistics of Production and Inventory, vol. 4, pp. 445–522. Elsevier, Amsterdam (1993)

15. Leung, J.Y.T.: Minimizing total weighted error for imprecise computation tasks. In: Leung, J.Y.T. (ed.) Handbook of Scheduling: Algorithms, Models and Performance Analysis, pp. 34-1–34-16. Chapman & Hall/CRC, Boca Raton (2004)

16. Leung, J.Y.-T., Yu, V.K.M., Wei, W.-D.: Minimizing the weighted number of tardy task units. Discr. Appl. Math. **51**, 307–316 (1994)

17. McCormick, S.T.: Fast algorithms for parametric scheduling come from extensions to parametric maximum flow. Oper. Res. **47**, 744–756 (1999)

18. McNaughton, R.: Scheduling with deadlines and loss functions. Manage. Sci. **12**, 1–12 (1959)

19. Nemhauser, G.L., Wolsey, L.A.: Integer and Combinatorial Optimization. Wiley, New York (1988)

20. Nowicki, E., Zdrzałka, S.: A survey of results for sequencing problems with controllable processing times. Discr. Appl. Math. **26**, 271–287 (1990)

21. Nowicki, E., Zdrzałka, S.: A bicriterion approach to preemptive scheduling of parallel machines with controllable job processing times. Discr. Appl. Math. **63**, 237–256 (1995)

22. Sahni, S.: Preemptive scheduling with due dates. Oper. Res. **27**, 925–934 (1979)

23. Sahni, S., Cho, Y.: Scheduling independent tasks with due times on a uniform processor system. J. ACM **27**, 550–563 (1980)

24. Schrijver, A.: A combinatorial algorithm minimizing submodular functions in strongly polynomial time. J. Comb. Theory B **80**, 346–355 (2000)

25. Schrijver, A.: Combinatorial Optimization: Polyhedra and Efficiency, p. 1879. Springer, Heidelberg (2003)

26. Shabtay, D., Steiner, G.: A survey of scheduling with controllable processing times. Discr. Appl. Math. **155**, 1643–1666 (2007)

27. Shakhlevich, N.V., Strusevich, V.A.: Pre-emptive scheduling problems with controllable processing times. J. Sched. **8**, 233–253 (2005)

28. Shakhlevich, N.V., Strusevich, V.A.: Preemptive scheduling on uniform parallel machines with controllable job processing times. Algorithmica **51**, 451–473 (2008)

29. Shakhlevich, N.V., Shioura, A., Strusevich, V.A.: Single machine scheduling with controllable processing times by submodular optimization. Int. J. Found. Comput. Sci. **20**, 247–269 (2009)

30. Shih, W.-K., Lee, C.-R., Tang, C.H.: A fast algorithm for scheduling imprecise computations with timing constraints to minimize weighted error. In: Proceedings of 21th IEEE Real-Time Systems Symposium (RTSS 2000), pp. 305–310 (2000)

31. Shioura, A., Shakhlevich, N.V., Strusevich, V.A.: A submodular optimization approach to bicriteria scheduling problems with controllable processing times on parallel machines. SIAM J. Discr. Math. **27**, 186–204 (2013)

32. Shioura, A., Shakhlevich, N.V., Strusevich, V.A.: Decomposition algorithms for submodular optimization with applications to parallel machine scheduling with controllable processing times. Math. Progr. A **153**, 495–534 (2015)

33. Shioura, A., Shakhlevich, N.V., Strusevich, V.A.: Scheduling imprecise computation tasks on parallel machines to minimize linear and non-linear error penalties: reviews, links and improvements. University of Greenwich, London, Report SORG-04-2015 (2015)

34. Shioura, A., Shakhlevich, N.V., Strusevich, V.A.: Application of submodular optimization to single machine scheduling with controllable processing times subject to release dates and deadlines. INFORMS J. Comput. **28**, 148–161 (2016)

Discrete Optimization

Constant-Factor Approximations for Cycle Cover Problems

Alexander Ageev[✉]

Sobolev Institute of Mathematics, pr. Koptyuga 4, Novosibirsk, Russia
ageev@math.nsc.ru

Abstract. A cycle cover of a graph is a set of cycles such that every vertex lies in exactly one cycle. We consider the following two cycle cover problems. Problem A: given a complete undirected graph $G = (V, E)$, edge costs $c : E \to R_+$ and positive integers $s_1, \ldots, s_m$, find a cycle cover $C_1, C_2, \ldots, C_m$ of minimum total cost subject to C_i has length s_i for each $i = 1, \ldots m$. Problem B: given a complete undirected graph $G = (V, E)$, edge costs $c : E \to R_+$, special vertices (depots) $w_1 \ldots, w_m \in V$ and positive integers $s_1, \ldots, s_m$, find a cycle cover $C_1, C_2, \ldots, C_m$ of minimum total cost subject to C_i has length s_i and contains vertex w_i for each $i = 1, \ldots m$. Problem B is a version of vehicle routing problem with m vehicles and routings of given lengths. Both problems include the TSP as a special case and so do not admit constant-factor approximations unless P=NP. We consider the metric case. Goemans and Williamson established that a case of Problem A when all cycles have length k is approximable within a factor of 4. In this paper we present the following results. Problem B can be solved by a 4-approximation algorithm in $O(n^2 \log n)$ time for $m = 2$ ($n = |V|$). Problem A can be solved by a 4-approximation algorithm in $O(n^3 \log n)$ time for $m = 2$. Problem A can be solved by a 8-approximation algorithm in $O(n^{m+1} \log n)$ time for any $m \geq 3$.

Keywords: Undirected graph · Cycle cover · Approximation algorithm · Worst-case analysis · Running time

1 Introduction

A cycle cover of a graph is a set of cycles such that every vertex lies in exactly one cycle. Cycle covers are important in designing approximation algorithms for various versions of the traveling salesman problem [2,3,5–7,13], for the shortest superstring problem [4,18], and for vehicle routing problems [11]. Hamiltonian cycles are special cases of cycle covers. Another classical example are cycle covers of minimum total weight which can be computed in polynomial time. This fact is exploited in the above mentioned algorithms, which commonly start by computing a cycle cover and then tie cycles to obtain a Hamiltonian cycle. Short cycles restrict the approximation ratios achieved by such algorithms: the longer the cycles in the initial cycle cover, the better the approximation ratio. Thus, we

© Springer International Publishing Switzerland 2016
Y. Kochetov et al. (Eds.): DOOR 2016, LNCS 9869, pp. 93–104, 2016.
DOI: 10.1007/978-3-319-44914-2_8

arrive at the problem of computing cycle covers without short cycles. Moreover, some vehicle routing problems [11] require covering vertices with cycles of bounded length. This motivates us to consider restricted cycle covers, where cycles have prescribed lengths.

Beyond applications in designing approximation algorithms, cycle covers are interesting on their own account. The classical matching problem is the problem of finding 1-factors, i.e., spanning subgraphs in which every vertex is incident to exactly one edge. Cycle covers of undirected graphs are also called 2-factors since every vertex is incident to exactly two edges in a cycle cover. Both structural properties of graph factors and the complexity of computing graph factors have been the topic of a considerable amount of research (see [14,16]).

1.1   Problem Formulations

Throughout the paper n stands for the number of vertices of a graph $G = (V, E)$. For a subgraph H of G, denote by $V(H)$ the set of its vertices.

We consider the following two problems.

Problem A

INSTANCE: A complete undirected graph $G = (V, E)$, edge costs $c : E \to R_+$ and positive integers $s_1, \ldots, s_m$ such that $s_i \geq 3$, $i = 1, \ldots, m$, and $\sum_{i=1}^{m} s_i = n$.

GOAL: Find a cycle cover $\mathcal{C} = (C_1, C_2, \ldots, C_m)$ of minimum total cost provided that the cycle C_i has length s_i for each $i = 1, \ldots m$.

Problem B

INSTANCE: A complete undirected graph $G = (V, E)$, edge costs $c : E \to R_+$, m special vertices (depots) $w_1, \ldots, w_m \in V$ and positive integers $s_1, \ldots, s_m$ such that $s_i \geq 3$, $i = 1, \ldots, m$ and $\sum_{i=1}^{m} s_i = n$.

GOAL: Find a cycle cover $\mathcal{C} = (C_1, C_2, \ldots, C_k')$ of minimum total cost provided that the cycle C_i has length s_i and contains vertex w_i for each $i = 1, \ldots m$.

Note that Problem B is a version of vehicle routing problem with m vehicles and routings of given lengths.

In the case when m is fixed Problem A polynomially reduces to problem B by fixing any vertex as one depot and examining all possible different $m-1$ vertices as the remaining depots. Thus Problem A can be solved in time $O(n^{m-1}Q)$ if Problem B can be solved in time Q. This reduction clearly preserves approximation factor. Both problems include the Traveling Salesman Problem (TSP) as a special case. Hence both problems are NP-hard and do not admit constant-factor approximations when the edge costs are arbitrary. In this paper we assume that the cost function c is a metric, i.e., the edge costs satisfy the triangle inequality.

1.2   Related Results

We give a short survey of the results concerning the metric case. The cycle cover problem without any restrictions on sizes of cycles can be solved in polynomial time by using Tutte's reduction to the classical perfect matching problem [14].

By using a modification of an algorithm due to Hartvigsen [10], a minimum-weight cycle cover with cycle lengths at least 4 in graphs with edge weights one and two (a special case of metric) can be computed in polynomial time. A cycle cover in which each cycle has size k can be found approximately with a factor of $7/6$ in the case when edge weights have values one and two [1]. Goemans and Williamson [8] show that the problems of finding minimum-weight cycle cover where each cycle has size k or at least k can be approximated within a factor of 4. Manthey [15] designs a constant-factor approximation algorithm for the problem of finding minimum-weight cycle cover provided that each cycle has size in a given subset $L \subseteq \{3, 4, 5, \ldots\}$. Manthey also shows that the problem cannot be approximated within a factor of $2 - \varepsilon$ for general L. Khachay and Neznakhina [12] give a simple 2-approximation algorithm for the problem where it is required to cover a graph by a given number of disjoint subgraphs of minimum weight. These subgraphs are allowed to be cycles, single edges and single vertices.

1.3 Our Results

In this paper we present the following results.

- Problem B can be solved by a 4-approximation algorithm in $O(n^2 \log n)$ time for $m = 2$.
- Problem A can be solved by a 4-approximation algorithm in $O(n^3 \log n)$ time for $m = 2$.
- Problem A can be solved by a 8-approximation algorithm in $O(n^{m+1} \log n)$ time for any m.

2 Goemans-Williamson Method

Goemans and Williamson [8] developed a powerful technique of designing algorithms with constant approximation factors applicable to quite a number of graph optimization problems. The following problem plays a key part in their approach.

Constrained Forest Problem
  Given a graph $G = (V, E)$ and a function $f : 2^{\mathbf{V}} \to \{0, 1\}$

$$\text{Min} \quad \sum_{e \in E} c_e x_e$$

subject to:

$$\sum_{e \in \delta(S)} x_e \geq f(S), \quad \emptyset \neq S \subset V, \tag{1}$$

$$x_e \in \{0, 1\}, \quad e \in E. \tag{2}$$

where c_e is a cost of edge e and $\delta(X)$ denotes the set of edges having exactly one endpoint in S (i.e., $\delta(X)$ is a cut). This integer program (IP) can be interpreted as a very special case of covering problem in which we need to find a minimum-cost set of edges that cover all cuts $\delta(S)$ corresponding to S with $f(S) = 1$. The minimal solutions to IP are incidence vectors of forests which justifies the name of the corresponding graph problem. Note that the classical minimum weight spanning tree is a special case of Constrained Forest Problem (it corresponds to the case when $f(S) \equiv 1$).

A function $f : \mathbf{2^V} \to \{0, 1\}$ is called *proper* if

(i)   (symmetry) $f(S) = f(V \setminus S)$ for all $S \subseteq V$,
(ii) (disjointness) If A and B are disjoint, then $f(A) = f(B) = 0$ implies $f(A \cup B) = 0$,
(iii) $f(\emptyset) = 0$.

Quite a number of interesting families of forests can be modeled by (IP) with proper functions. In particular, minimum-cost spanning tree, shortest path, Steiner tree, and T-join problems can be stated as (IP) with a proper function f or, equivalently, as a *Proper* Constrained Forest Problem. Though many proper constrained forest problems are NP-hard the general problem admits an approximation with a factor of 2.

Theorem 1 (Goemans-Williamson). *Proper constrained forest problem is approximable within a factor of 2 in time* $O(n^2 \log n)$.

The original theorem of Goemans and Williamson (Theorem 2.4 in [8]) is formulated in a slightly stronger form.

In the technical part of the paper we shall exploit some properties of the Goemans-Williamson algorithm. It is of a primal-dual type and together with a feasible solution $(\overline{x}_e)$ to (IP) finds a feasible solution $(\overline{y}_S)$ to the dual of linear programming relaxation of (IP) that is obtained from (IP) by replacing (2) with $x_e \geq 0$:

$$\text{Max} \quad \sum_{S \subset V} f(S) \cdot y_S$$

subject to:

$$\sum_{S : e \in \delta(S)} y_S \leq c_e, \quad e \in E,$$

$$y_S \geq 0, \quad \emptyset \neq S \subset V.$$

Moreover, these solutions satisfy the inequality

$$\sum_{e \in E} c_e \overline{x}_e \leq 2 \sum_S \overline{y}_S. \tag{3}$$

Goemans and Williamson [8] give quite a number of applications of Theorem 1. In particular, they show that the problems of finding minimum-weight cycle cover, where each cycle has size k (the case of Problem A where $s_i \equiv k$) or at least k for $i = 1, \ldots m$, can be approximated within a factor of 4 in time $O(n^2 \log n)$. However the trick they use is not straightforwardly applicable to Problem A in the general case. Fortunately, some advance is possible if instead of Problem A to try to develop approximations for the more general problem B. This is the key idea of our approach.

3   Problem B: The Case $m = 2$

This case of Problem B can be formulated in the following way:

Given a complete undirected graph $G = (V, E)$, a cost function $c : E \to R_+$, depots $w_1, w_2 \in V$ and positive integers $s_1, s_2 \geq 3$, $s_1 + s_2 = n$, find a cycle cover (C_1, C_2) of minimum total weight provided that for $i = 1, 2$ cycle C_i has size s_i and contains depot w_i.

The problem is a version of the vehicle routing problem. Each vehicle starts from its depot, visits a prescribed number of cities and gets back. Each city is visited by exactly one vehicle. We need to find routings for each vehicle of minimum total cost.

Let I be an instance of Problem B with $m = 2$. Consider the instance of Constrained Forest Problem with the following function f:

$f(S) = 0 \Leftrightarrow$ either $S = V$, or $S = \emptyset$, or S contains exactly one depot w_i and $|S| = s_i$, $i = 1, 2$.

Lemma 1

(i) The function f is proper.

(ii) Let $\mathcal{F}$ be a minimal feasible solution to the instance of Constrained Forest Problem with the function f. Then $\mathcal{F}$ is a spanning forest consisting of at most two components. If $\mathcal{F}$ consists of exactly two components (trees), then each component contains exactly one depot.

Proof. (i) Symmetry and disjointness follow straightforwardly from the definition of f. (ii) Let T be a component of $\mathcal{F}$. Assume that T is not a spanning tree. Then by the definition of f and in view of (1), T contains exactly one depot. Moreover, $S = V \setminus V(T)$ is a vertex set of the other component of T and this component contains the other depot.   □

Lemma 2. *The cost of the optimal solution of Constrained Forest Problem with the function f defined above provides a lower bound for the optimum of the instance I.*

Proof. Let C_1, C_2 be an optimal solution to I. By deleting a single edge in each cycle C_i, $i = 1, 2$ we obtain a forest that is a feasible solution to the instance of Constrained Forest Problem with the function f defined above.   □

Let G be a complete graph with edge costs satisfying the triangle inequality and let H be a connected subgraph of G. One of the ingredients of our algorithms will be the well-known procedure of generating a cycle of G by duplicating edges of H.

First, we replace each edge e of H by two parallel edges each having the cost of e. This gives an Eulerian multigraph $\widetilde{H}$. Then we find an Eulerian tour

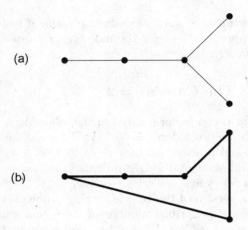

Fig. 1. (a) a connected graph H; (b) a cycle C obtained by shortcutting duplicated edges of H.

T in $\widetilde{H}$ and finally, transform T into a cycle C in G in the following way (this procedure is often called shortcutting). Starting at an arbitrary vertex v_0, follow the tour T, but mark vertices when you visit them as already visited. Later, when encountering a vertex previously visited, skip over it and go directly to the next vertex on the tour. If that vertex has also already been seen, go on to the next, etc. Keep going until encountering the vertex v_0 again. The resulting tour is a cycle C. For an example, see Fig. 1.

By the triangle inequality the cost of the cycle C is at most twice the cost of the subgraph H. In what follows we refer to this procedure as SHORTCUT. Note that SHORTCUT is basic in the folklore 2-approximation algorithm and an important part of the classical 3/2-approximation algorithm of Christofides-Serdyukov [9,17] for the metric TSP.

Now we are ready to present a description of the algorithm.

Algorithm ALGB2

The input is an instance I of Problem B with $m = 2$.

Step 1. Generate the instance of Constrained Forest Problem corresponding to I and solve it approximately by the Goemans–Williamson algorithm. We get a forest $\mathcal{F}$. Let r be the number of components of $\mathcal{F}$.

Step 2. By applying SHORTCUT to the components of $\mathcal{F}$ get cycles C_i, $i = 1, r$ $(r \leq 2)$.

Step 3. If $r = 2$, output the cycles C_1, C_2. Otherwise (i.e., $r = 1$) go to the next step.

Step 4. We have a single cycle C containing both depots. Split C into two paths $\mathcal{P}_1$ and $\mathcal{P}_2$ in such a way that $\mathcal{P}_i$ has size s_i and contains exactly one depot w_i, $i = 1, 2$. (the details in the Correctness, see also Fig. 2)

Step 5. By applying SHORTCUT to $\mathcal{P}_1 \mathcal{P}_2$ get two cycles C_1 and C_2 and declare them the output.

Correctness. We need to justify the splitting in Step 4. If the cycle contains two depots w_1 and w_2, we first construct the path $\mathcal{P}_1$ by adding to the depot w_1 $s_1 - 1$ pairwise adjacent edges not incident to w_2. Then we delete the edges incident to the endpoints of $\mathcal{P}_1$. The remaining part of the cycle C is also a path and contains s_2 vertices including the depot w_2 (see Fig. 2) as required.

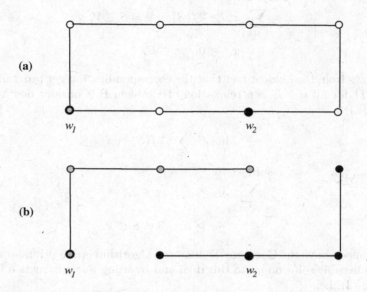

Fig. 2. (a) The single cycle C of length 8 at start of Step 4, $s_1 = 4$, $s_2 = 4$; (b) the paths $\mathcal{P}_1$ and $\mathcal{P}_2$ colored by grey and black, respectively.

Approximation ratio.

Let $cost(\cdot)$ denote the total cost of edges in a subgraph and let OPT denote the optimum of the instance I of Problem B where $m = 2$.

Lemma 3. *The cost of the solution returned by* ALGB2 *is within a factor of* 4 *of the cost of an optimal solution, i.e.,*

$$cost(C_1) + cost(C_2) \leq 4 \cdot OPT.$$

Proof. If ALGB2 finishes at Step 3 (i.e., $r = 2$) then it applies SHORTCUT one time only. By Theorem 1 and Lemma 2

$$cost(C_1) + cost(C_2) \leq 2 \cdot cost(F) \leq 4 \cdot OPT,$$

as required.

Therefore it remains to examine the case when $r = 1$, i.e., when $\mathcal{F}$ is a tree. In this case ALGB2 applies SHORTCUT twice and a similar argument results in a factor of 8. However, a refined argument gives the required factor of 4.

First, observe that the following linear program (LP) is a relaxation of Problem B:

$$\text{Min} \quad \sum_{e \in E} c_e x_e$$

subject to:

$$\sum_{e \in \delta(S)} x_e \geq 2f(S), \quad \emptyset \neq S \subset V,$$

$$x_e \geq 0, \quad e \in E.$$

This follows from the evident fact that the corresponding integer program where $x_e \in \{0, 1\}$ for all $e \in E$ is a relaxation of Problem B. Consider now the dual (DP) to (LP):

$$\text{Max} \quad \sum_{S \subset V} 2f(S) \cdot y_S$$

subject to:

$$\sum_{S : e \in \delta(S)} y_S \leq c_e, \quad e \in E,$$

$$y_S \geq 0, \quad \emptyset \neq S \subset V.$$

As mentioned above the Goemans-Williamson algorithm is of a primal-dual type. It finds a feasible solution y_S to this dual and by using y_S constructs a forest $\mathcal{F}$ satisfying (3), i.e.,

$$\sum_{e \in \mathcal{F}} c_e \leq 2 \sum_S y_S.$$

In the case of $r = 1$ $\mathcal{F}$ is a tree and ALGB2 applies SHORTCUT to transform it into a single cycle D satisfying

$$\sum_{e \in D} c_e \leq 4 \sum_S y_S.$$

The cycle D then splits into paths $\mathcal{P}_1$ and $\mathcal{P}_2$ and at Step 5 ALGB2 applies SHORTCUT to get the output cycles C_1 and C_2. Thus we have

$$\sum_{e \in C_1 \cup C_2} c_e \leq 8 \sum_S y_S.$$

However, the objective function of the dual (DP) is equal to $2 \sum y_S$ (since $y_S = 0$ for all S such that $f(S) = 0$) and so $2 \sum y_S$ is a lower bound for OPT. Thus

$$\sum_{e \in C_1 \cup C_2} c_e \leq 4 \cdot OPT. \qquad \square$$

It is clear that the most time-consuming part of ALGB2 is Step 1. Thus we arrive at the following theorem.

Theorem 2. *Problem B when $m = 2$ is approximable within a factor of 4 in time $O(n^2 \log n)$.* □

The theorem implies

Corollary 1. *Problem A when $m = 2$ is approximable within a factor of 4 in time $O(n^3 \log n)$.* □

4   Problem A: The General Case

To construct an approximation algorithm for Problem A we shall use Problem B as an auxiliary problem.

Let I_B be an instance of Problem B. Consider an instance I_{CFP} of Constrained Forest Problem with the following function:

$f(S) = 0 \Leftrightarrow$ either $S = V$, or $S = \emptyset$, or S contains depots $w_{i_1}, \ldots, w_{i_q}$ and $|S| = s_{i_1} + \ldots + s_{i_q}$, $i_1, \ldots, i_q \in \{1, \ldots m\}$, $i_1 < i_2 < \ldots i_q$, $q \geq 1$.

Lemma 4. *The cost of the optimal solution to the instance I_{CFP} provides a lower bound for the optimum of the instance I_B.*

Proof. Let $(C_1, \ldots, C_m)$ be an optimal cycle cover for the instance I_B. By deleting an edge in each cycle C_i, $i = 1, \ldots m$, we obtain a feasible solution to the instance I_{CFP}. □

Lemma 5

(i) *The function f is proper.*
(ii) *Let $\mathcal{F}$ be a minimal forest that is a feasible solution to I_{CFP}. Let T be a component of $\mathcal{F}$ and $w_{i_1}, \ldots w_{i_q}$ be the depots in $V(T)$. Then $V(T)$ contains at least one depot and*

$$|V(T)| = s_{i_1} + \ldots + s_{i_q}. \tag{4}$$

Proof. (i) Symmetry and disjointness straightforwardly follow from the definition of f. (ii) If T contains no depot or (4) is not valid, then by the definition of f (1) yields

$$\sum_{e \in \delta(V(T))} x_e \geq 1,$$

which contradicts the assumption that T is a component of $\mathcal{F}$. □

We first present an (auxiliary) algorithm for Problem B. Given an instance I_B of Problem B, it outputs a cycle cover which is a feasible solution to the corresponding instance I_A of Problem A (however, this solution is not necessarily feasible for I_B).

Algorithm AuxB
The input is an instance I of Problem B.

Step 1. By using Goemans-Williamson algorithm solve the Constrained Forest Problem with function f. The result is a forest $\mathcal{F}$.

Step 2. Apply SHORTCUT to the components of $\mathcal{F}$. The result is a cycle cover $D_1, \ldots D_r$.

Step 3. Split the cycles $D_1, \ldots D_r$ into paths $P_1, \ldots P_m$ such that P_i has s_i vertices, $i = 1, \ldots, m$ (for justification see Correctness below). Apply SHORTCUT to these paths. The result is a cycle cover $C_1, \ldots, C_m$. Output it.

Correctness. We need to justify Step 3. At start of Step 2 we have a forest $\mathcal{F}$ consisting of components $T_1, \ldots, T_r$. By the statement (ii) of Lemma 5 each component of $\mathcal{F}$ satisfies property (4). At Step 2 $T_1, \ldots, T_r$ transform into the cycles $D_1, \ldots D_r$ in such a way that $V(T_i) = V(D_i)$ for all $i = 1, \ldots, r$. Therefore the cycles $D_1, \ldots D_r$ also satisfy property (4), which makes possible Step 3. □

Approximation ratio.

As in the previous section let $cost(\cdot)$ denote the total cost of edges in a subgraph.

Lemma 6. AUXB *finds a feasible solution to the instance I_A whose cost is within a factor of 8 of the cost of an optimal solution to I_B.*

Proof. Let OPT_B denote the optimum of the instance I_B. By Theorem 1 and Lemma 4

$$cost(\mathcal{F}) \leq 2 \cdot OPT_B. \tag{5}$$

At Step 2 AUXB applies procedure SHORTCUT to the components of $\mathcal{F}$. Therefore

$$\sum_{i=1}^{r} cost(D_k) \leq 2cost(\mathcal{F}). \tag{6}$$

Since

$$\sum_{i=1}^{m} cost(P_i) \leq \sum_{i=1}^{r} cost(D_k),$$

the application of procedure SHORTCUT at Step 3 yields

$$\sum_{i=1}^{m} cost(C_i) \leq 2 \sum_{i=1}^{r} cost(D_k).$$

Together with (5) and (6) this gives

$$\sum_{i=1}^{m} cost(C_i) \leq 8 \cdot OPT_B,$$

as required.

Lemma 7. *Given an instance I_A, there exists an (ordered) collection of m depots $\{w_1, \ldots . w_m\} \subseteq V$ such that the optimum of I_B coincides with the optimum of I_A.*

Proof. Let $(C_1^*, \ldots, C_m^*)$ be an optimal cycle cover of I_A. Select a vertex w_i in each C_i, $i = 1, \ldots m$. Then the optimum of I_B with the set of depots $\{w_1, \ldots w_m\}$ coincides with the optimum of I_A. □

Algorithm ALGA

The input is an instance I of Problem A.

Fix an arbitrary vertex w_1 as the first depot. Browse through all possible (ordered) collections of $m-1$ depots $\{w_2, \ldots, w_m\} \subset V$ and apply AUXB to the instance of Problem B corresponding to each collection. Output the cycle cover of minimum total cost.

Theorem 3. ALGA *finds a solution to Problem A whose cost is within a factor of* 8 *of the cost of the optimal solution in* $O(n^{m+1} \log n)$ *time.*

Proof. The approximation bound follows from Lemmas 6 and 7. The time complexity bound follows from Theorem 1 and necessity to solve $O(n^{m-1})$ instances of Constrained Forest Problem. □

Acknowledgments. The author would like to thank the anonymous referees for their valuable comments and suggestions.

References

1. Bläser, M., Siebert, B.: Computing cycle covers without short cycles. In: Meyer auf der Heide, F. (ed.) ESA 2001. LNCS, vol. 2161, pp. 368–379. Springer, Heidelberg (2001)
2. Bläser, M.: A 3/4-approximation algorithm for maximum ATSP with weights zero and one. In: Jansen, K., Khanna, S., Rolim, J.D.P., Ron, D. (eds.) RANDOM 2004 and APPROX 2004. LNCS, vol. 3122, pp. 61–71. Springer, Heidelberg (2004)
3. Bläser, M., Manthey, B., Sgall, J.: An improved approximation algorithm for the asymmetric TSP with strengthened triangle inequality. J. Discrete Algorithms 4(4), 623–632 (2006)
4. Blum, A.L., Jiang, T., Li, M., Tromp, J., Yannakakis, M.: Linear approximation of shortest superstrings. J. ACM 41(4), 630–647 (1994)
5. Böckenhauer, H.-J., Hromkovič, J., Klasing, R., Seibert, S., Unger, W.: Approximation algorithms for the TSP with sharpened triangle inequality. Inf. Process. Lett. 75(3), 133–138 (2000)
6. Chandran, L.S., Ram, L.S.: On the relationship between ATSP and the cycle cover problem. Theor. Comput. Sci. 370(1–3), 218–228 (2007)
7. Chen, Z.-Z., Okamoto, Y., Wang, L.: Improved deterministic approximation algorithms for Max TSP. Inf. Process. Lett. 95(2), 333–342 (2005)
8. Goemans, M.X., Williamson, D.P.: A general approximation technique for constrained forest problems. SIAM J. Comput. 24(2), 296–317 (1995)
9. Christofides, N.: Worst-case analysis of a new heuristic for the travelling salesman problem. Report 388, Graduate School of Industrial Administration, Carnegie Mellon University (1976)
10. Hartvigsen, D.: An extension of matching theory. PhD thesis, Department of Mathematics, Carnegie Mellon University, Pittsburgh, Pennsylvania, USA (1984)

11. Hassin, R., Rubinstein, S.: On the complexity of the k-customer vehicle routing problem. Oper. Res. Lett. **33**(1), 71–76 (2005)
12. Khachay, M., Neznakhina, K.: Approximability of the minimum-weight k-size cycle cover problem. J. Glob. Optim. Online First, 1–18 (2015). http://link.springer.com/article/10.1007/s10898-015-0391-3
13. Kaplan, H., Lewenstein, M., Shafrir, N., Sviridenko, M.I.: Approximation algorithms for asymmetric TSP by decomposing directed regular multigraphs. J. ACM **52**(4), 602–626 (2005)
14. Lovász, L., Plummer, M.D.: Matching Theory. North-Holland Mathematics Studies, vol. 121. Elsevier, New York (1986)
15. Manthey, B.: Minimum-weight cycle covers and their approximability. Discrete Appl. Math. **157**(7), 1470–1480 (2009)
16. Schrijver, A.: Combinatorial Optimization: Polyhedra and Efficiency. Algorithms and Combinatorics, vol. 24. Springer, Heidelberg (2003)
17. Serdyukov, A.I.: On some extremal routs in graphs. Upravlyaemye Sistemy **17**, 76–79 (1978). (in Russian)
18. Sweedyk, Z.: A $2\frac{1}{2}$-approximation algorithm for shortest superstring. SIAM J. Comput. **29**(3), 954–986 (1999)

Precedence-Constrained Scheduling Problems Parameterized by Partial Order Width

René van Bevern[1]($\boxtimes$), Robert Bredereck[2], Laurent Bulteau[3], Christian Komusiewicz[4], Nimrod Talmon[5], and Gerhard J. Woeginger[6]

[1] Novosibirsk State University, Novosibirsk, Russian Federation
rvb@nsu.ru
[2] TU Berlin, Berlin, Germany
robert.bredereck@tu-berlin.de
[3] Université Paris-Est Marne-la-Vallée, Champs-sur-Marne, France
l.bulteau@gmail.com
[4] Friedrich-Schiller-Universität Jena, Jena, Germany
christian.komusiewicz@uni-jena.de
[5] Weizmann Institute of Science, Rehovot, Israel
nimrodtalmon77@gmail.com
[6] TU Eindhoven, Eindhoven, The Netherlands
gwoegi@win.tue.nl

Abstract. Negatively answering a question posed by Mnich and Wiese (Math. Program. 154(1–2):533–562), we show that P2|prec, $p_j \in \{1, 2\}|C_{max}$, the problem of finding a non-preemptive minimum-makespan schedule for precedence-constrained jobs of lengths 1 and 2 on two parallel identical machines, is W[2]-hard parameterized by the width of the partial order giving the precedence constraints. To this end, we show that SHUFFLE PRODUCT, the problem of deciding whether a given word can be obtained by interleaving the letters of k other given words, is W[2]-hard parameterized by k, thus additionally answering a question posed by Rizzi and Vialette (CSR 2013). Finally, refining a geometric algorithm due to Servakh (Diskretn. Anal. Issled. Oper. 7(1):75–82), we show that the more general RESOURCE-CONSTRAINED PROJECT SCHEDULING problem is fixed-parameter tractable parameterized by the partial order width combined with the maximum allowed difference between the earliest possible and factual starting time of a job.

Keywords: Resource-constrained project scheduling · Parallel identical machines · Makespan minimization · Parameterized complexity · Shuffle product

R. van Bevern—Supported by project 16-31-60007 mol_a_dk of the Russian Foundation for Basic Research.
C. Komusiewicz—Supported by the DFG, project MAGZ (KO 3669/4-1).
N. Talmon—Supported by a postdoctoral fellowship from I-CORE ALGO.

Y. Kochetov et al. (Eds.): DOOR 2016, LNCS 9869, pp. 105–120, 2016.
DOI: 10.1007/978-3-319-44914-2_9

1   Introduction

We study the parameterized complexity of the following NP-hard problem and various special cases [13,15] with respect to the width of the given partial order.

Problem 1.1. (Resource-constrained project scheduling (RCPSP))

Input: A set J of jobs, a partial order $\preceq$ on J, a set R of renewable resources, for each resource $\rho \in R$ the available amount R_ρ, and for each $j \in J$ a processing time $p_j \in \mathbb{N}$ and the amount $r_{j\rho} \leq R_\rho$ of resource $\rho \in R$ that it consumes.

Find: A *schedule* $(s_j)_{j\in J}$, that is, a *starting time* $s_j \in \mathbb{N}$ of each job j, such that

1. for $i \prec j$, job i finishes before job j starts, that is, $s_i + p_i \leq s_j$,
2. at any time t, at most R_ρ units of each resource ρ are used, that is, $\sum_{j\in s(t)} r_{j\rho} \leq R_\rho$, where $s(t) := \{j \in J \mid t \in [s_j, s_j + p_j)\}$, and
3. the maximum completion time $C_{\max} := \max_{j\in J}(s_j + p_j)$ is minimum.

A schedule satisfying (1)–(2) is *feasible*; a schedule satisfying (1)–(3) is *optimal*.

Intuitively, a schedule $(s_j)_{j\in J}$ processes each job $j \in J$ non-preemptively in the half-open real-valued interval $[s_j, s_j + p_j)$, which costs $r_{j\rho}$ units of resource ρ during that time. After finishing, jobs free their resources for later jobs. If there is only one resource and each job j requires one unit of it, then RCPSP is equivalent to P|prec|$C_{\max}$, the NP-hard problem of non-preemptively scheduling precedence-constrained jobs on a given number m of parallel identical machines to minimize the maximum completion time [15].

Mnich and Wiese [11] asked whether P|prec|$C_{\max}$ is solvable in $f(p_{\max}, w) \cdot$ poly(n) time, where $p_{\max}$ is the maximum processing time, w is the width of the given partial order $\preceq$, n is the input size, and f is a computable function independent of the input size. In other words, the question is whether P|prec|$C_{\max}$ is *fixed-parameter tractable* parameterized by $p_{\max}$ and w. Motivated by this question, which we answer negatively, we strengthen hardness results for P|prec|$C_{\max}$ and refine algorithms for RCPSP with small partial order width.

Due to space constraints, some details are deferred to a full version of this paper.

Stronger hardness results. We obtain new hardness results for the following special cases of P|prec|$C_{\max}$ (for basic definitions of parameterized complexity terminology, see the end of this section and recent textbooks [4,5]):

(1) P2|chains|$C_{\max}$, the case with two machines and precedence constraints given by a disjoint union of total orders, remains weakly NP-hard for width 3.
(2) P2|prec, $p_j \in \{1, 2\}$|$C_{\max}$, the case with two machines and processing times 1 and 2, is W[2]-hard parameterized by the partial order width w.
(3) P3|prec, $p_j = 1$, size$_j \in \{1, 2\}$|$C_{\max}$, the case with three machines, unit processing times, but where each job may require one or two machines, is also W[2]-hard parameterized by the partial order width w.

Towards showing (2) and (3), we show that SHUFFLE PRODUCT, the problem of deciding whether a given word can be obtained by interleaving the letters of k other given words, is W[2]-hard parameterized by k. This answers a question of Rizzi and Vialette [12]. We put these results into context in the following.

Result (1) complements the fact that $P|\text{prec}|C_{\max}$ with constant width w is solvable in pseudo-polynomial time using dynamic programming [14] and that $P2|\text{chains}|C_{\max}$ is *strongly* NP-hard for unbounded width [6].

Result (2) complements the NP-hardness result for $P2|\text{prec}, p_j\in\{1,2\}|C_{\max}$ due to Ullman [15] and the W[2]-hardness result for $P|\text{prec},p_j=1|C_{\max}$ parameterized by the number m of machines due to Bodlaender and Fellows [3]. While not made explicit, one can observe that Bodlaender and Fellows' reduction creates hard instances with $w = m + 1$. This is remarkable since $P|\text{prec}|C_{\max}$ is trivially polynomial-time solvable if $w \leq m$, and also since the result negatively answered Mnich and Wiese's question [11] twenty years before it was posed. Our result (2), however, gives a stronger negative answer: unless $W[2] = \text{FPT}$, not even $P2|\text{prec}, p_j\in\{1,2\}|C_{\max}$ allows for the desired $f(w) \cdot \text{poly}(n)$-time algorithm.

Refined algorithms. Servakh [14] gave a geometric pseudo-polynomial-time algorithm for RCPSP with constant partial order width w. The degree of the polynomial depends on w and, by (1) above, the algorithm cannot be turned into a true polynomial-time algorithm unless P = NP even for constant w. We refine this algorithm to solve RCPSP in $(2\lambda+1)^w \cdot 2^w \cdot \text{poly}(n)$ time, where λ is the maximum allowed difference between earliest possible and factual starting time of a job. The degree of the polynomial depends neither on w nor λ and is indeed a polynomial of the *input size* n. This does not contradict (1) since the factor $(2\lambda+1)^w$ might be superpolynomial in n. We note that fixed-parameter tractability for w or λ alone is ruled out by (2) and by Lenstra and Rinnooy Kan [10], respectively.

Preliminaries. A reflexive, symmetric, and transitive relation $\preceq$ on a set X is a *partial order*. We write $x \prec y$ if $x \preceq y$ and $x \neq y$. A subset $X' \subseteq X$ is a *chain* if $\preceq$ is a total order on X'; it is an *antichain* if the elements of X' are mutually incomparable by $\preceq$. The *width* of $\preceq$ is the size of largest antichain in X. A *chain decomposition* of X is a partition $X = X_1 \uplus \cdots \uplus X_k$ such that each X_i is a chain.

Recently, the *parameterized complexity* of scheduling problems attracted increased interest [2]. The idea is to accept exponential running times for solving NP-hard problems, but to restrict them to a small *parameter* [4,5]. Instances (x, k) of a *parameterized problem* $\Pi \subseteq \Sigma^* \times \mathbb{N}$ consist of an input x and a parameter k. A parameterized problem Π is *fixed-parameter tractable* if it is solvable in $f(k) \cdot \text{poly}(|x|)$ time for some computable function f. Note that the degree of the polynomial must not depend on k. FPT is the class of fixed-parameter tractable parameterized problems. There is a hierarchy of parameterized complexity classes $\text{FPT} \subseteq W[1] \subseteq W[2] \subseteq \cdots \subseteq W[P]$, where all inclusions are conjectured to be strict. A parameterized problem Π_2 is $W[t]$-*hard* if there is a *parameterized reduction* from each problem $\Pi_1 \in W[t]$ to Π_2, that is, an algorithm that maps an instance (x, k) of Π_1 to an instance (x', k') of Π_2 in time $f(k) \cdot \text{poly}(|x|)$ such that $k' \leq g(k)$ and $(x, k) \in \Pi_1 \Leftrightarrow (x', k') \in \Pi_2$, where f and g are arbitrary computable functions. No W[t]-hard problem is fixed-parameter tractable unless FPT = W[t].

2   Parallel Identical Machines and Shuffle Products

This section presents our hardness results for special cases of P|prec|$C_{\max}$. In Sect. 2.1, we show weak NP-hardness of P2|chains|$C_{\max}$ for three chains. In Sect. 2.2, we show W[2]-hardness of SHUFFLE PRODUCT as a stepping stone towards showing W[2]-hardness of P3|prec, $p_j{=}1$, size$_j{\in}\{1, 2\}$|$C_{\max}$ and P2|prec, $p_j{\in}\{1, 2\}$|$C_{\max}$ parameterized by the partial order width in Sect. 2.3.

2.1   Weak NP-hardness for Two Machines and Three Chains

Du et al. [6] showed that P2|chains|$C_{\max}$ is strongly NP-hard. We complement this result by the following theorem.

Theorem 2.1. P2|chains|$C_{\max}$ is weakly NP-hard even for precedence constraints of width three, that is, consisting of three chains.

Proof (sketch). We reduce from the weakly NP-hard PARTITION problem [8, SP12]: Given a multiset of positive integers $A = \{a_1, \ldots, a_t\}$, decide whether there is a subset $A' \subseteq A$ such that $\sum_{a_i \in A'} a_i = \sum_{a_i \in A \setminus A'} a_i$. Let $A = \{a_1, \ldots, a_t\}$ be a PARTITION instance. If $b := (\sum_{a_i \in A} a_i)/2$ is not an integer, then we are facing a no-instance. Otherwise, we construct a P2|chains|$C_{\max}$ instance as follows. Create three chains $J^0 := \{j_1^0 \prec \cdots \prec j_t^0\}$, $J^1 := \{j_1^1 \prec \cdots \prec j_{t+1}^1\}$, and $J^2 := \{j_1^2 \prec \cdots \prec j_{t+1}^2\}$ of jobs. For each $i \in \{1, \ldots, t\}$, job j_i^0 gets processing time a_i. The jobs in $J^1 \cup J^2$ get processing time $2b$ each. This construction can be performed in polynomial time and one can show that the input PARTITION instance is a yes-instance if and only if the created P2|chains|$C_{\max}$ instance allows for a schedule with makespan $T := (2t + 3)b$: in such a schedule, each machine must perform exactly $t + 1$ jobs from $J^1 \cup J^2$ and has b time for jobs from J^0.   □

2.2   W[2]-hardness for Shuffle Product

In this section, we show a W[2]-hardness result for SHUFFLE PRODUCT that we transfer to P2|prec, $p_j{\in}\{1, 2\}$|$C_{\max}$ and P3|prec, $p_j{=}1$, size$_j{\in}\{1, 2\}$|$C_{\max}$ in Sect. 2.3. We first formally introduce the problem (cf. Fig. 1).

Definition 2.2 (shuffle product). By $s[i]$, we denote the ith letter in a word s. A word t is said to be in the *shuffle product* of words s_1 and s_2, denoted by $t \in s_1 \shuffle s_2$, if t can be obtained by interleaving the letters of s_1 and s_2. Formally, $t \in s_1 \shuffle s_2$ if there are increasing functions $f_1 \colon \{1, \ldots, |s_1|\} \to \{1, \ldots, |t|\}$ and $f_2 \colon \{1, \ldots, |s_2|\} \to \{1, \ldots, |t|\}$ mapping positions of s_1 and s_2 to positions of t such that, for all $i \in \{1, \ldots, |s_1|\}$ and $j \in \{1, \ldots, |s_2|\}$, one has

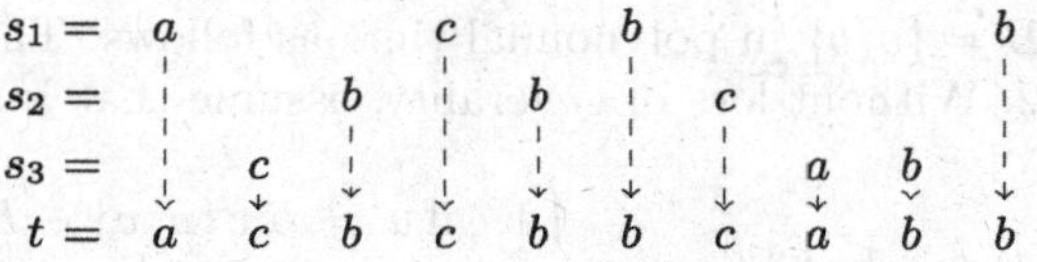

$$
\begin{array}{llllllllll}
s_1 = & a & & c & & b & & & b \\
s_2 = & & b & & b & & c & & \\
s_3 = & c & & & & & & a & b \\
t = & a & c & b & c & b & b & c & a & b & b
\end{array}
$$

Fig. 1. Illustration of a shuffle product: for $s_1 = acbb$, $s_2 = bbc$, and $s_3 = cab$, one has $t = acbcbbcabb \in s_1 \shuffle s_2 \shuffle s_3$. Dashed arcs show how the letters of each s_i map into t.

$t[f_1(i)] = s_1[i]$, $t[f_2(j)] = s_2[j]$, and $f_1(i) \neq f_2(j)$. This product is associative and commutative, which implies that the shuffle product of any set of words is well-defined.

Problem 2.3 ((Binary) Shuffle Product)

Input: Words $s_1, \ldots, s_k$, and t over a (binary) alphabet Σ.
Parameter: k.
Question: Is $t \in s_1 \shuffle s_2 \shuffle \cdots \shuffle s_k$?

BINARY SHUFFLE PRODUCT is NP-hard for unbounded k [16, Lemma 3.2], whereas SHUFFLE PRODUCT is polynomial-time solvable for constant k using dynamic programming. Rizzi and Vialette [12] asked about the parameterized complexity of SHUFFLE PRODUCT. We answer the question by the following theorem.

Theorem 2.4 BINARY SHUFFLE PRODUCT is W[2]-hard.

Our proof uses a parameterized reduction from the W[2]-hard DOMINATING SET problem [4,5] and is inspired by Bodlaender and Fellows's proof that $P|prec, p_j{=}1|C_{\max}$ is W[2]-hard parameterized by the number m of machines [3].

Problem 2.5 (Dominating Set)

Input: A graph $G = (V, E)$ and a natural number k.
Parameter: k.
Question: Is there a size-k *dominating set* D, that is, $V \subseteq N[D]$?

Herein, $N[D]$ is the set of vertices in D and their neighbors. In order to describe the construction, we introduce some notation.

Definition 2.6. We denote the concatenation of words $s_1, \ldots, s_k$ as $\prod_{i=1}^{k} s_i := s_1 s_2 \ldots s_k$ and denote k repetitions of a word s by s^k. The number of occurrences of a letter a in a word s is $|s|_a$.

Construction 2.7. Given a DOMINATING SET instance (G, k) with a graph $G = (V, E)$, we construct an instance of BINARY SHUFFLE PRODUCT with

$k + 3$ words over $\Sigma = \{a, b\}$ in polynomial time as follows. The construction is illustrated in Fig. 2. Without loss of generality, assume that $V = \{1, \ldots, n\}$.

$$\text{For } u, v \in V, \text{ let } \ell_{u,v} := \begin{cases} 1 & \text{if } u = v \text{ or } \{u, v\} \in E, \\ 2 & \text{otherwise.} \end{cases} \tag{2.1}$$

Moreover, define two words

$$A := \prod_{u=1}^{n} \prod_{v=1}^{n} ab^{\ell_{u,v}} \qquad \text{and} \qquad B := \left((a^k b^{2k})^{n-1} a^k b^{2k-1}\right)^n.$$

Finally, let $N := 2k(n-1) + 1$ and output an instance of SHUFFLE PRODUCT with the following $k + 3$ words:

$$s_i := A^N \text{ for each } i \in \{1, \ldots, k\}, \qquad\qquad t := B^N (a^k b^{2k})^{n-1},$$
$$s_{k+1} := a^{|t|_a - \sum_{i=1}^{k} |s_i|_a}, \text{ and} \qquad\qquad s_{k+2} := b^{|t|_b - \sum_{i=1}^{k} |s_i|_b}.$$

Note that A is simply the word that one obtains by concatenating the rows of the adjacency matrix of G and replacing ones by ab and zeroes by abb.

Before showing the correctness of Construction 2.7, we make some basic observations about the words it creates, for which we introduce some terminology.

Definition 2.8 (long and short blocks, positions). A *block* in a word s is a maximal consecutive subword using only one letter. A *c-block* is a block containing only the letter c. A block has *position* i in s if it is the ith successive

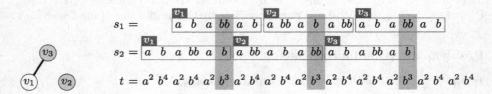

Fig. 2. Left: A DOMINATING SET instance with $k = 2$ and a solution $\{v_2, v_3\}$ (the gray nodes). Right: The "base pattern" of the corresponding SHUFFLE PRODUCT instance (only one repetition of A in s_1 and s_2 and only one repetition of B in t is shown). Blocks of s_1 and s_2 are mapped into the blocks of t displayed in the same column. The horizontal (blue) rectangles reflect that each s_i is built as the concatenation of the rows of the adjacency matrix, where zeroes are replaced by abb and ones by ab. The amount of horizontal offset of each s_i corresponds to the selection of a vertex as dominator (v_2 for s_1 and v_3 for s_2). The dark columns (red) correspond to the short b-blocks of t: they ensure that, in each row of the adjacency matrix, at least one selected vertex dominates the vertex corresponding to that row. The base pattern is repeated N times to ensure that at least one occurrence of the pattern is mapped to t without unwanted gaps. Additional words s_{k+1} and s_{k+2} are added to match the remaining letters from t. (Color figure online)

block in s. We call b-blocks of length $2k - 1$ in t *short* and b-blocks of length $2k$ *long*.

Observation 2.9. The words $s_1, \ldots, s_k$ and t created by Construction 2.7 from a DOMINATING SET instance (G, k) have the following properties:

(i) Each s_i for $i \in \{1, \ldots, k\}$ contains $2Nn^2$ blocks.
(ii) The word t contains $2Nn^2 + 2(n - 1)$ blocks.
(iii) For $i \in \{1, \ldots, k\}$, all a-blocks in s_i have length 1. All a-blocks of t have length k.
(iv) For $h \in \{1, \ldots, Nn\}$, the b-blocks at position $2hn$ in t are short. All other b-blocks in t are long.
(v) For each $i \in \{1, \ldots, k\}$, $p \in \{0, \ldots, N-1\}$, and $u, v \in \{1, \ldots, n\}$, the b-block at position $2pn^2 + 2n(u - 1) + 2v$ in s_i has length $\ell_{u,v}$: it corresponds to the entry in the uth row and vth column of the adjacency matrix of G.

Since Construction 2.7 runs in polynomial time and the number of words in the created SHUFFLE PRODUCT instance only depends on the size of the sought dominating set, for Theorem 2.4, it remains to prove the following lemma.

Lemma 2.10. *Let $s_1, \ldots, s_{k+2}$ and t be the words created by Construction 2.7 from a* DOMINATING SET *instance (G, k). Then G has a dominating set of size k if and only if $t \in s_1 \sqcup\!\sqcup s_2 \sqcup\!\sqcup \cdots \sqcup\!\sqcup s_{k+2}$.*

Proof. ($\Rightarrow$) Assume first that $G = (V, E)$ has a dominating set $D = \{d_1, \ldots, d_k\}$. We describe t as a shuffle product of the words s_i as follows. For each $i \in \{1, \ldots, k\}$, map all letters from the block at position x of s_i into block $x + 2(n - d_i)$ of t, that is, consecutive blocks of s_i are mapped into consecutive blocks of t with a small offset depending on d_i. So far, at most k letters are mapped into each a-block of t and at most $2k$ letters are mapped into each b-block of t. Hence, all a-blocks and all long b-blocks of t are long enough to accommodate all their designated letters. It remains to show that at most $2k - 1$ letters are mapped into each short b-block β of t. By Observation 2.9(iv), β is at position $2hn$ for some $h \in \{1, \ldots, Nn\}$. Thus, there are $p \in \{0, \ldots, N - 1\}$ and $u \in \{1, \ldots, n\}$ such that $2hn = 2(pn + u)n = 2pn^2 + 2un$. For each s_i, the block α_i of s_i mapped into β has position $(2pn^2 + 2un) - 2(n - d_i) = 2pn^2 + 2(u - 1)n + 2d_i$. Hence, α_i has length ℓ_{u,d_i} by Observation 2.9(v). Since D is a dominating set, it contains a vertex d_{i^*} such that $d_{i^*} = u$ or $\{d_{i^*}, u\} \in E$. Thus, by (2.1), α_{i^*} has length $\ell_{u,d_{i^*}} = 1$. Overall, at most k b-blocks of $\{s_1, \ldots, s_k\}$ are mapped into β. We have shown that at least one of them, namely α_{i^*}, has length one. Since the others have length at most two, at most $2k - 1$ letters are mapped into block β.

We have seen a mapping of the words s_i with $i \in \{1, \ldots, k\}$ to t. Thus, we have $|t|_a \geq \sum_{i=1}^{k} |s_i|_a$ and $|t|_b \geq \sum_{i=1}^{k} |s_i|_b$ and the words s_{k+1} and s_{k+2} are well-defined. It remains to map s_{k+1} and s_{k+2} to t. Since s_{k+1} consists only of a and s_{k+2} only of b, we only have to check that t contains as many letters a and b as all words s_i together, which is true by the definition of s_{k+1} and s_{k+2}. We conclude that $t \in s_1 \sqcup\!\sqcup s_2 \sqcup\!\sqcup \cdots \sqcup\!\sqcup s_{k+2}$ if G has a dominating set of size k.

($\Leftarrow$) Assume that $t \in s_1 \uplus s_2 \uplus \cdots \uplus s_{k+2}$. We show that G has a dominating set of size k. To this end, for $i \in \{1, \ldots, k\}$, let $y_i(x)$ be the position of the block in t into which the last letter of the block at position x of s_i is mapped and let $\delta_i(x) = y_i(x) - x$. We will see that, intuitively, one can think of $\delta_i(x)$ as the shift of the xth block of s_i in t. To show that G has a dominating set of size k, we use the following two facts about δ_i, which we will prove afterwards.

(i) For $i \in \{1, \ldots, k\}$ and $x \in \{1, \ldots, 2Nn^2\}$, one has $\delta_i(x) \in \{0, \ldots, 2(n-1)\}$.
(ii) There is a $p \in \{0, \ldots, N-1\}$ such that, for all $i \in \{1, \ldots, k\}$, δ_i is constant over the interval $I_p = \{2pn^2 + 1, \ldots, 2(p+1)n^2 + 1\}$.

We now focus on a $p \in \{0, \ldots, N-1\}$ as in (ii) and write δ_i for the value $\delta_i(x)$ taken for all $x \in I_p$. We show that $D := \{d_i = n - \delta_i/2 \mid k \in \{1, \ldots, k\}\}$ is a dominating set of size k for G, that is, we show $D \subseteq V$ and $V \subseteq N[D]$.

To this end, consider a vertex $u \in V$ and the block β of t at position $2pn^2 + 2un = 2hn$ for $h = pn + u \in \{1, \ldots, Nn\}$. By Observation 2.9(iv), β is a short b-block. For any $i \in \{1, \ldots, k\}$, let α_i be the block at position $2pn^2 + 2un - \delta_i$ in s_i. Because of (i), this position is in I_p. By definition of δ_i, the last letter of α_i is mapped into β. Thus, α_i is a b-block. Note that this implies that δ_i is even since a-blocks and b-blocks are alternating in t and s_i. Moreover, by (i), $d_i = n - \delta_i/2 \in \{1, \ldots, n\} = V$. It follows that $D \subseteq V$. We show that $u \in N[D]$. To this end, note that the a-block in s_i at position $2pn^2 + 2un - \delta_i - 1 \in I_p$ directly preceding α_i is mapped into the a-block of t at position $2pn^2 + 2un - 1$ directly preceding β. Thus, all letters of α_i are mapped into β and one has

$$\sum_{i=1}^{k} |\alpha_i| \leq |\beta|. \tag{2.2}$$

By Observation 2.9(v), α_i has length $\ell_{u,(n-\delta_i/2)} = \ell_{u,d_i}$. Since β is a short b-block, it has length $2k - 1$. From (2.2), we get $\sum_{i=1}^{k} \ell_{u,d_i} \leq 2k - 1$. Thus, there is some $i^* \in \{1, \ldots, k\}$ with $\ell_{u,d_{i^*}} = 1$. By (2.1), that means $d_{i^*} = u$ or $\{u, d_{i^*}\}$ is an edge in G. Hence, $u \in N[D]$ and D is a dominating set of size k for of G.

It remains to prove (i) and (ii). For (i), note that $y_i(1) \geq 1$ and $y_i(x+1) \geq y_i(x) + 1$. Hence, δ_i is non-decreasing with all values being non-negative. Furthermore, for $x = 2Nn^2$, $y_i(x) \leq 2Nn^2 + 2(n-1)$ since t has only so many blocks by Observation 2.9(ii). Thus, the maximum possible value of δ_i is $2(n-1)$. Towards (ii), we say that a value of $p \in \{0, \ldots, N-1\}$ is *bad for* i if δ_i is not constant over I_p. For such a p, one has $\delta_i(2pn^2 + 1) < \delta_i(2(p+1)n^2 + 1)$. Hence, there can be at most $2(n-1)$ values of p that are bad for i. Overall, there are at most $2k(n-1) < N$ values of p that are bad for some $i \in \{1, \ldots, k\}$. Thus, at least one value is not bad for any i. For this value of p, every δ_i is constant over the interval I_p. $\qquad\square$

2.3 W[2]-hardness of Scheduling Problems Parameterized by Width

In the previous section, we showed W[2]-hardness of SHUFFLE PRODUCT. We now transfer this result to scheduling problems on parallel identical machines.

Theorem 2.11. The following two problems are W[2]-hard parameterized by the width of the partial order giving the precedence constraints.

(i) $P2|\text{prec}, p_j \in \{1,2\}|C_{\max}$,
(ii) $P3|\text{prec}, p_j = 1, \text{size}_j \in \{1,2\}|C_{\max}$.

We prove (i) using the following parameterized reduction from SHUFFLE PRODUCT with $k + 1$ words to $P2|\text{prec}, p_j \in \{1,2\}|C_{\max}$ with $k + 2$ chains.

Construction 2.12. Let $(s_1, \ldots, s_k, t)$ be a SHUFFLE PRODUCT instance over the alphabet $\Sigma = \{1, 2\}$. Assume that $|t|_1 = \sum_{i=1}^{k} |s_i|_1$ and $|t|_2 = \sum_{i=1}^{k} |s_i|_2$ (otherwise, it is a no-instance). We create an instance of $P2|\text{prec}, p_j \in \{1,2\}|C_{\max}$:

(1) For each $i \in \{1, \ldots, k\}$, create a chain of *worker* jobs $j_{i1} \prec j_{i2} \prec \cdots \prec j_{i|s_i|}$, where $j_{i,x}$ has length $s_i[x]$.
(2) For each $x \in \{1, \ldots, |t|\}$, create three *floor* jobs $z_{x,1}, z_{x,2}, z_{x,3}$ with $z_{x,1} \prec z_{x,2}$ and $z_{x,1} \prec z_{x,3}$, where $z_{x,1}$ has length $t[x]$, and $z_{x,2}$ and $z_{x,3}$ have length 1. If $x < |t|$, then also add the precedence constraints $z_{x,2} \prec z_{x+1,1}$ and $z_{x,3} \prec z_{x+1,1}$.

Observe that $\{z_{x,1}, z_{x,2} \mid 1 \leq x \leq |t|\}$ is a chain. Thus, the makespan of any schedule is at least $T := \sum_{x=1}^{|t|}(t[x] + 1)$. For $x \in \{1, \ldots, n\}$, let $\tau(x) := \sum_{i=1}^{x-1}(t[x] + 1)$.

Observation 2.13. A schedule with makespan T must schedule job $z_{x,1}$ at time $\tau(x)$, and jobs $z_{x,2}$ and $z_{x,3}$ at time $\tau(x) + t[x]$. Thus, for $x \in \{1, \ldots, |t|\}$, both machines are used by floor jobs from $\tau(x) + t[x]$ to $\tau(x) + t[x] + 1$ and one machine is free of floor jobs between $\tau(x)$ and $\tau(x) + t[x]$ for $t[x]$ time units. We call these *available time slots*.

Construction 2.12 runs in polynomial time. Moreover, from $k + 1$ input words, it creates instances of width $k + 2$: there are k chains of worker jobs and the floor decomposes into two chains $\{z_{x,1}, z_{x,2} \mid 1 \leq x \leq |t|\}$ and $\{z_{x,1}, z_{x,3} \mid 1 \leq x \leq |t|\}$. To prove Theorem 2.11(i), one can thus show that $t \in s_1 \sqcup \cdots \sqcup s_k$ if and only if the created $P2|\text{prec}, p_j \in \{1,2\}|C_{\max}$ instance allows for a schedule of makespan T. By Observation 2.13, any such schedule has available time slots of lengths corresponding to the letters in t, each of which can accommodate a worker job corresponding to a letter of $s_1, \ldots, s_k$. The precedence constraints ensure that these worker jobs get placed into the time slots corresponding to letters of t in increasing order.

The proof of Theorem 2.11(ii) works analogously: one simply replaces worker jobs of length two by worker jobs of length one that require two machines and modifies the floor jobs so that they do not create time slots of length one or two, but so that each created time slot is available on only one or on two machines. To achieve this, the construction uses three machines.

3   Resource-Constrained Project Scheduling

In Sect. 2.3, we have seen that $P3|\text{prec}, p_j=1, \text{size}_j \in \{1,2\}|C_{\max}$ is W[2]-hard parameterized by the partial order width. It follows that also RCPSP (cf. Problem 1.1) is W[2]-hard for this parameter, even if the number of resources and the maximal resource usage are bounded by two and all jobs have unit processing times. In this section, we additionally consider the *lag* parameter:

Definition 3.1 (earliest possible starting time, lag). Let $J_0 \subseteq J$ be the jobs that are minimal elements in the partial order $\preceq$. The *earliest possible starting time* σ_j is 0 for a job $j \in J_0$ and, inductively, $\max_{i \prec j}(\sigma_i + p_i)$ for a job $j \in J \setminus J_0$. The *lag* of a feasible schedule $(s_j)_{j \in J}$ is $\lambda := \max_{j \in J}(s_j - \sigma_j)$.

Lenstra and Rinnooy Kan's NP-hardness proof for $P|\text{prec}, p_j=1|C_{\max}$ [10] shows that it is even NP-hard to decide whether there is a schedule of makespan at most three and lag at most one. Thus, the lag λ alone cannot lead to a fixed-parameter algorithm for RCPSP, just as the width w alone cannot. We show a fixed-parameter algorithm for the parameter $\lambda + w$.

Theorem 3.2. An optimal schedule with lag at most λ for RCPSP is computable in $(2\lambda + 1)^w \cdot 2^w \cdot \text{poly}(n)$ time if it exists, where w is the partial order width.

Our algorithm is a refinement of Servakh's pseudo-polynomial-time algorithm for RCPSP with constant width [14], which is based on graphical optimization methods introduced by Akers [1] and Hardgrave and Nemhauser [9] for hand-optimizing JOB SHOP schedules for two jobs. We provide a concise translation of Servakh's algorithm in Sect. 3.1 before we prove Theorem 3.2 in Sect. 3.2.

3.1   Geometric Interpretation of RCPSP

Given an RCPSP instance with precedence constraints $\preceq$ of width w, by Dilworth's theorem, we can decompose our set J of jobs into w pairwise disjoint chains. More specifically, these chains are efficiently computable [7]. For $\ell \in \{1, \ldots, w\}$, denote the jobs in chain ℓ by a sequence $(j_{\ell k})_{k=1}^{n_\ell}$ such that $j_{\ell k} \prec j_{\ell k+1}$ and let

$L_\ell^i := \sum_{k=1}^{i} p_{j_{\ell k}}$ be the sum of processing times of the first i jobs on chain ℓ,
$L_\ell := L_\ell^{n_\ell}$ be the sum of processing times of all jobs on chain ℓ.

Let $\mathbf{0} := (0, \ldots, 0) \in \mathbb{R}^w$ and $\mathbf{L} := (L_1, \ldots, L_w)$. Each point in the w-dimensional orthotope $X := \{\mathbf{x} \in \mathbb{R}^w \mid \mathbf{0} \leq \mathbf{x} \leq \mathbf{L}\}$ describes a *state* as follows.

Definition 3.3 (running, completed, feasibility). Let $\mathbf{x} = (x_1, \ldots, x_w) \in X$. For each chain $\ell \in \{1, \ldots, w\}$, if $x_\ell \in [L_\ell^{i-1}, L_\ell^i)$, then the jobs $(j_{\ell k})_{k=1}^{i-1}$ of chain ℓ are *completed* and job $j_{\ell i}$ has been *processed* for $x_\ell - L_\ell^{i-1}$ time. We call job $j_{\ell i}$ *running* if $L_\ell^{i-1} < x_\ell < L_\ell^i$. We denote by

$J(\boldsymbol{x}) \subseteq J$ the set of jobs running in state $\boldsymbol{x}$ and by
$C(\boldsymbol{x}) \subseteq J$ the set of jobs completed in state $\boldsymbol{x}$.

A point $\boldsymbol{x} \in X$ is *feasible* if it holds that both

(F1) the jobs $J(\boldsymbol{x})$ comply with resource constraints, that is, $\sum_{j \in J(\boldsymbol{x})} r_{j\rho} \leq R_\rho$
   for each resource $\rho \in R$, and

(F2) if there are two jobs $i \prec j$ such that $j \in J(\boldsymbol{x})$, then $i \in C(\boldsymbol{x})$.

Note that points $\boldsymbol{x} \in X$ may indeed violate (F2): there are not only prece-dence constraints between jobs on one chain, but also between jobs on different chains.

Each feasible schedule now yields a path of feasible points in the orthotope X from the point $\boldsymbol{0}$, where no job has started, to the point $\boldsymbol{L}$, where all jobs are completed. Each such path consists of (linear) segments of the form $[\boldsymbol{x}, \boldsymbol{x} + t\boldsymbol{\delta}]$ for some $\boldsymbol{\delta} = (\delta_1, \ldots, \delta_w) \in \{0, 1\}^w$, which corresponds to running exactly the jobs on the chains ℓ with $\delta_\ell = 1$ for t units of time. Since all processing times and starting times are integers (cf. Problem 1.1), we can assume $t \in \mathbb{N}$.

Definition 3.4 (feasibility of segments and their lengths). The *length* of a segment $[\boldsymbol{x}, \boldsymbol{x} + t\boldsymbol{\delta}]$ is t. The *length* of a path is the sum of the lengths of its segments. A segment $[\boldsymbol{x}, \boldsymbol{x} + t\boldsymbol{\delta}]$ is *feasible* if it contains only feasible points and interrupts no jobs; that is, if there is a job $j \in J(\boldsymbol{x})$ on chain ℓ, then $\delta_\ell = 1$.

There is now a one-to-one correspondence between feasible schedules and paths from $\boldsymbol{0}$ to $\boldsymbol{L}$ consisting only of feasible segments and between the shortest of these paths and optimal schedules. This leads to the following algorithm.

Algorithm 3.5 (Servakh [14]). Compute a shortest feasible path from $\boldsymbol{0}$ to $\boldsymbol{L}$ using dynamic programming: for each feasible point $\boldsymbol{x} \in X \cap \mathbb{N}^w$ in lexicographi-cally increasing order, compute the length $P(\boldsymbol{x})$ of a shortest feasible path from $\boldsymbol{0}$ to $\boldsymbol{x}$ using the recurrence relation

$$P(\boldsymbol{0}) = 0, P(\boldsymbol{x}) = \min_{\boldsymbol{\delta} \in \Delta_{\boldsymbol{x}}} P(\boldsymbol{x} - \boldsymbol{\delta}) + 1 \text{ for feasible } \boldsymbol{x} \in X \cap \mathbb{N}^w \setminus \{\boldsymbol{0}\}, \quad (3.1)$$

where $\Delta_{\boldsymbol{x}}$ is the set of vectors $\boldsymbol{\delta} \in \{0, 1\}^w$ such that segment $[\boldsymbol{x} - \boldsymbol{\delta}, \boldsymbol{x}]$ is feasible.

To compute $P(\boldsymbol{L})$, one thus iterates over at most $\prod_{\ell=1}^{w}(L_\ell + 1)$ points $\boldsymbol{x} \in X \cap \mathbb{N}^w$, for each of them over 2^w vectors $\boldsymbol{\delta} \in \{0, 1\}^w$, and, for each, decides whether $[\boldsymbol{x} - \boldsymbol{\delta}, \boldsymbol{x}]$ is feasible. Since the set of running jobs is the same for all interior points of the segment, it is enough to check the feasibility of its end points and one interior point, which can be done in polynomial time. Thus, the algorithm runs in $\prod_{\ell=1}^{w}(L_\ell + 1) \cdot 2^w \cdot \text{poly}(n)$ time, which is pseudo-polynomial for constant w.

3.2   Fixed-Parameter Algorithm for Arbitrary Processing Times

The bottleneck of Algorithm 3.5 is that it searches for a shortest path from $\mathbf{0}$ to $\mathbf{L}$ in the whole orthotope X. For the case where we are only accepting schedules of maximum lag λ, we will shrink the search space significantly: we show that we only have to search for paths within a tight corridor around the path corresponding to the schedule $(\sigma_j)_{j \in J}$ that starts jobs at the earliest possible time.

Definition 3.6 (point at time t on a path). Let p be the path from $\mathbf{0}$ to $\mathbf{L}$ corresponding to a not necessarily feasible schedule $(s_j)_{j \in J}$ that, however, respects precedence constraints. Let $t \geq 0$ and T be the length of p.

Then, $\boldsymbol{p}(t)$ is the endpoint of the subpath of length t of p starting in $\mathbf{0}$ for $t \leq T$, and $\boldsymbol{p}(t) := \mathbf{L}$ for $t > T$.

Since the definition requires $(s_j)_{j \in J}$ to respect precedence constraints, $\boldsymbol{p}(t)$ determines the state (cf. Definition 3.3) at time t according to schedule $(s_j)_{j \in J}$.

Definition 3.7 (λ-corridored). Let p be the path corresponding to the schedule $(\sigma_j)_{j \in J}$ that starts jobs at the earliest possible time (cf. Definition 3.1).

$$\Gamma_{\lambda}(t) := \{\boldsymbol{x} \in X \mid \boldsymbol{p}(t) - \boldsymbol{\lambda} \leq \boldsymbol{x} \leq \boldsymbol{p}(t)\}, \quad \text{where } \boldsymbol{\lambda} = (\lambda, \ldots, \lambda) \in \mathbb{N}^w.$$

We call a path q λ-*corridored* if $\boldsymbol{q}(t) \in \Gamma_{\lambda}(t)$ for all $t \geq 0$.

Note that points on the path p in Definition 3.7 may violate Definition 3.3(F1), but not (F2). One can show the following relation between λ-corridored paths and schedules of lag λ.

Lemma 3.8. A feasible schedule $(s_j)_{j \in J}$ has lag at most λ if and only if its corresponding path q is λ-corridored.

Lemma 3.8 allows us to compute a shortest feasible path from $\mathbf{0}$ to $\mathbf{L}$ using only points in $\Gamma_{\lambda}(t)$ for some t. Herein, we will exploit the following condition for checking whether a path segment can be part of a λ-corridored path.

Lemma 3.9. Let $[\boldsymbol{x}, \boldsymbol{x} + t\boldsymbol{\delta}]$ for $\boldsymbol{\delta} \in \{0, 1\}^w$. If $\boldsymbol{x} \in \Gamma_{\lambda}(t_0)$ and $\boldsymbol{x} + t\boldsymbol{\delta} \in \Gamma_{\lambda}(t_0 + t)$ for some $t_0 \geq 0$, then $\boldsymbol{x} + \tau\boldsymbol{\delta} \in \Gamma_{\lambda}(t_0 + \tau)$ for all $0 \leq \tau \leq t$.

Proof. Let p be the path corresponding to schedule $(\sigma_j)_{j \in J}$ as in Definition 3.7 and let $\boldsymbol{\delta} = (\delta_1, \ldots, \delta_w) \in \{0, 1\}^w$. For any $\tau \in [0, t]$, consider

$$\boldsymbol{x}^{\tau} = (x_1^{\tau}, \ldots, x_w^{\tau}) := \boldsymbol{x} + \tau\boldsymbol{\delta} \quad \text{and} \quad \boldsymbol{y}^{\tau} = (y_1^{\tau}, \ldots, y_{\ell}^{\tau}) := \boldsymbol{p}(t_0 + \tau).$$

By the prerequisites of the lemma, we have $\boldsymbol{y}^0 - \boldsymbol{\lambda} \leq \boldsymbol{x}^0 \leq \boldsymbol{y}^0$ and $\boldsymbol{y}^t - \boldsymbol{\lambda} \leq \boldsymbol{x}^t \leq \boldsymbol{y}^t$. We show $\boldsymbol{y}^{\tau} - \boldsymbol{\lambda} \leq \boldsymbol{x}^{\tau} \leq \boldsymbol{y}^{\tau}$ for any $\tau \in [0, t]$.

We start with $\boldsymbol{x}^\tau \leq \boldsymbol{y}^\tau$. For the sake of contradiction, assume that there is some chain ℓ and a $\tau \in [0, t]$ such that $x_\ell^\tau > y_\ell^\tau$. Then, $x_\ell^\tau > y_\ell^\tau \geq y_\ell^0 \geq x_\ell^0$. It follows that $\delta_\ell = 1$, which contradicts $\boldsymbol{x}^t \leq \boldsymbol{y}^t$ because, then,

$$x_\ell^t = x_\ell^0 + t = x_\ell^0 + \tau + (t - \tau) = x_\ell^\tau + (t - \tau) > y_\ell^\tau + (t - \tau) \geq y_\ell^{\tau + (t - \tau)} = y_\ell^t.$$

Now, we show $\boldsymbol{y}^\tau - \boldsymbol{\lambda} \leq \boldsymbol{x}^\tau$. Consider some chain ℓ. If $\delta_\ell = 1$, then we have $y_\ell^\tau - \lambda \leq y_\ell^0 + \tau - \lambda \leq x_\ell^0 + \tau = x_\ell^\tau$ and we are fine. If $\delta_\ell = 0$ and there is a $\tau \in [0, t]$ such that $y_\ell^\tau - \lambda > x_\ell^\tau$, then $y_\ell^t - \lambda \geq y_\ell^\tau - \lambda > x_\ell^\tau = x_\ell^t$, contradicting $\boldsymbol{y}^t - \boldsymbol{\lambda} \leq \boldsymbol{x}^t$. $\square$

We can now prove the following result by computing recurrence (3.1) for each of the $(\lambda + 1)^w$ feasible points $\boldsymbol{x} \in \Gamma_\lambda(t) \cap \mathbb{Z}^w$ for all $t \in \{0, \ldots, L\}$.

Proposition 3.10. An optimal schedule of lag at most λ for RCPSP if it exists is computable in $(\lambda + 1)^w \cdot 2^w \cdot \text{poly}(L)$ time, where L is the sum of all processing times and w is the partial order width.

However, note that this is a fixed-parameter algorithm only for polynomial processing times, which is why we skip the proof and go on towards proving Theorem 3.2—a fixed-parameter algorithm that works for arbitrarily large processing times. To this end, we prove that all maximal segments of a path corresponding to a schedule with lag at most λ start and end in one of $2 \cdot |J|$ hypercubes with edge length $2\lambda + 1$.

Lemma 3.11. Let q be the path of a feasible schedule $(s_j)_{j \in S}$ of lag at most λ and let $t_2 \leq t_1 \leq t_2 + \lambda$. Then, $\boldsymbol{q}(t_1) \in \Gamma_{2\lambda}(t_2 + \lambda)$ (cf. Definition 3.7).

Proof. Consider the schedule $(\sigma_j)_{j \in J}$ that starts each job at the earliest possible time and its path p. Our aim is to show

$$\boldsymbol{p}(t_2 + \lambda) - 2\boldsymbol{\lambda} \leq \boldsymbol{q}(t_1) \leq \boldsymbol{p}(t_2 + \lambda),$$

where $\boldsymbol{\lambda} = (\lambda, \ldots, \lambda) \in \mathbb{N}^w$. By Lemma 3.8, q is λ-corridored. Thus,

$$\boldsymbol{p}(t_1) - \boldsymbol{\lambda} \leq \boldsymbol{q}(t_1) \leq \boldsymbol{p}(t_1) \quad \text{and} \quad \boldsymbol{p}(t_2 + \lambda) - \boldsymbol{\lambda} \leq \boldsymbol{q}(t_2 + \lambda) \leq \boldsymbol{p}(t_2 + \lambda).$$

From this, one easily gets $\boldsymbol{q}(t_1) \leq \boldsymbol{p}(t_1) \leq \boldsymbol{p}(t_2 + \lambda)$. Moreover, one has

$$\boldsymbol{p}(t_2 + \lambda) - 2\boldsymbol{\lambda} \leq \boldsymbol{q}(t_2 + \lambda) - \boldsymbol{\lambda} \leq \boldsymbol{q}(t_2) + \boldsymbol{\lambda} - \boldsymbol{\lambda} = \boldsymbol{q}(t_2) \leq \boldsymbol{q}(t_1). \quad \square$$

Lemma 3.12. Let q be the path of a feasible schedule $(s_j)_{j \in S}$ of lag at most λ and let $[\boldsymbol{x}, \boldsymbol{x} + t\boldsymbol{\delta}]$ be a maximal segment of q such that the set $J(\boldsymbol{x} + \tau\boldsymbol{\delta})$ of running jobs (cf. Definition 3.3) is the same for all $\tau \in (0, t)$. Then,

$$\{\boldsymbol{x}, \boldsymbol{x} + t\boldsymbol{\delta}\} \subseteq \Gamma := \bigcup_{j \in J} \Gamma_{2\lambda}(\sigma_j + \lambda) \cup \bigcup_{j \in J} \Gamma_{2\lambda}(\sigma_j + p_j + \lambda),$$

where $(\sigma_j)_{j \in J}$ is the schedule that starts each job at the earliest possible time.

Proof. Let t_0 be chosen arbitrarily such that $q(t_0) \in \{x, x + t\delta\}$. By maximality of the segment, some job $j \in J$ is starting or ending at time t_0, that is, $t_0 = s_j$ or $t_0 = s_j + p_j$. Then, $\{x, x + t\delta\} \subseteq \Gamma_{2\lambda}(\sigma_j + \lambda) \cup \Gamma_{2\lambda}(\sigma_j + p_j + \lambda)$ follows from $\sigma_j \leq s_j \leq \sigma_j + \lambda$ and Lemma 3.11.   □

We are now ready to show a fixed-parameter algorithm for RCPSP parameterized by length and maximum lag. That is, we prove Theorem 3.2.

Proof (of Theorem 3.2). We compute the shortest feasible λ-corridored path from the state $\mathbf{0}$, were no job has started, to the state L, where all jobs have been completed (cf. Lemma 3.8). We use dynamic programming similarly to Algorithm 3.5. By Lemma 3.12, it is enough to consider those paths whose segments start and end in Γ. Thus, for each $x \in \Gamma \cap \mathbb{N}^w$ in lexicographically increasing order, we compute the length $P(x)$ of a shortest λ-corridored path from $\mathbf{0}$ to x with segments starting and ending in Γ. To this end, for an $x \in \Gamma \cap \mathbb{N}^w$, let Δ_x be the set of vectors $\delta \in \{0,1\}^w$ such that,

(i) there is a smallest integer $t_\delta \geq 1$ such that $x - t_\delta \cdot \delta \in \Gamma$ and such that
(ii) the segment $[x - t_\delta \cdot \delta, x]$ is feasible.

Then, $P(\mathbf{0}) = 0$ and, for feasible $x \in \Gamma \cap \mathbb{N}^w \setminus \{\mathbf{0}\}$, one has

$$P(x) = \min\{P(x - t_\delta \cdot \delta) + t_\delta \mid \delta \in \Delta_x \text{ and } x \in \Gamma_\lambda(P(x - t_\delta \cdot \delta) + t_\delta)\},$$

where $\min \emptyset = \infty$ and the last condition on x uses Lemma 3.9 to ensure that we are indeed computing the length $P(x)$ of a λ-*corridored* path (cf. Definition 3.7) to x: by induction, we know that $P(x - t_\delta \cdot \delta)$ is the length of a shortest λ-corridored path to $x - t_\delta \cdot \delta$, and thus $x - t_\delta \cdot \delta \in \Gamma_\lambda(P(x - t_\delta \cdot \delta))$.

We have to discuss how to check (i) and (ii). One can check (ii) in polynomial time since it is enough to check feasibility at the end points and one interior point of the segment since the set of jobs running at the interior points of $[x - t_\delta \cdot \delta, x]$ does not change: otherwise, since jobs are started or finished only at integer times, there is a maximal subsegment $[x, x - t \cdot \delta]$ with $t \leq t_\delta - 1$ where the set of running jobs does not change. Then $x - t \cdot \delta \in \Gamma$ by Lemma 3.12, contradicting the minimality of t_δ.

Towards (i), we search for the minimum $t_\delta \geq 1$ such that $x - t_\delta \cdot \delta \in \Gamma$. Consider the schedule $(\sigma_j)_{j \in J}$ that schedules each job at the earliest possible time (cf. Definition 3.1). It is computable in polynomial time. By Lemma 3.12, we search for the minimum $t_\delta \geq 1$ such that $x - t_\delta \cdot \delta \in \Gamma_{2\lambda}(\sigma_j + \lambda)$ or $x - t_\delta \cdot \delta \in \Gamma_{2\lambda}(\sigma_j + p_j + \lambda)$ for some job $j \in J$. That is, by Definition 3.7, for each job j, we find the minimum $t_j \geq 1$ that solves a system of linear inequalities of the form $y - 2\lambda \leq x - t_j \cdot \delta \leq y$, where $\delta \Rightarrow (\delta_1, \dots, \delta_w) \in \{0,1\}^w$. Writing $y = (y_1, \dots, y_w)$ and $x = (x_1, \dots, x_w)$, either $t_j = \max(\{1\} \cup \{x_\ell - y_\ell \mid \delta_\ell = 1\})$ is the minimum such t_j or there is no solution for job j. Note that t_j is an integer since x and y are integer vectors. Thus, $t_\delta = \min_{j \in J} t_j$ is computable in polynomial time.

We conclude that we process each $x \in \Gamma \cap \mathbb{N}^w$ in $2^w \cdot \text{poly}(n)$ time. Moreover, Γ contains at most $2 \cdot |J| \cdot (2\lambda + 1)^w$ integer points since each job $j \in J$ contributes at most $(2\lambda + 1)^w$ points in $\Gamma_{2\lambda}(\sigma_j + \lambda)$ and at most $(2\lambda + 1)^w$ points in $\Gamma_{2\lambda}(\sigma_j + p_j + \lambda)$. A total running time of $(2\lambda + 1)^w \cdot 2^w \cdot \text{poly}(n)$ follows.   □

4   Conclusion

Our algorithm for RCPSP shows, in particular, that $P3|\text{prec},p_j{=}1|C_{\max}$ is fixed-parameter tractable parameterized by the partial order width w and allowed lag λ. Since the NP-hardness of this problem is a long-standing open question [8, OPEN8], it would be surprising to show W[1]-hardness of this problem for *any* parameter: this would exclude polynomial-time solvability unless $\text{FPT} = \text{W}[1]$. Thus, it makes sense to search for a fixed-parameter algorithm for $P3|\text{prec},p_j{=}1|C_{\max}$ parameterized by w, whereas we showed that already $P2|\text{prec},p_j{\in}\{1,2\}|C_{\max}$ and $P3|\text{prec},p_j{=}1,\text{size}_j{\in}\{1,2\}|C_{\max}$ are W[2]-hard parameterized by w.

Acknowledgments. The authors are thankful to Sergey Sevastyanov for pointing out the work of Akers [1] and Servakh [14]. This research was initiated at the annual research retreat of the algorithms and complexity group of TU Berlin, April 3–9, 2016, Krölpa, Germany.

References

1. Akers Jr., S.B.: A graphical approach to production scheduling problems. Oper. Res. **4**(2), 244–245 (1956)
2. van Bevern, R.: Fixed-parameter algorithms in operations research: opportunities and challenges. Parameterized Complex. News **12**(1), 4–6 (2016)
3. Bodlaender, H.L., Fellows, M.R.: W[2]-hardness of precedence constrained k-processor scheduling. Oper. Res. Lett. **18**(2), 93–97 (1995)
4. Cygan, M., et al.: Parameterized Algorithms. Springer, Amsterdam (2015)
5. Downey, R.G., Fellows, M.R.: Fundamentals of Parameterized Complexity. Springer, Heidelberg (2013)
6. Du, J., Leung, J.Y.T., Young, G.H.: Scheduling chain-structured tasks to minimize makespan and mean flow time. Inform. Comput. **92**(2), 219–236 (1991)
7. Felsner, S., Raghavan, V., Spinrad, J.: Recognition algorithms for orders of small width and graphs of small Dilworth number. Order **20**(4), 351–364 (2003)
8. Garey, M.R., Johnson, D.S.: Computers and Intractability: A Guide to the Theory of NP-Completeness. Freeman, New York (1979)
9. Hardgrave, W.W., Nemhauser, G.L.: A geometric model and a graphical algorithm for a sequencing problem. Oper. Res. **11**(6), 889–900 (1963)
10. Lenstra, J.K., Rinnooy Kan, A.H.G.: Complexity of scheduling under precedence constraints. Oper. Res. **26**(1), 22–35 (1978)
11. Mnich, M., Wiese, A.: Scheduling and fixed-parameter tractability. Math. Program. **154**(1–2), 533–562 (2015)
12. Rizzi, Romeo, Vialette, Stéphane: On recognizing words that are squares for the shuffle product. In: Bulatov, Andrei A., Shur, Arseny M. (eds.) CSR 2013. LNCS, vol. 7913, pp. 235–245. Springer, Heidelberg (2013)
13. Schwindt, C., Zimmermann, J. (eds.): Handbook on Project Management and Scheduling. International Handbooks on Information Systems. Springer, Heidelberg (2015)

14. Servakh, V.V.: Effektivno razreshimy sluchaj zadachi kalendarnogo planirovaniya s vozobnovimymi resursami. Diskretn. Anal. Issled. Oper. **7**(1), 75–82 (2000)
15. Ullman, J.: NP-complete scheduling problems. J. Comput. Syst. Sci. **10**(3), 384–393 (1975)
16. Warmuth, M.K., Haussler, D.: On the complexity of iterated shuffle. J. Comput. Syst. Sci. **28**(3), 345–358 (1984)

A Scheme of Independent Calculations in a Precedence Constrained Routing Problem

Alexander G. Chentsov[1,2] and Alexey M. Grigoryev[1(✉)]

[1] Krasovskii Institute of Mathematics and Mechanics, Ekaterinburg, Russia
chentsov@imm.uran.ru, ag@uran.ru
[2] Ural Federal University, Ekaterinburg, Russia

Abstract. We consider a routing problem with constraints. To solve this problem, we employ a variant of the dynamic programming method, where the significant part (that is, the part that matters in view of precedence constraints) of the Bellman function is calculated by means of an independent calculations scheme. We propose a parallel implementation of the algorithm for a supercomputer, where the construction of position space layers for the hypothetical processors is conducted with use of discrete dynamic systems' apparatus.

Keywords: Dynamic programming · Routing problem · Precedence constraints · Sequential ordering problem · Parallel algorithms

1 Introduction

In this paper, we present a solution of a constrained routing problem, where the cost functions depend on the set of pending tasks. We investigate the dynamic programming method (DPM) and implement its variant that avoids the construction of unnecessary values of the Bellman function. For simplicity, we consider a problem of visiting a system of points; we call these points the cities (in [1–3], a more general variant of this problem connected with visiting megalopolises, or clusters, was investigated). Our variant of DPM follows the scheme of independent calculation of layers of the Bellman function; based on it, we construct a parallel algorithm for a supercomputer (a new contribution) and offer the results of a computational experiment.

A natural prototype of our problem is the well-known intractable Traveling Salesman Problem (TSP); in this connection, see [4–6]. Dynamic programming solutions for TSP were first proposed in [7,8]; for Precedence Constrained TSP (TSP-PC), similar techniques were proposed in [14,15]. The complexity of dynamic programming for TCP-PC was studied in, for example, [16,17]. The problem we consider in the present paper has several important qualitative singularities, which include precedence constraints and complicated cost functions.

This research was supported by Russian Science Foundation (project no. 14-11-00109).

Y. Kochetov et al. (Eds.): DOOR 2016, LNCS 9869, pp. 121–135, 2016.
DOI: 10.1007/978-3-319-44914-2_10

Namely, we consider cost functions depending on the set of pending tasks. Such a complicated variant of problem corresponds to [1–3, 9–11]. These cost functions are relevant in view of the problem of minimizing the radiation exposure of a team tasked with dismantling, say, a decommissioned fission power generating unit, see [18, Chap. 2], [19]. Indeed, each worker is subjected to the radiation produced by the radiation sources that are not dismantled at the time of travel. The set of such sources depends on the workers' route, hence the dependence of cost functions on the set of pending tasks (sequence dependence). In this connection, we recall [20], where a similar dependence for TSP was considired (in [20], a heuristic algorithm was proposed).

In our paper, we propose a parallel DPM precedure.

2   General Notation

In the following, we employ traditional set-theoretical notation. The symbol $\stackrel{\triangle}{=}$ denotes equality by definition. We call a family every set all elements of which are sets themselves. For arbitrary nonempty set T, we denote by $\mathcal{R}_+[T]$ the set of all real-valued functions from T into the infinite interval $[0, \infty[\stackrel{\triangle}{=} \{\xi \in \mathbb{R} | 0 \leq \xi\}$, where $\mathbb{R}$ is the real line. Let $\mathbb{N} \stackrel{\triangle}{=} \{1; 2; ...\}$ and $\mathbb{N}_0 \stackrel{\triangle}{=} \{0\} \cup \mathbb{N} = \{0; 1; 2; ...\}$. If $p \in \mathbb{N}_0$ and $q \in \mathbb{N}_0$, set $\overline{p, q} \stackrel{\triangle}{=} \{t \in \mathbb{N}_0 | (p \leq t) \& (t \leq q)\}$; for $p \in \mathbb{N}$ and $q \in \mathbb{N}$, we obtain $\overline{p, q} \subset \mathbb{N}$. If z is an ordered pair of arbitrary objects a and b (that is, $z = (a, b)$), then, by $\mathrm{pr}_1(z)$ and $\mathrm{pr}_2(z)$, we denote the first and second elements of that identically are defined by condition $z = (\mathrm{pr}_1(z), \mathrm{pr}_2(z))$. For every object x, by $\{x\}$ we denote the singleton containing x (so, $x \in \{x\}$). Moreover, for arbitrary objects a, b, and c, we define their triplet as follows: $(a, b, c) \stackrel{\triangle}{=} ((a, b), c)$. In addition, $A \times B \times C \stackrel{\triangle}{=} (A \times B) \times C$ for all nonempty sets A, B, and C.

If K is a nonempty finite set, then $|K| \in \mathbb{N}$ is the cardinality of K; we assume that $(\mathrm{bi})[K]$ is the set of all bijections from the "interval" $\overline{1, |K|}$ onto K. Of course, $|\emptyset| \stackrel{\triangle}{=} 0$. We recall that a permutation of a nonempty set is a bijection of this set onto itself. For every set H, we denote by $\mathcal{P}(H)$ the family of all subsets of H; let us also introduce the notation $\mathcal{P}'(H) \stackrel{\triangle}{=} \mathcal{P}(H) \setminus \{\emptyset\}$ (the family of all nonempty subsets of H).

3   Special Notation and Problem Statement

We fix $N \in \mathbb{N}$ such that $N \geq 2$. Moreover, fix

$$\mathbf{c} \in \mathcal{R}_+[\overline{O, N} \times \overline{O, N} \times \mathfrak{N}], \tag{3.1}$$

where $\mathfrak{N} \stackrel{\triangle}{=} \mathcal{P}'(\overline{1, N})$; we call elements of $\mathfrak{N}$ (nonempty subsets of $\overline{1, N}$) task lists. Function (3.1) specifies the intercity transportation costs. Finally, we introduce the function

$$f \in \mathcal{R}_+[\overline{O, N}] \tag{3.2}$$

that defines the cost of the terminal state. We use $\mathbf{c}$ and $\mathbf{f}$ (see (3.1), (3.2)) to define an additive criterion.

To define precedence constraints, we fix a set $\mathbf{K} \in \mathcal{P}(\overline{1,N} \times \overline{1,N})$. Thus, $\mathbf{K}$ is a set such that $\mathbf{K} \subset \overline{1,N} \times \overline{1,N}$. If $z \in \mathbf{K}$, then z is an ordered pair, for which $\mathrm{pr}_1(z) \in \overline{1,N}$ and $\mathrm{pr}_2(z) \in \overline{1,N}$; in addition, we call $\mathrm{pr}_1(z)$ its sender and $\mathrm{pr}_2(z)$ its recipient. For each such ordered pair, its sender $\mathrm{pr}_1(z)$ must be visited before its recipient $\mathrm{pr}_2(z)$. Let $\mathbf{P} \stackrel{\triangle}{=} (\mathrm{bi})[\overline{1,N}]$. Permutations from $\mathbf{P}$ are called routes. Precedence constraints are taken into account by restricting the choice of a route from $\mathbf{P}$ to its subset $\mathbf{A}$. Namely,

$$\mathbf{A} \stackrel{\triangle}{=} \{\alpha \in \mathbf{P} |\ \forall z \in \mathbf{K}\ \forall t_1 \in \overline{1,N}\ \forall t_2 \in \overline{1,N}\ (z = (\alpha(t_1), \alpha(t_2))) \Rightarrow (t_1 < t_2)\}$$

$$= \{\alpha \in \mathbf{P} |\ \alpha^{-1}(\mathrm{pr}_1(z)) < \alpha^{-1}(\mathrm{pr}_2(z))\ \forall z \in \mathbf{K}\} \quad (3.3)$$

is the set of all feasible (in the sense of precedence constraints) routes. Thus, every route from $\mathbf{A}$ visits the sender $\mathrm{pr}_1(z)$ before the corresponding recipient $\mathrm{pr}_2(z)$ for each ordered pair $z \in \mathbf{K}$. We assume that $\forall \mathbf{K}_0 \in \mathcal{P}'(\mathbf{K})\ \exists z_0 \in \mathbf{K}_0 :$ $\mathrm{pr}_1(z_0) \neq \mathrm{pr}_2(z)\ \forall z \in \mathbf{K}_0$ (in [12, Chap. 2], several specific classes of problems satisfying this condition are pointed out). Then, $\mathbf{A} \neq \emptyset$ (see [12, Chap. 2]). Thus, $\mathbf{A}$ is a nonempty finite set. If $\alpha \in \mathbf{A}$, then the value

$$\mathfrak{C}_\alpha \stackrel{\triangle}{=} \mathbf{c}\left(0, \alpha(1), \overline{1,N}\right) + \sum_{t=1}^{N-1} \mathbf{c}\left(\alpha(t), \alpha(t+1), \{\alpha(j) : j \in \overline{t+1,N}\}\right)$$

$$+ f(\alpha(N)) \in [0, \infty[\quad (3.4)$$

is well-defined. We consider the following problem:

$$\mathfrak{C}_\alpha \to \min,\ \ \alpha \in \mathbf{A}. \quad (3.5)$$

For this problem, its value (extremum)

$$V \stackrel{\triangle}{=} \min_{\alpha \in \mathbf{A}} \mathfrak{C}_\alpha \in [0, \infty[\quad (3.6)$$

and the set $\mathbf{A}_{\mathrm{opt}} \stackrel{\triangle}{=} \{\alpha_0 \in \mathbf{A} |\ \mathfrak{C}_{\alpha_0} = V\} \in \mathcal{P}'(\mathbf{A})$ are defined. Our goal consists in determining V and finding a route $\alpha_0 \in \mathbf{A}_{\mathrm{opt}}$, i.e., it is required to find a feasible route with the least value of additive criterion (3.4).

4 Dynamic Programming

We use the procedure on the base of DPM from [6, 7, 9, 11]. This procedure is a development of [12, Sect. 4.9]. But, for simplicity, we consider this procedure for the "point" statement (we do not use the model with megalopolises in order to concentrate on issues connected with parallelization of our procedure). Now, we consider a natural scheme of constraints transformation: we replace the feasibility in the sense of precedence constraints with crossing-out feasibility. To this end, we use the mapping $\mathbf{I}$ of [12, Chap. 2] operating in the family $\mathfrak{N}$ of all (nonempty) lists of tasks. Namely, this mapping $\mathbf{I}$ is defined by the following rule: for $K \in \mathfrak{N}$, $\mathbf{I}(K) \stackrel{\triangle}{=} K \setminus \{\mathrm{pr}_2(z) : z \in \Xi[K]\}$, where $\Xi[K] \stackrel{\triangle}{=} \{z \in \mathbf{K} | (\mathrm{pr}_1(z) \in K)\&(\mathrm{pr}_2(z) \in K)\}$. On this base, we introduce crossing-out feasible partial routes: for $K \in \mathfrak{N}$,

$$(\mathbf{I} - \mathrm{bi})[K] \stackrel{\triangle}{=} \{\alpha \in (\mathrm{bi})[K] | \alpha(s) \in \mathbf{I}(\{\alpha(t) : t \in \overline{s, |K|}\}) \ \forall s \in \overline{1, |K|}\}; \quad (4.1)$$

in addition, (4.1) is a nonempty set (see [12, Propositions 2.2.2 and 2.2.3]). Of course, the case $K = \overline{1, N}$ is possible. Moreover, in this case,

$$\mathbf{A} = (\mathbf{I} - \mathrm{bi})[\overline{1, N}] = \{\alpha \in \mathbf{P} | (\alpha(1) \in \mathbf{I}(\overline{1, N})) \quad\quad\quad (4.2)$$
$$\& (\alpha(k) \in \mathbf{I}(\overline{1, N} \setminus \{\alpha(t) : t \in \overline{1, k-1}\}) \ \forall k \in \overline{2, N})\}.$$

Using (4.1), we introduce partial routing problems: if $s \in \overline{0, N}$ and $K \in \mathfrak{N}$, then, for $|K| \geq 2$, we consider the following problem

$$\mathbf{c}(s, \alpha(1), K) + \sum_{t=1}^{|K|-1} \mathbf{c}(\alpha(t), \alpha(t+1), \{\alpha(j) : j \in \overline{t+1, |K|}\}) \quad\quad (4.3)$$
$$+ f(\alpha(|K|)) \to \min, \alpha \in (\mathbf{I} - \mathrm{bi})[K];$$

for problem (4.3), we introduce its corresponding value,

$$v(s, K) \stackrel{\triangle}{=} \min_{\alpha \in (\mathbf{I} - \mathrm{bi})[K]} [c(s, \alpha(1), K) + \sum_{t=1}^{|K|-1} c(\alpha(t), \alpha(t+1), \quad\quad (4.4)$$
$$\{\alpha(j) : j \in \overline{t+1, |K|}\}) + f(\alpha(|K|))] \in [0, \infty[.$$

For $s \in \overline{0, N}$ and $r \in \overline{1, N}$, we assume

$$v(s, \{r\}) \stackrel{\triangle}{=} c(s, r, \{r\}) + f(r); \quad\quad\quad (4.5)$$

obviously, $v(s, \{r\}) \in [0, \infty[$. In connection with (4.5), note that $\mathbf{I}(\{l\}) = \{l\}$ for $l \in \overline{1, N}$; see [3, Remark 3.2]. Evidently, (4.5) defines $v(s, K)$ for $s \in \overline{0, N}$, $K \in \mathfrak{N}$, and $|K| = 1$. Finally, set $v(s, \emptyset) \stackrel{\triangle}{=} f(s) \ \forall s \in \overline{0, N}$. Thus,

$$v(s, K) \in [0, \infty[\ \forall s \in \overline{0, N} \ \forall K \in \mathcal{P}(\overline{1, N}). \quad\quad (4.6)$$

In other words, we obtain the function $v \in \mathcal{R}_+[\overline{0, N} \times \mathcal{P}(\overline{1, N})]$. In addition, for $K = \overline{1, N}$, we have the equality $|K| = N$ and, by (3.4) and (4.2),

$$\mathfrak{C}_\alpha = c(0, \alpha(1), K) + \sum_{t=1}^{|K|-1} c(\alpha(t), \alpha(t+1), \{\alpha(j) : j \in \overline{t+1, |K|}\})$$
$$+ f(\alpha(|K|)) \ \forall \alpha \in \mathbf{A}.$$

Therefore, from (3.6) and (4.4), we obtain

$$V = v(0, \overline{1, N}). \quad\quad\quad (4.7)$$

Consequently, (4.6) provides a natural representation of initial problem (3.5). We view v as a Bellman function. By general statements of [1–3,13],

$$v(s, K) = \min_{j \in \mathbf{I}(K)} [c(s, j, K) + v(j, K \setminus \{j\})] \quad \forall s \in \overline{0, N} \quad \forall K \in \mathfrak{N}. \tag{4.8}$$

Thus, (4.8) is the Bellman equation for a backward procedure. From (4.7) and (4.8), we obtain

$$V = \min_{j \in \mathbf{I}(\overline{1, N})} [c(0, j, \overline{1, N}) + v(j, \overline{1, N} \setminus \{j\})]. \tag{4.9}$$

5   Constructing Layers of the Bellman Function

The procedure based on (4.8) can be used to solve problems of small dimensionality. To partially overcome the computational difficulties, we consider a modification of procedure from [12, Sect. 4.9] (in addition, see [1–3,11]) connected with construction of layers of the Bellman function. We again note that not all values of this function are constructed.

Thus, we decrease the computational complexity to a certain extent. To this end, we consider below the construction of of feasible task lists. Namely, we set

$$\mathbf{G} \triangleq \{K \in \mathfrak{N} | \forall z \in \mathbf{K} \quad (\mathrm{pr}_1(z) \in K) \Rightarrow (\mathrm{pr}_2(z) \in K)\}. \tag{5.1}$$

Of course, $\overline{1, N} \in \mathbf{G}$; thus, $\mathbf{G} \neq \emptyset$. Moreover, we assume that $\mathbf{G}_s \triangleq \{K \in \mathbf{G} | s = |K|\} \quad \forall s \in \overline{1, N}$. Then, the family $\{\mathbf{G}_1; ...; \mathbf{G}_N\}$ is a finite partition of $\mathbf{G}$. In addition, $\mathbf{G}_N = \{\overline{1, N}\}$ is the singleton containing the set $\overline{1, N}$. Moreover, for $\mathbf{K}_1 \triangleq \{\mathrm{pr}_1(z) : z \in \mathbf{K}\}$, we obtain

$$\mathbf{G}_1 \triangleq \{\{t\} : t \in \overline{1, N} \setminus \mathbf{K}_1\}. \tag{5.2}$$

Finally, we realize the procedure

$$\mathbf{G}_N \to \mathbf{G}_{N-1} \to ... \to \mathbf{G}_1 \tag{5.3}$$

using the following transformation [1,2]

$$\mathbf{G}_{s-1} = \{K \setminus \{j\} : K \in \mathbf{G}_s, \quad j \in \mathbf{I}(K)\} \quad \forall s \in \overline{2, N}. \tag{5.4}$$

Thus, in (5.3), we begin with $\mathbf{G}_N$ and then apply (5.4). Using (5.2) and (5.3), we construct the layers $D_0, D_1, ..., D_N$ of the space of positions. Let

$$D_0 \triangleq \{(s, \emptyset) : s \in \overline{1, N} \setminus \mathbf{K}_1\}, \quad D_N \triangleq \{(0, \overline{1, N})\}. \tag{5.5}$$

In (5.5), we introduce the simplest layers of the position space. Now, we introduce intermediate layers. To this end, for $s \in \overline{1, N-1}$ and $K \in \mathbf{G}_s$, we sequentially introduce the following sets:

$$J_s(K) \triangleq \{j \in \overline{1, N} \setminus K | K \cup \{j\} \in \mathbf{G}_{s+1}\}, \quad \mathbb{D}_s[K] \triangleq \{(j, K) : j \in J_s(K)\}. \tag{5.6}$$

In (5.6), the latter sets are mutually disjoint. Then, we set

$$D_s \triangleq \bigcup_{K \in \mathbf{G}_s} \mathbb{D}_s[K] \in \mathcal{P}'(\overline{1,N} \times \mathbf{G}_s) \quad \forall s \in \overline{1, N-1}. \tag{5.7}$$

Obviously, by means of (5.6) and (5.7), for $s \in \overline{1, N-1}$, we obtain a finite partition of D_s into the system of cells $\mathbb{D}_s[K], K \in \mathbf{G}_s$:

$$D_s \triangleq \bigsqcup_{K \in \mathbf{G}_s} \mathbb{D}_s[K].$$

We recall that $(t, K \setminus \{t\}) \in D_{s-1} \ \forall s \in \overline{1,N} \ \forall (l, K) \in D_s \ \forall t \in \mathbf{I}(K)$; see [12, Proposition 4.9.4]. We note that, by (4.6) and (5.1), for every $l \in \overline{0, N}$ and $(j, K) \in D_l$, the value $v(j, K) \in [0, \infty[$ is defined. Therefore, for $l \in \overline{0, N}$, we assume that $v_l \in \mathcal{R}_+[D_l]$ is defined by the following rule:

$$v_l(j, K) \triangleq v(j, K) \ \forall (j, K) \in D_l. \tag{5.8}$$

From (5.7), we see that, for $s \in \overline{1, N}$, $(l, K) \in D_s$, and $t \in \mathbf{I}(K)$, the value $v_{s-1}(t, K \setminus \{t\}) \in [0, \infty[$ is defined. Moreover, by (4.8) and (5.8),

$$v_s(l, K) = \min_{t \in \mathbf{I}(K)} [c(l, t, K) + v_{s-1}(t, K \setminus \{t\})] \ \forall s \in \overline{1, N} \ \forall (l, K) \in D_s. \tag{5.9}$$

Thus, (5.9) defines the rule of transformation $v_{s-1} \to v_s$ for $s \in \overline{1, N}$. This results into the following recurrence procedure:

$$v_0 \to v_1 \to \ldots \to v_N, \tag{5.10}$$

where v_0 is defined by the rule $v_0(j, \emptyset) = f(j) \ \forall j \in \overline{1, N} \setminus \mathbf{K}_1$. For $s \in \overline{1, N}$, the transformation $v_{s-1} \to v_s$ is defined by (5.9). We note that it is very simple to obtain v_1, namely,

$$v_1(j, \{s\}) = c(j, s, \{s\}) + f(s) \ \forall s \in \overline{1, N} \setminus \mathbf{K}_1 \ \forall j \in J_1(\{s\}); \tag{5.11}$$

in (5.11), we use the equality $D_1 = \{(j, \{t\}) : \ t \in \overline{1, N} \setminus \mathbf{K}_1, \ j \in J_1(\{t\})\}$. From (4.7) and (5.8), we have the following equality:

$$V = v_N(0, \overline{1, N}) = \min_{t \in \mathbf{I}(\overline{1,N})} [c(0, t, \overline{1, N}) + v_{N-1}(t, \overline{1, N} \setminus \{t\})]. \tag{5.12}$$

We note that, to determine V, we only need the function v_{N-1}. A similar arrangement works for all the remaining functions $v_s, s \in \overline{1, N-1}$. Therefore, we can propose the following algorithm:

Algorithm of Value Construction. Actually, this algorithm is a modification of (5.10); see [11]. We know the function v_0. The values of v_0 are used in construction of v_1. Actually, we use (5.11) for construction of v_1. Let $s \in \overline{1, N}$ and let us have only the function v_s. If $s = N$, then, by (5.12), the desired value V is defined. If $s < N$, then we construct v_{s+1} by the rule

$$v_{s+1}(l, K) = \min_{t \in \mathbf{I}(K)} [c(l, t, K) + v_s(t, K \setminus \{t\})] \quad \forall (l, K) \in D_{s+1}. \qquad (5.13)$$

If $s + 1 = N$, then we obtain V; see (5.12). If $s + 1 < N$, then we replace the array of values of v_s with the array of values of v_{s+1} (the array of values of v_s is erased) and consider the construction of v_{s+2} by the rule similar to (5.9). In this scheme, we keep only one function of $\{v_1; \ldots v_{N-1}\}$ in the memory of the computer. A more detailed account is given in [11]. A similar idea was proposed in [14].

Construction of Optimal Routes. To construct an optimal route, all functions $\{v_1; \ldots; v_N\}$ are required. Therefore, we assume procedure (5.10) was carried out to its fullest extent: all functions $v_1, \ldots, v_N$ are known.

Using (5.12), we choose an index $\mathbf{j}_1 \in \mathbf{I}(\overline{1, N})$ with the property

$$V = \mathbf{c}(0, \mathbf{j}_1, \overline{1, N}) + v_{N-1}(\mathbf{j}_1, \overline{1, N} \setminus \{\mathbf{j}_1\}). \qquad (5.14)$$

We use the equality $V = v_N(0, \overline{1, N})$ (see (5.12)). In addition, by (5.5), $(\mathbf{j}_1, \overline{1, N} \setminus \{\mathbf{j}_1\}) \in D_{N-1}$. As a corollary, from (5.9), we obtain $v_{N-1}(\mathbf{j}_1, \overline{1, N} \setminus \{\mathbf{j}_1\}) = \min_{t \in \mathbf{I}(\overline{1, N} \setminus \{\mathbf{j}_1\})} [\mathbf{c}(\mathbf{j}_1, t, \overline{1, N} \setminus \{\mathbf{j}_1\}) + v_{N-2}(t, \overline{1, N} \setminus \{\mathbf{j}_1; t\})]$. Then, we choose $\mathbf{j}_2 \in \mathbf{I}(\overline{1, N} \setminus \{\mathbf{j}_1\})$ such that

$$v_{N-1}(\mathbf{j}_1, \overline{1, N} \setminus \{\mathbf{j}_1\}) = \mathbf{c}(\mathbf{j}_1, \mathbf{j}_2, \overline{1, N} \setminus \{\mathbf{j}_1\}) + v_{N-2}(\mathbf{j}_2, \overline{1, N} \setminus \{\mathbf{j}_1; \mathbf{j}_2\}). \qquad (5.15)$$

In addition, $(\mathbf{j}_2, \overline{1, N} \setminus \{\mathbf{j}_1; \mathbf{j}_2\}) \in D_{N-2}$ and the value $v_{N-2}(\mathbf{j}_2, \overline{1, N} \setminus \{\mathbf{j}_1; \mathbf{j}_2\}) \in [0, \infty[$ is defined. If $N = 2$, then, from (5.14), we obtain the optimal route (see the definition of v_0). If $N > 2$, we must continue the procedure of solution of local extremal problems similar to (5.14) and (5.15) until we exhaust the whole list $\overline{1, N}$. Thus we will come to a "full" route $J = (\mathbf{j}_t)_{t \in \overline{1, N}} \in \mathbf{A}$ such that $\mathfrak{C}_J = V$. Obviously, this route J is optimal in problem (3.5), $J \in \mathbf{A}_{\mathrm{opt}}$.

6   Scheme of Independent Computations: General Constructions

In this section, we consider a simplified variant of the corresponding procedure of [1–3] (in addition, see [13]). We recall that

$$\mathbf{G}_{N-1} = \{\overline{1, N} \setminus \{j\} : \ j \in \mathbf{I}(\overline{1, N})\} \neq \emptyset. \qquad (6.1)$$

Thus, task sets from $\mathbf{G}_{N-1}$ (6.1) are sets $\overline{1, N} \setminus \{j\}$, $j \in \mathbf{I}(\overline{1, N})$, and nothing else. Evidently, $|K| = N - 1 \ \forall K \in \mathbf{G}_{N-1}$. The values $v_{N-1}(t, \overline{1, N} \setminus \{t\})$, $t \in \mathbf{I}(\overline{1, N})$, are sufficient for determination of V; see (5.12). In the following, with every set $K \in \mathbf{G}_{N-1}$, we connect one hypothetical processor. This approach (see [1–3,13]) is the basis of the parallel structure. In addition, this structure must reproduce all functions $v_0, v_1, ..., v_{N-1}$. Therefore, for every processor, we construct its specific layer of position space and the corresponding layer of the Bellman function. To construct these layers, we use auxiliary discrete dynamic models.

Discrete Dynamic Systems. In the following, we assume $N \geq 3$. In this item, for simplicity, fix an arbitrary task set $K \in \mathbf{G}_{N-1}$. Denote by $\mathbb{T}[K]$ the set of all tuples

$$(K_t)_{t \in \overline{0,N-2}} : \overline{0, N-2} \to \mathbf{G} \tag{6.2}$$

such that

$$(K_0 = K) \& (\forall \tau \in \overline{1, N-2} \; \exists s \in \mathbf{I}(K_{\tau-1}) : K_\tau = K_{\tau-1} \setminus \{s\}). \tag{6.3}$$

Items (6.2) and (6.3) define the bundle of trajectories with the initial state K. It is possible to view trajectories (6.2) and (6.3) as motions of some dynamic system. This system is defined by the mapping $\mathbf{I}$ of Sect. 4. In addition, we consider motions in $\mathbf{G}$. Assume that, for every $P \in \mathbf{G}$,

$$\mathcal{I}[P] \stackrel{\triangle}{=} \{P \setminus \{s\} : \; s \in \mathbf{I}(P)\}; \tag{6.4}$$

then, $P \to \mathcal{I}[P]$ defines elementary permutations of our dynamic system connected with the mapping $\mathcal{I}[\cdot]$. Thus, (6.3) transforms into

$$(K_0 = K) \& (K_\tau \in \mathcal{I}[K_{\tau-1}] \; \forall \tau \in \overline{1, N-2}). \tag{6.5}$$

In addition, $\mathbb{T}[K]$ is the bundle of all tuples (6.2) with property (6.5); clearly, this bundle corresponds to our initial state K. Using (6.5), we can introduce attainability domains for our discrete system: for $t \in \overline{0, N-2}$, assume

$$\widetilde{\mathbb{T}}[K;t] \stackrel{\triangle}{=} \{K_t : (K_i)_{i \in \overline{0,N-2}} \in \mathbb{T}[K]\} \in \mathcal{P}'(\mathbf{G}_{N-(t+1)}); \tag{6.6}$$

see [1, Proposition 14]. In addition, we have the following statement:

Proposition 6.1. *For $t \in \overline{0, N-3}$, the following equality holds:* $\widetilde{\mathbb{T}}[K; t+1] = \bigcup_{P \in \widetilde{\mathbb{T}}[K;t]} \mathcal{I}[P]$.

The proof follows from (6.4) and [1, Proposition 16]. In view of Proposition 6.1, we obtain the following recurrence procedure: starting with $\widetilde{\mathbb{T}}[K; 0] = \{K\}$, we can transform $\widetilde{\mathbb{T}}[K; t]$ into $\widetilde{\mathbb{T}}[K; t+1]$, where $t \in \overline{0, N-3}$ is that of Proposition 6.1. Thus, $\widetilde{\mathbb{T}}[K; 1] = \mathcal{I}[K], \widetilde{\mathbb{T}}[K; 2] = \bigcup_{P \in \widetilde{\mathbb{T}}[K;t]} \mathcal{I}[P]$ (for $N > 3$) and so on. As a result, we obtain the procedure

$$(\widetilde{\mathbb{T}}[K; 0] = \{K\}) \to \widetilde{\mathbb{T}}[K; 1] \to ... \to \widetilde{\mathbb{T}}[K; N-2]. \tag{6.7}$$

Thus, we can construct all attainability domains without using bundles (6.6).

7   Representations of Feasible Task Sets and Layers of Positions

In the previous section, we considered the construction of attainability domains for a fixed initial state. Now, we consider the totality of procedures (6.7). We note that

$$\mathbf{G}_{N-(t+1)} = \bigcup_{K \in \mathbf{G}_{N-1}} \widetilde{\mathbb{T}}[K; t] = \bigcup_{j \in \mathbf{I}(\overline{1,N})} \widetilde{\mathbb{T}}[\overline{1,N} \setminus \{j\}; t] \quad \forall t \in \overline{0, N-2}. \quad (7.1)$$

This property was established in [1, Proposition 12]. From (6.7) and (7.1), we see that (in particular)

$$\mathbf{G}_1 = \bigcup_{K \in \mathbf{G}_{N-1}} \widetilde{\mathbb{T}}[K; N-2].$$

Moreover, from (7.1), we obtain

$$\mathbf{G}_s = \bigcup_{K \in \mathbf{G}_{N-1}} \widetilde{\mathbb{T}}[K; N-(s+1)] \quad \forall s \in \overline{1, N-1}. \quad (7.2)$$

Clearly, (7.2) is a case of "distributed" representation of families used in (5.3). Now, we introduce certain new layers in the position space. In this construction, we use the cells defined in (5.6). Set

$$\mathfrak{D}_s[K] \overset{\triangle}{=} \bigcup_{H \in \widetilde{\mathbb{T}}[K; N-(s+1)]} \mathbb{D}_s[H] \in \mathcal{P}'(D_s) \quad \forall s \in \overline{1, N-1} \quad \forall K \in \mathbf{G}_{N-1}. \quad (7.3)$$

From [1, Proposition 17], the next representation of layers (5.7) follows:

$$D_s = \bigcup_{K \in \mathbf{G}_{N-1}} \mathfrak{D}_s[K] = \bigcup_{j \in \mathbf{I}(\overline{1,N})} \mathfrak{D}_s[\overline{1,N} \setminus \{j\}] \quad \forall s \in \overline{1, N-1}. \quad (7.4)$$

To consider the simplest case of $s = 1$, we introduce the sets

$$\mathbf{M}_0[K] \overset{\triangle}{=} \{h \in \overline{1,N} \setminus \mathbf{K}_1 | \{h\} \in \widetilde{\mathbb{T}}[K; N-2]\} \quad \forall K \in \mathbf{G}_{N-1}.$$

Then, in view of (7.2), for $K \in \mathbf{G}_{N-1}$, we obtain

$$\mathfrak{D}_1[K] = \bigcup_{h \in \mathbf{M}_0[K]} \mathbb{D}_1[\{h\}] \quad (7.5)$$

since $\widetilde{\mathbb{T}}[K; N-2] = \{\{h\} : h \in \mathbf{M}_0[K]\}$. Using (5.5) and (7.5), we have that

$$\mathfrak{D}_1[K] = \{(j, \{h\}) : h \in \mathbf{M}_0[K], j \in J_1(\{h\})\} \quad \forall K \in \mathbf{G}_{N-1}. \quad (7.6)$$

From (7.4) and (7.6), we obtain the following representation for D_1:

$$D_1 = \bigcup_{K \in \mathbf{G}_{N-1}} \mathfrak{D}_1[K] = \bigcup_{j \in \mathbf{I}(\overline{1,N})} \mathfrak{D}_1[\overline{1,N} \setminus \{j\}]$$

$$= \{(l, \{h\}) : h \in \bigcup_{j \in \mathbf{I}(\overline{1,N})} \mathbf{M}_0[\overline{1,N} \setminus \{j\}], \ l \in J_1(\{h\})\}.$$

For our goals, relation (7.6) is of greater importance. Clearly, in view of (6.1) and (7.6),

$$\mathfrak{D}_1[\overline{1,N} \setminus \{j\}] = \{(l, \{h\}) : h \in \mathbf{M}_0[\overline{1,N} \setminus \{j\}], \ l \in J_1(\{h\})\} \ \ \forall j \in \mathbf{I}(\overline{1,N}).$$

Returning to (7.2), we note the following statement:

Proposition 7.1. *If $K \in \mathbf{G}_{N-1}, t \in \overline{1,N-2}, \ (x,Q) \in \mathcal{D}_{t+1}[K]$, and $s \in \mathbf{I}(Q)$, then $(s, Q \setminus \{s\}) \in \mathfrak{D}_t[K]$.*

This proposition is a particular case of [3, Proposition 9.1]. Moreover, see [13, p. 236]. The property noted in Proposition 7.1 is similar to the analogous property of layers $D_0, D_1, ..., D_N$ in Sect. 5.

8 The Restriction of the Bellman Function Layers and Their Construction by a Recurrence Procedure

Now, we consider a decomposition procedure for layers of the Bellman function. In this decomposition, we follow [1–3]. We recall that, by (7.2), $\mathfrak{D}_s[K]$ is a nonempty subset of D_s for $s \in \overline{1, N-1}$ and $K \in \mathbf{G}_{N-1}$. We obtain

$$\mathcal{W}_s[K] \triangleq (v_s(j,P))_{(j,P) \in \mathfrak{D}_s[K]} \in \mathcal{R}_+[\mathfrak{D}_s[K]] \ \ \forall s \in \overline{1, N-1} \ \ \forall K \in \mathbf{G}_{N-1}. \quad (8.1)$$

Thus, we introduce the restrictions of the functions used in (5.10). By (5.8) and (7.2),

$$\mathcal{W}_s[K](j,P) = v(j,P) \ \forall s \in \overline{1, N-1} \ \ \forall K \in \mathbf{G}_{N-1} \ \ \forall (j,P) \in \mathfrak{D}_s[K]. \quad (8.2)$$

It (8.1) and (8.2), we have a new arrangement of the initial Bellman function. It is easily proved that

$$\mathcal{W}_1[K](j, \{h\}) = \mathbf{c}(j, h, \{h\}) + f(h) \ \ \forall K \in \mathbf{G}_{N-1} \ \ \forall h \in \mathbf{M}_0[K] \ \ \forall j \in J_1(\{h\}) \quad (8.3)$$

(we use (7.6)). By (7.6) and (8.3), all the functions $\mathcal{W}_1[K], K \in \mathbf{G}_{N-1}$, are defined. Consider a specific transformation $\mathcal{W}_t[K] \rightarrow \mathcal{W}_{t+1}[K]$ for $K \in \mathbf{G}_{N-1}$ and $t \in \overline{1, N-2}$. Namely, by Proposition 7.1, the value $\mathcal{W}_t[K](s, Q \setminus \{s\}) \in [0, \infty[$ is defined for $K \in \mathbf{G}_{N-1}, t \in \overline{1, N-2}, (\nu, Q) \in \mathfrak{D}_{t+1}[K]]$, and $s \in \mathbf{I}(Q)$.

Proposition 8.1. *If $K \in \mathbf{G}_{N-1}, \ t \in \overline{1, N-2}$, and $(\nu, Q) \in \mathfrak{D}_{t+1}[K]$, then*

$$\mathcal{W}_{t+1}[K](\nu, Q) = \min_{s \in \mathbf{I}(Q)} [\mathbf{c}(\nu, s, Q) + \mathcal{W}_t[K](s, Q \setminus \{s\})] \ \ \forall (\nu, Q) \in \mathfrak{D}_{t+1}[K].$$

This proposition is a variant of [3, (10.17)]; see also [13, Proposition 6.3]. Thus, with the aid of (8.3) and Proposition 8.1, we obtain the recurrence procedure

$$\mathcal{W}_1[K] \to \mathcal{W}_2[K] \to ... \to \mathcal{W}_{N-1}[K] \qquad (8.4)$$

for $K \in \mathbf{G}_{N-1}$; procedure (8.4) is carried out by a single processor. Let us now return to the construction of functions used in (5.9). In this connection, we use (7.3). Of course, we assume that all functions

$$\mathcal{W}_s[K], \ K \in \mathbf{G}_{N-1}, s \in \overline{1, N-1}, \qquad (8.5)$$

were constructed (all functions (8.4) are known). Consider the construction of v_s, where $s \in \overline{1, N-1}$.

Let $(j, Q) \in D_s$. Then, with use of (7.4), we choose an arbitrary $\mathbb{K} \in \mathbf{G}_{N-1}$ such that

$$(j, Q) \in \mathfrak{D}_s[\mathbb{K}]. \qquad (8.6)$$

Such a set $\mathbb{K}$ exists by (7.4). Then, by (8.1), we obtain

$$v_s(j, Q) = \mathcal{W}_s[\mathbb{K}](j, Q). \qquad (8.7)$$

Thus, using the functions $\mathcal{W}_s[K], \ K \in \mathbf{G}_{N-1}$, we construct v_s, where $s \in \overline{1, N-1}$; see (8.6). As a result, using relations similar to (8.6), we determine $v_1, ..., v_{N-1}$ (recall that v_0 is known; see Sect. 5).

Parallel Algorithm of Value Construction. We recall the procedure of Sect. 5, which is a particular case of the analogous procedure of [11]. For this procedure of Sect. 5, we consider a variant of parallel realization. To this end, let us recall (along with [11]) relations (5.9) and (5.12).

(1) Fix $\mathbb{K} \in \mathbf{G}_{N-1}$ and consider a sequential computation of the functions

$$\mathcal{W}_1[\mathbb{K}], \mathcal{W}_2[\mathbb{K}], \ldots, \mathcal{W}_{N-1}[\mathbb{K}].$$

This procedure is intended to be carried out by a processer dedicated to $\mathbb{K}$.

For representation of $\mathcal{W}_1[\mathbb{K}]$, we use (8.3). Namely, this function is completely defined by the following rule:

$$\mathcal{W}_1[\mathbb{K}](j, \{h\}) = \mathbf{c}(j, h, \{h\}) + f(h) \ \ \forall h \in \mathbf{M}_0[\mathbb{K}] \ \ \forall j \in J_1(\{h\}) \qquad (8.8)$$

(see (7.5)). Thus, we know all the values $\mathcal{W}_1[\mathbb{K}](j, \{h\})$, where $h \in \mathbf{M}_0[K]$ and $j \in J_1(\{h\})$. In other words, we know all the values $\mathcal{W}_1[\mathbb{K}], (j, K) \in \mathfrak{D}_1[\mathbb{K}]$.

For $\Theta \in \overline{1, N-2}$, assume we know the function $\mathcal{W}_\Theta[\mathbb{K}]$. That is, we know all the values $\mathcal{W}_\Theta[\mathbb{K}](j, K), (j, K) \in \mathfrak{D}_\Theta[\mathbb{K}]$. Moreover, by (8.3) and Proposition 7.1, we obtain $(s, Q \setminus \{s\}) \in \mathfrak{D}_\Theta[\mathbb{K}] \ \forall (j, Q) \in \mathfrak{D}_{\Theta+1}[\mathbb{K}] \ \ \forall s \in \mathbf{I}(Q)$. Now, we use Proposition 8.1: we compute the function $\mathcal{W}_{\Theta+1}[\mathbb{K}] \in \mathcal{R}_+[\mathfrak{D}_{\Theta+1}[\mathbb{K}]]$ by the rule

$$\mathcal{W}_{\Theta+1}[\mathbb{K}](\nu, Q) = \min_{s \in \mathbf{I}(Q)}[\mathbf{c}(\nu, s, Q) + \mathcal{W}_\Theta[\mathbb{K}](s, Q \setminus \{s\})] \ \ \forall (\nu, Q) \in \mathfrak{D}_{\Theta+1}[\mathbb{K}]. \qquad (8.9)$$

Of course, values (8.9) completely determine the function $\mathcal{W}_{\Theta+1}$.

If $\Theta = N-2$, then $\mathcal{W}_{\Theta+1}[\mathbb{K}] = \mathcal{W}_{N-1}[\mathbb{K}]$; in this case, we stop our procedure. If $\Theta < N - 2$, we replace the array of values of $\mathcal{W}_\Theta[\mathbb{K}]$ with the new array of values of the function $\mathcal{W}_{\Theta+1}$. In addition, the array of $\mathcal{W}_\Theta[\mathbb{K}]$ is destroyed. The new array is used (for $\Theta < N-2$) for construction of $\mathcal{W}_{\Theta+2}[\mathbb{K}]$ instead of $\mathcal{W}_\Theta[\mathbb{K}]$. Thus, we obtain the sequential procedure

$$\mathcal{W}_1[\mathbb{K}] \to ... \to \mathcal{W}_{N-1}[\mathbb{K}] \tag{8.10}$$

for which, in the memory of the processor, only one function from $\{\mathcal{W}_1[\mathbb{K}]; ...; \mathcal{W}_{N-1}[\mathbb{K}]\}$ is stored each time.

(2) For every processor, assume a procedure similar to (8.10) to be carried out. As a result, we have the functions $\mathcal{W}_{N-1}[K], K \in \mathbf{G}_{N-1}$. Now, we construct v_{N-1} by sewing them together on the basis of (7.4):

$$D_{N-1} = \bigcup_{K \in \mathbf{G}_{N-1}} \mathfrak{D}_{N-1}[K]. \tag{8.11}$$

From (8.2) and (8.11), for $(j, Q) \in D_{N-1}$, we choose an arbitrary $P \in \mathbf{G}_{N-1}$ such that $(j, Q) \in \mathfrak{D}_{N-1}[P]$. Then, in view of (8.1) we assume that

$$v_{N-1}(j, Q) = \mathcal{W}_{N-1}[P](j, Q). \tag{8.12}$$

Using (8.10) and (8.11), we completely construct the function v_{N-1}: the array of the values of v_{N-1} is defined by (8.12) under the corresponding choice of the set P. For final computation of V, we use relation (5.12).

9   Implementation of the Construction Procedure for the Bellman Function with Independent Computations

In this section, we consider a natural scheme of realization of the procedure with use of independent computations. In addition, we modify certain aspects of our theoretical construction. Namely, in the following, we assume that each layer connected with $K \in \mathbf{G}_{N-1}$ is examined by several processors with shared memory (RAM).

We call the arising totality of processors a node. A system of nodes is a computational cluster. Such system (that is, a node) is used in lieu of each single processor of the previous section.

The Data Storage. Let us consider a variant of the data storage for a single cluster node. In addition, we fix $K \in \mathbf{G}_{N-1}$ (of course, we could as well say we fix $j \in \mathbf{I}(\overline{1,N})$ for which $K = \overline{1,N} \setminus \{j\}$). In our data storage, the system $(\mathcal{W}_1[K], ..., \mathcal{W}_{N-1}[K])$ is to be stored.

We use a hash table as a container for the above-mentioned system (compared with array, this container uses less memory since there is no need to allocate memory for the storage of infeasible task sets). In our hash table, the keys are bit masks of subsets; essentially, a 1 bit at the position **i** means that the point numbered **i** is present in this subset.

The Algorithm

1. The main processor forms the family $\mathbf{G}_{N-1}$ (6.1), every element of which is a set of the cardinality $N-1$. The sets of $\mathbf{G}_{N-1}$ are distributed among the nodes by the MPI protocol. Every node contains **k** processors with shared memory. In the node connected with the set $K \in \mathbf{G}_{N-1}$, the layers $\mathcal{W}_s[K]$, $s \in \overline{1, N-1}$, are equally distributed between the processors. For resulting fragments of the above-mentioned layers, the values of the Bellman function are calculated. Since each node's memory is shared among its processors, no data exchange is necessary. In addition, operations with fragments of the positions' layers and Bellman function's layers are conducted with the aid of the OpenMP API (see also [21]).
2. The layers of the Bellman function values obtained on the first stage 1 and connected with the corresponding set $K \in \mathbb{G}_{N-1}$ are collected by the main node. Later on, the layer v_{N-1} is computed by means of (8.11) and (8.12). Finally, the value V is determined from (5.12).
3. The main processor constructs the optimal route by solving the local extremal problems similar to (5.14) and (5.15).

Computational Experiment. In the experiment, we considered a problem with $N = 46$ cities and 35 address pairs **K** that define precedence constraints ($|\mathbf{K}| = 35$). The experiment was conducted on the Uran supercomputer; 22 cluster nodes were used, each of which had 8 computational cores, thus, our implementation used 176 cores. The cost function was defined as Euclidean distance (for simplicity). The following results were obtained: $V = 773.091003$ (the extremum of the problem), the computation time was 6312 s, and the maximum memory usage per node was 142.17 GB. The sequential procedure was 22 times slower than the parallel (exactly the number of nodes in the cluster).

10   Conclusion

In this paper, we explore a rather general parallel solution procedure [1,2] for a constrained routing problem with complicated cost functions. This procedure was implemented in the form of a parallel algorithm on the Uran supercomputer. Its possible applications are connected with the engineering problems of fission power generation and (in the future) the problem of routing the cutting tools in CNC sheet cutting machines (in this connection, see [22–26]).

References

1. Chentsov, A.G.: One parallel procedure for the construction of the Bellman function in the generalized problem of the courier with the inner workings. Vestnik YuUrGU. Gos. Univ. Ser. Mat. Model Program. **12**, 53–76 (2012)
2. Chentsov, A.G.: One parallel procedure for the construction of the Bellman function in the generalized problem of the courier with the inner workings. Autom. Remote Control **3**, 34–149 (2012)

3. Chentsov, A.G.: To question of routing of works complexes: Vestnik Udmurtskogo Universiteta. Matematika. Mekhanika. Komp'yuternye Nauki, (1), 59–82 (2013)
4. Gutin, G., Punnen, A.P.: The Traveling Salesman Problem and Its Variations. Springer, New York (2002)
5. Cook, W.J.: In Pursuit of the Traveling Salesman. Mathematics at the limits of computation, p. 248. Princeton University Press, New Jersey (2012)
6. Melamed, I.I., Sergeev, S.I., Sigal, I.: The traveling salesman problem. I. Issues in Theory; II Exact Methods; III Approximate Algorithms: Automation and Remote Control 50(9), 1147–1173; 50(10), 1303–1324; 50(11), 1459–1479 (1989)
7. Held, M., Karp, R.M.: A dynamic programming approach to sequencing problems. J. Soc. Ind. Appl. Math. 10(1), 196–210 (1962)
8. Bellman, R.: Dynamic programming treatment of the travelling salesman problem. J. Assoc. Comput. Mach. 9, 61–63 (1962)
9. Chentsov, A.G., Chentsov, A.A.: Route problem with constraints depending on a list of tasks. Doklady Math. 92(3), 685–688 (2015)
10. Chentsov, A.A., Chentsov, A.G., Chentsov, P.A.: Elements of dynamic programming in extremal route problems. Autom. Remote Control 75(3), 537–550 (2014)
11. Chentsov, A.G., Chentsov, A.A.: On the problem of obtaining the value of routing problem with constraints. J. Autom. Inf. Sci. 6, 41–54 (2016)
12. Chentsov, A.G.: Extreme Problems of Routing and Tasks Distribution: Regular and Chaotic Dynamics. Izhevsk Institute of Computer Research, 240 p. (2008). (in Russian)
13. Chentsov, A.G., Kosheleva, M.S.: Dynamic programming in the precedence constrained TSP with the costs depending on a list of tasks. Mechatron. Autom. Control 16(4), 232–244 (2015). (in Russian)
14. Lawler, E.L.: Efficient implementation of dynamic programming algorithms for sequencing problems. CWI Technical report. Stichting Mathematisch Centrum. Mathematische Besliskunde-BW 106/79, pp. 1–16 (1979). http://oai.cwi.nl/oai/asset/9663/9663A.pdf
15. Bianco, L., Mingozzi, A., Ricciardelli, S., Spadoni, M.: The traveling salesman problem with precedence constraints. In: Bühler, W., Feichtinger, G., Hartl, R.F., Radermacher, F.J., Stähly, P. (eds.) Papers of the 19th Annual Meeting / Vorträge der 19. Jahrestagung. Operations Research Proceedings, vol. 1990, pp. 299–306. Springer, Heidelberg (1992)
16. Steiner, G.: On the complexity of dynamic programming for sequencing problems with precedence constraints. Ann. Oper. Res. 26(1), 103–123 (1990)
17. Salii, Y.V.: On the effect of precedence constraints on computational complexity of dynamic programming method for routing problems. Vestnik Udmurtskogo Universiteta. Matematika. Mekhanika. Komp'yuternye Nauki, (1), 76–86 (2014). (in Russian)
18. Korobkin, V.V., Sesekin, A.N., Tashlykov, O.L., Chentsov, A.G.: Routing Methods and Their Applications to the Enhancement of Safety and Efficiency of Nuclear Plant Operation. Novye Tekhnologii, Moscow (2012)
19. Tashlykov, O.L.: Personnel Dose Costs in the Nuclear Industry. Analysis. Ways to Decrease. Optimization. LAP LAMBERT Academic Publishing GmbH & Co. RG., Saarbruke (2011)
20. Alkaya, A.F., Duman, E.: A new generalization of the traveling salesman problem. Appl. Comput. Math. 9(2), 162–175 (2010)

21. Salii, Y.V.: Restricted dynamic programming heuristic for precedence constrained bottleneck generalized TSP. In: Proceedings of the 1st Ural Workshop on Parallel, Distributed, and Cloud Computing for Young Scientists, pp. 85–108 (2015). http://ceur-ws.org/Vol-1513/#paper-10

22. Wang, G.G., Xie, S.Q.: Optimal process planning for a combined punch-and-laser cutting machine using ant colony optimization. Int. J. Prod. Res. **43**(11), 2195–2216 (2005)

23. Lee, M.-K., Kwon, K.-B.: Cutting path optimization in CNC cutting processes using a two-step genetic algorithm. Int. J. Prod. Res. **44**(24), 5307–5326 (2006)

24. Jing, Y., Zhige, C.: An optimized algorithm of numerical cutting-path control in garment manufacturing. Adv. Mater. Res. **796**, 454–457 (2013)

25. Dewil, R., Vansteenwegen, P., Cattrysse, D: Construction heuristics for generating tool paths for laser cutters. Int. J. Prod. Res. pp. 1–20 (2014)

26. Petunin, A.A.: Modelling of tool path for the CNC sheet cutting machines. In: AIP Conference Proceedings, 41st International Conference Applications of Mathematics in Engineering and Economics (AMEE 2015), vol. 1690, pp. 060002(1)–060002(7) (2015)

On Asymptotically Optimal Approach to the m-Peripatetic Salesman Problem on Random Inputs

Edward Kh. Gimadi[1,2], Alexey M. Istomin[1], and Oxana Yu. Tsidulko[1,2(✉)]

[1] Sobolev Institute of Mathematics, 4 Acad. Koptyug avenue,
630090 Novosibirsk, Russia
gimadi@math.nsc.ru, alexeyistomin@gmail.com, tsidulko.ox@gmail.com
[2] Novosibirsk State University, 2 Pirogova Str., 630090 Novosibirsk, Russia

Abstract. We study the m-Peripatetic Salesman Problem on random inputs. In earlier papers we proposed a polynomial asymptotically optimal algorithm for the m-PSP with different weight functions on random inputs. The probabilistic analysis carried out for that algorithm is not suitable in the case of the m-PSP with identical weight functions.

In this paper we present an approach which under certain conditions gives polynomial asymptotically optimal algorithms for the m-PSP on random inputs with identical weight functions and for the m-PSP with different weight functions, as well. We describe in detail the cases of uniform and shifted exponential distributions of random inputs.

Keywords: m-PSP · Asymptotically optimal algorithm · Performance guarantees · Random inputs · Uniform distribution · Shifted exponential distribution

1 Introduction

The m-Peripatetic Salesman Problem (m-PSP) is a natural generalization of the classical Traveling Salesman Problem (TSP). It is formulated as follows: given a complete undirected n-vertex graph $G = (V, E)$ and weight functions $w_i : E \to \mathbb{R}_+$, $i = 1, \ldots, m$, the problem is to find m edge-disjoint Hamiltonian cycles $H_1, \ldots, H_m \subset E$ such that their total weight is minimum or maximum:

$$\sum_{i=1}^{m} w_i(H_i) = \sum_{i=1}^{m} \sum_{e \in H_i} w_i(e).$$

In the literature, the m-PSP is usually referred to as the case of the problem, where all the weight functions w_i are identical: $w_1 = \ldots = w_m = w$. In this paper we will sometimes specify this case as the m-PSP *with identical weight functions*, while the general case of the problem where $w_i \neq w_j$ for $1 \leq i \neq j \leq m$ will be referred to as the m-PSP *with different weight functions*.

© Springer International Publishing Switzerland 2016
Y. Kochetov et al. (Eds.): DOOR 2016, LNCS 9869, pp. 136–147, 2016.
DOI: 10.1007/978-3-319-44914-2_11

The m-PSP was introduced in [24] by Krarup. De Kort [9] showed that the problem of finding two edge-disjoint Hamiltonian cycles is NP-complete. This implies that the 2-PSP with identical weight functions is NP-hard both in the maximization and minimization variants. These results can be extended to the general m-PSP. The problem is also NP-hard, in the case of different weight functions [5].

The most studied variant of the problem is the 2-PSP. De Brey [8] presented several polynomially solvable cases of the 2-PSP. Papers [9–11] provide upper and lower bounds for the minimum 2-PSP and use them in the branch-and-bound algorithm. For the metric minimum 2-PSP algorithms with approximation ratios 9/4 [5] and 2 [2] are known. For the symmetric maximum 2-PSP algorithms with approximation ratios 3/4 and 7/9 were designed in [1,20]. For the asymmetric maximum 2-PSP an algorithm with approximation ratio 2/3 is given in [16]. The results for the 2-PSP, where edge weights belong to a given interval or a finite set of numbers, can be found in [14,18,21].

Baburin and Gimadi in [4] developed an asymptotically optimal algorithm with running-time $O(n^3)$ for the maximum Euclidean m-PSP.

Let's now introduce the main definitions we need to explore a problem on random inputs. By $F_A(I)$ and $OPT(I)$ we denote the approximate (obtained by some approximation algorithm A) and the optimum value of the objective function of the problem on the input I, respectively. An algorithm A is said to have *performance guarantees* $\left(\varepsilon_A(n), \delta_A(n)\right)$ on the set of random inputs of the problem of size n, if

$$\mathbf{Pr}\Big\{F_A(I) > \big(1 + \varepsilon_A(n)\big)OPT(I)\Big\} \leq \delta_A(n), \tag{1}$$

where $\varepsilon_A(n)$ is an assessment of *the relative error* of the solution obtained by algorithm A, $\delta_A(n)$ is an estimation of *the failure probability* of the algorithm, which is equal to the proportion of cases when the algorithm fails, i.e. it does not hold the relative error $\varepsilon_A(n)$ or does not produce any answer at all.

It is often important to understand the behavior of $\delta_A(n)$ and $\varepsilon_A(n)$ as the size $n = 2, 3, \ldots$ of the problem increases. An algorithm A is called *asymptotically optimal* on the class of instances of the problem, if there exist performance guarantees such that $\varepsilon_A(n) \to 0$ and $\delta_A(n) \to 0$ as $n \to \infty$.

In papers [15,17] we studied the m-PSP with *different weight functions* on random inputs. We assumed the weights of the edges to be independent and identically distributed random reals with uniform distribution on $[a_n, b_n], 0 < a_n \leq b_n$, or shifted exponential distribution on $[a_n, \infty), 0 < a_n$. We presented an $O(mn^2)$ running-time algorithm solving the problem, and showed that it is asymptotically optimal, if $m = o(n)$ and a few other conditions on the parameters of the distribution function are satisfied.

The algorithm from [15,17] constructs m Hamiltonian cycles in series. Building the i-th Hamiltonian cycle it applies the principle "go to the nearest not visited vertex" $n - 4i$ times, then converts the resulting path into the Hamiltonian cycle via the extension-rotation procedure.

In solving the m-PSP with different weight functions, the weights of the edges chosen by the algorithm are independent random variables $\xi_{is}, 1 \le i \le m, 1 \le s \le n$. To prove that the algorithm is asymptotically optimal we estimated the probability

$$\mathbf{Pr}\left\{\sum_{i=1}^{m}\sum_{s=1}^{n}\xi_{is} > (1+\varepsilon_n)OPT\right\} \le \delta_n,$$

using the theorem from [25, Chap. 2.2], which holds for the sum of independent random variables. If we apply this algorithm to the m-PSP with identical weight functions, the weights ξ_{is} of the chosen edges will be dependent. The dependent random variables are less studied in general, and we could not find any results similar to those from [25] for such cases. Thus we could not carry out the analogous analysis.

In this paper we present an approach which under certain conditions gives asymptotically optimal algorithms for the m-PSP with identical weight functions on random inputs. This approach will be also correct for the m-PSP with different weight functions.

The paper is organized as follows. In Sect. 2 we describe our approach, which involves an algorithm for finding a Hamiltonian cycle in a random graph. A short review on the algorithms that find a Hamiltonian cycle in a random sparse graph can be found in Sect. 2.1. Sections 2.2 and 2.3 briefly describe algorithms from [19] and [3]. In Sect. 3 we give the probabilistic analysis of the approach assuming that the weights of edges in the input graph are i.i.d. real numbers with distribution function defined on $[a_n, b_n]$ or $[a_n, \infty)$, $0 < a_n$. Section 3.1 provides the preliminary analysis of the approach. Finally, in Sect. 3.2 we present the performance guarantees of our approach for the uniform and shifted exponential distributions of inputs.

2   The Asymptotically Optimal Approach

Given a complete undirected weighted n-vertex graph $G = (V, E)$, the problem is to find $m \le n/2$ edge-disjoint Hamiltonian cycles $H_1, \ldots, H_m \subset E$ such that their total weight is minimum. We assume that the weights of the edges are independent and identically distributed random reals, with distribution function $f(x)$ defined on $[a_n, b_n]$ or $[a_n, \infty)$, $0 < a_n \le b_n$.

The approach consists of the following three steps.

Step 1. We uniformly split the initial complete n-vertex graph G into subgraphs $G_1, \ldots G_m$, so that each G_i has n vertices and about $\frac{n(n-1)}{2m}$ edges.

procedure SPLIT(G):
 begin
  for $1 \le i \le m$ set $V(G_i) = V(G)$, $E(G_i) = \emptyset$;
  for each $e \in E(G)$:
   select at random with equal probabilities one of the sets
   $E(G_1), \ldots, E(G_m)$, let it be $E(G_i)$, add the edge e to $E(G_i)$.
 end.

Step 2. Construct subgraphs $\widetilde{G}_1, \ldots, \widetilde{G}_m$ deleting all edges in G_i, $1 \le i \le m$, which are heavier than w^*. Later we will select w^* so as to retain only light edges in subgraphs, though still providing enough edges in each $\widetilde{G}_i$ for Step 3.

Step 3. In each subgraph $\widetilde{G}_i$ build a Hamiltonian cycle, using polynomial randomized algorithms, that *with high probability* or *w.h.p.* (with probability $\to 1$ as $n \to \infty$) find a Hamiltonian cycle in a sparse random graph. In this paper we will try algorithms by Gimadi and Perepelitsa (1973) [19] and Angluin and Valiant (1979) [3].

Steps 1 and 2 take $O(n^2)$ time, at Step 3 the chosen algorithm with time complexity $T(n)$ runs m times. So the total time complexity of the approach is $O(n^2 + mT(n))$.

2.1 Algorithms Finding a Hamiltonian Cycle in a Random Graph

The Hamiltonian cycle problem is a well-known NP-complete problem. A series of papers study the problem of finding a Hamiltonian cycle in random graphs.

Researchers in this field use two concepts of a random graph. The first is an n-vertex graph G_p, where each edge exists with probability p, independently of other edges. This concept is convenient for proving statements. The second concept is a graph G_N with n vertices and exactly N edges, chosen uniformly from the set of all such graphs. This concept is commonly used for making statements. In [3] it was shown that these two concepts are interchangeable for appropriate values N and p.

Erdos and Renyi [12] in 1959 obtained the threshold condition of existence of a Hamiltonian cycle: for any $\varepsilon > 0$ if the number of edges $N < (1/2 - \varepsilon)n \log n$, then w.h.p. the graph contains isolated vertices. Posa in 1976 gave a non-algorithmic proof of the fact that almost all undirected graphs with $cn \log n$ edges contain a Hamiltonian cycle. Komlos and Szemeredi [22] and independently Korshunov [23] stated that the required density may be reduced to $1/2(n \log n + n \log \log n + Q(n))$ edges, where $Q(n)$ is any function such that $Q(n) \to \infty$ as $n \to \infty$ (i.e. $Q(n)$ grows slowly as the $\log \log n$).

In 1973 Gimadi and Perepelitsa [19] presented an algorithm with a running-time of $O(n^2 / \ln n)$ which w.h.p. finds a Hamiltonian cycle in directed and undirected graphs with $N \ge n\sqrt{n \ln n}$ edges. Angluin and Valiant in 1979 proposed a randomized algorithm, which in $O(n \ln^2 n)$ time finds with probability $1 - O(n^{-\alpha})$ a Hamiltonian cycle in a directed or undirected random graph with number of edges $N \ge c(\alpha)n \ln n$, where $c(\alpha)$ is sufficiently large constant.

Bollobas, Fenner and Frieze in 1987 [6] gave a deterministic polynomial time algorithm that works w.h.p. at the exact threshold for Hamiltonicity, and runs in $O(n^{3+o(1)})$ time.

In 2015 Frieze and Haber [13] presented an algorithm, which in almost linear time $O(n^{1+o(n)})$ w.h.p finds a Hamiltonian cycle in random graph with minimum degree ≥ 3 and number of edges $N = cn$, where c is sufficiently large. Unfortunately, we cannot apply this algorithm to our approach, as long as we can not guarantee (even w.h.p.) minimum degree ≥ 3 in subgraphs $\widetilde{G}_i$, if each $\widetilde{G}_i$ has exactly cn edges.

Now we are going to briefly describe algorithm A_{GP} by Gimadi and Pere-pelitsa [19] and algorithm A_{AV} by Angluin and Valiant [3].

2.2   Gimadi-Perepelitsa Algorithm 1973

Paper [19] presents one of the first algorithms, that solves the problem of finding a Hamiltonian cycle in undirected sparse random graphs. The algorithm suc-ceeds w.h.p. for random n-vertex graphs with at least $N = n\sqrt{n}\ln n$ edges.

Algorithm A_{GP} with parameters k, τ, ρ. Algorithm attempts to find a Hamil-tonian cycle in a graph $G = (V, E)$. It consists of 5 Stages (Fig. 1). If it cannot execute a step, it stops and returns "failure".

Let $v = \lfloor \frac{n-\rho}{k} \rfloor$, $v' = \lfloor \frac{n-\rho}{k}\tau \rfloor$. The values of the parameters k, τ, ρ will be defined later.

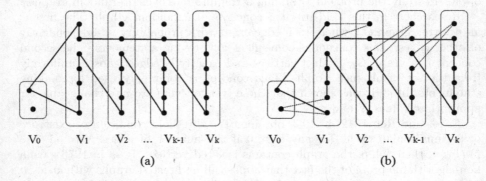

$$V_0 \quad V_1 \quad V_2 \quad ... \quad V_{k\text{-}1} \quad V_k \qquad V_0 \quad V_1 \quad V_2 \quad ... \quad V_{k\text{-}1} \quad V_k$$

$$(a) \hspace{6cm} (b)$$

Fig. 1. Algorithm A_{GP}: (a) Stages 0–3, (b) Stage 4.

Stage 0. Fix an arbitrary vertex i_1. Create subset $V_0 \subset V$ of $(n - kv)$ vertices such that $(u, i_1) \in E$ for all $u \in V_0$. Split the remaining kv vertices of the graph into k disjoint sets V_λ, $|V_\lambda| = v, 1 \le \lambda \le k$, where $i_1 \in V_1$.

Stage 1. Start building partial path $P = \{i_1\}$. Set $\lambda = 1$. While $\lambda \le k$ do:

Repeat $(v - v')$ times: Take the last vertex i_s of P and find an edge $(i_s, i_{s+1}) \in E \setminus P$ such that $i_{s+1} \in V_\lambda$. Add the edge (i_s, i_{s+1}) to P.

If $\lambda < k$, find an edge from the last vertex of P to some vertex in $V_{\lambda+1}$, add it to the path and set $\lambda = \lambda + 1$.

At this stage we have created a partial path P with $k(v - v')$ vertices.

Stage 2. Repeat $(k - 1)v'$ times:

Take the last vertex i_s of P, find an edge $(i_s, i_{s+1}) \in E$ such that $i_{s+1} \in V \setminus (P \cup V_0)$. Add this edge to P.

Now we have a partial path P with $kv - v'$ vertices.

Stage 3. Convert the path $P = \{i_1, \ldots, i_s\}$ into a cycle: find a pair of vertices $i_\alpha \in V \setminus (P \cup V_0)$ and $i_{\alpha+1} \in V_0$, such that there exist edges (i_s, i_α) and $(i_\alpha, i_{\alpha+1})$. Add these edges and the edge $(i_{\alpha+1}, i_1)$ to P.

Now we have a cycle P with $kv - v' + 2$ vertices.

Stage 4. For each $\lambda = 0, 1, \ldots, k$ we will try to add the remaining vertices of V_λ to P. First we cover the set $V_\lambda \setminus P$ with paths. For each path $\{u_0, \ldots, u_s\}, 0 \leq s$, find an edge (u_1, u_2) in P, such that $u_1, u_2 \in P \setminus V_\lambda$ and there exist edges (u_0, u_1) and (u_s, u_2). Delete the edge (u_1, u_2) from P and add the edges (u_0, u_1) and (u_s, u_2).

Theorem 1. *Let* $p \geq 2\sqrt{\frac{\ln n}{n}}$. *Set the parameters of the algorithm as follows:*

$$k = \frac{\ln n}{2}, \ \rho = 0.3np \ and \ \tau = \frac{kp}{1 + pk}.$$

Then algorithm A_{GP} *w.h.p. builds a Hamiltonian cycle in a random n-vertex graph with at least* $N = n\sqrt{n \ln n}$ *edges. The failure probability of the algorithm is*

$$\delta_{GP} = O\left(\frac{\sqrt{\ln n}}{n^{1.5 - o(1)}}\right) = O\left(\frac{\sqrt{\ln n}}{n^{0.8}}\right).$$

Algorithm runs in $O(n^2 / \ln n)$ *time.*

To prove this theorem the authors of [19] computed the conditional probabilities $P_0, \ldots, P_4$ that each stage of the algorithm has successfully completed, assuming that all previous stages were successful. From these computations the failure probability δ_{GP} can be derived.

2.3   Angluin-Valiant Algorithm 1979

In 1952 Dirac proved that a simple graph with n vertices ($n \geq 3$) is Hamiltonian, if every vertex has degree $n/2$ or greater. From Dirac's Theorem it easily follows an $O(n^2)$ time algorithm for finding a Hamiltonian cycle in an n-vertex graphs with minimum degree $\geq n/2$. Angluin and Valiant [3] showed that almost the same algorithm with probability $1 - O(n^{-\alpha})$ finds a Hamiltonian cycle in a random graph with number of edges $N \geq c(\alpha)n \ln n$ (minimum degree w.h.p. is $\geq c_1 \ln n$), where $c(\alpha)$ and c_1 are some constants.

This is a simple fast classic algorithm. Nevertheless, it should be noted that its weakness is a fairly large constant $c(\alpha)$ in the definition of N. If n is not large enough, the restriction on N implies an almost complete graph.

procedure SELECT(u):
begin
    if $\forall v \in V(G)$ $(u, v) \notin E(G)$, return "*";
    else select at random with equal probabilities one of the edges
    $(u, v) \in E(G)$, delete (u, v) from $E(G)$, and return the value v;
  end.

AlgorithmA_{AV}(G,v,t):

Algorithm attempts to find a Hamiltonian path P in a graph G from vertex s to t, returning "success" if it succeeds and "failure" otherwise. If $s = t$ it will search for a Hamiltonian cycle.

Stage 1. Set $P := \emptyset$, $ndp := s$, where ndp is the last vertex of P

Stage 2.(a) If P includes every node of G (except t, if $s \neq t$), and if we previously deleted the edge (ndp, t) from G, then we add (ndp, t) to P and return "success".

(b) Set v=SELECT(ndp).

(b.1) If $v = $ "*", return "failure".

(b.2) If $v \notin P$ and $v \neq t$, add (ndp, v) to P, $ndp := v$, and go to Stage 2.

(b.3) If $v \in P$ and $v \neq s$, then apply what is called the rotation procedure (Fig. 2). Set $u = $ the neighbor of v in P which is closer to ndp, delete (u, v) from P, add (ndp, v) to P, set $ndp = u$, go to Stage 2.

(b.4) Otherwise (i.e. if $v = t$), go to Stage 2.

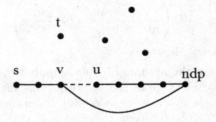

Fig. 2. The rotation procedure.

Theorem 2. *For all $\alpha > 0$ there exist M, K such that if the number of edges in a random graph is $N > c(\alpha)n \ln n$, where $c(\alpha)$ is a sufficiently large constant, then the probability that algorithm A_{AV} returns "success" before SELECT has been called $Mn \ln n$ times is $1 - O(n^{-\alpha})$.*

The proof [3] of this theorem consists in analyzing all the cases in which the algorithm returns "failure" or calls SELECT more than $Mn \ln n$ times, and showing that the probability of such events is negligible $O(n^{-\alpha})$.

3　Probabilistic Analysis of the Approach

3.1　Preliminary Analysis

Further we will need the following inequality:

Proposition 1. *For all n, p, β, with n integer, $0 \leq p \leq 1$, $0 \leq \beta \leq 1$*

$$\sum_{k=0}^{\lfloor (1-\beta)np \rfloor} \binom{n}{k} p^k (1-p)^{n-k} \leq \exp\{-\beta^2 np/2\}.$$

Theorem 3. *Let the weights of the input graph be i.i.d. random reals with a distribution function $f(x)$ defined on $[a_n, b_n]$ or $[a_n, \infty)$, $0 < a_n \leq b_n$. Let the algorithm chosen at Step 3 require N edges to succeed w.h.p., and let it fail with negligible probability δ.*

Then our approach gives the following performance guarantees. The relative error is

$$\varepsilon_A < \frac{w^* - a_n}{a_n},$$

where w^* is the upper bound of the weights of edges left in subgraphs at Step 2,

$$w^* = f^{-1}\left(\frac{4m(N+n)}{n(n-1)}\right).$$

The failure probability is

$$\delta_A \leq m\delta + me^{-n}.$$

Proof. At Step 1 we create random graphs $G_1, \ldots, G_m$, where each edge is present with probability $1/m$, independently of other edges. At Step 2 we delete heavy edges from $G_1, \ldots, G_m$. The probability that an edge is light is

$$\mathbf{Pr}\{\text{weight of the edge} < w^*\} = f(w^*).$$

Thus, $\widetilde{G}_1, \ldots, \widetilde{G}_m$ are random graphs, where each edge exists with probability

$$p = f(w^*)/m, \tag{2}$$

independently of other edges.

We estimate the optimum value of the objective function as $OPT \geq mna_n$. If the algorithm used at Step 3 does not fail for each subgraph $\widetilde{G}_i$, $1 \leq i \leq m$, then the value of the objective function obtained by our approach is $F_A < w^*mn$. So for the relative error we have:

$$\varepsilon_A = \frac{F_A - OPT}{OPT} < \frac{w^*mn}{mna_n} - 1 = \frac{w^* - a_n}{a_n}. \tag{3}$$

At Step 3 we apply one of two algorithms to random graphs $\widetilde{G}_1, \ldots, \widetilde{G}_m$. Both algorithms have requirements of the form "the graph should have at least N edges for the algorithm to succeed w.h.p.". Using Proposition 1, let's estimate the probability that there will not be enough edges in $\widetilde{G}_i$, $1 \leq i \leq m$, in the beginning of Step 3:

$$\delta' = \mathbf{Pr}\{\text{there are} < N \text{edges in} \widetilde{G}_i\} = \sum_{i=0}^{N-1} \binom{\frac{n(n-1)}{2}}{i} p^i (1-p)^{\frac{n(n-1)}{2} - i}$$

$$\leq \exp\left(-\frac{n(n-1)}{4}p + N\right) \leq e^{-n},$$

if

$$p \geq \frac{4(N + n)}{n(n - 1)}. \tag{4}$$

In other words, if the probability p of existence of an edge satisfies (4), δ' is negligible. Combining (2) and (4) we get the following upper bound w^* on the weight of edges left in subgraphs:

$$w^* = f^{-1}\left(\frac{4m(N + n)}{n(n - 1)}\right). \tag{5}$$

Finally, the failure probability δ_A of the approach consists of the probabilities δ' that there were not enough edges in $\widetilde{G}_i$, and probabilities δ that even though there were enough edges in $\widetilde{G}_i$ the algorithm at Step 3 has failed.

$$\delta_A \leq m\delta + m\delta' \leq m\delta + me^{-n}. \tag{6}$$

3.2   The Cases of Uniform and Exponential Distributions of Inputs

Theorem 4. *Let the weights of the input graph be i.i.d. random reals with uniform distribution function $UNI(x)$ defined on $[a_n, \beta_n]$, $0 < a_n \leq \beta_n$, or with shifted exponential distribution function $Exp(x)$ defined on $[a_n, \infty)$, $0 < a_n$, with parameters a_n, β_n:*

$$Exp(x) = 1 - \exp\left(-\frac{x - a_n}{\beta_n}\right), \ 0 < a_n \leq x.$$

Our approach will give asymptotically optimal solutions for the m-PSP, if:

— $m = O(n^{0.5-\theta})$, $0 < \theta < 0.5$, and $\beta_n/a_n = o\left(\frac{n^\theta}{\sqrt{\ln n}}\right)$, if we use algorithm A_{GP} at Step 3.

— $m = O(n^{1-\theta})$, $0 < \theta < 1$, and $\beta_n/a_n = o\left(\frac{n^\theta}{\ln n}\right)$, if we use algorithm A_{AV} at Step 3.

Proof. For the inputs with $UNI(x)$ distribution function according to (5) the weight of the most heavy edge left in subgraphs is less than

$$w^* = (\beta_n - a_n)\frac{4m(N + n)}{n(n - 1)} + a_n.$$

Thereby, according to Theorem 3 the relative error of the approach is

$$\varepsilon_n < \frac{\beta_n}{a_n}\frac{4m(N + n)}{n(n - 1)}. \tag{7}$$

For the inputs with shifted exponential distribution function $Exp(x)$ according to (5) the weight of the most heavy edge left in subgraphs is less than

$$w^* = a_n - \beta_n \ln\left(1 - \frac{4m(N + n)}{n(n - 1)}\right).$$

Thereby, using Theorem 3, for the relative error of the approach we have:

$$\varepsilon_n < -\frac{\beta_n}{a_n} \ln\left(1 - \frac{4m(N+n)}{n(n-1)}\right) \leq \frac{\beta_n}{a_n} \frac{4m(N+n)}{n(n-1)},$$

which is equivalent to (7).

Further analysis depends on the algorithm chosen at Step 3.

(1) If we apply algorithm A_{GP}, the required number of edges $N = n\sqrt{n \ln n}$, so for the relative error we have:

$$\varepsilon_A < \frac{\beta_n}{a_n} \frac{4m}{n(n-1)}\left(n\sqrt{n \ln n} + n\right) = \frac{\beta_n}{a_n} \frac{4m(\sqrt{n \ln n} + 1)}{(n-1)} = O\left(\frac{\beta_n}{a_n} \frac{\sqrt{\ln n}}{n^\theta}\right) = o(1),$$

$$\text{if } m = O(n^{0.5-\theta}),\, 0 < \theta < 0.5 \text{ and } \frac{\beta_n}{a_n} = o\left(\frac{n^\theta}{\sqrt{\ln n}}\right).$$

According to Theorem 3, Theorem 1 and the value of m chosen above, for the failure probability we have:

$$\delta_A \leq me^{-n} + mO\left(\frac{\sqrt{\ln n}}{n^{0.8}}\right) = O\left(\frac{\sqrt{\ln n}}{n^{0.3+\theta}}\right) = o(1).$$

(2) If we use algorithm A_{AV}, then the required number of edges $N = c(\alpha)n \ln n$.

$$\varepsilon_A < \frac{\beta_n}{a_n} \frac{4m}{(n-1)}(c\ln n + 1) = O\left(\frac{\beta_n}{a_n} \frac{\ln n}{n^\theta}\right) = o(1),$$

$$\text{if } m = O(n^{1-\theta}),\, 0 < \theta < 1, \text{ and } \frac{\beta_n}{a_n} = o\left(\frac{n^\theta}{\ln n}\right).$$

According to Theorems 2 and 3 and the value of m chosen above:

$$\delta_A \leq me^{-n} + mO(n^{-\alpha}) = O(n^{1-\theta-\alpha}) = o(1),\, \text{if } \alpha > 1.$$

Remark 1. *Considering (3) and (6), it is easy to prove that the results obtained in this paper can be extended to the case of any distribution function that dominates the considered distributions (distribution of majorizing type): $\widehat{f}(x) \geq f(x)$ for all x, and $\widehat{f}(x)$ has the same domain as $f(x)$.*

For example, the truncated normal distribution $\mathcal{N}_{a_n, \sigma_n^2}(x)$ with parameters $a_n, \sigma_n = \beta_n/2$:

$$\mathcal{N}_{a_n, \sigma_n^2}(x) = \frac{2}{\sqrt{2\pi\sigma_n^2}} \int_{a_n}^x \exp\left(-\frac{(t-a_n)^2}{2\sigma_n^2}\right) dt,\, 0 < a_n \leq x,$$

dominates the shifted exponential distribution function with parameters a_n, β_n. Thus, the same performance guarantees of the approach hold for the inputs with truncated normal distribution on $[a_n, \infty]$.

Remark 2. *The described algorithms A_{AV} and A_{GP} have counterparts for the case of directed graphs* [3, 19] *with failure probability of the same exponent. Thus the results obtained in this section are also true for the directed m-PSP.*

Remark 3. *The approach will also work well for any distribution function defined on* $[a_n, b_n]$ *or* $[a_n, \infty)$, $0 < a_n$, *which has large enough probability density near* a_n. *Then with increasing n the number of edges of almost a_n weight increases, and for large n we would be able to build Hamiltonian cycles using only the lightest edges.*

Acknowledgments. The authors are supported by the RSCF grant 16-11-10041.

References

1. Ageev, A.A., Baburin, A.E., Gimadi, E.K.: A 3/4 approximation algorithms for finding two disjoint Hamiltonian cycles of maximum weight. J. Appl. Ind. Math. **1**(2), 142–147 (2007)
2. Ageev, A.A., Pyatkin, A.V.: A 2-approximation algorithm for the metric 2-peripatetic salesman problem. Diskretn. Anal. Issled. Oper. **16**(4), 3–20 (2009)
3. Angluin, D., Valiant, L.G.: Fast probabilistic algorithms for Hamiltonian circuits and matchings. J. Comput. Syst. Sci. **18**(2), 155–193 (1979)
4. Baburin, A.E., Gimadi, E.K.: On the asymptotic optimality of an algorithm for solving the maximum m-PSP in a multidimensional Euclidean space. Proc. Steklov Inst. Math. **272**(1), 1–13 (2011)
5. Baburin, A.E., Gimadi, E.K., Korkishko, N.M.: Approximation algorithms for finding two edge-disjoint Hamiltonian cycles of minimal total weight (in Russian). J. Discr. Anal. Oper. Res. **2**(11), 11–25 (2004)
6. Bollobas, B., Fenner, T.I., Frieze, A.M.: An algorithm for finding Hamilton paths and cycles in random graphs. Combinatorica **7**, 327–341 (1987)
7. Chebolu, P., Frieze, A.M., Melsted, P.: Finding a maximum matching in a sparse random graph in O(n) expected time. JACM **57**, 24 (2010)
8. De Brey, M.J.D., Volgenant, A.: Well-solved cases of the 2-peripatetic salesman problem. Optimization **39**(3), 275–293 (1997)
9. De Kort, J.B.J.M.: Upper bounds and lower bounds for the symmetric K-Peripatetic Salesman Problem. Optimization **23**(4), 357–367 (1992)
10. DeKort, J.B.J.M.: Lower bounds for symmetric K-peripatetic salesman problems. Optimization **22**(1), 113–122 (1991)
11. De Kort, J.B.J.M.: A branch and bound algorithm for symmetric 2-peripatetic salesman problems. Eur. J. Oper. Res. **70**, 229–243 (1993)
12. Erdos, P., Renyi, A.: On random graphs I. Publ. Math. Debrecen **6**, 290–297 (1959)
13. Frieze, A., Haber, S.: An almost linear time algorithm for finding Hamilton cycles in sparse random graphs with minimum degree at least three. J. Random Struct. Algorithms **47**(1), 73–98 (2015)
14. Gimadi, E.K., Glazkov, Y.V., Glebov, A.N.: Approximation algorithms for sollving the 2-peripatetic salesman problem on a complete graph with edge weights 1 and 2. J. Appl. Ind. Math. **3**(1), 46–60 (2009)
15. Gimadi, E.K., Glazkov, Y.V., Tsidulko, O.Y.: The probabilistic analysis of an algorithm for solving the m-planar 3-dimensional assignment problem on one-cycle permutations. J. Appl. Ind. Math. **8**(2), 208–217 (2014)

16. Gimadi, E.K., Glebov, A.N., Skretneva, A.A., Tsidulko, O.Y., Zambalaeva, D.Z.: Combinatorial algorithms with performance guarantees for finding several Hamiltonian circuits in a complete directed weighted graph. Discrete Appl. Math. **196**(11), 54–61 (2015)

17. Gimadi, E.K., Istomin, A.M., Rykov, I.A., Tsidulko, O.Y.: Probabilistic analysis of an approximation algorithm for the m-peripatetic salesman problem on random instances unbounded from above. Proc. Steklov Inst. Math. **289**(1), 77–87 (2015)

18. Gimadi, E.K., Ivonina, E.V.: Approximation algorithms for the maximum 2-peripatetic salesman problem. J. Appl. Ind. Math. **6**(3), 295–305 (2012)

19. Gimadi, E.K., Perepelitsa, V.A.: A statistically effective algorithm for the selection of a Hamiltonian contour or cycle (in Russian). J. Diskret. Anal. **22**, 15–28 (1973)

20. Glebov, A.N., Zambalaeva, D.Z.: A polynomial algorithm with approximation ratio 7/9 for the maximum two peripatetic salesmen problem. J. Appl. Ind. Math. **6**(1), 69–89 (2012)

21. Glebov, A.N., Zambalaeva, D.Z.: An approximation algorithm for the minimum two peripatetic salesmen problem with different weight functions. J. Appl. Ind. Math. **6**(2), 167–183 (2012)

22. Komlos, J., Szemeredi, E.: Limit distributions for the existence of Hamilton circuits in a random graph. Discrete Math. **43**, 5563 (1983)

23. Korshunov, A.D.: On the power of some classes of graphs. Sov. Math. Dokl. **11**, 1100–1104 (1970)

24. Krarup, J.: The peripatetic salesman and some related unsolved problems. In: Roy, B. (ed.) Combinatorial Programming, Methods and Applications. NATO Advanced Study Institutes Series, vol. 19, pp. 173–178. Reidel, Dordrecht (1975)

25. Petrov, V.V.: Limit Theorems of Probability Theory. Sequences of Independent Random Variables. Clarendon Press, Oxford (1995)

Efficient Randomized Algorithm
for a Vector Subset Problem

Edward Gimadi[1,2] and Ivan Rykov[1,2(✉)]

[1] Sobolev Institute of Mathematics, Koptuga Avenue 4, 630090 Novosibirsk, Russia
rykovweb@gmail.com
[2] Novosibirsk State University, 2 Pirogova Street, 630090 Novosibirsk, Russia
gimadi@math.nsc.ru

Abstract. We introduce a randomized approximation algorithm for
NP-hard problem of finding a subset of m vectors chosen from n given
vectors in multidimensional Euclidean space $\mathbb{R}^k$ such that the norm of the
corresponding sum-vector is maximum. We derive the relation between
algorithm's time complexity, relative error and failure probability parameters. We show that the algorithm implements Polynomial-time Randomized Approximation Scheme (PRAS) for the general case of the problem. Choosing particular parameters of the algorithm one can obtain
asymptotically exact algorithm with significantly lower time complexity
compared to known exact algorithm. Another set of parameters provides
polynomial-time 1/2-approximation algorithm for the problem. We also
show that the algorithm is applicable for the related (minimization) clustering problem allowing to obtain better performance guarantees than
existing algorithms.

Keywords: Vector subset · Randomized algorithm · Asymptotically
exact algorithm · Performance guarantees

1   Introduction

We consider the following Maximization Vector Subset problem (VSP).

Problem 1. Given a set of n vectors $(v_1, \ldots v_n)$ in Euclidean space $\mathbb{R}^k$ and a
number $m < n$ choose a subset of m vectors $(v_{i_1}, \ldots v_{i_m})$ such that the norm
of the sum-vector $\sum_{j=1}^{m} v_{i_j}$ is maximum (we consider the Euclidean norm in k-
dimensional space $\mathbb{R}^k$; i. e. $\|\boldsymbol{x}\| = \sqrt{x_1^2 + \ldots + x_k^2}$).

This problem was considered in [5] in context of solving an applied problem
of determining a quasiperiodic segment in a noisy numerical sequence provided
that the number of repetitions of the segment is known. This problem typically arises in various applications such as electronic prospecting, radiolocation,
telecommunication, geophysics, voice signals processing, medical and technical
diagnostics, etc.

© Springer International Publishing Switzerland 2016
Y. Kochetov et al. (Eds.): DOOR 2016, LNCS 9869, pp. 148–158, 2016.
DOI: 10.1007/978-3-319-44914-2_12

The Vector Subset problem was thoroughly studied in [2]. It was shown that its general case of the problem is NP-hard. An approximation scheme was also proposed which allows to find a solution with relative error $\varepsilon > 0$ in $O(nk^{(k+3)/2}(1/\varepsilon)^{(k-1)/2})$ time. Thus, a FPTAS for any fixed dimension k was obtained.

In [2,4] two exact pseudopolynomial (for any fixed k) algorithms were proposed for the integer case of the vector subset problem.

Observe that the one-dimensional case of the problem (when $k = 1$ and v_i are real numbers) is trivial: the optimal solution either consists of m maximal or m minimal numbers. The complexity status of the particular case when the dimension is a fixed number $k > 1$ remained opened until Gimadi et al. [6] proposed an algorithm which finds an optimal solution in running-time $O(k^2 n^{2k})$.

Thus, the Problem 1 is polynomially solvable for any fixed dimension k. However, the complexity of the proposed algorithm quickly grows with the increase of dimension k. In current paper we describe the randomized algorithm having significantly lower time complexity while remaining asymptotically exact "**whp**" (with high probability). Hereinafter this abbreviation means "with probability tending to 1 when $n \to \infty$".

It's worth noting that all three mentioned algorithms (approximation, exact and asymptotically exact) use the same idea of solving a set of one-dimensional problems (of finding m best projections on a given ray) and choosing the best set of vectors among provided by the solutions of these problems.

In the second section we briefly describe an application of this idea to constructing the first two algorithms. In the third section we describe a new randomized algorithm in more details.

2 Previous Results

Theorem 1 [2]. *VSP is NP-hard.*

This theorem is proved by the reduction of the Clique problem to VSP.

Theorem 2 [2]. *There exists an algorithm A_1 that solves any instance of VSP with relative error less than $(k-1)/(8L^2)$ in $O(nk^2(2L+1)^{k-1})$ time.*

It's worth to describe briefly the idea of the algorithm A_1, since the randomized algorithm introduced in this paper shares the common approach with it.

It is based on the idea that if we know the direction of the optimal sum vector, then the corresponding m vectors can be found as those providing m largest projections on this direction. This is complemented by the note that if two directions are close (angle between them is small), then the projections of any vector on them have close values. Thus, the algorithm A_1 tries to cover the space with a set of directions such that any direction (including unknown optimal direction) would be close to some direction from the chosen set. For each chosen direction one solves the 1-dimensional problem of finding m vectors with largest projections on this direction.

In the case of algorithm A_1 the directions are taken as vectors starting at the origin and going to all the integer points on the faces of hypercube $\{x \in \mathbb{R}^k | -L \le x_i \le L, i = 1 \ldots k\}$. It is proved that this set provides the stated complexity and error estimates. Taking $\varepsilon = (k-1)/(8L^2)$ we easily see that algorithm A_1 provides FPTAS (for each fixed k) for VSP.

Theorem 3 [6]. *There exists an algorithm A_2 that optimally solves VSP in $O(k^2 n^{2k})$ time.*

The algorithm A_2 is based on more precise analysis of how the projections of vectors to the directions depend on the direction. It was shown that the set of vectors with largest projections remains the same within each of $O(kn^{2k-2})$ regions separated by hyperplane perpendiculars to vector differences. Thus, it is enough to solve the 1-dimensional problem for one direction from each of these regions to guarantee the optimal set of vectors will be presented in the set of obtained solutions. By applying the lemma from [3] it is shown in [6] that this can be done in $O(k^2 n^{2k})$ time.

3   Randomized Algorithm

The algorithm A' proposed in this section uses the same idea as the approximation algorithm A_1. The difference between them is that the algorithm A' is randomized; i.e. it carries out random decisions during its execution. Thus, we apply a probabilistic analysis to this algorithm; i.e. we estimate both the relative error and the failure probability.

We say that a randomized algorithm A has $(\varepsilon_n^A, \delta_n^A)$ estimates (called relative error and failure probability respectively) over the set $\mathfrak{I_n}$ of all the instances with n elements in the input data if

$$\mathbb{P} \left\{ \frac{f^*(I) - f_A(I)}{f^*(I)} < \varepsilon_n^A \ \middle| \ I \in \mathfrak{I_n} \right\} \ge 1 - \delta_n^A$$

The probability distribution is often set over the instances of the same size; i.e. values ε_n^A and δ_n^A depend on n. Algorithm is called asymptotically exact if both ε_n^A and δ_n^A tend to zero when n tends to infinity.

When constructing randomized algorithm, we can control the relation between approximation error, failure probability and time complexity. For the algorithm A' we choose the values of parameters such that the algorithm has much lower complexity, than the exact algorithm, while remaining asymptotically exact.

Let's describe A' stepwise.

Algorithm A'.

1. Randomly and uniformly choose L points on the unit sphere in $\mathbb{R}^k$;
2. for each of the directions, defined by these points, find the set of m vectors with maximal projections on the direction;
3. among the sets of vectors found at step 2 choose the set with the maximum norm of sum-vector.

Thus, the algorithm A' differs from the algorithm A_1 by the rule of directions generation. Instead of enumerating all the integer points on faces of the hypercube we randomly and uniformly choose the fixed number of points on the unit sphere.

We define the following criterion of algorithm's successful execution: if the smallest angle between optimal sum vector and one of the generated directions is less than φ, the algorithm is executed successfully; otherwise, algorithm fails.

Note that L and φ are the parameters of the algorithm that we can choose arbitrary and that affect relative error, failure probability and time complexity values.

The main theorem of the paper establishes the relationship between parameters L and φ and the algorithm performance guarantees.

Theorem 4. *Algorithm A' with parameters L and φ has the following performance guarantees: approximation error*

$$\varepsilon^{A'} \leq 2\sin^2\frac{\varphi}{2},$$

failure probability

$$\delta^{A'} = \exp\left(-\frac{\sin^{k-1}\varphi}{3\sqrt{k-1}}L\right)$$

and time complexity

$$T^{A'} = O(nkL).$$

3.1   Proof of Theorem 4

Let us show that all three estimates stated by the theorem hold.

Time Complexity Estimate. The time complexity estimate follows from the complexity of solving one-dimensional problem of choosing m vectors with maximum projections for each of L generated points. The complexity of each one-dimensional problem is $O(nk)$. Indeed, we need $O(nk)$ operations to calculate all projection values (n scalar products in k-dimensional space). Then, choosing m largest numbers among n numbers is done in $O(n)$ time (see [1]).

Relative Error Estimate. We prove the stated estimate for ε in assumption of algorithm's success; i.e. assuming that the angle between the optimal direction and the closest chosen point is less than φ.

Denote by $s(X)$ the sum vector of any arbitrary set X of vectors, denote by X^A the set of m vectors found by the algorithm A', denote by X^* the optimal solution of the initial problem, denote by $\widetilde{X}$ the solution of one-dimensional problem for the direction which is closest to the direction of the optimal sum vector $s(X^*)$. Denote by v^A and $\widetilde{v}$ the points on the sphere (unit vectors) chosen from L generated points that produce solutions X^A and $\widetilde{X}$ respectively. The following inequalities hold:

1. $\|s(X^A)\| \geq \|s(\widetilde{X})\|$ since algorithm has chosen the solution with the best value of the initial objective function;
2. $\|s(\widetilde{X})\| \geq \sum_{i=1}^{m} \widetilde{X}_i * \widetilde{v}$ since the length of the vector is greater or equal to its projection on any direction and the projection of sum is the sum of projections;
3. $\sum_{i=1}^{m} \widetilde{X}_i * \widetilde{v} \geq \sum_{i=1}^{m} X_i^* * \widetilde{v}$ since m vectors from $\widetilde{X}$ give the largest projections on the direction $\widetilde{v}$;
4. $\sum_{i=1}^{m} X_i^* * \widetilde{v} = \|s(X^*)\| * \cos \widetilde{\varphi} \geq \|s(X^*)\| * \cos \varphi$ since the angle $\widetilde{\varphi}$ between $s(X^*)$ and $\widetilde{v}$ is less then φ (in case, when algorithm did not fail).

Thus,

$$\varepsilon = \frac{\|s(X^*)\| - \|s(X^A)\|}{\|s(X^*)\|} = 1 - \frac{\|s(X^A)\|}{\|s(X^*)\|} \leq 1 - \cos \varphi = 2\sin^2 \frac{\varphi}{2}.$$

Failure Probability Estimate. Denote by p_0 the probability for the random uniformly chosen unit vector to form an angle less than φ with some fixed direction (direction of the optimal sum vector in our case).

It is quite obvious that algorithm's failure probability is estimated by

$$\delta \leq (1 - p_0)^L$$

which is a probability of the event "none of directions of L points chosen on the sphere forms angle less than φ with optimal direction".

Hence,

$$\delta \leq e^{-p_0 L}$$

So, to prove the stated estimate for δ it suffices to show that

$$p_0 \geq \frac{\sin^{k-1} \varphi}{3\sqrt{k-1}}. \tag{1}$$

Assume that φ is a half angle of the spherical cone with its apex in the center of the unit ball in k-dimensional space. The intersection of this cone with the boundary of the ball (i.e. with a unit sphere) forms a "cap" which we denote as $S_k^{cap}(\varphi)$. It's easy to see that p_0 equals to the ratio of the area of the cap $S_k^{cap}(\varphi)$

to the area of the unit sphere S_k. Thus, we can use the results obtained for the area of the cap in k-dimensional space to estimate the success probability.

E.g., the paper [8] provides the following formula for the area of the unit sphere and the area of a cap on the unit sphere:

$$S_k = \frac{2\pi^{k/2}}{\Gamma(\frac{k}{2})},$$

$$S_k^{cap}(\varphi) = \frac{2\pi^{(k-1)/2}}{\Gamma(\frac{k-1}{2})} \int_0^\varphi \sin^{k-2}\theta d\theta$$

where Γ is the Gamma function.

Thus, p_0 is estimated as

$$p_0 = \frac{S_k^{cap}(\varphi)}{S_k} = \frac{\Gamma(\frac{k}{2})}{\sqrt{\pi}\,\Gamma(\frac{k-1}{2})} \int_0^\varphi \sin^{k-2}\theta d\theta. \tag{2}$$

In order to obtain the required estimate (1) we need two following arithmetic lemmas.

Lemma 1.

$$\int_{\theta=0}^\varphi \sin^{k-2}\theta d\theta \geq \frac{\sin^{k-1}\varphi}{k-1}. \tag{3}$$

Proof. Using substitution $t = \sin\theta$ we get

$$\int_{\theta=0}^\varphi \sin^{k-2}\theta d\theta = \int_{\theta=0}^{\sin\varphi} \frac{t^{k-2}}{(1-t^2)^{1/2}} dt \geq \int_{\theta=0}^{\sin\varphi} t^{k-2} dt = \frac{\sin^{k-1}\varphi}{k-1}.$$

Lemma 2. *For any fixed $k > 1$ the following inequality holds:*

$$\frac{\Gamma(\frac{k}{2})}{2\sqrt{\pi}\,\Gamma(\frac{k+1}{2})} \geq \frac{1}{3\sqrt{k-1}}. \tag{4}$$

Proof. Case 1. k is even; i.e. $k = 2j$, $j > 0$:

$$\frac{\Gamma(\frac{k}{2})}{\Gamma(\frac{k+1}{2})} = \frac{\Gamma(j)}{\Gamma(j+\frac{1}{2})} = \frac{(j-1)!\,2^j}{\sqrt{\pi}(2j-1)!!}$$

$$= \frac{1\cdot 2\cdots(j-1)\cdot 2^j}{\sqrt{\pi}\cdot 1\cdot 3\cdots(2j-1)} = \frac{2}{\sqrt{\pi}}\cdot\frac{2\cdot 4\cdots(2j-2)}{3\cdot 5\cdots(2j-1)}$$

$$= \frac{2}{\sqrt{\pi}}\sqrt{\frac{2\cdot 2}{3}\cdot\frac{4\cdot 4}{3\cdot 5}\cdot\frac{6\cdot 6}{5\cdot 7}\cdots\frac{(2j-2)(2j-2)}{(2j-3)(2j-1)}\cdot\frac{1}{(2j-1)}}.$$

Noting that $s^2 > (s-1)(s+1)$ for any natural $s > 1$ we obtain for the even k:

$$\frac{\Gamma(\frac{k}{2})}{2\sqrt{\pi}\Gamma(\frac{k+1}{2})} > \frac{1}{2\sqrt{\pi}}\cdot\frac{4}{\sqrt{3\pi(2j-1)}} > \frac{1}{3\sqrt{k-1}}.$$

Case 2. k is odd; i.e. $k = 2j + 1$, $j \geq 0$:

$$\frac{\Gamma(\frac{k}{2})}{\Gamma(\frac{k+1}{2})} = \frac{\Gamma(j+\frac{1}{2})}{\Gamma(j+1)} = \frac{\sqrt{\pi}(2j-1)!!}{j! \, 2^j} = \frac{\sqrt{\pi} \cdot 1 \cdot 3 \cdots (2j-1)}{1 \cdot 2 \cdots j \cdot 2^j}$$

$$= \frac{\sqrt{\pi} \cdot 3 \cdot 5 \cdots (2j-1)}{2 \cdot 4 \cdots 2j} = \left(\pi \cdot \frac{1}{2} \cdot \frac{3 \cdot 3}{2 \cdot 4} \cdot \frac{5 \cdot 5}{4 \cdot 6} \cdots \frac{(2j-1)(2j-1)}{(2j-2)2j} \cdot \frac{1}{2j} \right)^{1/2}.$$

Similarly to the previous case we have

$$\frac{\Gamma(\frac{k}{2})}{2\sqrt{\pi}\Gamma(\frac{k+1}{2})} \geq \frac{1}{2\sqrt{\pi}} \cdot \frac{3\sqrt{\pi}}{4\sqrt{2j}} \geq \frac{3}{8\sqrt{k-1}} > \frac{1}{3\sqrt{k-1}},$$

hence the inequality (4) holds for odd k as well.

The proof of Lemma is complete.

Now, combining the relations (2), (3), (4) and the relation

$$\frac{k-1}{2}\Gamma(\frac{k-1}{2}) = \Gamma(\frac{k+1}{2}),$$

we immediately get the desired estimate (1).

This completes the proof of the Theorem 4.

3.2   Polynomial-Time Approximation Scheme

Assigning values to the parameters L and φ we can control relations between time complexity T, relative error ε and failure probability δ. Taking into account that the value of φ is unknown while the algorithm is executed, it makes sense to provide a statement where this parameter is excluded from all the expressions. The following theorem estimates δ and T in terms of expressions depending on ε and L where required relative error ε can be considered as a given parameter while L is a parameter which is directly chosen in the algorithm input.

Theorem 5. *Algorithm A' with a given parameter L finds a $(1-\varepsilon)$-approximate solution with failure probability*

$$\delta_{A'} \leq \exp\left(-\beta_k(\varepsilon)\frac{L}{3\sqrt{k-1}}\right),$$

in $T_{A'} = O(nkL)$ time, where $\beta_k(\varepsilon) = (\varepsilon(2-\varepsilon))^{\frac{k-1}{2}}$.

Proof. Since

$$\varepsilon = 1 - \cos\varphi = \sqrt{1 - \sin^2\varphi},$$

we have

$$\sin^2\varphi = 1 - (1-\varepsilon)^2 = \varepsilon(2-\varepsilon),$$

hence

$$\sin^{k-1}\varphi = (\varepsilon(2-\varepsilon))^{(k-1)/2} = \beta_k(\varepsilon).$$

Thus, the stated inequality is implied by Theorem 4.

Let's assume that we are given a target relative error ε. A natural goal would be to find the least possible time complexity such that the algorithm would be successfully executed **whp**.

Corollary 1. *Algorithm A' with*

$$L = \frac{3\sqrt{k-1}}{\beta_k(\varepsilon)} \ln \ln n$$

whp *finds $(1 - \varepsilon)$-approximate solution for the problem in*

$$O\left(\frac{3k\sqrt{k-1}}{\beta_k(\varepsilon)} n \ln \ln n \right)$$

time.

Proof. Indeed,

$$\delta \le \exp\left(-\frac{\beta_k(\varepsilon) L}{3\sqrt{k-1}} \right) = e^{-\ln \ln n} = \frac{1}{\ln n} \to 0,$$

which means that the algorithm is executed successfully **whp**.

Since $\varepsilon < 1$, we can obtain more explicit dependence of time complexity on ε.

Corollary 2. *Algorithm A' with*

$$L = \frac{3\sqrt{k-1}}{\beta_k(\varepsilon)} \ln \ln n$$

whp *finds $(1 - \varepsilon)$-approximate solution in*

$$T_{A'} = O\left(k^{\frac{3}{2}} \varepsilon^{-\frac{k-1}{2}} n \ln \ln n \right)$$

time; i.e. it implements Polynomial-time Randomized Approximation Scheme (PRAS) for the general case of the problem.

It can be also noted that the algorithm implements FPRAS for the case of any fixed k.

3.3   Polynomial Algorithm with Performance Guarantee

Analysing relations between L, T, ε, and δ it is also possible to obtain a significantly less complexity for the general case of the problem compared to the exact and to known approximation algorithms reaching constant estimate for the relative error ε.

Corollary 3. *Algorithm A' with*

$$L = \frac{3\sqrt{k-1}}{\beta_k(\frac{1}{2})} \ln \ln n$$

whp *finds $\frac{1}{2}$-approximate solution for the problem in $T_{A'} = O(k^{\frac{3}{2}}(4/3)^{\frac{k-1}{2}} n \ln \ln n)$ time.*

4   Application to Minimization Problem

In [7] the related minimization problem of splitting a set of vectors into two clusters was considered.

Problem 2. Given a set of n vectors $V = (v_1, \ldots v_n)$ in Euclidean space $\mathbb{R}^k$ and a number $m < n$ choose the subset of m vectors $C \subseteq V$ such that the objective function

$$\sum_{v \in C} \|v - v(C)\|^2 + \sum_{v \in V \setminus C} \|v\|^2$$

is minimized where $v(C) = \frac{1}{m} \sum_{v \in C} \|v\|$ is a geometrical center (centroid) of a chosen set.

In [7] a 2-approximation algorithm was constructed for this minimization problem with time complexity $O(kn^2)$ polynomial in both n and k. It was also noted that the values of objective functions of Problems 1 and 2 are related by the formula

$$S = K - \frac{F^2}{m} \tag{5}$$

where S is the value of the objective function of Problem 2, F is the value of the objective function of Problem 1, $K = \sum_{v \in V} \|v\|^2$ is a constant value. Correspondingly,

$$S^* = K - \frac{(F^*)^2}{m} \tag{6}$$

for the optimal values.

Since the solutions for both problems are the sets of m vectors, one can evaluate the algorithm for one problem to be applied to another problem. In [7] it was pointed, though, that the approximation results for the minimization problem can't be transferred to obtain polynomial algorithms with performance guarantees for the maximization problem. We show that the randomized algorithm for the maximization problem can be used to obtain better estimations for relative error of minimization problem than the one obtained in [7]. Of course, the algorithm A' is not polynomial in k, however, the complexity in k is moderate, especially for the case of constant approximation ratio (see Corollary 3).

From Eqs. (5) and (6) we derive:

$$\frac{S}{S^*} = \frac{K - \frac{F^2}{m}}{K - \frac{(F^*)^2}{m}} \leq \frac{K - (1 - \varepsilon_{A'})^2 \frac{(F^*)^2}{m}}{K - \frac{(F^*)^2}{m}} \leq 1 + \varepsilon_S$$

where

$$\varepsilon_S = \frac{(2 - \varepsilon_{A'})\varepsilon_{A'}}{\frac{mK}{(F^*)^2} - 1}.$$

Thus, the larger is the ratio $\frac{mK}{(F^*)^2}$ for the problem (which depends only on the input), the better is the relative error ε_S for Problem 2 compared to the relative error $\varepsilon_{A'}$ of the algorithm for the Problem 1.

Let us estimate this relation as the ratio between largest vectors in the set compared to all vectors. Without loss of generality assume $||v_1||^2 \geq ||v_2||^2 \geq \ldots \geq ||v_n||^2$. Set

$$K_r = \sum_{j=1}^{r} ||v_j||^2,$$

$r = 1, \ldots, n;\ K = K_n.$

Lemma 3.

$$\frac{mK}{(F^*)^2} \geq \frac{K}{K_m} \tag{7}$$

Hence

$$\varepsilon_S \leq \frac{(2 - \varepsilon_{A'})\varepsilon_{A'}}{\frac{K}{K_m} - 1} \tag{8}$$

Proof. First we prove that for each m-element subset $C \subseteq V$.

$$\left(\sum_{v \in C} v\right)^2 \leq m \sum_{v \in C} v^2. \tag{9}$$

Indeed,

$$\left(\sum_{v \in C} v\right)^2 = \sum_{v \in C} v^2 + 2 \sum_{v \neq w \in C} v \cdot w$$

$$= m \sum_{v \in C} v^2 - \sum_{v \neq w}(v^2 + w^2 - 2v \cdot w) = m \sum_{v \in C} v^2 - \sum_{v \neq w}(v - w)^2 \leq m \sum_{v \in C} v^2.$$

Next, we show that for any m-element subset $C \subseteq V$.

$$\sum_{v \in C} v^2 \leq K_m. \tag{10}$$

Indeed,

$$\sum_{v \in C} v^2 \leq \max_{C' \subseteq V} \sum_{v \in C'} v^2 = \sum_{j=1}^{m} v_j^2 = K_m.$$

Considering Eqs. (9) and (10) we get

$$(F^*)^2 = \left(\sum_{v \in C^*} v\right)^2 \leq m \sum_{v \in C^*} v^2 \leq mK_m,$$

which completes the proof of the Lemma.

Now we can use ratio $\frac{K}{K_m}$ to analyze the relation between the values ε_S and $\varepsilon_{A'}$, using (8).

Theorem 6. *The solution of the Problem 1 obtained by algorithm A' (in $O(nkL)$ time) is at least $(1 + \varepsilon_{A'})$-approximate if $K_m \leq K/3$. In case when $K_m = o(K)$ the algorithm A' is asymptotically exact for any fixed $\varepsilon_{A'}$.*

5   Conclusion

In this paper we considered the problem of finding a m-subset of a given set with n vectors in multidimensional Euclidean space $\mathbb{R}^k$ such that the norm of the corresponding sum-vector is maximized. We introduced a randomized approximation algorithm for solving the problem. It was shown that the algorithm implements Polynomial-time Randomized Approximation Scheme (PRAS) for the general case of the problem. Choosing particular parameters of the algorithm, we obtain asymptotically exact algorithm with significantly lower time complexity compared to known exact algorithm. Another set of parameters provides polynomial-time 1/2-approximation algorithm for the problem. It is shown that the algorithm is applicable for the related (minimization) clustering problem, allowing to obtain better performance guarantees than existing algorithms.

Acknowledgments. The authors are supported by the RSCF grant 16-11-10041.

References

1. Aho, A.V., Hopcroft, J.E., Ullman, J.D.: The Design and Analysis of Computer Algorithms. Addison-Wesley, Reading (1974)
2. Baburin, A.E., Gimadi, E.K., Glebov, N.I., Pyatkin, A.V.: The problem of finding a subset of vectors with the maximum total weight. J. Appl. Ind. Math. **2**(1), 32–38 (2008)
3. Baburin, A.E., Pyatkin, A.V.: Polynomial algorithms for solving the vector sum problem. J. Appl. Ind. Math. **1**(3), 268–272 (2007)
4. Gimadi, E.K., Glazkov, Y.V., Rykov, I.A.: On two problems of choosing some subset of vectors with integer coordinates that has maximum norm of the sum of elements in Euclidean space. J. Appl. Ind. Math. **3**(3), 343–352 (2009)
5. Gimadi, E.K., Kel'manov, A.V., Kel'manova, M.A., Khamidullin, S.A.: A posteriori detection of a quasiperiodic fragment in a numerical sequence. Pattern Recogn. Image Anal. **18**(1), 30–42 (2008)
6. Gimadi, E.K., Pyatkin, A.V., Rykov, I.A.: On polynomial solvability of some problems of a vector subset choice in a Euclidean space of fixed dimension. J. Appl. Ind. Math. **4**(1), 48–53 (2010)
7. Dolgushev, A.V., Kel'manov, A.V.: An approximation algorithm for solving a problem of cluster analysis. J. Appl. Ind. Math. **5**(4), 551–558 (2011)
8. Li, S.: Concise formulas for the area and volume of a hyperspherical cap. asian J. Math. Stat. **4**(1), 66–70 (2011)

An Algorithm with Approximation Ratio 5/6 for the Metric Maximum m-PSP

Aleksey N. Glebov[1,2]([✉]) and Anastasiya V. Gordeeva[2]

[1] Sobolev Institute of Mathematics, 4 Acad. Koptyug Avenue,
630090 Novosibirsk, Russia
anglemob1973@mail.ru

[2] Novosibirsk State University, 2 Pirogova Street, 630090 Novosibirsk, Russia
anast.gordeeva@gmail.com

Abstract. We present a polynomial algorithm with guaranteed approximation ratio 5/6 for the metric maximization version of the m-PSP with identical weight functions. This result extends the well-known algorithm by Kostochka and Serdyukov for the metric TSP (1985) to the case of several Hamiltonian cycles and improves the approximation ratio of the algorithm by Gordeeva (2010) for the metric 2-PSP.

Keywords: Metric TSP · Metric m-PSP · Maximization version of the TSP · Maximization version of the m-PSP · Approximation algorithm

1 Introduction

The m-Peripatetic Salesman Problem (m-PSP) was introduced by Krarup [23] as a natural generalization of the well-known Traveling Salesman Problem (TSP). The input of the m-PSP is a complete (directed or undirected) n-vertex graph $G = (V, E)$ and weight functions $w_i : E \to \mathbf{R}_+$, $i = 1, \ldots, m$. The problem is to find m edge disjoint Hamiltonian cycles $H_1, \ldots, H_m \subset E$ such that their total weight

$$\sum_{i=1}^{m} w_i(H_i) = \sum_{i=1}^{m} \sum_{e \in H_i} w_i(e)$$

is maximum or minimum. Most of the research deals with the case when all the weight functions w_i are the same (identical): $w_1 = \ldots = w_m = w$.

De Brey and Volgenant [7] investigated some easy cases of the 2-PSP. De Kort developed lower and upper bounds for the 2-PSP, through branch-and-bound algorithms, and showed that the problem of finding two edge disjoint Hamiltonian circuits is NP-complete [8–10]. This result implies that the 2-PSP with identical weight functions is NP-hard both in the maximization and minimization variants. These results can be extended for the general m-PSP. The problem remains NP-hard in the case of different weight functions [4].

A.N. Glebov—The work is supported by RFBR (projects 15-01-00976 and 15-01-05867).

© Springer International Publishing Switzerland 2016
Y. Kochetov et al. (Eds.): DOOR 2016, LNCS 9869, pp. 159–170, 2016.
DOI: 10.1007/978-3-319-44914-2_13

So, the efforts of most researchers are concentrated on finding cases when the problem can be solved in polynomial time and developing polynomial approximation algorithms for the TSP and the m-PSP. A review of the most significant results in this area can be found in [13,25].

At the moment, the problem is studied both for deterministic and random instances, for arbitrary [1,18], metric [2,4,6,20–22,24] and Euclidean [3,14] weight functions which may be identical or different for all cycles, and for special cases of the problem where edge weights belong to a given interval or a finite set of numbers [15–17,19].

In particular, for the symmetric version of the problem the following results were established. In [1,18] two polynomial algorithms for the maximization version of the 2-PSP with approximation ratios 3/4 and 7/9 were designed. The authors of [16] present a series of polynomial approximation algorithms for the minimization version of the 2-PSP with edge weights 1 and 2, where the weight function is common for both Hamiltonian cycles, while in [17,19] the same problem was studied for two different weight functions. In this case two polynomial algorithms with approximation ratios 7/5 and 4/3 were developed. For the Euclidean maximum-weight m-PSP an asymptotically exact algorithm with time complexity $O(n^3)$ was designed [3].

Applications include the design of patrol tours [5] where it is often important to assign a set of edge disjoint tours to the watchman in order to avoid constant repetition of the same tour and thus enhance security. De Kort [10] cites a network design application where, in order to protect the network from link failure, several edges-disjoint cycles must be determined. He also mentions a scheduling application of the 2-PSP where each job must be processed twice by the same machine but technological constraints prevent the repetition of identical job sequences.

Here we concentrate on the metric variant of the m-PSP (with identical weight functions) where the edge weights satisfy *the triangle inequality*. For the metric minimization version of the TSP, the best known result is the algorithm with approximation ratio 3/2 which was independently designed by Christofides [6] and Serdyukov [24]. For the minimization version of the 2-PSP, Ageev and Pyatkin [2] developed an algorithm with approximation ratio 2. What concerns the metric maximization version of the TSP, Kostochka and Serdyukov [21] in 1985 proposed a 5/6-approximation algorithm, while the best known approximation ratio for this problem is 7/8 by Kowalik and Mucha [22]. For the maximization version of the 2-PSP, Gordeeva [20] presented two polynomial algorithms with asymptotical approximation ratios 5/6 and 11/16 for the cases of identical and different weight functions respectively.

In this paper we extend the result of [21] and improve the first result in [20] by constructing a polynomial algorithm for the metric maximization version of the m-PSP with identical weight functions $w_1 = \ldots = w_m = w$. The algorithm has an exact (non-asymptotical) approximation ratio 5/6 and deals with arbitrary many Hamiltonian cycles, $m \geq 2$, provided that the number of vertices of G is sufficiently large ($n \geq 36m - 28$).

2   Preliminary Definitions and Notation

We consider a complete n-vertex graph $G = G(V, E)$ with the vertex set $V = V(G)$ and the edge set $E = E(G)$. Let $w : E \to \mathbf{R}_+$ be an arbitrary non-negative weight function of the edges of G. By $w(OPT)$ we denote the weight of the optimal solution of the maximization version of the m-PSP for G.

Suppose $K = K(V_K, E_K)$ is an arbitrary graph (which can be a subgraph of G); $v \in V_K$ is a vertex in K. We use the following notation:

• $d_K(v) = d(v)$ is *the degree of v* in K (the number of edges of K incident with v).

• $\Delta(K)$ is *the maximum degree of K*;

• $\delta(K)$ is *the minimum degree of K*;

• A 2-*factor (cycle cover)* of K is a collection of vertex-disjoint cycles covering all vertices of K.

• A *partial tour* in K is a collection of vertex-disjoint paths covering all vertices of K.

For a partial tour T we introduce the following notation:

• $P(T)$ is the set of all paths of T;

• $p(T) = |P(T)|$ is the number of paths in T;

• $|T|$ is the number of edges in T;

• $L(T)$ is the set of all edges in K joining end-points of paths in T.

Clearly, $|T| + p(T) = |V_K|$ for any partial tour T in K. A path of T is *trivial* if it consists of just one vertex and *non-trivial* otherwise.

Suppose F is a 2-factor in K. Let R be a subset of edges in F which contains at least one edge from each cycle of F. Then $T = F - R$ is a partial tour in K such that $p(T) = |R|$ and $w(T) = w(F) - w(R)$. Moreover, if R is a matching in F, then T consists of non-trivial paths. On the other hand, if T is a partial tour in K, then by adding an appropriate set of edges $A \subset L(T)$ with $|A| = p(T)$ we produce a Hamiltonian cycle $H = T \cup A$ in K whose total weight is $w(H) = w(T) + w(A)$.

In what follows we assume that some edges of the graph G can be *bad*. By B we denote the set of all bad edges in G and by $d_B(v)$ denote the 'bad degree' of a vertex v, i.e. the number of bad edges incident with v. We assume that all edges in $E(G) \setminus B$ are *good*.

3   Some Algorithms for Constructing Hamiltonian Cycles

We start with describing three auxiliary procedures for finding Hamiltonian cycles in graphs with high minimum degree. All the procedures are based on the following well-known

Theorem 1 *(Dirac's Theorem [11]). Every graph K with $k \geq 3$ vertices and with minimum degree at least $k/2$ contains a Hamiltonian cycle.*

Let $\bar{K}$ be the complement of K. We say that edges of K are *good* while edges of $\bar{K}$ are supposed to be *bad*. The condition of Dirac's Theorem implies that every vertex v of K satisfies $d_B(v) \leq k/2 - 1$. For every graph K meeting this requirement, the Hamiltonian cycle can be found by the following procedure which can be derived from the proof of Dirac's Theorem.

Procedure *Dirac's HC*:

Step 0. Start with an arbitrary path $P = v_0v_1 \ldots v_t$ in K.

Step 1. If P is a Hamiltonian path, go to Step 3, otherwise — to Step 2.

Step 2. If v_0 or v_t is joined by a good edge to a vertex $x \notin P$, add this edge to P and go to Step 1. Otherwise, proceed to Step 3.

Step 3. Transform P to a cycle C applying the following method. If v_0v_t is a good edge, then set $C := P \cup \{v_0v_t\}$. Otherwise, find an index $i \in [2, t-1]$ such that both edges v_0v_i and $v_{i-1}v_t$ are good and set $C := P \cup \{v_0v_i, v_{i-1}v_t\} \setminus \{v_{i-1}v_i\}$. If C is a Hamiltonian cycle in K, then present C as a solution. Otherwise, proceed to Step 4.

Step 4. Suppose $C = (v_0v_1 \ldots v_tv_0)$. Find a vertex $v_j \in C$ which is joined by a good edge to a vertex $x \notin C$. Construct a path $P = xv_jv_{j+1} \ldots v_{j-1}$ and go to Step 1.

The existence of the index i at Step 3 and of the good edge v_jx at Step 4 follows from Dirac's degree condition $\delta(K) \geq k/2$ (which is equivalent to $d_B(v) \leq k/2 - 1$ for every vertex v). Clearly, the running-time of the procedure 'Dirac's HC' is $O(k^2)$.

Our next goal is to develop a modification of this procedure for the case when we want to produce a Hamiltonian cycle by joining paths of a given partial tour. Suppose T is a partial tour in a complete graph G and T consists of k non-trivial paths. Assume that $L(T)$ contains a subset B of bad edges and all edges in $D = L(T) \setminus B$ are good.

The following procedure 'THC' transforms T to a Hamiltonian cycle H by adding good edges, i.e. it finds a subset $A \subset D$ such that $H = T \cup A$ is a Hamiltonian cycle in G. At every stage the procedure constructs a path or a cycle of the type $Q = a_1P_1b_1a_2P_2b_2a_3 \ldots b_{t-1}a_tP_tb_t$, where $P_1 = a_1 \ldots b_1$, $P_2 = a_2 \ldots b_2, \ldots, P_t = a_t \ldots b_t$ are pairwise distinct paths of T and b_1a_2, $b_2a_3, \ldots, b_{t-1}a_t$ are good edges (if Q is a cycle, then the edge b_ta_1 is also good). The procedure runs correctly if every end-vertex v of a path in T satisfies $d_B(v) \leq k/2$.

Procedure *THC*:

Step 0. Start with an arbitrary path $Q = a_1P_1b_1a_2P_2b_2a_3 \ldots b_{t-1}a_tP_tb_t$, where $P_1, P_2, \ldots, P_t \in P(T)$; $b_1a_2, b_2a_3, \ldots, b_{t-1}a_t \in D$. (For example, set $Q = P_1 = a_1 \ldots b_1 \in P(T)$, $t = 1$.)

Step 1. If Q is a Hamiltonian path, go to Step 3, otherwise — to Step 2.

Step 2. If b_t or a_1 is joined by a good edge to an end-vertex a of a path $P = a \ldots b \in P(T)$ which is not contained in Q, then add the edge b_ta (or a_1a) and the path P to Q and go to Step 1. Otherwise, proceed to Step 3.

Step 3. Transform Q to a cycle C in the following way. If a_1b_t is a good edge, then set $C := Q \cup \{a_1b_t\}$. Otherwise, find an index $i \in [2, t]$ such that both edges a_1a_i and $b_{i-1}b_t$ are good and set $C := Q \cup \{a_1a_i, b_{i-1}b_t\} \setminus \{b_{i-1}a_i\}$. If C is a Hamiltonian cycle in G, then present C as a solution. Otherwise, proceed to Step 4.

Step 4. Suppose $C = (a_1 P_1 b_1 a_2 P_2 b_2 a_3 \ldots b_{t-1} a_t P_t b_t a_1)$. Find a vertex $x \in \{a_1, b_1, a_2, b_2, \ldots, a_t, b_t\}$ which is joined by a good edge to an end-vertex b of a path $P = a \ldots b \in P(T)$, where P is not contained in C. W.l.o.g. assume that $x = a_i$ for some $i \in \{1, \ldots, t\}$. Construct a path $Q = a P b a_i P_i b_i a_{i+1} \ldots a_{i-1} P_{i-1} b_{i-1}$ and go to Step 1.

Similarly to the procedure Dirac's *HC*, the existence of the index i at Step 3 and of the good edge ab at Step 4 follows from the 'bad degree condition' $d_B(v) \le k/2$ for end-vertices of T. The running-time of the procedure '*THC*' is $O(k^2)$.

We conclude this section by showing that if bad degrees of end-vertices in T satisfy stronger restrictions than in procedure '*THC*', then we can construct a cyclic ordering $P_1, P_2, \ldots, P_k$ of the paths in T such that for any two consecutive paths P_i, P_{i+1} (and for P_k, P_1) the end-vertices of these two paths are joined only by good edges. The following procedure runs correctly provided that $k \ge 2$ and $d_B(v) \le (k-2)/4$ for any end-vertex v of a path in T.

Procedure *Path Order T*:

Step 1. Define a graph K with the vertex set $P(T)$, where vertices P and Q are adjacent iff the end-points of the paths P and Q are joined only by good edges in G.

Step 2. By applying the procedure Dirac's *HC* to K produce a Hamiltonian cycle $H = (P_1, P_2, \ldots, P_k, P_1)$ in K. The vertex ordering $P_1, P_2, \ldots, P_k$ of H is the desired path ordering of T.

In order to establish the correctness of applying Dirac's *HC* at Step 2 notice that each vertex $P = a \ldots b$ of K satisfies

$$d_B(P) \le d_B(a) + d_B(b) \le 2 \times (k-2)/4 = k/2 - 1.$$

The procedure 'Path Order T' will be a helpful tool for us in constructing our algorithm since it allows to implement the main procedure from the algorithm by Kostochka and Serdyukov.

4   The Key Procedure by Kostochka and Serdyukov

In this section we represent the main procedure of the algorithm by Kostochka and Serdyukov for the metric maximization version of the TSP [21].

Suppose $G(V, E)$ is a complete n-vertex graph; $w : E \to R^+$ is a weight function of its edges satisfying the triangle inequality. Let T be a partial tour in G consisting of non-trivial paths $P_1 = a_1 \ldots b_1$, $P_2 = a_2 \ldots b_2$, $\ldots, P_k = a_k \ldots b_k$. The following procedure '*KS*' transforms T to a Hamiltonian cycle $H = T \cup A$ by adding a set of edges $A \subset L(T)$ with the property $w(A) \ge \frac{1}{2} \sum_{i=1}^{k} w(a_i b_i)$.

Procedure KS:

Step 1. If $k = 1$, set $H := T \cup \{a_1b_1\}$. If $k = 2$, then set $A_1 = \{a_1b_2, b_1a_2\}$, $A_2 = \{a_1a_2, b_1b_2\}$. Define A to be a set with the maximum weight among A_1, A_2 and construct $H := T \cup A$. If $k \geq 3$ and k is even, go to Step 2; and if k is odd — to Step 3.

Step 2. Set

$$A_1 = A_2 = \{b_1a_2, b_2a_3, \ldots, b_ka_1\},$$

$$A_3 = \{a_1a_2, a_3a_4, \ldots, a_{k-1}a_k, b_2b_3, b_4b_5, \ldots, b_kb_1\},$$

$$A_4 = \{a_2a_3, a_4a_5, \ldots, a_ka_1, b_1b_2, b_3b_4, \ldots, b_{k-1}b_k\},$$

and go to Step 4.

Step 3. Set

$$A_1 = \{b_1a_2, b_2a_3, \ldots, b_ka_1\},$$

$$A_2 = A_1 \setminus \{b_1a_2, b_ka_1\} \cup \{a_1a_2, b_kb_1\},$$

$$A_3 = \{b_1a_2, a_3a_4, a_5a_6, \ldots, a_ka_1, b_2b_3, b_4b_5, \ldots, b_{k-1}b_k\},$$

$$A_4 = \{a_2a_3, a_4a_5, \ldots, a_{k-1}a_k, b_1b_2, b_3b_4, \ldots, b_{k-2}b_{k-1}, b_ka_1\}$$

and go to Step 4.

Step 4. Define A to be a set with the maximum weight among A_1, A_2, A_3, A_4. Set $H := T \cup A$.

It was proved in [21] that the Hamiltonian cycle H constructed by the procedure 'KS' satisfies $w(H) \geq w(T) + \frac{1}{2} \sum_{i=1}^{k} w(a_ib_i)$.

Another important observation concerning the procedure 'KS' is that for any $k \geq 2$ the set A constructed by it consists only of edges joining the end-vertices of consecutive paths P_i, P_{i+1} (or P_k, P_1) in T. This implies that if the ordering $P_1, P_2, \ldots, P_k$ is produced by the procedure 'Path Order T' from the previous section, then A consists of good edges only.

5   Processing Long and Short Cycles

Before we present our main algorithm, we introduce two auxiliary procedures. The first procedure removes some light edges from a given 2-factor F^L consisting of 'long' cycles and produces a set T^L consisting of non-trivial paths. The second procedure converts a 2-factor F^S consisting of 'short' cycles to path collections T' and T'' with special properties by removing edges with the minimum weight from cycles of F^S.

Suppose F is a 2-factor with cycles $C_1, C_2, \ldots, C_t$ in a complete weighted n-vertex graph G. By $l_j = l(C_j)$ denote the *length* of a cycle C_j, i.e. the number of edges in C_j. We say that C_j is *short* if $l_j \leq 5$, and *long* otherwise. Denote by F^S and F^L the sets of all short and long cycles in F, respectively. W.l.o.g. assume that $F^L = \{C_1, \ldots, C_q\}$, $F^S = \{C_{q+1}, \ldots, C_t\}$ (where $q = 0$ if F consists only of short cycles). Set $l^L = \sum_{j=1}^{q} l_j$ and $l^S = \sum_{j=q+1}^{t} l_j$. Observe that $l^L + l^S = n$.

The following procedure 'Remove M' removes a matching M with at least $\lfloor l^L/6 \rfloor$ edges having total weight $w(M) \le \frac{1}{6}w(F^L)$ from F^L. The important property of M is that it contains at least one edge from each cycle of F^L. Therefore, $T^L = F^L \setminus M$ is a collection of $|M|$ non-trivial paths in G whose total weight is at least $\frac{5}{6}w(F^L)$.

The procedure 'Remove M' starts by colouring edges of F^L with colours from the set $\{1, 2, \ldots, 6\}$. The colouring is *periodic* in the sense that the order of colours used is $1, 2, \ldots, 6, 1, 2 \ldots, 6, 1, 2 \ldots$ without changing it after finishing to colour the edges of a cycle C_j and proceeding to C_{j+1}.

Procedure *Remove M*:

Step 1. Colour periodically the edges of C_1 in a cyclic order. If the colours of e_1 and e_{l_1} coincide, switch the colours of e_1 and e_2. After that proceed by colouring the edges of $C_2, \ldots, C_q$ in the similar way.

Step 2. Let M^i, $i = 1, \ldots 6$, be the set of all edges of colour i in F^L. Define M to be a set with the minimum weight among $M^1, M^2, \ldots, M^6$.

Step 3. Remove M from F^L. Denote the resulting collection of paths by $T^L = F^L \setminus M$.

By the definition of a periodic colouring, it follows that the sets $M^1, M^2, \ldots, M^6$ are edge disjoint matchings forming the partition of $E(F^L)$. Thus, $w(M) \le \frac{1}{6}w(F^L)$. Furthermore, it follows that every M^i contains at least one edge from each cycle C_j of F^L since at Step 1 we use the periodic 6-colouring and $l_j \ge 6$ for $j = 1, \ldots, q$. One can easily observe that $|M^i| \in \{\lfloor l^L/6 \rfloor, \lfloor l^L/6 \rfloor + 1\}$ for $i = 1, \ldots, 6$. So, we have $|M| \ge \lfloor l^L/6 \rfloor$ as desired.

Next we present a procedure which transforms a partial 2-factor F^S consisting of short cycles $C_{q+1}, \ldots, C_t$ to collections of paths T' and T''. Let $s = t - q$ be the number of cycles in F^S and let e_j be the edge of the minimum weight in C_j. Denote by P_j the non-trivial path $C_j \setminus e_j$, $j = q+1, \ldots, t$.

Procedure *Split F^S*:

Step 1. Order the cycles of F^S so that $w(e_{q+1}) \le w(e_{q+2}) \le \ldots \le w(e_t)$.

Step 2. Set $s' = \lfloor l^S/3 \rfloor - s$; $T' = \{P_{q+1}, \ldots, P_{q+s'}\}$; $T'' = \{P_{q+s'+1}, \ldots, P_t\}$.

The main property of the procedure 'Split F^S' is stated by the following

Lemma 1. $\displaystyle\sum_{j=q+1}^{q+s'} w(e_j) + \frac{1}{2} \sum_{j=q+s'+1}^{t} w(e_j) \le \frac{1}{6}w(F^S)$.

Proof. The idea of the proof is to define pairwise disjoint subsets of edges $S_{q+1}, \ldots, S_t$ in $E(F^S)$ such that $|S_j| = 6$ if $j \le q + s'$, $|S_j| = 3$ if $j > q + s'$ and the weight of every edge in S_j is at least $w(e_j)$ for all $j = q+1, \ldots, t$. This would immediately imply

$$w(F^S) \ge \sum_{j=q+1}^{q+s'} w(S_j) + \sum_{j=q+s'+1}^{t} w(S_j) \ge 6 \sum_{j=q+1}^{q+s'} w(e_j) + 3 \sum_{j=q+s'+1}^{t} w(e_j)$$

which gives the desired inequality.

For $j = q + s' + 1, \ldots, t$, define S_j as an arbitrary subset of cardinality 3 in $E(C_j)$. For $j = q + 1, \ldots, q + s'$, set $S_j = E(C_j) \cup S'_j$, where S'_j is an arbitrary subset of cardinality $6 - l_j$ in $\bigcup_{k=q+s'+1}^{t} E(C_k) \setminus S_k$. By this definition and the condition $w(e_{q+1}) \leq w(e_{q+2}) \leq \ldots \leq w(e_t)$ it follows that for every $j = q + 1, \ldots, t$ each edge $e \in S_j$ is contained in some $E(C_k)$ with $k \geq j$, and hence we have $w(e) \geq w(e_k) \geq w(e_j)$.

It remains to check that we can choose edge disjoint subsets $S'_{q+1}, \ldots, S'_{q+s'}$ in $\bigcup_{k=q+s'+1}^{t} E(C_k) \setminus S_k$, i.e. that $\sum_{j=q+1}^{q+s'} (6 - l_j) \leq \sum_{j=q+s'+1}^{t} (l_j - 3)$. This follows by the choice of s':

$$\sum_{j=q+s'+1}^{t} (l_j - 3) - \sum_{j=q+1}^{q+s'} (6 - l_j) = l^S - 3(s - s') - 6s' = l^S - 3(s + s') = l^S - 3\lfloor l^S/3 \rfloor \geq 0.$$

6 Algorithm $A_{5/6}$ for the Metric Maximum m-PSP

Now we are ready to formulate and analyze the Algorithm $A_{5/6}$ for the metric maximization version of the m-PSP.

First, we give a brief outline of the algorithm. At the preliminary stage, by successively applying Gabow's algorithm [12] and a 2-factorization algorithm, we construct a collection of edge disjoint 2-factors $F_1, F_2, \ldots, F_m$ in G whose total weight is at least $w(OPT)$. After that, for each $i = 1, \ldots, m$, we transform F_i to a Hamiltonian cycle H_i with the total weight $w(H_i) \geq \frac{5}{6} w(F_i)$ such that H_i has no edge in common with already constructed $H_1, \ldots, H_{i-1}$ nor with remaining 2-factors $F_{i+1}, \ldots, F_m$. So we treat the edges of the cycles $H_1, \ldots, H_{i-1}$ and 2-factors $F_{i+1}, \ldots, F_m$ as bad (in the sense that they cannot be added to H_i). Hence at every moment each vertex $v \in V$ is incident with $2(m - 1)$ bad edges, i.e. it has $d_B(v) = 2(m - 1)$.

We subdivide each F_i to subsets F_i^S and F_i^L of short and long cycles respectively. We start processing F_i by applying 'Remove M' to F_i^L which gives a collection of paths T_i^L and by applying 'Split F^S' to F_i^S which produces path collections T_i' and T_i''. For F_i, F_i^L, F_i^S, T_i^L, T_i' and T_i'' we introduce the parameters t, q, l^L, l^S, s and s' as described in Sect. 5 (omitting the subscript i for simplicity).

If the path collection $\widetilde{T}_i = T_i^L \cup T_i'$ contains at least $4(m - 1)$ paths, then we proceed by applying the procedure by Gordeeva [20] which appends all paths in T_i'' to paths of $\widetilde{T}_i$ (Step 5). After that we construct H_i by applying the procedure 'THC' to the modified $\widetilde{T}_i$ (Step 6). If $\widetilde{T}_i$ initially contains less than $4(m - 1)$ paths, we construct H_i by successively applying the procedures 'Order T' and 'KS' to the partial tour $T_i = T_i^L \cup T_i' \cup T_i''$ (Step 7).

Now we give a detailed description of the algorithm.

Algorithm $A_{5/6}$:

Step 1. Applying Gabow's algorithm [12] find a $2m$-regular subgraph G_{2m} in G having maximum weight. Note that $w(G_{2m}) \geq w(OPT)$. The running-time of Step 1 is $O(mn^3)$.

Step 2. Split the edge set of G_{2m} to edge disjoint 2-factors $F_1, F_2, \ldots, F_m$ using the well-known 2-factorization algorithm with running-time $O(m^2n^2)$.

Successively, for each $i = 1, \ldots, m$, transform F_i to a Hamiltonian cycle H_i applying the method described below. Start from Step 3.

Step 3. Define the set of bad edges $B_i := E(H_1 \cup \ldots \cup H_{i-1} \cup F_{i+1} \cup \ldots \cup F_m)$. Subdivide F_i to sets $F_i^L = \{C_1, \ldots, C_q\}$ and $F_i^S = \{C_{q+1}, \ldots, C_t\}$ consisting of long and short cycles respectively. For each $j = q + 1, \ldots, t$, denote by $e_j = a_j b_j$ the edge in C_j with the minimum weight.

Step 4. By applying 'Remove M' to F_i^L produce a path collection T_i^L consisting of at least $\lfloor l^L/6 \rfloor$ non-trivial paths. By applying 'Split F^S' to F_i^S produce path collections T_i' and T_i''. Set $\widetilde{T}_i = T_i^L \cup T_i'$, $\nu = p(\widetilde{T}_i)$. If $\nu \geq 4(m-1)$, go to Step 5, otherwise — to Step 7.

Step 5. Successively, for each $j = q + s' + 1, \ldots, t$, append the path P_j to $\widetilde{T}_i$ using the following method. Find a path $P = a \ldots b$ in $\widetilde{T}_i$ and its end-vertex a such that both aa_j and ab_j are good edges. Let e_j' be an edge with the maximum weight among aa_j, ab_j. Append the path P_j to P by inserting the edge e_j'. Observe that $w(e_j') \geq \frac{1}{2}w(e_j)$ due to the choice of e_j' and the triangle inequality.

Step 6. Construct the Hamiltonian cycle H_i by applying the procedure 'THC' to the path collection $\widetilde{T}_i$ obtained at the end of Step 5.

Step 7. Define the partial tour $T_i = T_i^L \cup T_i' \cup T_i''$. Apply 'Path Order T' to T_i. Finally, construct H_i by applying the procedure 'KS' to T_i.

Our main result is the following

Theorem 2. *Let $G(V, E)$ be a complete n-vertex graph and $w : E \to R^+$ be a weight function satisfying the triangle inequality. Suppose $n \geq 36m - 28$. Then algorithm $A_{5/6}$ constructs m edge disjoint Hamiltonian cycles $H_1, H_2, \ldots, H_m$ in G such that $w(H_1) + w(H_2) + \ldots + w(H_m) \geq \frac{5}{6}w(OPT)$. The running-time of the algorithm is $O(mn^3)$.*

Proof. First we examine that all the procedures involved in the algorithm are implemented correctly.

The existence of the path $P = a \ldots b$ and its end-vertex a with desired properties at Step 5 follows from the inequality $\nu \geq 4(m-1)$ since $\widetilde{T}_i$ contains ν paths and $2\nu \geq 8(m-1)$ end-vertices while a_j and b_j are incident only with $d_B(a_j) + d_B(b_j) = 2 \times 2(m-1) = 4(m-1) < 8(m-1)$ bad edges. The procedure 'THC' at Step 6 runs correctly because the bad degree of each vertex v satisfies $d_B(v) = 2(m-1) = \frac{1}{2} \times 4(m-1) \leq \nu/2$.

Finally, we have to establish the correctness of applying 'Path Order T' at Step 7. According to the condition formulated in Sect. 3 and taking into account the equality $d_B(v) = 2(m-1)$, we need to prove that $2(m-1) \leq (k-2)/4$, i.e. that $k \geq 8m - 6$, where $k = p(T_i)$. Assume that $8m - 7 \geq k = p(T_i^L) + s \geq \lfloor l^L/6 \rfloor + s$.

By the conditions of Step 7, we have $4m - 5 \geq \nu = p(T_i^L) + s' \geq \lfloor l^L/6 \rfloor + s'$. Combining the last two inequalities we get:

$$12m - 12 \geq 2\lfloor l^L/6 \rfloor + s + s' = 2\lfloor l^L/6 \rfloor + \lfloor l^S/3 \rfloor \geq 2 \cdot \frac{l^L - 5}{6} + \frac{l^S - 2}{3},$$

which implies $36m - 29 \geq l^S + l^L = n$ yielding a contradiction.

Next we prove the inequality $w(H_1) + w(H_2) + \ldots + w(H_m) \geq \frac{5}{6}w(OPT)$. Notice that $w(F_1) + w(F_2) + \ldots + w(F_m) \geq w(OPT)$ by the result of Steps 1 and 2. So, it suffices to check that $w(H_i) \geq \frac{5}{6}w(F_i)$ for every $i = 1, \ldots, m$.

Clearly, $w(F_i) = w(F_i^S) + w(F_i^L)$. Observe that $w(T_i') + w(T_i'') = w(F_i^S) - \sum_{j=q+1}^{t} w(e_j)$ and $w(T_i^L) \geq \frac{5}{6}w(F_i^L)$ by the main property of 'Remove M'.

If the Hamiltonian cycle H_i is constructed at Steps 5 and 6, then since $w(e_j') \geq \frac{1}{2}w(e_j)$ for $j = q + s' + 1, \ldots, t$ and by Lemma 1, after Step 5 we have

$$w(\widetilde{T}_i) = w(T_i^L) + w(T_i') + w(T_i'') + \sum_{j=q+s'+1}^{t} w(e_j') \geq \frac{5}{6}w(F_i^L) + w(T_i') + w(T_i'')$$

$$+ \frac{1}{2}\sum_{j=q+s'+1}^{t} w(e_j) = \frac{5}{6}w(F_i^L) + w(F_i^S) - \sum_{j=q+1}^{q+s'} w(e_j) - \frac{1}{2}\sum_{j=q+s'+1}^{t} w(e_j)$$

$$\geq \frac{5}{6}w(F_i^L) + \frac{5}{6}w(F_i^S) = \frac{5}{6}w(F_i).$$

Hence, after Step 6 we have $w(H_i) \geq w(\widetilde{T}_i) \geq \frac{5}{6}w(F_i)$.

Suppose H_i is constructed at Step 7. In this case we have $w(T_i) = w(T_i^L) + w(T_i') + w(T_i'')$. Taking into account the main property of the procedure 'KS', similarly to the previous case, we obtain

$$w(H_i) \geq w(T_i^L) + w(T_i') + w(T_i'') + \frac{1}{2}\sum_{j=q+1}^{t} w(e_j) \geq \frac{5}{6}w(F_i^L) + w(T_i')$$

$$+ w(T_i'') + \frac{1}{2}\sum_{j=q+s'+1}^{t} w(e_j) \geq \frac{5}{6}w(F_i).$$

It remains to notice that the time complexity of the algorithm $A_{5/6}$ is determined by the implementation of Gabow's algorithm which finds a $2m$-regular subgraph G_{2m} at Step 1. So the running-time of $A_{5/6}$ is $O(mn^3)$. This completes the proof of Theorem 2.

The authors want to thank Prof. E. Kh. Gimadi for fruitful discussions and interesting ideas concerning the m-PSP.

References

1. Ageev, A.A., Baburin, A.E., Gimadi, E.: A 3/4 approximation algorithm for finding two disjoint Hamiltonian cycles of maximum weight. Diskretn. Anal. Issled. Oper. Ser. 1 **13**(2), 11–20 (2006). (in Russian)
2. Ageev, A.A., Pyatkin, A.V.: A 2-approximation algorithm for the metric 2-peripatetic salesman problem. Diskretn. Anal. Issled. Oper. **16**(4), 3–20 (2009). (in Russian)
3. Baburin, A.E., Gimadi, E.Kh.: On the asymptotic optimality of an algorithm for solving the maximization version of the m-PSP in a multidimensional euclidean space. Proc. Steklov Inst. Math. **272**(1), 1–13 (2011)
4. Baburin, A.E., Gimadi, E.Kh., Korkishko, N.M.: Approximation algorithms for finding two edge-disjoint hamiltonian cycles of minimal total weight. Diskretn. Anal. Issled. Oper. Ser 2 **11**(1), 11–25 (2004). (in Russian)
5. Wolfter Calvo, R., Cordone, R.: A heuristic approach to the overnight security service problem. Comput. Oper. Res. **30**, 1269–1287 (2003)
6. Christofides, N.: Worst-case analysis of a new heuristic for the traveling salesman problem. Technical report CS-93-13, Carnegie Mellon University (1976)
7. De Brey, M.J.D., Volgenant, A.: Well-solved cases of the 2-peripatetic salesman problem. Optimization **39**(3), 275–293 (1997)
8. De Kort, J.B.J.M.: Upper bounds for the symmetric 2-PSP. Optimization **23**(4), 357–367 (1992)
9. De Kort, J.B.J.M.: Lower bounds for symmetric K-PSP. Optimization **22**(1), 113–122 (1991)
10. De Kort, J.B.J.M.: A branch and bound algorithm for symmetric 2-PSP. EJOR **70**, 229–243 (1993)
11. Dirac, G.A.: Some theorems on abstract graphs. Proc. London Math. Soc. **3**(2), 69–81 (1952)
12. Gabow, H.N.: An efficient reduction technique for degree-restricted subgraph and bidirected network flow problems. In: Proceedings of 15th Annual ACM Symposium on Theory of Computing, pp. 448–456. ACM, New York (1983)
13. Gimadi, E.Kh.: Approximation efficient algorithms with performance guarantees for some hard routing problems. In: Proceedings of II International Conference 'Optimzation and Applications' OPTIMA-2011, pp. 98–101. Petrovac, Montenegro (2011)
14. Gimadi, E.Kh.: Asymptotically optimal algorithm for finding one and two edge-disjoint traveling salesman routes of maximal weight in euclidean space. Proc. Steklov Inst. Math. **263**(2), 56–67 (2008)
15. Gimadi, E.Kh., Ivonina, E.V.: Approximation algorithms for the maximum 2-peripatetic salesman problem. Diskretn. Anal. Issled. Oper. **19**(1), 17–32 (2012). (in Russian)
16. Gimadi, E.Kh., Glazkov, Yu.V., Glebov, A.N.: Approximation algorithms for solving the 2-peripatetic salesman problem on a complete graph with edge weights 1 and 2. Diskretn. Anal. Issled. Oper. Ser 2. **14**(2), 41–61 (2007). (in Russian)
17. Glebov, A.N., Gordeeva, A.V., Zambalaeva, D.Zh.: An algorithm with approximation ratio 7/5 for the minimum two peripatetic salesmen problem with different weight functions. Sib. Electron. Math. Izv. **8**, 296–309 (2011). (in Russian)
18. Glebov, A.N., Zambalaeva, D.Zh.: A polynomial algorithm with approximation ratio 7/9 for the maximum two peripatetic salesmen problem. Diskretn. Anal. Issled. Oper. **18**(4), 17–48 (2011). (in Russian)

19. Glebov, A.N., Zambalaeva, D.Zh.: An approximation algorithm for the minimum two peripatetic salesmen problem with different weight functions. Diskretn. Anal. Issled. Oper. **18**(5), 11–37 (2011). (in Russian)
20. Gordeeva, A.V.: Polynomial algorithms with performance garauntees for the metric maximization version of the 2-PSP. Specialist's Thesis, 22 p. Novosibirsk State University, Novosibirsk (2010). (in Russian)
21. Kostochka, A.V., Serdyukov, A.I.: Polynomial algorithms with approximation ratios 3/4 and 5/6 for the maximization version of the TSP. Upravlyaemye Systemy. Sobolev Institute of Mathematics, Novosiirsk, vol. 26, pp. 55–59 (1985). (in Russian)
22. Kowalik, L., Mucha, M.: Deterministic 7/8-approximation for the metric maximum TSP. Theor. Comput. Sci. **410**(47–49), 5000–5009 (2009)
23. Krarup, J.: The peripatetic salesman and some related unsolved problems. In: Combinatorial Programming: Methods and Applications (Proceedings of NATO Advanced Study Inst., Versailles, 1974), pp. 173–178. Reidel, Dordrecht (1975)
24. Serdyukov, A.I.: On some extremal tours in graphs. Upravlyaemye Systemy. Sobolev Institute of Mathematics, Novosiirsk, vol. 17, pp. 76–79 (1984). (in Russian)
25. Gutin, G., Punnen, A.P. (eds.): The Traveling Salesman Problem and its Variations, 830 p. Kluver Academic Publishers, Dordrecht (2002)

An Approximation Algorithm for a Problem of Partitioning a Sequence into Clusters with Restrictions on Their Cardinalities

Alexander Kel'manov[1,2], Ludmila Mikhailova[1], Sergey Khamidullin[1], and Vladimir Khandeev[1,2(✉)]

[1] Sobolev Institute of Mathematics, 4 Koptyug Ave., 630090 Novosibirsk, Russia
{kelm,mikh,kham,khandeev}@math.nsc.ru
[2] Novosibirsk State University, 2 Pirogova St., 630090 Novosibirsk, Russia

Abstract. We consider the problem of partitioning a finite sequence of points in Euclidean space into a given number of clusters (subsequences) minimizing the sum of squared distances between cluster elements and the corresponding cluster centers. It is assumed that the center of one of the desired clusters is the origin, while the centers of the other clusters are unknown and determined as the mean values over clusters elements. Additionally, there are a few structural restrictions on the elements of clusters with unknown centers: (1) clusters form non-overlapping subsequences of the input sequence, (2) the difference between two consecutive indices is bounded from below and above by prescribed constants, and (3) the total number of elements in these clusters is given as an input. It is shown that the problem is strongly NP-hard. A 2-approximation algorithm which runs in polynomial time for a fixed number of clusters is proposed for this problem.

Keywords: Clustering · Structural constraints · Euclidean space · Minimum sum-of-squared distances · NP-hardness · Guaranteed approximation factor

1 Introduction

The subject of this study is a problem of partitioning a finite sequence of points in Euclidean space into subsequences. Our goal is to find out the computational complexity of the problem and to provide a polynomial-time factor-2 approximation algorithm.

The research is motivated by insufficient study of the problem and its relevance, in particularly, to problems of approximation, clustering, sequence (time series) analysis as well as to many natural science and engineering applications that require classification of results of chronologically sorted numerical experiments and observations on the state of some objects (see, for example, [1–4] and references therein). Some applications (sources) of the problem are presented in the next section.

© Springer International Publishing Switzerland 2016
Y. Kochetov et al. (Eds.): DOOR 2016, LNCS 9869, pp. 171–181, 2016.
DOI: 10.1007/978-3-319-44914-2_14

This is the incremental work to the results previously obtained in [5–7]. Each of the cited works is an essential building-block in the algorithm presented in this work — the first algorithm with a guaranteed approximation factor.

2 Problem Formulation, Complexity, and Related Problems

Everywhere below $\mathbb{R}$ denotes the set of real numbers, $\|\cdot\|$ denotes the Euclidean norm, and $\langle\cdot,\cdot\rangle$ denotes the scalar product.

Formally, we consider the following problem.

Problem 1. *Given* a sequence $\mathcal{Y} = (y_1, \ldots, y_N)$ of points from $\mathbb{R}^q$ and some positive integers $T_{\min}, T_{\max}, L$, and M. *Find* nonempty disjoint subsets $\mathcal{M}_1, \ldots, \mathcal{M}_L$ of $\mathcal{N} = \{1, \ldots, N\}$, i.e. subsets of indices of the elements from the sequence $\mathcal{Y}$, such that

$$F(\mathcal{M}_1, \ldots, \mathcal{M}_L) = \sum_{l=1}^{L} \sum_{j \in \mathcal{M}_l} \|y_j - \overline{y}(\mathcal{M}_l)\|^2 + \sum_{i \in \mathcal{N} \setminus \mathcal{M}} \|y_i\|^2 \longrightarrow \min, \qquad (1)$$

where $\mathcal{M} = \bigcup_{l=1}^{L} \mathcal{M}_l$, and $\overline{y}(\mathcal{M}_l) = \frac{1}{|\mathcal{M}_l|} \sum_{j \in \mathcal{M}_l} y_j$ is the centroid of subset $\{y_j \,|\, j \in \mathcal{M}_l\}$, under the following constraints: (i) the cardinality of $\mathcal{M}$ is equal to M, (ii) concatenation of elements of subsets $\mathcal{M}_1, \ldots, \mathcal{M}_L$ is an increasing sequence, provided that the elements of each subset are in ascending order, (iii) the following inequalities for the elements of $\mathcal{M} = \{n_1, \ldots, n_M\}$ are satisfied:

$$T_{\min} \le n_m - n_{m-1} \le T_{\max} \le N, \quad m = 2, \ldots, M. \qquad (2)$$

From the above formulation, it is clear that Problem 1 belongs to the class of clustering problems with a quadratic criterion. Clusters are the unknown index subsets $\mathcal{M}_1, \ldots, \mathcal{M}_L, \mathcal{N} \setminus \mathcal{M}$ and the corresponding subsequences of the input sequence.

One of the sources of Problem 1 is the next problem which is typical for many natural science and technical applications, in particular, for noise-proof remote monitoring, electronic intelligence, analysis and recognition of biomedical and speech signals, data mining, machine learning, and others.

There is a series of N chronologically ordered measurements $y_1, \ldots, y_N$ of a q-tuple y of numerical characteristics of some object. The object has $L+1$ states. Among them L states are active and one state is passive. In the passive state all the numerical characteristics in the tuple equal zero, while, in each active state the value of at least one characteristic is nonzero. The data contains some measurement errors. It is known that for some time the object is located in one of the active states, and then switches to a different active state. At that all the active states of the object are accompanied by a switching into the passive state for some unknown time interval which is bounded from above and below. In addition we are given the natural numbers $T_{\min}$ and $T_{\max}$, which correspond to the minimum and maximum time interval between any two successive active states of the object. The correspondence of the sequence element to some state of the object is not known in advance. It is required to find the sequence of active states of the object and to estimate the characteristics of the object in each of the active states (which correspond to the respective cluster centers).

Formalization of this problem with respect to the criterion of the minimum sum of squared deviations induces the following approximation problem. Given a sequence $\mathcal{Y} = (y_1, \ldots, y_N)$ of points from $\mathbb{R}^q$ and some positive integers $T_{\min}$, $T_{\max}$, L, and M. Find an approximating sequence $z_1, \ldots, z_N$ having the following structure

$$z_n = \begin{cases} x_1, & n \in \mathcal{M}_1, \\ \ldots \\ x_L, & n \in \mathcal{M}_L, \\ 0, & n \in \mathcal{N} \backslash \mathcal{M}, \end{cases} \tag{3}$$

where $x_1, \ldots, x_L$ are unknown points from $\mathbb{R}^q$, such that

$$\sum_{i \in \mathcal{N}} \|y_i - z_i\|^2 \longrightarrow \min, \tag{4}$$

under the same constraints on the numbers from subsets $\mathcal{M}_l, \ldots, \mathcal{M}_L$, and $\mathcal{M}$ as in Problem 1.

Schematically, the segment of sequence z_n, $n \in \mathcal{N}$, has the following structure

$$\ldots 0 x_{l-1} 0 \ldots 0 x_{l-1} 0 \ldots \ldots 0 x_l 0 \ldots 0 x_l 0 \ldots \tag{5}$$

Here $x_{l-1}, x_l \in \mathbb{R}^q$ are unknown nonzero points corresponding to the $(l-1)$-th and l-th active states of the object, 0 corresponds to the passive state of the object. The number of zero points between the nonzero points is unknown and lies within the admissible range from $T_{\min} - 1$ to $T_{\max} - 1$ in accordance with the constraints (2).

Relying on (3), expanding the sum (4) and grouping the terms, it is easy to verify by differentiation that the values $x_l = \overline{y}(\mathcal{M}_l)$, $l = 1, \ldots, L$, are optimal in the sense of (4), and thus the formulated approximation problem induces Problem 1. Herein in the optimal approximating sequence, the segment (5) has the following form

$$\ldots 0 \overline{y}(\mathcal{M}_{l-1}) 0 \ldots 0 \overline{y}(\mathcal{M}_{l-1}) 0 \ldots \ldots 0 \overline{y}(\mathcal{M}_l) 0 \ldots 0 \overline{y}(\mathcal{M}_l) 0 \ldots$$

For all $l = 1, \ldots, L$ in this sequence, the indices from the set $\mathcal{M}_l$, the cluster $\{y_j \mid j \in \mathcal{M}_l\}$, and its centroid $\overline{y}(\mathcal{M}_l)$ are determined as the result of solving Problem 1. Centroid $\overline{y}(\mathcal{M}_l)$ is an estimate for the point x_l.

From the above mentioned schematic record of sequences in the string form, it is evident that each of them can be interpreted as a sequence containing the segments with some quasiperiodic (because of the constraints (2)) repetitions. If we define the boundaries of the series on the first or the last repetition, then one can interpret all of the above problems as problems of partitioning a sequence into segments with quasiperiodic repetitions of a priori unknown points, estimating these points, and finding their positions in the sequence.

The next statement establishes the complexity status of Problem 1.

Proposition 1. *The Problem 1 is strongly NP-hard.*

Proposition 1 follows from the fact that the special case of Problem 1 with $L = 1$ is strongly NP-hard [5]. Thus, Problem 1 belongs to the class of computationally intractable problems.

3   Known and Obtained Results

Problem 1 is among the poorly studied discrete optimization problems. It is closely related to the problem (see [7]) in which the input sequence $\mathcal{Y}$ is one-dimensional, i.e. $q = 1$. The points from tuple $(x_1, \ldots, x_L)$ belong to $\mathbb{R}^d$, where $d \geq 1$, and they are given at the problem input, at that $T_{\min} \geq d$ in the restrictions (2). In the objective function of the problem instead of the centroids $\overline{y}(\mathcal{M}_1), \ldots, \overline{y}(\mathcal{M}_L)$ of the desired subsets appear the elements from the given tuple $(x_1, \ldots, x_L)$. The unknown variables are the sets $\mathcal{M}_1, \ldots, \mathcal{M}_L$. This problem can be interpreted as a problem searching a sequence for non-overlapping segments with quasiperiodic repetitions of points from the tuple together with the positions of these points in the sequence. It was shown in [7] that this problem is solvable in polynomial time using dynamic programming. Below we apply a simplification of this dynamic program in our algorithm.

Except for the special case with $L = 1$ in Eq. (1), no algorithms with guaranteed approximation factor are known at the moment for Problem 1. For this special case, the following results were obtained.

In [5], the variant of Problem 1 in which $T_{\min}$ and $T_{\max}$ are the parameters was analyzed. In the cited work it was shown that in the case when $L = 1$, this parameterized variant is strongly NP-hard for any $T_{\min} < T_{\max}$. In the trivial case when $T_{\min} = T_{\max}$, the problem is solvable in polynomial time.

In [6], for the same case of Problem 1, when $L = 1$, a 2-approximation polynomial-time algorithm running in $\mathcal{O}(N^2(MN + q))$ time was presented.

In addition, in [8,9], two special cases of the case $L = 1$ were studied. In both subcases the dimension q of the space is fixed. For the subcase with integer inputs in [8] an exact pseudopolynomial algorithm was constructed. The time complexity of this algorithm is $\mathcal{O}(MN^2(MD)^q)$, where D is the maximum absolute in any coordinate of the input points. For the subcase with real inputs in [9] a fully polynomial-time approximation scheme was proposed which, given a relative error ε, finds a $(1 + \varepsilon)$-approximate solution of Problem 1 in $\mathcal{O}(MN^3(1/\varepsilon)^{q/2})$ time.

The main result of this paper is an algorithm that allows to find a 2-approximate solution of Problem 1 in $\mathcal{O}(LN^{L+1}(MN + q))$ time, which is polynomial if the number L of clusters is fixed.

4   Fundamentals of Algorithm

To construct the algorithm we need a few basic assertions, an auxiliary problem and an exact polynomial algorithm for its solution.

The geometrical foundations of the algorithm are given by the following lemmas.

Lemma 1. *For any point $u \in \mathbb{R}^q$ and any finite nonempty set $\mathcal{Z} \subset \mathbb{R}^q$ the following equality holds*

$$\sum_{z \in \mathcal{Z}} \|z - u\|^2 = \sum_{z \in \mathcal{Z}} \|z - \overline{z}\|^2 + |\mathcal{Z}| \cdot \|u - \overline{z}\|^2 , \qquad (6)$$

where $\overline{z} = \frac{1}{|\mathcal{Z}|} \sum_{z \in \mathcal{Z}} z$ is the centroid of $\mathcal{Z}$.

Lemma 1 has quite simple proof and is well-known. Its proof has been given in several publications (for example, in [10]).

Lemma 2. *Assume that the conditions of Lemma 1 hold. Then, if some point $u \in \mathbb{R}^q$ is closer (with respect to the Euclidean distance) to the centroid $\overline{z}$ of $\mathcal{Z}$ than all points in $\mathcal{Z}$, then*

$$\sum_{z \in \mathcal{Z}} \|z - u\|^2 \le 2 \sum_{z \in \mathcal{Z}} \|z - \overline{z}\|^2 .$$

Lemma 2 follows from (6), because by the assumption for every point $z \in \mathcal{Z}$ we have the inequality $\|u - \overline{z}\| \le \|z - \overline{z}\|$.

From now on we use $f^x(y)$ to denote a function $f(x, y)$ for which x is fixed.

Lemma 3. *Let*

$$S(\mathcal{M}_1, \ldots, \mathcal{M}_L, x_1, \ldots, x_L) = \sum_{l=1}^{L} \sum_{j \in \mathcal{M}_l} \|y_j - x_l\|^2 + \sum_{i \in \mathcal{N} \setminus \mathcal{M}} \|y_i\|^2 , \qquad (7)$$

$$G(\mathcal{M}_1, \ldots, \mathcal{M}_L, x_1, \ldots, x_L) = \sum_{l=1}^{L} \sum_{j \in \mathcal{M}_l} \left(2\langle y_j, x_l \rangle - \|x_l\|^2 \right) ,$$

where $x_1, \ldots, x_L$ are points from $\mathbb{R}^q$, and elements of the sets $\mathcal{M}_l, \ldots, \mathcal{M}_L$, and $\mathcal{M}$ satisfy restrictions of Problem 1. Then the following statements are true:

(1) for any nonempty fixed subsets $\mathcal{M}_1, \ldots, \mathcal{M}_L$ the minimum of function (7) over $x_1, \ldots, x_L$ is reached at the points $x_l = \overline{y}(\mathcal{M}_l)$, $l = 1, \ldots, L$, and is equal to $F(\mathcal{M}_1, \ldots, \mathcal{M}_L)$;

(2) for any tuple $x = (x_1, \ldots, x_L)$ of fixed points from $\mathbb{R}^q$ the minimum of function $S^x(\mathcal{M}_1, \ldots, \mathcal{M}_L)$ over $\mathcal{M}_1, \ldots, \mathcal{M}_L$ is reached at the subsets $\mathcal{M}_1^x, \ldots, \mathcal{M}_L^x$ that maximize function $G^x(\mathcal{M}_1, \ldots, \mathcal{M}_L)$.

Proof. The first statement of this lemma is easily verified by differentiation and also follows from Lemma 1. To prove the second statement it is sufficient to note that the following equality holds

$$S^x(\mathcal{M}_1, \ldots, \mathcal{M}_L) = \sum_{j \in \mathcal{N}} \|y_j\|^2 - G^x(\mathcal{M}_1, \ldots, \mathcal{M}_L), \quad (8)$$

where the sum on the right-hand side is independent of $\mathcal{M}_1, \ldots, \mathcal{M}_L$. □

The main ingredient to our algorithm is an exact polynomial-time algorithm for solving the following auxiliary problem.

Problem 2. Given a sequence $\mathcal{Y} = (y_1, \ldots, y_N)$ and a tuple $x = (x_1, \ldots, x_L)$ of points from $\mathbb{R}^q$, and some positive integers $T_{\min}$, $T_{\max}$, and M. *Find* nonempty disjoint subsets $\mathcal{M}_1, \ldots, \mathcal{M}_L$ of $\mathcal{N} = \{1, \ldots, N\}$ that maximize the objective function $G^x(\mathcal{M}_1, \ldots, \mathcal{M}_L)$, under the same constraints on the optimized variables as in Problem 1.

To explain the algorithm for solving this auxiliary problem, we define the function

$$g_l^x(n) = 2\langle y_n, x_l \rangle - \|x_l\|^2, \quad n \in \mathcal{N}, \quad l = 1, \ldots, L, \quad (9)$$

where x_l is a point from tuple x, and y_n is an element of sequence $\mathcal{Y}$.

In accordance with the definition (9), for the objective function $G^x(\mathcal{M}_1, \ldots, \mathcal{M}_L)$ we have

$$G^x(\mathcal{M}_1, \ldots, \mathcal{M}_L) = \sum_{l=1}^{L} \sum_{n \in \mathcal{M}_l} g_l^x(n).$$

In addition, we note that Lemma 3 yields the following equalities

$$(\mathcal{M}_1^x, \ldots, \mathcal{M}_L^x) = \arg \min_{\mathcal{M}_1, \ldots, \mathcal{M}_L} S^x(\mathcal{M}_1, \ldots, \mathcal{M}_L)$$

$$= \arg \max_{\mathcal{M}_1, \ldots, \mathcal{M}_L} G^x(\mathcal{M}_1, \ldots, \mathcal{M}_L). \quad (10)$$

In the next lemma and its corollary we give a dynamic programming scheme. This scheme guarantees finding the optimal solution $\mathcal{M}_1^x, \ldots, \mathcal{M}_L^x$ of Problem 2 and (according to the Eq. (10)) the optimal solution of the problem of minimizing the function $S^x(\mathcal{M}_1, \ldots, \mathcal{M}_L)$. The presented scheme follows from the results obtained in [7] and is given here for completeness.

Lemma 4. *Let the conditions of Problem 2 hold. Then for any positive integers L and M such that $(M-1)T_{\min} < N$ and $L \leq M$, the optimal value $G_{\max}^x$ of the objective function of Problem 2 is given by the formula*

$$G_{\max}^x = \max_{n \in \{1+(M-1)T_{\min}, \ldots, N\}} G_{L,M}^x(n); \quad (11)$$

here, the values of $G_{L,M}^x(n)$ *are calculated using the recurrence formula*

$$G_{l,m}^x(n) = g_l^x(n)$$

$$+ \begin{cases} 0, & if\ l = 1, m = 1, \\ \max\limits_{j \in \gamma_{m-1}(n)} G_{1,m-1}^x(j), & \\ & if\ l = 1, m = 2, \ldots, M - (L-1), \\ \max\limits_{j \in \gamma_{m-1}(n)} G_{l-1,m-1}^x(j), & \\ & if\ l = 2, \ldots, L,\ m = l, \\ \max\{\max\limits_{j \in \gamma_{m-1}(n)} G_{l,m-1}^x(j), \max\limits_{j \in \gamma_{m-1}(n)} G_{l-1,m-1}^x(j)\}, & \\ & if\ l = 2, \ldots, L,\ m = l+1, \ldots, M - (L-l), \end{cases} \qquad (12)$$

where

$$\gamma_{m-1}(n) = \{j \mid \max\{1 + (m-2)T_{\min}, n - T_{\max}\} \le j \le n - T_{\min}\}, \\ m = 2, \ldots, M, \qquad (13)$$

for every $n = 1 + (m-1)T_{\min}, \ldots, N - (M-m)T_{\min}$.

Corollary 1. *Let the conditions of Lemma 4 hold. In addition, let*

$$r_{l,m}^x(n) = \begin{cases} 1, & if\ l = 1,\ m = 2, \ldots, M - (L-1), \\ l - 1, & if\ l = 2, \ldots, L,\ m = l, \\ l - 1, & if\ \max\limits_{j \in \gamma_{m-1}(n)} G_{l,m-1}^x(j) < \max\limits_{j \in \gamma_{m-1}(n)} G_{l-1,m-1}^x(j), \\ & \qquad l = 2, \ldots, L,\ m = l+1, \ldots, M - (L-l), \\ l, & if\ \max\limits_{j \in \gamma_{m-1}(n)} G_{l,m-1}^x(j) \ge \max\limits_{j \in \gamma_{m-1}(n)} G_{l-1,m-1}^x(j), \\ & \qquad l = 2, \ldots, L,\ m = l+1, \ldots, M - (L-l), \end{cases}$$

$$I_{l,m}^x(n) = \arg \max\limits_{j \in \gamma_{m-1}(n)} G_{l,m-1}^x(j),\ l = 1, \ldots, L,\ m = l+1, \ldots, M - (L-l),$$

for every $n = 1 + (m-1)T_{\min}, \ldots, N - (M-m)T_{\min}$;

$$n^x(m) = \begin{cases} \arg \max\limits_{n \in \{1+(M-1)T_{\min}, \ldots, N\}} G_{L,M}^x(n), & if\ m = M, \\ I_{k^x(m),m+1}^x(n^x(m+1)), & if\ m = M - 1, \ldots, 1, \end{cases}$$

$$k^x(m) = \begin{cases} L, & if\ m = M, \\ r_{k^x(m+1),m+1}^x(n^x(m+1)), & if\ m = M - 1, \ldots, 1; \end{cases}$$

$$J^x(l) = \begin{cases} 0, & if\ l = 0, \\ \left| \left\{ m \in \{1, \ldots, M\} \mid k^x(m) \le l \right\} \right|, & if\ l = 1, \ldots, L. \end{cases}$$

Then the sets $\mathcal{M}_1^x, \ldots, \mathcal{M}_L^x$ *are given by the formula*

$$\mathcal{M}_l^x = \{n \mid n = n^x(m),\ m = J^x(l-1) + 1, \ldots, J^x(l)\} \qquad (14)$$

for every $l = 1, \ldots, L$.

A step-by-step description of the algorithm implementing the above scheme is given in the following.

Algorithm $\mathcal{A}_1$.

Input: sequence $\mathcal{Y}$, tuple $(x_1, \ldots, x_L)$ of points, numbers $T_{\min}$, $T_{\max}$, and M.

Step 1. Compute the values $g_l^x(n)$ for $l = 1, \ldots, L$, and $n = 1 + (l-1)T_{\min}$, $\ldots, N - (L-l)T_{\min}$ using Formula (9).

Step 2. Using Formulae (12) and (13), compute the values $G_{l,m}^x(n)$ for each $l = 1, \ldots, L$, $m = l, \ldots, M - (L-l)$, $n = 1 + (m-1)T_{\min}, \ldots, N - (M-m)T_{\min}$.

Step 3. Find the maximum $G_{\max}^x$ of the objective function G^x by Formula (11), and the optimal subsets $\mathcal{M}_l^x$ by Formula (14).

Output: the family $\{\mathcal{M}_1^x, \ldots, \mathcal{M}_L^x\}$ of subsets.

Remark 1. Before the start of the algorithm it is required to verify the two conditions of Lemma 4. These necessary conditions provide the consistency of the constraints in Problems 1 and 2, as well as the correctness of the input data of the algorithm.

Remark 2. In [7], it was found that Algorithm $\mathcal{A}_1$ finds the optimal solution of Problem 2 in $\mathcal{O}(LN(M(T_{\max} - T_{\min} + 1) + q))$ time. In this expression, the value of $T_{\max} - T_{\min} + 1$ is at most N. Therefore, the algorithm running time can be estimated as $\mathcal{O}(LN(MN + q))$.

5   Approximation Algorithm

Our approach to Problem 1 is as follows. For each ordered set (tuple) containing L elements of the sequence $\mathcal{Y}$, we find an exact solution of the auxiliary Problem 2, i.e. a family containing disjoint subsets of indices of the input sequence, which is a feasible solution of the original Problem 1.

The found family of subsets we declare a solution candidate for Problem 1 and include this family in the set of solution candidates.

From the obtained set as the final solution we choose a family of subsets which yields the largest value for the objective function of Problem 2.

Let us formulate an algorithm that implements the described approach. Below, in the step-by-step description, it is assumed that the input positive integers satisfy the conditions of Lemma 4 (see Remark 1).

Algorithm $\mathcal{A}$.

Input: sequence $\mathcal{Y}$, numbers $T_{\min}$, $T_{\max}$, M, and L.

Step 1. For every tuple $x = (x_1, \ldots, x_L) \in \mathcal{Y}^L$ of elements of the sequence $\mathcal{Y}$, using Algorithm $\mathcal{A}_1$, find the optimal solution $\{\mathcal{M}_1^x, \ldots, \mathcal{M}_L^x\}$ of Problem 2.

Step 2. Find a tuple $x(A) = \arg\max_{x \in \mathcal{Y}^L} G^x(\mathcal{M}_1^x, \ldots, \mathcal{M}_L^x)$ and a family $\{\mathcal{M}_1^A, \ldots, \mathcal{M}_L^A\} = \{\mathcal{M}_1^{x(A)}, \ldots, \mathcal{M}_L^{x(A)}\}$. If the optimum is taken by several tuples, we choose any of them.

Output: the family $\{\mathcal{M}_1^A, \ldots, \mathcal{M}_L^A\}$ of subsets.

Lemma 5. *Let $\{\mathcal{M}_1^*, \ldots, \mathcal{M}_L^*\}$ be the optimal solution of Problem 1, and $\{\mathcal{M}_1^A, \ldots, \mathcal{M}_L^A\}$ be the solution found by Algorithm $\mathcal{A}$. Then*

$$F(\mathcal{M}_1^A, \ldots, \mathcal{M}_L^A) \leq 2F(\mathcal{M}_1^*, \ldots, \mathcal{M}_L^*).$$

Proof. The optimal solution $\{\mathcal{M}_1^*, \ldots, \mathcal{M}_L^*\}$ of Problem 1 corresponds to the tuple $(\overline{y}(\mathcal{M}_1^*), \ldots, \overline{y}(\mathcal{M}_L^*))$ of centroids, where $\overline{y}(\mathcal{M}_l^*) = \frac{1}{|\overline{y}(\mathcal{M}_l^*)|} \sum_{y \in \mathcal{M}_l^*} y$, $l = 1, \ldots, L$. Let us consider the point $t_l = \arg \min_{y \in \mathcal{M}_l^*} \|y - \overline{y}(\mathcal{M}_l^*)\|$, $l = 1, \ldots, L$, from the subset $\mathcal{M}_l^*$, closest to the centroid of this subset. This point in the set $\mathcal{M}_l^*$ and the set $\mathcal{M}_l^*$ itself satisfy the conditions of Lemma 2. Therefore, by applying the inequality of Lemma 2 to every subset $\mathcal{M}_l^*$, $l = 1, \ldots, L$, we can estimate the sum

$$
\begin{aligned}
S(\mathcal{M}_1^*, \ldots, \mathcal{M}_L^*, t_1, \ldots, t_L) &= \sum_{l=1}^{L} \sum_{y \in \mathcal{M}_l^*} \|y - t_l\|^2 + \sum_{i \in \mathcal{N} \setminus \mathcal{M}^*} \|y_i\|^2 \\
&\leq 2 \sum_{l=1}^{L} \sum_{y \in \mathcal{M}_l^*} \|y - \overline{y}(\mathcal{M}_l^*)\|^2 + \sum_{i \in \mathcal{N} \setminus \mathcal{M}^*} \|y_i\|^2 \\
&\leq 2 \sum_{l=1}^{L} \sum_{y \in \mathcal{M}_l^*} \|y - \overline{y}(\mathcal{M}_l^*)\|^2 + 2 \sum_{i \in \mathcal{N} \setminus \mathcal{M}^*} \|y_i\|^2 = 2 F(\mathcal{M}_1^*, \ldots, \mathcal{M}_L^*) ,
\end{aligned}
\tag{15}
$$

where $\mathcal{M}^* = \cup_{l=1}^{L} \mathcal{M}_l^*$.

On the other hand, we notice that the tuple $t = (t_1, \ldots, t_L)$ is among the tuples from $\mathcal{Y}^L$ that have been examined at Step 1 of Algorithm $\mathcal{A}$. Let $\{\mathcal{M}_1^t, \ldots, \mathcal{M}_L^t\}$ be the optimal solution found at Step 2 of Algorithm $\mathcal{A}$ for Problem 2 at $x = t$. Then according to statement 2 of Lemma 3, i.e. according to (10), the family $\{\mathcal{M}_1^t, \ldots, \mathcal{M}_L^t\}$ supplies the minima to the function $S^x(\mathcal{M}_1, \ldots, \mathcal{M}_L)$ at $x = t$. Consequently the bound

$$
S(\mathcal{M}_1^t, \ldots, \mathcal{M}_L^t, t_1, \ldots, t_L) \leq S(\mathcal{M}_1^*, \ldots, \mathcal{M}_L^*, t_1, \ldots, t_L)
\tag{16}
$$

is valid for the left-hand side of (15).

Furthermore, by the definition of Step 2 and according to (8) we have the bound

$$
S(\mathcal{M}_1^A, \ldots, \mathcal{M}_L^A, x_1^A, \ldots, x_L^A) \leq S(\mathcal{M}_1^t, \ldots, \mathcal{M}_L^t, t_1, \ldots, t_L) ,
\tag{17}
$$

where $(x_1^A, \ldots, x_L^A) = x(A)$. Additionally, from the first statement of Lemma 3 we have the inequality

$$
F(\mathcal{M}_1^A, \ldots, \mathcal{M}_L^A) \leq S(\mathcal{M}_1^A, \ldots, \mathcal{M}_L^A, x_1^A, \ldots, x_L^A) .
\tag{18}
$$

Finally, by combining (15)–(18) we get the chain of estimation inequalities

$$
\begin{aligned}
F(\mathcal{M}_1^A, \ldots, \mathcal{M}_L^A) &\leq S(\mathcal{M}_1^A, \ldots, \mathcal{M}_L^A, x_1^A, \ldots, x_L^A) \\
&\leq S(\mathcal{M}_1^t, \ldots, \mathcal{M}_L^t, t_1, \ldots, t_L) \leq S(\mathcal{M}_1^*, \ldots, \mathcal{M}_L^*, t_1, \ldots, t_L) \\
&\leq 2 F(\mathcal{M}_1^*, \ldots, \mathcal{M}_L^*) ,
\end{aligned}
$$

which proves Lemma 5. □

We finally prove the running time of the algorithm and that the bound of 2 on its approximation factor is tight.

Theorem 1. *Algorithm $\mathcal{A}$ finds a 2-approximate solution of Problem 1 in $\mathcal{O}(LN^{L+1}(M(T_{max} - T_{min} + 1) + q))$ time. The performance guarantee 2 of the algorithm is tight.*

Proof. The 2-accuracy bound of the algorithm follows from Lemma 5. We bound the time complexity of the algorithm using its step-by-step description.

The computation time is determined by the time complexity of Step 1, at which Problem 2 is solved $\mathcal{O}(N^L)$ times by applying Algorithm $\mathcal{A}_1$, whose time complexity is $\mathcal{O}(LN(M(T_{max} - T_{min} + 1) + q))$ (see Remark 2). In addition, it needs $\mathcal{O}(N^L)$ comparisons for searching a largest value of the objective function of Problem 2 at Step 2. By summing all these times we obtain the final bound for the algorithm time complexity.

The tightness of the performance guarantee of Algorithm $\mathcal{A}$ follows from the tightness of the performance guarantee of the 2-approximation algorithm for the case of Problem 1 when $L = 1$ (see [6]).   □

Remark 3. According to Remark 2, the running time of Algorithm $\mathcal{A}$ is $\mathcal{O}(LN^{L+1}(MN + q))$, which is polynomial if the number L of clusters is fixed.

6   Conclusion

In this paper we have shown the strong NP-hardness of one problem of partitioning a finite sequence of points of Euclidean space into clusters with restrictions on their cardinalities. We also have shown an approximation algorithm for this problem. The proposed algorithm allows to find a 2-approximate solution of the problem in a polynomial time if the number of clusters is fixed.

In our opinion, the presented algorithm would be useful as one of the tools for solving problems in applications related to data mining, and analysis and recognition of time series (signals).

Of considerable interest is the development of faster polynomial-time approximation algorithms for the case when the number of clusters is not fixed. An important direction of study is searching subclasses of this problem for which faster polynomial-time approximation algorithms can be constructed.

Acknowledgments. This work was supported by Russian Science Foundation, project no. 16-11-10041.

References

1. Fu, T.: A review on time series data mining. Eng. Appl. Artif. Intell. **24**(1), 164–181 (2011)
2. Kuenzer, C., Dech, S., Wagner, W.: Remote Sensing Time Series. Remote Sensing and Digital Image Processing, vol. 22. Springer, Switzerland (2015)
3. Warren Liao, T.: Clustering of time series data – a survey. Pattern Recogn. **38**(11), 1857–1874 (2005)
4. Aggarwal, C.C.: Data Mining: The Textbook. Springer, Switzerland (2015)

5. Kel'manov, A.V., Pyatkin, A.V.: On complexity of some problems of cluster analysis of vector sequences. J. Appl. Ind. Math. **7**(3), 363–369 (2013)
6. Kel'manov, A.V., Khamidullin, S.A.: An approximating polynomial algorithm for a sequence partitioning problem. J. Appl. Ind. Math. **8**(2), 236–244 (2014)
7. Kel'manov, A.V., Mikhailova, L.V.: Joint detection of a given number of reference fragments in a quasi-periodic sequence and its partition into segments containing series of identical fragments. Comput. Math. Math. Phys. **46**(1), 165–181 (2006)
8. Kel'manov, A.V., Khamidullin, S.A., Khandeev, V.I.: An exact pseudopolynomial algorithm for a sequence bi-clustering problem (in Russian). In: Book of Abstract of the XVth Russian Conference "Mathematical Programming and Applications", pp. 139–140. Inst. Mat. Mekh. UrO RAN, Ekaterinburg (2015)
9. Kel'manov, A.V., Khamidullin, S.A., Khandeev, V.I.: A fully polynomial-time approximation scheme for a sequence 2-cluster partitioning problem. J. Appl. Indust. Math. **10**(2), 209–219 (2016)
10. Kel'manov, A.V., Romanchenko, S.M.: An FPTAS for a vector subset search problem. J. Appl. Indust. Math. **8**(3), 329–336 (2014)

A Fully Polynomial-Time Approximation Scheme for a Special Case of a Balanced 2-Clustering Problem

Alexander Kel'manov[1,2(✉)] and Anna Motkova[2(✉)]

[1] Sobolev Institute of Mathematics, 4 Koptyug Avenue, 630090 Novosibirsk, Russia
kelm@math.nsc.ru
[2] Novosibirsk State University, 2 Pirogova Street, 630090 Novosibirsk, Russia
anitamo@mail.ru

Abstract. We consider the strongly NP-hard problem of partitioning a set of Euclidean points into two clusters so as to minimize the sum (over both clusters) of the weighted sum of the squared intracluster distances from the elements of the clusters to their centers. The weights of sums are the cardinalities of the clusters. The center of one of the clusters is given as input, while the center of the other cluster is unknown and determined as the geometric center (centroid), i.e. the average value over all points in the cluster. We analyze the variant of the problem with cardinality constraints. We present an approximation algorithm for the problem and prove that it is a fully polynomial-time approximation scheme when the space dimension is bounded by a constant.

Keywords: NP-hardness · Euclidian space · Fixed dimension · FPTAS

1 Introduction

The subject of this study is a strongly NP-hard quadratic Euclidean problem of partitioning a finite set of points into two clusters. We will show a fully polynomial-time approximation scheme (FPTAS) for a special case of the problem.

Our research is motivated by insufficient study of the problem from an algorithmic direction and its importance in some applications including geometry, cluster analysis, statistical problems of joint evaluation and hypotheses testing with heterogeneous samples, data interpretation problem, etc.

The paper has the following structure. Section 2 contains the problem formulation, some applications, and some closely related problems. Additionally, known and our new results are discussed. In Sect. 3 we formulate and prove some basic properties exploited by our algorithm. In Sect. 4, an approximation algorithm is presented. Finally, also in Sect. 4 we show that our algorithm is a fully polynomial-time approximation scheme when the space dimension is fixed.

© Springer International Publishing Switzerland 2016
Y. Kochetov et al. (Eds.): DOOR 2016, LNCS 9869, pp. 182–192, 2016.
DOI: 10.1007/978-3-319-44914-2_15

2 Problem Formulation, Its Origin, Related Problems, known and New Results

Everywhere below we use the standard notations, namely: $\mathbb{R}$ is the set of the real numbers, $\mathbb{R}_+$ is the set of positive real numbers, $\mathbb{Z}$ is the set of integers, $\|\cdot\|$ is the Euclidean norm, and $\langle\cdot,\cdot\rangle$ is the scalar product.

The problem under consideration is formulated as follows (see also [1,2]).

Problem 1 (*Balanced Variance-based 2-Clustering with given center*). *Given a set* $\mathcal{Y} = \{y_1,\dots,y_N\}$ *of points from* $\mathbb{R}^q$ *and a positive integer* M. *Find* a partition of $\mathcal{Y}$ into two non-empty clusters $\mathcal{C}$ and $\mathcal{Y} \setminus \mathcal{C}$ such that

$$F(\mathcal{C}) = |\mathcal{C}| \sum_{y \in \mathcal{C}} \|y - \overline{y}(\mathcal{C})\|^2 + |\mathcal{Y} \setminus \mathcal{C}| \sum_{y \in \mathcal{Y} \setminus \mathcal{C}} \|y\|^2 \longrightarrow \min, \qquad (1)$$

where $\overline{y}(\mathcal{C}) = \frac{1}{|\mathcal{C}|} \sum_{y \in \mathcal{C}} y$ is the geometric center (centroid) of $\mathcal{C}$ and such that $|\mathcal{C}| = M$.

The problem has an obvious geometrical interpretation. It is a partition of a finite set of points in Euclidean space into two geometrical structures minimizing (1). In formula (1) the weights of the sums are the cardinalities of the desired clusters. So, Problem 1 can be interpreted as the problem of optimal weighted (by the cardinalities of the clusters) summing and also as a problem of balanced partitioning (or clustering).

In addition, the problem has applications in Data mining problem (see, for example, [3–5]). The essence of this multifaceted problem is the approximation of data by some mathematical model that allows to plausibly explain the origin of the data in terms of the model. In particular, the next statistical hypothesis can be used as such mathematical model: it is true that the input data $\mathcal{Y}$ is the inhomogeneous sample from two distributions, and that one of these distributions has zero mean while another mean is unknown and non-equal to zero. To test this hypothesis, first we need to find an optimal solution to Problem 1, and only then we will be able to use the classical results in the field of statistical hypothesis testing.

It is widely known that applied researchers, who study and analyze data, use algorithms as the basic mathematical tools for solving a variety of clustering problems in which clusters consist of similar or related by certain criteria objects. Creating such mathematical tools for solving data mining problems causes the development of new algorithms with guaranteed performance estimates of accuracy and time complexity.

The strong NP-hardness of Problem 1 was proved in [1,2]. This fact implies that, unless P=NP, there are neither exact polynomial-time nor exact pseudopolynomial-time algorithms for it [6]. In addition, in [1,2], the nonexistence of an FPTAS was shown (unless P=NP) for Problem 1. So, finding subclasses of this problem for which there exists an FPTAS is a question of topical interest.

Note that there is only one algorithmic result for Problem 1, i.e. an exact algorithm [7] for the case of integer components of the input points. The time complexity of this algorithm is $\mathcal{O}(qN(2MB+1)^q)$, where B is the maximum absolute value of the components of the input points. If the dimension q of the space is bounded by a constant, then the time complexity of the algorithm is $\mathcal{O}(N(MB)^q)$. So, in this case the algorithm is pseudopolynomial.

At the same time, there are a lot of results for problems closely related to Problem 1. Properties of algorithms for these problems can be found in the papers cited below.

The NP-hard *Balanced variance-based* 2-*clustering* problem is one of the most closely related to Problem 1. The objective function in this problem is different from (1) in that the center of cluster $\mathcal{Y} \setminus \mathcal{C}$ is not fixed:

$$|\mathcal{C}| \sum_{y \in \mathcal{C}} \|y - \overline{y}(\mathcal{C})\|^2 + |\mathcal{Y} \setminus \mathcal{C}| \sum_{y \in \mathcal{Y} \setminus \mathcal{C}} \|y - \overline{y}(\mathcal{Y} \setminus \mathcal{C})\|^2 \longrightarrow \min. \tag{2}$$

In problem (2) the centroids $\overline{y}(\mathcal{C})$ and $\overline{y}(\mathcal{Y} \setminus \mathcal{C})$ of both clusters $\mathcal{C}$ and $\mathcal{Y} \setminus \mathcal{C}$ are the functions of $\mathcal{C}$. It is well-known that this problem is equivalent to *Min-sum all-pairs* 2-*clustering* problem in which it is required to find a partition such that

$$\sum_{x \in \mathcal{C}} \sum_{z \in \mathcal{C}} \|x - z\|^2 + \sum_{x \in \mathcal{Y} \setminus \mathcal{C}} \sum_{z \in \mathcal{Y} \setminus \mathcal{C}} \|x - z\|^2 \longrightarrow \min. \tag{3}$$

Algorithmic questions for problems (2) and (3) were studied, for example, in [1,2,8–13].

The well-known NP-hard [14] *Minimum sum-of-squares* 2-*clustering* problem is close to Problem 1. In this problem (related to classical work by Fisher [15] and also called 2-*Means* [16]), we need to find two clusters $\mathcal{C}$ and $\mathcal{Y} \setminus \mathcal{C}$ such that

$$\sum_{y \in \mathcal{C}} \|y - \overline{y}(\mathcal{C})\|^2 + \sum_{y \in \mathcal{Y} \setminus \mathcal{C}} \|y - \overline{y}(\mathcal{Y} \setminus \mathcal{C})\|^2 \longrightarrow \min. \tag{4}$$

In problem (4) as well as in problem (2) the centroids of both clusters are the functions of $\mathcal{C}$, but in problem (4) the sums are not weighted by the cluster cardinalities. Thousands of publications are dedicated to problem (4) and its applications.

The strongly NP-hard problem *Minimum sum-of-squares 2-clustering with given center* has been actively studied in the last decade. In this problem we need to find a 2-partition such that

$$\sum_{y \in \mathcal{C}} \|y - \overline{y}(\mathcal{C})\|^2 + \sum_{y \in \mathcal{Y} \setminus \mathcal{C}} \|y\|^2 \longrightarrow \min. \tag{5}$$

Problem (5) differs from Problem 1 in that the sums are not weighted by the cardinalities of the desired clusters. The algorithmic results for this problem can be found in [17–24].

In the considered Problem 1, the centroid $\overline{y}(\mathcal{C})$ of the cluster $\mathcal{C}$ is unknown and the center of the cluster $\mathcal{Y} \setminus \mathcal{C}$ is given at the origin as in the problem (5).

Since Problem 1 is neither equivalent nor a special case of the problems (2)–(5), the previous algorithmic results for these closely related problems do not apply to Problem 1. We need new explorations for this problem.

In this work we present an approximation algorithm for Problem 1. Given a relative error ε, the algorithm finds a $(1 + \varepsilon)$-approximate solution in $\mathcal{O}\left(qN^2(\sqrt{\frac{2q}{\varepsilon}} + 1)^q\right)$ time. In the case of a fixed space dimension q the running time of the algorithm is equal to $\mathcal{O}\left(N^2\left(\frac{1}{\varepsilon}\right)^{q/2}\right)$ and so, it implements a fully polynomial-time approximation scheme.

3   Foundations of the Algorithm

In this section, we provide some basic statements exploited by our algorithm.

The following two lemmas are well known. Their proofs are presented in many publications (see, for example, [25, 26]).

Lemma 1. *For an arbitrary point* $x \in \mathbb{R}^q$ *and a finite set* $\mathcal{Z} \subset \mathbb{R}^q$, *it is true that*

$$\sum_{z \in \mathcal{Z}} \|z - x\|^2 = \sum_{z \in \mathcal{Z}} \|z - \overline{z}\|^2 + |\mathcal{Z}| \cdot \|x - \overline{z}\|^2,$$

where $\overline{z}$ *is the centroid of* $\mathcal{Z}$.

Lemma 2. *Let the conditions of Lemma 1 hold. If a point* $u \in \mathbb{R}^q$ *is closer (in terms of distance) to the centroid* $\overline{z}$ *of* $\mathcal{Z}$ *than any point in* $\mathcal{Z}$, *then*

$$\sum_{z \in \mathcal{Z}} \|z - u\|^2 \leq 2 \sum_{z \in \mathcal{Z}} \|z - \overline{z}\|^2.$$

Lemma 3. *Let*

$$S(\mathcal{C}, x) = |\mathcal{C}| \sum_{y \in \mathcal{C}} \|y - x\|^2 + |\mathcal{Y} \setminus \mathcal{C}| \sum_{y \in \mathcal{Y} \setminus \mathcal{C}} \|y\|^2, \ \mathcal{C} \subseteq \mathcal{Y}, \ x \in \mathbb{R}^q, \qquad (6)$$

where $\mathcal{Y}$ *is the input set of Problem 1. Then it is true that*

$$S(\mathcal{C}, x) = F(\mathcal{C}) + |\mathcal{C}|^2 \|x - \overline{y}(\mathcal{C})\|^2.$$

Proof. Applying Lemma 1 to the set $\mathcal{C}$ and its centroid, we have

$$\sum_{y \in \mathcal{C}} \|y - x\|^2 = \sum_{y \in \mathcal{C}} \|y - \overline{y}(\mathcal{C})\|^2 + |\mathcal{C}| \cdot \|x - \overline{y}(\mathcal{C})\|^2. \qquad (7)$$

After the substitution of (7) in the definition (6), we obtain

$$S(\mathcal{C}, x) = |\mathcal{C}| \sum_{y \in \mathcal{C}} \|y - x\|^2 + |\mathcal{Y} \setminus \mathcal{C}| \sum_{y \in \mathcal{Y} \setminus \mathcal{C}} \|y\|^2$$

$$= |\mathcal{C}| \sum_{y \in \mathcal{C}} \|y - \overline{y}(\mathcal{C})\|^2 + |\mathcal{C}|^2 \|x - \overline{y}(\mathcal{C})\|^2 + |\mathcal{Y} \setminus \mathcal{C}| \sum_{y \in \mathcal{Y} \setminus \mathcal{C}} \|y\|^2$$

$$= F(\mathcal{C}) + |\mathcal{C}|^2 \|x - \overline{y}(\mathcal{C})\|^2.$$

$\square$

For any function $f(x, y)$, we denote by $f^x(y)$ the function when the argument x is fixed and by $f^y(x)$ the function when the argument y is fixed.

Lemma 4. *For the conditional minimums of the function* (6) *the next statements are true:*

(1) *for any nonempty fixed set $\mathcal{C} \subseteq \mathcal{Y}$ the minimum of the function $S^{\mathcal{C}}(x)$ over $x \in \mathbb{R}^q$ is reached at the point $x = \overline{y}(\mathcal{C}) = \frac{1}{|\mathcal{C}|} \sum\limits_{y \in \mathcal{C}} y$ and is equal to $F(\mathcal{C})$;*

(2) *if $|\mathcal{C}| = M = \mathrm{const}$, then, for any fixed point $x \in \mathbb{R}^q$, the minimum of function $S^x(\mathcal{C})$ over $\mathcal{C} \subseteq \mathcal{Y}$ satisfies*

$$\arg \min_{\mathcal{C} \subseteq \mathcal{Y}} S^x(\mathcal{C}) = \arg \min_{\mathcal{C} \subseteq \mathcal{Y}} G^x(\mathcal{C}),$$

where

$$G^x(\mathcal{C}) = \sum_{y \in \mathcal{C}} g^x(y), \tag{8}$$

$$g^x(y) = (2M - N)\|y\|^2 - 2M \langle y, x \rangle, \quad y \in \mathcal{Y}, \tag{9}$$

and

$$\min_{\mathcal{C} \subseteq \mathcal{Y}} G^x(\mathcal{C}) = \sum_{y \in \mathcal{B}^x} g^x(y), \tag{10}$$

where the set $\mathcal{B}^x$ consists of M points of the set $\mathcal{Y}$, at which the function $g^x(y)$ has the smallest values.

Proof. The first statement follows from Lemma 3.

Since $|\mathcal{Y}| = N$ and $|\mathcal{C}| = M$, the second statement follows from the next chain of equalities:

$$S^x(\mathcal{C}) = M \sum_{y \in \mathcal{C}} \|y - x\|^2 + (N - M) \sum_{y \in \mathcal{Y} \setminus \mathcal{C}} \|y\|^2$$

$$= M \sum_{y \in \mathcal{C}} \|y\|^2 + M^2 \|x\|^2 - 2M \sum_{y \in \mathcal{C}} \langle y, x \rangle + (N - M) \sum_{y \in \mathcal{Y} \setminus \mathcal{C}} \|y\|^2$$

$$= (N - M) \sum_{y \in \mathcal{Y}} \|y\|^2 + M^2 \|x\|^2 + (2M - N) \sum_{y \in \mathcal{C}} \|y\|^2 - 2M \sum_{y \in \mathcal{C}} \langle y, x \rangle$$

$$= (N - M) \sum_{y \in \mathcal{Y}} \|y\|^2 + M^2 \|x\|^2 + \sum_{y \in \mathcal{C}} g^x(y) = (N - M) \sum_{y \in \mathcal{Y}} \|y\|^2 + M^2 \|x\|^2 + G^x(\mathcal{C}).$$

It remains to note that in the last two equalities the first two addends do not depend on $\mathcal{C}$. The formula (10) is obvious. □

Lemma 5. *Let the conditions of Lemma 4 hold and $\mathcal{C}^*$ be the optimal solution of Problem 1. Then, for a fixed point $x \in \mathbb{R}^q$, the following inequality is true*

$$F(\mathcal{B}^x) \leq F(\mathcal{C}^*) + M^2 \|x - \overline{y}(\mathcal{C}^*)\|^2.$$

Proof. The definitions (1) and (6), and Lemma 4 imply

$$F(\mathcal{B}^x) = S^{\overline{y}(\mathcal{B}^x)}(\mathcal{B}^x) \leq S^x(\mathcal{B}^x) \leq S^x(\mathcal{C}^*). \tag{11}$$

Applying Lemma 3 to the right-hand side of (11), we obtain

$$S^x(\mathcal{C}^*) = F(\mathcal{C}^*) + M^2\|x - \overline{y}(\mathcal{C}^*)\|^2. \tag{12}$$

Combining (11) and (12) yields the statement of the lemma. $\square$

Lemma 6. *Let the conditions of Lemma 5 hold and* $t = \arg\min_{y\in\mathcal{C}^*} \|y - \overline{y}(\mathcal{C}^*)\|^2$ *be the point from the subset* $\mathcal{C}^*$ *closest to its centroid. Then the following inequality is true*

$$\|t - \overline{y}(\mathcal{C}^*)\|^2 \leq \frac{1}{M^2}F(\mathcal{B}^t), \tag{13}$$

where $\mathcal{B}^t$ *is the set defined in Lemma 4 (for* $x = t$*).*

Proof. By the definition of point t we have

$$\|t - \overline{y}(\mathcal{C}^*)\|^2 \leq \|y - \overline{y}(\mathcal{C}^*)\|^2$$

for each $y \in \mathcal{C}^*$. Summing up both sides of this inequality over all $y \in \mathcal{C}^*$, we obtain

$$M\|t - \overline{y}(\mathcal{C}^*)\|^2 \leq \sum_{y\in\mathcal{C}^*} \|y - \overline{y}(\mathcal{C}^*)\|^2. \tag{14}$$

Since $\mathcal{C}^*$ is the optimal solution,

$$F(\mathcal{C}^*) \leq F(\mathcal{B}^t). \tag{15}$$

Then (14), (1) and (15) imply

$$M\|t - \overline{y}(\mathcal{C}^*)\|^2 \leq \sum_{y\in\mathcal{C}^*} \|y - \overline{y}(\mathcal{C}^*)\|^2 \leq \frac{1}{M}F(\mathcal{C}^*) \leq \frac{1}{M}F(\mathcal{B}^t).$$

$\square$

Lemma 7. *Let the conditions of Lemma 6 hold. Let*

$$\|x - \overline{y}(\mathcal{C}^*)\|^2 \leq \frac{\varepsilon}{2M^2}F(\mathcal{B}^t) \tag{16}$$

for some $\varepsilon > 0$ *and* $x \in \mathbb{R}^q$. *Then the subset* $\mathcal{B}^x$ *(defined in Lemma 4) is a* $(1 + \varepsilon)$*-approximate solution of Problem 1.*

Proof. From (1), Lemma 4 and the definition of the point t we have

$$F(\mathcal{B}^t) = S^{\overline{y}(\mathcal{B}^t)}(\mathcal{B}^t) \leq S^t(\mathcal{B}^t) \leq S^t(\mathcal{C}^*). \tag{17}$$

Applying Lemma 2 to the set $\mathcal{C}^*$ and the point t, we have

$$\sum_{y\in\mathcal{C}^*} \|y - t\|^2 \leq 2\sum_{y\in\mathcal{C}^*} \|y - \overline{y}(\mathcal{C}^*)\|^2.$$

Therefore, definition (6) yields

$$S^t(\mathcal{C}^*) = M \sum_{y \in \mathcal{C}^*} \|y - t\|^2 + (N - M) \sum_{y \in \mathcal{Y} \setminus \mathcal{C}^*} \|y\|^2$$

$$\le 2M \sum_{y \in \mathcal{C}^*} \|y - \overline{y}(\mathcal{C}^*)\|^2 + (N - M) \sum_{y \in \mathcal{Y} \setminus \mathcal{C}^*} \|y\|^2 \le 2F(\mathcal{C}^*). \quad (18)$$

Combining (16), (17), and (18) we obtain

$$\|x - \overline{y}(\mathcal{C}^*)\|^2 \le \frac{\varepsilon}{2M^2} F(\mathcal{B}^t) \le \frac{\varepsilon}{2M^2} S^t(\mathcal{C}^*) \le \frac{\varepsilon}{M^2} F(\mathcal{C}^*). \quad (19)$$

Finally, from Lemma 5 and (19) for the subset $\mathcal{B}^x$ we obtain the following estimate of the value of the objective function

$$F(\mathcal{B}^x) \le F(\mathcal{C}^*) + M^2 \|x - \overline{y}(\mathcal{C}^*)\|^2 \le (1 + \varepsilon) F(\mathcal{C}^*).$$

This estimate means that the subset $\mathcal{B}^x$ is a $(1 + \varepsilon)$-approximate solution for Problem 1. $\qquad \square$

4   Approximation Algorithm

In this section, we present our approximation algorithm for Problem 1. Its main idea is as follows. For each point of the input set a domain (cube) is constructed so that the center of the desired subset necessarily belongs to one of these domains. Given (as input) the prescribed relative error ε of the solution, a lattice (a grid) is generated that discretizes the cube with a uniform step in all coordinates. For each lattice node, a subset of M points from the input set that have the smallest values of the function (9) is formed (the minimum of (8) is reached at that subset). The resulting set is declared as a solution candidate. The candidate that minimizes the objective function is chosen to be the final solution.

For an arbitrary point $x \in \mathbb{R}^q$ and positive numbers h and H, we define the set of points

$$\mathcal{D}(x, h, H) = \{d \in \mathbb{R}^q \,|\, d = x + h \cdot (i_1, \ldots, i_q), i_k \in \mathbb{Z}, |hi_k| \le H, k \in \{1, \ldots, q\}\} \quad (20)$$

which is a cubic lattice of size $2H$ centered at the point x with node spacing h.

For any point $x \in \mathbb{R}^q$ the number of nodes in this lattice is

$$|\mathcal{D}(x, h, H)| \le \left(2 \left\lfloor \frac{H}{h} \right\rfloor + 1 \right)^q \le \left(2 \frac{H}{h} + 1 \right)^q. \quad (21)$$

Remark 1. If some point z from $\mathbb{R}^q$ and some node x from the lattice $\mathcal{D}(x, h, H)$ satisfy the inequality $\|z - x\| \le H$ then the distance from z to the nearest node of the lattice obviously does not exceed $\frac{h\sqrt{q}}{2}$.

For constructing an algotithmic solution we need to determine adaptively the size H of the lattice and its node spacing h for each point y of the input set $\mathcal{Y}$ so that the domain of the lattice contains the centroid of the desired subset. The node spacing is defined by the relative error ε. To this end we define the functions:

$$H(y) = \frac{1}{M}\sqrt{F(\mathcal{B}^y)}, \ y \in \mathcal{Y}, \tag{22}$$

$$h(y, \varepsilon) = \frac{1}{M}\sqrt{\frac{2\varepsilon}{q}F(\mathcal{B}^y)}, \ y \in \mathcal{Y}, \ \varepsilon \in \mathbb{R}_+, \tag{23}$$

where $\mathcal{B}^y$ is a set determined in Lemma 4, if $x = y$.

Note that all calculations in the algorithm described below are based on constructing candidate (approximate) solutions of Problem 1 as a subset $\mathcal{B}^x$ (defined in Lemma 4) for any point x from the support set of points. In this way we use two support sets. The first of them is the input set $\mathcal{Y}$ and the second one is the set of nodes of the lattice $\mathcal{D}(y, h, H)$ centered at y. The lattice is adaptively calculated by formulae (22) and (23) for each input point $y \in \mathcal{Y}$. The approximation factor is finally bounded using the basic statements in Sect. 3.

Remark 2. For any point $y \in \mathcal{Y}$ the cardinality $|\mathcal{D}(y, h, H)|$ of the lattice does not exceed the value

$$L = \left(\sqrt{\frac{2q}{\varepsilon}} + 1\right)^q$$

due to (21), (22), and (23).

Below is the step-by-step description of the algorithm.

Algorithm $\mathcal{A}$.
Input: a set $\mathcal{Y}$ and numbers M and ε.
For each point $y \in \mathcal{Y}$ Steps 1–6 are executed.
Step 1. Compute the values $g^y(z)$, $z \in \mathcal{Y}$, using formula (9); find a subset $\mathcal{B}^y \subseteq \mathcal{Y}$ with M smallest values $g^y(z)$, compute $F(\mathcal{B}^y)$ using formula (1).
Step 2. If $F(\mathcal{B}^y) = 0$, then put $\mathcal{C}_\mathcal{A} = \mathcal{B}^y$; exit.
Step 3. Compute H and h using formulae (22) and (23).
Step 4. Construct the lattice $\mathcal{D}(y, h, H)$ using formula (20).
Step 5. For each node x of the lattice $\mathcal{D}(y, h, H)$ compute the values $g^x(y)$, $y \in \mathcal{Y}$, using formula (9) and find a subset $\mathcal{B}^x \subseteq \mathcal{Y}$ with M smallest values $g^x(y)$. Compute $F(\mathcal{B}^x)$ using formula (1), remember this value and the set $\mathcal{B}^x$.
Step 6. If $F(\mathcal{B}^x) = 0$, then put $\mathcal{C}_\mathcal{A} = \mathcal{B}^x$; exit.
Step 7. In the family $\{\mathcal{B}^x \,|\, x \in \mathcal{D}(y, h, H), y \in \mathcal{Y}\}$ of candidate sets that have been constructed in Steps 1–6, choose as a solution $\mathcal{C}_\mathcal{A}$ the set $\mathcal{B}^x$ for which $F(\mathcal{B}^x)$ is minimal.
Output: the set $\mathcal{C}_\mathcal{A}$.

Theorem 1. *For any fixed $\varepsilon > 0$ Algorithm $\mathcal{A}$ finds a $(1 + \varepsilon)$-approximate solution of Problem 1 in $\mathcal{O}\left(qN^2\left(\sqrt{\frac{2q}{\varepsilon}} + 1\right)^q\right)$ time.*

Proof. Let us bound the approximation factor of the algorithm. If the equality $F(\mathcal{B}^y) = 0$ holds at Step 2 for some point $y \in \mathcal{Y}$, then the subset $\mathcal{B}^y \subseteq \mathcal{Y}$ is an optimal solution of Problem 1, since, for any set $\mathcal{C} \subseteq \mathcal{Y}$, it is true that $F(\mathcal{C}) \geq 0$. We get an optimal solution at Step 6 in the same way.

Consider the case when the condition $F(\mathcal{B}^y) = 0$ at Step 2 does not hold. Obviously, there exists a point $t \in \mathcal{Y}$ such that $t = \arg\min_{y \in C^*} \|y - \overline{y}(C^*)\|$ and the algorithm meets it at least once in the set $\mathcal{Y}$ while running. By Lemma 6, inequality (13) holds for this point. This inequality and (22) mean that $\|t - \overline{y}(C^*)\| \leq H(t)$, so the centroid of the optimal subset lies within the lattice $\mathcal{D}(t, h, H)$ of the size $H = H(t)$ and the node spacing $h = h(t, \varepsilon)$.

Let $x^* = \arg\min_{x \in \mathcal{D}(t,h,H)} \|x - \overline{y}(C^*)\|$ be a node of the grid $\mathcal{D}(t, h, H)$, the nearest to the centroid of the optimal subset. Since the squared distance from the optimal centroid $\overline{y}(C^*)$ to the nearest node x^* of the lattice does not exceed $\frac{h^2 q}{4}$ (by remark 1), we have the estimate

$$\|x^* - \overline{y}(C^*)\|^2 \leq \frac{h^2 q}{4} = \frac{\varepsilon}{2M^2} F(\mathcal{B}^t).$$

Therefore, the point x^* satisfies the conditions of Lemma 7 and, hence, the set $\mathcal{B}^{x^*}$ is a $(1 + \varepsilon)$-approximate solution of Problem 1.

It is clear, that any subset $\mathcal{B}^x$ in the family of candidate solutions on Step 7 constructed for node x such that $\|x - \overline{y}(C^*)\|^2 \leq \|x^* - \overline{y}(C^*)\|^2$ guarantees a $(1 + \varepsilon)$-approximation also.

Let us evaluate the time complexity of the algorithm.

At Step 1 calculation of $g^y(z)$ requires at most $\mathcal{O}(qN)$-time. Finding the M smallest elements in the set of N elements is performed in $\mathcal{O}(N)$ operations (for example, using the algorithm of finding the n-th smallest value in an unordered array [27]). Computation of the value $F(\mathcal{B}^y)$ takes $\mathcal{O}(qN)$ time.

Steps 2, 3 and 6 are executed in $\mathcal{O}(1)$ operations. It requires $\mathcal{O}(qL)$ operations for generating the lattice at Step 4 (by remark 2).

At Step 5, computation of the elements of the set $\mathcal{B}^x$ for each node of the grid requires $\mathcal{O}(qN)$ time, and the same is true for the computation of $F(\mathcal{B}^x)$ (as computations at Step 1). Thus, at this step the computational time for all nodes of the grid is $\mathcal{O}(qNL)$.

Since Steps 1–6 are performed N times, the time complexity of these steps is $\mathcal{O}(qN^2 L)$. The time complexity of Step 7 is bounded by $\mathcal{O}(NL)$, and the total time complexity of all Steps is $\mathcal{O}(qN^2 L)$. Therefore, the time complexity of Algorithm $\mathcal{A}$ is $\mathcal{O}\left(qN^2 \left(\sqrt{\frac{2q}{\varepsilon}} + 1\right)^q\right)$.   □

Remark 3. In the case when the dimension q of space is bounded by a constant value and $\varepsilon < 2q$, we have

$$qN^2 \left(1 + \sqrt{\frac{2q}{\varepsilon}}\right)^q \leq qN^2 2^q \left(\frac{2q}{\varepsilon}\right)^{q/2} = \mathcal{O}\left(N^2 \left(\frac{1}{\varepsilon}\right)^{q/2}\right),$$

and it means that Algorithm $\mathcal{A}$ is an FPTAS.

Remark 4. It is clear that the constructed algorithm can be applied for solving a problem in which the cardinalities of the clusters are the optimized variables. For this purpose, it is sufficient to solve Problem 1 N times with the help of Algorithm $\mathcal{A}$ for each $M = 1, \ldots, N$, and then choose the best of these solutions in the sense of minimizing the objective function. The time complexity of this algorithm obviously equals $\mathcal{O}\left(N^3 \left(\frac{1}{\varepsilon}\right)^{q/2}\right)$. But it is interesting to construct algorithms with less time complexity without searching for such candidate solutions.

5 Conclusion

In this paper we presented an approximation algorithm for one strongly NP-hard quadratic Euclidian problem of balanced partitioning a finite set of points into two clusters. It was proved that our algorithm is a fully polynomial-time approximation scheme if the space dimension is bounded by a constant.

In the algorithmical sense, the considered problem is poorly studied. Therefore, it seems important to continue studying the questions on algorithmical approximability of the problem.

Acknowledgments. This work was supported by the RFBR, projects 15-01-00462, 16-31-00186 and 16-07-00168.

References

1. Kel'manov, A.V., Pyatkin, A.V.: NP-hardness of some quadratic euclidean 2-clustering problems. Doklady Math. **92**(2), 634–637 (2015)
2. Kel'manov, A.V., Pyatkin, A.V.: On the complexity of some quadratic euclidean 2-clustering problems. Comput. Math. Math. Phys. **56**(3), 491–497 (2016)
3. Aggarwal, C.C.: Data Mining: The Textbook. Springer International Publishing, Switzerland (2015)
4. Bishop, C.M.: Pattern Recognition and Machine Learning. Springer Science+Business Media, LLC, New York (2006)
5. Hastie, T., Tibshirani, R., Friedman, J.: The Elements of Statistical Learning: Data Mining, Inference, and Prediction. Springer-Verlag, New York (2001)
6. Garey, M.R., Johnson, D.S.: Computers and Intractability: A Guide to the Theory of NP-Completeness. Freeman, San Francisco (1979)
7. Kel'manov, A.V., Motkova, A.V.: An exact pseudopolynomial algorithm for a special case of a euclidean balanced variance-based 2-clustering problem. In: Abstracts of the VI International Conference "Optimization and Applications" (OPTIMA-2015), P. 98. Petrovac, Montenegro (2015)
8. Sahni, S., Gonzalez, T.: P-complete approximation problems. J. ACM **23**, 555–566 (1976)
9. Brucker, P.: On the complexity of clustering problems. Lect. Notes Econ. Math. Syst. **157**, 45–54 (1978)
10. Inaba, M., Katoh, N., Imai, H.: Applications of Weighted Voronoi Diagrams and Randomization toVariance-Based k-Clustering: (extended abstract). Stony Brook, NY, USA, pp. 332–339 (1994)

11. Hasegawa, S., Imai, H., Inaba, M., Katoh, N., Nakano, J.: Efficient algorithms for variance-based k-clustering. In: Proceedings of the 1st Pacific Conference on Computer Graphics andApplications (Pacific Graphics 1993, Seoul, Korea),World Scientific, River Edge, NJ. 1, pp. 75–89 (1993)

12. de la Vega, F., Kenyon, C.: A randomized approximation scheme for metric max-cut. J. Comput. Syst. Sci. **63**, 531–541 (2001)

13. de la Vega, F., Karpinski, M., Kenyon, C., Rabani, Y.: Polynomial Time Approximation Schemes for Metric Min-Sum Clustering. Electronic Colloquium on Computational Complexity (ECCC), 25 (2002)

14. Aloise, D., Deshpande, A., Hansen, P., Popat, P.: NP-hardness of euclidean sum-of-squares clustering. Mach. Learn. **75**(2), 245–248 (2009)

15. Fisher, R.A.: Statistical Methods and Scientific Inference. Hafner Press, New York (1956)

16. Rao, M.: Cluster analysis and mathematical programming. J. Amer. Statist. Assoc. **66**, 626–662 (1971)

17. Gimadi, E.K., Kel'manov, A.V., Kel'manova, M.A., Khamidullin, S.A.: Aposteriori finding of a quasiperiodic fragment with a given number of repetitions in a numerical sequence (in Russian). Sib. Zh. Ind. Mat. **9**(25), 55–74 (2006)

18. Gimadi, E.K., Kel'manov, A.V., Kel'manova, M.A., Khamidullin, S.A.: A posteriori detecting a quasiperiodic fragment in a numerical sequence. Pattern Recogn. Image Anal. **18**(1), 30–42 (2008)

19. Dolgushev, A.V., Kel'manov, A.V.: An approximation algorithm for solving a problem of cluster analysis. J. Appl. Indust. Math. **5**(4), 551–558 (2011)

20. Dolgushev, A.V., Kel'manov, A.V., Shenmaier, V.V.: A polynomial-time approximation scheme for a problem of partitioning a finite set into two clusters (in Russian). Trudy Inst. Mat. i Mekh. UrO. RAN. **21**(3), 100–109 (2015)

21. Kel'manov, A.V., Khandeev, V.I.: A 2-approximation polynomial algorithm for a clustering problem. J. Appl. Indust. Math. **7**(4), 515–521 (2013)

22. Kel'manov, A.V., Khandeev, V.I.: A randomized algorithm for two-cluster partition of a set of vectors. Comput. Math. Math. Phys. **55**(2), 330–339 (2015)

23. Kel'manov, A.V., Khandeev, V.I.: An exact pseudopolynomial algorithm for a problem of the two-cluster partitioning of a set of vectors. J. Appl. Indust. Math. **9**(4), 497–502 (2015)

24. Kel'manov, A.V., Khandeev, V.I.: Fully polynomial-time approximation scheme for a special case of a quadratic euclidean 2-clustering problem. Comput. Math. Math. Phys. **56**(2), 334–341 (2016)

25. Kel'manov, A.V., Romanchenko, S.M.: An approximation algorithm for solving a problem of search for a vector subset. J. Appl. Ind. Math. **6**(1), 90–96 (2012)

26. Kel'manov, A.V., Romanchenko, S.M.: An FPTAS for a vector subset search problem. J. Appl. Indust. Math. **8**(3), 329–336 (2014)

27. Wirth, N.: Algorithms + Data Structures = Programs. Prentice Hall, New Jersey (1976)

PTAS for the Euclidean Capacitated Vehicle Routing Problem in R^d

Michael Khachay[1,2,3]([✉]) and Roman Dubinin[2]

[1] Krasovskii Institute of Mathematics and Mechanics, Ekaterinburg, Russia
mkhachay@imm.uran.ru
[2] Ural Federal University, Ekaterinburg, Russia
romandubinin94@gmail.com
[3] Omsk State Technical University, Omsk, Russia

Abstract. Capacitated Vehicle Routing Problem (CVRP) is the well-known combinatorial optimization problem remaining NP-hard even in the Euclidean spaces of fixed dimension. Thirty years ago, in their celebrated paper, M. Haimovich and A. Rinnoy Kan proposed the first PTAS for the Planar Single Depot CVRP based on their Iterated Tour Partition heuristic. For decades, this result was extended by many authors to numerous useful modifications of the problem taking into account multiple depots, pick up and delivery options, time window restrictions, etc. But, to the best of our knowledge, almost none of these results go beyond the Euclidean plane. In this paper, we try to bridge this gap and propose an EPTAS for the Euclidean CVRP for any fixed dimension.

Keywords: Vehicle routing · Euclidean space · EPTAS

1 Introduction

We consider the Capacitated Vehicle Routing Problem, which is the well-known special case of Vehicle Routing Problem [17] belonging to the class of combinatorial optimization models widely adopted in operations research. It is generally believed that, for the first time, as an optimization problem, the VRP was introduced by Dantzig and Ramser in their seminal paper [5]. They considered a routing problem for a fleet of gasoline delivery trucks servicing a number of gas stations supplied by a unique bulk terminal. Demands of serviced gas stations and distances between any two locations were specified. The goal was to find the least cost set of truck routes visiting all the stations.

In its simplest setting, the VRP can be defined as the combinatorial optimization problem aiming at designing the cheapest collection of delivery routes from some dedicated point (*depot*) to a set of customers (*clients*) given by their spatial locations. This problem has many known modifications [9,14] taking into account different additional features and constraints, e.g. depots multiplicity, heterogeneity of customer demand, vehicle capacity, time windows, etc.

In this paper, we suppose that all clients have the same one unit demand and all vehicles have the same *capacity*, equal to some predefined number q.

© Springer International Publishing Switzerland 2016
Y. Kochetov et al. (Eds.): DOOR 2016, LNCS 9869, pp. 193–205, 2016.
DOI: 10.1007/978-3-319-44914-2_16

This specific problem is called [10] Capacitated Vehicle Routing Problem or CVRP. Complexity of this problem is determined by its closeness to some well-known intractable combinatorial problems. For instance, Traveling Salesman Problem (TSP) is just a special case of the CVRP such that a depot is col-located with one of clients and $q \geq n$. Therefore, the CVRP is strongly NP-hard even in the Euclidean plane, since the same results are proven for the TSP [15]. Almost all known special cases of the CVRP (except the case when $q \leq 2$) (see, e.g. [14]) are also NP-hard even in the Euclidean spaces of finite dimension.

For these reasons, research on the CVRP is mostly focused on design of approximation algorithms and heuristics. For a general metric, the CVRP is shown to be APX-complete [2] for any fixed $q \geq 3$, i.e. there exists $\varepsilon > 0$ such that the existence of a polynomial time $(1 + \varepsilon)$-approximation algorithm implies $P = NP$.

Most positive approximation results for CVRP are obtained for the Euclidean plane. One of the first studies of two-dimensional Euclidean CVRP has been due to Haimovich and Rinnooy Kan [10], who presented several heuristics for this problem leading to the first PTAS for $q = O(\log \log n)$. Asano et al. [2] substantially improved this result by designing a PTAS for $q = O(\log n / \log \log n)$. Also they construct a PTAS for the case of $\Omega(n)$ based on the famous Arora's PTAS [1] for the two-dimensional Euclidean TSP. Recently, Das and Mathieu [6,7] proposed a quasi-polynomial time approximation scheme (QPTAS) for the two-dimensional Euclidean CVRP for every q with time complexity of $n^{(\log n)^{O(1/\varepsilon)}}$. Khachay and Zaytseva [13] applied the approach proposed in [10] to the construction a PTAS for the Single Depot CVRP in three-dimensional Euclidean space.

The extension of the latter result to the case of any fixed number m of depots and any fixed dimension $d > 1$ is the main contribution of this paper. Actually, on the basis of recent geometric results describing the structure of finite ε-nets on the surface of the unit Euclidean sphere S^{d-1}, we propose a new Efficient Polynomial Time Approximation Scheme[1] (EPTAS) for the Euclidean CVRP, for which capacity q, the number of depots m and dimension $d > 1$ are fixed. The algorithm proposed remains PTAS for the problem with fixed m and d and $q = O(\log \log n)^{1/d}$.

The rest of the paper is organized as follows. In Sect. 2, we recall the general statement of the CVRP along with its metric and Euclidean settings. In Sect. 3, we describe our EPTAS based on the famous Iterated Tour Partition [10] for the case of single depot. Further, in Sect. 4 we extend this result to the case of an arbitrary fixed number m of depots. Section 5 contains summarizing remarks and a short overview of future work.

2   Problem Statement

Recall the necessary definitions and notation.
1. $X = \{x_1, \ldots, x_n\}$ is a set of clients, $Y = \{y_1, \ldots, y_m\}$ is a set of depots. Denote by $G^0(X \cup Y, E, w)$ a complete weighted digraph, whose weight function

---

[1] A PTAS with time complexity $f(1/\varepsilon)p(n)$ for some polynomial p.

$w : E \to \mathbb{R}_+$ defines transportation costs for any pair of locations. Hereinafter, the function w is supposed to be symmetric. For any route R consisting of arcs $e_1, \ldots, e_p$, its *cost* $w(R)$ is defined by the equation $w(R) = \sum_{i=1}^{p} w(e_i)$. Along with the digraph G^0, we consider its subgraph $G = G^0\langle X \rangle$ induced by the vertex subset X.

2. To any client x_i, assign a number $r_i = \min\{w(y_j, x_i) : j = 1, \ldots, m\}$ defining the least direct transportation cost among the depots. Breaking ties arbitrarily, define a partition $X_1 \cup \ldots \cup X_m = X$ into subsets

$$X_j = \{x_i \in X : r_i = w(x_i, y_j)\}, \tag{1}$$

such that any client x_i is assigned to the nearest depot y_j.

3. Any feasible route has a form $y_{j_s}, x_{i_1}, \ldots, x_{i_t}, y_{j_f}$, where y_{j_s} and y_{j_f} are depots[2], $x_{i_1}, \ldots, x_{i_t}$ are distinct clients visited by this route, and $t \leq q$.

If $m = 1$ the problem in question is called the Single Depot Capacitated Vehicle Routing Problem (SDCVRP). In this case, all feasible routes are simple circuits. Otherwise, the problem is called the Multiple Depot CVRP (MDCVRP). We distinguish two special settings of this problem. In the first one, we denote it MDCVRP1, any feasible route can start and terminate at separate depots. In the second one, MDCVRP2, for any feasible route, its start and finish depots should be identical ($y_{j_s} = y_{j_f}$).

For any aforementioned setting, the goal is, for a given digraph $G^0(X \cup Y, E, w)$ and a capacity q, to find a cheapest set of routes visiting each client exactly once.

Along with the general setting of the SDCVRP and MDCVRP we consider two important their special cases defined in terms of the weight function w.

Metric CVRP. In this case, the graph G^0 is supposed to be undirected and w meets the triangle inequality $w(z_1, z_2) \leq w(z_1, z_3) + w(z_3, z_2)$ for each $z_1, z_2, z_3 \in X \cup Y$.

Euclidean CVRP. Here, $X \cup Y \subset \mathbb{R}^d$ and $w(z_1, z_2) = \|z_1 - z_2\|_2$.

3 Approximability of SDCVRP

The main idea of our approach stems from the famous Iterated Tour Partition (ITP) heuristic introduced by Haimovich and Rinnooy Kan [10] and presented below as Algorithm 1. Using ITP in combination with approximation algorithms for the metric TSP, we construct polynomial time algorithms with asymptotically fixed performance guarantees for the metric SDCVRP and polynomial time approximation scheme for the d-dimensional Euclidean SDCVRP for any fixed $d > 1$. Further, in Sect. 4, we extend this approach to the case of multiple depots.

For our constructions, we need the following technical claims proved for the first time in [10]. Although, all the proofs in that paper were carried out for the

---

[2] Not necessarily distinct.

Algorithm 1. ITP heuristic

Input: a complete weighted digraph $G^0(X \cup \{y\}, E, w)$ of order n, a natural number q and an arbitrary Hamiltonian circuit H in G.
Output: an approximate solution S_{ITP} of SDCVRP.

1: **for all** $x \in H$ **do**
2:     starting from the vertex x, partition the circuit H into $l = \lceil n/q \rceil$ chains, each of them, except maybe one, spans q vertexes;
3:     connecting endpoints of each chain with the depot y directly, construct a set $S(x)$ of l routes;
4: **end for**
5: output the set $S_{\text{ITP}} = \arg\min\{w(S(x)) \colon x \in H\}$.

Euclidean plane only, it is easy to verify that they remain true in the much more general setting of the CVRP as well. For the sake of brevity, we skip the proofs (see [13] for details).

Lemma 1. *For* $\bar{r} = 1/n \sum_{i=1}^{n} r_i$, *the following equation*

$$w(S_{\text{ITP}}) \le 2 \lceil n/q \rceil \bar{r} + \left(1 - \frac{\lceil n/q \rceil}{n} \right) w(H) \le 2 \lceil n/q \rceil \bar{r} + (1 - 1/q) w(H) \quad (2)$$

is valid.

It should be noted that the upper bound claimed in Lemma 1 is valid for the most general setting of the SDCVRP. In the metric case, for any feasible solution S of the SDCVRP, we can obtain also a lower bound on $w(S)$ by means of the cost of the Hamiltonian circle H_S induced by S in the graph G.

Indeed, let S consists of routes $C_1, \ldots, C_t$. Excluding the depot y from each route C_i and connecting arbitrarily the chains obtained to produce a single (Hamiltonian) circle H_S, we obtain the following bound.

Lemma 2.
$$w(S) \ge \max\left\{ 2n\bar{r}/q, w(H_S) \right\}. \quad (3)$$

Combining the bounds given by Lemmas 1 and 2, we obtain the following equation relating the optimum value $\text{VRP}^*(X, \{y\})$ of the metric SDCVRP and the weight $\text{TSP}^*(X)$ of an optimal Hamiltonian circle in the corresponding TSP instance.

Theorem 1.

$$\min\left\{ 2\bar{r}n/q, \text{TSP}^*(X) \right\} \le \text{VRP}^*(X, \{y\}) \le 2\lceil n/q \rceil \bar{r} + (1 - 1/q) \text{TSP}^*(X).$$

The results above give us an ability to represent a performance guarantee of any ITP-based approximation algorithm for metric SDCVRP in terms of heuristics used for obtaining approximate solutions of the inner TSP. Indeed, suppose, we obtain a Hamiltonian cycle H in the graph G, whose cost $TSP^* \leq w(H) \leq \rho TSP^*$ for some $\rho \geq 1$. Using (2) and (3), we obtain

$$\frac{w(S)}{VRP^*(X, \{y\})} \leq \frac{2 \lceil n/q \rceil \bar{r} + (1 - 1/q) \rho TSP^*(X)}{\max \{2\bar{r}n/q, TSP^*(X)\}} \leq \frac{q}{n} + 1 + \rho. \qquad (4)$$

Since the RHS of Eq. (4) tends to $1 + \rho$ any time when $q = o(n)$, an arbitrary ρ-approximation algorithm for the metric TSP produces asymptotically $(1 + \rho)$-approximation algorithm for the metric SDCVRP.

Further, since the running time of the ITP is at most $O(n^2)$, the overall time complexity of any based-on-ITP approximation algorithm is defined by the running time of the initial approximation algorithm for the metric TSP. For instance, the famous Christofides' 3/2-approximation algorithm [4] with the running time of $O(n^3)$ produces asymptotically 5/2-approximation algorithm with the same time complexity bound, and the well-known Arora's PTAS [1] for the d-dimensional Euclidean TSP, for any fixed $d > 1$ and any $\varepsilon \in (0, 1)$, produces asymptotically $(2 + \varepsilon)$-approximation algorithm with the time complexity of $(n(\log n)^{(O(\sqrt{d}/\varepsilon))^{d-1}}$.

To proceed with PTAS for the Euclidean SDCVRP, we recall Algorithm 2 proposed in [10].

Algorithm 2. Combined ITP scheme (CITP)

Input: a complete weighted graph $G^0(X \cup \{y\}, E, w)$ of order n, natural number q, and an upper relative error bound $\varepsilon > 0$.
Output: an approximate solution S_{CITP} of the SDCVRP.
 1: relabel the clients so that $r_1 \geq r_2 \geq \ldots \geq r_n$;
 2: for some value $k = k(\varepsilon)$ (which will be specified later), partition X into subsets $X(k) = \{x_1, \ldots, x_{k-1}\}$ of *inner* and $X \setminus X(k)$ *outer* clients;
 3: find an exact solution $S^*(X(k))$ of the instance of SDCVRP specified by the subgraph $G^0 \langle X(k) \cup \{y\} \rangle$;
 4: apply Algorithm 1 for construction of an approximate solution $S_{ITP}(X \setminus X(k))$ of the SDCVRP defined by the subgraph $G^0 \langle X \setminus X(k) \cup \{y\} \rangle$;
 5: output $S_{CITP} = S^*(X(k)) \cup S_{ITP}(X \setminus X(k))$.

Algorithm 2 makes a decomposition of the initial instance into two smaller instances of the SDCVRP. The first subproblem, for the outer clients, is supposed to be solved to optimality, while for the second one describing the inner clients, an approximate solution is found by Algorithm 1. Lemma 3 helps us to relate optimum values of these two subproblems.

Lemma 3. *For an arbitrary $k \in \{1, \ldots, n\}$ the equation*

$$VRP^*(X, \{y\}) \leq VRP^*(X(k), \{y\}) + VRP^*(X \setminus X(k), \{y\})$$
$$\leq VRP^*(X, \{y\}) + 4(k - 1)r_k.$$

is valid.

As previous results, Lemma 3 remains true for an arbitrary metric.

Further, we restrict ourselves to finite dimensional Euclidean spaces. All remaining assertions of this section are based on the following existence lemma for a finite ε-net on the surface of the unit Euclidean sphere S^{d-1} in terms of the angular distance

$$dist(x_1, x_2) = \arccos(x_1, x_2), \ (x_1, x_2 \in S^{d-1})$$

(see, e.g. [11], Lemma 3.1). According to the classic definition (see, e.g. [16]), we call some finite subset $N \subset S^{d-1}$ a finite ε-net (on the sphere S^{d-1}) if, for any $x \in S^{d-1}$, there exists $\xi \in N$ such that $dist(\xi, x) \leq \varepsilon$.

Lemma 4. *For an arbitrary* $h \in (0, h_0)$, $h_0 = \pi/(6\sqrt{d-1})$, *on the sphere* S^{d-1} *there exists an* $h\sqrt{d-1}$-*net* $N = N(d, h)$ *such that* $|N| = Ch^{-(d-1)}$ *for some constant* $C = C(d)$.

Using the claim of Lemma 4, we obtain an upper bound for the optimum value $\text{TSP}^*(X)$ of an instance of the Euclidean TSP in terms of the radius of an enclosing sphere. Suppose, a TCP instance is specified by a set $X = \{x_1, \ldots, r_n\}$ contained within the Euclidean ball $B(y, R) \subset \mathbb{R}^d$ of radius R centered at y. As above, we assume that all the clients are numbered in non-increasing order of their distances $r_i = \|x_i - y\|_2$ from the depot y.

Lemma 5. *For an arbitrary* $d > 1$ *and a finite* $X \subset B(y, R)$ *the following bounds*

$$\text{TSP}^*(X) \leq \begin{cases} C_1 R^{1/d} (\sum_{i=1}^n r_i)^{(d-1)/d}, \text{ if } \sum_{i=1}^n r_i > RC(\pi/6)^{-d}(d-1)^{(d+1)/2}, \\ C_2 R, \text{ otherwise,} \end{cases}$$

are valid, where

$$C_1 = 2dC^{1/d}(d-1)^{(d-1)/2d} \text{ and } C_2 = 2dC(\pi/6)^{-(d-1)}(d-1)^{(d-1)/2}.$$

Proof. By Lemma 4, for any $h \in (0, h_0)$, $h_0 = \pi/(6\sqrt{d-1})$ on the surface of the ball $B(y, R)$ there exists a finite $h\sqrt{d-1}$-net N of $Ch^{-(d-1)}$ elements. Connect any $\xi_j \in N$ with the center y by radial segment, after that connect each client x_i with the nearest radius $[y, \xi_j]$ (by the appropriate orthogonal line segment). Further, we construct a salesman tour by the well-known edge-doubling technique for the tree obtained.

Let $\Phi(h)$ be the length of the tour constructed. Again, by Lemma 4, for any $h \in (0, h_0)$,

$$\text{TSP}^*(X) \leq \Phi(h) = 2h\sqrt{d-1}\sum_{i=1}^n r_i + 2RCh^{-(d-1)}. \tag{5}$$

Minimizing the RHS of Eq. (5) subject to $0 < h < \pi/(6\sqrt{d-1})$, we obtain the claimed bounds.

Indeed, $\inf \Phi(h)$ on this range coincides either with $\Phi(h_{min})$, where

$$h_{min} = \left(\frac{RC}{\sum_{i=1}^{n} r_i} \sqrt{d-1} \right)^{1/d},$$

if

$$h_{min} < h_0, \text{ i.e. } \sum_{i=1}^{n} r_i > RC(d-1)^{(d+1)/2}(\pi/6)^{-d},$$

or with $\Phi(h_0)$, otherwise. In the first case, we obtain

$$\Phi(h_{min}) = 2 \left(\frac{RC}{\sum_{i=1}^{n} r_i} \sqrt{d-1} \right)^{1/d} \sqrt{d-1} \sum_{i=1}^{n} r_i + 2RC \left(\frac{RC}{\sum_{i=1}^{n} r_i} \sqrt{d-1} \right)^{-(d-1)/d}$$

$$= \underbrace{2dC^{1/d}(d-1)^{-(d-1)/(2d)}}_{C_1} \cdot R^{1/d} (\sum_{i=1}^{n} r_i)^{(d-1)/d}$$

If, on the other hand,

$$\sum_{i=1}^{n} r_i \leq RC(d-1)^{(d+1)/2}(\pi/6)^{-d}, \tag{6}$$

then

$$\Phi(h_0) = 2\sqrt{d-1} \frac{\pi}{6\sqrt{d-1}} \sum_{i=1}^{n} r_i + 2RC \left(\frac{\pi}{6} \right)^{d-1} (\sqrt{d-1})^{(d-1)}$$

$$\leq \frac{\pi}{3} RC(d-1)^{(d+1)/2} \left(\frac{\pi}{6} \right)^{-d} + 2RC \left(\frac{\pi}{6} \right)^{-(d-1)} (d-1)^{(d-1)/2}$$

$$= RC \left(\frac{\pi}{6} \right)^{-d} \left(\frac{\pi}{3}(d-1) + 2 \cdot \frac{\pi}{6} \right) (d-1)^{(d-1)/2}$$

$$= \underbrace{2Cd \left(\frac{\pi}{6} \right)^{-(d-1)} (d-1)^{(d-1)/2}}_{C_2} \cdot R.$$

Lemma 5 is proved.

Further, for some $d > 1$, consider an instance of the SDCVRP in the d-dimension Euclidean space. By

$$e(k) = \frac{w(S_{\text{CITP}}(X)) - \text{VRP}^*(X)}{\text{VRP}^*(X)}$$

$$= \frac{\text{VRP}^*(X(k)) + w(S_{\text{ITP}}(X \setminus X(k))) - \text{VRP}^*(X)}{\text{VRP}^*(X)},$$

denote a relative error of the approximate solution produced by Algorithm 2, using for some ρ a ρ-approximation algorithm for finding an approximate solution of the inner TSP.

Lemma 6. *For an arbitrary $\rho \geq 1$ and $\varepsilon > 0$ there exists $k = k(\varepsilon) \in \mathbb{N}$ such that $e(k) \leq \varepsilon$.*

Proof. Applying the claims of Lemmas 1–3 and introducing the notation

$$\bar{r}_k = \frac{\sum_{i=k}^{n} r_i}{n - k + 1},$$

we obtain

$$e(k) \leq \frac{4(k-1)r_k + 2\lceil (n-k+1)/q \rceil \bar{r}_k + \rho \mathrm{TSP}^*(X \setminus X(k)) - 2\bar{r}_k(n-k+1)/q}{2n\bar{r}/q}$$

$$\leq q(2k-1)\frac{r_k}{\sum_{i=1}^{n} r_i} + \frac{q\rho}{2\sum_{i=1}^{n} r_i}\mathrm{TSP}^*(X \setminus X(k)).$$

By Lemma 5,

$$e(k) \leq q(2k-1)\frac{r_k}{\sum_{i=1}^{n} r_i} + \frac{q\rho}{2}\max\left\{ C_1 \left(\frac{r_k}{\sum_{i=1}^{n} r_i} \right)^{1/d}, C_2 \frac{r_k}{\sum_{i=1}^{n} r_i} \right\}$$

$$\leq q(2k-1)\frac{r_k}{\sum_{i=1}^{n} r_i} + \frac{q\rho}{2}\max\{C_1, C_2\} \left(\frac{r_k}{\sum_{i=1}^{n} r_i} \right)^{1/d},$$

since $r_k \leq \sum_{i=1}^{n} r_i$.

Further, denote $(r_k/\sum_{i=1}^{n} r_i)^{1/d}$ by s_k. Suppose that, for any $t \in \{1, \ldots, k\}$,

$$q(2t-1)s_t^d + \frac{q\rho}{2}C^* s_t > \varepsilon \tag{7}$$

is valid, where $C^* = \max\{C_1, C_2\}$ depends on d ultimately.

There exist two options. In the first option, $s_t \geq \varepsilon/(q\rho C^*)$ for each t. Then,

$$1 \geq \sum_{t=1}^{k} s_t^d \geq k \left(\frac{\varepsilon}{q\rho C^*} \right)^d,$$

therefore,

$$k \leq \left(\frac{q\rho C^*}{\varepsilon} \right)^d. \tag{8}$$

Consider the other option. Let t_0 be the smallest number, for which

$$s_{t_0} < \varepsilon/(q\rho C^*).$$

By construction, the same inequality is valid also for each $t_0 \le t \le k$, and, by (7), $s_t^d > \varepsilon/(2q(2t-1))$. Combining the bounds obtained, we get

$$1 \ge \sum_{t=1}^{k} s_t^d \ge (t_0 - 1) \left(\frac{\varepsilon}{q\rho\,C^*}\right)^d + \frac{\varepsilon}{2q} \sum_{t=t_0}^{k} \frac{1}{2t-1}$$

$$\ge (t_0 - 1) \left(\frac{\varepsilon}{q\rho\,C^*}\right)^d + \frac{\varepsilon}{2q} \int_{t_0}^{k+1} \frac{dt}{2t-1}$$

$$= (t_0 - 1) \left(\frac{\varepsilon}{q\rho\,C^*}\right)^d + \frac{\varepsilon}{4q}(\ln(2k+1) - \ln(2t_0 - 1)). \quad (9)$$

Without loss of generality suppose that $\varepsilon \le 4q\rho$. This equation together with the obvious (for $d > 1$) condition $C^* \ge 4$ implies

$$\left(\frac{\varepsilon}{q\rho\,C^*}\right)^d \le \frac{\varepsilon}{4q},$$

and

$$\left(\frac{\varepsilon}{q\rho\,C^*}\right)^{-d} \ge t_0 - 1 + \ln(2k+1) - \ln(2t_0 - 1). \quad (10)$$

Minimizing the RHS of (10) subject to $t_0 \in \{1, \ldots, k\}$, we obtain

$$k \le \frac{1}{2} e^{\left(\frac{q\rho\,C^*}{\varepsilon}\right)^d}. \quad (11)$$

Comparing bounds (8) and (11), come to the decision that the segment

$$\left[1, \frac{1}{2} e^{\left(\frac{q\rho\,C^*}{\varepsilon}\right)^d} + 1\right] \quad (12)$$

definitely contains the required number $k = k(\varepsilon)$. Lemma is proved.

Theorem 2. *Suppose that ρ-approximation algorithm with the running time of $O(n^c)$ is used for the inner TSP, then, for any fixed q, $\rho \ge 1$, and $d \ge 2$, Algorithm 2 is an Efficient Polynomial Time Approximation Scheme (EPTAS) for the SDCVRP.*

Proof. Indeed, for a given $\varepsilon > 0$, we can find $k(\varepsilon)$ such that $e(k) \le \varepsilon$ by Lemma 6. An exact solution $S^*(X(k(\varepsilon))$ can be found by dynamic programming (see, e.g. [3]) in time $O(K^q 2^K)$, where K is the upper end of the segment (12). The rest of Algorithm 2 requires $O(n^c) + O(n^2)$ time. Therefore, the overall time complexity of Algorithm 2 can be bounded from above by a polynomial function of n, whose order and all the coefficients except the constant term does not depend on ε. That is, Algorithm 2 is an EPTAS for the SDCVRP for any fixed q, $\rho \ge 1$, and $d \ge 2$. Theorem is proved.

It should be noted that Algorithm 2 remains a PTAS for the SDCVRP even under slightly relaxed restrictions on its parameters, e.g. for any fixed d, ρ, and $q = O((\log\log(n))^{1/d})$.

4   PTAS for the MDCVRP

Further, we extend the results of Sect. 3 to the case of multiple depots. The main idea of such an extension for the Euclidean plane is proposed in [3]. The authors proposed to partition the client set X according to equation (1), after that the initial MDCVRP can be decomposed into a collection of the appropriate SDCVRP instances for subsets $X_j \setminus X(k) \cup \{y_j\}$. Below, we give a short overview of this technique.

Algorithm 3. Combined ITP (the case of multiple depots)

Input: complete weighted graph $G^0(X \cup Y, E, w)$ of order n, a natural number q, and and an upper relative error bound $\varepsilon > 0$.
Output: an approximate solution S_{CITP} of the MDCVRP.
1: relabel the clients according to their distances $r_1 \geq r_2 \geq \ldots \geq r_n$ from the set Y;
2: find a value $k = k(\varepsilon)$, specifying the partition of the client set X onto subsets $X(k) = \{x_1, \ldots, x_{k-1}\}$ and $X \setminus X(k)$;
3: find an exact solution $S^*(X(k))$ for MDCVRP, defined by the subgraph $G^0 \langle X(k) \cup Y \rangle$;
4: using Algoritm 1, construct an approximate solution $S_{\mathrm{ITP}}(X_j \setminus X(k))$ for any subgraph $G^0 \langle X_j \setminus X(k) \cup Y \rangle$;
5: output $S_{\mathrm{CITP}} = S^*(X(k)) \cup S_{\mathrm{ITP}}(X_1 \setminus X(k)) \cup \ldots \cup S_{\mathrm{ITP}}(X_m \setminus X(k))$.

Similarly to Sect. 3, denote the relative error of Algorithm 3 by

$$e(k) = \frac{w(S_{\mathrm{CITP}}(X)) - \mathrm{VRP}^*(X)}{\mathrm{VRP}^*(X)} \tag{13}$$
$$= \frac{\mathrm{VRP}^*(X(k)) + \sum_{j=1}^m w(S_{\mathrm{ITP}}(X_j \setminus X(k))) - \mathrm{VRP}^*(X)}{\mathrm{VRP}^*(X)}.$$

Lemma 7. *In the MDCVRP1, for an arbitrary $m > 1$, $\rho \geq 1$, and $\varepsilon > 0$, there exists a number $k = k(\varepsilon) \in \mathbb{N}$ such that $e(k) \leq \varepsilon$.*

Proof. Let $X'_j = X_j \setminus X(k)$, $n_j = |X'_j|$ and $\bar{r}_{jk} = \sum_{x_i \in X'_j} r_i / n_j$. Following the proof idea of Lemma 6,

$$e(k) \leq \frac{4(k-1)r_k + \sum_{j=1}^m (2\lceil n_j/q \rceil \bar{r}_{jk} + \rho \mathrm{TSP}^*(X'_j) - 2\bar{r}_{jk} n_j/q)}{2n\bar{r}/q}$$

$$\leq q(2k - 2 + m)\frac{r_k}{\sum_{i=1}^n r_i} + \frac{q\rho}{2\sum_{i=1}^n r_i} \sum_{j=1}^m \mathrm{TSP}^*(X'_j)$$

$$\leq q(2k - 2 + m)\frac{r_k}{\sum_{i=1}^n r_i} + \frac{mq\rho}{2}C^* \left(\frac{r_k}{\sum_{i=1}^n r_i}\right)^{1/d}.$$

Suppose that, for an arbitrary $t \in \{1, \ldots, k\}$

$$q(2t - 2 + m)\frac{r_t}{\sum_{i=1}^n r_i} + \frac{mq\rho}{2}C^*\left(\frac{r_t}{\sum_{i=1}^n r_i}\right)^{1/d} > \varepsilon \qquad (14)$$

and simultaneously

$$s_t^d = \frac{r_t}{\sum_{i=1}^n r_i} \geq \left(\frac{\varepsilon}{mq\rho C^*}\right)^d, \qquad (15)$$

we get the bound

$$k \leq \left(\frac{mq\rho C^*}{\varepsilon}\right)^d,$$

similar to Eq. (8). On the other hand, if the system (14) does not imply (15) and t_0 the smallest number, for which the opposite inequality holds. Then, similarly to (9), we obtain

$$1 \geq \sum_{t=1}^k s_t^d \geq (t_0 - 1)\left(\frac{\varepsilon}{mq\rho C^*}\right)^d + \frac{\varepsilon}{2q}\sum_{t=t_0}^k \frac{1}{2t - 2 + m}$$

$$\geq (t_0 - 1)\left(\frac{\varepsilon}{mq\rho C^*}\right)^d + \frac{\varepsilon}{2q}\int_{t_0}^{k+1}\frac{dt}{2t - 2 + m}$$

$$\geq \left(\frac{\varepsilon}{mq\rho C^*}\right)^d\left((t_0 - 1) + (\ln(2k + m) - \ln(2t_0 - 2 + m))\right)$$

$$\geq \left(\frac{\varepsilon}{mq\rho C^*}\right)^d \ln((2k + m)/m),$$

and

$$k \leq \frac{m}{2}e^{\left(\frac{mq\rho C^*}{\varepsilon}\right)^d}.$$

Therefore, the range

$$\left[1, \frac{m}{2}e^{\left(\frac{mq\rho C^*}{\varepsilon}\right)^d} + 1\right] \qquad (16)$$

definitely contains the required number $k(\varepsilon)$. Lemma is proved.

Lemma 7 implies that Algorithm 3 is an EPTAS for MDCVRP1. To prove that Algorithm 3 is an EPTAS for MDCVRP2 as well, we need a version of Lemma 3 proven in [3].

Lemma 8. *For the MDCVRP2 and for an arbitrary $k \in \{1, \ldots, n\}$, the equation*

$$\mathrm{VRP}^*(X, Y) \leq \mathrm{VRP}^*(X(k), Y) + \mathrm{VRP}^*(X \setminus X(k), Y)$$
$$\leq \mathrm{VRP}^*(X, Y) + 2(q - 1)(k - 1)r_k$$

is valid.

Lemma 9. *For the MDCVRP2 and for an arbitrary $m > 1$, $\rho \geq 1$, and $\varepsilon > 0$ there exists a number $k = k(\varepsilon) \in \mathbb{N}$ such that $e(k) \leq \varepsilon$.*

Proof. Almost the same way that the proof of Lemma 7 was obtained from the claims of Lemmas 1–3 and Lemma 5 one can show that Lemmas 9 follows from Lemmas 1–2, 5, and 8. Finally, we obtain that the range

$$\left[1, (m+1)e^{\left(\frac{mq\rho\, C^*}{\varepsilon}\right)^d}\right] \tag{17}$$

definitely contains the required $k = k(\varepsilon)$, for which $e(k) \leq \varepsilon$. Lemma is proved.

Our final results follows from Lemmas 7 and 9 and can be proved in the same way as Theorem 2.

Theorem 3. *Under the conditions of Theorem 2, for any $\varepsilon > 0$, for any fixed $m, d > 1$, $\rho \geq 1$, and q, Algorithm 3 is an EPTAS for the MDCVRP1 and MDCVRP2, whose time complexity is $O(n^c + n^2 + mK^q 2^K)$, where $K = K(\varepsilon)$ coincides with right-hand ends of ranges (16) and (17), respectively.*

5 Conclusion

In the paper, we show that Algorithms 2 and 3 together with an arbitrary polynomial time fixed-guarantee approximation algorithm for the TSP induce an EPTAS for the SDCVRP and MDCVRP in any fixed-dimension Euclidean spaces, respectively. Furthermore, the time complexity bounds found remain polynomial with respect to n even for less accurate[3] but maybe much more fast algorithms for the inner TSP, which can be useful for tackling *Big Data*.

Future work can be focused on combining the results obtained with recent results on cycle covers of graphs (see, e.g. [8,12]).

Acknowledgments. This research was supported by Russian Science Foundation, project no. 14-11-00109.

References

1. Arora, S.: Polynomial time approximation schemes for euclidean traveling salesman and other geometric problems. J. ACM **45**, 753–782 (1998)
2. Asano, T., Katoh, N., Tamaki, H., Tokuyama, T.: Covering points in the plane by k-tours: a polynomial time approximation scheme for fixed k. IBM Tokyo Research (1996)
3. Cardon, S., Dommers, S., Eksin, C., Sitters, R., Stougie, A., Stougie, L.: A PTAS for the multiple depot vehicle routing problem. Technical Report 2008.03, Eindhoven Univ. of Technology, March 2008. http://www.win.tue.nl/bs/spor/2008-03.pdf

---

[3] e.g. for $\rho = O((\log \log n)^{1/d})$.

4. Christofides, N.: Worst-case analysis of a new heuristic for the traveling salesman problem. In: Symposium on New Directions and Recent Results in Algorithms and Complexity, p. 441 (1975)
5. Dantzig, G.B., Ramser, J.H.: The truck dispatching problem. Manage. Sci. **6**(1), 80–91 (1959)
6. Das, A., Mathieu, C.: A quasi-polynomial time approximation scheme for Euclidean capacitated vehicle routing. In: Proceedings of the Twenty-first Annual ACM-SIAM Symposium on Discrete Algorithms. pp. 390–403. SODA 2010, Society for Industrial and Applied Mathematics, Philadelphia, PA, USA (2010)
7. Das, A., Mathieu, C.: A quasipolynomial time approximation scheme for Euclidean capacitated vehicle routing. Algorithmica **73**, 115–142 (2015)
8. Gimadi, E.K., Rykov, I.A.: On the asymptotic optimality of a solution of the euclidean problem of covering a graph by m nonadjacent cycles of maximum total weight. Dokl. Math. **93**(1), 117–120 (2016)
9. Golden, B.L., Raghavan, S., Wasil, E.A. (eds.): The Vehicle Routing Problem: Latest Advances and New Challenges. Operations Research/Computer Science Interfaces Series, vol. 43. Springer, Heidelberg (2008)
10. Haimovich, M., Rinnooy Kan, A.H.G.: Bounds and heuristics for capacitated routing problems. Math. Oper. Res. **10**(4), 527–542 (1985)
11. Hubbert, S., Gia, Q.T.L., Morton, T.M.: Spherical Radial Basis Functions, Theory and Applications. SpringerBriefs in Mathematics, 1st edn. Springer International Publishing, Heidelberg (2015)
12. Khachay, M., Neznakhina, K.: Approximability of the minimum-weight k-size cycle cover problem. J. Global Optim. (2015). http://dx.doi.org/10.1007/s10898-015-0391-3
13. Khachay, M., Zaytseva, H.: Polynomial time approximation scheme for single-depot euclidean capacitated vehicle routing problem. In: Lu, Z., Kim, D., Wu, W., Li, W., Du, D.-Z. (eds.) COCOA 2015. LNCS, vol. 9486, pp. 178–190. Springer, Cham (2015). http://dx.doi.org/10.1007/978-3-319-26626-8_14
14. Kumar, S., Panneerselvam, R.: A survey on the vehicle routing problem and its variants. Intell. Inf. Manage. **4**, 66–74 (2012)
15. Papadimitriou, C.: Euclidean TSP is NP-complete. Theor. Comput. Sci. **4**, 237–244 (1997)
16. Sutherland, W.A.: Introduction to Metric and Topological Spaces, 2nd edn. Oxford University Press, Oxford (2009). [Oxford Mathematics]
17. Toth, P., Vigo, D. (eds.): The Vehicle Routing Problem. Society for Industrial and Applied Mathematics, Philadelphia, PA, USA (2001)

On Integer Recognition over Some Boolean Quadric Polytope Extension

Andrei Nikolaev[✉]

Department of Discrete Analysis, P.G. Demidov Yaroslavl State University,
Yaroslavl, Russia
werdan.nik@gmail.com

Abstract. The problem of integer recognition is to determine whether the maximum of a linear objective function achieved at an integral vertex of a polytope. We consider integer recognition over polytope $SATP$ and its LP relaxation $SATP_{LP}$. These polytopes are natural extensions of the well-known Boolean quadric polytope BQP and its rooted semimetric relaxation BQP_{LP}.

Integer recognition over $SATP_{LP}$ is NP-complete, since various special instances of 3-SAT problem like NAE-3SAT and X3SAT are transformed to it. We describe polynomially solvable subproblems of integer recognition over $SATP_{LP}$ with constrained objective functions. Based on that, we solve some cases of edge constrained bipartite graph coloring.

1 Introduction

We consider the well-known Boolean quadric polytope $BQP(n)$ [10], constructed from the NP-hard problem of unconstrained Boolean quadratic programming:

$$Q(x) = x^T Q x \to \max,$$

where vector $x \in \{0,1\}^n$, and Q is an upper triangular matrix, by introducing new variables $x_{i,j} = x_i x_j$.

In the standard form $BQP(n)$ can be defined as the convex hull of all integral solutions of the system

$$x_{i,j}^{1,1} + x_{i,j}^{1,2} + x_{i,j}^{2,1} + x_{i,j}^{2,2} = 1, \tag{1}$$

$$x_{i,j}^{1,1} + x_{i,j}^{1,2} = x_{k,j}^{1,1} + x_{k,j}^{1,2}, \tag{2}$$

$$x_{i,j}^{1,1} + x_{i,j}^{2,1} = x_{i,l}^{1,1} + x_{i,l}^{2,1}, \tag{3}$$

$$x_{i,i}^{1,2} = x_{i,i}^{2,1} = 0, \tag{4}$$

$$x_{i,j}^{1,1} \geq 0, \ x_{i,j}^{1,2} \geq 0, \ x_{i,j}^{2,1} \geq 0, \ x_{i,j}^{2,2} \geq 0, \tag{5}$$

where $1 \leq k \leq i \leq j \leq l \leq n$ [4] (see also [9]).

System (1)–(5) itself describe the Boolean quadric polytope LP relaxation $BQP_{LP}(n)$. Since $BQP_{LP}(n)$ and $BQP(n)$ have the same integral vertices, Boolean quadratic programming is reduced to integer programming over $BQP_{LP}(n)$.

Y. Kochetov et al. (Eds.): DOOR 2016, LNCS 9869, pp. 206–219, 2016.
DOI: 10.1007/978-3-319-44914-2_17

Theorem 1. *Integer programming over $BQP_{LP}(n)$ is NP-hard.*

Boolean quadric polytope arises in many fields of mathematics and physics. Sometime it is called the correlation polytope, since its members can be interpreted as joint correlations of events in some probability space. Besides, $BQP(n)$ is in one-to-one correspondence via the covariance linear mapping with the cut polytope $CUT(n+1)$ of the complete graph on $n+1$ vertices [5] (see also [2]). Cut polytope LP relaxation, corresponding to $BQP_{LP}(n)$, is known as the rooted semimetric polytope [6].

In recent years, the Boolean quadric polytope has been under the close attention in connection with the problem of estimating the extension complexity. An extension of the polytope P is another polytope Q such that P is the image of Q under a linear map. The number of facets of Q is called the size of an extension. Extension complexity of P is defined as the minimum size of all possible extensions. Fiorini et al. proved that the extension complexity of the Boolean quadric polytope is exponential [7] (see also [8]).

Theorem 2. *The extension complexity of $BQP(n)$ and $CUT(n)$ is $2^{\Omega(n)}$.*

Since polytopes of many combinatorial problems, including stable set, knapsack, 3-dimensional matching, and traveling salesman, contain a face that is an extension of $BQP(n)$, those polytopes also have an exponential extension complexity. Thus, such problems cannot be solved effectively by linear programming, as any LP formulation will have an exponential number of inequalities.

We consider a problem of integer recognition: for a given linear objective function $f(x)$ and a polytope P determine whether $\max\{f(x)|x \in P\}$ achieved at an integral vertex of P. It is similar to the integer feasibility problem and NP-complete in general case. In [4] integer recognition over $BQP_{LP}(n)$ was solved by linear programming over $BQP_{LP}(n)$ and the metric polytope $MET(n)$, obtained by augmenting the system (1)–(5) by the triangle inequalities that define the $BQP(3)$ facets [10].

Lemma 3 *(see [4]). If for some linear objective function $f(x)$ we have*

$$\max_{x \in BQP_{LP}(n)} f(x) = \max_{x \in MET(n)} f(x),$$

then the maximum is achieved at an integral vertex of $BQP_{LP}(n)$. Otherwise,

$$\max_{x \in BQP_{LP}(n)} f(x) > \max_{x \in MET(n)} f(x),$$

and $f(x)$ reaches its maximum at the face containing only fractional vertices.

Hence, we have

Theorem 4. *Integer recognition over $BQP_{LP}(n)$ is polynomially solvable.*

Metric polytope $MET(n)$ itself is also important, since it is the most simple and natural relaxation of the $CUT(n)$ polytope, and has many practical applications, such as being a compact LP formulation for the max-cut problem on graphs not contractible to K_5 [1]. Integer recognition over metric polytope is examined in [3].

Note that integer programming and integer recognition problems over polytope $BQP_{LP}(n)$ differ greatly in their complexity.

2   3-SAT Relaxation Polytope

We consider a more general polytope $SATP(m, n) \subset \mathbb{R}^{6mn}$ (see [4]), obtained as the convex hull of all integral solutions of the system

$$\sum_{k,l} x_{i,j}^{k,l} = 1, \tag{6}$$

$$x_{i,j}^{1,1} + x_{i,j}^{2,1} + x_{i,j}^{3,1} = x_{i,t}^{1,1} + x_{i,t}^{2,1} + x_{i,t}^{3,1}, \tag{7}$$

$$x_{i,j}^{k,1} + x_{i,j}^{k,2} = x_{s,j}^{k,1} + x_{s,j}^{k,2}, \tag{8}$$

$$x_{i,j}^{k,l} \geq 0, \tag{9}$$

where $k = 1, 2, 3$; $l = 1, 2$; $i, s = 1, \ldots m$; $j, t = 1, \ldots n$.

Inequalities (6)–(9) without the integrality constraint define LP relaxation $SATP_{LP}(m, n)$. Points that satisfy the system can be conveniently represented as a block matrix (Table 1).

Table 1. Fragment of the $SATP_{LP}(m, n)$ block matrix.

$x_{i,j}^{1,1}$	$x_{i,j}^{1,2}$	$x_{i,t}^{1,1}$	$x_{i,t}^{1,2}$
$x_{i,j}^{2,1}$	$x_{i,j}^{2,2}$	$x_{i,t}^{2,1}$	$x_{i,t}^{2,2}$
$x_{i,j}^{3,1}$	$x_{i,j}^{3,2}$	$x_{i,t}^{3,1}$	$x_{i,t}^{3,2}$
$x_{s,j}^{1,1}$	$x_{s,j}^{1,2}$	$x_{s,t}^{1,1}$	$x_{s,t}^{1,2}$
$x_{s,j}^{2,1}$	$x_{s,j}^{2,2}$	$x_{s,t}^{2,1}$	$x_{s,t}^{2,2}$
$x_{s,j}^{3,1}$	$x_{s,j}^{3,2}$	$x_{s,t}^{3,1}$	$x_{s,t}^{3,2}$

If we consider a face of the $SATP(n, n)$ polytope, constructed as follows:

$$\forall i, j : x_{i,j}^{3,1} = x_{i,j}^{3,2} = 0,$$

$$\forall i : x_{i,i}^{1,2} = x_{i,i}^{2,1} = 0,$$

and discard all the coordinates for $i < j$ (orthogonal projection), we get the $BQP(n)$ polytope. As a result, we have

Theorem 5. *The extension complexity of the $SATP(m,n)$ polytope is* $2^{\Omega(\min\{m,n\})}$.

In [4] by reduction from 3-SAT it was shown that

Theorem 6. *Integer recognition over $SATP_{LP}(m,n)$ is NP-complete.*

We prove that the polytope $SATP_{LP}(m,n)$ can be seen as a LP relaxation of various special instances of 3-SAT problem as well.

Lemma 7. *Let z be the vertex of the $SATP(m,n)$ polytope, then its coordinates are determined by the vectors $\boldsymbol{row}(z) \in \{0,1\}^m$ and $\boldsymbol{col}(z) \in \{0,1,2\}^n$ by the following formulas:*

$$x_{i,j}^{1,1} = \frac{1}{2}(1 - \boldsymbol{row}_i(z))(2 - \boldsymbol{col}_j(z))(1 - \boldsymbol{col}_j(z)), \qquad (10)$$

$$x_{i,j}^{1,2} = \frac{1}{2}\boldsymbol{row}_i(z)(2 - \boldsymbol{col}_j(z))(1 - \boldsymbol{col}_j(z)), \qquad (11)$$

$$x_{i,j}^{2,1} = (1 - \boldsymbol{row}_i(z))\boldsymbol{col}_j(z)(2 - \boldsymbol{col}_j(z)), \qquad (12)$$

$$x_{i,j}^{2,2} = \boldsymbol{row}_i(z)\boldsymbol{col}_j(z)(2 - \boldsymbol{col}_j(z)), \qquad (13)$$

$$x_{i,j}^{3,1} = \frac{1}{2}(1 - \boldsymbol{row}_i(z))\boldsymbol{col}_j(z)(1 - \boldsymbol{col}_j(z)), \qquad (14)$$

$$x_{i,j}^{3,2} = \frac{1}{2}\boldsymbol{row}_i(z)\boldsymbol{col}_j(z)(1 - \boldsymbol{col}_j(z)). \qquad (15)$$

Proof. From constraints (6)–(9) it follows that vertices of $SATP(m,n)$ polytope are zero-one points with exactly one unit per block. For any vertex z of $SATP(m,n)$ we define $\boldsymbol{row}(z) \in \{0,1\}^m$ and $\boldsymbol{col}(z) \in \{0,1,2\}^n$ vectors by the following rules:

$$\boldsymbol{row}_i(z) = \begin{bmatrix} 0, & \text{if } x_{i,1}^{1,1} + x_{i,1}^{2,1} + x_{i,1}^{3,1} = 1, \\ 1, & \text{otherwise.} \end{bmatrix}$$

$$\boldsymbol{col}_j(z) = \begin{bmatrix} 0, & \text{if } x_{1,j}^{1,1} + x_{1,j}^{1,2} = 1, \\ 1, & \text{if } x_{1,j}^{2,1} + x_{1,j}^{2,2} = 1, \\ 2, & \text{otherwise.} \end{bmatrix}$$

All the vertex coordinates are uniquely determined by the first row and first column of blocks from the system (6)–(8). Equations (10)–(15) correspond to them for zero-one points. Thus, polytope $SATP(m,n)$ has exactly $2^m 3^n$ vertices.

We consider one-in-three 3-satisfiability or exactly-1 3-satisfiability (X3SAT): given a set $U = \{u_1, \ldots, u_m\}$ of variables and collection $C = \{c_1, \ldots, c_n\}$ of 3-literal clauses over U, the problem is to determine whether there exists a truth assignment to the variables so that each clause has exactly one true literal [11].

With each instance of the problem we associate an objective vector $w \in \mathbb{R}^{6mn}$:

– if clause c_j has literal u_i at the place k, then

$$\forall s \in \{1,2,3\} \backslash k : \ w_{i,j}^{k,1} = w_{i,j}^{s,2} = 1,$$

– if clause c_j has literal $\bar{u}_i$ at the place k, then

$$\forall s \in \{1,2,3\}\backslash k: \ w_{i,j}^{k,2} = w_{i,j}^{s,1} = 1,$$

– all the remaining coordinates of vector w equal to 0.

An example of vector w construction for a formula

$$(x \vee y \vee \bar{z}) \wedge (\bar{x} \vee z \vee t) \wedge (\bar{y} \vee z \vee \bar{t}) \tag{16}$$

is shown in Table 2(a).

Table 2. Examples of objective vectors for X3SAT and NAE-3SAT problems

1	0	0	1	0	0
0	1	1	0	0	0
0	1	1	0	0	0
0	1	0	0	0	1
1	0	0	0	1	0
0	1	0	0	1	0
1	0	0	1	0	1
1	0	1	0	1	0
0	1	0	1	0	1
0	0	0	1	1	0
0	0	0	1	1	0
0	0	1	0	0	1

(a) X3SAT

1	0	0	1	0	0
0	1	1	0	0	0
1	1	1	1	0	0
1	1	0	0	0	1
1	0	0	0	1	0
0	1	0	0	1	1
1	0	1	1	1	1
1	1	1	0	1	0
0	1	0	1	0	1
0	0	0	1	1	0
0	0	1	1	1	1
0	0	1	0	0	1

(b) NAE-3SAT

With each integral vertex z we associate a truth assignment u by the following rule: $u_i = 1 - \boldsymbol{row}_i(z)$.

Now we consider linear objective function $f_w(x) = \langle w, x \rangle$.

Theorem 8. *There exists a truth assignment for X3SAT problem with exactly one true literal per clause if and only if*

$$\max_{x \in SATP_{LP}(m,n)} f_w(x) = \max_{z \in SATP(m,n)} f_w(z) = 3n.$$

Proof. It suffices to verify that for any integral vertex $z \in SATP(m,n)$ on the corresponding truth assignment we have

$$f_w(z_j) = \langle w_j, z_j \rangle = \begin{cases} 3, & \text{if the clause } c_j \text{ has exactly one true literal,} \\ 1, & \text{if all three literals in } c_j \text{ are true,} \\ 2, & \text{otherwise.} \end{cases}$$

Thus, X3SAT problem is transformed to the integer recognition over polytope $SATP_{LP}(m,n)$. Another popular variant of 3-SAT problem is Not-All-Equal 3-SAT (NAE-3SAT): given a set $U = \{u_1, \ldots, u_m\}$ of variables and collection $C = \{c_1, \ldots, c_n\}$ of 3-literal clauses over U, the problem is to determine whether there exists a truth assignment so that each clause has at least one true literal and at least one false literal [11].

With each instance of the problem we associate an objective vector $y \in \mathbb{R}^{6mn}$:

– if clause c_j has literal u_i at the place k, then

$$y_{i,j}^{k,1} = y_{i,j}^{(k+1)mod3,2} = y_{i,j}^{(k+2)mod3,1} = y_{i,j}^{(k+2)mod3,2} = 1,$$

– if clause c_j has literal $\overline{u}_i$ at the place k, then

$$y_{i,j}^{k,2} = y_{i,j}^{(k+1)mod3,1} = y_{i,j}^{(k+2)mod3,1} = y_{i,j}^{(k+2)mod3,2} = 1,$$

– all the remaining coordinates of vector y equal to 0.

An example of vector y for the formula (16) is shown in Table 2(b).

We consider linear objective function $f_y(x) = \langle y, x \rangle$.

Theorem 9. *There exists a truth assignment for NAE-3SAT problem with at least one true literal and at least one false literal per clause if and only if*

$$\max_{x \in SATP_{LP}(m,n)} f_y(x) = \max_{z \in SATP(m,n)} f_y(z) = 3n.$$

Proof. Again it suffices to verify that for any integral vertex $z \in SATP(m,n)$ on the corresponding truth assignment we have

$$f_w(z_j) = \langle w_j, z_j \rangle = \begin{bmatrix} 2, \text{ if the clause } c_j \text{ has three true literals or three false literals,} \\ 3, \text{ otherwise.} \end{bmatrix}$$

3   Polynomially Solvable Subproblems

Integer recognition is NP-complete over entire $SATP_{LP}(m,n)$ polytope, but polynomially solvable over its face $BQP_{LP}(n)$. However, for some objective functions, other than those specified above, integer recognition over $SATP_{LP}(m,n)$ can be efficiently solved. In this main section we examine one of such polynomially solvable cases.

We consider a vector $c \in \mathbb{R}^{6mn}$, such that

$$\forall j \in \mathbb{N}_n, \ \exists a, b \in \{1,2,3\} \ (a \neq b), \forall i \in \mathbb{N}_m :$$
$$c_{i,j}^{a,1} + c_{i,j}^{b,2} = c_{i,j}^{a,2} + c_{i,j}^{b,1}, \tag{17}$$

and a corresponding linear objective function $f_c(x) = \langle c, x \rangle$.

Theorem 10. *For objective functions of the form $f_c(x)$ the problem of integer recognition over $SATP_{LP}(m,n)$ polytope is polynomially solvable.*

Proof. Without loss of generality, we assume that the vector c has the form:

$$\forall i, j : c_{i,j}^{2,1} + c_{i,j}^{3,2} = c_{i,j}^{2,2} + c_{i,j}^{3,1}. \tag{18}$$

For any other choices of restrictions on vector c following proof can be modified by just renaming the coordinates.

To make room for superscripts we introduce a new notation for the coordinates of the polytope:

$$x_{i,j}^{1,1} = x_{i,j}, \quad x_{i,j}^{1,2} = y_{i,j}, \quad x_{i,j}^{2,1} = z_{i,j},$$
$$x_{i,j}^{2,2} = t_{i,j}, \quad x_{i,j}^{3,1} = u_{i,j}, \quad x_{i,j}^{3,2} = v_{i,j}.$$

We construct a new polytope $SATP_{LP}^2(m, n)$, satisfying the system (6)–(9) and the additional constraints:

$$y_{i,j} + z_{i,j} + u_{i,j} + x_{i,l} + t_{i,l} + v_{i,l} + x_{k,j} + t_{k,j} + v_{k,j} + x_{k,l} + t_{k,l} + v_{k,l} \leq 3, \tag{19}$$

$$y_{i,j} + z_{i,j} + u_{i,j} + y_{i,l} + z_{i,l} + u_{i,l} + x_{k,j} + y_{k,j} + z_{k,j} + y_{k,l} + z_{k,l} + u_{k,l} \leq 3, \tag{20}$$

for all $i, k \in \mathbb{N}_m$ $(i \neq k)$ and $j, l \in \mathbb{N}_n$ $(j \neq l)$.

All integral vertices of $SATP(m, n)$ satisfy the inequalities (19)–(20), therefore $SATP_{LP}^2(m, n)$ is another LP relaxation of $SATP(m, n)$ polytope. Note that the total number of additional constraints is polynomially bounded above by $O(m^2 n^2)$.

Table 3. Rearrange of the columns in the i-th row of the block matrix at Step 1.

0	$y_{i,j}^w$	0	$y_{i,l}^w$		$x_{i,j}^{w^*} = y_{i,j}^w$	0	$x_{i,l}^{w^*} = y_{i,l}^w$	0
0	$t_{i,j}^w$	0	$t_{i,l}^w$	$\Rightarrow$	$z_{i,j}^{w^*} = t_{i,j}^w$	0	$z_{i,l}^{w^*} = t_{i,l}^w$	0
0	$v_{i,j}^w$	0	$v_{i,l}^w$		$u_{i,j}^{w^*} = v_{i,j}^w$	0	$u_{i,l}^{w^*} = v_{i,l}^w$	0

Let w be the point that maximize the function $f_c(x)$ over $SATP_{LP}^2(m, n)$. We claim that there exists a point $w^* \in SATP_{LP}^2(m, n)$ with $f_c(w) = f_c(w^*)$ and $\forall i, j : x_{i,j}^{w^*} > 0$, up to renaming the coordinates. We construct point w^* from w in a few steps.

1. If there exists some i that

$$x_{i,j}^w + z_{i,j}^w + u_{i,j}^w = 0,$$

then we change the columns in all blocks of the i-th row (Table 3). Due to the symmetry of the system (6)–(9), (19), (20) and the constraints (18), new point belongs to $SATP_{LP}^2(m, n)$ polytope and has the same value of the objective function. In fact, we simply rename some coordinates. Thus, we can now consider w^* simply as w and continue the procedure.

Table 4. Rearrange of the rows in the j-th column of the block matrix at Step 2.

-	-
0	0
$u_{i,j}^w$	$v_{i,j}^w$
-	-
0	0
$u_{k,j}^w$	$v_{k,j}^w$

$\Rightarrow$

-	-
$z_{i,j}^{w^*} = u_{i,j}^w$	$t_{i,j}^{w^*} = v_{i,j}^w$
0	0
-	-
$z_{k,j}^{w^*} = u_{k,j}^w$	$t_{k,j}^{w^*} = v_{k,j}^w$
0	0

Table 5. Construction of the block i, j of the point w^*.

-	-
0	$t_{i,j}^w$
$u_{i,j}^w$	$v_{i,j}^w$

$\Rightarrow$

0	0
$z^{w^*} = \epsilon$	$t_{i,j}^{w^*} = t_{i,j}^w - \epsilon$
$u_{i,j}^{w^*} = u_{i,j}^w - \epsilon$	$v_{i,j}^{w^*} = v_{i,j}^w + \epsilon$

Table 6. Rearrange of the rows in the j-th column of the block matrix.

0	0
$z_{i,j}^w$	$t_{i,j}^w$
$u_{i,j}^w$	$v_{i,j}^w$

$\Rightarrow$

$x^{w^*} = z_{i,j}^w$	$y^{w^*} = t_{i,j}^w$
$z^{w^*} = u_{i,j}^w$	$t^{w^*} = v_{i,j}^w$
0	0

Table 7. Fragment of the point w block matrix.

$x_{i,s}$	0	$x_{i,l}$	0	0	$y_{i,j}$
$z_{i,s}$	0	-	$t_{i,l}$	$z_{i,j}$	-
-	$v_{i,s}$	-	-	-	-
$x_{k,s}$	-	$x_{k,l}$	-	$x_{k,j}$	-
-	-	-	-	0	$t_{k,j}$
-	-	-	-	0	-

2. If there exists some j that

$$z_{i,j}^w + t_{i,j}^w = 0 \text{ and } u_{i,j}^w + v_{i,j}^w > 0,$$

then we change the second and third rows in all blocks of the j-th column as at Step 1 (Table 4). Again, point w^* belongs to $SATP_{LP}^2(m,n)$ and has the same value of the objective function.

3. There exists some j that $x_{i,j} + y_{i,j} = 0$. As a result of Steps 1 and 2 we have

$$\forall i : z_{i,j}^w + t_{i,j}^w > 0, \ z_{i,j}^w + u_{i,j}^w > 0.$$

Hence, if for some i: $z_{i,j}^w = 0$, then $t_{i,j}^w > 0$ and $u_{i,j}^w > 0$. We construct the point w^* as it's shown in Table 5. Since the coordinates $t_{i,j}^w$ and $u_{i,j}^w$ are nonnegative and $v_{i,j}^w < 1$, we can choose a sufficiently small value of ϵ so that w^* satisfy the system (6)–(9), (19), (20). We estimate the value of the objective function

$$f_c(w^*) = f_c(w) + \epsilon c_{i,j}^{2,1} + \epsilon c_{i,j}^{3,2} - \epsilon c_{i,j}^{2,2} - \epsilon c_{i,j}^{3,1},$$
$$f_c(w^*) = f_c(w) + \epsilon(c_{i,j}^{2,1} + c_{i,j}^{3,2} - c_{i,j}^{2,2} - c_{i,j}^{3,1}) = f_c(w),$$

by Eq. (18). Thus, we can assume that $z_{i,j}^w > 0$.

We change the rows in all blocks of the j-th column as it's shown in Table 6. Now in the j-th column we have $x_{i,j}^{w^*} > 0$ for all i. Without loss of generality, we assume that the Step 3 was applied to the first d columns. Here comes the tricky part: we can't just rearrange rows in such way, as w^* may not belong to $SATP_{LP}^2(m,n)$ polytope or has a different value of the objective function. Therefore, we simply rename the coordinates of the point w. Thus, for the first d columns constraints (18) and inequalities (19), (20) are modified accordingly.

4. We find the leftmost and uppermost block i, j with $x_{i,j}^w = 0$. As a result of the previous steps, $y_{i,j}^w$ is nonnegative, and if $z_{i,j}^w = 0$, then both $t_{i,j}^w$ and $u_{i,j}^w$ are nonnegative. Therefore, we can repeat the ϵ-procedure from Step 3 (Table 3) and achieve $z_{i,j}^{w^*} > 0$.

The idea behind the following procedure is to construct point w^* with the same value of objective function while preventing x coordinates in all the blocks to the up and to the left of i, j from becoming zero.

5. The next step depends on the form of the i-th row.
   (a) If for all $l < j : y_{i,l}^w > 0$, then we can rearrange the columns in the i-th row as at Step 1 (Table 3) to get $x_{i,l} > 0$ for all $l \leq j$.
   (b) There exists some l ($d < l < j$) that $y_{i,l}^w = 0$. Hence, $x_{i,l}^w$ is nonnegative, and if $t_{i,l}^w = 0$, then both $z_{i,j}^w$ and $v_{i,j}^w$ are nonnegative, hence, we can construct a point w^* with $t_{i,l}^{w^*} > 0$ by the similar ϵ-procedure. Thus, we assume that $t_{i,l}^w$ is nonnegative (Table 7).

(c) There exists some s $(s \le d < j)$ that $y_{i,s}^w = 0$. Since $s \le d$, the coordinates in the s-th column were renamed at Step 3. Thus, if $t_{i,s}^w > 0$, then we can construct the point w^* of $SATP_{LP}^2(m,n)$ by the ϵ-procedure with $y_{i,s}^{w^*} > 0$. Therefore, we assume that $t_{i,s}^w = 0$, and, due to that, $z_{i,s}^w$ and $v_{i,s}^w$ are nonnegative (Table 7).

6. Now we examine the j-th column.

(a) If for all k: $z_{k,j}^w > 0$, then we can rearrange the rows in j-th column as at Step 3 (Table 6) and achieve $x_{i,j}^{w^*} > 0$ for all i. Next, we rename the coordinates for the j-th column to become the $(d+1)$-th and increase the value of d by one.

(b) There exists some k that $z_{k,j}^w = 0$. Then $t_{k,j}^w$ is nonnegative, since $z_{i,j}^w > 0$. We assume $u_{k,j}^w = 0$ (Table 7), otherwise by ϵ-procedure we can achieve $z_{k,j}^{w^*}$ being nonnegative. In this case we can't make $x_{i,j}^{w^*}$ nonnegative. Let's verify if such point w belongs $SATP_{LP}^2(m,n)$ and check the inequality (19):

$$(*) = y_{i,j} + z_{i,j} + u_{i,j} + x_{i,l} + t_{i,l} + v_{i,l}$$
$$+ x_{k,j} + t_{k,j} + v_{k,j} + x_{k,l} + t_{k,l} + v_{k,l} \le 3,$$
$$(y_{i,j} = x_{k,j} + y_{k,j}, \quad x_{k,j} + y_{k,j} + t_{k,j} + v_{k,j} = 1),$$
$$(*) = 1 + z_{i,j} + u_{i,j} + x_{i,l} + t_{i,l} + v_{i,l} + x_{k,j} + x_{k,l} + t_{k,l} + v_{k,l} \le 3,$$
$$(z_{i,j} + u_{i,j} = x_{i,l} + z_{i,l} + u_{i,j}, \quad x_{i,l} + z_{i,l} + t_{i,l} + u_{i,l} + v_{i,l} = 1),$$
$$(*) = 2 + x_{i,l} + x_{k,j} + x_{k,l} + t_{k,l} + v_{k,l} \le 3,$$
$$(x_{i,l} = x_{k,l} + y_{k,l}, \quad x_{k,j} = x_{k,l} + z_{k,l} + u_{k,l},$$
$$x_{k,l} + y_{k,l} + z_{k,l} + t_{k,l} + u_{k,l} + v_{k,l} = 1),$$
$$(*) = 3 + 2x_{k,l} \le 3.$$

By construction, for all $l < j$ we have $x_{k,l}^w > 0$, hence, point w with such blocks i, j, k, l does not belong to the polytope $SATP_{LP}^2(m,n)$.

Now we check the inequality (20) for blocks i, j, k, s. Note that $s \le d$, and the s-th column was modified at Step 3. Therefore, the inequality has the form

$$(**) = y_{i,j} + z_{i,j} + u_{i,j} + x_{i,s} + z_{i,s} + v_{i,s}$$
$$+ x_{k,j} + t_{k,j} + v_{k,j} + x_{k,s} + z_{k,s} + v_{k,s} \le 3,$$
$$(y_{i,j} = x_{k,j} + y_{k,j}, \quad x_{k,j} + y_{k,j} + t_{k,j} + v_{k,j} = 1),$$
$$(**) = 1 + z_{i,j} + u_{i,j} + x_{i,s} + z_{i,s} + v_{i,s} + x_{k,j} + x_{k,s} + z_{k,s} + v_{k,s} \le 3,$$
$$(v_{i,s} = y_{i,j} + t_{i,j} + v_{i,j}, \quad y_{i,j} + z_{i,j} + t_{i,j} + u_{i,j} + v_{i,j} = 1),$$
$$(**) = 2 + x_{i,s} + z_{i,s} + x_{k,j} + x_{k,s} + z_{k,s} + v_{k,s} \le 3,$$
$$(x_{i,s} = x_{k,s} + y_{k,s}, \quad z_{i,s} = z_{k,s} + t_{k,s}, \quad x_{k,j} = x_{k,s} + z_{k,s} + u_{k,s},$$
$$x_{k,s} + y_{k,s} + z_{k,s} + t_{k,s} + u_{k,s} + v_{k,s} = 1),$$
$$(**) = 3 + 2(x_{k,s} + z_{k,s}) \le 3.$$

Since $x_{k,s}^w > 0$, point w with such blocks i, j, k, s does not belong to the polytope $SATP_{LP}^2(m, n)$.

Thereby, the combination of 5 (b or c) and 6 (b) is impossible, and we can repeat the Steps 4–6, until for all i, j we have $x_{i,j}^{w^*} > 0$.

Thus, for any point w that maximize the objective function $f_c(x)$ over polytope $SATP_{LP}^2(m, n)$ we can construct such point $w^* \in SATP_{LP}^2(m, n)$ that $x_{i,j}^{w^*} > 0$ for all i, j, up to renaming the coordinates, and $f_c(w) = f_c(w^*)$.

The point w^* can be decomposed into a convex combination

$$w^* = \epsilon q + (1 - \epsilon)h,$$

where $0 < \epsilon \leq 1$, q is an integral vertex of $SATP_{LP}(m, n)$ with $x_{i,j}^q = 1$ for all i, j, and h has the following coordinates:

$$x_{i,j}(h) = \frac{x_{i,j}(w^*) - \epsilon}{1 - \epsilon}, \quad y_{i,j}(h) = \frac{y_{i,j}(w^*)}{1 - \epsilon}, \quad z_{i,j}(h) = \frac{z_{i,j}(w^*)}{1 - \epsilon},$$

$$t_{i,j}(h) = \frac{t_{i,j}(w^*)}{1 - \epsilon}, \quad u_{i,j}(h) = \frac{u_{i,j}(w^*)}{1 - \epsilon}, \quad v_{i,j}(h) = \frac{v_{i,j}(w^*)}{1 - \epsilon}.$$

The point h satisfies the system (6)–(9), hence, both q and h belongs to $SATP_{LP}(m, n)$.

Our algorithm for integer recognition over $SATP_{LP}(m, n)$ polytope is similar to the one in Lemma 3: if

$$\max_{x \in SATP_{LP}(m,n)} f_c(x) > \max_{x \in SATP_{LP}^2(m,n)} f_c(x),$$

then, clearly, the maximum is not achieved at an integral vertex, since polytopes $SATP_{LP}(m, n)$ and $SATP_{LP}^2(m, n)$ have the same set of integral vertices, and if

$$\max_{x \in SATP_{LP}(m,n)} f_c(x) = \max_{x \in SATP_{LP}^2(m,n)} f_c(x),$$

then for a point w that maximize the objective function we can construct such point $w^* \in SATP_{LP}^2(m, n)$ that

$$f_c(w) = f_c(w^*) = f_c(q),$$

where q is an integral vertex, hence,

$$\max_{x \in SATP_{LP}(m,n)} f_c(x) = \max_{z \in SATP(m,n)} f_c(z)$$

and the integer recognition problem has a positive answer.

Polytope $SATP_{LP}^2(m, n)$ has a polynomial number of additional constraints, therefore, LP over it is polynomially solvable, and the entire algorithm is polynomial. Note, that the construction of w^* and integral vertex q also requires polynomial time ($O(mn(m + n))$), as in the worst case for each block i, j we have to check all the blocks in the row i and column j.

4   Edge Constrained Bipartite Graph Coloring

In this section we construct a special problem to show how the constraints and the algorithm from Theorem 10 may be used.

We consider a problem of 2-3 *edge constrained bipartite graph coloring* (2-3-ECBGC): for a given bipartite graph $G = (U, V, E)$ and a function of permitted color combinations for every edge

$$pc : E \times \{1,2\} \times \{1,2,3\} \rightarrow \{+,-\},$$

it is required to determine if it's possible to assign the vertex colors in such way

$$color : U \rightarrow \{1,2\} \text{ and } color : V \rightarrow \{1,2,3\},$$

that they satisfy the constraints of all the edges in the graph.

Theorem 11. *2-3-ECBGC problem is NP-complete.*

Proof. The problem obviously belongs to the class NP, as solution can be verified in $O(|E|)$ time.

We transform exactly-1 3-satisfiability problem to 2-3-ECBGC. Let m be the number of variables and n the number of clauses. First, we construct an instance of integer recognition over $SATP_{LP}(m,n)$ with an objective vector $w \in \mathbb{R}^{6mn}$ as shown in Theorem 8. Then we create a bipartite graph G_w with m vertices in U and n vertices in V. Graph G_w has an edge (i,j) if and only if clause c_j has literal u_i or $\bar{u}_i$. Permitted color combinations are defined as follows:

$$pc(i,j,k,s) = \begin{cases} +, & \text{if } w_{i,j}^{k,s} = 1, \\ -, & \text{otherwise.} \end{cases}$$

There is a bijection between possible color assignments and integral vertices of $SATP(m,n)$:

$$\forall i \in U : color(i) = \boldsymbol{row}_i(z) + 1,$$
$$\forall j \in V : color(j) = \boldsymbol{col}_j(z) + 1.$$

By Theorem 8, truth assignment for X3SAT exists if and only if there exists such integral vertex z of $SATP(m,n)$ that $f_w(z) = 3n$. Since some color assignment satisfy the permitted color constraints of the edge i,j if and only if $f_w(z_{i,j}) = 1$, and there are exactly $3n$ edges in the graph G, we have

$$\text{X3SAT} \leq_p \text{2-3-ECBGC}.$$

Using Theorem 10, we construct a special polynomially solvable subproblem of 2-3-edge constrained bipartite graph coloring.

Theorem 12. *2-3-ECBGC problem is polynomially solvable if the permitted colors function satisfy the following constraints*

$$\forall j \in V, \exists a_j, b_j \in \{1,2,3\} \ (a \neq b), \forall i \in U :$$
$$pc(i,j,a_j,1) = pc(i,j,b_j,2) = \text{``} + \text{''} \Leftrightarrow pc(i,j,a_j,2) = pc(i,j,b_j,1) = \text{``} + \text{''}.$$

$$(21)$$

Proof. Let $|U| = m$ and $|V| = n$. We reduce 2-3-ECBGC problem to integer recognition over $SATP_{LP}(m, n)$ by constructing the objective vector $c \in \mathbb{R}^{6mn}$ from the permitted colors function as follows:

$$c_{i,j}^{k,s} = \begin{cases} 1, & \text{if } pc(i,j,k,s) = \text{``}+\text{''}, \\ -1, & \text{in the case of } \textit{zero balancing}, \\ 0, & \text{otherwise}. \end{cases}$$

We have a *zero balancing* case if an edge i, j out of four color combinations $(a_j, 1)$, $(a_j, 2)$, $(b_j, 1)$, and $(b_j, 2)$ has only one permitted. Assume, without loss of generality, that it is $(a_j, 1)$, then we assign $c_{i,j}^{b_j,2} = -1$ to achieve zero balance:

$$c_{i,j}^{a_j,1} + c_{i,j}^{b_j,2} = c_{i,j}^{a_j,2} + c_{i,j}^{b_j,1} = 0.$$

An example of objective vector c construction for $a_1 = a_2 = 1$ and $b_1 = b_2 = 2$ is shown in Table 8.

Table 8. Example of objective vector c construction.

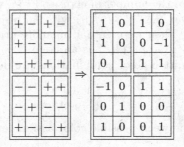

For every edge i, j we have 4 pairs of colors a_j and b_j, and 16 possible combinations of constraints. Six of them are forbidden by (21). Others transform into a block of vector c of the form (17). We again use the bijection between integral vertices and possible color assignments as in Theorem 11.

Thus, permitted color assignment for 2-3-ECBGC exists if and only if

$$\max_{x \in SATP_{LP}(m,n)} f_c(x) = \max_{z \in SATP(m,n)} f_c(z) = |E|.$$

Integer recognition over $SATP_{LP}(m, n)$ with the objective function $f_c(x)$ is polynomially solvable (Theorem 10), therefore, such instance of 2-3-ECBGC problem is polynomially solvable as well.

Not that the constrained objective function (17) is far more flexible than we used for 2-3-ECBGC problem, since it is not limited to the $\{-1, 0, 1\}$ values. For example, we can add the weight for permitted color combinations that will satisfy the constraints (17) and solve the problem by integer recognition algorithm.

5   Conclusions

We studied the problem of integer recognition over $SATP(m, n)$ polytope that is a simple extension of the well-known and important Boolean quadric polytope $BQP(n)$, constructed by adding two additional coordinates per block. In general case integer recognition over $SATP(m, n)$ is NP-complete, since several special instances of 3-SAT like X3SAT and NAE-3SAT are reduced to it.

We considered possible constraints on the objective function for which integer recognition over $SATP_{LP}(m, n)$ is polynomially solvable. We introduced a problem of 2-3 edge constrained bipartite graph coloring that is NP-complete in general case, and design a polynomial time algorithm for its special subproblem, based on $SATP(m, n)$ properties. This example shows how the polytope $SATP(m, n)$ may be used, and why it is of interest for further studying.

Acknowledgments. The research was partially supported by the Russian Foundation for Basic Research, Project 14-01-00333, and the President of Russian Federation Grant MK-5400.2015.1.

References

1. Barahona, F.: On cuts and matchings in planar graphs. Math. Program. **60**, 53–68 (1993)
2. Barahona, F., Mahjoub, A.R.: On the cut polytope. Math. Program. **36**, 157–173 (1986)
3. Bondarenko, V.A., Nikolaev, A.V., Symanovich, M.E., Shemyakin, R.O.: On a recognition problem on cut polytope relaxations. Autom. Remote Control **75**, 1626–1636 (2014)
4. Bondarenko, V.A., Uryvaev, B.V.: On one problem of integer optimization. Autom. Remote Control **68**, 948–953 (2007)
5. De Simone, C.: The cut polytope and the Boolean quadric polytope. Discrete Math. **79**, 71–75 (1990)
6. Deza, M.M., Laurent, M.: Geometry of Cuts and Metrics (Algorithms and Combinatorics). Springer, Heidelberg (1997)
7. Fiorini, S., Massar, S., Pokutta, S., Tiwary, H.R., De Wolf, R.: Exponential lower bounds for polytopes in combinatorial optimization. J. ACM **62**, 1–17 (2015)
8. Kaibel, V., Weltge, S.: A short proof that the extension complexity of the correlation polytope grows exponentially. Discrete Comput. Geom. **53**, 397–401 (2015)
9. Nikolaev, A.: On vertex denominators of the Boolean quadric polytope relaxation. Appl. Math. Sci. **9**, 5583–5591 (2015)
10. Padberg, M.: The Boolean quadric polytope: some characteristics, facets and relatives. Math. Program. **45**, 139–172 (1989)
11. Schaefer, T.J.: The complexity of satisfiability problems. In: Proceedings of the Tenth Annual ACM Symposium on Theory of Computing, (STOC 1978), pp. 216–226. ACM, New York (1978)

Variable Neighborhood Search-Based Heuristics for Min-Power Symmetric Connectivity Problem in Wireless Networks

Roman Plotnikov[1]([✉]), Adil Erzin[1,2], and Nenad Mladenovic[3]

[1] Sobolev Institute of Mathematics, Novosibirsk, Russia
nomad87@ngs.ru
[2] Novosibirsk State University, Novosibirsk, Russia
[3] University of Valenciennes and Hainaut-Cambresis, Famars, France

Abstract. We investigate the well-known NP-hard problem of finding an optimal communication subgraph in a given edge-weighted graph. This problem appears in different distributed wireless communication networks, e.g., in wireless sensor networks, when it is necessary to minimize transmission energy consumption. We propose new heuristic algorithms based on variable neighborhood search metaheuristic. Our results have been compared with the best known results, and the numerical experiment showed that, on a large number of instances, our algorithms outperform the previous ones, especially in a case of large dimensions.

Keywords: Wireless sensor networks · Energy efficiency · NP-hard problem · Variable neighborhood search

1 Introduction

In recent years different issues related to the wireless communication networks have been actively researched (see, e.g., [1,16]). Mainly, the problem is to minimize energy consumption of network elements per time unit in order to prolong the lifetime of the network. Since often the exact positions of the network elements and the topology of the network cannot be predefined, modern sensors have ability to adjust their transmission ranges in order to minimize energy consumption without breaking the connectivity of the network. Herewith, usually the energy consumption of a network's element is assumed to be proportional to d^s, where $s \geq 2$ and d is the transmission range [15]. But in the general case this condition may not be satisfied because of the inhomogeneity of the environment, radio interference and peculiar properties of network elements (e.g., the signal may not be spread equally in all directions). Thus, the communication energy consumption for each connection could be arbitrary.

We assume that the communication network is represented as a connected graph $G = (V, E)$. In this paper we consider the symmetric case: an edge between two vertices means that the both of them can send a message to each other and the energy consumption for this communication is the same for both of

© Springer International Publishing Switzerland 2016
Y. Kochetov et al. (Eds.): DOOR 2016, LNCS 9869, pp. 220–232, 2016.
DOI: 10.1007/978-3-319-44914-2_18

them. If $c_{ij} \geq 0$ is a transmission-related energy consumption needed for sending data from $i \in V$ to $j \in V$ (as well as from j to i), then in the connected subgraph $T = (V, E')$, $E' \subseteq E$ the energy consumption of node $i \in V$ equals to $E_i(T) = \max\limits_{j:(i,j) \in E'} c_{ij}$. The goal of this paper is the development of algorithms for the construction of a spanning subgraph T that minimizes $\sum\limits_{i \in V} E_i(T)$. Without loss of generality, we assume that subgraph T is a spanning tree.

In this paper we propose new heuristic algorithms which use the variable neighborhood search (VNS) metaheuristic, different local searches and two variants of shaking algorithm. We compare solutions obtained by these algorithms with optimal solutions in small dimension cases and with solutions obtained by other algorithms when dimension is large.

The rest of the paper is organized as follows. Section 2 contains the formulation of the problem. In Sect. 3 the related papers are described. The new heuristics are proposed in Sect. 4. In Sect. 5 the results of numerical experiments are presented. Section 6 concludes the paper.

2 Problem Statement

Mathematically, the considered problem can be formulated as follows. Given a simple connected weighted graph $G = (V, E)$ with a vertex set V, $|V| = n$, and an edge set E, find such spanning tree T^* of G, which is the solution to the following problem:

$$W(T) = \sum_{i \in V} \max_{j \in V_i(T)} c_{ij} \rightarrow \min_{T}, \tag{1}$$

where $V_i(T)$ is the set of vertices adjacent to the vertex i in the tree T and $c_{ij} \geq 0$ is the weight of the edge $(i, j) \in E$.

In literature, this problem is called the Minimum Power Symmetric Connectivity Problem (MPSCP) [2]. Any feasible solution of (1), i.e., a spanning tree of G, will be called a *communication tree* (subgraph). It is known that (1) is strongly NP-hard [1,7,8,12], and if P $\neq$ NP, then the problem is inapproximable within $1 + \frac{1}{260}$ [8]. Therefore, the construction and analysis of efficient approximation algorithms are some of the most important issues regarding the research on this problem.

3 Related Works

The more general Range Assignment Problem, where the goal is to find a strong connected subgraph in a given oriented graph, has been considered in [5,12]. Its subproblem, MPSCP, was first studied in [2]. The authors proved that Minimum Spanning Tree (MST) is 2-approximation for this problem. Also they proposed a polynomial-time approximation scheme with performance ratio of $1 + \ln 2 + \varepsilon \approx$ 1.69 and 15/8-approximate polynomial algorithm. In [3] a greedy heuristic, later called Incremental Power: Prim (IPP), was proposed. IPP is similar to the Prim's algorithm of finding of MST. A Kruscal-like heuristic, later called Incremental Power: Kruscal, was studied in [4]. Both of these so called incremental power heuristics have been proposed for the Minimum Power Asymmetric Broadcast Problem, but they are suitable for MPSCP too. It is proved in [14] that they both have an approximation ratio 2, and it was shown in the same paper that in practice they yield significantly more accurate solution than MST.

Authors of [1] proposed two local search heuristics. The first one is edge-switching (ES) which iteratively performs the best possible replacement of a tree edge and a non-tree edge until a local optimum is reached. The other local search algorithm is edge and fork switching (EFS), where at each step an attempt to replace one or two edges of a tree by an edge or a fork (two adjacent edges) in the best way. In [14] two ES-like heuristics were proposed. In the first one, ES1a, each non-tree edge is added at first and then an edge which belongs to the appeared cycle and causes the maximum power costs is removed. In the second heuristic, ES1b, each edge from a tree is considered to be replaced by the non-tree edge in such way that decrease of objective is maximum. It should be noticed that instead of finding a local optimum ES1a and ES1b perform a single loop on a fixed list of edges (i.e., once added or removed edges are never considered again). Also, they propose a faster sweep method (SW) and the most time-consuming double edge switching (ES2), which is said to be the generalization of EFS: it performs replacements of two edges from a tree and two non-tree edges while it leads to reduction of the objective. Their numerical experiments demonstrate the weakest results of SW (4–5 % improvement over MST for 50–100 nodes), better results of ES1a and ES1b (about 5.5 %), and incredibly high results of ES2 (12–14 %). However, we, as well as authors of [17], could not achieve even 7 % improvement over MST for these dimensions on random instances using our algorithms. It seems like the optimal solution, on average, does not outperform MST by more than 7 % on our random instances. Anyway, ES2 is not applicable for large dimensions because of the high time complexity ($O(|V|^3|E|^2)$). Also, the another two local searches should be mentioned: ST from [17] and LI from [7]. They are very similar because they use the same idea: at each step an edge is removed from a tree and the root of obtained subtree is reconnected with some vertex from another subtree in such way that the decrease of the objective is maximum. The difference between ST and LI is the following: in ST the best replacement is performed at each step, but in LI all edges are sequently considered to be removed and replaced by another edge in the best way, and this loop is repeated while the solution is improved at least at one its iteration.

In [13] a way to filter the edges without impairment of the optimal solution was proposed. This method allows to significantly simplify the initial communication graph and to reduce the computation time. Authors of [17] presented a new iterated local search (ILS) which uses ES and EFS at local search phase, filtration technique from [13] and two different mutation operators. Their numerical experiment results demonstrate that, on average, the best solution within acceptable time can be obtained by ILS with ES, filtration and so-called random increase mutation.

In [6] a hybrid genetic algorithm (GA), which uses variable neighborhood descent (VND) as mutation, was proposed. This algorithm is well parallelized and very fast.

Since we don't know any better heuristics proposed by other authors, we have implemented the best variant of ILS, hybrid GA with VND and compared our algorithms with them in Sect. 5.

4 Heuristics

As mentioned earlier, we use the VNS metaheuristic idea to get an approximate solution of (1). We use two well-known schemas: basic VNS and general VNS. Detailed descriptions of both these methods can be found in [9,10]. For the reader's convenience the pseudo codes of these algorithms are presented in Figs. 1 and 2. These metaheuristics consist of the local search and shaking phases. As a stopping criteria of VNS-based algorithms we used the following rule: if there were no any improvements in last 3 iterations then algorithm stops.

1: Select the set of neighborhood structures $\mathcal{N}_k$, for $k = 1, ..., k_{max}$ that will be used
    for the shaking phase, and let $\mathcal{N}_0 = \{x\}$; find an initial solution x; set $k = 0$;
2: **while** the stopping criteria is not met **do**
3:    **while** $k \leq k_{max}$ **do**
4:        Perform *Shaking*: generate a point x' at random from $\mathcal{N}_k(x)$;
5:        Perform a *Local search*. Let x'' be an obtained local optimum;
6:        **if** x'' is better than x **then**
7:            $x = x''$; $k = 1$
8:        **else**
9:            $k = k + 1$
10:      **end if**
11:    **end while**
12: **end while**

Fig. 1. Basic VNS

In order to reduce the computational complexity we use the filtration of edges presented in [13]. The idea of this method is the following. If the lower bound of the objectives of (1) on all communication trees, which contain the edge e, exceeds the objective on another known feasible solution then the edge e is removed from the communication graph. This filtration is applied to the communication graph as soon as new record solution has been obtained. For the first approximation of our heuristic we generate two trees: one by MST and another by IPP, and then we take the better of them.

1: Select the set of neighborhood structures $\mathcal{N}_k$, for $k = 1, ..., k_{max}$ that will be used
   for the shaking phase, and let $\mathcal{N}_0 = \{x\}$; select the set of neighborhood structures
   N_l, for $l = 1, ..., l_{max}$ that will be used for local search; find an initial solution x;
   set $k = 0$;
2: **while** the stopping criteria is not met **do**
3:     **while** $k \leq k_{max}$ **do**
4:         Perform *Shaking*: generate a point x' at random from $\mathcal{N}_k(x)$;
5:         **while** $l \leq l_{max}$ **do**
6:             Find the best solution $x'' \in N_l(x')$.
7:             **if** x'' is better than x' **then**
8:                 $x' = x''$; $l = 1$
9:             **else**
10:                 $l = l + 1$
11:             **end if**
12:         **end while**
13:         **if** x' is better than x **then**
14:             $x = x'$; $k = 1$
15:         **else**
16:             $k = k + 1$
17:         **end if**
18:     **end while**
19: **end while**

Fig. 2. General VNS

Local Searches. Each neighborhood structure of the local search phase of VNS-based heuristics is used only for local search. Therefore, the descriptions of the local search procedures are sufficient for the definition of the corresponding neighborhood structures, and there is no necessity for explicit formulation of the neighborhood structures.

We propose two local search heuristics which perform edge switchings, but, as opposed to the known ES-like heuristics, they do not perform each edge switching in the best way, but instead of this they iteratively consider a list of edges and perform the best switching for each considered edge. The procedure stops if at some iteration there was no any improvements in all steps of the loop over the edges. There are two possible variants of this approach, we called them Adding and Best Removing (ABR) and Removing and Best Adding (RBA). The pseudo-codes of these local searches can be found, respectively, in Fig. 3 and in Fig. 4. Note that these local searches are similar to ES1a and ES1b from [14], but, as opposed to ES1a and ES1b, ABR and RBA guarantee that the obtained solution is a local optimum.

Shaking. For the shaking procedure, which is used in basic VNS and general VNS, we propose two algorithms. The first one is random shaking which is described in Fig. 5. It consists of sequence of random edge additions and random edge removings. The second algorithm is intensified shaking (Fig. 6), which adds a random edge at first and then removes an edge from the cycle whose deletion

1: Input: $G = (V, E)$ - communication graph, $T = (V, F)$ — spanning tree;
2: improved = true;
3: **while** improved **do**
4:     improved = false;
5:     $G = FilterEdges(G, W(T))$;
6:     $D = E \setminus F$;
7:     **for** each edge $e \in D$ **do**
8:         Find the such edge f in a cycle of $F \cup \{e\}$, whose removing leads to the maximum decrease of the objective;
9:         $T' = (V, F \cup \{e\} \setminus \{f\})$;
10:         **if** $W(T') < W(T)$ **then**
11:             $T = T'$;
12:             improved = true;
13:         **end if**
14:     **end for**
15: **end while**

Fig. 3. ABR local search

1: Input: $G = (V, E)$ - communication graph, $T = (V, F)$ — spanning tree;
2: $G = FilterEdges(G, W(T))$;
3: improved = true;
4: **while** improved **do**
5:     improved = false;
6:     **for** each edge $e \in F$ **do**
7:         Let A and B be the edges of connected components obtained after removing of e from T;
8:         Find such edge $f \in E$ which connects A and B and whose adding to $A \cup B$ leads to the minimum increase of the objective;
9:         $T' = (V, A \cup B \cup \{f\})$;
10:         **if** $W(T') < W(T)$ **then**
11:             $T = T'$;
12:             $G = FilterEdges(G, W(T))$;
13:             improved = true;
14:         **end if**
15:     **end for**
16: **end while**

Fig. 4. RBA local search

reduces the objective at most. In both algorithms replacing of edges is repeated k times. Let $k_{\max}$ be the maximum number of edge replacements by the shaking procedure. Note that k_{max} is a free parameter in the considered VNS heuristics, and its best value of this parameter is estimated experimentally.

1: Input: $G = (V, E)$ - communication graph, $T = (V, F)$ – spanning tree, k – neighborhood index;
2: $i = 1$;
3: **while** $i \leq k$ **do**
4:     Select an edge $e_1 \in E \setminus F$ at random;
5:     $F' = F \cup \{e_1\}$;
6:     Let $C \subseteq F'$ be a cycle containing e_1; select at random an edge $e_2 \in C$;
7:     $F' = F' \setminus \{e_2\}$;
8:     $T = (V, F')$;
9: **end while**

Fig. 5. Random shaking

1: Input: $G = (V, E)$ - communication graph, $T = (V, F)$ — spanning tree, k — neighborhood index;
2: $i = 1$;
3: **while** $i \leq k$ **do**
4:     Select an edge $e_1 \in E \setminus F$ at random;
5:     $F' = F \cup \{e_1\}$;
6:     Let $C \subseteq F'$ be a cycle containing e_1; select an edge $e_2 \in C$ whose deletion reduces the objective at most;
7:     $F' = F' \setminus \{e_2\}$;
8:     $T = (V, F')$;
9: **end while**

Fig. 6. Intensified shaking

5   Simulation

All the proposed algorithms have been implemented in C++ using the Visual Studio 2010 Integrated Development Environment. A simulation was executed for $n = 10, 30, 50, 250$, and in some cases for $n = 500$. For the same dimension, 100 different instances were randomly generated. For each instance, a required number of points was uniformly scattered on a square area with a side of 10 units. After this, a complete edge-weighted graph whose vertices correspond to the points and whose edge weights were equal to the squared distances between the points was defined. Then the calculation of MST and IPP were run on the complete graph, and the best of the obtained two trees was chosen as the first approximation solution for the heuristics. The experiment was performed on an Intel Core i5-4460 (3.2 GHz) 8 Gb machine, and only one thread was used at the same time for all algorithms except CPLEX and GA.

In order to compare the algorithms for the large dimensions, when an optimal solution cannot be found in acceptable time, we calculated the average improvement compared to MST. This estimate was often used for these purposes in the related papers [1,14,17]. For the small dimensions ($n \leq 30$), we defined the parameters of the problem formulation as an integer linear programming problem (ILP), as proposed in [7], and then we obtained the optimal solution using the IBM ILOG CPLEX package. In Table 1 the improvement of optimal solution over MST and the average CPLEX CPU time are presented. Currently neither of known packages and ILP formulations allow to obtain an optimal solution for $n \geq 40$ in acceptable time [1,7,13]. Note that we have parallelized CPLEX on 4 threads to speed-up calculations.

Table 1. CPLEX (optimal solution). Improvement over MST and CPU time

n	Impr. to MST	CPU time
10	3.98 %	0.33 s
30	5.78 %	93.53 s

For the VNS-based heuristics, it is necessary to define the parameter k_{max}. For this goal, each algorithm was run on the same instances with different values of k_{max}. It appeared that, beginning from $k_{max} = 30$, on average, the objective of the obtained solution did not decrease significantly, whereas the runtime grown fast. Moreover, on average, the runtime of all the algorithms remained acceptable for $k_{max} = 30$. Therefore, in all the VNS-based algorithms, we set $k_{max} = 30$.

In the Table 2 the effect of filtration from [13] is presented. The first column represents the percentage of edges removed after applying the filtration procedure to the complete graph when the results of MST and IPP are known. In the other columns the speed-ups of some of the heuristics are reflected. One can see that filtration significantly simplifies the initial graph and speeds up the algorithms. In all further results all heuristics use filtration procedure.

Table 2. Filtration effect

n	Filtered edges	Speed-up of ES	Speed-up of ABR	Speed-up of RBA
30	53.02 %	53.08 %	58.96 %	46.15 %
50	55.81 %	59.62 %	63.36 %	53.74 %
100	59.71 %	66.38 %	64.3 %	59.57 %
250	60.84 %	70.59 %	66.53 %	62.73 %

In Table 3 the local search algorithms are compared. The best values are marked bold. The best solutions were always obtained by EFS, but its running time increases very fast with growing n, and it works more than 1000 s already on 200 nodes.

Table 4 represents CPU time and improvement over MST of the basic VNS with different local search procedures and random shaking. In Table 5 the results obtained by the same algorithms but with intensified shaking are presented. Since, on average, intensified shaking works slightly better than the random shaking, in the further tables intensified shaking is used in VNS-based heuristics.

Table 3. Local search heuristics. Improvement over MST and CPU time.

n	ABR		RBA		EFS		ES		LI	
	Impr. to MST	CPU time	Impr. to MST	CPU time	Impr. to MST	CPU time	Impr. to MST	CPU time	Impr. to MST	CPU time
10	3.76 %	0.00 s	3.72	0.00 s	**3.96** %	0.00 s	3.75 %	0.00 s	3.04 %	0.00 s
30	5.03 %	0.00 s	5.05 %	0.00 s	**5.58** %	0.09 s	5.07 %	0.00 s	3.56 %	0.00 s
50	5.35 %	0.00 s	5.33 %	0.00 s	**6.08** %	0.92 s	5.45 %	0.00 s	3.98 %	0.00 s
100	5.52 %	0.02 s	5.5 %	0.02 s	**6.17** %	29.39 s	5.59 %	0.05 s	3.77 %	0.00 s
250	5.61 %	0.26 s	5.6 %	0.23 s	–	–	5.71 %	1.09 s	3.94 %	0.01 s

Table 4. Basic VNS with random shaking. Improvement over MST and CPU time.

n	B_ABR		B_RBA		B_ES		B_LI	
	Impr. to MST	CPU time	Impr. to MST	CPU time	Impr. to MST	CPU time	Impr. to MST	CPU time
10	**3.98** %	0.00 s	3.96 %	0.00	**3.98** %	0.01 s	3.93 %	0.00 s
30	5.74 %	0.06 s	**5.78** %	0.06	5.76 %	0.08 s	4.40 %	0.00 s
50	6.21 %	0.24 s	6.29 %	0.26	**6.30** %	0.32 s	3.96 %	0.00 s
100	6.08 %	1.39 s	**6.23** %	1.71	6.11 %	1.81 s	3.50 %	0.01 s
250	6.12 %	17.97 s	**6.27** %	22.52	6.00 %	19.27 s	3.62 %	0.02 s

Table 5. Basic VNS with intensified shaking. Improvement over MST and CPU time.

n	B_ABR		B_RBA		B_ES		B_LI	
	Impr. to MST	CPU time	Impr. to MST	CPU time	Impr. to MST	CPU time	Impr. to MST	CPU time
10	3.94 %	0.00 s	**3.98** %	0.00 s	**3.98** %	0.00 s	3.77 %	0.00 s
30	5.71 %	0.05 s	**5.77** %	0.05 s	5.76 %	0.06 s	4.14 %	0.00 s
50	6.16 %	0.23 s	6.26 %	0.21 s	**6.27** %	0.21 s	3.82 %	0.00 s
100	6.02 %	1.44 s	**6.24** %	1.44 s	6.12 %	1.21 s	3.42 %	0.01 s
250	6.01 %	15.41 s	**6.27** %	21.46 s	5.96 %	12.09 s	3.45 %	0.02 s

The general VNS-based heuristics results presented in Table 6. We have run two variants of general VNS. Both of them used ABR and RBA as local searches and intensified shaking. G_AR is general VNS where in each iteration of the local search phase ABR was run at first and RBA was run next. G_RA is general VNS in each iteration of the local search phase RBA was run at first and ABR was run next.

In Table 7 the results obtained by ILS with different local searches are presented. The random increase mutation was used in ILS and, as well as it was done in [17], 200 iterations were run before stop.

Table 6. General VNS with intensified shaking. Improvement over MST and CPU time.

n	G_AR		G_RA	
	Impr. to MST	CPU time	Impr. to MST	CPU time
10	**3.98 %**	0.00 s	3.96 %	0.00 s
30	5.74 %	0.07 s	**5.78 %**	0.08 s
50	6.15 %	0.28 s	**6.29 %**	0.29 s
100	6.03 %	1.74 s	**6.20 %**	1.86 s
250	6.05 %	21.27 s	**6.30 %**	31.14 s

Table 7. ILS-based heuristics. Improvement over MST and CPU time.

n	ILS_ABR		ILS_RBA		ILS_ES		ILS_LI	
	Impr. to MST	CPU time	Impr. to MST	CPU time	Impr. to MST	CPU time	Impr. to MST	CPU time
10	**3.98 %**	0.02 s	3.9 %	0.02 s	**3.98 %**	0.02 s	3.17 %	0.01 s
30	5.72 %	0.32 s	5.75 %	0.33 s	**5.78 %**	0.43 s	2.929 %	0.02 s
50	6.23 %	1.111 s	6.28 %	1.23 s	**6.33 %**	1.73 s	3.187 %	0.04 s
100	6.12 %	6.353 s	6.23 %	7.96 s	**6.31 %**	13.33 s	3.048 %	0.09 s
250	6.06 %	65 s	6.16 %	107.5 s	**6.4 %**	250 s	3.21 %	0.29 s

Although, on average, ILS outperforms the VNS-based heuristics, it requires significantly more time, and the average excesses of the best of ILS-based heuristic ILS_ES over the best of VNS-based metaheuristics B_RBA and G_RA are not so significant — they never exceed 0.1 %. Therefore, we compared the solution obtained by one of the best VNS-based metaheuristics, basic VNS with RBA and intensified shaking (B_RBA), with the solution obtained by ILS_ES in the same running time as B_RBA. These results are presented in Table 8. In the same manner, we have compared the G_RA (which appeared to be the best of general VNS-based heuristics) with ILS_ES, see Table 9. Except the improvement over MST, for each of two heuristics B_RBA and G_RA, we calculated the percentage of cases when its solution is better than ILS and the percentage of cases when it is worse than ILS_ES. One can see that, on average, B_RBA and G_RA both outperform ILS_ES, especially on large dimensions. The advantages of the both VNS-based heuristics are most strongly shown when $n = 500$. In this case B_RBA yielded more accurate solution than ILS_ES in 99 % of cases, the average improvement of B_RBA over MST exceeds the same estimation of LI_ES by 0.44 % which is about 7.5 % of the improvement, and the maximum improvement over to MST exceeds the same estimation of LI_ES by 0.82 % which is 10.16 % of the improvement. The results obtained by G_RA in the case of $n = 500$ are very impressive as well: G_RA yields better solution than ILS_ES in 94 % of cases, its average excess of improvement over MST is 0.38 %, which is 6.37 %

Table 8. Comparison of the results for the best of the basic VNS-based heuristics B_RBA and for the best of the iterated local search-based algorithms ILS_ES.

n	B_RBA is better	ILS_ES is better	B_RBA: Impr. to MST			ILS_ES: Impr. to MST			CPU time
			Min	Avg	Max	Min	Avg	Max	
10	2 %	0 %	0 %	**3.98** %	19.82 %	0 %	3.95 %	19.82 %	0.00 s
30	11 %	2 %	0.82 %	**5.78** %	15 %	0.82 %	5.7 %	14.58 %	0.05 s
50	30 %	10 %	1.20 %	**6.28** %	13.56 %	1.20 %	6.2 %	13.41 %	0.20 s
100	54 %	26 %	2.48 %	**6.23** %	10.42 %	2.38 %	6.15 %	10.86 %	1.43 s
250	85 %	15 %	3.64 %	**6.29** %	9.50 %	3.46 %	6.04 %	9.36 %	22.63 s
500	99 %	1 %	4.10 %	**6.34** %	8.89 %	3.72 %	5.90 %	8.07 %	208.2 s

Table 9. Comparison of the results for the best of the general VNS-based heuristics G_RA and for the best of the iterated local search-based algorithms ILS_ES.

n	G_RA is better	ILS_ES is better	G_RA: Impr. to MST			ILS_ES: Impr. to MST			CPU time
			Min	Avg	Max	Min	Avg	Max	
10	0 %	0 %	0 %	**3.98** %	19.82 %	0 %	**3.98** %	19.82 %	0.01 s
30	6 %	6 %	0.82 %	**5.77** %	15 %	0.82 %	5.74 %	14.58 %	0.08 s
50	17 %	14 %	1.2 %	**6.29** %	13.66 %	1.2 %	**6.29** %	13.56 %	0.30 s
100	47 %	36 %	1.87 %	6.21 %	10.34 %	2.42 %	**6.23** %	10.86 %	2.11 s
250	72 %	28 %	3.69 %	**6.29** %	9.49 %	3.65 %	6.16 %	9.60 %	31.42 s
500	94 %	6 %	4.30 %	**6.35** %	8.6 %	3.75 %	5.97 %	8.29 %	279.7 s

of the improvement. It should be noted, that LI_ES had appeared to be too time-consuming in a case of $n = 500$. Its average running time on 10 instances exceeded 1200 s.

In [6] two hybrid genetic algorithms for the MPSCP were proposed. The best results had been obtained by the genetic algorithm which used VND-based heuristic as mutation. In Table 10 the results of this hybrid genetic algorithm GA_VND are compared with the best VNS-based heuristics: B_RBA and G_RA. One can see that GA_VND solved the problem significantly faster, but it should be taken into account that it was well parallelized and used four parallel threads. However, VNS-based heuristics yield more accurate solutions, especially on large

Table 10. Comparison of the results for the best of VNS-based heuristics and hybrid genetic algorithm GA_VND.

n	B_RBA		G_RA		GA_VND	
	Impr. to MST	CPU time	Impr. to MST	CPU time	Impr. to MST	CPU time
10	**3.98** %	0.00 s	**3.98** %	0.01 s	**3.98** %	0.06 s
30	**5.78** %	0.05 s	5.77 %	0.08 s	5.75 %	0.14 s
50	6.28 %	0.20 s	**6.29** %	0.3 s	6.20 %	0.31 s
100	**6.23** %	1.427 s	6.21 %	2.11 s	5.96 %	1.12 s
250	**6.29** %	22.63 s	**6.29** %	31.42 s	5.87 %	6.35 s
500	6.34 %	208.2 s	**6.35** %	279.7 s	5.71 %	31.8 s

instances: in a case of $n = 500$ their average improvement over MST exceeds the same estimation for the GA_VND by more than 0.6 %, which is 10.5 % of the improvement. Since GA_VND was stopped after stabilization (when the quality of solutions was not changed during the last 20 iterations), we did not expect that its solutions would become significantly better after the longer work. Therefore, GA_VND was not compared with the proposed VNS-based heuristics by time limit, as it was done for ILS_ES.

6 Conclusion

In this paper we have presented new variable neighborhood search-based heuristics for the Minimum Power Symmetric Connectivity Problem. We used two known variants of the VNS metaheuristic: basic VNS and general VNS. As local search we used already known heuristics ES, EFS and LI as well as two new heuristics: ABR and RBA. We also used filtration of edges of the communication graph inside our algorithms in order to reduce the computation time. The numerical experiment has shown that the best of the proposed VNS-based heuristics (namely, B_RBA and G_RA) are more suitable to use in practice than the best of known algorithms (iterated local search-based algorithm proposed in [17] and hybrid genetic algorithm proposed in [6]): on average, our heuristics obtain significantly more accurate solutions in short time and allow to successfully get solutions very close to optimal in large dimension cases. In future we plan to implement the variable neighborhood decomposition search [11] for this problem in order to solve it in larger dimensions in acceptable time.

Acknowledgments. The research of A. Erzin is partly supported by the Russian Foundation for Basic Research (grant no. 16-07-00552). The research of N. Mladenovic is partly supported by the Ministry of Education and Science, Republic of Kazakhstan (Institute of Information and Computer Technologies) (project no. 0115PK00546). The research of R. Plotnikov is supported by the Russian Foundation for Basic Research (grant no. 16-37-60006).

References

1. Althaus, E., Calinescu, G., Mandoiu, I.I., Prasad, S.K., Tchervenski, N., Zelikovsky, A.: Power efficient range assignment for symmetric connectivity in static ad hoc wireless networks. Wireless Netw. **12**(3), 287–299 (2006)
2. Calinescu, G., Mandoiu, I.I., Zelikovsky, A.: Symmetric connectivity with minimum power consumption in radio networks. In: Baeza-Yates, R.A., Montanari, U., Santoro, N. (eds.) Proceedings of the 2nd IFIP International Conference on Theoretical Computer Science. IFIP Conference Proceedings, vol. 223, pp. 119–130. Kluwer, Dordrecht (2002)
3. Cheng, X., Narahari, B., Simha, R., Cheng, M.X., Liu, D.: Strong minimum energy topology in wireless sensor networks: NP-completeness and heuristics. IEEE Trans. Mob. Comput. **2**(3), 248–256 (2003)
4. Chu, T., Nikolaidis, I.: Energy efficient broadcast in mobile ad hoc networks. In: Proceedings of AD-HOC Networks and Wireless (2002)

5. Clementi, A.E.F., Penna, P., Silvestri, R.: Hardness results for the power range assignment problem in packet radio networks. In: Hochbaum, D.S., Jansen, K., Rolim, J.D.P., Sinclair, A. (eds.) RANDOM 1999 and APPROX 1999. LNCS, vol. 1671, pp. 197–208. Springer, Heidelberg (1999)

6. Erzin, A., Plotnikov, R.: Using VNS for the optimal synthesis of the communication tree in wireless sensor networks. Electron. Notes Discrete Math. **47**, 21–28 (2015)

7. Erzin, A., Plotnikov, R., Shamardin, Y.: On some polynomially solvable cases and approximate algorithms in the optimal communication tree construction problem. J. Appl. Indust. Math. **7**, 142–152 (2013)

8. Fuchs, B.: On the hardness of range assignment problems. Technical report. TR05-113, Electronic Colloquium on Computational Complexity (2005)

9. Hanafi, S., Lazic, J., Mladenovic, N., Wilbaut, C., Crevits, I.: New variable neighbourhood search based 0–1 MIP heuristics. Yugoslav J. Oper. Res. (2015). doi:10. 2298/YJOR140219014H

10. Hansen, P., Mladenovic, N.: Variable neighborhood search: principles and applications. Eur. J. Oper. Res. **130**, 449–467 (2001)

11. Hansen, P., Mladenovic, N., Perez-Britos, D.: Variable neighborhood decomposition search. J. Heuristics **7**(4), 335–350 (2001)

12. Kirousis, L.M., Kranakis, E., Krizanc, D., Pelc, A.: Power consumption in packet radio networks. Theoret. Comput. Sci. **243**(1–2), 289–305 (2000)

13. Montemanni, R., Gambardella, L.M.: Exact algorithms for the minimum power symmetric connectivity problem in wireless networks. Comput. Oper. Res. **32**(11), 2891–2904 (2005)

14. Park, J., Sahni, S.: Power assignment for symmetric communication in wireless networks. In: Proceedings of the 11th IEEE Symposium on Computers and Communications (ISCC), Washington, pp. 591–596. IEEE Computer Society, Los Alamitos (2006)

15. Rappaport, T.S.: Wireless Communications: Principles and Practices. Prentice Hall, Upper Saddle River (1996)

16. Santi, P.: Topology Control in Wireless Ad Hoc and Sensor Networks. Wiley, Chichester (2005)

17. Wolf, S., Merz, P.: Iterated local search for minimum power symmetric connectivity in wireless networks. In: Cotta, C., Cowling, P. (eds.) EvoCOP 2009. LNCS, vol. 5482, pp. 192–203. Springer, Heidelberg (2009)

On the Facets of Combinatorial Polytopes

Ruslan Simanchev[1,2(✉)] and Inna Urazova[2]

[1] Omsk Scientific Center of SB RAS, 15 Marksa Avenue, 644024 Omsk, Russia
osiman@rambler.ru
[2] Omsk State University, 55a Mira Avenue, 644077 Omsk, Russia
urazovainn@mail.ru

Abstract. One of the central questions of polyhedral combinatorics is the question of the algorithmic relationship between vertex and facet descriptions of convex polytopes. In the sense of combinatorial optimization the reason for the relevance of this issue is the possibility of application of convex analysis methods to the decision combinatorial problems [6,10,15]. In this paper we consider combinatorial polytopes sufficiently general form. A number of necessary conditions and sufficient conditions for support inequality of polytope to be facet inequality are obtained, an illustration of the use of the developed technology to the connected k-factors polytope are given. Also we discuss the use of facet inequalities in cutting plane algorithms.

Keywords: Polytope · Facet inequality · Separation problem

1 Preliminaries

Let E is a finite set. With E we associate the $|E|$-dimensional Euclidean space R^E by one-to-one correspondence between the elements of set E, and the coordinate axes of space R^E. In other words, R^E is a set of column vectors whose components are indexed by elements of set E. The polytope in space R^E is a convex hull of a finite number of points in R^E. The affine hull of polytope P is set $affP$ of all affine combinations of points of P. The affine hull $affP$ is an affine subspace of R^E. Consequently, there is a system of linear equations that

$$affP = \{x \in R^E \mid A^T x = \alpha\},$$

where $A - (|E| \times n)$-matrix and without loss of generality $rankA = n$. The dimension of polytope P (denote $dimP$) is the cardinality of max to include affine independent family of points of P minus 1.

Let $a \in R^E$ and $a_0 \in R$. Linear inequality $a^T x \leq a_0$ is called valid to polytope P, if it holds for all points of P. A valid inequality is called a support inequality if there are $x', x'' \in P$, such that $a^T x' < a_0$ and $a^T x'' = a_0$. Any support inequality generates a face $\{x \in P \mid a^T x = a_0\}$ of P. A vertex of P is a face with the dimension equal to 0. A facet of P is a face with the dimension equal to $dimP - 1$ and corresponding inequality is called facet inequality. Facet inequalities play a special role

© Springer International Publishing Switzerland 2016
Y. Kochetov et al. (Eds.): DOOR 2016, LNCS 9869, pp. 233–243, 2016.
DOI: 10.1007/978-3-319-44914-2_19

in polyhedral combinatorics. According to the theorem of Weyl-Minkowski [15], to describe a polytope as a set of solutions of linear systems it is necessary and sufficient to know all its facet inequalities (up to equivalence). In addition, facet inequalities are widely used as cutting planes for the decision of combinatorial optimization problems of large dimension [1,2,4,9].

Let us turn to the description of the object, which is the subject of this article. For each $R \subseteq E$ we define its incidence vector $x^R \in R^E$ as a vector with coordinates $x_e^R = 1$ if $e \in R$ and $x_e^R = 0$ if $e \notin R$. Thus, a set of all subsets of E is placed in one-to-one correspondence with a set of all vertices of unit cube in R^E. Based on this correspondence the $(0,1)$-vector $x \in R^E$ will be understood as a subset of E too.

Let $\mathcal{H} \subseteq 2^E$ be a family of subsets of E. A combinatorial polytope associated with $\mathcal{H}$ is the set

$$P_\mathcal{H} = conv\{x^H \in R^E \mid H \in \mathcal{H}\},$$

where "$conv$" means "convex hull". Here are some of obvious, but important properties of $P_\mathcal{H}$.

(1) Each vertex of $P_\mathcal{H}$ is a $(0,1)$-vector.
(2) Vertices and only they correspond to sets of the family $\mathcal{H}$.
(3) Polytope $P_\mathcal{H}$ has no integer points other than the vertices.

In this article we justify an approach to the proof of the facetness of an inequality support to $P_\mathcal{H}$. Traditionally, to prove the facetness two approaches are used. The first one is based on finding a sufficient number of affinely independent vertices of an edge generated by the inequality. The second approach is based on the use of the following fact which is wide-known in convex analysis

Theorem 1 [10]. *Let $P \subset R^E$ be a polytope, $affP = \{x \in R^E \mid A^T x = \alpha\}$, and the matrix A have full rank. The support inequality $a^T x \leq a_0$ is a facet inequality if and only if for any support inequality $c^T x \leq c_0$ satisfying the condition*

$$\{x \in P \mid a^T x = a_0\} \subseteq \{x \in P \mid c^T x = c_0\},$$

there is a combination $c = \mu a + A\lambda$, $c_0 = \mu a_0 + \alpha^T \lambda$, where μ is a non-negative number and $\lambda \in R^{rankA}$.

This approach, as a rule, is connected with a large amount of algebraic calculations. Our results follow from Theorem 1 and consist in the presentation of linear combinations of constraints of affine hull of the face being studied using symmetrical differences of sets of family $\mathcal{H}$. It enables providing a combinatorial nature to the proof procedure. We introduce the concept of $b\mathcal{H}$-basis which links support inequality $b^T x \leq b_0$, family $\mathcal{H}$ and affine hull of $P_\mathcal{H}$ polytope. $b\mathcal{H}$-basis is a subset of set E and corresponds to the elements of set E which make up the basis of the matrix defining the affine hull of $P_\mathcal{H}$ polytope face being studied.

In this paper a number of necessary conditions (Sect. 2) and sufficient conditions (Sect. 3) for support inequality of the polytope $P_\mathcal{H}$ to be a facet inequality are obtained. The technique described in Sects. 2 and 3 is a generalization of the method from [11], that was used for a connected k-factors polytope. In Sect. 4, we present three facet inequality classes from [11] and discuss the separation problem for one of them.

We consider the so-called combinatorially full families $\mathcal{H} \subseteq 2^E$, i.e. such families, that satisfy the axiom:

for any $e_1, e_2 \in E$ there is such $H \in \mathcal{H}$ that $e_1 \in H$ and $e_2 \notin H$.

For example, on the complete graph with more than four nodes the matchings, Hamiltonian cycles, M-graphs [12] are combinatorially full families.

2 Necessary Conditions

In this section some of the properties of facet inequalities for $P_{\mathcal{H}}$ on combinatorially full families $\mathcal{H}$ will be discussed.

Theorem 2. *Let $a^T x \leq a_0$ be a support inequality for $P_{\mathcal{H}}$ and $\{x^1, x^2, \ldots, x^t\}$ be a set of all vertices of the corresponding face. If any of following conditions*

(1) $|\cap_{i=1}^{t} x^i| \geq 2$ or $|E \setminus \cup_{i=1}^{t} x^i| \geq 2$,

(2) $\cap_{i=1}^{t} x^i = \{e_0\}$ (or $E \setminus \cup_{i=1}^{t} x^i = \{e_0\}$) and exist suth $\overline{x} \in \mathcal{H}$ that $e_0 \in \overline{x}$ (or $e_0 \notin \overline{x}$ respectively) and $a^T \overline{x} < a_0$,

holds, then the inequality $a^T x \leq a_0$ is not facet inequality for $P_{\mathcal{H}}$.

Proof. Let F_a be a face of $P_{\mathcal{H}}$, that is generated by inequality $a^T x \leq a_0$.

(1) Let $e_1, e_2 \in \cap_{i=1}^{t} x^i$, $e_1 \neq e_2$ and $F_j = \{x \in P_{\mathcal{H}} \mid x_{e_j} = 1\}$, $j = 1, 2$ be faces that are generated by support inequalities $x_{e_j} \leq 1$, $j = 1, 2$. Then all points $x^1, x^2, \ldots, x^t$ lie on faces F_j, $j = 1, 2$. Consequently, face F_a belongs to the intersection of faces F_1 and F_2.

Since the family $\mathcal{H}$ is combinatorially full then F_1 and F_2 are proper faces of $P_{\mathcal{H}}$. Furthermore, $F_1 \neq F_2$. In fact if a set $H \in \mathcal{H}$ belongs to one of the elements e_1 and e_2 then H belongs to the second element. This is contradicts to combinatorial fullness of $\mathcal{H}$. Thus

$$dim F_a \leq dim(F_1 \cap F_2) < min\{dim F_1, dim F_2\} \leq dim P_{\mathcal{H}} - 1.$$

Hence F_a is not a facet.

For $e_1, e_2 \in E \setminus \cup_{i=1}^{t} x^i$ the arguments are completely analogous if we assume that $F_j = \{x \in P_{\mathcal{H}} \mid x_{e_j} = 0\}$, $j = 1, 2$.

(2) Let $F_0 = \{x \in P_{\mathcal{H}} \mid x_{e_0} = 1\}$. Since the family $\mathcal{H}$ is combinatorially full then F_0 is a proper face of $P_{\mathcal{H}}$. Moreover, from the conditions of the theorem it is obvious that $F_a \subseteq F_0$. If $F_a = F_0$ then for any $x' \in \mathcal{H}$ that contains e_0 inclusion $x' \in F_a$ holds. This contradicts to the condition. Hence $F_a \neq F_0$. Now we can write

$$dim F_a < dim F_0 \leq dim P_{\mathcal{H}} - 1.$$

Hence F_a is not a facet again.

For $E \setminus \cup_{i=1}^{t} x^i = \{e_0\}$ the arguments are analogous, if we assume $F_0 = \{x \in P_{\mathcal{H}} \mid x_{e_0} = 0\}$.

The theorem is proved. $\square$

The next necessary condition holds for support inequalities with coefficients 0 and 1 in the left side. The class of such inequalities is induced by the set of all subsets of E as follows. Let $W \subseteq E$. The number

$$r_{\mathcal{H}}(W) = max\{|W \cap H| \mid H \in \mathcal{H}\}$$

is called rank of W (with respect to $\mathcal{H}$). Accordingly, inequality

$$\sum_{e \in W} x_e \leq r_{\mathcal{H}}(W)$$

is called a rank inequality induced by W. Due to the definition of $r_{\mathcal{H}}$ value, any rank inequality is a support inequality for $P_{\mathcal{H}}$.

Theorem 3. *Let $W \subseteq E$ and $|W| \geq 2$. If there exists $H \in \mathcal{H}$ such that $W \subset H$ then rank inequality $\sum_{e \in W} x_e \leq r_{\mathcal{H}}(W)$ is not facet inequality for $P_{\mathcal{H}}$.*

In fact in this case $r_{\mathcal{H}}(W) = |W|$. Hence if for some $\overline{x} \in \mathcal{H}$ the equality $\sum_{e \in W} \overline{x}_e = r_{\mathcal{H}}(W)$ holds, then $W \subseteq \overline{x}$. Now the required follows from Theorem 2.

3   Sufficient Conditions

As before, let $\mathcal{H} \subseteq 2^E$ be combinatorially full family of subsets of E,

$$aff P_{\mathcal{H}} = \{x \in R^E \mid A^T x = \alpha\}$$

and the matrix A have full rank. Each row of matrix A corresponds to exactly one element $e \in E$ and vice versa. Therefore the set of rows of matrix A is denoted by E. The set of columns is denoted by the letter V and let $|V| = n$. Clearly $rank A = |V| \leq |E|$. For matrix coefficient of A, that is in line $e \in E$ and column $u \in V$ we will write a_{eu}. If $c \in R^E$ then by $(c|A)$ (or $(A|c)$) we denote matrix that obtained by ascribing to the matrix A column c on the left (respectively, right). Through $A(c, \widetilde{E})$ we denote the submatrix of $(c|A)$, formed lines $\widetilde{E} \subseteq E$. If the matrix A is empty, then by $(c|A)$ (or $(A|c)$) we mean a column c.

Let $b^T x \leq b_0$ be support inequality for $P_{\mathcal{H}}$. We need the following definitions.

Definition 1. *Non-empty set $S \subset E$ will be called $b\mathcal{H}$-switching if there are $H_1, H_2 \in \mathcal{H}$ such that*
    (1) $S = H_1 \triangle H_2$,
    (2) $b^T x^{H_1} = b^T x^{H_2} = b_0$.
    Here $H_1 \triangle H_2 = (H_1 \setminus H_2) \cup (H_2 \setminus H_1)$ is the symmetric difference of sets H_1 and H_2.

Definition 2. *Subset $\widetilde{E} \subset E$ called a $b\mathcal{H}$-basis if the following conditions hold*
(i) $|\widetilde{E}| = n + 1$;
(ii) matrix $A(b, \widetilde{E})$ has a full rank;
(iii) for any $e \in E \setminus \widetilde{E}$ there is an ordered sequence of $e_1, e_2, \ldots, e_t = e$ from E that for each $i \in \{1, 2, \ldots, t\}$ the element e_i belongs to some $b\mathcal{H}$-switching, lying in the $\widetilde{E} \cup \{e_1, e_2, \ldots, e_i\}$.

First, we prove an auxiliary lemma.

Lemma 1. *Let* $affP_{\mathcal{H}} = \{x \in R^E \mid A^T x = \alpha\}$, $H_1, H_2 \in \mathcal{H}$ – *pair of different sets and* $S = H_1 \bigtriangleup H_2$. *Then for each* $u \in V$ *holds that*

$$\sum_{e \in S \setminus H_1} a_{eu} = \sum_{e \in S \setminus H_2} a_{eu}.$$

Proof. Let $a^T x = \alpha_u$ is an equation from $A^T x = \alpha$ that corresponds to node $u \in V$. It is clear that vectors x^{H_1} and x^{H_2} satisfy this equation. Because $S \setminus H_2 = H_1 \setminus H_2$ and $S \setminus H_1 = H_2 \setminus H_1$ then

$$0 = a^T x^{H_1} - a^T x^{H_2} = a^T (x^{H_1} - x^{H_2}) = a^T (x^{H_1 \setminus H_2} - x^{H_2 \setminus H_1})$$
$$= a^T (x^{S \setminus H_2} - x^{S \setminus H_1}) = \sum_{e \in S \setminus H_2} a_{eu} - \sum_{e \in S \setminus H_1} a_{eu}.$$

The lemma is proved. $\qquad\qquad\qquad\qquad\qquad\qquad\qquad\qquad\qquad\qquad\quad$ □

Theorem 4. *Support inequality* $b^T x \le b_0$ *is a facet inequality for* $P_{\mathcal{H}}$ *if* $b\mathcal{H}$-*basis* $\widetilde{E} \subset E$ *exist.*

Proof. The proof is based on Theorem 1. Let $c^T x \le c_0$ be a support inequality for $P_{\mathcal{H}}$ satisfying the condition

$$\{x \in P \mid b^T x = b_0\} \subseteq \{x \in P \mid c^T x = c_0\}. \tag{1}$$

We show that the system of linear equations

$$\mu b + A\lambda = c \tag{2}$$

is relatively compatible to $\mu \in R$, $\lambda \in R^n$ and $\mu \ge 0$.

Any equation of the system (2) corresponds to a single $e \in E$. Let us denote the equations of system (2) across $\gamma(e)$, $e \in E$. Wherein we will have in mind right and left parts

$$\gamma(e): \quad b_e \mu + \sum_{u \in V} a_{eu} \lambda_u = c_e.$$

Let $S = H_1 \bigtriangleup H_2$ is the $b\mathcal{H}$-switching from the corresponding definition. Then $b^T x^{H_1} = b^T x^{H_1} = b_0$. Accordingly

$$0 = b^T x^{H_1} - b^T x^{H_2} = b^T (x^{H_1} - x^{H_2}) = b^T (x^{H_1 \setminus H_2} - x^{H_2 \setminus H_1})$$
$$= b^T (x^{S \setminus H_2} - x^{S \setminus H_1}) = \sum_{e \in S \setminus H_2} b_e - \sum_{e \in S \setminus H_1} b_e. \tag{3}$$

As by condition (1) we have $c^T x^{H_1} = c^T x^{H_1} = c_0$ then we get from similar considerations

$$0 = \sum_{e \in S \setminus H_2} c_e - \sum_{e \in S \setminus H_1} c_e. \tag{4}$$

Note that in Lemma 1 the same combination of elements of other columns of the system (2) as in (3) and (4) are used. Consequently, in matrix $(b|A|c)$ the sum of lines with the names from $S \setminus H_2$ minus the sum of lines with names from $S \setminus H_1$ gives the zero line. In other words, equations $\gamma(e)$, $e \in S$ are connected by the following linear equation:

$$0 = \sum_{e \in S \setminus H_2} \gamma(e) - \sum_{e \in S \setminus H_1} \gamma(e). \tag{5}$$

This means their linear dependence. Thus, if a set $S \subset E$ is $b\mathcal{H}$-switching then any one equation of a family $\{\gamma(e), e \in S\}$ can be discarded from the system (2) without prejudice to its compatibility.

We show that this property allows us to drop all of the equations $\gamma(e)$ with names $e \in E \setminus \widetilde{E}$ out of the system (2). Let $e \in E \setminus \widetilde{E}$ and $e_1, e_2, \ldots, e_t = e$ is corresponding ordered sequence from the condition (3). Note that the condition (3) is performed for each element of the sequence. Suppose that $S_t \subseteq \widetilde{E} \cup \{e_1, e_2, \ldots, e_t\}$ is a $b\mathcal{H}$-switch and $e_t \in S_t$. Then by (5) $\gamma(e_t)$ is a linear combination of equations from the set $\{\gamma(e), e \in \widetilde{E} \cup \{e_1, e_2, \ldots, e_{t-1}\}\}$. Similarly, we can easily see that $\gamma(e_{t-1})$ is a linear combination of the equations from $\{\gamma(e), e \in \widetilde{E} \cup \{e_1, e_2, \ldots, e_{t-2}\}\}$. Therefore, $\gamma(e_t)$ is a linear combination of equations $\{\gamma(e), e \in \widetilde{E} \cup \{e_1, e_2, \ldots, e_{t-2}\}\}$. Continuing these arguments in order of decrease of the numbers e_i, we conclude that the equation $\gamma(e_t) = \gamma(e)$ is a linear combination of equations $\{\gamma(e), e \in \widetilde{E}\}$. Hence the equation $\gamma(e)$ can be dropped out of the system (2) without prejudice to its compatibility.

Thus the system (2) is equivalent to system

$$A(b, \widetilde{E})\overline{\lambda} = \widetilde{c}, \tag{6}$$

where $\widetilde{c} = (c_e, e \in \widetilde{E}, \overline{\lambda} = (\mu, \lambda^T)^T \in R^{n+1}$. By (1) and (2) we have $rank A(b, \widetilde{E}) = n + 1$. Hence, system (6) and consequently the system (2) are compatible. The solution of system (2) is nontrivial because otherwise $c = 0$.

It remains to show that $\mu \geq 0$. Since $b^T x \leq b_0$ is a support inequality for $P_{\mathcal{H}}$ then there are $x^1, x^2 \in \mathcal{H}$ such that $b^T x^1 = b_0$ and $b^T x^2 < b_0$. Then by (1) $c^T x^1 = c_0$ and $c^T x^2 \leq c_0$. So we have

$$0 \leq c^T(x^1 - x^2) = (\mu b^T + \lambda^T A^T)(x^1 - x^2) = \mu(b^T x^1 - b^T x^2) + \lambda^T \alpha - \lambda^T \alpha.$$

Since $b^T x^1 - b^T x^2 > 0$ then $\mu \geq 0$.

The theorem is proved.   □

4 Connected k-factors Polytope: Facets and Separation Problem

The technique that described in Sects. 2 and 3 is quite cumbersome. Nevertheless, if we take a specific $\mathcal{H}$, affine hull of the $P_{\mathcal{H}}$ and a support inequality, constructive results are possible. Thus in [11] by this technique three facet inequality

classes for the connected k-factors polytope are described. In other words the
results of Sects. 2 and 3 are a generalization of the approach from [11] to an arbitrary polytope of the form $P_{\mathcal{H}}$. Later in [12] this technique was used for graph
approximation problem polytope.

In this section the three facet inequality classes from [11] are described. For
one of this classes the polynomial solvability of the separation problem with even
k will be proved.

Let $K_n = (V, E)$ is complete undirected graph without loops and multiple
edges with the vertex set V and edge set E. Let kn is even. The subgraph
$H \subseteq K_n$ is called k-factor if the degree of each vertex from V relative H equal
to k. In [11] a connected k-factors polytope was considered. Note that when
$k = 2$ a connected k-factors set is a Hamiltonian cycles set. The connected k-
factors polytope is constructed in the way described in Sect. 1. Here we believe
that the base set is E and the connected k-factors set is $\mathcal{H}$. The connected k-
factors polytope will be denoted by $P_{k,n}$. In [11] three facet inequalities classes
for the connected k-factors polytope were described:

- trivial facets: for every $e \in E$ inequalities $x_e \geq 0$ for $k < n - 2$ and $x_e \leq 1$
for $k \leq n - 2$ are facet inequalities;

- subtour eliminations inequalities: for clique $K = (VK, EK) \subset K_n$ inequality

$$\sum_{e \in EK} x_e \leq \lceil \frac{k \mid VK \mid}{2} - 1 \rceil \qquad (7)$$

is facet inequality if and only if $k < \mid VK \mid < n - k$;

- 2-matching inequalities: let $K = (VK, EK)$ is clique and $R = (VR, ER)$
is set of pairwise non-adjacent edges such that for each edge exactly one node
belongs VK. Then inequality

$$\sum_{e \in EK} x_e + \sum_{e \in ER} x_e \leq \lfloor \frac{k \mid VK \mid + \mid ER \mid}{2} \rfloor$$

is facet inequality if and only if $k < |VK| < n - k$ and $k|VK| \neq |ER|(mod 2)$. At
that if $k+1 = |VK|$ (or $k+1 = |V \setminus VK|$) then $|VK| - |ER| \leq 1$ (or respectively
$|V \setminus VK| - |ER| \leq 1$).

These results are generalized to the case $k = 2$.

Currently facet inequalities are quite actively used to solve combinatorial
optimization problems. Their application in the algorithms enabled finding exact
solutions to a large number of large-dimension problems [1, 2, 7, 9, 13, 14]. Facet
inequalities are most efficient as cutting planes. The interest to the facet inequalities is, in our opinion, due to the following considerations. First, the finiteness
of a polytope facet number ensures the finiteness of the cutting plane algorithm
based on the facet inequalities. Second, hyperplanes generating facets are in a
certain sense the strongest cuttings. Figuratively speaking, a hyperplane generating the face with the dimension smaller than $dim P_{\mathcal{H}} - 1$ may be "corrected",
placed on a larger dimension face. Third, in any linear system describing a polytope all of its facet inequalities are contained with the equivalent accuracy.

At the stage of using different inequality classes in the cutting plane algorithms the following algorithmic problem referred to as separation problem goes to the foreground. Let us speak that inequality $b^T x \leq b_0$ valid relative to $P_{\mathcal{H}}$ cuts off point $\bar{x} \in R^E$ if $b^T \bar{x} > b_0$. The separation problem consists in the following. Let us take point $\bar{x} \in R^E$ and family $\mathcal{L}$ of linear inequalities valid relative to $P_{\mathcal{H}}$. It is required to find an inequality cutting point $\bar{x} \in R^E$ among the family $\mathcal{L}$ or prove that $\mathcal{L}$ has no such inequality. Let us point out that there are no "reasonable" methods for the separation problem solution for polytope $P_{\mathcal{H}}$ in general. Moreover for different optimization problems and different inequality classes the separation problem has different complexity status. Thus, for example for a connected 2-factor polytope the separation problem for subtour eliminations inequalities and 2-matching inequalities are polynomially solvable and the separation problems for comb inequalities and clique tree inequalities are NP-hard [8] (see also [5,13]).

To prove the polynomial solvability of separation problem for subtour eliminations inequality relative to connected 2-factors [8] suggested reduction of this problem to the problem of minimum cut in the edge-weighted graph. This approach proved to be applicable for the case of arbitrary even k. Let us consider this situation in more details.

Two support inequalities to polytope are called equivalent if they generate the same polytope face. Let us denote a set of all edges incident to $u \in V$ through $\delta(u)$. In [11] the following results necessary for further discussion were obtained.

Lemma 2 [11]. *Affine hull of $P_{k,n}$ is the set of solutions of the system of linear equations*

$$\sum_{e \in \delta(u)} x_e = k, \quad u \in V. \tag{8}$$

From Lemma 2, particularly, follows that $dim P_{\mathcal{H}} = \frac{n^2 - n}{2} - n$.

Lemma 3 [11]. *Let cliques K and $\bar{K}$ satisfy the condition $VK = V \setminus VK$. Then inequalities of (3) type generated by cliques K and $\bar{K}$ are equivalent relative to $P_{k,n}$.*

For two non-overlapping sets $U, W \subset V$ we will denote a set of edges with one end in U and the other in W as $\gamma[U, W]$. The cut in K_n is defined by the set $U \subset V$. Hereby the sets U and $V \setminus U$ are called shores of cut whereas $\gamma[U, V \setminus U]$–edges set of cut.

Theorem 5. *Let $\bar{x} \in aff P_{k,n}$. In the class of subtour eliminations inequalities generating the facets of $P_{k,n}$ there will be an inequality cutting point $\bar{x}$ if and only if in K_n there exists such cut $\gamma[U, V \setminus U]$ that*

$$\sum_{e \in \gamma[U, V \setminus U]} \bar{x}_e < 2 + \lfloor \frac{k \mid U \mid}{2} \rfloor - \lceil \frac{k \mid U \mid}{2} \rceil.$$

Hereby the cutting inequality is generated by the clique on the set U.

Proof. Let

$$\sum_{e \in EK} \bar{x}_e > \lceil \frac{k \mid VK \mid}{2} - 1 \rceil \qquad (9)$$

be a facet inequality of the type (7) cutting point $\bar{x}$. Let $\bar{K}$ be clique on the set $V \setminus VK$. By virtue of Lemmas 2 and 3 for clique $\bar{K}$ a similar inequality is performed

$$\sum_{e \in E\bar{K}} \bar{x}_e > \lceil \frac{k \mid V \setminus VK \mid}{2} - 1 \rceil. \qquad (10)$$

It is obvious that

$$E = EK \cup E\bar{K} \cup \gamma[VK, V\bar{K}].$$

Then, by virtue of Lemma 2 and the fact that these three sets are pairwise disjoint, we have

$$\frac{kn}{2} = \sum_{e \in E} \bar{x}_e = \sum_{e \in EK} \bar{x}_e + \sum_{e \in E\bar{K}} \bar{x}_e + \sum_{e \in \gamma[VK, V\bar{K}]} \bar{x}_e.$$

Consequently,

$$\sum_{e \in \gamma[VK, V\bar{K}]} \bar{x}_e = \frac{kn}{2} - \sum_{e \in EK} \bar{x}_e - \sum_{e \in E\bar{K}} \bar{x}_e <$$

$$< \frac{kn}{2} - \lceil \frac{k \mid VK \mid}{2} - 1 \rceil - \lceil \frac{k \mid V \setminus VK \mid}{2} - 1 \rceil$$

$$= \frac{kn}{2} - \lceil \frac{k \mid VK \mid}{2} - 1 \rceil - (\frac{kn}{2} - \lfloor \frac{k \mid VK \mid}{2} + 1 \rfloor)$$

$$= \lfloor \frac{k \mid VK \mid}{2} + 1 \rfloor - \lceil \frac{k \mid VK \mid}{2} - 1 \rceil$$

$$= 2 + \lfloor \frac{k \mid VK \mid}{2} \rfloor - \lceil \frac{k \mid VK \mid}{2} \rceil.$$

The reverse statement follows from the same chain of equalities and the fact that inequalities (8) and (9) are satisfied simultaneously.

The theorem is proved.   □

Let $\bar{x} \in affP_{k,n}$. Let us associate weight $\bar{x}_e$ with each edge $e \in E$. Let us call the value $\sum_{e \in \gamma[U, V \setminus U]} \bar{x}_e$ the weight of cut $\gamma[U, V \setminus U]$. Therefore, at an even k the separation problem for subtour eliminations inequalities on the polytope of connected k-factors is reduced to the problem of the minimum weight cut in the edge-weighted graph and, consequently, it is polynomially solvable. At an odd k the consequences of Theorem 5 are:

Corollary 1. *At an odd k in the class of facet subtour eliminations inequalities there will be an inequality cutting the point $\bar{x} \in affP_{k,n}$, if and only if in K_n with the edge weights $\bar{x}_e, e \in E$, among the cuts with even shores the minimum cut has the weight smaller than 2.*

Corollary 2. *At an odd k in the class of facet subtour eliminations inequalities there will be an inequality cutting the point $\bar{x} \in affP_{k,n}$, if in K_n with the edge weights $\bar{x}_e$, $e \in E$, the minimum cut has the weight smaller than 1.*

5   Conclusion

This article describes the procedure of proving the facet nature of the support inequalities applicable to a wide class of combinatorial polytopes. The technique suggested is rather cumbersome but the specification of the set of combinatorial admissible objects, affine hull of the relevant polytope and the support inequality itself enables obtaining constructive results [11,12]. Besides, the article considers separation problem for the subtour eliminations inequality class [11] relative to the polytope of connected k-factors. Its polynomial solvability for even k is demonstrated. Let us add that for the problem of the minimum connected k-factor we implemented the cutting plane algorithm using the said subtour eliminations inequalities. The algorithm iteration consists in the following. For the current continuous optimum, separation problem for the subtour eliminations inequalities is solved. If separation problem had positive result, the subtour eliminations inequality found were used as the cutting plane. Otherwise Gomory's cutting plane was used. 85 problems with the n parameter from 20 to 90 and k parameter from 4 to $\lfloor \frac{n}{2} \rfloor$ were solved. On average, only 10 % of the algorithm iterations resulted in the subtour eliminations cutting planes. This is an evidence of the fact that the facets generated by subtour eliminations inequalities have rather a small "area" among all the polytope facets.

References

1. Crowder, H., Johnson, E.L., Padberg, M.W.: Solving large-scale zero-one linear programming problems. J. Oper. Res. **31**(5), 803–834 (1983)
2. Grötschel, M., Holland, O.: Solution of large-scale symmetric traveling salesman problems. J. Math. Program. **51**, 141–202 (1991)
3. Grötschel, M., Pulleyblank, W.R.: Clique tree inequalities and the symmetric traveling salesman problem. J. Math. Program. **11**, 537–569 (1989)
4. Gottlieb, E.S., Rao, M.R.: The generalized assignment problem: valid inequalities and facets. J. Math. Program. **46**, 31–52 (1990)
5. Kononov, A.V., Simanchev, R., Urazova, I.V.: On separation problem of same support inequalities class for M-graphs polytope. In: 17-th National Conference Mathematical Methods for Pattern Recognition, pp. 78–79, Moscow (2015)
6. Korte, B., Vygen, J.: Combinatorial Optimization. Theory and Algorithms. Springer, Heidelberg (2006)
7. Padberg, M.W.: $(1, k)$ Configurations and facets for packing problems. J. Math. Program. **18**, 94–99 (1980)
8. Padberg, M.W., Rinaldi, G.: Facet identification for the symmetric traveling salesman polytope. J. Math. Program. **47**, 219–257 (1990)
9. Padberg, M.W., Rinaldi, G.: A branch and cut algorithm for the resolution of large-scale symmetric traveling salesman problems. J. SIAM Rev. **33**, 60–100 (1991)

10. Schrijver, A.: Combinatorial Optimization. Polyhedra and Efficiency. Springer, Heidelberg (2004)
11. Simanchev, R.Y.: On rank inequalities generating facets of a polytope of connected k-factors. J. Diskretn. Anal. Issled. Oper. **3**(3), 84–110 (1996). (Russian)
12. Simanchev, R.Y., Urazova, I.V.: On the polytope faces of the graph approximation problem. J. Appl. Ind. Math. **9**(2), 283–291 (2015)
13. Simanchev, R.Y., Urazova, I.V.: An integer-valued model for the problem of minimizing the total servicing time of unit claims with parallel devices with precedences. J. Autom. Remote Control. **71**(10), 2102–2108 (2010)
14. Wolsey, L.A.: Valid inequalities for 0–1 Knapsacks and MIPS with generalised upper bound constraints. J. Discrete Appl. Math. **29**, 251–261 (1990)
15. Yemelichev, V., Kovalev, M., Kravtsov, M.: Polytopes, Graphs and Optimisation. Cambridge University Press, Cambridge (1984)

A Branch and Bound Algorithm for a Fractional 0-1 Programming Problem

Irina Utkina, Mikhail Batsyn[(✉)], and Ekaterina Batsyna

Department of Applied Mathematics and Informatics,
Laboratory of Algorithms and Technologies for Network Analysis,
National Research University Higher School of Economics,
136 Rodionova Street, Niznhy Novgorod, Russia
{iutkina,mbatsyn,batcyna}@hse.ru

Abstract. We consider a fractional 0-1 programming problem arising in manufacturing. The problem consists in clustering of machines together with parts processed on these machines into manufacturing cells so that intra-cell processing of parts is maximized and inter-cell movement is minimized. This problem is called Cell Formation Problem (CFP) and it is an NP-hard optimization problem with Boolean variables and constraints and with a fractional objective function. Because of its high computational complexity there are a lot of heuristics developed for it. In this paper we suggest a branch and bound algorithm which provides exact solutions for the CFP with a variable number of cells and grouping efficacy objective function. This algorithm finds optimal solutions for 21 of the 35 popular benchmark instances from literature and for the remaining 14 instances it finds good solutions close to the best known.

Keywords: Cell formation · Biclustering · Branch and bound · Upper bound · Exact solution

1   Introduction

The first work on the Group Technology in manufacturing was written by Flanders (1925). In Russia the Group Technology was introduced by Mitrofanov (1933). The main problem in the Group Technology (GT) is to find an optimal partitioning of machines and parts into manufacturing cells, in order to maximize intra-cell processing and minimize inter-cell movement of parts. Maximization of the so-called grouping efficacy is accepted in literature as a good objective combining these two goals (Kumar and Chandrasekharan 1990). This problem is called the Cell Formation Problem (CFP) (Goldengorin et al. 2013). CFP with grouping efficacy objective function is a fractional 0-1 programming problem.

Burbidge developed Product Flow Analysis (PFA) approach to this problem and described the GT and the CFP in his book (Burbidge 1961). Ballakur and Steudel (1987) have shown that the CFP is an NP-hard problem for different objective functions. That is why there have been developed a lot of heuristic approaches

Y. Kochetov et al. (Eds.): DOOR 2016, LNCS 9869, pp. 244–255, 2016.
DOI: 10.1007/978-3-319-44914-2_20

(Goncalves and Resende 2004; James et al. 2007; Bychkov et al. 2013, Paydar and Saidi-Mehrabad 2013) and almost no exact ones for the CFP with a variable number of cells and grouping efficacy objective function.

Kusiak et al. (1993) consider one of the most simple variants of the CFP called the machine partitioning problem in which it is necessary to partition only machines into the specified number of cells minimizing the total Hamming distance between machines inside the cells. The authors present an exact A* algorithm for this variant of the CFP. They also develop a branch and bound algorithm for the CFP with a variable number of cells, a limit on the number of machines inside each cell, and maximization of the size of so-called mutually separable cells as an objective function. Spiliopoulos and Sofianopoulou (1998) and Arket et al. (2012) also present branch and bound algorithms for the machine partitioning problem.

One of the recent exact approaches for the CFP with the grouping efficacy objective function is suggested by Elbenani and Ferland (2012). These authors suggest to reduce the fractional programming CFP problem to a number of ILP problems by means of Dinkelbach approach and to solve each ILP problem with CPLEX solver. Unfortunately they consider the CFP with a fixed number of cells which is much easier. They solve 27 of the 35 popular benchmark instances, but only for a fixed number of cells. The same simplified formulation of the CFP is considered by Brusco (2015). The author develops a branch and bound algorithm and solves 31 of the 35 instances, but again only for some fixed numbers of cells. For example problem 26 is solved only for 7 cells and it requires more than 15 days of computational time.

To the best of our knowledge the only existing exact approach to the CFP with a variable number of cells and grouping efficacy objective function is by Bychkov et al. (2014) who suggested a new approach to reduce the CFP problem to a small number of ILP problems and for the first time solved to optimality 14 of the 35 popular benchmark instances from literature using CPLEX software. Zilinskas et al. (2015) considered the CFP with a variable number of cells as a bi-objective optimization problem and developed an exact algorithm which finds Pareto frontier.

In this paper we suggest an efficient branch and bound algorithm for the CFP with a variable number of cells and grouping efficacy objective function. We are able to find optimal solutions for 21 of the 35 benchmark instances. Note also that the CFP is a biclustering problem in which we simultaneously cluster machines and parts into cells. So the suggested approach can be also applied to biclustering problems arising in data mining (Busygin et al. 2008).

2   Formulation

The objective of the CFP is to find an optimal partitioning of machines and parts into groups (production cells, or shops) in order to minimize the inter-cell movement of parts from one cell to another and to maximize intra-cell processing operations. The input data for this problem is matrix A which contains zeroes and ones. The size of this matrix is $m \times p$ which means that it has m machines and

p parts. The element a_{ij} of the input matrix is equal to one if part j should be processed on machine i. The objective is to minimize the number of zeroes inside cells and the number of ones outside cells. There have been suggested several objective functions which combine these two goals. The objective function which provides a good combination of these goals and is widely accepted in literature is the grouping efficacy suggested by Kumar and Chandrasekharan (1990):

$$f = \frac{n_1^{in}}{n_1 + n_0^{in}} \rightarrow \max, \tag{1}$$

where n_1 is the number of ones in the input matrix, n_1^{in} is the number of ones inside cells, n_0^{in} is the number of zeroes inside cells.

The mathematical programming model for the CFP is the following (see also Bychkov et al. (2014)).

Decision variables:

$$x_{ik} = \begin{cases} 1 & \text{if machine } i \text{ is assigned to cell } k \\ 0 & \text{otherwise} \end{cases} \tag{2}$$

$$y_{jk} = \begin{cases} 1 & \text{if part } j \text{ is assigned to cell } k \\ 0 & \text{otherwise} \end{cases} \tag{3}$$

Objective function:

$$\max \frac{n_1^{in}}{n_1 + n_0^{in}} \tag{4}$$

Constraints:

$$n_1^{in} = \sum_{k=1}^{c} \sum_{i=1}^{m} \sum_{j=1}^{p} a_{ij} x_{ik} y_{jk} \tag{5}$$

$$n_0^{in} = \sum_{k=1}^{c} \sum_{i=1}^{m} \sum_{j=1}^{p} (1 - a_{ij}) x_{ik} y_{jk} \tag{6}$$

$$\sum_{k=1}^{c} x_{ik} = 1 \quad \forall i = 1, \ldots, m \tag{7}$$

$$\sum_{k=1}^{c} y_{jk} = 1 \quad \forall j = 1, \ldots, p \tag{8}$$

$$\sum_{i=1}^{m} \sum_{j=1}^{p} x_{ik} y_{jk} \geq \sum_{i=1}^{m} x_{ik} \quad \forall k = 1, \ldots, c \tag{9}$$

$$\sum_{i=1}^{m} \sum_{j=1}^{p} x_{ik} y_{jk} \geq \sum_{j=1}^{p} y_{jk} \quad \forall k = 1, \ldots, c \tag{10}$$

Here $c = \min(m, p)$ is the maximum possible number of cells. Constrains (7) and (8) require that every machine and every part is assigned to exactly one cell. Constrains (9) and (10) require that there are no cells having only machines without parts or only parts without machines.

3 Branch and Bound Algorithm

3.1 Branching

Because of the biclustering structure of the CFP our branching goes by two parameters. The suggested algorithm has branching on machines and parts sequentially changing each other: machines-parts-machines-... We use vectors $M(1 \times m)$ and $P(1 \times p)$ for this purpose. Element M_i contains the cell to which machine i is assigned and element P_j contains the cell to which part j is assigned. For example M = [1231] and P = [11321] mean that cell 1 contains machines 1, 4 and parts 1, 2, 5, cell 2 contains machine 2 and part 4, and cell 3 contains machine 3 and part 3.

Branching on machines makes changes in vector M. It starts from assigning the first machine to cell 1. Let k be the number of cells in the current partial solution. When the algorithm branches on machines, it takes the first machine which is not assigned to any cell and tries to assign it to the existing cells with numbers from 1 to k or creates a new cell $(k + 1)$ for this machine.

Branching on parts makes changes in vector P. It starts with all zeroes inside P which means that no parts are assigned to any cell. When the algorithm branches on parts it takes the first part which is not assigned to any cell and tries to assign it to the existing cells from 1 to k or to a new cell $(k + 1)*$ (star means that the number of the cell can be $k + 1$ or greater) if there are some unassigned machines which can be also added later to this new cell. We assume that the number of parts is greater than the number of machines.

The algorithm branches on parts and machines successively. It starts with $M = [100 \dots 0]$ and $P = [00 \dots 0]$. Next it changes vector P, then - vector M and so on. This way the algorithm builds the search tree. The leaves of the search tree contain complete solutions and other nodes contain partial solutions. The complete search tree depends only on the number of machines and parts. It contains all feasible solutions as its leaves.

To provide an efficient branching, before choosing a branch we calculate an upper bound for each branch and choose the branch with the greatest value of the upper bound. This branching strategy allows us to find good solutions earlier.

3.2 Upper Bound

To obtain an upper bound for a given partial solution we relax the original CFP problem and suggest a polynomial algorithm to calculate an optimal solution or an upper bound for the relaxed problem. The relaxed problem is formulated as follows. We are given a partial solution in which some of the machines and parts are already assigned to some cells. For example in Table 1 machines 1, 2 with parts 1, 2, 3, 4 are assigned to cell 1, and machine 3 with part 5 is assigned to cell 2. The objective is to assign the remaining machines independently on each other to the existing cells or to a new cell, and assign the remaining parts to the existing cells taking into account only the rows already assigned in the given partial solution. In the relaxed problem we allow an independent assignment of machines and parts to cells. In this case the best assignment for machine 4 will be to put it to cell 1 with parts 1, 2, 3, 4, 7. This will bring 4 ones and 1 zero inside cells. The best assignment for machine 5 will be to put it to a new cell 3 with parts 7, 8. This will bring 2 ones and 0 zeroes inside cells. The best assignment for parts 6 and 7 which takes into account only rows 1, 2, 3 will be to put it to cell 2 (with machine 3). The best assignment for part 8 which takes into account only rows 1, 2, 3 will be to put it to cell 1 (with machines 1, 2). This optimal solution for the relaxed problem is shown in Table 2. This solution is infeasible for the original CFP problem because independent assignment of machines and parts is allowed and as a result we obtain non-rectangular cells which can also intersect by columns. Since it is an optimal solution to the relaxed problem it provides an upper bound to the original problem. In our example for the partial solution we have $f = \frac{8}{21+1} \approx 0.36$ and the solution of the relaxed problem gives us an upper bound to the complete solution of the CFP equal to $UB = \frac{8+10}{21+1} \approx 0.82$.

In our example from Table 1 it is not obvious whether the chosen alternative $(a_1, b_1) = (4, 1)$ (putting 4 ones and 1 zero inside cells) is better than alternative $(a_2, b_2) = (1, 0)$ for machine 3. To choose between two alternatives we use the following theorem.

Theorem 1. *If the unknown maximum value of the objective function $\frac{a}{b}$ for the relaxed CFP problem without assignment of machine i (considering all its ones and zeroes to be outside cells) can be estimated as $\frac{a}{b} \in [l, u]$, $b \in [b_l, b_u]$, then alternative (a_1, b_1) for machine i is better than alternative (a_2, b_2) if:*

$$b_l \left(l - \frac{\Delta a}{\Delta b} \right) \geq b_1 \frac{\Delta a}{\Delta b} - a_1 \tag{11}$$

Table 1. A partial solution for the CFP

	1	2	3	4	5	6	7	8
1	1	1	1	1	1	0	0	1
2	1	1	0	1	0	0	0	1
3	0	0	1	0	1	1	1	0
4	1	0	1	1	1	0	1	0
5	0	0	0	0	0	0	1	1

Table 2. Optimal solution for the relaxed CFP

	1	2	3	4	5	6	7	8
1	1	1	1	1	1	0	0	1
2	1	1	0	1	0	0	0	1
3	0	0	1	0	1	1	1	0
4	1	0	1	1	1	0	1	0
5	0	0	0	0	0	0	1	1

and is worse than (a_2, b_2) *if:*

$$b_u \left(u - \frac{\Delta a}{\Delta b} \right) \leq b_1 \frac{\Delta a}{\Delta b} - a_1 \tag{12}$$

Here $\Delta a = a_2 - a_1, \Delta b = b_2 - b_1 > 0$ *(if $\Delta b < 0$ we can always swap the alternatives).*

Proof. Alternative (a_1, b_1) is better than alternative (a_2, b_2) if:

$$\frac{a + a_1}{b + b_1} \geq \frac{a + a_2}{b + b_2} \tag{13}$$

Multiplying this inequality by the positive denominators and making simple transformations (here we need $\Delta b > 0$) we get the equivalent inequality:

$$b \left(\frac{a}{b} - \frac{\Delta a}{\Delta b} \right) \geq b_1 \frac{\Delta a}{\Delta b} - a_1 \tag{14}$$

Using the given estimations for the unknown maximum value of the objective function $\frac{a}{b}$ we have the following bounds for the left-hand side of this inequality:

$$b_l \left(l - \frac{\Delta a}{\Delta b} \right) \leq b \left(\frac{a}{b} - \frac{\Delta a}{\Delta b} \right) \leq b_u \left(u - \frac{\Delta a}{\Delta b} \right)$$

So if inequality (11) is true then we immediately have that inequalities (14) and (13) are true, which means that alternative (a_1, b_1) is better than alternative (a_2, b_2). Otherwise if inequality (12) is true then we immediately have that the opposite inequalities are true:

$$b \left(\frac{a}{b} - \frac{\Delta a}{\Delta b} \right) \leq b_1 \frac{\Delta a}{\Delta b} - a_1, \quad \frac{a + a_1}{b + b_1} \leq \frac{a + a_2}{b + b_2}$$

This means that alternative (a_2, b_2) is better in this case.   □

Note that in case $\Delta b = 0$ it is obvious which of the two alternatives is better. It is also not difficult to estimate the unknown maximum value of the objective function $\frac{a}{b}$ for the relaxed CFP problem without assignment of machine i (considering all its ones and zeroes to be outside cells). Let a_c, b_c be the current values of the objective function numerator $a_c = n_1^{in}$ and denominator $b_c = n_1 + n_0^{in}$ for the current partial solution. Then in the worst case we can get $\frac{a}{b} = \frac{a_c}{b_c}$ because every unassigned machine can be put to a new cell putting some ones and no zeroes inside it, and every part can be left unassigned. So the lower bound is $l = \frac{a_c}{b_c}$. In the best case we can put inside cells no zeroes and all the ones except $\bar{n}_1^{out}$ ones which lie in the already assigned area and cannot get inside cells (see gray area (the darkest area on black-and-white printing) in Table 1) and except n_1^i ones which lie in row i. So the upper bound is $u = (n_1 - \bar{n}_1^{out} - n_1^i)/b_c$. The value of the denominator b can be estimated as: $b_c \leq b \leq n_1 + n_0 - \bar{n}_0^{out} - n_0^i$. Here n_1 and n_0 are the number of ones and zeroes in the input matrix, $\bar{n}_0^{out}$ is the number of zeroes which lie in the already assigned area (gray area in Table 1) and n_0^i is the number of zeroes in row i.

Below we present a polynomial algorithm of calculating the suggested upper bound as an optimal solution of the relaxed CFP problem, if we can always choose the best alternative for every machine and part, or as an upper bound to this solution otherwise. We illustrate the algorithm on the instance shown in Table 3.

Algorithm 1. Algorithm to choose between two alternatives

---

function COMPAREALTERNATIVES($a_1, b_1, a_2, b_2, n_1, n_0, n_1^{in}, n_0^{in}, \bar{n}_1^{out}, \bar{n}_0^{out}, n_1^i, n_0^i$)
    $\Delta a \leftarrow a_2 - a_1, \Delta b \leftarrow b_2 - b_1$ $\triangleright$ Δb should be non-negative
    if ($\Delta b = 0$) **then**
        if ($\Delta a < 0$) **then**
            return 1
        else if ($\Delta a > 0$) **then**
            return 2
        else
            return 0
    $a_c \leftarrow n_1^{in}, b_c \leftarrow n_1 + n_0^{in}, b_l \leftarrow b_c, b_u \leftarrow n_1 + n_0 - \bar{n}_0^{out} - n_0^i, l \leftarrow \frac{a_c}{b_c}, u \leftarrow \frac{n_1 - \bar{n}_1^{out} - n_1^i}{b_c}$
    if $b_l \left(l - \frac{\Delta a}{\Delta b}\right) \geq b_1 \frac{\Delta a}{\Delta b} - a_1$ **then**
        return 1
    if $b_u \left(u - \frac{\Delta a}{\Delta b}\right) \leq b_1 \frac{\Delta a}{\Delta b} - a_1$ **then**
        return 2
    return -1

Table 3. Example for upper bound calculation

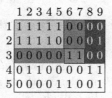

1. Calculate the number of ones n_1^{in} and zeroes n_0^{in} inside the cells of the given partial solution and the number of ones $\bar{n}_1^{out}$ and zeroes $\bar{n}_0^{out}$ which cannot get inside cells in any solution (see gray area (the gray area with black zeroes on black-and-white printing) in Table 3). The total number of ones n_1 and zeroes n_0 are constant. From these values we get the numerator $a_c = n_1^{in}$ and the denominator $b_c = n_1 + n_0^{in}$ for the efficacy $f = a_c/b_c$ of the current partial solution. For the example in Table 3 we have: $n_1^{in} = 11, n_0^{in} = 1, \bar{n}_1^{out} = 0, \bar{n}_0^{out} = 9, n_1 = 19, n_0 = 26, a_c = 11, b_c = 20$.

Table 4. Solution providing the upper bound

	1	2	3	4	5	6	7	8	9
1	1	1	1	1	1	0	0	0	0
2	1	1	1	1	0	0	0	0	1
3	0	0	0	0	0	1	1	0	0
4	0	1	1	0	0	0	0	1	1
5	0	0	0	0	1	1	0	0	1

2. For every unassigned machine (row) using Algorithm 1 we compare all possible alternatives of adding it to one of the existing cells or to a new cell. For our example we have 3 alternatives for machine $i = 4$: 1) $(4,3)$ - add it to cell 1 with parts 1, 2, 3, 4, 5, 8, 9 putting 4 ones and 3 zeroes inside this cell; 2) $(2,2)$ - add it to cell 2 with parts 6, 7, 8, 9 putting 2 ones and 2 zeroes inside; 3) $(2,0)$ - add it to a new cell 3 with parts 8, 9 putting 2 ones and 0 zeroes inside. Obviously alternative 3 is better than alternative 2. So we need to compare only two alternatives $(a_1, b_1) = (2, 0)$ and $(a_2, b_2) = (4, 3)$. We have: $n_1^i = 4, n_0^i = 5, \Delta a = 2, \Delta b = 3, l = a_c/b_c = 11/20, u = (n_1 - \bar{n}_1^{out} - n_1^i)/b_c = 15/20, b_l = 20, b_u = n_1 + n_0 - \bar{n}_0^{out} - n_0^i = 31$. And the values we need to apply Algorithm 1 are:

$$b_1 \frac{\Delta a}{\Delta b} - a_1 = -2, \quad b_l \left(l - \frac{\Delta a}{\Delta b}\right) = -\frac{7}{3}, \quad b_u \left(u - \frac{\Delta a}{\Delta b}\right) = \frac{31}{12}$$

So neither of the conditions in Algorithm 1 is satisfied and we cannot determine which alternative is better (Algorithm 1 returns -1). In this case we build an alternative $(\max(a_1, a_2), \min(b_1, b_2))$, which is better than both incomparable alternatives, and use it to obtain an upper bound on the solution of the relaxed CFP problem. In our example it is alternative $(4, 0)$.
Now for machine $i = 5$ we have: $n_1^i = 3, n_0^i = 6, l = 11/20, u = 16/20, b_l = 20, b_u = 30$. There are 3 alternatives $(2, 4)$, $(2, 1)$, and $(1, 0)$. It is clear that alternative $(2, 4)$ is worse than $(2, 1)$. For $(a_1, b_1) = (1, 0)$ and $(a_2, b_2) = (2, 1)$ we have:

$$b_1 \frac{\Delta a}{\Delta b} - a_1 = -1, \quad b_l \left(l - \frac{\Delta a}{\Delta b}\right) = -9, \quad b_u \left(u - \frac{\Delta a}{\Delta b}\right) = -6$$

So $b_u \left(u - \frac{\Delta a}{\Delta b}\right) \le b_1 \frac{\Delta a}{\Delta b} - a_1$ and Algorithm 1 returns alternative $(a_2, b_2) = (2, 1)$. Thus $(2, 1)$ is the best alternative for machine 5.
3. For every unassigned part (column) in the same way using Algorithm 1 we compare all possible alternatives of adding it to one of the existing cells or leaving it unassigned. However in this case we take into account only ones and zeroes which lie in the rows already assigned in the given partial solution (blue area (the darkest area with white digits on black-and-white printing) in Table 3).
In our example part 8 has only zeroes in this area and so it is better not to add it to any cell. For part 9 $(j = 9)$ we have 3 alternatives: (1) $(1, 1)$ - add

it to cell 1 putting 1 one and 1 zero inside; (2) $(0,1)$ - add it to cell 2 putting 0 ones and 1 zero inside; (3) $(0,0)$ - do not add it to any cell. It is clear that $(0,1)$ is a bad alternative and we need to compare only $(a_1, b_1) = (0,0)$ and $(a_2, b_2) = (1,1)$. We have $n_1^j = 1, n_0^j = 2, l = a_c/b_c = 11/20, u = (n_1 - \bar{n}_1^{out} - n_1^j)/b_c = 18/20, b_l = 20, b_u = n_1 + n_0 - \bar{n}_0^{out} - n_0^j = 34$. The values needed to apply Algorithm 1 are:

$$b_1 \frac{\Delta a}{\Delta b} - a_1 = 0, \quad b_l \left(l - \frac{\Delta a}{\Delta b} \right) = -9, \quad b_u \left(u - \frac{\Delta a}{\Delta b} \right) = -3.4$$

So $b_u \left(u - \frac{\Delta a}{\Delta b} \right) \leq b_1 \frac{\Delta a}{\Delta b} - a_1$ and Algorithm 1 returns alternative $(a_2, b_2) = (1,1)$ as the best alternative for part 9.
4. We calculate the upper bound by putting inside all the ones and zeroes corresponding to the best alternatives chosen for all unassigned machines and parts. For our example the corresponding solution which gives an upper bound to the relaxed CFP problem (and thus to the original CFP problem also) is shown in Table 4. For this example we have $UB = \frac{11+4+2+1}{19+0+1+1} = \frac{18}{21} \approx 0.86$.

4   Results

The suggested branch and bound algorithm has been able to solve 21 of 35 popular benchmark instances from the literature exactly and to find good solutions for the remaining 14 instances. The results are presented in Table 5. All computations were run on Intel Core i7 with 16 Gb RAM. Note that the algorithm was run without any initial solution while in the results reported by Bychkov et al. (2014) the best-known solutions were used as initial.

The results show that the developed algorithm is more efficient than the approach suggested by Bychkov et al. (2014). Bychkov et al. (2014) approach is able to solve to optimality only instances 1–13, where it is much slower (up to 1000 times) than our new algorithm, and also instance 22 which is very simple and is solved withing 0.01 s by both approaches. The suggested algorithm is able to solve to optimality 7 instances more (instances 14, 15, 16, 17, 20, 23, 24) which are very hard and have never been solved before to the best of our knowledge. All instances marked with footnote 'b' in Table 5 (21 instance) have not been solved to optimality by Bychkov et al. (2014). As it is noted in this footnote in this approach the problem is divided into a number of IP subproblems and a time limit of 300 s is set for every subproblem. That is why all these instances have different times after which the computation has been stopped without reaching an optimal solution. All instances marked with footnote 'a' in Table 5 (14 instances) have not been solved to optimality by the new approach. The time limit was set to 100000 s for all these instances.

Table 5. Results

#	Name	Size	Best-known solution	f	Time, s	Bychkov et al. (2014), Time, s
1	King and Nakornchai (1982)	5 × 7	0.8235	0.8235	0.00	0.63
2	Waghodekar and Sahu (1984)	5 × 7	0.6957	0.6957	0.00	2.29
3	Seifoddini (1989)	5 × 18	0.7959	0.7959	0.00	5.69
4	Kusiak (1987)	6 × 8	0.7692	0.7692	0.00	1.86
5	Kusiak and Chow (1987)	7 × 11	0.6087	0.6087	0.00	9.14
6	Boctor (1991)	7 × 11	0.7083	0.7083	0.00	5.15
7	Seifoddini and Wolfe (1986)	8 × 12	0.6944	0.6944	0.00	13.37
8	Chandrasekharan and Rajagopalan (1986a)	8 × 20	0.8525	0.8525	0.00	18.33
9	Chandrasekharan and Rajagopalan (1986b)	10 × 10	0.5872	0.5872	0.19	208.36
10	Mosier and Taube (1985a)	10 × 15	0.7500	0.7500	0.00	6.25
11	Chan and Milner (1982)	10 × 15	0.9200	0.9200	0.00	2.93
12	Askin and Subramanian (1987)	14 × 24	0.7206	0.7206	2.89	259.19
13	Stanfel (1985)	14 × 24	0.7183	0.7183	5.51	259.19
14	McCormick et al. (1972)	16 × 24	0.5326	0.5326	97117.43	[b]20829.38
15	Srinivasan et al. (1990)	16 × 30	[c]0.6899	0.6899	837.93	[b]13719.99
16	King (1980)	16 × 43	0.5753	0.5753	7045.64	[b]24930.93
17	Carrie (1973)	18 × 24	0.5773	0.5773	5668.25	[b]13250.01
18	Mosier and Taube (1985b)	20 × 20	0.4345	[a]0.4211	100000.00	[b]43531.77
19	Kumar et al. (1986)	20 × 23	0.5081	[a]0.4697	100000.00	[b]33020.13
20	Carrie (1973)	20 × 35	0.7791	0.7791	88.62	[b]11626.98
21	Boe and Cheng (1991)	20 × 35	0.5798	[a]0.5615	100000.00	[b]33322.08
22	Chandrasekharan and Rajagopalan (1989)	24 × 40	1.0000	1.0000	0.00	0.00
23	Chandrasekharan and Rajagopalan (1989)	24 × 40	0.8511	0.8511	33.70	[b]6916.24
24	Chandrasekharan and Rajagopalan (1989)	24 × 40	0.7351	0.7351	86007.93	[b]14408.88
25	Chandrasekharan and Rajagopalan (1989)	24 × 40	0.5329	[a]0.5185	100000.00	[b]34524.47
26	Chandrasekharan and Rajagopalan (1989)	24 × 40	0.4895	[a]0.4648	100000.00	[b]41140.94
27	Chandrasekharan and Rajagopalan (1989)	24 × 40	0.4726	[a]0.4468	100000.00	[b]44126.76
28	McCormick et al. (1972)	27 × 27	0.5482	[a]0.5017	100000.00	[b]22627.28
29	Carrie (1973)	28 × 46	0.4706	[a]0.4569	100000.00	[b]71671.08
30	Kumar and Vannelli (1987)	30 × 41	0.6331	[a]0.5942	100000.00	[b]22594.20
31	Stanfel (1985)	30 × 50	0.6012	[a]0.5789	100000.00	[b]31080.82
32	Stanfel (1985)	30 × 50	0.5083	[a]0.4860	100000.00	[b]48977.01
33	King and Nakornchai (1982)	30 × 90	0.4775	[a]0.4684	100000.00	[b]99435.64
34	McCormick et al. (1972)	37 × 53	0.6064	[a]0.5680	100000.00	[b]47744.04
35	Chandrasekharan and Rajagopalan (1987)	40 × 100	0.8403	[a]0.8403	100000.00	[b]24167.76

[a] The problem is not solved to optimality by our algorithm within the time limit of 100000 s.

[b] The problem is not solved to optimality by Bychkov et al. (2014). In this approach the problem is divided into a number of IP subproblems and a time limit of 300 s is set for every subproblem.

[c] A greater value is reported in some papers on heuristics probably due to an incorrect input matrix.

Acknowledgment. This research is supported by Laboratory of Algorithms and Technologies for Network Analysis, NRU HSE.

References

Arkat, J., Abdollahzadeh, H., Ghahve, H.: A new branch and bound algorithm for cell formation problem. Appl. Math. Model. **36**, 5091–5100 (2012)

Askin, R.G., Subramanian, S.P.: A cost-based heuristic for group technology configuration. Int. J. Prod. Res. **25**(1), 101–113 (1987)

Ballakur, A., Steudel, H.J.: A within cell utilization based heuristic for designing cellular manufacturing systems. Int. J. Prod. Res. **25**, 639–655 (1987)

Boctor, F.F.: A linear formulation of the machine-part cell formation problem. Int. J. Prod. Res. **29**(2), 343–356 (1991)

Boe, W., Cheng, C.H.: A close neighbor algorithm for designing cellular manufacturing systems. Int. J. Prod. Res. **29**(10), 2097–2116 (1991)

Brusco, J.M.: An exact algorithm for maximizing grouping efficacy in part machine clustering. IIE Transactions **47**(6), 653–671 (2015)

Burbidge, J.L.: The new approach to production. Prod. Eng. **40**(12), 3–19 (1961)

Busygin, S., Prokopyev, O., Pardalos, P.M.: Biclustering in data mining. Comput. Oper. Res. **35**(9), 2964–2987 (2008)

Bychkov, I., Batsyn, M., Sukhov, P., Pardalos, P.M.: Heuristic algorithm for the cell formation problem. In: Goldengorin, B.I., Kalyagin, V.A., Pardalos, P.M. (eds.) Models, Algorithms, and Technologies for Network Analysis. Springer Proceedings in Mathematics & Statistics, vol. 59, pp. 43–69. Springer, New York (2013)

Bychkov, I., Batsyn, M., Pardalos, P.: Exact model for the cell formation problem. Optimization Letters **8**(8), 2203–2210 (2014)

Carrie, S.: Numerical taxonomy applied to group technology and plant layout. Int. J. Prod. Res. **11**, 399–416 (1973)

Chan, H.M., Milner, D.A.: Direct clustering algorithm for group formation in cellular manufacture. J. Manuf. Syst. **1**(1), 64–76 (1982)

Chandrasekharan, M.P., Rajagopalan, R.: MODROC: an extension of rank order clustering for group technology. Int. J. Prod. Res. **24**(5), 1221–1233 (1986a)

Chandrasekharan, M.P., Rajagopalan, R.: An ideal seed non-hierarchical clustering algorithm for cellular manufacturing. Int. J. Prod. Res. **24**(2), 451–464 (1986b)

Chandrasekharan, M.P., Rajagopalan, R.: ZODIAC: an algorithm for concurrent formation of part families and machine cells. Int. J. Prod. Res. **25**(6), 835–850 (1987)

Chandrasekharan, M.P., Rajagopalan, R.: Groupability: analysis of the properties of binary data matrices for group technology. Int. J. Prod. Res. **27**(6), 1035–1052 (1989)

Elbenani, B., Ferland, J. A. Cell Formation Problem Solved Exactly with the Dinkelbach Algorithm. Montreal, Quebec. CIRRELT-2012-07, 1–14(2012)

Flanders, R.E.: Design manufacture and production control of a standard machine. Trans. ASME **46**, 691–738 (1925)

Goldengorin, B., Krushinsky, D., Pardalos, P.M.: Cell Formation in Industrial Engineering. Theory, Algorithms and Experiments. Springer Optimization and Its Applications, vol. 79, p. 206. Springer, New York (2013)

Goncalves, J.F., Resende, M.G.C.: An evolutionary algorithm for manufacturing cell formation. Comput. Ind. Eng. **47**, 247–273 (2004)

James, T.L., Brown, E.C., Keeling, K.B.: A hybrid grouping genetic algorithm for the cell formation problem. Comput. Oper. Res. **34**(7), 2059–2079 (2007)

King, J.R.: Machine-component grouping in production flow analysis: an approach using a rank order clustering algorithm. Int. J. Prod. Res. **18**(2), 213–232 (1980)

King, J.R., Nakornchai, V.: Machine-component group formation in group technology: review and extension. Int. J. Prod. Res. **20**(2), 117–133 (1982)

Kumar, K.R., Kusiak, A., Vannelli, A.: Grouping of parts and components in flexible manufacturing systems. Eur. J. Oper. Res. **24**, 387–397 (1986)

Kumar, K.R., Chandrasekharan, M.P.: Grouping efficacy: a quantitative criterion for goodness of block diagonal forms of binary matrices in group technology. Int. J. Prod. Res. **28**(2), 233–243 (1990)

Kumar, K.R., Vannelli, A.: Strategic subcontracting for efficient disaggregated manufacturing. Int. J. Prod. Res. **25**(12), 1715–1728 (1987)

Kusiak, A.: The generalized group technology concept. Int. J. Prod. Res. **25**(4), 561–569 (1987)

Kusiak, A., Chow, W.S.: Efficient solving of the group technology problem. J. Manuf. Syst. **6**(2), 117–124 (1987)

Kusiak, A., Boe, J.W., Cheng, C.: Designing cellular manufacturing systems: branch-and-bound and A* approaches. IIE Transactions **25**(4), 46–56 (1993)

McCormick, W.T., Schweitzer, P.J., White, T.W.: Problem decomposition and data reorganization by a clustering technique. Oper. Res. **20**(5), 993–1009 (1972)

Mitrofanov, S.P.: Nauchnye osnovy gruppovoy tekhnologii. Lenizdat, Leningrad, Russia, 435 pages (1933). (in Russian)

Mosier, C.T., Taube, L.: The facets of group technology and their impact on implementation. OMEGA **13**(6), 381–391 (1985a)

Mosier, C.T., Taube, L.: Weighted similarity measure heuristics for the group technology machine clustering problem. OMEGA **13**(6), 577–583 (1985b)

Paydar, M.M., Saidi-Mehrabad, M.: A hybrid genetic-variable neighborhood search algorithm for the cell formation problem based on grouping efficacy. Comput. Oper. Res. **40**(4), 980–990 (2013)

Seifoddini, H.: A note on the similarity coefficient method and the problem of improper machine assignment in group technology applications. Int. J. Prod. Res. **27**(7), 1161–1165 (1989)

Seifoddini, H., Wolfe, P.M.: Application of the similarity coefficient method in group technology. IIE Transactions **18**(3), 271–277 (1986)

Spiliopoulos, K., Sofianopoulou, S.: An optimal tree search method for the manufacturing systems cell formation problem. Eur. J. Oper. Res. **105**, 537–551 (1998)

Srinivasan, G., Narendran, T.T., Mahadevan, B.: An assignment model for the part-families problem in group technology. Int. J. Prod. Res. **28**(1), 145–152 (1990)

Stanfel, L.: Machine clustering for economic production. Eng. Costs Prod. Econ. **9**, 73–81 (1985)

Waghodekar, P.H., Sahu, S.: Machine-component cell formation in group technology MACE. Int. J. Prod. Res. **22**, 937–948 (1984)

Zilinskas, J., Goldengorin, B., Pardalos, P.M.: Pareto-optimal front of cell formation problem in group technology. J. Global Optim. **61**(1), 91–108 (2015)

Scheduling Problems

Approximating Coupled-Task Scheduling Problems with Equal Exact Delays

Alexander Ageev[✉] and Mikhail Ivanov

Sobolev Institute of Mathematics, pr. Koptyuga 4, Novosibirsk, Russia
ageev@math.nsc.ru

Abstract. We consider a coupled-task single machine scheduling problem with equal exact delays and makespan as the objective function. We design a 3-approximation algorithm for the general case of this problem. We also prove that the existence of a $(1.25 - \varepsilon)$-approximation algorithm implies $P = NP$. The inapproximability result remains valid for the case when the processing times of the two operations of each job are equal. We prove that this case is approximable within a factor of 1.5.

Keywords: Couple-task scheduling · Inapproximability lower bound · Approximation algorithm · Worst-case analysis

1 Introduction

We consider the single-machine coupled-task scheduling problem with exact delays. In the problem a set $J = \{1, \ldots, n\}$ of independent jobs is given. Each job $j \in J$ consists of two operations with processing times a_j and b_j separated by a given *exact* delay l_j, which means that the second operation of job j must start processing exactly l_j time units after the first operation of job j has been completed. It is assumed that at any time the machine can process at most one operation and no preemptions are allowed. The objective is to minimize the makespan (the schedule length). In the standard three-field notation scheme introduced by Graham et al. [11] (see also [2]) the single machine problem is denoted by $1 \mid \text{exact } l_j \mid C_{\max}$.

In this paper, we consider the case of the single machine problem when all delays are equal, i.e., $l_j = L$ for all $j \in J$. We refer to this case as $1 \mid \text{exact } l_j = L \mid C_{\max}$.

The scheduling problems with exact delays arise in command-and-control applications where a commander distributes a set of orders (associated with the first operations) and must wait to receive responses (corresponding to the second operations) that do not conflict with any other (for more detailed discussion on the subject, see [9,14]). Research papers on problem $1 \mid \text{exact } l_j \mid C_{\max}$ are mostly motivated by applications in pulsed radar systems, where the machine is a multifunctional radar whose purpose is to simultaneously track various targets by emitting a pulse and receiving its reflection some time later [7,9,10,13,14].

© Springer International Publishing Switzerland 2016
Y. Kochetov et al. (Eds.): DOOR 2016, LNCS 9869, pp. 259–271, 2016.
DOI: 10.1007/978-3-319-44914-2_21

Coupled-task scheduling problems with exact delays also arise in chemistry manufacturing where there often may be an exact technological delay between the completion time of some operation and the starting time of the next operation.

1.1 Related Work

Coupled-task scheduling problems have been investigated for decades. Quite a few various results related to these problem are surveyed by Blazewicz et al. in [5] (for later results see [6,12]). We cite here only previously known approximation results as well as those related to the case of equal delays.

Orman and Potts [13] establish that the problem is strongly NP-hard even in some special cases. In particular, they prove it for $1 \mid$ exact $l_j = L, b_j = b \mid C_{\max}$, i.e., in the case when $l_j = L, b_j = b$ for all $j \in J$. Baptiste [3] presents an algorithm with running time $O(\log n)$ for the very special case when $a_j = a, b_j = b, l_j = L$ for all jobs j provided that a, b, and L are fixed. The complexity status of the case, when a, b, and L are not fixed, remains open [4,13].

Ageev and Baburin [1] present non-trivial constant-factor approximation algorithms for both the single and two machine problems under the assumption of unit processing times. More specifically, in [1] it is shown that problem $1 \mid$ exact $l_j, a_j = b_j = 1 \mid C_{\max}$ is approximable within a factor of 7/4.

Ageev and Kononov [2] present a 3.5-approximation algorithm for the general case of $1 \mid$ exact $l_j \mid C_{\max}$ and a 3-approximation algorithms for the cases when $a_j \leq b_j$, or $a_j \geq b_j$ for all $j \in J$. They also show that the latter algorithms provide a 2.5-approximation for the case when $a_j = b_j$ for all $j \in J$. Moreover, they prove that problem $1 \mid$ exact $l_j \mid C_{\max}$ is not $(2 - \varepsilon)$-approximable unless $P = NP$ even in the case of $a_j = b_j$ for all $j \in J$.

Table 1. A summary of the approximability results.

Problem	Appr. factor	Inappr. bound	Ref.
$1 \mid$ exact $l_j, \mid C_{\max}$	3.5	$2 - \varepsilon$	[2]
$1 \mid$ exact $l_j, a_j \leq b_j \mid C_{\max}$	3	$2 - \varepsilon$	[2]
$1 \mid$ exact $l_j, a_j = b_j \mid C_{\max}$	2.5	$2 - \varepsilon$	[2]
$1 \mid$ exact $l_j, a_j = b_j = 1 \mid C_{\max}$	1.75		[1]
$1 \mid$ exact $l_j = L \mid C_{\max}$	3	$1.25 - \varepsilon$	this paper
$1 \mid$ exact $l_j = L, a_j \leq b_j \mid C_{\max}$	2	$1.25 - \varepsilon$	this paper
$1 \mid$ exact $l_j = L, a_j = b_j \mid C_{\max}$	1.5	$1.25 - \varepsilon$	this paper

1.2 Our Results

We show that the existence of a $(1.25 - \varepsilon)$-approximation for $1 \mid$ exact $l_j = L \mid C_{\max}$ even in the case $a_j = b_j$ for all jobs $j = 1, \ldots n$ implies $P = NP$.

On the positive side, we design a 3-approximation for $1 \mid \text{exact } l_j = L \mid C_{\max}$. For the cases of $1 \mid \text{exact } l_j = L, \mid C_{\max}$ when $a_j \leq b_j$ or $a_j = b_j$ for all jobs $j = 1, \ldots n$ we present 2- and 1.5-approximations, respectively. The 1.5-approximation algorithm has a remarkable property: its approximation factor tends to the inapproximability lower bound of 1.25 when the number of blocks it constructs tends to infinity.

Our results compared with the previously known approximation results are shown in Table 1.

1.3   Basic Notation

For both problems an instance will be represented as a collection of triples $\{(a_j, l_j, b_j) : j \in J\}$ where $J = \{1, \ldots, n\}$ is the set of jobs, a_j and b_j are the lengths of the first and the second operations of job j, respectively and l_j is the given delay between these operations. As usual, we assume that all input numbers are nonnegative integers. For a schedule σ and any $j \in J$, denote by $\sigma(j)$ the starting time of the first operation of job j. As the starting times of the first operations uniquely determine the starting times of the second operations, any feasible schedule is uniquely specified by the collection of starting times of the first operations $\{\sigma(1), \ldots, \sigma(n)\}$. For a schedule σ and any $j \in J$, denote by $C_j(\sigma)$ the completion time of job j in σ; note that $C_j(\sigma) = \sigma(j) + l_j + a_j + b_j$ for all $j \in J$. The length of a schedule σ is denoted by $C_{\max}(\sigma)$ and thus $C_{\max}(\sigma) = \max_{j \in J} C_j(\sigma)$. The length of a shortest schedule is denoted by $C_{\max}^*$.

2   Inapproximability Lower Bound

In this section we establish an inapproximability lower bound for the problem $1 \mid \text{exact } l_j = L, a_j = b_j \mid C_{\max}$. This bound holds for all cases of $1 \mid \text{exact } l_j = L \mid C_{\max}$ we consider in this paper.

The only related complexity result we aware of is that of Orman and Potts [13]: they prove that the case when $l_j = L, b_j = b$ for all j is strongly NP-hard.

To obtain our bound we construct a specific polynomial-time reduction from the following well-known NP-complete problem [8]:

PARTITION

Instance: Nonnegative integers $w_1, \ldots, w_m$ such that $\sum_{k=1}^{m} w_k = 2S$.

Question: Does there exist a subset $X \subseteq \{1, \ldots, m\}$ such that $\sum_{k \in X} w_k = S$?

Consider an instance I of PARTITION. Construct an instance of $1 \mid \text{exact } l_j = L, a_j = b_j \mid C_{\max}$.

Let $J = \{1, \ldots, m+2\}$ and

$$a_j = b_j = w_j, \quad l_j = R + S \quad \text{for } j = 1, \ldots m,$$

$$a_j = b_j = R, \quad l_j = R + S \quad \text{for } j = m+1, m+2$$

where $R > 6S$.

We call the jobs $m + 1$ and $m + 2$ *big* and the remaining jobs, *small*.

Fig. 1. The jobs in $\{1, \ldots, m\}$ are executed within the shaded intervals.

Lemma 1. (i) *If $\sum_{k \in X} w_k = S$ for some subset $X \subseteq \{1, \ldots m\}$, then there exists a feasible schedule σ such that $C_{\max}(\sigma) \leq 4R + 6S$.*
(ii) *If there exists a feasible schedule σ with $C_{\max}(\sigma) \leq 4R + 6S$, then*

$$\sum_{k \in X} w_k = S$$

for some set $X \subseteq \{1, \ldots m\}$.
(iii) *If there does not exists a feasible schedule σ with $C_{\max}(\sigma) \leq 4R + 6S$, then $C_{\max}^* > 5R$.*

Proof.   (i) We construct a schedule σ such that $C_{\max}(\sigma) \leq 4R + 6S$. For the big jobs $m + 1$ and $m + 2$ set $\sigma(m + 2) = \sigma(m + 1) + R + S$. Then the first operation of job $m + 1$ and the first operation of job $m + 2$ are separated by the interval $A = [\sigma(m + 1) + R, \sigma(m + 1) + R + S]$ of size S. The same is true for the second operations of these jobs: they are separated by the interval $B = [\sigma(m + 2) + 2R, \sigma(m + 2) + 2R + S]$ of size S (see Fig. 1).

The small jobs in X are scheduled in the following way. The second operations of these jobs are executed within the interval A one by one without idles in the nondecreasing order of w_j (see Fig. 1). The first operations of the jobs $j \in \{1, \ldots, m\} \setminus X$ are executed within the interval B one by one without idles in the nonincreasing order of w_j. Since $l_j = R + S$ and $w_j \leq S$ for all $j \in \{1, \ldots, m\}$, the constructed schedule σ has length at most $4R + 6S$.

(ii) W.l.o.g. we may assume that $\sigma(m+1) < \sigma(m+2)$. Then the first operation of job $m + 2$ must be executed before the second operation of job $m + 1$ since otherwise $C_{\max}(\sigma) > 6R$. Moreover, by the definitions of jobs no operation of jobs in $j \in \{1, \ldots, m\}$ can be executed between the first operation of job $m + 2$ and the second operation of job $m+1$. On the other hand, if some job $j \in \{1, \ldots, m\}$ is scheduled outside the big jobs, i.e., both operations of j are not in the interval $[\sigma(m + 1), \sigma(m + 1) + 4R + 3S]$, then $C_{\max}(\sigma) > 5R > 4R + 6S$. It follows that exactly one operation of each small job $j \in \{1, \ldots, m\}$ is executed either between the first operations of the big jobs $m + 1$ and $m + 2$, or between the second operations of these jobs. Since $\sum_{k=1}^{m} w_k = 2S$, the interval between the first operation of job $m + 2$ and the second operation of job $m + 1$ is empty, i.e., $\sigma(m + 2) = \sigma(m + 1) + R + S$. Moreover, $\sum_{k=1}^{m} w_k = 2S$ also implies that the machine has no idles within time intervals $[\sigma(m + 1) + R, \sigma(m + 1) + R + S]$ and $[\sigma(m+1)+3R+S, \sigma(m+1)+3R+2S]$. Thus the schedule σ has the form shown in Fig. 1 and for the set of small jobs $X \subseteq \{1, \ldots, m\}$ whose second operations are executed within the time interval $[\sigma(m + 1) + R, \sigma(m + 1) + R + S]$ we have $\sum_{k \in X} w_k = S$ as required.

(iii) In (ii) we actually proved that if for a schedule σ $C_{\max}(\sigma) \leq 4R + 6S$, then exactly one operation of each small job $j \in \{1, \ldots, m\}$ is executed either between the first operations of the "big" jobs $m + 1$ and $m + 2$, or between the second operations of these jobs. Otherwise at least one job $j \in \{1, \ldots, m\}$ is scheduled outside the big jobs, i.e., both operations of j are executed either earlier, or later than both operations of the big jobs. However, then $C_{\max}(\sigma) > 5R$.   □

Set $R = qS$ where $q \geq 6$. Lemma 1 implies that the existence of an α-approximation algorithm with $\alpha < 5R/(4R + 6S) = 5q/(4q + 6)$ implies that $P = NP$. Thus we have the following result.

Theorem 1. *The existence of a* $(5/4 - \varepsilon)$-*approximation algorithm for problem* $1 \mid exact\ l_j = L,\ a_j = b_j \mid C_{\max}$ *implies* $P = NP$.   □

3   Approximation Algorithms

We give approximations for three special cases of $1 \mid exact\ l_j = L \mid C_{\max}$.

3.1   Algorithm for $1 \mid exact\ l_j = L, a_j \leq b_j \mid C_{\max}$

The algorithm of this section is essentially Algorithm 1M$\leq$ in [2] described for the special case when $l_j = L, j = 1, \ldots n$. We give a refined analysis of the algorithm for this case.

Informally, the algorithm does the following. First it arranges the jobs in order of the non-increasing lengths of the first operations. Then it looks over the jobs in this order and successively constructs blocks B_s ($s = 1, \ldots r$) which are some bundles of jobs $j_s, \ldots, j_{s+1} - 1$ ($j_1 = 1$). In each block B_s, the second operations of job $j_s = 1, \ldots, j_s - 1$ are processed one after the other (see Fig. 2(a)). A block becomes complete when it cannot be augmented in this way by the current job. Then the algorithm starts constructing the next block. Finally, the algorithm outputs a schedule which consists in the successive execution of blocks $B_1, \ldots B_r$ (see Fig. 2(b) with $r = 4$).

Now we give a formal description of the algorithm. Recall that we consider the case when $a_j \leq b_j\ j = 1, \ldots, n$.

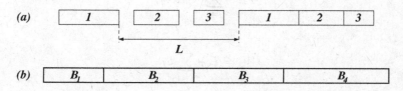

Fig. 2. (a) a block consisting of three jobs; (b) a schedule consisting of four blocks.

ALGORITHM $A^{\leq}$.

PHASE I *(jobs ordering)*. Number the jobs in the following way:

$$a_1 \geq a_2 \geq \ldots \geq a_n .$$

PHASE II *(constructing blocks $1, \ldots, r$)*. By examining the set of jobs in the order $j = 1, \ldots, n$ compute the indices $j_1 < j_2 < \ldots < j_r \leq n$ in the following way.

Step 1. Set $j_1 = 1$. If $\sum_{s=1}^{t-1} b_s \leq L$ for all $t = 2, \ldots, n$, then set $r = 1$, otherwise go to Step 2.

Step $k(k \geq 2)$. Set j_k to be equal to the minimum index among indices $t > j_{k-1}$ such that $\sum_{s=j_{k-1}}^{t-1} b_s > L$. If $j_k = n$ or $\sum_{s=j_k}^{t-1} b_s \leq L$ for all $t = j_k+1, \ldots, n$, then set $r = k$, otherwise go to Step $k + 1$.

PHASE III *(constructing the schedule)*. Set $\sigma(j_1) = \sigma(1) = 0$. If $r > 1$, then for $s = 2, \ldots, r$ set

$$\sigma(j_s) = \sigma(j_{s-1}) + a_{j_{s-1}} + L + \sum_{k=j_{s-1}}^{j_s - 1} b_k .$$

For every $j \in J \setminus \{j_1, \ldots, j_r\}$, set

$$\sigma(j) = \sigma(j_s) + a_{j_s} - a_j + \sum_{k=j_s}^{j-1} b_k$$

where s is the maximum index such that $j_s < j$.

The correctness of the algorithm is established by Lemma 1 in [2]. The running time is $O(n \log n)$.

Approximation Ratio

Note that $C_{\max}^* \geq \max\{W_1, W_2\}$ where

$$W_1 = \sum_{j=1}^{n} (a_j + b_j) \tag{1}$$

and

$$W_2 = L + \sum_{i=1}^{n} b_j.$$

The lower bound W_1 is trivial. The bound W_2 follows from the fact that in any feasible schedule all second operations are executed outside the delay ($= L$) of the first executed job.

For every $s = 1, \ldots, r$ define

$$H_s = a_{j_s} + L + \sum_{k=j_s}^{j_{s+1}-1} b_k, \tag{2}$$

i.e., H_s is the length of the block B_s.

Since the blocks are executed one after the other without idles,

$$C_{\max}(\sigma) = \sum_{s=1}^{r} H_s = \sum_{s=1}^{r} (a_{j_s} + L) + \sum_{j=1}^{n} b_j.$$

By the description of the algorithm (see Step k) for any $s = 1, \ldots r - 1$

$$\sum_{k=j_s}^{j_{s+1}-1} b_k > L. \tag{3}$$

It follows that

$$
\begin{aligned}
C_{\max}(\sigma) &\leq \sum_{s=1}^{r} a_{j_s} + L + \sum_{s=1}^{r-1} \sum_{k=j_s}^{j_{s+1}-1} b_k + \sum_{j=1}^{n} b_j \\
&\leq \left(\sum_{s=1}^{r} a_{j_s} + \sum_{j=1}^{n} b_j \right) + \sum_{j=1}^{n} b_j + L \\
&\leq W_1 + W_2 \\
&\leq C_{\max}^*.
\end{aligned}
\tag{4}
$$

Theorem 2. *Algorithm $A^{\leq}$ finds a schedule of length within factor of 2 of the length of a shortest schedule.*   □

3.2   Algorithm A^{gen} for $1 \mid$ exact $l_j = L \mid C_{\max}$

To construct an 3-approximation for the case of arbitrary a_j and b_j we use a trick in [2]. Then we give a refined worst-case approximation bound analysis.

Let $I = \{(a_j, l_j, b_j) : j \in J\}$ be an instance of $1 \mid$ exact $l_j = L \mid C_{\max}$.

ALGORITHM A^{gen}.

1. If $\sum_{j=1}^{n} a_j > \sum_{j=1}^{n} b_j$, replace $I = \{(a_j, L, b_j) : j \in J\}$ by the symmetrical instance $\{(b_j, L, a_j) : j \in J\}$ (which is equivalent to the inverse of the time axis).
2. Form a new instance $I^* = \{(a_j, l_j = L, \bar{b}_j) : j \in J\}$ where $\bar{b}_j = \max\{a_j, b_j\}$ (note that I^* is an instance of $1 \mid$ exact $l_j = L, a_j \leq b_j \mid C_{\max}$.)
3. By applying Algorithm $A^{\leq}$ to I^* find a schedule σ.
4. If $\sum_{j=1}^{n} a_j \leq \sum_{j=1}^{n} b_j$ output σ; otherwise output the inverse of σ.

It is clear that the running time of Algorithm A^{gen} is $O(n \log n)$.

Approximation Ratio

We may evidently assume that

$$\sum_{j=1}^{n} a_j \leq \sum_{j=1}^{n} b_j. \tag{5}$$

Now by (5) and the definition of $\bar{b}_j$ we have

$$C_{\max}(\sigma) \leq \Big(\sum_{s=1}^{r} a_{j_s} + \sum_{j=1}^{n} \bar{b}_j\Big) + \sum_{j=1}^{n} \bar{b}_j + L$$

$$\leq \sum_{j=1}^{n} a_j + \sum_{j=1}^{n} \max\{a_j, b_j\} + \sum_{j=1}^{n} \max\{a_j, b_j\} + L$$

$$\leq \sum_{j=1}^{n} a_j + \sum_{j=1}^{n} (a_j + b_j) + \sum_{j=1}^{n} (a_j + b_j) + L$$

$$= \Big(\sum_{j=1}^{n} a_j + \sum_{j=1}^{n} b_j\Big) + \Big(\sum_{j=1}^{n} b_j + L\Big) + 2\sum_{j=1}^{n} a_j$$

$$\leq W_1 + W_2 + 2\sum_{j=1}^{n} a_j$$

$$\leq 2W1 + W_2 (\text{by (5)})$$

$$\leq 3C_{\max}^{*}.$$

Theorem 3. *Algorithm A^{gen} finds a schedule of length within factor of 3 of the length of a shortest schedule.* □

3.3   Improved Analysis of Algorithm $A^{\leq}$ for $1 \mid$ exact $l_j = L, a_j = b_j \mid C_{\max}$

In this subsection we give an improved worst-case analysis of Algorithm $A^{\leq}$ for the case when $a_j = b_j$ for all $j \in J$.

If for some $j \in J$, $a_j > L$ then no operation of any other job can be executed within the delay of job j. Hence such jobs are executed independently of the other jobs in any feasible schedule and so the problem reduces to the case when $a_j \leq L$ for all $j \in J$. In this subsection we assume that

$$a_j \leq L \text{ for all } j \in J. \tag{6}$$

In the analysis, we use the following lower bounds for $C_{\max}^*$:

$$W_1 = 2 \sum_{j=1}^{n} a_j$$

and

$$W_3 = 2a_1 + L + \sum_{j=2}^{n} a_j.$$

The first bound is a special case of (1). The bound W_3 follows from the fact that in any feasible schedule at least one operation of any job $j \geq 2$ will be executed outside the delay of job 1.

Recall that by (2) H_s denotes the length of block B_s, $s = 1, \ldots, r$. We call a block *complete* if it satisfies (3). By the description of Algorithm $A^{\leq}$ all blocks except, possibly B_r are complete.

Note first that in the case $r = 1$ (of one block) Algorithm $A^{\leq}$ outputs an optimal schedule since $C_{\max}(\sigma) = W_3$.

Case $r = 2$.

In this case the first block is complete and so satisfies (3), i.e., $\sum_{j=1}^{j_2-1} a_j > L$. Thus

$$\begin{aligned}
C_{\max}(\sigma) &= H_1 + H_2 \\
&= \Big(a_1 + L + \sum_{j=1}^{j_2-1} a_j\Big) + \Big(a_{j_2} + L + \sum_{j=j_2}^{n} a_j\Big) \\
&= 2a_1 + L + \sum_{j=2}^{n} a_j + L + a_{j_2} \\
&= W_3 + L + a_{j_2}.
\end{aligned}$$

Since B_1 is complete,

$$L + a_{j_2} \leq \sum_{j=1}^{j_2-1} a_j + a_{j_2} \leq \sum_{j=1}^{n} a_j = 1/2 W_1.$$

Thus

$$C_{\max}(\sigma) \leq W_3 + 1/2 W_1 \leq 3/2 C_{\max}^*.$$

Case $r = s + 1$, $s \geq 2$.

For $k = 1, \ldots r$, let x_k denote the idle time within the delay of the first job in the block B_k, i.e., the idle time within B_k.

Lemma 2. For $k = 1, \ldots s$, $x_k < a_{j_k}$.

Proof. Let $k \in \{1, \ldots s\}$. First, we have

$$H_k = a_{j_k} + L + \sum_{j=j_k}^{j_{k+1}-1} a_j.$$

On the hand, by the definition of x_k

$$H_k = 2 \sum_{j=j_k}^{j_{k+1}-1} a_j + x_k.$$

It follows that

$$a_{j_k} - x_k = \sum_{j=j_k}^{j_{k+1}-1} a_j - L. \tag{7}$$

Since B_k is complete,

$$\sum_{j=j_k}^{j_{k+1}-1} a_j > L.$$

Together with (7) this gives the required inequality $x_k < a_{j_k}$.   □

Lemma 3. *For any blocks B_k and B_l such that $1 \le k < l \le s$*

$$x_k + x_l < L. \tag{8}$$

Proof. Set

$$S_t = \sum_{j=j_t+1}^{j_{t+1}-1} a_t$$

for $t = 1, \ldots, s$. Note that $x_t = L - S_t$. Moreover, since B_t for $t = 1, \ldots, s$ is complete the assumption (6) implies $S_t > 0$. Recall that

$$a_1 \ge a_2 \ge \ldots \ge a_n.$$

Since $l > k$ this implies $a_{j_l} \le S_k$. By Lemma 2 it follows that

$$x_k + x_l = (L - S_k) + x_l < (L - S_k) + a_{j_l} \le L.$$

□

Since $H_{s+1} \le a_{s+1} + L + \sum_{j=s+1}^{n} a_j$ we have

$$C_{\max}(\sigma) = \sum_{k=1}^{s} H_k + H_{s+1}$$

$$= \left(2 \sum_{j=1}^{j_2-1} a_j + x_1\right) + \ldots + \left(2 \sum_{j=j_s}^{j_{s+1}-1} a_j + x_s\right) + H_{s+1}$$

$$\le 2 \sum_{j=1}^{n} a_j + \sum_{k=1}^{s} x_k + L.$$

By the inequality (8) it follows that

$$C_{\max}(\sigma) \leq 2\sum_{j=1}^{n} a_j + G + L,$$

where G is the optimum of the following linear program

$$\max\{\sum_{k=1}^{s}\lambda_k : \lambda_p + \lambda_q \leq L, 1 \leq p < q \leq s\}.$$

It is easy to see that $\lambda^* = 1/2$ is an optimal solution of this problem and so $G = \frac{sL}{2}$. Thus

$$C_{\max}(\sigma) \leq 2\sum_{j=1}^{n} a_j + \frac{sL}{2} + L. \qquad (9)$$

Moreover, since the blocks B_k, $k = 1, \ldots s$, are complete,

$$\sum_{j=j_k}^{j_{k+1}-1} a_j > L.$$

It follows that

$$W_1 = 2\sum_{j=1}^{n} a_j \geq 2sL. \qquad (10)$$

Now by (9) and (10)

$$C_{\max}(\sigma) \leq W_1 + \frac{1}{4}W_1 + L.$$

Since by (10) $L \leq \frac{1}{2s}W_1$, it follows that

$$C_{\max}(\sigma) \leq W_1 + \frac{1}{4}W_1 + \frac{W_1}{2s} \leq (1 + \frac{1}{4} + \frac{1}{2s}) C_{\max}^*. \qquad (11)$$

Finally, since we assume that $r = s + 1 > 2$, i.e., $s \geq 2$

$$C_{\max}(\sigma) \leq \frac{3}{2}C_{\max}^*.$$

Moreover, the inequality (11) shows that the approximation ratio tends to $5/4$ when the number of blocks tends to infinity. It is a remarkable fact since by Theorem 1 $5/4$ is the inapproximability lower bound for this problem.

Finally, we collect the results of the subsection in the following

Theorem 4. *Let $a_j = b_j$ for all $j \in J$. Then in the worst case Algorithm $A^{\leq}$ finds a schedule of length within factor of $3/2$ of the length of a shortest schedule. Moreover, the approximation factor is equal to $\frac{5}{4} + \frac{1}{2(r-1)}$ when $r \geq 3$.* □

4   Concluding Remarks

We conclude the paper by pointing out possible directions for further work. First, we suppose that there exists a bit more sophisticated algorithm for $1 \mid$ exact $l_j = L \mid C_{\max}$ with a better approximation factor. Second, as was mention above even the complexity status of $1 \mid$ exact $l_j = L, a_j = a, b_j = b \mid C_{\max}$ remains open [4, 13] though Algorithm 1M$\leq$ provides a 2-approximation for this case. Therefore it seems reasonable to ask if there exists a better approximation for it. Third, since PARTITION is weakly NP-hard, Theorem 1 leaves open the question of the existence of a full $(1.25 + \varepsilon)$-approximation scheme for the problem $1 \mid$ exact $l_j = L, a_j = b_j \mid C_{\max}$.

Acknowledgments. The authors thank the anonymous referees for their helpful comments and suggestions.

References

1. Ageev, A.A., Baburin, A.E.: Approximation algorithms for UET scheduling problems with Exact Delays. Oper. Res. Lett. **35**, 533–540 (2007)
2. Ageev, A.A., Kononov, A.V.: Approximation algorithms for scheduling problems with exact delays. In: Erlebach, T., Kaklamanis, C. (eds.) WAOA 2006. LNCS, vol. 4368, pp. 1–14. Springer, Heidelberg (2007)
3. Baptiste, P.: A note on scheduling identical coupled tasks in logarithmic time. Disc. Appl. Math. **158**, 583–587 (2010)
4. Bekesi, J., Galambos, G., Jung, M.N., Oswald, M., Reinelt, G.: A branch-and-bound algorithm for the coupled task problem. Math. Methods Oper. Res. **80**, 47–81 (2014)
5. Blazewicz, J., Pawlak, G., Tanas, M., Wojciechowicz, W.: New algorithms for coupled tasks scheduling – a survey. RAIRO - Oper. Res. - Recherche Operationnelle **46**, 335–353 (2012)
6. Condotta, A., Shakhlevich, N.V.: Scheduling coupled-operation jobs with exact time-lags. Discrete Appl. Math. **160**, 2370–2388 (2012)
7. Farina, A., Neri, P.: Multitarget interleaved tracking for phased array radar. IEEE Proc. Part F: Comm. Radar Signal Process. **127**, 312–318 (1980)
8. Garey, M.R., Johnson, D.S.: Computers and Intractability: A Guide to the Theory of NP-completeness. Freeman, San Francisco (1979)
9. Elshafei, M., Sherali, H.D., Smith, J.C.: Radar pulse interleaving for multi-target tracking. Naval Res. Logist. **51**, 79–94 (2004)
10. Izquierdo-Fuente, A., Casar-Corredera, J.R.: Optimal radar pulse scheduling using neural networks. In: IEEE International Conference on Neural Networks, vol. 7, pp. 4588–4591 (1994)
11. Graham, R.L., Lawler, E.L., Lenstra, J.K., Rinnooy Kan, A.H.G.: Optimization and approximation in deterministic sequencing and scheduling: a survey. Ann. Discret. Math. **5**, 287–326 (1979)

12. Hwang, F.J., Lin, B.M.T.: Coupled-task scheduling on a single machine subject to a fixed-job-sequence. J. Comput. Indust. Eng. **60**, 690–698 (2011)
13. Orman, A.J., Potts, C.N.: On the complexity of coupled-task scheduling. Discrete Appl. Math. **72**, 141–154 (1997)
14. Sherali, H.D., Smith, J.C.: Interleaving two-phased jobs on a single machine. Discrete Optim. **2**, 348–361 (2005)

Routing Open Shop
with Unrelated Travel Times

Ilya Chernykh[1,2(✉)]

[1] Sobolev Institute of Mathematics, Novosibirsk, Russia
idchern@math.nsc.ru
[2] Novosibirsk State University, Novosibirsk, Russia

Abstract. The routing open shop problem is a generalization of scheduling open shop problem and metric TSP. The jobs are located at nodes
of some transportation network while the machines travel on the network to execute the jobs in the open shop environment. The machines
are initially located at the same node (the depot) and have to return
to the depot after completing all the jobs. The goal is to construct a
feasible schedule minimizing the makespan. The problem is known to be
NP-hard even for the trivial case with two machines on a link.

We discuss the generalization of that problem in which each machine
has individual travel times between the nodes of the network. For this
model with two machines on a tree we suggest a linear time algorithm
for a case when the depot is not predefined and has to be chosen.

Keywords: Scheduling · Routing open shop · Unrelated travel times

1 Introduction

We consider the routing open shop problem which was introduced in [1,2]. This
problem is a generalization of two classic discrete optimization problems: Open
Shop scheduling problem and metric Travelling Salesman Problem. Note that
both problems are strongly NP-hard.

The Open Shop problem [7] can be described as follows. Given a set of
machines $\mathcal{M} = \{M_1, \ldots, M_m\}$ and a set of *jobs* $\mathcal{J} = \{J_1, \ldots, J_n\}$ one has
to process each job by each machine, this operation takes p_{ji} time. Preemption
is not allowed. Each machine can process at most one job and each job can
be processed by at most one machine at any time moment. The objective is to
construct feasible schedule minimizing the *makespan* (or the length of the schedule). Note that for the Open Shop problem the makespan coincided with the
maximum operation's completion time $C_{\max}$. The input of this problem can be
described by the matrix of processing times $\mathcal{P} = (p_{ji})_{m \times n}$.

The Open Shop is known to be solvable in linear time for the case of two
machines and is NP-hard for the three-machine case [7]. It is strongly NP-hard
when the number of machines is the part of input. For this case for any $\rho < \frac{5}{4}$
no ρ-approximation algorithm exist (unless P = NP) [11].

© Springer International Publishing Switzerland 2016
Y. Kochetov et al. (Eds.): DOOR 2016, LNCS 9869, pp. 272–283, 2016.
DOI: 10.1007/978-3-319-44914-2_22

In the Metric Travelling Salesman problem we have an undirected edge-weighted graph $G = \langle V, E \rangle$. The weight of edge $e_{pq} = [v_p, v_q]$ is a nonnegative integer representing a distance between nodes v_p and v_q. Distances satisfy the triangle unequality. The goal is to find a hamiltonian tour in G with minimal total weight of its edges. The problem is strongly NP-hard [6]. The input of this problem can be described by the matrix of distances $\mathcal{D} = (\tau_{pq})_{k \times k}$, there $k = |V|$.

The Routing Open Shop problem is the combination of the two problems mentioned. The jobs are located at the nodes of transportation network represented by an edge-weighted connected graph G. The machines have to travel between the nodes to execute all the jobs. Thus each machine spends time not only to process its operation but also to travel.

It is assumed that each machine travels with unit speed and travel paths are not bottle-neck: any number of machines can travel the same edge in the same time. All the machines are initially located at the same node (referred to as the *depot*) and have to return back after completing all their operations. The makespan of a feasible schedule is the length of the time interval between the instant when the machines start working or moving and the instant when the last machine returns to the depot after completing all its operations. The goal is to minimize the makespan $R_{\max}$ (note that it differs from $C_{\max}$). Following standard three-field notation [9] we'll denote this problem as $RO||R_{\max}$ (or $ROm||R_{\max}$ in case number of machines m is fixed).

The graph G is not necessary the complete one but the routing problem on G is equivalent to the metric TSP on its transitive closure. Thus we'll denote the shortest distances between nodes v_p and v_q of G by τ_{pq}. The input of the Routing Open Shop problem consists of the matrix of processing times $\mathcal{P}$, matrix of distances $\mathcal{D}$, the location function V ($V(j)$ stands for the index of node containing the job J_j) and the index of the depot (without lost of generality let it be 0).

The Routing Open Shop problem is strongly NP-hard as it contains the metric TSP as a subcase. Moreover is remains NP-hard even on a 2-node network with only two machines [2]. For the latter case a FPTAS exists [8].

Several approximation algorithms are known for the Routing Open Shop problem. Most of them can be found in [3] and references within. Two most relevant to the present paper algorithms are described in [3]: a $\frac{13}{8}$-approximation algorithm for $RO2||R_{\max}$ and a $\frac{4}{3}$-approximation for $RO2|easy-TSP|R_{\max}$ (for the case when the optimal solution of the TSP on the graph G is known or can be found in polynomial time due to the graph structure or special properties of the distance matrix).

In the Routing Open Shop problem all the machines travel with the same unit speed hence distances τ_{pq} coincide with travel times. In this paper we introduce the generalization of the Routing Open Shop problem in which travel times are specific for different machines. Let $\tau_{pq}^{(i)}$ stand for the travel time between nodes v_p and v_q for machine M_i. We can consider the following hierarchy of the travel time models (inspired by classic scheduling models with parallel machines, see [9] for example):

- $RO||R_{\max}$: $\tau_{pq}^{(i)} = \tau_{pq}$ (*identical* travel times);
- $RO|Qtt|R_{\max}$: $\tau_{pq}^{(i)} = \dfrac{\tau_{pq}}{s_i}$ (*uniform* travel times, s_i represents the travel speed of machine M_i);
- $RO|Rtt|R_{\max}$: $\tau_{pq}^{(i)}$ are individual for each machine (*unRelated* travel times).

Obviously the last model is the hardest because unlike the previous ones in $RO|Rtt|R_{\max}$ each machine has its own individual optimal route over G. This model has some similarity to the Routing Open Shop with Missed Operations, in which each machine has it's own subset of jobs to process and therefore doesn't have to visit each node of the transportation network.

In this paper we consider a special case of $RO2|Rtt|R_{\max}$ in which G is a tree and the depot node is not predefined but has to be chosen by the scheduler. We will denote the problem by $RO2|Rtt, tree, variable - depot|R_{\max}$. For this problem we present a linear time algorithm.

The remainder of the paper is organized as follows. Section 2 provides necessary definitions, notation and preliminary results. Section 3 contains a simple approximation algorithm for $RO2|Rtt|R_{\max}$. Section 4 thoroughly describes the main tools for achieving our result — jobs' aggregation procedure and its properties. Section 5 contains the description of the main algorithm and the proof of its optimality. In Sect. 6 we discuss some open questions and future research plans.

2 Main Definitions and Notation

Let I be an instance of the Routing Open Shop Problem, $\mathcal{P}(I) = (p_{ji})_{m \times n}$ is its matrix of processing times, $\mathcal{D}(I) = (\tau_{pq})_{k \times k}$ is the matrix of distances on the graph $G = \langle V, E \rangle$, $V = \{v_0, \ldots, v_{k-1}\}$, v_0 is the depot and job J_j is located at the node $v_{V(j)}$. Let $\mathcal{J}^t = \{J_j \in \mathcal{J}|V(j) = t\}$ be the set of all jobs from node v_t. We assume that each non-depot node contains at least one job (i.e. $\forall t \in \{1, \ldots, k-1\}$ $\mathcal{J}^t \neq \emptyset$).

We will also use p_{ji} to denote the operation of machine M_i and job J_j (in addition to the notation of its processing time). In the case of two machines we will use a_j and b_j instead of p_{j1} and p_{j2}.

Let $s_{ji}(\sigma)$ and $c_{ji}(\sigma) \doteq s_{ji}(\sigma) + p_{ji}$ denote the starting and the completion times of operation of job J_j and machine M_i in schedule σ. The feasibility requirements for schedule σ are the following. If job J_a is processed by machine M_i before job J_b, then

$$s_{bi}(\sigma) \geqslant c_{ai}(\sigma) + \tau_{V(a)V(b)}.$$

Also for any job $J_j \in \mathcal{J}$

$$s_{ji}(\sigma) \geqslant \tau_{0V(j)}.$$

The *release moment* of machine M_i in schedule σ is the time moment when machine M_i returns to the depot after completing all its operations and is equal to $R_i(\sigma) \doteq \max\limits_j(c_{ji}(\sigma) + \tau_{V(j)0})$. The makespan of schedule σ is the maximal release moment $R_{\max}(\sigma) \doteq \max\limits_i R_i(\sigma)$. The goal is to find a feasible schedule minimizing the makespan $R_{\max}$. $R_{\max}^*$ stands for the optimal makespan.

Let $\ell_i \doteq \sum\limits_{j=1}^{n} p_{ji}$ denote the total processing time of the operations of machine M_i, referred to as the *load* of M_i. The total processing time of operations of the job J_j is called the *length* of the job and is denoted by $d_j \doteq \sum\limits_{i=1}^{m} p_{ji}$. $\ell_{\max} \doteq \max \ell_i$ and $d_{\max} \doteq \max d_j$ are the maximal machine load and maximal job length respectively.

Note that for the Open Shop problem with the input $\mathcal{P}$ the value

$$\bar{C} \doteq \max\{\ell_{\max}, d_{\max}\} \tag{1}$$

is the lower bound for the optimum. In the two-machine case optimum always coincides with $\bar{C}$ [7].

Let T^* denotes the length of the optimal tour over graph G with matrix of distances $\mathcal{D}$. Due to the fact that each machine has to visit each node at least once, we have the following simple lower bound for the optimum of the $RO||R_{\max}$ problem:

$$\hat{R} \doteq \max\{\ell_{\max} + T^*, d_{\max}\} \leqslant R_{\max}^*. \tag{2}$$

Although it is easy to observe that some job J_j cannot be completed in time less than $d_j + 2\tau_{0V(j)}$, therefore

$$\bar{R} \doteq \max\left\{\ell_{\max} + T^*, \max\{d_j + 2\tau_{0V(j)}\}\right\} \leqslant R_{\max}^* \tag{3}$$

is also a more precise lower bound to the optimum.

Now let us formulate the Routing Open Shop problem with Unrelated Travel Times $(RO|Rtt|R_{\max})$. The input of the problem with unrelated travel times is similar to the input of the original Routing Open Shop problem only instead of one matrix of distances $\mathcal{D}$ we have m such matrices $\mathcal{D}^{(i)} = \left(\tau_{pq}^{(i)}\right)_{k\times k}$ representing travel times for each machine independently. Let T_i^* denote the length of the optimal tour for machine M_i.

Adapting the lower bound $\hat{R}$ to this case we obtain

$$\hat{R}^{Rtt} \doteq \max\left\{\max\limits_i(\ell_i + T_i^*), d_{\max}\right\}. \tag{4}$$

The adaptation of the lower bound $\bar{R}$ is not that easy since scheduling of a single job even in Qtt case generally is an NP-hard problem even on a two-node network (link).

Lemma 1. *The problem $RO|link, n = 1, Qtt|R_{\max}$ is NP-hard.*

Proof. We will reduce the PARTITION problem to the decision-making version of our problem. Let $\mathcal{S} = \{a_1, \ldots, a_m\}$ be the input of the PARTITION problem, $\sum a_i = 2K$. Consider the following instance $I_{\mathcal{S}}$ of the $RO|link, n = 1, Qtt|R_{\max}$. This instance has $m + 1$ machines, two nodes, v_0 being a depot, and a single job J with processing times $(a_1, \ldots, a_m, 0)$ located at v_1. Distance between nodes is equal to $K + 1$, $s_{m+1} = 1$, $s_i = K + 1$ for $i = 1, \ldots, m$, therefore the travel times for machines M_1 to M_m equal to 1 while $\tau^{(m+1)} = K + 1$. Let us proof that the schedule with makespan not exceeding $2K + 2$ for instance $I_{\mathcal{S}}$ exists iff the partition of set $\mathcal{S}$ exists.

The sufficiency is obvious. Let us have a partition $\sum_{i=1}^{l} a_i = \sum_{i=l+1}^{m} a_i = K$. Then the schedule with makespan $2K + 2$ looks as follows: operations of machines $M_1, \ldots, M_m$ are performed without idles in that sequence in interval $[1, 2K + 1]$, while the zero operation of machine M_{m+1} starts and completes at time moment $K + 1$ (see Fig. 1). Concave arcs represent travel intervals of the machines.

Now consider we have a schedule with makespan $2K + 2$. Then the zero operation of machine M_{m+1} has to be scheduled at moment $K + 1$ and operations of other machines are performed without idles in interval $[1, K + 1]$. As soon as preemption is not allowed, some operation's completion time should coincide with the moment $K + 1$, hence we have a partition.   □

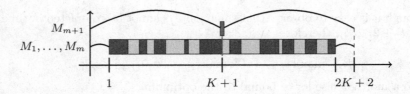

Fig. 1. Illustration of an optimal schedule for instance $I_{\mathcal{S}}$ in case the answer to the PARTITION is positive.

3   Simple Heuristic for $RO2|Rtt|R_{\max}$

Consider the following algorithm for $RO2|Rtt|R_{\max}$.
    Algorithm S.

1. For each machine M_i use a polytime approximation algorithm with the best approximation guarantee available to find the near-optimal tour ρ_i over G. Create permutations of jobs π_i consistent with route ρ_i, $i = 1, 2$.
2. Let each machine M_i process its operations in order π_i, idling if some job is busy being processed by the other machine. If two machines arrive at some job at the same time, machine M_1 goes first.

Theorem 1. *Let σ be the schedule built by Algorithm S, T_i stands for the length of tour ρ_i, $T_2 = \kappa T_1$, without lost of generality $\kappa \geqslant 1$. Then*

1. if $T_i = T_i^*$ (easy − TSP case) then $R_{\max}(\sigma) \leqslant 2\hat{R}^{Rtt}$;
2. else $R_{\max}(\sigma) \leqslant \min\left\{\dfrac{5}{2}, 1 + \max\left\{1, \dfrac{\kappa}{2}\right\}\right\} \hat{R}^{Rtt}$.

Proof. Note that idle time of any machine M_i in schedule σ doesn't exceed the load ℓ_{3-i} of the other machine. For case 1 it means

$$R_i \leqslant \ell_i + T_i^* + \ell_{3-i} = \ell_1 + \ell_2 + T_i^* \leqslant 2\hat{R}^{Rtt}, \ i = 1,2.$$

Consider case 2. Let us use the Christofides-Serdyukov algorithm for finding near-optimal tours over G [5,10]. Then $T_i \leqslant \dfrac{3}{2}T_i^*$.

Note that

$$R_i \leqslant \ell_i + \frac{3}{2}T_i^* + \ell_{3-i} = \ell_1 + \ell_2 + \frac{3}{2}T_i^* \leqslant \frac{5}{2}\hat{R}^{Rtt}, \ i = 1,2.$$

From the other hand

$$R_1 \leqslant \ell_1 + \frac{3}{2}T_1^* + \ell_2 \leqslant (\ell_1 + T_1^*) + (\ell_2 + \frac{1}{2}T_2^*) \leqslant 2\hat{R}^{Rtt},$$

$$R_2 \leqslant \ell_2 + \frac{3}{2}T_2^* + \ell_1 = (\ell_2 + T_2^*) + (\ell_1 + \frac{\kappa}{2}T_1^*) \leqslant \hat{R}^{Rtt} + \max\left\{1, \frac{\kappa}{2}\right\} \hat{R}^{Rtt},$$

therefore $R_{\max}(\sigma) \leqslant \min\left\{\dfrac{5}{2}, 1 + \max\left\{1, \dfrac{\kappa}{2}\right\}\right\} \hat{R}^{Rtt}$. $\square$

4   Jobs' Aggregation

Consider the following *jobs' aggregation operation*.

Definition 1. *Let I be an instance of the Routing Open Shop problem with set of jobs $\mathcal{J} = \{J_1, \ldots, J_n\}$ and $\mathcal{K} \subseteq \mathcal{J}^t$ is a subset of jobs from node v_t. By aggregation of set $\mathcal{K}$ we understand the following transformation of I into new instance I' in which*

$$\mathcal{J}'^t = \mathcal{J}^t \setminus \mathcal{K} \cup \{J_{\mathcal{K}}\}, \ p_{\mathcal{K}i} = \sum_{J_j \in \mathcal{K}} p_{ji}.$$

This way a set of jobs $\mathcal{K}$ is replaced by a new *aggregated* job for which operations' processing times are the total processing times of corresponding operations of jobs replaced. Note that any feasible schedule of instance I' can be treated as a feasible schedule of the initial instance I with the same makespan. Our goal is to reduce the number of jobs using the aggregation operations providing that the lower bound $\hat{R}^{Rtt}$ wouldn't change.

From now on we consider a two-machine case of the Routing Open Shop problem.

Definition 2. *Let* $\Delta^t \doteq \sum\limits_{J_j \in \mathcal{J}^t} d_j$ *be the* load *of the node* v_t. *We will refer to the node* v_t *as* overloaded *if* $\Delta^t > \hat{R}^{Rtt}$.

Lemma 2. *For any instance of the* $RO2|Rtt|R_{\max}$ *there is at most one overloaded node.*

Proof. Note that from (4) we have

$$\sum_{j=1}^{n} d_j = \sum_{t=0}^{k-1} \Delta^t = \ell_1 + \ell_2 \leqslant 2\hat{R}^{Rtt} - T_1^* - T_2^*. \tag{5}$$

It follows that for any two different nodes v_p and v_q

$$\Delta^p + \Delta^q \leqslant 2\hat{R}^{Rtt},$$

therefore at least one of v_p and v_q is not overloaded. $\qquad\square$

Lemma 3. *For any instance* I *of the problem* $RO2|Rtt|R_{\max}$ *there exists another instance* I' *obtained from* I *by series of jobs' aggregation operations such that*

1. $\hat{R}^{Rtt}(I') = \hat{R}^{Rtt}(I)$,
2. *all the nodes of* I' *excluding at most one contain exactly one job. The one "exclusive" node contains at most three jobs.*

Proof. In order to preserve the lower bound $\hat{R}^{Rtt}$ we may only apply jobs' aggregation operation to sets $\mathcal{K} \in \mathcal{J}^t$ such that

$$\sum_{J_j \in \mathcal{K}} d_j \leqslant \hat{R}^{Rtt}.$$

Thus for any not overloaded node v_t we may safely apply the aggregation operation to the whole set $\mathcal{J}^t$. According to Lemma 2 there is at most one overloaded node v_p in instance I. Let $\mathcal{J}^p = \{J_1, \ldots, J_l\}$. Let us perform the following aggregation operations.

Let f be the maximal index such that $\sum\limits_{j=1}^{f} d_j \leqslant \hat{R}^{Rtt}$. First apply the aggregation operation to the set $\mathcal{K} = \{J_1, \ldots, J_f\}$. Now we have jobs $\{J_\mathcal{K}, J_{f+1}, \ldots, J_l\}$ in the node v_p and due to the choice of f

$$d_\mathcal{K} + d_{f+1} > \hat{R}^{Rtt}.$$

Consider the case when $f + 1 < l$ (otherwise we already have only two jobs $J_\mathcal{K}$ and J_{f+1} in node v_l and the proof is complete). Then from (5) we have

$$\sum_{j=f+2}^{l} d_j = \Delta^p - d_\mathcal{K} - d_{f+1} < 2\hat{R}^{Rtt} - T_1^* - T_2^* - \hat{R}^{Rtt} \leqslant \hat{R}^{Rtt},$$

hence we may apply the aggregation operation to the remaining jobs $\{J_{f+2}, \ldots, J_l\}$ in v_p, which completes the proof. $\qquad\square$

We will refer to the transformation described in the proof above as *jobs' aggregation procedure*. Note that the jobs' aggregation procedure can be performed in $O(n)$ time.

The idea of the algorithm described in the following section is based on the reduction of the number of jobs using the jobs' aggregation procedure, constructing a schedule of the reduced instance and considering that schedule as one for the initial problem instance.

5   The Main Result

In this section we consider the $RO2|Rtt, tree, variable - depot|R_{\max}$ problem. In this model graph G is a tree and depot node is not predefined. One has to choose the location of the depot and find an optimal schedule relatively the choice made.

We will prove that this problem is solvable in linear time to the optimum. Note that $RO2|Rtt, tree|R_{\max}$ is NP-hard as it contains a known NP-hard problem $RO2|link|R_{\max}$ as a special case [1].

Also note that $\hat{R}^{Rtt}$ is still a lower bound for a $variable - depot$ problem.

5.1   The Idea of the Algorithm A

The basic idea behind the Algorithm A is the job's aggregation procedure. We will use it to transform the initial instance of the problem into the one with small number of jobs preserving the lower bound $\hat{R}^{Rtt}$ (which is possible due to the Lemma 3). Note that any feasible schedule for the aggregated instance can be easily interpreted as a feasible schedule for the initial instance: we only need to treat each aggregated operation as a continuous block of operations aggregated performed in any order without idles.

We will also use a subroutine REMOVE thoroughly described in the next subsection. The idea of that subroutine is the following. Consider some terminal node $u \in G$ containing a single job J_j with operations' processing times a_j and b_j. Let node w be adjacent to u and travel times of machines M_1 and M_2 over the edge $[u, w]$ are $\tau^{(1)}$ and $\tau^{(2)}$ respectively. The subroutine translates job J_j from node u to node w, increasing its operations' processing times by $2\tau^{(1)}$ and $2\tau^{(2)}$ respectively while removing node u from G. This way the processing of operations of transformed job J_j can be interpreted as a combination of travelling from w to u, performing the initial operation and travelling back to w.

5.2   Description of the Algorithm A

Subroutine REMOVE(u)

1. Let u be a terminal node containing a single job J_j with processing times a_j and b_j, and weights of its incident edge $e = [u, w]$ are $\tau^{(1)}$ and $\tau^{(2)}$ for machines M_1 and M_2 respectively.
2. Remove node u from G. Relocate job J_j to the node w. Set processing times for J_j to $a_j + 2\tau^{(1)}$ and $b_j + 2\tau^{(2)}$.

Algorithm A.

1. Perform the job's aggregation procedure. IF some node v is overloaded THEN GOTO Step 2 ELSE GOTO Step 4.
2. Choose v as the depot.
   FOR EACH terminal node $u \neq v \in G$ DO
   (a) REMOVE(u).
   (b) Perform a job's aggregation procedure for a modified graph G.

(Note that G is being modified inside the loop and so is the set of its terminal nodes.)

3. Build an optimal schedule for jobs from node v using Gonzalez-Sàhni algorithm [7]. GOTO Step 8.
4. FOR EACH terminal node $u \in G$ DO
   (a) Let u contain a single job J_j with processing times a_j and b_j, and weights of its incident edge $e = [u, v]$ are $\tau^{(1)}$ and $\tau^{(2)}$ for machines M_1 and M_2 respectively.
   (b) IF $d_j + 2\tau^{(1)} + 2\tau^{(2)} \leqslant \hat{R}^{Rtt}$ THEN
      i REMOVE(u).
      ii Perform a job's aggregation procedure for a modified graph G.
      iii. IF node v is overloaded THEN GOTO Step 2.
5. IF G contains single node v THEN set choose it as the depot and GOTO Step 3.
6. Modified graph G contains only two nodes, v_0 and v_1, each containing single job J_0 and J_1 respectively. Without lost of generality suppose $\Delta^0 \geqslant \Delta^1$. Choose node v_0 as the depot.
7. Denote operations of job J_j by a_j and b_j. Build a schedule according to the following order of operations:

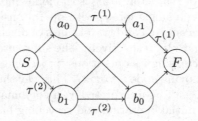

Vertices S and F denote start and completion moments of the schedule.
8. Interpret the resulting schedule as a solution for initial instance of the problem, treating the aggregated operations as continuous blocks of initial operations and treating processing of operations modified by the subroutine REMOVE(u) as a sequence of travelling from w to u, performing the operation and travelling back to w. END.

5.3   Optimality of the Algorithm A

Theorem 2. *For any instance of the problem $RO2|Rtt, tree, variable - depot|R_{\max}$ Algorithm A obtains a feasible schedule of length $\hat{R}^{Rtt}$. The running time of Algorithm A is $O(n)$.*

Proof. At first lets prove that instance transformations used in Algorithm A don't affect the lower bound $\hat{R}^{Rtt}$. Indeed the jobs' aggregation procedure doesn't change the lower bound due to the Lemma 3. The subroutine REMOVE(u) doesn't increase $\hat{R}^{Rtt}$ if the following condition holds:

$$d_j + 2\tau^{(1)} + 2\tau^{(2)} \leqslant \hat{R}^{Rtt}. \tag{6}$$

Here J_j is a single job from u, $\tau^{(1)}$ and $\tau^{(2)}$ are travel times of machines over edge incident to u. Note that at Step 4(b) condition (6) is explicitly verified before calling of the subroutine.

Let's prove that condition also holds at Step 2(a). In that case node v is overloaded and $u \neq v$. Denote the load of the node v by $\Delta > \hat{R}^{Rtt}$. Inequality (5) implies

$$\Delta + d_j + 2\tau^{(1)} + 2\tau^{(2)} \leqslant \ell_1 + \ell_2 + T_1^* + T_2^* \leqslant 2\hat{R}^{Rtt}$$

therefore (6) holds.

Now we have to prove that at Step 6 graph G contains only two nodes. Observe the IF clause at Step 4(b). Lets prove that its condition is false at most once (and that means that there is at most one not removed terminal edge).

Consider some terminal node v_1 with job J_1 and travel times over its incident edge are $\tau_1^{(1)}$ and $\tau_1^{(2)}$. Let the condition at Step 4(b) for node v_1 be false implying

$$d_1 + 2\tau_1^{(1)} + 2\tau_1^{(2)} > \hat{R}^{Rtt}. \tag{7}$$

Consider another terminal node v_2 (not adjacent to v_1, otherwise we have only two nodes already) with job J_2 and travel times $\tau_2^{(1)}$ and $\tau_2^{(2)}$. Note that $d_1 + 2\tau_1^{(1)} + 2\tau_1^{(2)} + d_2 + 2\tau_2^{(1)} + 2\tau_2^{(2)} \leqslant \ell_1 + \ell_2 + T_1^* + T_2^*$. Now it follows from (5) and (7) that $d_2 + 2\tau_2^{(1)} + 2\tau_2^{(2)} < \hat{R}^{Rtt}$.

To conclude the proof we need to show that Algorithm A builds a schedule of length $\hat{R}^{Rtt}$ for the transformed instance of the problem. For the case when the transformed graph G contains a single node (Step 3) that fact follows from the properties of the Gonzalez-Sàhni algorithm [7]. Consider Step 7.

The makespan of the schedule for jobs J_0 and J_1 built at Step 7 coincides with the length of some critical path from S to F in scheme from Step 7. The length of a path is a total weight of its nodes (operations processing times) and arcs (travel times). Those possible lengths are

1. $a_0 + a_1 + 2\tau^{(1)} = \ell_1 + T_1^* \leqslant \hat{R}^{Rtt}$,
2. $a_0 + b_0 = d_0 \leqslant \hat{R}^{Rtt}$,

3. $b_1 + b_0 + 2\tau^{(2)} = \ell_2 + T_2^* \leqslant \hat{R}^{Rtt}$,
4. $b_1 + a_1 + \tau^{(1)} + \tau^{(2)} = \Delta^1 + \tau^{(1)} + \tau^{(2)}$.

The only questionable variant is the fourth as from (4) each of the others doesn't exceed $\hat{R}^{Rtt}$. Let's prove that $\Delta^1 + \tau^{(1)} + \tau^{(2)} \leqslant \hat{R}^{Rtt}$. Suppose that's not true and $\Delta^1 + \tau^{(1)} + \tau^{(2)} > \hat{R}^{Rtt}$. Note that $\Delta^0 \geqslant \Delta^1$ (see Step 6 of the Algorithm A) and $\Delta^0 + \Delta^1 \leqslant 2\hat{R}^{Rtt} - T_1^* - T_2^*$ from (5). But from the other hand we have $\Delta^0 + \Delta^1 + T_1^* + T_2^* = \Delta^0 + \Delta^1 + 2\tau^{(1)} + 2\tau^{(2)} \geqslant 2\Delta^1 + 2\tau^{(1)} + 2\tau^{(2)} > 2\hat{R}^{Rtt}$. The contradiction obtained proves our assumption to be wrong.

Note that the running time of the Algorithm A is majorized by the running time of the jobs' aggregation procedure ($O(n)$). The enumeration of terminal nodes also can be done in $O(k)$ and $k \leqslant n + 1$.   □

6   Open Questions and Future Research

Consider the problem $RO2|Rtt|R_{\max}$. Although generally it is difficult to adapt lower bound $\bar{R}$ for the problem with unrelated (or uniform) travel times (as discussed in Sect. 2) this adaptation works fine for two machines. The following value

$$\bar{R}_2^{Rtt} \doteq \left\{ \max_i (\ell_i + T_i^*), \max_k \left(d_{\max}^k + \tau_{k0}^{(1)} + \tau_{k0}^{(2)} \right) \right\}$$

is a lower bound for the optimum of $RO2|Rtt|R_{\max}$ problem. Here $d_{\max}^k \doteq \max_{J_j \in \mathcal{J}_k} d_j$ stands for the maximal job length at node v_k. The question is, how good can we approximate the optimum of $RO2|Rtt|R_{\max}$ in comparison with the lower bound $\bar{R}_2^{Rtt}$? More precisely, we are talking about the following *optima localization problem*: what is the smallest possible value ρ such that optimum of any instance of $RO2|Rtt|R_{\max}$ is guaranteed to belong to the interval $[\bar{R}_2^{Rtt}, \rho\bar{R}_2^{Rtt}]$?

Note that for $RO2||R_{\max}$ this problem is solved only partially. It is known that optima of $RO2|link|R_{\max}$ belong to the interval $\left[\bar{R}, \frac{6}{5}\bar{R}\right]$ and the bounds of the interval are tight [1]. This result was resently generalized to the case of triangular transportation network $RO2|triangle|R_{\max}$ [4]. From the other hand, there is an approximation algorithm which builds a schedule with makespan not exceeding $\frac{4}{3}\bar{R}$ for $RO2|easy-TSP|R_{\max}$ [3] an therefore optimum of $RO2||R_{\max}$ doesn't exceed $\frac{4}{3}\bar{R}$. One of the priorities of the future research of this problem is to close the gap between $\frac{6}{5}$ and $\frac{4}{3}$ of the upper bound of the optima localization interval.

As for $RO2|Rtt|R_{\max}$, the analogues question is even more open. **Algorithm S** from Sect. 3 provides a schedule with makespan not greater than $2\bar{R}^{Rtt}$ for $RO2|Rtt, easy - TSP|R_{\max}$. One of the points of future research is to find a better approximation algorithm for that problem.

Now consider optima localization problem for $RO2|Rtt, tree|R_{\max}$. For this case we can simplify the structure of the tree using the REMOVE(u) subroutine from Subsect. 5.2. This subroutine would not increase the value of $\bar{R}^{Rtt}$ unless u is either the depot or the overloaded node. Since there is at most one overloaded node according to Lemma 2 we may transform any tree into the chain whose ends are the depot and the overloaded node. The optima localization interval for the case of such a chain is still has not been studied yet and also is of interest.

References

1. Averbach, I., Berman, O., Chernykh, I.: A $\frac{6}{5}$-approximation algorithm for the two-machine routing open shop problem on a 2-node network. Eur. J. Oper. Res. **166**(1), 3–24 (2005)
2. Averbach, I., Berman, O., Chernykh, I.: The routing open-shop problem on a network: complexity and approximation. Eur. J. Oper. Res. **173**(2), 521–539 (2006)
3. Chernykh, I., Kononov, A., Sevastyanov, S.: Efficient approximation algorithms for the routing open shop problem. Comput. Oper. Res. **40**(3), 841–847 (2013)
4. Chernykh, I., Lgotina, E.: The 2-machine routing open shop on a triangular transportation network. In: Kochetov, Y., Khachay, M., Beresnev, V., Nurminski, E., Pardalos, P. (eds.) DOOR 2016. LNCS, vol. 9869, pp. 284–297. Springer, Heidelberg (2016)
5. Christofides, N.: Worst-case analysis of a new heuristic for the travelling salesman problem. Report 388, Graduate School of Industrial Administration, Carnegie-Mellon University, Pittsburg, PA (1976)
6. Garey, M., Johnson, D.: Computers and Intractability, A Guide to the Theory of NP-Completeness. W.H. Freemann and Company, San Francisco (1979)
7. Gonzalez, T., Sahni, S.: Open shop scheduling to minimize finish time. J. Assoc. Comp. Math. **23**, 665–679 (1976)
8. Kononov, A.: On the two-machine routing open shop problem on a 2-node network. Discrete Anal. Oper. Res. **19**(2), 54–74 (2012). (in Russian)
9. Lawler, E.L., Lenstra, J.K., Rinnooy Kan, A.H.G., Shmoys, D.B., Sequencing, S.: Algorithms and Complexity. In: Graves, S.C., et al. (eds.) Handbooks in Operations Research and Management Science. Logistics of Production and Inventory, vol. 4, pp. 445–522. North Holland, Amsterdam (1993)
10. Serdyukov, A.: On some extremal routes in graphs. Upravlyaemye Sistemy **17**, 76–79 (1978). (in Russian)
11. Williamson, D.P., Hall, L.A., Hoogeveen, J.A., Hurkens, C.A.J., Lenstra, J.K., Sevast'janov, S.V., Shmoys, D.B.: Short shop schedules. Oper. Res. **45**(2), 288–294 (1997)

The 2-Machine Routing Open Shop on a Triangular Transportation Network

Ilya Chernykh[1,2(✉)] and Ekaterina Lgotina[2]

[1] Sobolev Institute of Mathematics, Novosibirsk, Russia
idchern@math.nsc.ru
[2] Novosibirsk State University, Novosibirsk, Russia
kate.lgotina@outlook.com

Abstract. The two machine routing open shop being a generalization of the metric TSP and two machine open shop scheduling problem is considered. It is known to be NP-hard even for the simplest case when the transportation network consists of two nodes only. For that simplest case it is known that the optimal makespan for any instance belongs to interval $[\bar{R}, \frac{6}{5}\bar{R}]$, there $\bar{R}$ is the standard lower bound. We generalize that classic result to the case of three-nodes transportation network and present a linear time $\frac{6}{5}$-approximation algorithm for that case.

Keywords: Scheduling · Routing open shop · Optima localization

1 Introduction

In classic open shop model there are given sets of jobs $\mathcal{J}$ and machines $\mathcal{M}$ and machines have to perform operations of each job in arbitrary order to minimize finish time [6]. The input consists of given processing times for each operation and can be described as $m \times n$ matrix, m and n being the numbers of machines and jobs respectively. It is supposed that after performing an operation of some job machine is immediately available for any successive job to process. However in real life environment the latter is not the case. Jobs usually represent some material objects and therefore some time lags between processing of different operations are often unavoidable.

Various ways to model such time lags are known. Detailed review can be found in [3] and references therein.

We consider the following *routing open shop* model. The jobs are supposed to represent some large immovable objects located at the nodes of some transportation network while machines are mobile and have to travel between the locations of jobs to perform their operations.

The routing open shop model was introduced in [1,2]. It generalizes two classic NP-hard discrete optimization problems: metric traveling salesman problem (TSP) and open shop scheduling problem. The routing open shop problem can be described as following. There is a transportation network represented by an undirected edge-weighted graph. Nodes represent some *locations* and weight of

© Springer International Publishing Switzerland 2016
Y. Kochetov et al. (Eds.): DOOR 2016, LNCS 9869, pp. 284–297, 2016.
DOI: 10.1007/978-3-319-44914-2_23

edge represents a distance between corresponding nodes. One of the nodes is given to be a *depot*. There is a number of *machines* initially located at the depot and a number of *jobs* distributed among all the nodes, each node contains at least one job. Machines have to travel between nodes with unit speed using shortest routes, processing operations of each job in an open shop environment. After performing all the operations machines have to return to the depot. The *makespan* of a schedule is the time moment of returning of the last machine to the depot after processing all the operations. The goal is to minimize the makespan. Following the traditional three-field notation (see [8] for example) we will denote the routing open shop problem as $RO||R_{\max}$.

The routing open shop with a single machine is equivalent to a metric TSP and therefore is well-known to be NP-hard in strong sense. A single-node routing open shop is just a plain open shop problem and is NP-hard for three and more machines while being polynomially solvable in the two-machine case [6]. The simplest non-trivial case of routing open shop is the two-machine problem on a link ($RO2|link|R_{\max}$). This case is shown to be NP-hard in [1]. A fully polynomial time approximation scheme and a few polynomially solvable subcases for $RO2|link|R_{\max}$ are described in [7].

Problem $RO2|link|R_{\max}$ is thoroughly investigated in [2]. It is shown that the optimal makespan for any instance doesn't exceed $\frac{6}{5}\bar{R}$, $\bar{R}$ stands for the standard lower bound (see Sect. 2), while reaching $\frac{6}{5}\bar{R}$ for some instances. The approximation algorithm described in [2] produces a schedule with makespan belonging to an interval $[\bar{R}, \frac{6}{5}\bar{R}]$, therefore that algorithm provides an $\frac{6}{5}$-approximation.

There are several approximation algorithms known for the general two-machine routing open shop ($RO2||R_{\max}$). An $\frac{7}{4}$-approximation algorithm is described in [1]. More precise $\frac{13}{8}$-approximation algorithm is given in [3]. Note that the $RO2||R_{\max}$ problem includes a metric TSP as a special case. Since the best known algorithm for the metric TSP is the $\frac{3}{2}$-approximation algorithm due to Christofides [5] and Serdyukov [10] we cannot hope to achieve better than $\frac{3}{2}$-approximation for the $RO2||R_{\max}$ problem until a better approximation for the metric TSP will be found. From the other hand the *easy-TSP* version of the $RO2||R_{\max}$ (the case when an optimal solution for the underlying TSP is known or the time complexity of its search is not taken into account) problem admits a $\frac{4}{3}$-approximation algorithm described in [3].

Note that all the approximation algorithms mentioned in the previous paragraph use the standard lower bound $\bar{R}$ to justify their performance guarantees: ρ-approximation algorithm actually obtains a schedule with makespan belonging to an interval $[\bar{R}, \rho\bar{R}]$. Therefore for any instance of the $RO2||R_{\max}$ problem its optimal makespan belongs to the interval $[\bar{R}, \frac{4}{3}\bar{R}]$, though for the case on a link this optima localization interval can be shrinked down to $[\bar{R}, \frac{6}{5}\bar{R}]$. That observations leads us to the question: what is the tightest interval of form $[\bar{R}, \rho\bar{R}]$ which contains optima for all the instances of the $RO2||R_{\max}$ problem? From the previous research we know that $\frac{6}{5} \leqslant \rho \leqslant \frac{4}{3}$.

This paper addresses that question for a case of triangular transportation network. For that $RO2|triangle|R_{\max}$ problem we show that optimum of any instance belongs to the interval $[\bar{R}, \frac{6}{5}\bar{R}]$ hence generalizing the known result for a link [2].

Previously the routing open shop on a triangular transportation network was addressed in [4] for the preemptive version of the problem. It was shown that for any instance of the $RO2|triangle, pmtn|R_{\max}$ problem its optimum belongs to inteval $[\bar{R}, \frac{11}{10}\bar{R}]$ while the algorithmic complexity of the problem is still unknown. As for the $RO2|link, pmtn|R_{\max}$ problem it is shown in [9] that the problem is polynomially solvable and optimum always coincides with the standard lower bound $\bar{R}$.

The structure of the paper is the following. Section 2 contains the formal description of the problem under consideration, necessary notation and preliminary results. In Sect. 3 we will provide the proof of the main result for three important special cases. The final proof and the description of the $\frac{6}{5}$-approximation algorithm for the general $RO2|triangle|R_{\max}$ problem as well as concluding remarks will be given in Sect. 4.

2 Preliminary Notes

2.1 Formal Description and Necessary Notation

Let us give a formal description of the $RO2||R_{\max}$ problem.

There are given sets $\mathcal{J} = \{J_1, \ldots, J_n\}$ of jobs and $\mathcal{M} = \{A, B\}$ of machines. Each job J_j consists of two operations a_j and b_j to be processed by machines A and B respectively in an arbitrary order. An undirected transportation network is represented by a connected edge-weighted graph $G = \langle V, E \rangle$, $V = \{v_0, \ldots, v_{c-1}\}$. The weight ω_{pq} of edge $e_{pq} = [v_p, v_q] \in E$ represents distance between nodes v_p and v_q. Distances are symmetric and satisfy the triangle inequality. Graph G is not necessary complete but we will use the notation ω_{pq} for distance between any two even nonadjacent nodes. Jobs from $\mathcal{J}$ are distributed between the nodes of transportation network and each node contains at least one job. Both machines are initially located at v_0 (the *depot*) and have to travel with unit speed between nodes to perform operations of the jobs. Machines have to return to the depot after completing all the jobs in some arbitrary sequence without preemption.

The goal is to construct a feasible schedule of processing all the operations and returning to the depot in minimal time.

Notation a_j (b_j) will also be used for the processing time of corresponding operation. The set of jobs located at node v_k will be denoted as $\mathcal{J}^k$.

As preemption is not allowed any schedule S can be described by specifying the starting times $s_{jA}(S)$ ($s_{jB}(S)$) for operations a_j (b_j) of each job J_j. Completion time $c_{jA}(S)$ can be defined as $s_{jA}(S) + a_j$, $c_{jB}(S) = s_{jB}(S) + b_j$.

For any feasible schedule S if machine $M \in \mathcal{M}$ processes job $J_j \in \mathcal{J}^k$ before job $J_i \in \mathcal{J}^l$ then the following condition should be carried out:

$$s_{iM}(S) \geqslant c_{jM}(S) + \omega_{kl}.$$

If job $J_j \in \mathcal{J}^k$ is the first job processed by machine M in schedule S then the following condition should hold:

$$s_{jM}(S) \geqslant \omega_{0k}.$$

Let job $J_j \in \mathcal{J}^k$ be the last job processed by machine $M \in \mathcal{M}$ in schedule S. Then machine M releases from duty at time moment

$$R_M(S) \doteq c_{jM}(S) + \omega_{k0}.$$

The makespan of a schedule S is defined as $R_{\max}(S) \doteq \max\{R_A(S), R_B(S)\}$.

We will omit the notation of schedule S in cases it is uniquely defined by the context.

For any problem instance I with weight function ω we will use the following additional notation:

- $\ell_A(I) = \sum\limits_{j=1}^{n} a_j$, $\ell_B(I) = \sum\limits_{j=1}^{n} b_j$ are the *loads of machines* A and B correspondingly;
- $d_j(I) = a_j + b_j$ is the *length of job* $J_j \in \mathcal{J}$;
- $\ell_{\max}(I) = \max\{\ell_A(I), \ell_B(I)\}$ is the maximum machine load;
- $d^k_{\max}(I) = \max\limits_{J_j \in \mathcal{J}^k} d_j(I)$ is the maximum job length at node v_k;
- T^* is the length of the minimal travel route;
- $R^*_{\max}(I)$ stands for the optimal makespan.

Now we can describe the standard lower bound for the optimal makespan introduced in [1]:

$$R^*_{\max}(I) \geqslant \bar{R}(I) \doteq \max\left\{\ell_{\max}(I) + T^*, \max_k\left(d^k_{\max}(I) + 2\omega_{0k}\right)\right\}. \qquad (1)$$

We will focus on a special case $RO2|triangle|R_{\max}$ in which graph G is triangular and $V = \{v_0, v_1, v_2\}$. Lets introduce notation specific to the triangular case:

- $\tau \doteq \omega_{01}$, $\nu \doteq \omega_{12}$, $\mu \doteq \omega_{02}$;
- $T^* = \tau + \mu + \nu$.

In this case the standard lower bound has the following simplified form:

$$\bar{R} = \max\left\{\ell_{\max} + T^*, d^0_{\max}, d^1_{\max} + 2\tau, d^2_{\max} + 2\mu\right\}. \qquad (2)$$

2.2 Jobs' Aggregation

We will use the following definition introduced in [4].

Definition 1. *The* load of node v_k *is the total processing time of all operations from that node:*

$$\Delta^k \doteq \sum_{J_j \in \mathcal{J}^k} d_j.$$

Node v_k is referred to as overloaded *if*

$$\Delta^k + 2\omega_{0k} > \bar{R},$$

otherwise the node is underloaded.

The following statement holds for any instance of $RO2||R_{\max}$.

Statement 1. *Let I be an instance of $RO2||R_{\max}$ with graph $G = \langle V, E \rangle$. Then V contains at most one overloaded node.*

Proof. Note that due to (1) the following unequality holds for the *total load*:

$$\Delta \doteq \sum_{k=0}^{c-1} \Delta^k = \ell_A + \ell_B \leqslant 2(\bar{R} - T^*). \tag{3}$$

Suppose we have an overloaded node v_k, $\Delta^k > \bar{R} - 2\omega_{0k} \geqslant \bar{R} - T^*$. Then for any other node v_l unequality (3) implies

$$\Delta^l \leqslant \Delta - \Delta^k < 2(\bar{R} - T^*) - \bar{R} + T^* = \bar{R} - T^*,$$

therefore

$$\Delta^l + 2\omega_{0l} \leqslant \Delta^l + T^* < \bar{R}.$$

$\square$

The algorithm we'll present is based on the following operation of jobs' aggregation.

Definition 2. *Let I be some instance of $RO2||R_{\max}$, $\mathcal{K} \subseteq \mathcal{J}^k$ for some v_k. Then we say that instance I' is obtained from I by* aggregation *of jobs from $\mathcal{K}$ if*

$$\mathcal{J}^k(I') = \mathcal{J}^k(I) \setminus \mathcal{K} \cup \{J_\mathcal{K}\}, \quad a_\mathcal{K} = \sum_{J_j \in \mathcal{K}} a_j, \ b_\mathcal{K} = \sum_{J_j \in \mathcal{K}} b_j,$$

$$\forall l \neq k \ \mathcal{J}^l(I') = \mathcal{J}^l(I).$$

The instance $\tilde{I}$ obtains from I by a series of job's aggregation will be referred to as a modification *of I.*

It is obvious that any feasible schedule for some modification $\tilde{I}$ of I can be treated as a feasible schedule for I with the same makespan, therefore the optimum of any modification of I is greater or equal to $R^*_{\max}(I)$. Note that machine loads and node loads are preserved by any job's aggregation operation.

Statement 2. *For any instance I of $RO2||R_{\max}$ there exists its modification $\tilde{I}$ such that*

1. $\bar{R}(\tilde{I}) = \bar{R}(I)$,
2. *every underloaded node in $\tilde{I}$ contains exactly one job, the only overloaded node (if any) contains at most three jobs.*

Proof. In order to preserve the standard lower bound (1) under the job's aggregation operation we may only choose such sets $\mathcal{K} \subseteq \mathcal{J}^k$ that

$$\sum_{J_j \in \mathcal{K}} d_j \leqslant \bar{R} - 2\omega_{0k}.$$

Therefore for any underloaded node v_k aggregation of jobs from set $\mathcal{J}^k$ doesn't increase the standard lower bound. From Statement 1 there is at most one overloaded node v_l in I. Let's prove that we can aggregate jobs from $\mathcal{J}^l$ into at most three jobs preserving $\bar{R}$.

Let $\mathcal{J}^l = \{J_1, \ldots, J_p\}$. Let j be the maximal index such that

$$\sum_{t=1}^{j} d_t \leqslant \bar{R} - 2\omega_{0l}.$$

Note that $j < p$ because v_l is overloaded. Perform the aggregation operation for the set $\mathcal{K} = \{J_1, \ldots, J_j\}$. Due to the choice of index j we have $d_{\mathcal{K}} + d_{j+1} > \bar{R} - 2\omega_{0l}$. Suppose $j + 1 < p$ (otherwise we have two jobs at v_l and statement is correct). Let $\mathcal{K}' = \{J_{j+2}, \ldots, J_p\}$. From (3) we have

$$\sum_{J_t \in \mathcal{K}'} d_j \leqslant \Delta - d_{\mathcal{K}} - d_{j+1} < 2(\bar{R} - T^*) - (\bar{R} - 2\omega_{0l}) \leqslant \bar{R} - T^* \leqslant \bar{R} - 2\omega_{0l},$$

therefore aggregation of set $\mathcal{K}'$ doesn't increase $\bar{R}$. Thus the modification claimed to exist is achieved by aggregation operations of all jobs at each underloaded node, of set $\mathcal{K}$ and of set $\mathcal{K}'$.   □

Note that for any instance I the modification $\tilde{I}$ described can be found in $O(n)$ time.

Let $\tilde{I}$ be a modification of I, $\bar{R}(\tilde{I}) = \bar{R}(I) = \bar{R}$. If there exists a schedule S for $\tilde{I}$ such that $R_{\max}(S) \leqslant \rho\bar{R}$ then $R_{\max}^*(I) \leqslant \rho\bar{R}$. Hence it is sufficient to establish the optima localization interval for an instance with small number of jobs which exists due to Statement 2. Such results are described in Sect. 3.

3   Three Important Subcases

We will use the branch-and-bounds method to prove the main result for special subcases with small number of jobs. Analogous approach was described in [11], and also used for obtaining similar optima localization results in [2,4,9]. The underlying idea is to describe a subset of instances by the choice of critical path in a digraph which represents a partial order of operations used to build an early schedule. Such digraphs determine an order of operations for each job and each machine and will be referred to as *schemes* of a schedule, weights of vertices are correspondent operations' processing times and weights of arcs are travel distances. By $S_{\mathcal{H}}$ we will denote the early schedule built according to the scheme $\mathcal{H}$ for some current instance.

Depending on actual instance the makespan of schedule $S_{\mathcal{H}}$ can be described by different complete paths in $\mathcal{H}$. Knowing which path is critical (i.e. has the maximal length) we can describe the makespan of a $S_{\mathcal{H}}$ by that path's length, i.e. by a sum of weights of nodes and arcs of that path. The enumeration of such complete (and potentially critical) paths lies underneath the branching procedure of the proof.

3.1   All Nodes are Underloaded

Lemma 1. *Let I be an instance of the $RO2|triangle|R_{\max}$ problem with single job at each node. Then there exists a feasible schedule S for I such that $R_{\max}(S) \leqslant \frac{6}{5}\bar{R}$.*

Proof. Let us have set of jobs $\mathcal{J}^0 = \{J_0\}$, $\mathcal{J}^1 = \{J_\alpha\}$ and $\mathcal{J}^2 = \{J_\beta\}$. Without lost of generality assume that

$$a_\alpha \geqslant b_\beta. \tag{4}$$

If that is not the case we can renumerate nodes and/or machines to achieve the condition above.

Now we will consider a series of schedules and prove that at least one of them satisfies the lemma's claim.

Consider the schedule $S_1 = S_{\mathcal{H}_1}$ (see Fig. 1). S and F mark the start and finish time moments respectively.

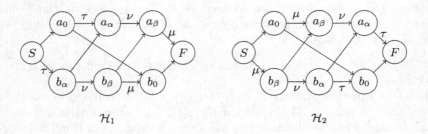

$$\mathcal{H}_1 \qquad\qquad\qquad \mathcal{H}_2$$

Fig. 1. Schemes $\mathcal{H}_1$ and $\mathcal{H}_2$.

Following a well-known fact from project scheduling the makespan of S_1 coincides with the weighted length of a critical path of $\mathcal{H}_1$. Therefore

$$R_{\max}(S_1) = \max\{ a_0 + \tau + a_\alpha + \nu + a_\beta + \mu, \tau + b_\alpha + \nu + b_\beta + \mu + b_0, a_0 + b_0,$$
$$\tau + b_\alpha + a_\alpha + \nu + a_\beta + \mu, \tau + b_\alpha + \nu + b_\beta + a_\beta + \mu\}.$$

From (2) the first three sums from the max clause above clearly don't exceed the lower bound $\bar{R}$. If one of the correspondent paths turns out to be critical then $R_{\max}(S_1) = \bar{R}$ and the claim of the lemma follows immediately. Further such to be called *trivial* paths will be excluded from consideration.

Using assumption (4) we can conclude that

$$R_{\max}(S_1) = \tau + b_\alpha + a_\alpha + \nu + a_\beta + \mu = T^* + b_\alpha + a_\alpha + a_\beta. \tag{5}$$

Now let $S_2 = S_{\mathcal{H}_2}$ (Fig. 1). Using similar reasoning and excluding trivial paths we conclude that

$$R_{\max}(S_2) = \max\{T^* + b_\beta + a_\beta + a_\alpha, T^* + b_\beta + b_\alpha + a_\alpha\} = T^* + b_\beta + a_\alpha + \max\{a_\beta, b_\alpha\}.$$

Note that due to the metric property of the distances the makespan of S_2 can be evaluated as

$$R_{\max}(S_2) \leqslant T^* + a_\alpha + \bar{R} - 2\mu \tag{6}$$

or as

$$R_{\max}(S_2) \leqslant T^* + b_\beta + \bar{R} - 2\tau. \tag{7}$$

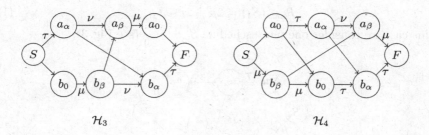

Fig. 2. Schemes $\mathcal{H}_3$ and $\mathcal{H}_4$.

Consider schedules $S_3 = S_{\mathcal{H}_3}$ and $S_4 = S_{\mathcal{H}_4}$ (see Fig. 2). There is the only non-trivial path in $\mathcal{H}_3$ therefore

$$R_{\max}(S_3) = b_0 + b_\beta + a_\beta + a_0 + 2\mu. \tag{8}$$

The scheme $\mathcal{H}_4$ contains three non-trivial paths:

1. $S \to a_0 \xrightarrow{\tau} a_\alpha \to b_\alpha \xrightarrow{\tau} F$;
2. $S \to a_0 \to b_0 \xrightarrow{\tau} b_\alpha \xrightarrow{\tau} F$;
3. $S \xrightarrow{\mu} b_\beta \xrightarrow{\mu} b_0 \xrightarrow{\tau} b_\alpha \xrightarrow{\tau} F$.

We will consider those cases one by one.

Case 1:

$$R_{\max}(S_4) = a_0 + a_\alpha + b_\alpha + 2\tau. \tag{9}$$

Let S be the best schedule among $S_1, \ldots, S_4$. Using (2), (5), (6), (8) and (9) we obtain

$$5R_{\max}(S) \leqslant R_{\max}(S_1) + R_{\max}(S_2) + 2R_{\max}(S_3) + R_{\max}(S_4) \leqslant$$
$$\leqslant (T^* + b_\alpha + a_\alpha + a_\beta) + (T^* + a_\alpha + \bar{R} - 2\mu) + 2(b_0 + b_\beta + a_\beta + a_0 + 2\mu) +$$
$$+ (a_0 + a_\alpha + b_\alpha + 2\tau) = \bar{R} + 2T^* + 2\mu + 2\tau + 3\ell_A + 2\ell_B \leqslant 6\bar{R},$$

therefore $R_{\max}(S) \leqslant \frac{6}{5}\bar{R}$.

Case 2:
$$R_{\max}(S_4) = a_0 + b_0 + b_\alpha + 2\tau. \tag{10}$$

Again, let S be the best schedule among $S_1, \ldots, S_4$. Using (2), (5)–(8) and (10) we obtain

$$5R_{\max}(S) \leqslant R_{\max}(S_1) + 2R_{\max}(S_2) + R_{\max}(S_3) + R_{\max}(S_4) \leqslant$$
$$\leqslant (T^* + b_\alpha + a_\alpha + a_\beta) + (T^* + b_\beta + \bar{R} - 2\tau) + (T^* + a_\alpha + \bar{R} - 2\mu) +$$
$$+ (b_0 + b_\beta + a_\beta + a_0 + 2\mu) + (a_0 + b_0 + b_\alpha + 2\tau) = 2\bar{R} + 3T^* + 2\ell_A + 2\ell_B \leqslant 6\bar{R},$$

therefore $R_{\max}(S) \leqslant \frac{6}{5}\bar{R}$.

Case 3:
$$R_{\max}(S_4) = 2\mu + 2\tau + \ell_B. \tag{11}$$

In this case we consider one more schedule $S_5 = S_{\mathcal{H}_5}$ (see Fig. 3).

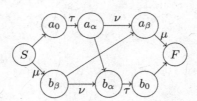

Fig. 3. Scheme $\mathcal{H}_5$.

There is the only non-trivial path in $\mathcal{H}_5$ therefore we may assume

$$R_{\max}(S_5) = a_0 + b_0 + 2\tau + a_\alpha + b_\alpha. \tag{12}$$

Let S be the best schedule among $S_1, \ldots, S_5$. Using (2), (5), (7), (8), (11) and (12) we obtain

$$5R_{\max}(S) \leqslant R_{\max}(S_1) + R_{\max}(S_2) + R_{\max}(S_3) + R_{\max}(S_4) + R_{\max}(S_5) \leqslant$$
$$\leqslant (T^* + b_\alpha + a_\alpha + a_\beta) + (T^* + b_\beta + \bar{R} - 2\tau) + (b_0 + b_\beta + a_\beta + a_0 + 2\mu) +$$
$$+ (2\mu + 2\tau + \ell_B) + (a_0 + b_0 + 2\tau + a_\alpha + b_\alpha) \leqslant \bar{R} + 5T^* + 2\ell_A + 3\ell_B \leqslant 6\bar{R},$$

therefore $R_{\max}(S) \leqslant \frac{6}{5}\bar{R}$.

This concludes the proof of Lemma 1.   □

3.2   The Depot is Overloaded

Lemma 2. *Let I be an instance of the $RO2|triangle|R_{\max}$ problem with single job at each node except the depot which is overloaded and contains at most three jobs. Then there exists a feasible schedule S for I such that $R_{\max}(S) \leqslant \frac{6}{5}\bar{R}$.*

Proof. Let us use the following notation for sets of jobs: $\mathcal{J}^0 = \{J_1, J_2, J_3\}$, $\mathcal{J}^1 = \{J_\alpha\}$ and $\mathcal{J}^2 = \{J_\beta\}$. If the depot contains only two jobs we will add dummy job J_3 with zero processing times.

Without lost of generality we can assume that

$$a_2 \geqslant b_1, \; a_3 \geqslant b_2. \tag{13}$$

Indeed, we can always achieve that condition by proper re-numeration of mahcines or/and jobs from $\mathcal{J}^0$ due to the following reasoning. Consider three pairs of operations: a_2 and b_1, a_3 and b_2, a_1 and b_3, and compare them pairwise. Without lost of generality due to possible re-numeration of machines at least for two of those pairs operation of machine A is greater or equal to the respective operation of machine B. Using proper numeration of jobs we can assure that (13) holds.

Note that as the depot v_0 is overloaded we have $d_1 + d_2 + d_3 > \bar{R}$. Since $\sum_j d_j = \ell_A + \ell_B \leqslant 2\bar{R} - 2T^*$ we have

$$d_\alpha + d_\beta + 2T^* < 2(\bar{R} - T^*) - \bar{R} + 2T^* \leqslant \bar{R}. \tag{14}$$

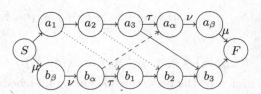

Fig. 4. Scheme $\mathcal{H}_1$.

Now consider a schedule $S_1 = S_{\mathcal{H}_1}$ (see Fig. 4). Note that complete paths containing dotted arcs cannot be critical due to the assumption (13). We will not consider such paths and omit reference to that assumption further. Also the length of the path containing dashed arc is at most $\bar{R}$ due to (14). Therefore we have to consider just one non-trivial path $S \to a_1 \to a_2 \to a_3 \to b_3 \to F$:

$$R_{\max}(S_1) = a_1 + a_2 + a_3 + b_3. \tag{15}$$

Consider a schedule $S_2 = S_{\mathcal{H}_2}$ (see Fig. 5). We need to consider just one path $S \xrightarrow{\mu} b_\beta \xrightarrow{\nu} b_\alpha \xrightarrow{\tau} b_3 \to a_3 \xrightarrow{\tau} a_\alpha \xrightarrow{\nu} a_\beta \xrightarrow{\mu} F$:

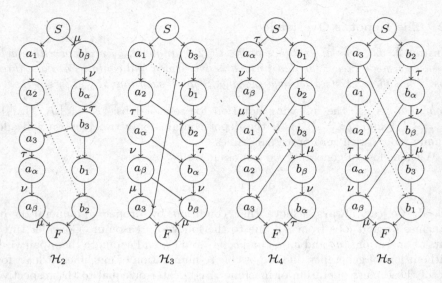

Fig. 5. Schemes $\mathcal{H}_2$, $\mathcal{H}_3$, $\mathcal{H}_4$ and $\mathcal{H}_5$.

$$R_{\max}(S_2) = 2T^* + b_\beta + b_\alpha + b_3 + a_3 + a_\alpha + a_\beta. \tag{16}$$

Now consider schedules $S_3 = S_{\mathcal{H}_3}$ and $S_4 = S_{\mathcal{H}_4}$ (Fig. 5). We need to consider three non-trivial paths in $\mathcal{H}_3$ and due to (14) only three paths in $\mathcal{H}_4$.

Case 1:
$$R_{\max}(S_3) = T^* + a_1 + a_2 + a_\alpha + b_\beta + \max\{a_\beta, b_\alpha\}. \tag{17}$$

Case 1.1:
$$R_{\max}(S_4) = b_1 + a_2 + a_3 + \max\{a_1, b_2\}. \tag{18}$$

Here and further let S be the best schedule among all schedules built in each case. Then from (15)–(18) we have

$$5R_{\max}(S) \leqslant R_{\max}(S_1) + R_{\max}(S_2) + R_{\max}(S_3) + 2R_{\max}(S_4) \leqslant 4\ell_A + 2\ell_B + 3T^* \leqslant 6\bar{R},$$

therefore $R_{\max}(S) \leqslant \frac{6}{5}\bar{R}$.

Case 1.2:
$$R_{\max}(S_4) = b_1 + b_2 + b_3 + a_3. \tag{19}$$

From (17) and (19) we have

$$2R_{\max}(S) \leqslant R_{\max}(S_3) + R_{\max}(S_4) \leqslant \ell_A + \ell_B + T^* \leqslant 2\bar{R},$$

therefore $R_{\max}(S) = \bar{R}$.

Case 2:
$$R_{\max}(S_3) = T^* + a_1 + a_2 + b_2 + b_\alpha + b_\beta. \tag{20}$$

Case 2.1:
$$R_{\max}(S_4) = b_1 + a_1 + a_2 + a_3. \tag{21}$$
In this case from (15), (16), (20) and (21) we have
$$5R_{\max}(S) \leqslant R_{\max}(S_1) + R_{\max}(S_2) + R_{\max}(S_3) + 2R_{\max}(S_4) \leqslant 4\ell_A + 2\ell_B + 3T^* \leqslant 6\bar{R},$$
therefore $R_{\max}(S) \leqslant \frac{6}{5}\bar{R}$.

Case 2.2:
$$R_{\max}(S_4) = b_1 + b_2 + b_3 + a_3. \tag{22}$$
In this case from (15), (16), (20) and (22) we have
$$5R_{\max}(S) \leqslant R_{\max}(S_1) + R_{\max}(S_2) + 2R_{\max}(S_3) + R_{\max}(S_4) \leqslant 3\ell_A + 3\ell_B + 3T^* \leqslant 6\bar{R},$$
therefore $R_{\max}(S) \leqslant \frac{6}{5}\bar{R}$.

Case 2.3:
$$R_{\max}(S_4) = b_1 + b_2 + a_2 + a_3. \tag{23}$$
In this case we build one last schedule $S_5 = S_{\mathcal{H}_5}$ (Fig. 5). Consider three non-trivial paths in $\mathcal{H}_5$.

Case 2.3.1:
$$R_{\max}(S_5) = a_3 + b_3 + b_1. \tag{24}$$
In this case from (20) and (24) we have
$$2R_{\max}(S) \leqslant R_{\max}(S_3) + R_{\max}(S_5) \leqslant \ell_A + \ell_B + 2T^* \leqslant 2\bar{R},$$
therefore $R_{\max}(S) = \bar{R}$.

Case 2.3.2:
$$R_{\max}(S_5) = b_2 + b_\alpha + a_\beta + \max\{a_\alpha, b_\beta\} + T^*. \tag{25}$$
In this case from (15) and (25) we have
$$2R_{\max}(S) \leqslant R_{\max}(S_1) + R_{\max}(S_5) \leqslant \ell_A + \ell_B + 2T^* \leqslant 2\bar{R},$$
therefore $R_{\max}(S) = \bar{R}$.

This concludes the proof of Lemma 2.   □

3.3   Some Distant Node is Overloaded

Lemma 3. *Let I be an instance of the $RO2|triangle|R_{\max}$ problem with single job at each node except one of distant nodes which is overloaded and contains at most three jobs. Then there exists a feasible schedule S for I such that $R_{\max}(S) \leqslant \frac{6}{5}\bar{R}$.*

The proof of Lemma 3 is similar to that of Lemma 2. We omit this proof due to space limitations of the issue.

4   Conclusion

Statement 2, Lemmas 1–3 imply the following

Theorem 1. *For any instance I of $RO2|triangle|R_{max}$ there exists a feasible schedule S with makespan from interval $[\bar{R}, \frac{6}{5}\bar{R}]$. Such schedule can be found in linear time.*

Indeed, we just need to perform the jobs' aggregation procedure described in proof of Statement 2 to obtain modification $\tilde{I}$, then use the proof of correspondent Lemma according to the existence and location of overloaded node to build a feasible schedule for $\tilde{I}$ with makespan from $[\bar{R}, \frac{6}{5}\bar{R}]$, and finally transform that schedule into the feasible schedule for initial instance, treating each aggregated operation as a block of initial operations performed without idles in arbitrary order.

Note that the interval $[\bar{R}, \frac{6}{5}\bar{R}]$ is a tight optima localization interval for $RO2|triangle|R_{max}$ problem as it is for its special case $RO2|link|R_{max}$ [2].

The most important question still to be investigated is the following

Open Question 1. *What is the smallest value ρ such that interval $[\bar{R}, \rho\bar{R}]$ is an optima localization interval for $RO2||R_{max}$ problem?*

The second interesting question concerns the possibility of generalizing results for $RO2|link|R_{max}$ from [7] to our case $RO2|triangle|R_{max}$. Those results (polynomially solvable subcases and an FPTAS) are based on the properties of the Gonzalez-Sáhni algorithm for two-machine open shop problem [6].

The Gonzalez-Sáhni algorithm consists of three main steps.

Step 1. Separate all jobs from $\mathcal{J}$ into two subsets:

$$\mathcal{J}_{\leqslant} \doteq \{J_j \in \mathcal{J}|a_j \leqslant b_j\} \text{ and } \mathcal{J}_{>} \doteq \{J_j \in \mathcal{J}|a_j > b_j\}.$$

Step 2. Choose the *diagonal* job J_r such that the maximum

$$\max\{\max\{a_j|J_j \in \mathcal{J}_{\leqslant}\}, \max\{b_j|J_j \in \mathcal{J}_{>}\}\}$$

is reached at J_r. Without lost of generality $J_r \in \mathcal{J}_{\leqslant}$.

Step 3. Sequence operations of machine A in an arbitrary order such that operations of jobs from $\mathcal{J}_{\leqslant} \setminus \{J_r\}$ precede operations of jobs from $\mathcal{J}_{>}$ and a_r is the last operation processed by A. Operations of machine B are sequenced in the same order except for b_r which is processed first.

Also note that if $d_r \geqslant \ell_{max}$ then sequence for machine A at Step 3 can be arbitrary providing that a_r is the last operation of A.

Two theorems from [7] state the following:

For any instance I of $RO2|link|R_{max}$

1. *if $J_r \in \mathcal{J}^0$ then optimum of I equals to $\bar{R}$ and can be found in linear time;*
2. *if $J_r \in \mathcal{J}^1$ and $d_r \geqslant \ell_{max}$ then optimum of I equals to $\bar{R}$ and can be found in linear time;*
3. *otherwise a schedule for I of makespan $\ell_{max} + 2T^*$ can be built in linear time.*

Similar technique can be used to prove the following

Lemma 4. *For any instance* I *of* $RO2|triangle|R_{\max}$

1. *if* $J_r \in \mathcal{J}^0$ *then a schedule for* I *of makespan* $\ell_{\max} + 2(\tau + \nu)$ *can be built in linear time;*
2. *if* $J_r \in \mathcal{J}^1$ *and* $d_r \geqslant \ell_{\max}$ *then optimum of* I *equals to* $\bar{R}$ *and can be found in linear time;*
3. *otherwise a schedule for* I *of makespan* $\ell_{\max} + 2T^*$ *can be built in linear time.*

As we see the first case $(J_r \in \mathcal{J}^0)$ resolves differently for those two problems. Therefore the technique from [7] will not help us to get an FPTAS for $RO2|triangle|R_{\max}$ that easily. This observation leads us to the following

Open Question 2. *Does an FPTAS for* $RO2|triangle|R_{\max}$ *exist?*

References

1. Averbakh, I., Berman, O., Chernykh, I.: The routing open-shop problem on a network: complexity and approximation. Eur. J. Oper. Res. **173**(2), 521–539 (2006)
2. Averbakh, I., Berman, O., Chernykh, I.: A 6/5-approximation algorithm for the two-machine routing open shop problem on a 2-node network. Eur. J. Oper. Res. **166**(1), 3–24 (2005)
3. Chernykh, I., Kononov, A., Sevastyanov, S.: Efficient approximation algorithms for the routing open shop problem. Comput. Oper. Res. **40**(3), 841–847 (2013)
4. Chernykh, I., Kuzevanov, M.: Sufficient condition of polynomial solvability of two-machine routing open shop with preemption allowed. Intellektual'nye sistemy **17**(1–4), 552–556 (2013). (in Russian)
5. Christofides, N.: Worst-case analysis of a new heuristic for the travelling salesman problem. Report 388, Graduate School of Industrial Administration, Carnegie-Mellon University, Pittsburg, PA (1976)
6. Gonzalez, T., Sahni, S.: Open shop scheduling to minimize finish time. J. Assoc. Comput. Mach. **23**, 665–679 (1976)
7. Kononov, A.V.: On the routing open shop problem with two machines on a two-vertex network. J. Appl. Ind. Math. **6**(3), 318–331 (2012). ISSN: 1990–4789
8. Lawler, E.L., Lenstra, J.K., Rinnooy Kan, A.H.G., Shmoys, G.B.: Sequencing and scheduling: algorithms and complexity. In: Logistics of Production and Inventory, pp. 445–452. Elsevier, Amsterdam (1993)
9. Pyatkin, A.V., Chernykh, I.D.: The open shop problem with routing at a two-node network and allowed preemption. J. Appl. Ind. Math. **6**(3), 346–354 (2012). ISSN: 1990–4789
10. Serdyukov, A.: On some extremal routes in graphs. Upravlyaemye Sistemy **17**, 76–79 (1978). (in Russian)
11. Sevastianov, S.V., Tchernykh, I.D.: Computer-aided way to prove theorems in scheduling. In: Bilardi, G., Pietracaprina, A., Italiano, G.F., Pucci, G. (eds.) ESA 1998. LNCS, vol. 1461, pp. 502–513. Springer, Heidelberg (1998)

Mixed Integer Programming Approach to Multiprocessor Job Scheduling with Setup Times

Anton V. Eremeev$^{(\boxtimes)}$ and Yulia V. Kovalenko

Sobolev Institute of Mathematics, 4, Akad. Koptyug Avenue,
630090 Novosibirsk, Russia
eremeev@ofim.oscsbras.ru, julia.kovalenko.ya@yandex.ru

Abstract. Multiprocessor jobs require more than one processor at the same moment of time. We consider two basic variants of scheduling multiprocessor jobs with various regular criteria. In the first variant, for each job the number of required processors is given and fixed, and the job can be processed on any subset of parallel processors of this size. In the second variant, the subset of dedicated processors required by a job is given and fixed. A sequence dependent setup time is needed between different jobs. We formulate mixed integer linear programming models based on a continuous time representation for the NP-hard scheduling problems under consideration. Using these models, we identify new polynomially solvable cases with the number of jobs bounded above by a constant.

Keywords: Multiprocessor job · Setup time · Integer linear programming · Polynomial solvability

1 Introduction

We consider the multiprocessor scheduling problem, where a set of k jobs $\mathcal{J} = \{1, \ldots, k\}$ has to be executed by m processors such that each processor can work on at most one job at a time, and each job must be processed simultaneously by several processors. Let $\mathcal{M} = \{1, \ldots, m\}$ and let p_j denote the processing time of job j for each $j \in \mathcal{J}$. Dedicated and parallel variants of the problem are studied here. In the first variant, there is a size $size_j$ associated with each job $j \in \mathcal{J}$ indicating that the job can be processed on any subset of parallel processors of the given size. In the second variant, each job $j \in \mathcal{J}$ requires a simultaneous use of a pre-specified subset (mode) fix_j of dedicated processors. Following the traditional definitions in scheduling theory [1] we consider rigid jobs in the first variant and single mode multiprocessor jobs in the second variant.

A sequence dependent setup time is required to switch a processor from one job to another. For the parallel model of the problem let $s_{jj'}$ be the non-negative setup time from job j to job j', where j, $j' \in \mathcal{J}$. For the dedicated variant of the problem let $s^l_{jj'}$ denote the setup time from job j to job j' on processor l,

© Springer International Publishing Switzerland 2016
Y. Kochetov et al. (Eds.): DOOR 2016, LNCS 9869, pp. 298–308, 2016.
DOI: 10.1007/978-3-319-44914-2_24

$l \in \mathcal{M}$, j, $j' \in F_l$, where $F_l = \{j \in \mathcal{J} : l \in fix_j\}$ is the set of jobs that use processor $l \in \mathcal{M}$.

Following the traditional thee-field notation for scheduling problems [1], we denote preemptive version of the problem with single mode multiprocessor jobs (rigid jobs) as $P|fix_j, pmtn, s^l_{jj'}|\gamma$ ($P|size_j, pmtn, s_{jj'}|\gamma$) and non-preemptive version of the problem is denoted by $P|fix_j, s^l_{jj'}|\gamma$ ($P|size_j, s_{jj'}|\gamma$). Here γ specifies an objective function. It is assumed that the subset of processors used by a rigid job can be changed at runtime in preemptive scheduling.

We consider four widely used objective functions to be minimized: the makespan, $C_{\max} = \max_{j \in \mathcal{J}} C_j$, the maximum lateness, $L_{\max} = \max_{j \in \mathcal{J}}(C_j - d_j)$, the sum of completion times, $C_{\sum} = \sum_{j \in \mathcal{J}} C_j$, the sum of latenesses, $L_{\sum} = \sum_{j \in \mathcal{J}}(C_j - d_j)$, where C_j denote the completion time of job $j \in \mathcal{J}$, and d_j is the due date of job $j \in \mathcal{J}$, i.e. the time by which job j should be completed.

In practice one often may assume that the setup times satisfy the triangle inequality:

$$s^l_{j'',j'} \leq s^l_{j'',j} + s^l_{j,j'}, \ j, \ j', \ j'' \in F_l, \ l \in \mathcal{M} \text{ (for dedicated processors)}, \quad (1)$$

$$s_{j'',j'} \leq s_{j'',j} + s_{j,j'}, \ j, \ j', \ j'' \in \mathcal{J} \text{ (for parallel processors)}. \quad (2)$$

We denote this special case by placing Δ in front of $s^l_{jj'}$ (or $s_{jj'}$) in the second field of the three-field notation. In the special case where the number of machines $m = 1$ is not a part of the input, there is no difference between rigid jobs and single mode multiprocessor jobs and the problem notation simplifies to $1|pmtn, \Delta s_{jj'}|\gamma$ and $1|\Delta s_{jj'}|\gamma$ for preemptive and non-preemptive jobs respectively.

All mentioned above problems $P|fix_j, pmtn, \Delta s^l_{jj'}|\gamma$, $P|fix_j, \Delta s^l_{jj'}|\gamma$, $P|size_j, pmtn, \Delta s_{jj'}|\gamma$ and $P|size_j, \Delta s_{jj'}|\gamma$ with $\gamma \in \{C_{\max}, L_{\max}, C_{\sum}, L_{\sum}\}$ are NP-hard even in the single-machine case as implied by the following proposition.

Proposition 1. *Problems $1|pmtn, \Delta s_{jj'}|\gamma$ and $1|\Delta s_{jj'}|\gamma$ with $\gamma \in \{C_{\max}, L_{\max}, C_{\sum}, L_{\sum}\}$ are strongly NP-hard.*

Proposition 1 may be attributed to the "folklore". However for the sake of completeness we provide its proof in the appendix.

In [2], the problem of scheduling multiprocessor jobs with sequence-dependent setup times was combined with a lot-sizing problem where it is required to produce a set of the products in demanded volumes. In this setting, the multiprocessor jobs were called *multi-machine technologies*, each technology engaging a number of machines simultaneously to produce a batch of some product. In [2], this problem with $C_{\max}$ criterion was shown to be hard to approximate and new NP-hard and polynomially solvable special cases were identified.

A MIP model was proposed by Shaik et al. in [10] for a problem of scheduling multi-machine technologies with sequence-dependent setup times for continuous production plants. Besides that, a decomposition method was developed in [10] to solve real-life problems from chemical industry where straightforward application of the MIP model was impractical.

A survey of results on multiprocessor jobs scheduling in the case of zero setup times is provided by Drozdowski in [1]. It is known that in this case problem $P|size_j, pmtn|C_{max}$ with the number of processors bounded above by a constant (denoted $Pm|size_j, pmtn|C_{max}$) and problem $P|fix_j, pmtn|C_{max}$ with only two-processor jobs are both polynomially solvable. This result is based on the fact that the considered special cases may be treated as linear programming (LP) problems using the so-called *configurations* introduced by Jansen and Porkolab [7,8] and resembling *patterns* proposed by Kantorovich and Zalgaller for the one-dimensional cutting-stock problem in the middle of 20-th century (see e.g. [9]). In [7,8] a configuration is defined as a set of jobs which may be processed simultaneously. The total number of configurations is $O(k^m)$ and the resulting LP problems contain $O(k^m)$ variables, each one representing the time of using the corresponding configuration in a schedule, and $O(k)$ constraints. These problems may be solved in polynomial time by considering the dual LP problems and applying the ellipsoid method of Grötschel, Lovász and Schrijver [4].

Unfortunately the configurations-based approach can not be extended to the case of sequence-dependent setup times because in the general case evaluation of objective function requires not only the durations but also the sequence of jobs on each machine. In the present paper, we develop the MIP models for multiprocessor jobs scheduling with sequence-dependent setup times using the notion of *event points*, which was originally proposed in the context of single-processor jobs by Ierapetritou and Floudas in [6]. In the case of multiprocessor jobs, an event point, as well as a configuration, corresponds to some set of compatible jobs, but in contrast to the set of jobs of a configuration (which is defined a priori) the set of jobs of an event point is defined by the values of Boolean variables of this event point. The presence of Boolean variables allows to account for the sequence dependent setup times. Besides that, unlike the jobs of a configuration, the jobs of an event point may have different starting and completion times.

The MIP models proposed on the basis of event points are described in Sect. 2. Using these models, the new polynomially solvable cases with sequence-dependent setup times are identified in Sect. 3. The concluding remarks are provided in Sect. 4.

2 Mixed Integer Linear Programming Models

2.1 Single Mode Multiprocessor Jobs

Let us define the notion of *event points* analogously to [2,6]. By event point we will mean a subset of variables in mixed integer linear programming (MIP) model, which characterize a selection of a certain set of jobs and their starting and completion times. In one event point each processor may be utilized in at most one job. The set of all event points will be denoted by $N = \{1, \ldots, n_{max}\}$, where the parameter n_{max} is chosen sufficiently large on the basis of a prior estimates or preliminary experiments.

The structure of the schedule is defined by the Boolean variables w_{jn} such that $w_{jn} = 1$ if job j is executed in event point n, and $w_{jn} = 0$ otherwise.

In case job j is executed in event point n, the staring time and the completion time of job j in this event point are given by the real-valued variables T_{jn}^{st} and T_{jn}^{f} accordingly.

Let H be an upper bound on the schedule length,

$$H := \sum_{j \in \mathcal{J}} p_j + (k-1) \cdot \max_{l \in \mathcal{M},\ j \neq j' \in F_l} \{s_{jj'}^l\}.$$

Then the set of feasible solutions for problem $P|fix_j, pmtn, s_{jj'}^l|\gamma$ is defined as follows

$$\sum_{j \in F_l} w_{jn} \leqslant 1, \ l \in \mathcal{M}, \ n \in N, \tag{3}$$

$$T_{jn}^f \geqslant T_{jn}^{st}, \ j \in \mathcal{J}, \ n \in N, \tag{4}$$

$$T_{jn}^{st} \geqslant T_{j'n'}^f + s_{j'j}^l - H(2 - w_{jn} - w_{j'n'} + \sum_{\tilde{j} \in F_l} \sum_{n' < \tilde{n} < n} w_{\tilde{j}\tilde{n}}), \tag{5}$$

$$l \in \mathcal{M}, \ j, j' \in F_l, \ n, n' \in N, \ n \neq 1, n' < n,$$

$$T_{jn}^f - T_{jn}^{st} \leqslant w_{jn} \cdot p_j, \ j \in \mathcal{J}, \ n \in N, \tag{6}$$

$$\sum_{n \in N} \frac{T_{jn}^f - T_{jn}^{st}}{p_j} \geqslant 1, \ j \in \mathcal{J}, \tag{7}$$

$$w_{jn} \in \{0, 1\}, \ T_{jn}^{st} \geqslant 0, \ j \in \mathcal{J}, \ n \in N. \tag{8}$$

Constraint (3) implies that in any event point on processor l at most one job may be executed. Constraint (5) indicates that the starting time of job j on processor l should not be less than the completion time of a preceding job on the same processor, plus the setup time. Constraint (4) guarantees that all jobs may be performed only for non-negative time. If a job j is not executed in the event point n (i.e. $w_{jn} = 0$) then its duration should be zero – this is ensured by inequality (6). Constraint (7) implies that each job $j \in \mathcal{J}$ is entirely executed. Constraints (8) give the area where the variables are defined.

The set of feasible solutions for problem $P|fix_j, s_{jj'}^l|\gamma$ may be obtained from (3)–(8) by adding the inequality

$$\sum_{n \in N} w_{jn} \leqslant 1, \ j \in \mathcal{J}, \tag{9}$$

which ensures each job is executed without preemptions.

The optimization criteria for presented models are formulated in the following form.

1. Makespan

$$C_{\max} \to \min,$$

$$C_{\max} \geqslant T_{jn}^f, \ j \in \mathcal{J}, \ n \in N.$$

2. Sum of completion times

$$\sum_{j \in \mathcal{J}} T_j^f \to \min,$$

$$T_j^f \geqslant T_{jn}^f, \; j \in \mathcal{J}, \; n \in N.$$

3. Maximum lateness

$$L_{\max} \to \min,$$

$$L_{\max} \geqslant T_{jn}^f - d_j, \; j \in \mathcal{J}, \; n \in N.$$

4. Sum of latenesses

$$\sum_{j \in \mathcal{J}} L_j \to \min,$$

$$L_j \geqslant T_{jn}^f - d_j, \; j \in \mathcal{J}, \; n \in N.$$

2.2 Rigid Jobs

MIP models for problems with rigid jobs are constructed based on the same principles as the previous models. However, in this case, if only the jobs are allocated in the event points, then a problem of assignment of the jobs to processors arises. Mainly, this assignment is needed to calculate the setup times between jobs on processors. The following proposition shows that such assignment problems with criteria $C_{\max}$ and $L_{\max}$ are NP-hard.

Proposition 2. *Suppose a family of subsets of jobs $\{\mathcal{J}_1, \ldots, \mathcal{J}_{n_{\max}}\}$ is given as a part of the problem input. Then problems $P2|size_j, pmtn, s_{jj'}|\gamma$ and $P2|size_j, s_{jj'}|\gamma$ with $\gamma \in \{C_{\max}, L_{\max}\}$, under additional constraint that n-th job processed on each machine belongs to $\mathcal{J}_n$, $n = 1, \ldots, n_{\max}$, are NP-hard.*

Proof. The hardness of considered problems can be shown by a polynomial reduction of ORDERED PARTITION problem, which is known to be NP-complete [3]. ORDERED PARTITION problem is formulated as follows: Let an ordered set $A = \{a_1, a_2, \ldots, a_{2k_0}\}$ be given. A positive integer e_i is associated with each element $a_i \in A$, $i = 1, \ldots, k$, such that $\sum_{a_i \in A} e_i = 2E$. ORDERED PARTITION problem asks if there exists a partition of A into two subsets A_1 and A_2 such that $\sum_{a_i \in A_1} e_i = \sum_{a_i \in A_2} e_i = E$, $|A_1| = |A_2| = k_0$ and set A_1 includes exactly one element from each pair $a_{2i-1}, a_{2i}, i = 1, \ldots, k_0$.

For brevity we will denote $P|size_j, pmtn, s_{jj'}|C_{\max}$ and $P|size_j, s_{jj'}|C_{\max}$ problems under additional constraint that n-th job processed on each machine belongs to $\mathcal{J}_n$, $n = 1, \ldots, n_{\max}$, by P1 and P2 respectively.

We reduce an ORDERED PARTITION instance to instances of problems P1 and P2 as follows. Put the number of jobs $k := 2k_0$; the number of processors $m := 2$; $size_j := 1$, $p_j := e_j$ and $s_{jj'} := 0$ for all $j \neq j' \in \mathcal{J}$. Besides that we define a family of subsets of jobs assuming $n_{\max} := k_0$ and $\mathcal{J}_n := \{2n - 1, 2n\}$ for $n = 1, \ldots, k_0$.

Consider the decision versions of problems P1 and P2 which ask if there is a schedule with makespan $C_{\max} \leq K$ for a given K.

Note that in the instances of P1 and P2 defined above, each job j belongs only to one subset $\mathcal{J}_n$ and $\sum_{j \in \mathcal{J}} p_j = 2E$. Hence, in a schedule with $C_{\max} \leq K$, assuming $K := E$, all jobs are executed without preemptions on one of the two processors, because overall available time on two processors does not exceed $2E$. Therefore, a positive answer to an instance of decision problem P1 or P2 implies a positive answer to the ORDERED PARTITION problem and vice versa.

In the case of criterion $L_{\max}$, the statement of the proposition holds because criteria $L_{\max}$ and $C_{\max}$ are equivalent when $d_j = 0$ for all $j \in \mathcal{J}$.   □

In view of Proposition 2, in the case of rigid jobs we formulate our MIP models in such a way that both jobs and processors, on which they are executed, are assigned in the event points.

The structure of the schedule is also defined by the Boolean variables w_{jn} and the real-valued variables T_{jn}^{st} and T_{jn}^{f}, which have the same meaning as in the case of single mode multiprocessor jobs. Moreover, we include additional Boolean variables z_{jln} such that $z_{jln} = 1$ if job j is executed in event point n and uses processor l, and $z_{jln} = 0$ otherwise.

Let H be an upper bound on schedule length. It suffices to put

$$H = \sum_{j \in \mathcal{J}} p_j + (k-1) \cdot \max_{j \neq j'} \{s_{jj'}\}.$$

Based on the above remarks and variables, the set of feasible solutions for problem $P|size_j, pmtn, s_{jj'}|\gamma$ is defined by the following constraints:

$$\sum_{j \in \mathcal{J}} z_{jln} \leqslant 1, \; l \in \mathcal{M}, \; n \in N, \tag{10}$$

$$\sum_{l \in \mathcal{M}} z_{jln} = size_j \cdot w_{jn}, \; j \in \mathcal{J}, \; n \in N, \tag{11}$$

$$T_{jn}^{f} \geqslant T_{jn}^{st}, \; j \in \mathcal{J}, \; n \in N, \tag{12}$$

$$T_{jn}^{st} \geqslant T_{j'n'}^{f} + s_{j'j} - H(2 - z_{jln} - z_{j'ln'} + \sum_{\tilde{j} \in \mathcal{J}} \sum_{n' < \tilde{n} < n} z_{\tilde{j}l\tilde{n}}), \tag{13}$$

$$l \in \mathcal{M}, \; j \neq \tilde{j} \in \mathcal{J}, \; n, n' \in N, \; n \neq 1, n' < n, \tag{14}$$
$$T_{jn}^{f} - T_{jn}^{st} \leqslant w_{jn} \cdot p_j, \; j \in \mathcal{J}, \; n \in N,$$

$$\sum_{n \in N} \frac{T_{jn}^{f} - T_{jn}^{st}}{p_j} \geqslant 1, \; j \in \mathcal{J}, \tag{15}$$

$$z_{jln} \in \{0,1\}, \; w_{jn} \in \{0,1\}, \; T_{jn}^{st} \geqslant 0, \; j \in \mathcal{J}, \; l \in \mathcal{M}, \; n \in N. \tag{16}$$

Constraints (10), (12)–(15) have the same interpretation as in the model for problem $P|fix_j, pmtn, s_{jj'}^l|\gamma$. Constraint (11) guarantees that job j uses exactly $size_j$ processors if it is executed in the event point n (i.e. $w_{jn} = 1$).

The set of feasible solutions for problem $P|size_j, s_{jj'}|\gamma$ may be obtained from (10)–(16) by adding inequality (9). The optimization criteria are modeled as in the case of dedicated processors.

3 Polynomially Solvable Cases

New polynomially solvable special cases with non-zero setup times are found using proposed MIP models, under assumption that the number of jobs is bounded by a constant. An instance of a multiprocessor job scheduling problem is reduced to a number of instances of a linear programming problem, obtained from the MIP model assigning some fixed values to Boolean variables.

3.1 Single Mode Multiprocessor Jobs

In order to find an optimal solution to $P|fix_j, s_{jj'}^l|\gamma$ using model (3)–(9), it is sufficient to assume $n_{\max} = k$ because the preemptions are not allowed. Denote $\mathcal{P}_{fix}$ the linear programming problem obtained by fixing all Boolean variables (w_{jn}) in model (3)–(9) supplemented by a linear programming formulation of optimization criterion γ. Here and below by *fixing* of the variables we assume assignment of some fixed values to them (which turns these variables into parameters). Problem $\mathcal{P}_{fix}$ with $n_{\max} = k$ involves a polynomially bounded number of variables and constraints, which means it is polynomially solvable (see e.g. [4]).

Let τ_{fix} be an upper bound on the time complexity of solving problem $\mathcal{P}_{fix}$. The problem $P|fix_j, s_{jj'}^l|\gamma$, where the number of jobs is bounded from above by a constant, we denote by $P|fix_j, s_{jj'}^l, k = \mathrm{const}|\gamma$. This problem reduces to $(n_{\max})^k$ problems of $\mathcal{P}_{fix}$ type with $n_{\max} = k$. Therefore the following theorem holds.

Theorem 1. *Problem* $P|fix_j, s_{jj'}^l, k = \mathrm{const}|\gamma$, $\gamma \in \{C_{\max}, C_{\sum}, L_{\max}, L_{\sum}\}$, *is polynomially solvable within* $O(\tau_{fix} \cdot k^k)$ *time.*

To find an optimal solution to $P|fix_j, pmtn, \Delta s_{jj'}^l|\gamma$ problem, it suffices to set $n_{\max} = k^m$ in model (3)–(8). Indeed, the number of different sets of jobs that may be executed simultaneously does not exceed k^m. Besides that, there exists an optimal solution to problem $P|fix_j, pmtn, \Delta s_{jj'}^l|\gamma$ where each of the above mentioned sets of jobs is executed simultaneously at most once. This fact follows by the lot shifting technique which is applicable here since the setup times obey the triangle inequality (see e.g. [11]).

Let $\mathcal{P}'_{fix}$ denote the linear programming problem obtained by fixing all Boolean variables (w_{jn}) in MIP model (3)–(8) supplemented by a linear programming formulation of optimization criterion γ. A problem $\mathcal{P}'_{fix}$ with $n_{\max} = k^m$ and the number of processors bounded above by a constant is polynomially solvable. Let τ'_{fix} denote an upper bound of the time complexity of solving $\mathcal{P}'_{fix}$. The problem $P|fix_j, pmtn, \Delta s^l_{jj'}|\gamma$, where the numbers of processors and jobs are bounded by a constant is denoted by $Pm|fix_j, pmtn, \Delta s^l_{jj'}, k = \text{const}|\gamma$ in what follows. This problem reduces to $2^{kn_{\max}}$ problems of $\mathcal{P}'_{fix}$ type, where $n_{\max} = km$. So the following result holds.

Theorem 2. *Problem $Pm|fix_j, pmtn, \Delta s^l_{jj'}, k = \text{const}|\gamma$, $\gamma \in \{C_{\max}, C_\Sigma, L_{\max}, L_\Sigma\}$, is polynomially solvable within $O\left(\tau'_{fix} \cdot 2^{k^{m+1}}\right)$ time.*

In some works it is assumed that a job has a number of alternative modes, where each processing mode is specified by a subset of processors and the execution time of the job on that particular processor set. Such jobs are called *multimode multiprocessor* jobs [1]. Our MIP models and polynomially solvable cases for single mode multiprocessor jobs may be extended to the scheduling problem with multimode multiprocessor jobs and various regular criteria, which can be formulated in terms of linear programming.

3.2 Rigid Jobs

In order to find an optimal solution to $P|size_j, s_{jj'}|\gamma$ using model (9)–(16), it is sufficient to set $n_{\max} = k$ because the preemptions are not allowed. Denote by $\mathcal{P}_{size}$ the linear programming problem obtained by fixing all Boolean variables (w_{jn}) and (z_{jln}) in model (9)–(16) with a linear formulation of optimization criterion γ. Problem $\mathcal{P}_{size}$ with $n_{\max} = k$ involves $O(k^2)$ variables and $O(k^4 m)$ constraints, then it is pseudopolynomially solvable, since m is the numerical parameter of the problem. Let τ_{size} be an upper bound on the time complexity of solving problem $\mathcal{P}_{size}$.

The problem $P|size_j, s_{jj'}|\gamma$, where the number of jobs is bounded by a constant from above, is denoted by $P|size_j, s_{jj'}, k = \text{const}|\gamma$. This problem reduces to $O\left(n^k_{\max} \prod_{j=1}^{k} C^{size_j}_m\right)$ problems of $\mathcal{P}_{size}$ type with $n_{\max} = k$. Problem $P|size_j, s_{jj'}, k = \text{const}|\gamma$ is polynomially solvable in $O\left(\tau_{size} \cdot k^k \prod_{j=1}^{k} C^{size_j}_m\right)$ time, if $m \leqslant \sum_{j=1}^{k} size_j$ and sizes $size_j$ are bounded by a constant for all $j \in \mathcal{J}$, and the problem is trivial, if $m > \sum_{j=1}^{k} size_j$. In the latter case, all jobs start at time moment $t = 0$ in the early schedule. Therefore the following theorem holds.

Theorem 3. *Problem $P|size_j, s_{jj'}, k = \text{const}|\gamma$, $\gamma \in \{C_{\max}, C_\Sigma, L_{\max}, L_\Sigma\}$, is polynomially solvable, when parameters $size_j$ are bounded by a constant for all $j \in \mathcal{J}$*

To find an optimal solution to $P|size_j, pmtn, \Delta s_{jj'}|\gamma$ problem, it suffices to set $n_{\max} = k^m$ in model (10)–(16) as in the case of dedicated processors (see

Sect. 3.1). Let $\mathcal{P}'_{size}$ be the linear programming problem obtained by fixing all Boolean variables (w_{jn}) and (z_{jln}) in MIP model (10)–(16) supplemented by a linear formulation of optimization criterion γ. Denote by τ'_{size} an upper bound of the time complexity of solving $\mathcal{P}'_{size}$. The problem $P|size_j, pmtn, \Delta s_{jj'}|\gamma$, where the numbers of processors and jobs are bounded by a constant, we denote by $Pm|size_j, pmtn, \Delta s_{jj'}, k = \text{const}|\gamma$. This problem reduces to $2^{kmn_{max}}$ problems of $\mathcal{P}'_{size}$ type, where $n_{max} = k^m$. Thus, we have

Theorem 4. *Problem* $Pm|size_j, pmtn, \Delta s_{jj'}, k = \text{const}|\gamma$, $\gamma \in \{C_{max}, C_\Sigma, L_{max}, L_\Sigma\}$, *is polynomially solvable within* $O\left(\tau'_{size} \cdot 2^{mk^{m+1}}\right)$ *time.*

Let us assume that there is a set of usable processor numbers for each job $j \in \mathcal{J}$. Then the jobs are called *moldable* jobs [1], if the number of required processors is chosen before starting a job and is not changed until the job termination. Jobs are called *malleable* [1], if the number of processors can be changed at runtime. The MIP models and polynomially solvable cases presented above for the case of rigid jobs may be generalized to the scheduling problems with moldable and malleable jobs.

4   Conclusions

The problem of multiprocessor job scheduling is studied in parallel and dedicated versions. MIP models are formulated for both versions of the problem using the event-points approach and continuous time representation. New polynomially solvable special cases of the problem are found using the MIP models, under assumption that the number of jobs is bounded by a constant.

Presented models and polynomially solvable cases are extended to the other (more general) scheduling problems with moldable jobs, malleable jobs, multimode multiprocessor jobs and various regular criteria, which can be formulated in linear form.

Acknowledgements. This research is supported by the Russian Science Foundation grant 15-11-10009. The authors are grateful to M.Y. Kovalyov for helpful remarks.

Appendix: The Proof of Proposition 1

Proposition 1. *Problems* $1|pmtn, \Delta s_{jj'}|\gamma$ *and* $1|\Delta s_{jj'}|\gamma$ *with* $\gamma \in \{C_{max}, L_{max}, C_\Sigma, L_\Sigma\}$ *are strongly NP-hard.*

Proof. We will consider only C_Σ criterion since the problems with other three criteria are treated analogously.

In [5] it is proven that recognition of grid graphs with a Hamiltonian path (the HAMILTON PATH PROBLEM) is NP-complete. Recall that a graph $G' = (V', E')$ with vertex set V' and edge set E' is called a *grid* graph, if its vertices are the integer vectors $v = (x_v, y_v) \in \mathbf{Z}^2$ on plane, i.e., $V' \subset \mathbf{Z}^2$, and a pair of vertices is connected by an edge iff the Euclidean distance between them is equal to 1. Here and below, $\mathbf{Z}$ denotes the set of integer numbers. We can assume that graph G' is connected since otherwise Hamiltonian path does not exist and this can be recognized in polynomial time.

Let us first reduce HAMILTON PATH problem to $1|\Delta s_{jj'}|C_{\sum}$, assuming that jobs correspond to vertices and the setup times are equal to Euclidean distances between the integer points where the corresponding vertices are located. All processing times $p_j = 1$.

Then minimal setup times are equal to one. The earliest completion times of the jobs are $1, 3, 5, 7, \ldots, 2k - 1$, where k is the number of jobs.

In the recognition version of $1|\Delta s_{jj'}|C_{\sum}$ it is required to answer the question: Is there a schedule with $C_{\sum}$ value not greater than a given value K?

Let us put $K := (1 + 3 + 5 + 7 + \ldots + 2k - 1) = k^2$.

On one hand, if a schedule with the value of $C_{\sum}$ at most K exists, then all setups of this schedule are equal to 1 and graph G' contains a Hamilton path. On the other hand, if graph G' contains a Hamilton path then ordering the jobs in the sequence of vertices of this path we obtain a schedule with the value of $C_{\sum} = k^2$.

This reduction is computable in polynomial time and all input parameters of $1|\Delta s_{jj'}|C_{\sum}$ instance are upper bonded by k, so we conclude that $1|\Delta s_{jj'}|C_{\sum}$ problem is strongly NP-hard.

In the case of $1|pmtn, \Delta s_{jj'}|C_{\sum}$ problem we construct the same reduction. Note that in a schedule with $C_{\sum} \leq k^2$ the preemptions are impossible. Indeed, suppose that job j' is the first job that has a preemption and all n preceding jobs have no preemptions. Let job j' be executed for a units of time and then preemption took place and let j'' be the first job, which finishes after this preemption.

In case $j'' \neq j'$, job j'' ends at time $t \geq 2(n+1) - 1 + a$. In this case even if all jobs after j'' finish in the earliest possible times $2(n+2)-1, 2(n+3)-1, \ldots, 2k-1$, then still $C_{\sum} > k^2$.

In case $j'' = j'$, job j' ends at time $t \geq 2(n + 1) - 1 + b$, where b is the total preemption time of job j'. Thus by the same reasoning as in the previous case, we conclude that $C_{\sum} > k^2$. $\square$

References

1. Drozdowski, M.: Scheduling for Parallel Processing, p. 386. Springer, London (2009)
2. Eremeev, A.V., Kovalenko, J.V.: On multi-product lot-sizing and scheduling with multi-machine technologies. In: Lübbecke, M., Koster, A., Letmathe, P., Peis, B., Walther, G. (eds.) Operations Research Proceedings 2014. Operations Research Proceedings, pp. 301–306. Springer, Heidelberg (2016)
3. Garey, M.R., Johnson, D.S.: Computers and Intractability: A Guide to the Theory of NP-Completeness. W.H. Freeman and Company, San Francisco (1979)
4. Grötschel, M., Lovász, L., Schrijver, A.: Geometric Algorithms and Combinatorial Optimizations, 2nd edn, p. 362. Springer, Heidelberg (1993)

5. Itai, A., Papadimitriou, C.H., Szwarcfiter, J.L.: Hamilton paths in grid graphs. SIAM J. Comput. **11**(4), 676–686 (1982)
6. Ierapetritou, M.G., Floudas, C.A.: Effective continuous-time formulation for short-term scheduling: I. multipurpose batch process. Ind. Eng. Chem. Res. **37**, 4341–4359 (1998)
7. Jansen, K., Porkolab, L.: Preemptive parallel task scheduling in $O(n) + poly(m)$ time. In: Lee, D.T., Teng, S.-H. (eds.) ISAAC 2000. LNCS, vol. 1969, pp. 398–409. Springer, Heidelberg (2000)
8. Jansen, K., Porkolab, L.: Preemptive scheduling with dedicated processors: applications of fractional graph coloring. J. Sched. **7**, 35–48 (2004)
9. Mukhacheva, E.A., Mukhacheva, A.S.: L.V. Kantorovich and cutting-packing problems: new approaches to combinatorial problems of linear cutting and rectangular packing. J. Math. Sci. **133**(4), 1504–1512 (2006)
10. Shaik, M.A., Floudas, C.A., Kallrath, J., Pitz, H.J.: Production scheduling of a large-scale industrial continuous plant: short-term and medium-term scheduling. Comput. Chem. Eng. **33**, 670–686 (2009)
11. Tanaev, V.S., Kovalyov, M.Y., Shafransky, Y.M.: Scheduling theory. Group Technologies, Minsk, Institute of Technical Cybernetics NAN of Belarus (1998). (in Russian)

On Speed Scaling Scheduling of Parallel Jobs with Preemption

Alexander Kononov and Yulia Kovalenko[(✉)]

Sobolev Institute of Mathematics,
4, Akad. Koptyug Avenue, 630090 Novosibirsk, Russia
alvenko@math.nsc.ru, julia.kovalenko.ya@yandex.ru

Abstract. Parallel jobs require more than one processor at the same time. We study speed scaling scheduling of parallel jobs with preemption. We propose "almost-exact" algorithms for problems with rigid jobs and single mode two-processor jobs. Based on configuration linear programs, our algorithms obtain an $OPT + \varepsilon$ solution for any fixed $\varepsilon > 0$.

Keywords: Speed scaling · Scheduling · Parallel jobs · NP-hardness · Approximation algorithm

1 Introduction

We are given a set of parallel jobs $\mathcal{J} = \{1, \dots, n\}$, each job $j \in \mathcal{J}$ is specified by its release date r_j, its deadline d_j, its processing volume (work) W_j, and a set of m speed-scalable processors. In our paper we consider two basic variants of scheduling multiprocessor jobs. In the first variant, processing of job j simultaneously requires precisely $size_j$ processors. In the second variant, execution of job j simultaneously requires a prespecified subset fix_j of dedicated processors. Note that the parallel execution of parts of the same job is not allowed. Moreover execution of each job can be interrupted and resumed without incurring any costs or delays. According to the definitions in the literature on scheduling theory we consider rigid tasks (jobs) and single mode multiprocessor tasks [8].

We consider the standart model in speed-scaling in which if a processor runs at a speed s then the energy consumption is s^{α} units of energy per time unit, where $\alpha > 1$ is a constant (practical studies show that $\alpha \leq 3$). We assume that if processors execute the same job simultaneously then all these processors run at the same speed. For each job $j \in \mathcal{J}$, we say that j is alive during the interval $[r_j, d_j]$. Since processors may change their speed, a job j may be completed faster (or slower) than the time W_j it needs to be executed at speed 1. It is supposed that a continuous spectrum of processor speeds is available. The goal is to find a feasible schedule respecting the release dates and deadlines of jobs so that the total energy consumption is minimized.

Minimization of energy consumption under deadline constraints is an important real-time problem for computational systems [9]. The motivation to consider speed scaling scheduling of parallel jobs consists in the fact that some

© Springer International Publishing Switzerland 2016
Y. Kochetov et al. (Eds.): DOOR 2016, LNCS 9869, pp. 309–321, 2016.
DOI: 10.1007/978-3-319-44914-2_25

jobs can not be performed asynchronously on modern computers. Such situation takes place in multi-processor graphics cards, where the memory capacity of one processor is not sufficient, or when CPU and a dedicated processor must run together.

1.1 Related Work

For the preemptive single-processor case, Yao et al. [14] proposed an optimal algorithm for finding a feasible schedule with minimum energy consumption. The case, where there are m parallel processors available and all jobs are of the single-processor type, has been solved optimally in polynomial time when both the preemption and the migration of jobs are allowed [1,3,5]. A schedule is called migratory if a job may be interrupted and resumed on the same or on another processor. We note that the migration of jobs is equivalent to the possibility to execute a parallel job in different modes.

Albers et al. [2] considered the problem on parallel processors where the preemption of jobs is allowed but not their migration. They proved that instances with agreeable deadlines and unit-work jobs are solvable in polynomial time. For general instances with unit-work jobs, they proved that the problem becomes strongly NP-hard and they proposed an $(\alpha^\alpha 2^{4\alpha})$-approximation algorithm. For the case where the jobs have arbitrary processing volumes, the problem was proved to be NP-hard even for instances with common release dates and common deadlines. Albers et al. [1] proposed a $2(2 - \frac{1}{m})^\alpha$-approximation algorithm for instances with common release dates, or common deadlines, and an $(\alpha^\alpha 2^{4\alpha})$-approximation algorithm for instances with agreeable deadlines. Greiner et al. [10] gave a generic reduction transforming an optimal schedule for the problem on parallel processors with migration to a $B_{\lceil \alpha \rceil}$-approximate solution for the problem on parallel processors with preemptions but without migration, where $B_{\lceil \alpha \rceil}$ is the $\lceil \alpha \rceil$-th Bell number. This result holds only when $\alpha \leq m$. Cohen-Addad et al. [6] showed that the problem without migration is APX-hard even for jobs with common life intervals and work volumes in $1, 3, 4$.

Bampis et al. [4] studied the heterogeneous preemptive problem on parallel processors where every processor i has a different speed-to power function, $s^{\alpha(i)}$, and both a life interval and a processing volume of each job are processor dependent. For the migratory variant they proposed a polynomial in problem size and in $\frac{1}{\varepsilon}$ algorithm returning a solution within an additive error ε. They also proposed an $(1 + \varepsilon)^\alpha \tilde{B}_\alpha$-approximation algorithm for the nonmigratory variant of the problem, where $\tilde{B}_\alpha$ is the generalized Bell number [4].

To the best of our knowledge no one considered the speed scaling scheduling of parallel jobs. For more information on scheduling problems with parallel jobs, we refer the reader to the survey book by Drozdowski [8]. Following the traditional three-field notation for scheduling problems [4,8] the speed scaling problem with rigid jobs is denoted by $P|size_j, pmtn, r_j, d_j|E$ and the speed scaling problem with single mode multiprocessor jobs is denoted by $P|fix_j, pmtn, r_j, d_j|E$.

1.2   Our Results

In this paper, we present two algorithms for problems with rigid jobs and single mode two-processor jobs. We formulate the problems as a *configuration linear program* (LP) with an exponential number of variables and a polynomial number of constraints. Surprisingly we use the same configuration LP for both problems. First, we prove that for any $\varepsilon > 0$ there is a feasible solution of the LP with energy consumption at most $OPT + \varepsilon$, where OPT is an optimal solution of the original problem. Then, we consider the dual LP and we show how to apply the Ellipsoid algorithm to it and obtain an optimal solution for the primal LP. For this purpose, we provide two separation oracles, one for the problem with rigid jobs and one for the problem with single mode two-processor jobs, i.e. we present two algorithms which given a solution for the dual LP decide if this solution is feasible or otherwise identify a violated constraint. More precisely, we get the following results. Given an instance of $P|size_j, pmtn, r_j, d_j|E$ with m processors, we can solve the Separation Problem for $P|size_j, pmtn, r_j, d_j|E$ in time polynomial in m, $1/\varepsilon$ and the size of the input. Thus we have a polynomial time algorithm if m is fixed and a pseudo-polynomial time algorithm if m is a part of the input. Given an instance of $P|fix_j, pmtn, r_j, d_j|E$ with $|fix_j| = 2$ for all jobs, we can solve the Separation Problem in time polynomial in $1/\varepsilon$ and the size of the input. As we can compute an optimal solution for the dual LP, we can also find an optimal solution for the primal LP by solving it with the variables corresponding to the constraints that were found to be violated during the run of the ellipsoid algorithm and setting all other primal variables to zero. Thus we get two our main results.

Theorem 1. *A schedule of energy consumption $OPT + \varepsilon$ can be found for the speed scaling problem with rigid jobs in time polynomial in m, $1/\varepsilon$ and the input size.*

Theorem 2. *A schedule of energy consumption $OPT + \varepsilon$ can be found for the speed scaling problem with single mode two-processor jobs in time polynomial in $1/\varepsilon$ and the input size.*

In the paper we assume that a continuous spectrum of processor speeds is available. If only a finite set of discrete speed levels is available, then our algorithm finds an optimal solution in time polynomial in size of the instance only.

We present our main results and their generalizations in Sect. 2. In Sect. 3, we discuss a connection between the computational complexity of the speed scaling problems and the computational complexity of the scheduling problems with $C_{\max}$ and $L_{\max}$ criteria.

2   "Almost-Exact" Algorithms

In this section, we present an algorithm for the speed scaling scheduling of rigid jobs or single mode two-processor jobs with preemption. Using the approach from [4], we formulate the problem as a configuration linear program, with an exponential number of variables and a polynomial number of constraints, and we show how to obtain an $OPT + \varepsilon$ solution with the Ellipsoid algorithm in time polynomial in the input size and $1/\varepsilon$, where OPT is an optimal solution of the problem and $\varepsilon > 0$.

A configuration c is a one-to-one feasible assignment of n_c, $1 \leq n_c \leq m$, jobs to the m processors joined with an assignment of a speed value for every processor. We denote by $\mathcal{C}$ the set of all possible configurations. A schedule for our problem has to specify exactly one configuration at each time t. The cardinality of $\mathcal{C}$ is unbounded, since the speeds of processors may accept any real values. At first, we discretize the possible speed values and consider only a finite number of speeds at which the processors can run.

To discretize the speeds, we define a lower and an upper bound on the speed of any processor in an optimal schedule. For the lower bound, consider a job $j \in \mathcal{J}$. Note that the processing time of j in any feasible schedule is at most $(d_j - r_j)$. The convexity of the speed-to-power function implies that a lower bound on the speed of every processor is greater than or equal to the minimum density among all the jobs, i.e., $S_{LB} = \min_{j \in \mathcal{J}} \frac{W_j}{d_j - r_j}$.

To compute an upper bound we assign all the jobs in the smallest alive interval. We obtain $S_{UB} = \frac{\sum_{j \in \mathcal{J}} W_j}{\min_{j \in \mathcal{J}} (d_j - r_j)}$. Without loss of generality we assume that all release dates and deadlines are integer and, hence, $S_{UB} \leq nW$, where $W = \max_{j \in \mathcal{J}} W_j$.

Given these lower and upper bounds and a small constant $\delta > 0$. We consider only the speeds from the set $S_\delta = \{S_{LB}, (1 + \delta)S_{LB}, (1 + 2\delta)S_{LB}, \ldots, (1 + k\delta)S_{LB}\}$, where k is the smallest integer such that $(1 + k\delta)S_{LB} \geqslant S_{UB}$. Hence, the number of speed values is bounded by $k \leq \frac{S_{UB}}{\delta S_{LB}}$, which is polynomial in $\frac{1}{\delta}$ and exponential in the size of the input.

Consider now an optimal schedule for some instance of our problem. We obtain the schedule σ from the optimal one by rounding up the speeds of processors to the closest discrete value. The ratio of the energy consumption of any processor i at any time t in σ to the energy consumption of the same processor i at time t in the optimal schedule is at most $(1 + \delta)^\alpha$. By summing up for all processors and all time intervals, we conclude that the energy consumption of σ is at most $(1+\delta)^\alpha OPT$. Finally, if we set a δ such that $\delta = \left(1 + \frac{\varepsilon}{OPT}\right)^{\frac{1}{\alpha}} - 1$, then the energy consumption of σ is at most $\varepsilon + OPT$. We note that this selection made the number of discrete speeds to be exponential in the size of the instance and polynomial in $1/\varepsilon$. Indeed, we have $k \leq \frac{S_{UB}}{\left((1+\varepsilon/OPT)^{\frac{1}{\alpha}} - 1\right)S_{LB}} \leq \frac{\alpha S_{UB} OPT}{\varepsilon S_{LB}}$.

Finally we get the following result.

Lemma 1. *For problems $P|size_j, pmtn, r_j, d_j|E$ and $P|fix_j, pmtn, r_j, d_j|E$ there exists a feasible schedule of energy consumption at most $OPT + \varepsilon$ that uses a finite (exponential in the size of the instance and polynomial in $1/\varepsilon$) number of discrete processors' speeds, for any $\varepsilon > 0$.*

In what follows in this paper, we deal with schedules that satisfy Lemma 1.

2.1 Configuration LP

Let $t_0 < t_1 < \cdots < t_l$ be the events corresponded to release dates and deadlines of jobs sorted according to the increasing time. We denote the set of all possible intervals of the form $(t_{i-1}, t_i]$, $i = 1, \ldots, l$, by $\mathcal{I}$. Let $|I|$ be the length of interval $I \in \mathcal{I}$. We introduce a variable $x_{I,c}$, for each $I \in \mathcal{I}$ and $c \in \mathcal{C}$, which corresponds to the total processing time during the interval $I \in \mathcal{I}$ where the processors run according to the configuration $c \in \mathcal{C}$. We denote by $E_{I,c}$ the instantaneous energy consumption of the processors if they run with respect to the configuration c during the interval I. Let $S_{j,c}$ be the speed of job j according to the configuration c, and let (I, c) be the set of jobs which are alive during the interval I and which are executed on some processors by the configuration c. We propose the following configuration LP:

$$\sum_{I \in \mathcal{I}, c \in \mathcal{C}} E_{I,c} \cdot x_{I,c} \to \min, \tag{1}$$

$$\sum_{c \in \mathcal{C}} x_{I,c} \leqslant |I|, \ I \in \mathcal{I}, \tag{2}$$

$$\sum_{I,c: j \in (I,c)} \frac{S_{j,c} \cdot x_{I,c}}{W_j} \geqslant 1, \ j \in \mathcal{J}, \tag{3}$$

$$x_{I,c} \geqslant 0, \ I \in \mathcal{I}, c \in \mathcal{C}. \tag{4}$$

The configurations, assigned to interval I, can be executed in an arbitrary order. Inequality (2) ensures that for each interval $I \in \mathcal{I}$ there is exactly one configuration for each time $t \in I$. Constraint (3) implies that each job $j \in \mathcal{J}$ is entirely executed.

The above LP has an exponential number of variables, but its number of constraints is linear in n. It is shown in [11] that if one can solve the *separation* problem for the dual LP in polynomial time then one can also solve the optimization problem of the primal LP in polynomial time. We consider the dual LP of (1)–(4):

$$\sum_{j \in \mathcal{J}} \lambda_j - \sum_{I \in \mathcal{I}} \mu_I |I| \to \max, \tag{5}$$

$$\sum_{j \in (I,c)} \frac{S_{j,c}}{W_j} \lambda_j - \mu_I \leqslant E_{I,c}, \ I \in \mathcal{I}, \ c \in \mathcal{C}, \tag{6}$$

$$\lambda_j \geqslant 0, \mu_I \geqslant 0, \ j \in \mathcal{J}, \ I \in \mathcal{I}. \tag{7}$$

The dual LP has a number of variables which is linear in n and an exponential number of constraints. In the next subsection we provide polynomial time separation oracles, i.e., we give polynomial-time algorithms which given a solution for the dual LP decide if this solution is feasible or otherwise identify a violated constraint. It follows that the strong optimization problem for the dual linear program can be solved by the Ellipsoid method in oracle polynomial time (see Theorem (6.4.9) in [11]). As we can compute an optimal solution for the dual

LP, we can also find an optimal solution for the primal LP by solving it with the variables corresponding to the constraints that were found to be violated during the run of the Ellipsoid method and setting all other primal variables to zero. The number of these violated constraints is bounded by the number of calls to the separation oracle. In turn, the number of calls to the separation oracle and the number of elementary arithmetic operations are bounded by a polynomial in the facet complexity ϕ of the corresponding polyhedron P (see Theorem (6.6.5) in [11]). A polyhedron P has a facet complexity at most ϕ if there exists a system of inequalities with rational coefficients that has the solution set P and such that the encoding length of each inequality of the system is at most ϕ. In our case the facet complexity ϕ can be bounded by polynomial in size of the instance and $log(1/\varepsilon)$.

2.2   Separation Problems

The separation oracle for the dual LP works as follows. For each $I \in \mathcal{I}$, we try to find if there is a violated constraint. Recall that there are $O(n)$ intervals in the set $\mathcal{I}$. For a given I, it suffices to check the minimum among the values $E_{I,c} - \sum_{J_j \in (I,c)} \frac{S_{j,c}}{W_j} \lambda_j$ over all possible configurations c. If this minimum value is less than $(-\mu_I)$, then we have a violated constraint. Otherwise, if we cannot find any violated constraint for all $I \in \mathcal{I}$, then the dual solution is feasible. Note that $E_{I,c} = \sum_{j \in (I,c)} (S_{j,c})^\alpha m_j$, where $m_j = size_j$ in the case of problem $P|size_j, pmtn, r_j, d_j|E$, and $m_j = |fix_j|$ in the case of problem $P|fix_j, pmtn, r_j, d_j|E$, for all $j \in (I,c)$. Hence, we want to find the minimum value of $\sum_{j \in (I,c)} (S_{j,c}^\alpha m_j - \frac{S_{j,c}}{W_j} \lambda_j)$.

For each job $j \in \mathcal{J}$ that is alive during I, the term $(S_{j,c}^\alpha m_j - \frac{S_{j,c}}{W_j} \lambda_j)$ is minimized at a discrete value $v_j \in S_\delta$ which is one of the two closest to the value $\left(\frac{\lambda_j}{W_j \cdot m_j \cdot \alpha} \right)^{1/(\alpha-1)}$ discrete speeds. To see this we just need to notice that we minimize a convex function of one variable over a set of possible discrete values. The value $\left(\frac{\lambda_j}{W_j \cdot m_j \cdot \alpha} \right)^{1/(\alpha-1)}$ is obtained by minimizing $(S_{j,c}^\alpha m_j - \frac{S_{j,c}}{W_j} \lambda_j)$ if there is no discretization of the speeds and it is obtained by setting the derivative of the last expression to zero. For job j, the discrete value $v_j \in S_\delta$ is calculated by Binary search in time polynomial in $1/\varepsilon$ and size of the instance. Hence, given an interval I, we want to find a configuration c that minimizes

$$\sum_{j \in (I,c)} \left(v_j^\alpha m_j - \frac{v_j}{W_j} \lambda_j \right). \tag{8}$$

Rigid Jobs. For the speed scaling scheduling with rigid jobs, the problem of minimizing expression (8) reduces to a maximum KNAPSACK problem and can be solved as follows.

We denote the subset of jobs, which are alive during interval I, by $\mathcal{J}' = \{j_1', \ldots, j_{n'}'\}$ and let $V(j) = (\frac{v_j}{W_j} \lambda_j - v_j^\alpha size_j)$ be the weight of job $j \in \mathcal{J}'$. Let F_{ik}

denote the set of all subsets of jobs from $\{j'_1, \ldots, j'_k\}$ that can be scheduled simultaneously on processors from $\{1, \ldots, i\}$ and let $F(i,k) = \max\{\sum_{j \in f_{ik}} V(j) : f_{ik} \in F_{ik}\}$. Then the problem of testing the feasibility is reduced to computing $\max\{F(i, n') : i = 1, \ldots, m\}$ for each interval $I \in \mathcal{I}$, which can be done by using a dynamic programming procedure based on the following recursions.

$$F(0, k) = F(i, 0) = 0, \; i = 1, \ldots, m, \; k = 0, 1, \ldots, n';$$

$$F(i, k) = F(i, k - 1), \; i = 1, \ldots, size_{j'_k} - 1, \; k = 1, \ldots, n';$$

$$F(i, k) = \max\{F(i, k - 1), \; F(i - size_{j'_k}, k - 1) + V(j'_k)\},$$

$$i = size_{j'_k}, \ldots, m, \; k = 1, \ldots, n'.$$

This method is based on the approach proposed in paper [12].

Hence, there is a separation oracle for the dual problem which runs in time polynomial in m, $1/\varepsilon$ and the size of the instance. Thus we have a polynomial time algorithm if m is fixed and a pseudo-polynomial time algorithm if m is a part of the input.

Theorem 1. *A schedule of energy consumption at most $OPT + \varepsilon$ can be found for problem $P|size_j, pmtn, r_j, d_j|E$ in time polynomial in m, $1/\varepsilon$ and the input size.*

Our algorithm can be generalized for problem $P|size_j, pmtn, r_j, d_j, \alpha_i|E$, where each processor has its own power function. The only change in the primal LP for this problem is the value $E_{I,c}$, namely $E_{I,c} = \sum_{j \in (I,c)} \left(\sum_{\{i(j,c)\}} S_{j,c}^{\alpha_{i(j,c)}} \right)$. Here $\{i(j,c)\}$ is the set of processors on which job j is assigned into configuration c, $|\{i(j,c)\}| = size_j$.

The separation oracle for the dual LP reduces to minimization of

$$\sum_{j \in (I,c)} \left(\sum_{\{i(j,c)\}} S_{j,c}^{\alpha_{i(j,c)}} - \frac{S_{j,c}}{W_j} \lambda_j \right). \tag{9}$$

For every job j, the "optimal" discrete speed $v_j \in S_\delta$ can also be calculated by Binary search, because each component of sum (9) is a one variable function with a single minimum over a set of positive values. Hence the following result holds.

Corollary 1. *A schedule of energy consumption at most $OPT + \varepsilon$ can be found for problem $P|size_j, pmtn, r_j, d_j, \alpha_i|E$ in time polynomial in $1/\varepsilon$ and the input size.*

Let us assume that the number of processors used by a job is chosen by the scheduler, and can be changed at runtime. Then these jobs are called malleable jobs [8], and the speed scaling problem with malleable jobs is denoted by $P|var, r_j, d_j|E$. Our algorithm can be adopted for problem $P|var, r_j, d_j|E$.

Let $var_j = \{size_j^1, \ldots, size_j^{l_j}\}$ denote the set of usable processor sizes for job $j \in \mathcal{J}$, where $size_j^h < size_j^{h+1}$, and let W_{jh} be the workload of job $j \in \mathcal{J}$ if it is executed on $size_j^h$ processors. For malleable jobs the set of discrete processors' speeds and the configuration LP are constructed the same way as for rigid jobs, but a configuration c specifies a feasible subset of jobs and an assignment of processor numbers to jobs as well as an assignment of a speed value for each processor.

The separation problem for the dual LP reduces to finding a configuration c that minimizes

$$\sum_{j \in (I,c)} \left(v_{j,h(j,c)}^{\alpha} size_j^{h(j,c)} - \frac{v_{j,h(j,c)}}{W_{j,h(j,c)}} \lambda_j \right) \tag{10}$$

for each interval $I \in \mathcal{I}$.

Here $h(j,c)$ is the number of processor size for job j in configuration c and $v_{j,h(j,c)}$ is the "optimal" discrete speed of job j, if it is executed on $size_j^{h(j,c)}$ processors.

We denote the subset of jobs, which are alive during interval I, by $\mathcal{J}' = \{j_1', \ldots, j_{n'}'\}$ and let $V(j,h) = (\frac{v_{j,h}}{W_{j,h}} \lambda_j - v_{j,h}^{\alpha} size_j^h)$ be the weight of job $j \in \mathcal{J}'$, if it is executed on $size_j^h$ processors. Then the problem of minimizing expression (10) is equivalent to maximization of $\sum_{j \in (I,c)} V(j,h(j,c))$. The latter problem is solved by a dynamic programming.

Let F_{ik} denote the set of all subsets of jobs from $\{j_1', \ldots, j_k'\}$ that can be scheduled simultaneously on processors from $\{1, \ldots, i\}$ (the number of processors is indicated for each job) and let $F(i,k) = \max\{\sum_{j \in f_{ik}} V(j,h(j,f_{ik})) : f_{ik} \in F_{ik}\}$. In our algorithm, the values $F(i,k)$ are recursively computed, and the recursion is given by

$$F(0,k) = F(i,0) = 0, \ i = 1, \ldots, m, \ k = 0, 1, \ldots, n';$$

$$F(i,k) = F(i,k-1), \ i = 1, \ldots, size_{j_k'}^1 - 1, \ k = 1, \ldots, n';$$

$$F(i,k) = \max\{\max_{size_{j_k'}^h \leq i} \{F(i - size_{j_k'}^h, k-1) + V(j_k', h)\}; \ F(i,k-1)\},$$

$$i = size_{j_k'}^1, \ldots, m, \ k = 1, \ldots, n'.$$

Therefore, there is a separation oracle for the dual problem which runs in time polynomial in m, $1/\varepsilon$ and size of the instance. Thus, we have

Corollary 2. *A schedule of energy consumption at most $OPT + \varepsilon$ can be found for problem $P|var, r_j, d_j|E$ in time polynomial in m, $1/\varepsilon$ and the input size.*

Single Mode Two-Processor Jobs. For the speed scaling scheduling with single mode two-processor jobs (i.e. $|fix_j| = 2$ for $j \in \mathcal{J}$), the problem of minimizing expression (8) reduces to a MAXIMUM WEIGHTED MATCHING problem on a multi-graph, where vertices correspond to processors and edges represent the jobs. There is an edge $\{i, i'\}$ with weight equal to $\left(\frac{v_j}{W_j} \lambda_j - 2v_j^{\alpha} \right)$ for each alive job $j \in \mathcal{J}$ in interval I, whose required processor set consists of i and i' (i.e. $fix_j = \{i, i'\}$). The maximum weighted matching in such a multi-graph can be found in time polynomial in n and defines a configuration c that minimizes (8).

Hence, there is a separation oracle for the dual problem which runs in time polynomial in $1/\varepsilon$ and the size of the instance.

Theorem 2. *A schedule of energy consumption at most $OPT + \varepsilon$ can be found for problem $P|fix_j, pmtn, r_j, d_j|E$ with $|fix_j| = 2$ in time polynomial in $1/\varepsilon$ and the input size.*

As in the case of rigid jobs, the presented approach can be generalized for problem $P|fix_j, pmtn, r_j, d_j, \alpha_i|E$ with $|fix_j| = 2$. Thus, we have

Corollary 3. *A schedule of energy consumption at most $OPT + \varepsilon$ can be found for problem $P|fix_j, pmtn, r_j, d_j, \alpha_i|E$ with $|fix_j| = 2$ in time polynomial in $1/\varepsilon$ and the input size.*

The complexity of the Separation Problem changes when there are three prespecified processors for each job. In this case, the problem of minimizing expression (8) is NP-hard, because the NP-complete THREE-DIMENSIONAL MATCHING problem can be reduced to it.

Theorem 3. *The problem of finding a configuration that minimizes expression (8), given an interval $I \in \mathcal{I}$, is NP-hard for problem $P|fix_j, pmtn, r_j, d_j|E$ with $|fix_j| = 3$.*

Proof. THREE-DIMENSIONAL MATCHING problem is formulated as follows: Let the set $M \subseteq X \times Y \times Z$ be given, where X, Y, and Z are disjoint sets having the same number of elements q. The question is, does M contains a matching, i.e., a subset $M' \subseteq M$ such that $|M'| = q$ and no two elements of M' agree in any coordinate.

Given an instance of THREE-DIMENSIONAL MATCHING, we construct the following instance of problem $P|fix_j, pmtn, r_j, d_j|E$. Set $n = |M|$, $m = 3q$, $W_j = 1$, $r_j = 0$, $d_j = 1$, and $|fix_j| = 3$ for all $j \in \mathcal{J}$. Suppose that each element $l = (x, y, z)$ from M corresponds to a job j and coordinates of l represent the required processors for job j, i.e. $fix_j = \{x, y, z\}$.

The separation oracle for the constructed instance is reduced to finding the maximum value of $\sum_{j \in (I,c)} (S_{j,c} \lambda_j - 3 S_{j,c}^{\alpha})$ over all possible configurations c for a single interval $I = (0, 1]$. Let λ_j be the same for each job $j \in \mathcal{J}$ and be equal to $\lambda > 0$, then the "optimal" discrete speeds $v_j \in S_\delta$ are identical for all jobs. Denote these speeds by v. Hence, we have to find a configuration c that maximizes

$$\sum_{j \in (I,c)} (v\lambda - 3v^{\alpha}). \tag{11}$$

Note that at each time moment at most q jobs can be executed. Therefore, a configuration c with $\sum_{j \in (I,c)} (v\lambda - 3v^{\alpha}) \geqslant q(v\lambda - 3v^{\alpha})$ exists and contains exactly q jobs if and only if the THREE-DIMENSIONAL MATCHING problem has a positive answer. Our reduction is polynomial and thus the theorem is proved. □

3   NP-Hardness Results

Suppose that the parameter α is the same for all processors and the number of required processors is given and fixed for each job. Under this assumption it is easy to see that in any optimal schedule, any job runs at a constant speed due to the convexity of the speed-to-power function. Let us assume that we know the optimal execution speed of each job. We still need to find a feasible schedule with respect to release dates and deadlines. Intuition suggests that speed-scaling problems should be harder than their counterparts with fixed processing times. Indeed, let us prove the NP-hardness of $P|size_j, pmtn, r_j, d_j|E$.

Theorem 4. $P|size_j, pmtn, r_j, d_j|E$ *is ordinary NP-hard even if all jobs have a common release date, a common deadline, and unit processing volumes.*

Proof. We show that the NP-complete PARTITION problem polynomially transforms to $P|size_j, pmtn, r_j, d_j|E$.

PARTITION: Given a set $A = \{a_1, a_2, \ldots, a_n\}$ of natural numbers. Is there a subset $A' \subset A$ such that $\sum\limits_{a_i \in A'} a_i = \sum\limits_{a_i \in A \setminus A'} a_i$?

So let $a_1, a_2, \ldots, a_n$ be an instance of PARTITION and let $\sum\limits_{a_i \in A} a_i = 2B$. We construct an instance of $P|size_j, pmtn, r_j, d_j|E$ with n jobs and B processors as follows. For every a_j we generate a job j. We set $size_j = a_j$, $W_j = 1$, $r_j = 0$, $d_j = 2$ for $j \in \mathcal{J}$. It is required to determine if there is a schedule of energy consumption at most $E = 2B$. Due to convexity of the speed-to-power function and the fact that $2m = \sum_{j \in \mathcal{J}} (W_j \cdot size_j)$, a feasible schedule of energy consumption at most $2B$ has no idle time and the speed of each processor is equal to 1 during the whole interval $[0, 2]$. Therefore, at each time moment t the sizes of jobs running in parallel must be equal to $B = \sum_{j \in \mathcal{J}_t} size_j$, where $\mathcal{J}_t$ is the set of jobs executed at time t. Therefore, a positive answer to $P|size_j, pmtn, r_j, d_j|E$ with $E \leq 2B$ implies a positive answer to PARTITION and vice versa. □

In the proof of Theorem 4 we use the common technique for proving NP-hardness results in scheduling. Moreover, our proof is almost a step by step reproduction of the NP-hardness proof for $P|size_j, pmtn, r_j, d_j|C_{\max}$ [7]. We claim that most of the NP-hardness proofs for scheduling problems with the minimization maximum lateness criterion L_{max} may be easily transformed to their speed scaling counterparts. We note that C_{max} is the special case of L_{max} when all jobs have due dates at time 0.

Consider the decision version Π of some scheduling problem with the criterion L_{max} : Given n jobs and m processors. Each processor has a constant speed. With each job j we associate a release date r_j, a processor dependent processing time p_{ij}, and a deadline d_j. A feasible schedule is an assignment of jobs to processors such that each job j starts no earlier than its release date r_j and completes no later than its deadline d_j. It is required to find a feasible schedule. We do not specify the scheduling environment here. We may consider a single processor, parallel identical or unrelated processors, open shop, job shop, e.t.c. We may consider models that allow preemption and those that do not, models in which jobs have precedence constraints, models with rigid jobs or multi-processor jobs, or other variations.

Let $r_{min} = \min_j r_j$ and $d_{max} = \max_j d_j$. We say that a schedule is called *non-idle* if no processor is idle during the interval $[r_{min}, d_{max}]$. We say an instance I has the non-idle property if there is no feasible schedule that is not non-idle. Let Π' be a subproblem of Π that contains only instances with the non-idle property.

First, we show that the problem Π' can be polynomially reduced to the corresponding speed scaling problem $\bar{\Pi}$. Indeed, let I be an instance of Π'. We consider an instance $\bar{I}$ of the speed scaling problem with the same set of jobs and processors. Each job j has the same release date r_j and the same deadline d_j as in I. The processing volume W_{ij} of job j on processor i is equal to the processing time p_{ij} of job j on processor i in the instance I. The instance $\bar{I}$ has the same processor and job environment as the instance I except one. The processors may change their speed. If a processor runs at speed s then the energy consumption is s^α units of energy per time unit, where $\alpha > 1$ is a constant. Thus, if processor i performs job j at a speed s then it spends $\frac{W_{ij}}{s}$ time units to complete the job. It is required to determine if there is a schedule of energy consumption at most $m(d_{max} - r_{min})$. Due to convexity of the speed-to-power function and the fact that all feasible schedules in the instance I (if any) should be non-idle, a schedule with the required energy consumption exists if and only if there exists a feasible non-idle, schedule in the instance I. It follows that the NP-completeness of Π' implies the NP-completeness of $\bar{\Pi}$.

Second, we observe that most of the NP-hardness proofs for scheduling problems with the minimization maximum lateness criterion are based on the polynomial reduction of a well-known NP-complete problem to the decision version of scheduling problem. In most cases either the obtained instance of decision problem has the non-idle property or we can easily modify the instance so that a new instance has the non-idle property and there exists a feasible schedule in it if and only if there exists a feasible schedule in the original instance.

To illustrate our observation let us consider the problem $P|fix_j, pmtn|C_{max}$. Jansen and Porkolab [13] proved that $P|fix_j, pmtn|C_{max}$ is strongly NP-hard even if all jobs have unit execution time and $|fix_j| \leq 3$ for each job $j \in \mathcal{J}$. They reduced the fractional coloring problem on a graph $G = (V, A)$ with fractional chromatic number $\chi_f(G) \leqslant 3$ to the decision version of $P|fix_j, pmtn|C_{max}$ in such a way that there are at most three prespecified processors for each job and two or three jobs on each processor. They proved that a schedule of length 3 for the corresponding instance I of $P|fix_j, pmtn|C_{max}$ exists if and only if the instance of the fractional coloring problem has a fractional chromatic number $\chi_f(G) \leqslant 3$. We add a "dummy" single-processor job for each processor with two jobs. It is obvious that a schedule of length 3 in a new instance exists if and only if there exists a schedule of length 3 in the instance I. Finally, we get the following result as a corollary of Theorem 3.1 in [13].

Theorem 5. $P|fix_j, pmtn, r_j, d_j|E$ with $|fix_j| \leq 3$ *is strongly NP-hard even if all jobs have a common release date, a common deadline, and unit processing volumes.*

4   Conclusion

The NP-hard speed scaling scheduling problems of parallel jobs with preemption are studied. "Almost-exact" algorithms are proposed for solving the problems with rigid jobs and single mode two-processor jobs. Based on configuration linear programs, our algorithms return a solution within an additive factor of $\varepsilon > 0$ from the optimal solution.

Further research appears to be appropriate in extending the obtained results to the speed scaling scheduling of moldable jobs $(P|any, pmtn, r_j, d_j|E)$ and multi mode multiprocessor jobs $(P|set_j, pmtn, r_j, d_j|E)$.

Acknowledgements. This research is supported by the Russian Science Foundation grant 15-11-10009.

References

1. Albers, S., Antoniadis, A., Greiner, G.: On multi-processor speed scaling with migration: extended abstract. In: 23rd ACM Symposium on Parallelism in Algorithms and Architectures (SPAA 2011), pp. 279–288. ACM (2011)
2. Albers, S., Müller, F., Schmelzer, S.: Speed scaling on parallel processors. In: 19th ACM Symposium on Parallelism in Algorithms and Architectures (SPAA 2007), pp. 289–298. ACM (2007)
3. Angel, E., Bampis, E., Kacem, F., Letsios, D.: Speed scaling on parallel processors with migration. In: Kaklamanis, C., Papatheodorou, T., Spirakis, P.G. (eds.) Euro-Par 2012. LNCS, vol. 7484, pp. 128–140. Springer, Heidelberg (2012)
4. Bampis, E., Kononov, A., Letsios, D., Lucarelli, G., Sviridenko, M.: Energy efficient scheduling and routing via randomized rounding. In: FSTTCS, pp. 449–460 (2013)
5. Bingham, B.D., Greenstreet, M.R.: Energy optimal scheduling on multiprocessors with migration. In: International Symposium on Parallel and Distributed Processing with Applications (ISPA 2008), pp. 153–161. IEEE (2008)
6. Cohen-Addad, V., Li, Z., Mathieu, C., Milis, I.: Energy-efficient algorithms for non-preemptive speed-scaling. In: Bampis, E., Svensson, O. (eds.) WAOA 2014. LNCS, vol. 8952, pp. 107–118. Springer, Heidelberg (2015)
7. Drozdowski, M.: On complexity of multiprocessor tasks scheduling. Bull. Pol. Acad. Sci. Tech. Sci. **43**(3), 381–392 (1995)
8. Drozdowski, M.: Scheduling for Parallel Processing, p. 386. Springer, London (2009)
9. Gerards, M.E.T., Hurink, J.L., Hölzenspies, P.K.F.: A survey of offline algorithms for energy minimization under deadline constraints. J. Sched. **19**, 3–19 (2016)
10. Greiner, G., Nonner, T., Souza, A.: The bell is ringing in speed-scaled multiprocessor scheduling. In: 21st ACM Symposium on Parallelism in Algorithms and Architectures (SPAA 2009), pp. 11–18. ACM (2009)
11. Grötschel, M., Lovász, L., Schrijver, A.: Geometric Algorithms and Combinatorial Optimizations, 2nd, p. 362. Springer, Heidelberg (1993)

12. Jansen, K., Porkolab, L.: Preemptive parallel task scheduling in $O(n) + \text{poly}(m)$ time. In: Lee, D.T., Teng, S.-H. (eds.) ISAAC 2000. LNCS, vol. 1969, pp. 398–409. Springer, Heidelberg (2000)
13. Jansen, K., Porkolab, L.: Preemptive scheduling with dedicated processors: applications of fractional graph coloring. J. Sched. **7**, 35–48 (2004)
14. Yao, F., Demers, A., Shenker, S.: A scheduling model for reduced CPU energy. In: 36th Annual Symposium on Foundation of Computer Science (FOCS 1995), pp. 374–382 (1995)

Facility Location

Facility Location in Unfair Competition

Vladimir Beresnev[1,2(✉)] and Andrey Melnikov[1,2]

[1] Sobolev Institute of Mathematics, Novosibirsk, Russia
{beresnev,melnikov}@math.nsc.ru
[2] Novosibirsk State University, Novosibirsk, Russia

Abstract. We consider a mathematical model belonging to the family of competitive location problems. In the model, there are two competing parties called Leader and Follower, which open their facilities with the goal to capture customers and maximize profit. In our model we assume that Follower is able to open own facilities as well as to close the Leader's ones. The model can be written as a pessimistic bilevel integer programming problem. We show that the problem of Leader's profit maximization can be represented as a problem of pseudo–Boolean function maximization. The number of variables the function depends on equals to the number of sites available for opening a facility. We suggest a method of calculation of an upper bound for the optimal value of the function based on strengthening of a bilevel model with valid inequalities and further relaxation of the model by removing the lower–level optimization problem.

Keywords: Stackelberg game · Upper bound · Competitive location

1 Introduction

In contrast to the classical location problem [9] models of competitive location consider several competing parties [4–6,10]. The parties simultaneously or sequentially open their facilities with the aim to optimize personal objective functions. The goals of the competitors are associated with the customers capture and satisfying their demands. There is a number of customer behavior models resulting from the characteristics of the demand and other factors [13]. We assume that the customer capture is based on his preferences. They are assumed to be known for both parties.

In our model, we consider the competition of two sides that open their facilities sequentially. The decision making process can be considered as a Stackelberg game [15]. The formalization of this kind of games can be done in a natural way in terms of bilevel programming [8]. According to the game terminology the party that opens its facilities first will be referred tó as Leader. The second party that opens its facilities knowing Leader's decision will be referred to as Follower.

In the present work, we deal with the model of competitive location where in contrast to models from [4–6] Follower is able to close Leader's facilities by using discrediting, black PR and other methods of unfair competition.

© Springer International Publishing Switzerland 2016
Y. Kochetov et al. (Eds.): DOOR 2016, LNCS 9869, pp. 325–335, 2016.
DOI: 10.1007/978-3-319-44914-2_26

The Leader's aim in this competition is to open such a set of facilities that brings maximum profit provided that Follower can close some of Leader's facilities and capture some customers. Follower's goal is to maximize profit as well. Follower decides which Leader's facilities are to be closed and where to open own facilities.

First publications considering bilevel location models with opportunity of closing or destructing of the facilities appeared in 2008. In [14] authors formulate the model of interdiction median problem with fortification (RIMF), where one party called a defender commits resources to protect facilities serving customers from the rational attack of another party. The authors investigate properties of the model and suggest an enumeration scheme to obtain an optimal defender's solution. Further developments of the model can be found in [1,11,16] where stochastic generalizations of the model are considered. Other models of protection against the rational attack are investigated in [2,3].

An important feature of our model is necessity of a revision of the feasibility definition. The most common concepts of feasibility for bilevel programming problems are optimistic and pessimistic solutions. In the present work we focus on the problem of finding a pessimistic optimal solution. The suggested approach is based on the ideas developed and approved in [4,6]. The first point is representation of the Leader's problem in the form of pseudo–Boolean function maximization. The number of variables the function depends on is equal to the number of places available for facility opening. The representation allows implementing inexact methods of search in a Boolean cube such as local search and its modifications. The second important point is calculation of an upper bound for the values the function takes on subsets specified by partial (0,1)–vectors. It allows developing an implicit enumeration scheme proved to be effective when applied to previously studied models.

In this paper, we show that given values of the Leader's location variables the problem of finding a pessimistic feasible solution is reduced to mixed–integer programming problem. This implies that the required pseudo–Boolean function can be constructed. By using the approach from [4,6] we define a modified system of evaluating subsets, which allow to formulate sufficient conditions of capturing the customer by Follower. This conditions written in a form of linear inequalities, are used as valid inequalities for strengthening the bilevel model. The relaxation of the strengthened model by removing the lower–level problem provides an upper bound for the optimal value of the pseudo–Boolean function.

The paper is organized as follows. In Sect. 1 we propose the model of the facility location in unfair competition in the form of a bilevel integer programming problem. Section 2 is devoted to the problem of finding a pessimistic feasible solution of the model. A reduction to a pseudo–Boolean function maximization problem is discussed as well. In Sect. 3 we construct an estimating problem providing an upper bound for the optimal value of the pseudo–Boolean function.

2   Mathematical Model

Let us introduce the necessary notations.

Index sets:
$I = \{1, \ldots, m\}$ is a set of locations (candidate sites for opening facilities);
$J = \{1, \ldots, n\}$ is a set of customers.

Parameters:
f_i is a fixed cost of opening a Leader's facility $i \in I$;
g_i is a fixed cost of opening a Follower's facility $i \in I$;
G_i is a cost of closing a Leader's facility $i \in I$;
p_{ij} is a profit of Leader's facility $i \in I$ obtained from a customer $j \in J$;
q_{ij} is a profit of Follower's facility $i \in I$ obtained from a customer $j \in J$;

Variables:
$$x_i = \begin{cases} 1, & \text{if Leader opens facility } i \\ 0, & \text{otherwise}, \end{cases}$$
$$z_i = \begin{cases} 1, & \text{if Follower opens facility } i \\ 0, & \text{otherwise}, \end{cases}$$
$$s_i = \begin{cases} 1, & \text{if Follower closes Leader's facility } i \\ 0, & \text{otherwise}, \end{cases}$$
$$x_{ij} = \begin{cases} 1, & \text{if Leader's facility } i \text{ serves the customer } j \\ 0, & \text{otherwise}, \end{cases}$$
$$z_{ij} = \begin{cases} 1, & \text{if Follower's facility } i \text{ serves the customer } j \\ 0, & \text{otherwise}. \end{cases}$$

We assume that the preferences of a customer $j \in J$ are represented with a linear order $\succeq_j$ on the set I. The relation $i_1 \succeq_j i_2$ shows that either facility i_1 is more preferable for j than i_2, or $i_1 = i_2$. If $i_1 \neq i_2$ and $i_1 \succeq_j i_2$, we use denotation $i_1 \succ_j i_2$.

Given $j \in J$, we denote the greatest element of a nonempty set $K \subseteq I$ according to the order $\succeq_j$ with $i_j(K)$. In other words, $i_j(K)$ is a $i \in K$ such that $i \succeq_j k$ for all $k \in K$. For a nonzero Boolean vector $x = (x_i)$, $i \in I$ we assume that $i_j(x) = i_j(\{i \in I | x_i = 1\})$.

It is assumed that a customer is captured by the party that opens the most preferable facility for him. Moreover, the party is able to serve the captured customer only with a facility that is more preferable for him than any competitor's facility. If Boolean vectors x and z correspond to Leader's and Follower's facilities locations respectively, then Leader's facility $i \in I$ can serve customer $j \in J$ iff $i \succ_j i_j(z)$. Similarly, Follower's facility $i \in I$ can serve a customer $j \in J$ iff $i \succ_j i_j(x)$.

Now we can formulate the model of facility location in unfair competition in terms of bilevel integer programming:

$$\max_{(x_i),(x_{ij})} \min_{(z_i),(z_{ij}),(s_i)} \left(-\sum_{i \in I} f_i x_i + \sum_{j \in J} \sum_{i \in I} p_{ij} x_{ij} \right), \tag{1}$$

$$\tilde{z}_i + \sum_{k|i\succeq_j k} x_{kj} \leq 1, \quad i \in I, \quad j \in J; \tag{2}$$

$$x_i - \tilde{s}_i \geq x_{ij}, \quad i \in I, \quad j \in J; \tag{3}$$

$$x_i, x_{ij} \in \{0,1\}, \quad i \in I, \quad j \in J; \tag{4}$$

$$\text{where } (\tilde{z}_i), (\tilde{z}_{ij}), (\tilde{s}_i) \text{ solves} \tag{5}$$

$$\max_{(z_i),(z_{ij}),(s_i)} \left\{ -\sum_{i \in I} G_i s_i - \sum_{i \in I} g_i z_i + \sum_{j \in J}\sum_{i \in I} q_{ij} z_{ij} \right\}, \tag{6}$$

$$x_i - s_i + z_i \leq 1, \quad i \in I; \tag{7}$$

$$x_i \geq s_i, \quad i \in I; \tag{8}$$

$$x_i - s_i + \sum_{k|i\succeq_j k} z_{kj} \leq 1, \quad i \in I, \quad j \in J; \tag{9}$$

$$z_i \geq z_{ij}, \quad i \in I, \quad j \in J; \tag{10}$$

$$z_i, z_{ij}, s_i \in \{0,1\}, \quad i \in I, \quad j \in J. \tag{11}$$

We denote the upper–level problem (1)–(5) with $\mathcal{L}$ and the lower–level problem (6)–(11) with $\mathcal{F}$. The problem (1)–(11) is denoted by $(\mathcal{L}, \mathcal{F})$.

Leader's objective function (1) expresses the value of his profit and consists of two components. The first one is the cost of facilities to be opened, and the second summand represents the income collected by them. We assume that in the cases when the problem $\mathcal{F}$ has several optimal solutions Follower plays against Leader and chooses the solution that minimizes (1). Constraints (2) ensure that Leader serves the customer with a facility which is more preferable for the customer than any Follower's facility. In addition, these constraints ensure that the customer is served with no more than one Leader's facility. Constraints (3) guarantee that customers are served with open facilities. Follower's problem $\mathcal{F}$ has a similar form. Additional term in Follower's objective function (6) equals to the cost of closing Leader's facilities. Constraints (7) ensures that Follower's facility can be opened only in a location without Leader's one, and constraints (8) allow to close only the Leader's facility which is open.

3   Pessimistic Feasible Solutions

A pair $(X, \tilde{Z})$ is called a *feasible solution* of the problem $(\mathcal{L}, \mathcal{F})$ if $X = ((x_i), (x_{ij}))$ is a feasible solution of the problem $\mathcal{L}$ with given $\tilde{z} = (\tilde{z}_i)$, $\tilde{s} = (\tilde{s}_i)$, and $\tilde{Z} = ((\tilde{z}_i), (\tilde{z}_{ij}), (\tilde{s}_i))$ is an optimal solution of the problem $\mathcal{F}$ with given $x = (x_i)$.

Denote the value of objective function (6) on a feasible solution Z of the problem $\mathcal{F}$ with $F(Z)$ and the value of objective function (1) on a feasible solution $(X, \tilde{Z})$ of the problem $(\mathcal{L}, \mathcal{F})$ with $L(X, \tilde{Z})$.

Given values of variables $x = (x_i)$, $i \in I$, let us select "good" Leader's solutions among all feasible solutions $(X, \tilde{Z})$ of the problem $(\mathcal{L}, \mathcal{F})$. We call a feasible solution $(\tilde{X}, \tilde{Z})$, $\tilde{X} = ((x_i), (\tilde{x}_{ij}))$ *strong* if $L(\tilde{X}, \tilde{Z}) \geq L(X, \tilde{Z})$ for every feasible solution $(X, \tilde{Z})$, where $X = ((x_i), (x_{ij}))$. It is clear that a feasible solution $(\tilde{X}, \tilde{Z})$, $\tilde{X} = ((x_i), (\tilde{x}_{ij}))$ is strong if for all $j \in J$ holds

$$\sum_{i \in I} p_{ij} \tilde{x}_{ij} = \max_{k | k \succ_j i_j(\tilde{z})} p_{kj}(x_k - \tilde{s}_k),$$

where maximum over an empty set is assumed to be equal to zero.

We say that a strong feasible solution $(\bar{X}, \bar{Z})$ of the problem $(\mathcal{L}, \mathcal{F})$, where $\bar{X} = ((x_i), (\bar{x}_{ij}))$, is *pessimistic*, if $L(\bar{X}, \bar{Z}) \leq L(\tilde{X}, \tilde{Z})$ for each strong feasible solution $(\tilde{X}, \tilde{Z})$, $\tilde{X} = ((x_i), (\tilde{x}_{ij}))$. A pessimistic feasible solution (X^*, Z^*) of the problem $(\mathcal{L}, \mathcal{F})$ is called a *pessimistic optimal solution* if $L(X^*, Z^*) \geq L(\bar{X}, \bar{Z})$ for each pessimistic feasible solution $(\bar{X}, \bar{Z})$.

Given a Boolean vector $x = (x_i)$, $i \in I$, consider the problem of finding a pessimistic feasible solution $(\bar{X}, \bar{Z})$, $\bar{X} = ((x_i), (\bar{x}_{ij}))$ of the problem $(\mathcal{L}, \mathcal{F})$. This solution can be computed in two steps.

At the first step given a vector x solve the problem $\mathcal{F}$ and get an optimal value F^* of its objective function. At the second step solve the following auxiliary problem. To formulate it we introduce new variables $u_j, j \in J$. The variable u_j takes the value of the maximum profit Leader gets from serving the customer j.

The aforementioned problem is formulated as follows:

$$\min_{(z_i),(z_{ij}),(s_i),(u_j)} \sum_{j \in J} u_j \tag{12}$$

$$x_i - s_i + z_i \leq 1, \quad i \in I; \tag{13}$$

$$x_i \geq s_i, \quad i \in I; \tag{14}$$

$$x_i - s_i + \sum_{k | i \succ_j k} z_{kj} \leq 1, \quad i \in I, \quad j \in J; \tag{15}$$

$$z_i \geq z_{ij}, \quad i \in I, \quad j \in J; \tag{16}$$

$$u_j \geq p_{ij}(x_i - s_i - \sum_{k | k \succ_j i} z_k), \quad i \in I, \quad j \in J; \tag{17}$$

$$-\sum_{i \in I} G_i s_i - \sum_{i \in I} g_i z_i + \sum_{j \in J} \sum_{i \in I} q_{ij} z_{ij} \geq F^*; \tag{18}$$

$$z_i, z_{ij}, s_i \in \{0, 1\}, \quad i \in I, \quad j \in J; \tag{19}$$

$$u_i \geq 0, \quad j \in J. \tag{20}$$

Let $(\bar{Z}, \bar{U})$, $\bar{Z} = ((\bar{z}_i), (\bar{z}_{ij}), (\bar{s}_i))$, $\bar{U} = (\bar{u}_j)$ be an optimal solution of the problem (12)–(20), and let $\bar{z} = (\bar{z}_i)$. Notice that for solution $(\bar{Z}, \bar{U})$ the following equality holds for each $j \in J$:

$$\bar{u}_j = \max_{i | i \succ_j i_j(\bar{z})} \{p_{ij}(x_i - \bar{s}_i)\}.$$

Now we are able to construct a strong feasible solution $(\bar{X}, \bar{Z})$, $\bar{X} = ((x_i), (\bar{x}_{ij}))$ of the problem $(\mathcal{L}, \mathcal{F})$ *corresponding* to $(\bar{Z}, \bar{U})$. For $j \in J$ such that $\bar{u}_j > 0$ let us denote by i_j the index $i \in I$ for which the constraint (17) is active. Then for $i \in I, j \in J$ we set

$$\bar{x}_{ij} = \begin{cases} 1, & \text{if } \bar{u}_j > 0 \text{ and } i = i_j \\ 0 & \text{otherwise} \end{cases}.$$

Notice that $(\bar{X}, \bar{Z})$ is a strong feasible solution of the problem $(\mathcal{L}, \mathcal{F})$. In addition, observe that $\bar{u}_j = \sum\limits_{i \in I} p_{ij} \bar{x}_{ij}, j \in J$.

Theorem 1. *Given (0,1)–vector* $x = (x_i)$, $i \in I$, *if* $(\bar{Z}, \bar{U})$, $\bar{Z} = ((\bar{z}_i), (\bar{z}_{ij}), (\bar{s}_i))$, $\bar{U} = (\bar{u}_j)$ *is an optimal solution of the problem (12)–(20), then the solution* $(\bar{X}, \bar{Z})$, $\bar{X} = ((x_i), (\bar{x}_{ij}))$ *of the problem* $(\mathcal{L}, \mathcal{F})$, *corresponding to* $(\bar{Z}, \bar{U})$ *is a pessimistic feasible solution of the problem* $(\mathcal{L}, \mathcal{F})$.

Proof. Let $(\tilde{X}, \tilde{Z})$, $\tilde{X} = ((x_i), (\tilde{x}_{ij}))$, $\tilde{Z} = ((\tilde{z}_i), (\tilde{z}_{ij}), (\tilde{s}_i))$ be a strong feasible solution of the problem $(\mathcal{L}, \mathcal{F})$. Set $\tilde{z} = (\tilde{z}_i)$ and

$$\tilde{u}_j = \sum_{i \in I} p_{ij} \tilde{x}_{ij}, \quad j \in J.$$

Since $(\tilde{X}, \tilde{Z})$ is a strong feasible solution, then

$$\tilde{u}_j = \max_{i | i \succ_j i_j(\tilde{z})} p_{ij}(x_i - \tilde{s}_i), \quad j \in J.$$

Consequently, $(\tilde{Z}, \tilde{U})$, $\tilde{U} = (\tilde{u}_i)$ is a feasible solution of the problem (12)–(20). We get

$$\sum_{j \in J} \sum_{i \in I} p_{ij} \tilde{x}_{ij} = \sum_{j \in J} \tilde{u}_j \geq \sum_{j \in J} \bar{u}_j = \sum_{j \in J} \sum_{i \in I} p_{ij} \bar{x}_{ij}.$$

It follows that $L(\bar{X}, \bar{Z}) \leq L(\tilde{X}, \tilde{Z})$, and the Theorem 1 is proved.

Since any (0,1)–vector x defines the value of objective function (1) on the corresponding pessimistic feasible solution, then the problem $(\mathcal{L}, \mathcal{F})$ can be considered as a pseudo–Boolean function maximization problem. This function f depends on m Boolean variables and for every vector of Leader's locations gives the value of Leader's profit.

4   Upper Bound

Consider the problem of computing an upper bound for values of the aforementioned pseudo–Boolean function $f(x)$, $x \in \{0,1\}^m$. Our goal is to modify the approach from [4,6] and apply it to the problem under investigation. The method consists in strengthening of the initial bilevel problem with some additional constraints satisfied by all pessimistic feasible solutions. The relaxation of the strengthened model by removing the lower–level problem provides a valid upper bound.

Valid inequalities for the problem $(\mathcal{L}, \mathcal{F})$ utilize a specially constructed system of subsets $\{I_j\}$, $j \in J$. Our goal is to form a nontrivial subset I_j for each $j \in J$ such that in the case, when the most preferable for $j_0 \in J$ Leader's facility is not in I_{j_0}, then j_0 does not bring profit to Leader. Given $j_0 \in J$, let us formulate the rule to determine if an arbitrary $i \in I$ is in the subset I_{j_0} or not.

Consider the set $N(i) = \{k \in I \mid k \succ_{j_0} i\}$ of facilities more preferable for j_0 than i and its superset $\bar{N}(i) = N(i) \cup \{i\}$. The set

$$J(i) = \{j \in J \mid i = i_j(I \backslash N(i))\}$$

contains customers for which all the facilities that are more preferable than i are contained in $N(i)$. Since $j_0 \in J(i)$, then $J(i) \neq \emptyset$.

For each $k \in N(i)$ denote the subset of $J(i)$ that can be captured by k by

$$J_1(i, k) = \{j \in J(i) \mid k = i_j((I \backslash N(i)) \cup \{k\})\},$$

and for each $k \in \bar{N}(i)$ the subset of $J(i)$ that can be captured by k after closing the facility i by

$$J_2(i, k) = \{j \in J(i) \mid k = i_j((I \backslash \bar{N}(i)) \cup \{k\})\}.$$

Suppose that $i \notin I_{j_0}$ if there exists $k \in N(i)$ such that

$$\sum_{j \in J_1(i,k)} q_{kj} \geq g_k,$$

or if there exists $k \in \bar{N}(i)$ such that $\sum_{j \in J_2(i,k)} q_{kj} \geq g_k + G_i$. Otherwise we assume that $i \in I_{j_0}$.

Lemma 1. *Let $(\bar{X}, \bar{Z})$, $\bar{X} = ((\bar{x}_i), (\bar{x}_{ij}))$, $\bar{Z} = ((\bar{z}_i), (\bar{z}_{ij}), (\bar{s}_i))$ be a pessimistic feasible solution of the problem $(\mathcal{L}, \mathcal{F})$ and $\{I_j\}$ be a system of estimating subsets. For each $j_0 \in J$ the following holds: if $i_{j_0}(\{i \in I \mid \bar{x}_i - \bar{s}_i = 1\}) \notin I_{j_0}$, then $\sum_{i \in I} p_{ij_0} \bar{x}_{ij_0} = 0$.*

Proof. Consider $(0,1)$–vectors $\bar{x} - \bar{s} = (\bar{x}_i - \bar{s}_i)$, $i \in I$ and $\bar{z} = (\bar{z}_i)$, $i \in I$. If $\bar{x} - \bar{s} = 0$, then from (3) we obtain the required. Otherwise, set $i_x = i_{j_0}(\bar{x} - \bar{s})$. Assume that $i_x \notin I_{j_0}$ and consider the set $N(i_x) = \{i \in I \mid i \succ_{j_0} i_x\}$. If $N(i_x) \neq \emptyset$ and $\sum_{i \in N(i_x)} \bar{z}_i > 0$, then from (2) and (3) we get $\sum_{i \in I} \bar{x}_{ij_0} = 0$.

Otherwise, consider the set $J(i_x) = \{j \in J \mid i_x = i_j(I \backslash N(i_x))\}$. Since $(\bar{x}_i - \bar{s}_i) = \bar{z}_i = 0$ for all $i \in N(i_x)$, then $i_j(\bar{x} - \bar{s}) \succ_j i_j(\bar{z})$ for each $j \in J(i_x)$. From $i_x \notin I_{j_0}$ we get two possibilities:

(1) there exists $k \in N(i_x)$ such that for $J_1(i_x, k)$ we have $\sum_{j \in J_1(i_x, k)} q_{kj} \geq g_k$;

(2) there exists $k \in \bar{N}(i_x)$ such that for $J_2(i_x, k)$ we have $\sum\limits_{j \in J_2(i_x,k)} q_{kj} \geq g_k + G_{i_x}$.

In the first case we can construct a new feasible solution $Z = ((z_i), (z_{ij}), (s_i))$ of the problem $\mathcal{F}$ which differs from the optimal solution $\bar{Z}$ only in that $z_k = 1$ and $z_{kj} = 1$ for $j \in J_1(i_x, k)$.

For solutions Z and $\bar{Z}$ the following inequality holds:

$$F(Z) - F(\bar{Z}) = -g_k + \sum_{j \in J_1(i_x,k)} q_{kj} \geq 0.$$

If this inequality is strict we have a contradiction with optimality of $\bar{Z}$. If $\sum\limits_{i \in I} p_{ij_0} \bar{x}_{ij_0} > 0$ the replacement of the optimal solution $\bar{Z}$ of the problem $\mathcal{F}$ with a feasible solution Z does not reduce the objective function of the lower–level problem but reduces the upper–level one. It contradicts with the fact that $(\bar{X}, \bar{Z})$ is a pessimistic feasible solution.

In the second case we construct a feasible solution $Z = ((z_i), (z_{ij}), (s_i))$ of the problem $\mathcal{F}$, which differs from $\bar{Z}$ only in that $z_k = 1$, $z_{kj} = 1$ for $j \in J_2(i_x, k)$, and $s_{i_x} = 1$. For the lower–level objective function, we have:

$$F(Z) - F(\bar{Z}) = -g_k - G_{i_x} + \sum_{j \in J_2(k,i_x)} q_{kj} \geq 0.$$

By repeating the argument for the first case we get the Lemma 1 proved.

Corollary 1. *Let $(\bar{X}, \bar{Z})$ be a pessimistic feasible solution of the problem $(\mathcal{L}, \mathcal{F})$ and $\{I_j\}$ is a system of estimating subsets. There exists a pessimistic feasible solution (X, Z), $X = ((x_i), (x_{ij}))$, $Z = ((z_i), (z_{ij}), (s_i))$ of the problem $(\mathcal{L}, \mathcal{F})$ such that $L(X, Z) = L(\bar{X}, \bar{Z})$ and for each $j \in J$ the following inequality holds:*

$$\sum_{i \in I} x_{ij} \leq \sum_{i \in I_j} x_i. \tag{21}$$

Proof. Set (X, Z) to be equal to $(\bar{X}, \bar{Z})$. If the right hand side of (21) is positive, then (21) results from the constraints (2).

If for some $j \in J$ we have $x_i = 0$ for all $i \in I_j$ then Lemma 1 can be applied. Indeed, in this case $i_j(\{i \in I | x_i - s_i = 1\}) \notin I_j$ and thus $\sum\limits_{i \in I} p_{ij} x_{ij} = 0$. By setting $x_{ij} = 0$ for all $i \in I$ we get the required.

Consider the following problem, which we refer to as *estimating problem* for $(\mathcal{L}, \mathcal{F})$. It is obtained from the problem $(\mathcal{L}, \mathcal{F})$ by adding the constraints (21) and removing the lower–level objective function. From Corollary 1 we conclude that the first modification does not change the optimal value of the objective function. The second modification increases the feasible region by relaxing the constraints on the lower–level variables to get values from the set of optimal solutions. Obviously, after this relaxation all lower–level variables can be set to be equal to zero. This allows us to remove them as well. Finally, the estimating problem is written as follows:

$$\max_{(x_i),(x_{ij})} \left\{ - \sum_{i \in I} f_i x_i + \sum_{j \in J} \sum_{i \in I} p_{ij} x_{ij} \right\},$$

$$\sum_{i \in I} x_{ij} \leq 1, \quad j \in J;$$

$$x_{ij} \leq x_i, \quad i \in I, j \in J;$$

$$\sum_{i \in I} x_{ij} \leq \sum_{i \in I_j} x_i, \quad j \in J;$$

$$x_i, x_{ij} \in \{0, 1\}, \quad i \in I, j \in J.$$

Denote the value of its objective function on the feasible solution $X = ((x_i), (x_{ij}))$ with $B(X)$. Let $X^0 = ((x_i^0), (x_{ij}^0))$ be an optimal solution of the estimating problem.

Theorem 2. *Let X^0 be an optimal solution of the estimating problem. For each pessimistic feasible solution of the problem $(\mathcal{L}, \mathcal{F})$ the following inequality holds:* $L(\bar{X}, \bar{Z}) \leq B(X^0)$.

Proof. Let $(\bar{X}^*, \bar{Z}^*)$ be a pessimistic optimal solution of the problem $(\mathcal{L}, \mathcal{F})$. From the Corollary 1 we conclude that there exists a pessimistic feasible solution (X^*, Z^*) satisfying (21) and such that $L(X^*, Z^*) = L(\bar{X}^*, \bar{Z}^*)$. Since the value $B(X^0)$ is an optimal value of the estimating problem, which is a relaxation of the problem $(\mathcal{L}, \mathcal{F})$ with additional constraint (21), we have $B(X^0) \geq L(X^*, Z^*)$. The Theorem 2 is proved.

Thus computing the upper bound for the pseudo–Boolean function $f(x)$ consists in solving a single–level mixed–integer programming problem.

5   Conclusions and Future Research

In this paper, we have introduced a new model of competitive facility location, which belongs to the class of Stackelberg games. Players called Leader and Follower maximizes their profit obtained from customers serving with deduction of the fixed costs of facilities opening. The model of customers' behavior assumes that customer is captured by the side which opens the most preferable facility for him or her. The novelty of the model consists in ability of Follower to close Leader's facility by paying some known price. It models the situation of unfair competition where discrediting and other forms of dishonest activities can be applied to Leader's facilities in order to force them to close.

We propose the method to construct a pessimistic feasible solution corresponding to Boolean vector, representing Leader's facilities location. Consequently, the Leader's problem can be represented in a form of pseudo–Boolean function maximization. It allows to construct local search based methods and apply a large pool of metaheuristic schemes [7,12] to obtain approximate solutions of the problem in reasonable time.

The proposed upper bound can be utilized in estimation of the inexact methods effectiveness. Valid inequalities presented by the Corollary 1 strengthen the formulation of the problem and can increase the convergence rate of bilevel solvers to come. Due to proximity of the estimating problem and the Leader's problem, the optimal solution of the first one can be taken as a starting point of the search.

The next step of our research is incorporation of the fixed values of Leader's location variables into the procedure of upper bound calculation. This modification is necessary for the implicit enumeration scheme development but coupled with difficulties caused by an uncertainty of the status of Leader's facilities, which are fixed to be open in branching procedure, but are able to be closed by Follower. Another direction of research can be associated with protection planning, where Leader is able to protect some of his facilities from closing.

Acknowledgments. The research is supported by Russian Foundation for Basic Research (project 15-01-01446) and Presidium of Russian Academy of Sciences (program 15, project 227). We deeply grateful to Alexander Ageev for his assistance in preparing the English text.

References

1. Aksen, D., Piyade, N., Aras, N.: The budget constrained r-interdiction median problem with capacity expansion. Cent. Eur. J. Oper. Res. **18**(3), 269–291 (2010)
2. Aksen, D., Aras, N.: A matheuristic for leader-follower games involving facility location-protection-interdiction decisions. In: Talbi, E.-G., Brotcorne, L. (eds.) Metaheuristics for Bi-level Optimization. SCI, vol. 482, pp. 115–152. Springer, Heidelberg (2013)
3. Aliakbarian, N., Dehghanian, F., Salari, M.: A bi-level programming model for protection of hierarchical facilities under imminent attacks. Comp. Oper. Res. **64**, 210–224 (2015)
4. Beresnev, V.: Branch-and-bound algorithm for competitive facility location problem. Comput. Oper. Res. **40**, 2062–2070 (2013)
5. Beresnev, V.: On the competitive facility location problem with a free choice of suppliers. Autom. Remote Control **75**, 668–676 (2014)
6. Beresnev, V., Mel'nikov, A.: The branch-and-bound algorithm for a competitive facility location problem with the prescribed choice of suppliers. J. Appl. Ind. Math. **8**, 177–189 (2014)
7. Davydov, I.A., Kochetov, Y.A., Mladenovic, N., Urosevic, D.: Fast metaheuristics for the discrete $(r|p)$-centroid problem. Autom. Remote Control **75**, 677–687 (2014)
8. Dempe, S.: Foundations of Bilevel Programming. Kluwer Acad. Publ, Dortrecht (2002)
9. Krarup, J., Pruzan, P.M.: The simple plant location problem: survey and synthesis. Eur. J. Oper. Res. **12**(1), 36–81 (1983)
10. Kress, D., Pesch, E.: Sequential competitive location on networks. Eur. J. Oper. Res. **217**, 483–499 (2012)

11. Liberatore, F., Scaparra, M.P., Daskin, M.S.: Analysis of facility protection strategies against an uncertain number of attacks: the stochastic r-interdiction median problem with fortification. Comp. Oper. Res. **38**, 357–366 (2011)
12. Mel'nikov, A.: Randomized local search for the discrete competitive facility location problem. Autom. Remote Control **75**, 700–714 (2014)
13. Santos-Penate, D.R., Suarez-Vega, R., Dorta-Gonzalez, P.: The leader follower location model. Netw. Spat. Econ. **7**, 45–61 (2007)
14. Scaparra, M.P., Church, R.L.: A bilevel mixed-integer program for critical infrastructure protection planning. Comp. Oper. Res. **35**, 1905–1923 (2008)
15. Stackelberg, H.: The Theory of the Market Economy. Oxford Univ. Press, Oxford (1952)
16. Zhu, Y., Zheng, Z., Zhang, X., Cai, K.Y.: The r-interdiction median problem with probabilistic protection and its solution algorithm. Comp. Oper. Res. **40**, 451–462 (2013)

Variable Neighborhood Descent
for the Capacitated Clustering Problem

Jack Brimberg[1], Nenad Mladenović[2,3], Raca Todosijević[2],
and Dragan Urošević[2(✉)]

[1] Royal Military College of Canada, Kingston, ON, Canada
Jack.Brimberg@rmc.ca
[2] Mathematical Institute SANU, Belgrade, Serbia
racatodosijevic@gmail.com, {nenad,draganu}@mi.sanu.ac.rs
[3] LAMIH-UVHC, Le Mont Houy Valenciennes, Cedex 9, France

Abstract. In this paper we propose a Variable neighborhood descent
based heuristic for the capacitated clustering problem and related handover minimization problem. The performance of the proposed approach
is assessed on benchmark instances from the literature. The obtained
results confirm that of our approach is highly competitive with the state-
of-the-art methods, significantly outperforming all of them on the set of
randomly-generated instances tested.

Keywords: Optimization · Variable neighborhood descent · Heuristic ·
Clustering

1 Introduction

Given a set P containing N elements, each of which has a weight w_i ($i \in \{1, 2, ..., N\}$), the goal of the Capacitated Clustering Problem (CCP) is to partition the set P into a required number G of disjoint groups (clusters) so that
the sum of diversities over each cluster is maximized and the sum of the weights
of the elements in each cluster is within some capacity limits. The diversity of
each cluster is expressed as the sum of the distances between the elements in the
cluster. The distance d_{ij} between any pair of elements i and j actually reflects
diversity between them.

Formally, using binary variables $x_{ig}, i = 1, 2, \ldots, N, g = 1, 2, \ldots, G$ such that
x_{ig} receives value 1 if the element i is assigned to the group g, and 0 otherwise,
CCP may be expressed as the following quadratic binary integer program:

$$\max \sum_{g=1}^{G} \sum_{i=1}^{N} \sum_{j=1}^{N} d_{ij} x_{ig} x_{jg} \tag{1}$$

s.t.

$$\sum_{g=1}^{G} x_{ig} = 1, \quad i = 1, 2, \ldots, N \tag{2}$$

© Springer International Publishing Switzerland 2016
Y. Kochetov et al. (Eds.): DOOR 2016, LNCS 9869, pp. 336–349, 2016.
DOI: 10.1007/978-3-319-44914-2_27

$$\sum_{i=1}^{N} w_i x_{ig} \geq a_g, \quad g = 1, 2, \ldots, G \tag{3}$$

$$\sum_{i=1}^{N} w_i x_{ig} \leq b_g, \quad g = 1, 2, \ldots, G \tag{4}$$

$$x_{ig} \in \{0, 1\}, \quad i = 1, 2, \ldots, N, \quad g = 1, 2, \ldots, G \tag{5}$$

The constraints (2) ensure that each element is assigned to exactly one group. Constraints (3) and (4) guarantee that the minimum capacity, a_g, and the maximum capacity, b_g, requirements of each group are fulfilled.

The applications of CCP arise in the context of facility planners at mail processing and distribution centers within the US Postal Service [2] as well as in the context of mobility networks [9].

CCP is an NP hard problem [4]. To tackle it several methods have been proposed. In [2], the authors developed greedy randomized adaptive search procedures (GRASP) coupled with variable neighborhood descent (VND) variants. In addition, they proposed a Path Relinking post-processing procedure which did not result in a significant improvement. In [9], the authors proposed several randomized heuristics for solving the handover minimization problem, an equivalent of CCP where the objective is to minimize the sum of diversities over clusters instead of maximize. In particular they designed GRASP with path-relinking for the generalized quadratic assignment problem, a GRASP with evolutionary path-relinking, and a biased random-key genetic algorithm. Recently, Martínez-Gavara et al. [6] proposed tabu search and several GRASP variants for solving CCP, which may be considered as state-of-the-art methods. Note that in the case $w_i = 1$ CCP becomes the Maximally Diverse Grouping Problem (MDGP). To solve MDGP several specialized heuristics have been proposed in the literature such as ones described in [1,3,5,10,11].

In this paper we develop a Variable neighborhood descent (VND) heuristic for solving the CCP problem. VND is the deterministic variant of Variable Neighborhood Search (VNS) [8] that uses $k_{max} > 1$ predefined neighborhood structures in the search for a better solution. Once several neighborhood types are defined $(N_1, \ldots, N_{k_{max}})$, the deterministic search using all of them may be organized in many different ways. Most common is the so-called Basic VND, where the order of neighborhoods is chosen, and then a search following that order is organized. If a better solution is not found in the complete neighborhood N_k, the search continues in the next one, i.e., N_{k+1}. If a better solution is found, then the move is performed and the search continues again from the first neighborhood (for a recent survey on VND, see [7]). Following this sequential framework, we propose a VND based heuristic for CCP, and assess its performance by extensive testing on the benchmark instances from the literature. The obtained computational results demonstrate the superiority of the proposed heuristic over the existing state-of-the-art.

The rest of the paper is organized as follows. In the next section we describe the main steps of VND for solving CCP. Section 3 presents numerical experiments devoted to assessing the quality of several improvement procedures and the proposed Multi-start VND. Finally, Sect. 4 concludes the paper.

2 VND for CCP

The VND we develop for CCP follows the steps presented in Algorithm 1. In what follows we provide a detailed description of its main ingredients i.e., solution representation, initial solution, neighborhood structures and local searches.

2.1 Solution Representation

The solution space of CCP consists of all feasible partitions of the set P into G disjoint groups such that sum of weights of elements in each group is within desired bounds. Each solution can be represented by an array x of length N where each value x_i represents the label of the group containing element i in the current solution. Along with array x, in order to make the local searches more efficient (as will be explained later), we maintain two auxiliary data structures:

- the array sw with G elements, where each element sw_g equals to the total weight of elements belonging to the group g in the corresponding solution, i.e.:

$$sw_g = \sum_{i:x_i=g} w_i \qquad (6)$$

- matrix sd whose dimensions are $N \times G$ so that each entry sd_{ig} equals to the sum of distances between the element i and all elements belonging to the group g in the corresponding solution:

$$sd_{ig} = \sum_{j:x_j=g} d_{ij} \qquad (7)$$

2.2 Initial Solution

The procedure for generating an initial solution includes three phases. In the first phase, G elements are chosen at random and put into different groups. After that, in the second phase, new elements are iteratively inserted into groups until the total weight of elements in each group reaches the lower limit. At each iteration of the second phase we choose a random element i currently not inserted and after that choose the group g for insertion in the following way:

$$g = \operatorname{argmax}\{\frac{sd_{ig}}{sw_g}|g = 1, 2, ..., G; sw_g < a_g\}. \qquad (8)$$

In the third phase we distribute the remaining elements by iteratively selecting the next element i and choosing the group g for insertion in a similar way as in the second phase:

$$g = \operatorname{argmax}\{\frac{sd_{ig}}{sw_g}|g = 1, 2, ..., G; sw_g + w_i \leq b_g\}. \qquad (9)$$

2.3   Neighborhood Structures for the CCP

Each neighborhood is defined by a move that relocates a certain number of elements from their current groups to another. In particular, the following three neighborhoods are distinguished.

Insertion neighborhood (N_{ins}). The insertion neighborhood of a current solution x consists of all solutions obtained by choosing one element i and moving it from the group $g = x_i$ to a group $g' \neq g$. Note that the change of the objective function value obtained by removing element i from group $g = x_i$ and inserting it into group $g' \neq g$ may be calculated, using previously defined data structures as:

$$\Delta f = sd_{ig'} - sd_{ig}. \tag{10}$$

However, the insertion neighborhood may contain also solutions that violate the capacity constraints. In order to quickly recognize feasible solutions the data structure sw is used. Namely, to check feasibility of a neighboring solution it suffices to check whether the capacity constraints of the two groups involved in the move (defining this neighboring solution) remain fulfilled or not. Using the data structure sw this may be accomplished in the way stated in the following two inequalities:

$$sw_g - w_i \geq a_g \quad \text{and} \quad sw_{g'} + w_i \leq b_{g'} \tag{11}$$

Swap neighborhood (N_{swap}). The swap neighborhood of the current solution x consists of all solutions obtained by selecting two elements i and j belonging to different groups in the current solution ($g' = x_i \neq x_j = g''$) and moving each of them to the group containing the other element i.e.,: the element i will be moved to the group g'', while the element j will be relocated to the group g'. In this case, the difference between the objective function value of the solution obtained by performing such a move and the current solution value may be calculated using the data structure sd:

$$\Delta f = (sd_{ig''} + sd_{jg'}) - (sd_{ig'} + sd_{jg''}) - 2d_{ij}. \tag{12}$$

Similarly, as in the case of the insertion move, some swap moves lead to infeasible solutions. So, in order to check the feasibility of a resulting solution, it is necessary to check whether capacity requirements of groups g' and g'' remain satisfied. This may be accomplished by verifying the next two inequalities that make use of the data structure sw:

$$a_{g'} \leq sw_{g'} - w_i + w_j \leq b_{g'} \quad \text{and} \quad a_{g''} \leq sw_{g''} - w_j + w_i \leq b_{g''} \tag{13}$$

Two Out − One In neighborhood ($N_{2out1in}$)). This neighborhood consists of all solutions obtained by selecting three elements i_1, i_2 and j such that elements i_1 and i_2 belong to the same group g' (i.e., $g' = x_{i_1} = x_{i_2}$) while the element j belongs to another group g'' (i.e., $g'' = x_j \neq g'$), and moving elements i_1 and i_2 to the group g'' and the element j to group g'. The change of the objective function value caused by executing such a move can be calculated in the following way using the data structure sd:

$$\Delta f = (sd_{i_1 g''} + sd_{i_2 g''} + sd_{jg'}) - (sd_{i_1 g'} + sd_{i_2 g'} + sd_{jg''}) + 2d_{i_1 i_2} - 2(d_{i_1 j} + d_{i_2 j}) \quad (14)$$

Following the same principle as in the case of the swap neighborhood, the feasibility of each solution in this neighborhood may be checked by verifying the capacity constraints of the two groups involved in the move as stated in the next two inequalities:

$$a_{g'} \le sw_{g'} - w_{i_1} - w_{i_2} + w_j \le b_{g'} \quad a_{g''} \le sw_{g''} - w_j + w_{i_1} + w_{i_2} \le b_{g''} \quad (15)$$

2.4   Variable neighborhood descent

The three neighborhood structures above are explored within a sequential Variable Neighborhood Descent (VND) procedure whose steps are given in Algorithm 1. In the algorithm, the local search procedure is called three times. Statement LocalSearch(x, N_i) means that the local search relative to the current neighborhood structure is performed using x as the initial solution. For local search within the Insertion and Swap neighborhoods, the local search continues until there is an improvement in the given neighborhood of the current solution. We use the so–called first–improvement strategy which means that each time an improved solution in the given neighborhood is found, we make the corresponding move and continue exploring the same neighborhood now centered at this new solution. On the other hand, the local search within the 2Out–1In neighborhood terminates after finding the first solution which is better than the current one. More precisely, all three local searches use the first improvement strategy, but they differ in the way they proceed after executing an improvement move: the first two continue the search w.r.t. the same neighborhood structure while the last one finishes its work. The reason for this strategy is the fact that the 2Out–1In neighborhood is significantly larger than the Insertion and Swap neighborhoods, and therefore the exploration of the entire neighborhood is a time consuming process.

The order of neighborhoods presented in Algorithm 1, i.e., *Insertion, Swap* and *2Out–1In*, is selected after exhaustive experimentation as will be shown in Sect. 3.

3   Computational Results

The computational experiments are devoted to an evaluation of the quality of different improvement procedures and the benefits of using sequential Basic VND instead of using the single neighborhoods. The second part assesses the quality of the proposed multi-start VND in comparison with the current state-of-the-art approaches. The proposed method as well as the improvement procedures are implemented in C++ and executed on an Intel Core 2 Duo CPU E6750 with 8Gb RAM. For testing purposes benchmark instances from the literature are used.

Algorithm 1. Variable Neighborhood Descent

Function VND(x);
$k \leftarrow 1$;
while $k \leq 3$ do
    if $k == 1$ then $x' \leftarrow$ LocalSearch(x, N_{ins});
    if $k == 2$ then $x' \leftarrow$ LocalSearch(x, N_{swap});
    if $k == 3$ then $x' \leftarrow$ LocalSearch($x, N_{2out1in}$);
    $k \leftarrow k + 1$;
    if $f(x') > f(x)$ then
        $x \leftarrow x'$; if $k > 2$ then $k \leftarrow 1$;
    end
end
return x

These instances constitute CCPLIB publicly available at http://www.optsicom. es/ccp/. Based on their characteristics, four data sets may be distinguished:

- RanReal 240 - 20 instances with $N = 240$, $G = 12$, $a_g = 75$, and $b_g = 125$;
- RanReal 480 - 20 instances with $N = 480$, $G = 20$, $a_g = 100$, and $b_g = 150$;
- Sparse 82 - 10 instances with $N = 82$, $G = 8$, $a_g = 25$, and $b_g = 75$;
- Handover - 83 synthetic instances introduced in [9] for the handover minimization problem.

3.1   Comparison of Improvement Procedures

The first set of experiments aims to compare different improvement procedures based on the exploration of one or more neighborhoods. Namely, we compare three local search procedures that use Insertion, Swap and 2Out – 1In neighborhood structures, VND obtained by exploring Insertion and Swap neighborhoods (VND2) in this order, and VND obtained by using all three introduced neighborhoods (VND3) whose steps are given in Algorithm 1. These methods are compared on only one large instance with 480 elements (RanReal480_01). Each Local Search / VND variant is executed 1000 times, starting each time from a different random solution. The summarized results are reported in Table 1 and Fig. 1.

In Table 1 columns 2, 3, and 4 give the minimum, the average, and the maximum % deviation from the best known solution, respectively, over 1000 runs. Columns 5, 6, and 7 report the minimum, average and maximum normalized distance between the generated local optima over 1000 runs and the best known solution. The (normalized) distance between solutions x and y is defined in the following way:

$$d(x,y) = \frac{|\{(i,j)|1 \leq i < j \leq N, ((x_i = x_j) \wedge (y_i \neq y_j)) \vee ((x_i \neq x_j) \wedge (y_i = y_j))\}|}{|\{(i,j)|1 \leq i < j \leq N, (x_i = x_j) \vee (y_i = y_j)\}|}$$

$$(16)$$

Table 1. Comparison of different local searches on instance RanReal480_10

Imp. procedure	% deviation			Norm. distance			Time
	Min.	Avg.	Max.	Min.	Avg.	Max.	
Insertion LS	32.537	34.880	37.214	0.947	0.956	0.963	0.001
Swap LS	32.537	34.880	37.214	0.943	0.957	0.964	0.038
2 Out – 1 In LS	3.021	4.762	7.253	0.598	0.674	0.761	9.901
VND2	21.133	24.359	26.963	0.925	0.939	0.949	0.050
VND3	0.787	1.924	5.712	0.530	0.623	0.783	1.417

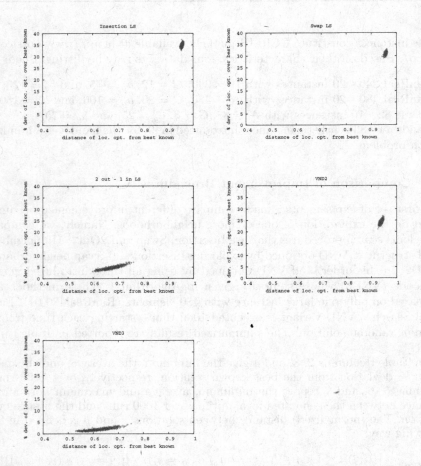

Fig. 1. Distribution of 1000 local maxima on distance–to–target diagram for different local search algorithms

which estimates the fraction of pairs belonging to the same group in one solution, but not to the same group in the other solution. The last column reports the average computing time spent to reach a local maximum (in seconds).

From the table we can make the following observations:

- There is a correlation between the percentage deviation and the normalized distance: for example
  - The percentage deviations for the local optima obtained by Local search with respect to the 2Out–1In neighborhood belong to the interval [3.021, 7.253] while the distances belong to the interval [0.598, 0.761]
  - The percentage deviations for the local optima obtained by VND2 belong to the interval [21.133, 26.963] while the distances belong to the interval [0.925, 0.949]

This observation is also confirmed by the so-called F-test. Namely, the F values of 324.852, 417.955, 4612.75, 923.036 and 2914.707 obtained for Insertion LS, Swap LS, 2Out–1In LS, VND2, and VND3, respectively, are greater than the corresponding critical F value 254.19. Hence, for each local search we accept the alternative hypothesis that there is a positive correlation between the percentage deviation and the normalized distance.

- The percentage deviations of local optima obtained by Local search with respect to the 2Out–1In neighborhood are significantly lower than with the other two neighborhoods, but execution time is significantly greater.
- Very efficient VND, which outperforms all other considered improvement procedures, is obtained by combining 2Out–1In, Insertion and Swap neighborhoods into VND3.

Distributions of local optima for all five variants are shown on distance-to-target diagrams in Fig. 1, where each local optimum is presented by the point (x, y) whose coordinates are:

- x - the normalized distance between the local optimum and the best known solution;
- y - the percentage deviation of value of the local optimum from the best known value

Based on these results we decided to use multi-start VND3 to compare with other methods found in the literature.

3.2   Main Computational Results

In this section we present the comparison of our multi-start VND with the state-of-the-art approaches from the literature: GRASP from [2] (referred to as Prev_GRASP); an adaptation of tabu search from [5] (AdTS_SO); GRASP, tabu search (TS) and GRASP+TS proposed in [6]; and GRASP with Path Relinking developed for handover minimization in [9] (PR-HMP).

In Tables 2 and 3 summarized results on each data set are presented. On each test instance multi-start VND has been executed with a time limit of 60 seconds on an Intel Core 2 Duo CPU E6750. Note that AdTS_SO, GRASP, TS and GRASP+TS, the best methods among the previous ones, were executed with the same time limit of 60 seconds on an Intel Core 2 Quad CPU Q8300 (the faster machine). In the first column of Table 2 the name of the data set is given. The next five columns contain results for already published methods (all these results are provided to us by Anna Martinez-Gavara). The last three columns contain summarized results for VND, i.e., the average of the best, average, and the worst solution values found on a certain data set in ten runs. Handover instances are grouped according to the number of elements and summarized results are given in Table 3. Again, the table provides summarized results of existing methods and our VND. Note that the objective on Handover instances is to minimize handover (the objective function in (1)). Detailed results on the entire set of instances are given in the Appendix.

Table 2. Summarized results for Random Real instances and Sparse instances

Data set	Prev_GRASP	AdTS_SO	TS	GRASP	GRASP+TS	VND		
						Best	Avg.	Worst
RanReal 240	182171.08	195945.10	201125.69	171073.71	200253.11	203616.70	203226.62	202899.21
RanReal 480	300192.92	476879.89	505615.95	413325.91	505202.05	517305.80	516168.78	514976.26
Sparse 82	1326.89	1330.98	1271.00	1330.17	1329.42	1331.20	1331.20	1331.20

From the presented results it follows that multi-start VND outperforms all existing state-of-the-art methods from the literature on Random Real instances. On the Handover set of instances, our VND is better than all other heuristics, except AdTS_SO. According to Table 3, AdTS_SO was better on average on the 2 largest instances, worse on the third largest, and tied on the remaining.

In Tables 4 and 5 we present the results of a non-parametric Wilcoxon-Mann-Whitney test for statistical significance applied on the results obtained by the methods in comparison. The level of significance is chosen to be $\alpha = 0.05$. From Table 4 we may conclude the following: on the RanReal240 data set, VND significantly outperforms all methods; on the RanReal480 data set, there is no significant difference between VND and GRASP+TS, while all the other methods are significantly outperformed by our VND; on the Sparse 82 data set, VND

Table 3. Summarized results for Handover instances

Instance size	PR-HMP 60 s	PR-HMP 900 s	AdTS_SO	GRASP+TS	CPLEX	VND		
						Best	Avg.	Worst
20	1056.67	1056.67	1056.67	1056.67	1056.67	1056.67	1056.67	1056.67
30	2295.71	2295.71	2295.71	2295.71	2295.71	2295.71	2295.71	2295.71
40	3090.27	3090.27	3090.27	3090.27	3090.27	3090.27	3090.27	3090.27
100	39167.33	38844.80	38725.87	38523.73	38430.13	38463.87	38581.27	38707.47
200	152867.07	150533.07	145984.93	146885.87	145980.27	146477.47	146973.13	147426.80
400	605507.87	594255.47	568690.40	574382.00	568690.40	570082.53	572243.44	574158.13

Table 4. Critical values are 52 if number of instances is 20 and 8 if number of instances is 10

Instance	# inst.	VND vs Prev_GRASP	VND vs AdTS_SO	VND vs TS	VND vs GRASP	VND vs GRASP+TS
RanReal 240	20	**210 : 0**	**210 : 0**	**210 : 0**	**210 : 0**	**210 : 0**
RanReal 480	20	**210 : 0**	**210 : 0**	**210 : 0**	**210 : 0**	147 : 63
Sparse 82	10	**54.5 : 0.5**	32.5 : 22.5	**55 : 0**	**50 : 5**	**47.5 : 7.5**

Table 5. Critical values are 5 if number of instances is 9, 21 if number of instances is 14, and 25 if number of instances is 15

Instance	# inst.	VND vs PR-HMP600	VND vs PR-HMP900	VND vs AdTS_SO	VND vs GRASP+TS	VND vs CPLEX
Handover - 20	9	22.5 : 22.5	22.5 : 22.5	22.5 : 22.5	22.5 : 22.5	22.5 : 22.5
Handover - 30	14	52.5 : 52.5	52.5 : 52.5	52.5 : 52.5	52.5 : 52.5	52.5 : 52.5
Handover - 40	15	60 : 60	60 : 60	60 : 60	60 : 60	60 : 60
Handover - 100	15	**119.5 : 0.5**	**106.5 : 13.5**	83 : 37	37 : 83	32 : 88
Handover - 200	15	**120 : 0**	**117.5 : 2.5**	*9.5 : 110.5*	**100.5 : 19.5**	*9.5 : 110.5*
Handover - 400	15	**120 : 0**	**120 : 0**	*13 : 107*	**119 : 1**	*13 : 107*

significantly outperforms all methods except AdTS_SO. On the other hand, from Table 5 we may conclude that on Handover instances with 200 and 400 elements, VND significantly outperforms PR-HMP and GRASP+TS while it is significantly outperformed by AdTS_SO and CPLEX. In addition it follows that on small size Handover instances there is no significant difference between VND and other methods.

4   Conclusions

In this paper we study the Capacitated Clustering Problem (CCP) which aims to partition a given set in a predefined number of groups so that the diversity among them is maximized and the weight of each group is within required limits. To solve this NP hard problem we propose three neighborhood structures and use them within a sequential VND heuristic. In order to assess the merit of this new approach, we have performed extensive testing on the benchmark instances from the literature. The obtained results confirm that of our approach is highly competitive with the state-of-the-art methods significantly outperforming all of them on randomly-generated instances.

Appendix

Here we present detailed results obtained by all considered methods. For RanRail instances, a larger value indicates a better solution, while for Handover instances the smaller values are better (Tables 6, 7, 8, 9 and 10).

Table 6. Detailed results for Random real instances with 240 elements

Instance	Prev_GRASP	AdTS_SO	TS	GRASP	GRASP+TS	VND		
						Best	Average	Worst
RanReal240_01	191036.02	214551.76	220226.18	176036.63	219783.86	223220.90	222931.65	222771.36
RanReal240_02	185798.62	195530.69	200838.30	172012.53	198460.55	202095.20	201691.73	201331.47
RanReal240_03	177679.46	188525.31	193859.64	166702.33	192114.51	196391.99	196199.95	196001.32
RanReal240_04	197891.69	212960.48	219449.77	178811.29	218197.83	222711.88	222455.67	222055.42
RanReal240_05	170257.12	186650.26	189761.81	165567.11	190303.48	193570.70	192892.46	192612.46
RanReal240_06	191333.75	205944.67	210283.10	174582.81	211133.78	214707.61	214256.87	213801.64
RanReal240_07	188621.39	199490.34	204135.95	177680.33	203220.69	207115.03	206830.64	206477.47
RanReal240_08	181243.99	194202.85	201800.33	171333.36	200695.50	202820.87	202576.92	202249.77
RanReal240_09	186776.58	200032.73	205216.80	173904.23	204184.97	207004.32	206739.99	206272.79
RanReal240_10	170689.20	185400.22	188445.01	165790.13	187605.60	190248.29	189846.25	189514.36
RanReal240_11	183728.50	197887.83	199946.79	170377.82	198975.29	202692.08	202179.11	201772.79
RanReal240_12	177213.43	191849.93	196681.44	168895.22	195151.19	199006.57	198691.78	198398.79
RanReal240_13	182626.67	189022.28	199299.16	170930.93	197380.00	199766.31	199484.53	199219.61
RanReal240_14	191464.86	217608.00	225673.50	175415.88	223312.09	226724.31	226584.34	226338.20
RanReal240_15	168569.52	181484.27	185752.72	163757.16	185090.80	189022.16	188426.05	188109.30
RanReal240_16	180009.20	191972.77	200223.47	170355.34	198040.51	202272.69	201554.40	201180.86
RanReal240_17	178388.69	188751.19	190064.90	168755.03	190517.19	192878.67	192373.26	191882.45
RanReal240_18	175509.44	186078.02	190713.89	165896.21	189637.71	192747.90	192334.36	192089.73
RanReal240_19	175657.91	189107.14	195055.37	168383.72	193445.34	197005.23	196657.15	196435.39
RanReal240_20	188925.60	201851.31	205085.62	176286.14	207811.23	210331.30	209825.28	209468.92
Average	182171.08	195945.10	201125.69	171073.71	200253.11	203616.70	203226.62	202899.21

Table 7. Detailed results for Random real instances with 480 elements

Instance	Prev_GRASP	AdTS_SO	TS	GRASP	GRASP+TS	VND		
						Best	Average	Worst
RanReal480_01	299619.14	497472.55	536382.59	420832.68	500223.84	549014.41	547892.82	546286.17
RanReal480_02	304227.71	470373.18	488909.30	411623.39	533603.96	503366.82	502322.88	501119.26
RanReal480_03	298926.33	454315.61	477286.03	404696.02	492730.51	487656.24	486391.16	485221.87
RanReal480_04	293139.02	485070.78	497608.91	416435.47	475803.24	514164.35	513016.49	512519.53
RanReal480_05	299281.24	449988.89	464195.50	390749.48	501165.11	475361.16	474143.57	472698.82
RanReal480_06	298472.76	477858.44	515459.84	425578.62	463642.73	524772.59	523003.55	520939.69
RanReal480_07	299188.50	492304.63	523172.66	414492.43	513597.68	537904.95	536212.13	534476.05
RanReal480_08	300385.03	490816.35	514573.17	419901.78	525983.27	524539.40	523419.36	522634.67
RanReal480_09	303649.83	497163.10	533724.82	424433.32	510734.97	549638.40	548621.16	547690.30
RanReal480_10	306294.40	464872.52	499690.31	408016.64	537376.71	512324.63	511338.69	510695.01
RanReal480_11	299633.46	488063.47	503977.89	414517.44	501140.16	517628.43	516058.83	514660.96
RanReal480_12	301473.40	452603.22	485198.38	404855.36	500631.77	493481.90	491899.13	490561.03
RanReal480_13	295120.81	486569.63	511113.37	423469.53	482841.68	524838.96	524255.44	523146.83
RanReal480_14	301714.49	460747.35	496735.29	396540.06	515433.77	506472.86	505311.16	504011.07
RanReal480_15	298579.03	465794.52	498214.01	410008.93	492767.37	509668.23	508386.57	507261.68
RanReal480_16	302554.13	491840.64	529684.49	420109.36	495708.62	542107.14	541827.80	541137.01
RanReal480_17	297147.32	485308.51	523209.02	421288.17	532540.46	530474.35	529671.66	528983.76
RanReal480_18	297464.34	472499.44	507540.96	417418.02	519121.34	517785.85	516230.55	514441.58
RanReal480_19	300848.84	482816.75	505461.40	412504.97	503347.70	515234.80	514385.62	513680.57
RanReal480_20	306138.70	471118.18	500181.05	409046.51	505646.05	509680.56	508986.95	507359.35
Average	300192.92	476879.89	505615.95	413325.91	505202.05	517305.80	516168.78	514976.26

Table 8. Detailed results for Sparse instances

Instance	Prev_GRASP	AdTS_SO	TS	GRASP	GRASP+TS	VND		
						Best	Average	Worst
Sparse82_01	1337.66	1342.17	1268.05	1338.03	1337.31	1342.17	1342.17	1342.17
Sparse82_02	1295.38	1304.48	1225.95	1306.64	1306.64	1306.64	1306.64	1306.64
Sparse82_03	1350.89	1353.94	1295.32	1352.00	1353.36	1353.94	1353.94	1353.94
Sparse82_04	1286.65	1291.22	1233.28	1288.45	1283.97	1291.22	1291.22	1291.22
Sparse82_05	1351.38	1352.35	1309.09	1352.35	1352.35	1352.35	1352.35	1352.35
Sparse82_06	1343.59	1354.61	1301.14	1354.61	1350.01	1354.61	1354.61	1354.61
Sparse82_07	1265.34	1266.94	1206.38	1266.70	1266.94	1266.94	1266.94	1266.94
Sparse82_08	1393.02	1393.02	1329.98	1393.02	1393.02	1393.02	1393.02	1393.02
Sparse82_09	1288.49	1294.12	1258.39	1293.39	1294.12	1294.12	1294.12	1294.12
Sparse82_10	1356.48	1356.98	1282.43	1356.48	1356.48	1356.98	1356.98	1356.98
Average	1326.89	1330.98	1271.00	1330.17	1329.42	1331.20	1331.20	1331.20

Table 9. Detailed results for Handover instances - Part I

Instance size	PR-HMP 60 s	PR-HMP 900 s	AdTS_SO	GRASP+TS	CPLEX	VND		
						Best	Avg.	Worst
20_5_270001	540	540	540	540	540	540	540	540
20_5_270002	54	54	54	54	54	54	54	54
20_5_270003	816	816	816	816	816	816	816	816
20_5_270004	126	126	126	126	126	126	126	126
20_5_270005	372	372	372	372	372	372	372	372
20_10_270001	2148	2148	2148	2148	2148	2148	2148	2148
20_10_270002	1426	1426	1426	1426	1426	1426	1426	1426
20_10_270003	2458	2458	2458	2458	2458	2458	2458	2458
20_10_270004	1570	1570	1570	1570	1570	1570	1570	1570
Average 20	1056.67	1056.67	1056.67	1056.67	1056.67	1056.67	1056.67	1056.67
30_5_270001	772	772	772	772	772	772	772	772
30_5_270002	136	136	136	136	136	136	136	136
30_5_270003	920	920	920	920	920	920	920	920
30_5_270004	52	52	52	52	52	52	52	52
30_5_270005	410	410	410	410	410	410	410	410
30_10_270001	3276	3276	3276	3276	3276	3276	3276	3276
30_10_270002	1404	1404	1404	1404	1404	1404	1404	1404
30_10_270003	2214	2214	2214	2214	2214	2214	2214	2214
30_10_270004	2150	2150	2150	2150	2150	2150	2150	2150
30_10_270005	2540	2540	2540	2540	2540	2540	2540	2540
30_15_270001	6178	6178	6178	6178	6178	6178	6178	6178
30_15_270002	4042	4042	4042	4042	4042	4042	4042	4042
30_15_270003	4126	4126	4126	4126	4126	4126	4126	4126
30_15_270004	3920	3920	3920	3920	3920	3920	3920	3920
Average 30	2295.71	2295.71	2295.71	2295.71	2295.71	2295.71	2295.71	2295.71
40_5_270001	610	610	610	610	610	610	610	610
40_5_270002	136	136	136	136	136	136	136	136
40_5_270003	234	234	234	234	234	234	234	234
40_5_270004	232	232	232	232	232	232	232	232
40_5_270005	774	774	774	774	774	774	774	774
40_10_270001	4544	4544	4544	4544	4544	4544	4544	4544
40_10_270002	2068	2068	2068	2068	2068	2068	2068	2068
40_10_270003	2090	2090	2090	2090	2090	2090	2090	2090
40_10_270004	1650	1650	1650	1650	1650	1650	1650	1650
40_10_270005	4316	4316	4316	4316	4316	4316	4316	4316
40_15_270001	8646	8646	8646	8646	8646	8646	8646	8646
40_15_270002	4586	4586	4586	4586	4586	4586	4586	4586
40_15_270003	5396	5396	5396	5396	5396	5396	5396	5396
40_15_270004	4800	4800	4800	4800	4800	4800	4800	4800
40_15_270005	6272	6272	6272	6272	6272	6272	6272	6272
Average 40	3090.27	3090.27	3090.27	3090.27	3090.27	3090.27	3090.27	3090.27

Table 10. Detailed results for Handover instances - Part II

Instance size	PR-HMP 60 s	PR-HMP 900 s	AdTS_SO	GRASP+TS	CPLEX	VND Best	Avg.	Worst
100_15_270001	19174	19174	19000	19000	19000	19000	19190.80	19298
100_15_270002	22686	22686	22686	22686	22686	22686	22768.40	22806
100_15_270003	14696	14638	14558	14558	14558	14558	14558.00	14558
100_15_270004	20172	19802	19700	19700	19700	19700	19700.00	19700
100_15_270005	23114	22892	22746	22746	22746	22782	22959.80	23020
100_25_270001	37412	36960	36412	36412	36412	36412	36455.20	36534
100_25_270002	40278	39886	38608	38608	38608	38608	38701.20	38768
100_25_270003	34110	33978	32692	32696	32692	32692	32716.00	32830
100_25_270004	37350	36958	35322	35322	35322	35322	35347.60	35470
100_25_270005	37956	37208	36690	36690	36690	36882	36994.80	37148
100_50_270001	61610	61474	61956	61410	61410	61476	61712.20	61858
100_50_270002	62596	62400	63128	62208	62208	62296	62515.40	62692
100_50_270003	55472	54868	55854	54846	54846	54928	55174.00	55688
100_50_270004	58326	58068	59072	57894	57894	58192	58336.00	58490
100_50_270005	62558	61680	62464	63080	61680	61424	61589.60	61752
Average	39167.33	38844.80	38725.87	38523.73	38430.13	38463.87	38581.27	38707.47
200_15_270001	86528	84210	81558	81558	81558	81558	81611.20	81824
200_15_270002	95840	92808	89564	90794	89564	90976	91577.00	92414
200_15_270003	79320	79232	79232	79232	79232	79232	79531.40	79956
200_15_270004	79464	78358	78428	79108	78358	79012	79607.40	80008
200_15_270005	99732	98070	96040	96088	96040	96248	96758.20	97324
200_25_270001	139492	134886	133578	134432	133578	134536	134793.40	135010
200_25_270002	143860	140994	134072	136392	134072	135556	136118.00	136670
200_25_270003	142976	142976	136892	138166	136892	137506	138179.60	138736
200_25_270004	137870	131316	128590	129628	128590	128966	129404.40	129830
200_25_270005	158736	154364	148312	149772	148312	148304	149445.40	150014
200_50_270001	226108	225080	216640	216766	216640	216410	216961.80	217400
200_50_270002	222116	220224	213208	213766	213208	213712	213992.40	214368
200_50_270003	225658	224916	215700	217168	215700	216530	216854.80	217118
200_50_270004	215834	215410	207162	208658	207162	207412	207810.60	208190
200_50_270005	239472	235152	230798	231760	230798	231204	231951.40	232540
Average	152867.07	150533.07	145984.93	146885.87	145980.27	146477.47	146973.13	147426.80
400_15_270001	415284	400450	373458	375106	373458	372802	373658.00	374872
400_15_270002	401710	380550	368686	374666	368686	372356	375841.80	377898
400_15_270003	375532	375130	356722	358994	356722	358846	360574.20	362852
400_15_270004	378590	353808	335374	338858	335374	333948	337383.60	339364
400_15_270005	401546	372506	361000	366556	361000	366650	368426.20	370752
400_25_270001	582280	571484	552576	556878	552576	552286	554673.80	557246
400_25_270002	571750	562430	534458	536658	534458	533928	536058.00	537828
400_25_270003	573036	563564	531228	537864	531228	532260	535742.80	537914
400_25_270004	515828	512258	487002	489922	487002	488546	489652.00	490870
400_25_270005	604928	596358	554798	561428	554798	556794	559350.80	562890
400_50_270001	871340	864784	831626	843754	831626	833104	834656.00	836418
400_50_270002	861396	859170	830384	838612	830384	831156	833272.40	834442
400_50_270003	844196	837952	810044	813798	810044	811828	813111.20	814900
400_50_270004	808748	797706	768104	776166	768104	769506	771372.00	772646
400_50_270005	876454	865682	834896	846470	834896	837228	839878.80	841480
Average	605507.87	594255.47	568690.40	574382.00	568690.40	570082.53	572243.44	574158.13

References

1. Brimberg, J., Mladenović, N., Urošević, D.: Solving the maximally diverse grouping problem by skewed general variable neighborhood search. Inf. Sci. **295**, 650–675 (2015)
2. Deng, Y., Bard, J.F.: A reactive grasp with path relinking for capacitated clustering. J. Heuristics **17**(2), 119–152 (2011)
3. Fan, Z., Chen, Y., Ma, J., Zeng, S.: A hybrid genetic algorithmic approach to the maximally diverse grouping problem. J. Oper. Res. Soc. **62**(1), 92–99 (2011)
4. Feo, T.A., Khellaf, M.: A class of bounded approximation algorithms for graph partitioning. Networks **20**(2), 181–195 (1990)
5. Gallego, M., Laguna, M., Martí, R., Duarte, A.: Tabu search with strategic oscillation for the maximally diverse grouping problem. J. Oper. Res. Soc. **64**(5), 724–734 (2012)
6. Martínez-Gavara, A., Campos, V., Gallego, M., Laguna, M., Martí, R.: Tabu search and grasp for the capacitated clustering problem. Comput. Optim. Appl. **62**(2), 589–607 (2015)
7. Mjirda, A., Todosijević, R., Hanafi, S., Hansen, P., Mladenović, N.: Sequential variable neighborhood descent variants: an empirical study on the traveling salesman problem. Int. Trans. Oper. Res. (2016). doi:10.1111/itor.12282
8. Mladenović, N., Hansen, P.: Variable neighborhood search. Comput. Oper. Res. **24**(11), 1097–1100 (1997)
9. Morán-Mirabal, L., González-Velarde, J., Resende, M.G., Silva, R.M.: Randomized heuristics for handover minimization in mobility networks. J. Heuristics **19**(6), 845–880 (2013)
10. Rodriguez, F., Lozano, M., García-Martínez, C., González-Barrera, J.: An artificial bee colony algorithm for the maximally diverse grouping problem. Inf. Sci. **230**, 183–196 (2013)
11. Urošević, D.: Variable neighborhood search for maximum diverse grouping problem. Yugoslav J. Oper. Res. **24**(1), 21–33 (2014)

A Leader-Follower Hub Location Problem Under Fixed Markups

Dimitrije D. Čvokić[1]([✉]), Yury A. Kochetov[2,3], and Aleksandr V. Plyasunov[2,3]

[1] University of Banja Luka, Mladena Stojanovića 2, 78000 Banja Luka,
Republika Srpska, Bosnia and Herzegovina
dimitriye.chwokitch@yahoo.com
[2] Sobolev Institute of Mathematics, 4, Akad. Koptyug Avenue,
630090 Novosibirsk, Russia
[3] Novosibirsk State University, 2, Pirogova Street, 630090 Novosibirsk, Russia

Abstract. Two competitors, a Leader and a Follower, are sequentially creating their hub and spoke networks to attract customers in a market where prices have fixed markups. Each competitor wants to maximize his profit, rather than a market share. Demand is split according to the logit model. The goal is to find the optimal hub and spoke topology for the Leader. We represent this Stackelberg game as a nonlinear mixed-integer bi-level optimisation problem and show how to reformulate the Follower's problem as a mixed-integer linear program. Exploiting this reformulation, we solve instances based on a synthetic data using the alternating heuristic as a solution approach. Computational results are thoroughly discussed, consequently providing some managerial insights.

Keywords: Hub location · Pricing · Fixed markup · Stackelberg competition · Linear reformulation · Matheuristic

1 Introduction

Competition between firms that use hub and spoke networks has been studied mainly from the sequential location approach. An existing firm, the Leader, serves the demand in some region, and a new one, the Follower, wants to enter. This topic of research is quite fresh, and the first paper on competitive hub location is attributed to Marianov, Serra and ReVelle [1]. Their approach was extended and followed by Eiselt and Marianov in [2], Gelareh, Nickel and Pisinger in [3], and by many others. Sasaki and Fukushima presented a (continuous) Stackelberg Hub Location Problem in [4], in which the incumbent competes with several entrants for profit maximisation. For every route, only one hub was allowed. Adler and Smilowitz introduced in [5] a framework to decide the convenience of merging airlines or creating alliances, using a game-theory-based approach. Later, Sasaki et al. in [6] proposed a problem in which two agents are locating hub-arcs to maximise their respective revenues under the Stackelberg framework, allowing more than one hub on a route.

© Springer International Publishing Switzerland 2016
Y. Kochetov et al. (Eds.): DOOR 2016, LNCS 9869, pp. 350–363, 2016.
DOI: 10.1007/978-3-319-44914-2_28

Here, we consider a sequential hub location and pricing problem in which two competitors, a Leader and a Follower, compete to attract the customers and aim to maximize their profits rather than a market share. The pricing is identified as an important service attribute that affects the client's choice [7–10], as expected. Therefore, we are interested in studying its impact to the optimal hub and spoke topology. In contrast to [11], we assume that the prices are regulated, moreover that markups are fixed.

Regulation is a legal norm intended to shape a conduct that is a by-product of imperfection. It may be used to prescribe or proscribe a conduct, to calibrate incentives, or to change preferences. Common examples of regulation includes control of market entries, *prices*, wages, development, approvals, pollution effects, employment for some people in certain industries, standards of production for some goods, the military forces and devices. For more information we refer the reader to [12–14]. The normative economic theories conclude that the regulations should *encourage competition* where feasible, minimize the cost of information asymmetries, *provide for price structures that improve economic efficiency*, establish regulatory processes that provide for regulation under the law and independence, transparency, *predictability*, legitimacy and credibility of the regulatory system (see [13, 15], for example). Price regulation refers to the policy of setting prices by a government agency, legal statute, or regulatory authority. Under such policy, fixed, minimum and maximum prices may be set. Referring to [15, 16], a decision may be based on costs, return on investments, or even markups.

As it was previously said, we are interested in a direct price setting as *a form of regulation*, particularly, a scenario where the markups are fixed. Fixing prices is not just a theoretical scenario. When it comes to the transportation industry, a famous example is the IATA (International Air Transport Association) price regulation. That is, several years ago, the price for a non-stop flight from an origin to a destination in a given passenger class was fixed for IATA airlines. The fact that one had or had not to change planes did not affect the price. A passenger could, in principle, use his Lufthansa ticket on a British Airways flight, because tickets were transferable within a fare class, as it was reported by Grammig et al. in [17]. Moreover, Lüer-Villagra and Marianov showed in [11] that if demand is non-elastic and logit model is used for calculating the discrete choice probability, the optimal prices for all routes connecting a particular origin-destination (OD) pair have the same markup.

We note that fixing markups does not mean that prices will be the same, that is they could vary if the routes are composed of several different lines that have different travel costs. As a matter of fact, this approach could be seen as a transition case to a Stackelberg competition in hub location where prices are not regulated. Nevertheless, a hub location or a route opening decision, or even an entrance into a market, can be very dependent on the revenues that a company can obtain using its network. Revenues, in turn, depend on the pricing structure and competitive context, as it was observed in [11].

Following the work of Lüer-Villagra and Marianov in [11], and because this research is still fresh, we will assume that the demand is non-elastic and customers patronize the route by price. Customers' decision process is modelled using a logit model, which is well validated in the transportation literature (see for example [11, 18, 20]).

Regarding the economies of scale, we use a model in which a constant (flow-independent) discount between hubs and no discount on spokes are considered. In the literature, modeling of the economies of scale in this fashion is addressed as the fundamental approach, which incidentally results in an entirely connected inter-hub network if the objective is the cost minimisation [19]. Most of the researchers use this method of discounting the flow between hubs [19], independent of its magnitude, mainly because of its computational attractiveness and the fact that the search for an entirely successful model is still open [11]. Therefore, we take the same approach in this paper, although we do not expect that hubs have to be completely inter-connected, as we are dealing with the profit maximization problems.

The proposed model is applied to the air passenger industry. However, with slight changes in the discrete choice model, they can be applied to any other industry that benefits from a hub and spoke network structure. We will call this problem a Leader-Follower Hub Location Problem under Fixed Markups (LFHLPuFM).

The contributions of this paper are as follows. Section 2 describes this Stackelberg game. In Sect. 3 we present the mixed-integer linear reformulation of the Follower's problem. After that, in Sect. 4, we describe our solution approach based on the alternating heuristic. In the end, we give some comments and managerial insights.

2   A Leader-Follower Hub Location Problem Under Fixed Markups

The problem is defined over a directed multi-graph $G = G(N, A)$, where N is the non-empty set of nodes and A is the set of arcs that are connecting every pair of nodes in the graph. We assume that for every arc $(i, j) \in A$, there is an opposite arc $(j, i) \in A$. Situations where this does not hold are quite rare and they do not make the problem computationally more attractive. Possible location for hubs are the nodes $i \in N$, and for each of them, there is a fixed cost f_i. The hubs can be shared. We note that the number of hubs to be located is not fixed. Its value is to be determined by the solution of a model. For every arc $(i, j) \in A$ there is a fixed (positive) cost g_{ij} for allocating it as a spoke and a (positive) travel cost per unit of flow c_{ij}. We assume that the travel cost is a non-decreasing function of distance. To model the inter-hub discounts, let $\aleph, \alpha, \psi$ be the discount factors due to flow consolidation in collection (origin to hub), transfer (between hubs), and distribution (hub to destination), respectively. At most two hubs are allowed to be on a single route. The travel cost $c_{ij/kl}$ over a route $i \to k \to l \to j$ is defined as $c_{ij/kl} = \aleph c_{ik} + \alpha c_{kl} + \psi c_{lj}$. It is assumed that pricing is regulated, and a form of regulation is a direct price setting, so that all markups are fixed.

In other words, for every route $i \rightarrow k \rightarrow l \rightarrow j$ there is a fixed markup $\mu_{ij/kl}$. The set of all routes is trimmed to avoid the ones which are impractical, i.e. those routes that have the second arrival point. We define it in a similar fashion as it was done by O'Kelly et al. in [21]

$$I = \{(i,j,k,l) \in N^4 \mid (i = l \wedge l \neq k \wedge k \neq j) \vee (j = k \wedge k \neq l \wedge l \neq i) \vee (i \neq l \wedge k \neq j)\}.$$

On the basis of this set we define the set of valid indices for our routes as $M = \{(i,j,k,l) \in N^4 \mid (i,j,k,l) \in N^4 \backslash I\}$. The sets of valid indices for the possible hubs between the OD pairs $(i,j) \in N^2$ are defined in a similar manner $M_{ij} = \{(k,l) \in N^2 \mid (i,j,k,l) \in N^4 \backslash I\}$. The demand w_{ij} for every OD pair $(i,j) \in N^2$ is assumed to be non-elastic and non-negative. The logit model has a sensitivity parameter Θ that corresponds to the pricing. It has an already known positive value assigned. Higher values of sensitivity parameters mean that the customers are very sensitive to the differences in prices. In other words, they will mostly choose the less expensive routes. Both competitors have a large amount of resources to cover the entire market with their networks. The goal is to maximize the profit, rather than a market share.

This Stackelberg game can be represented as a non-linear mix-integer bi-level mathematical program, where we have that:

- $u_{ij/kl}$ is the fraction of the flow going from $i \in N$ to $j \in N$ through the Leader's hubs located at $k,l \in N$
- $v_{ij/kl}$ is the fraction of the flow going from $i \in N$ to $j \in N$ through the Follower's hubs located at $k,l \in N$
- $x_k = 1$ if the Leader locates a hub at node $k \in N$ and 0 otherwise
- $y_k = 1$ if the Follower locates a hub at node $k \in N$, and 0 otherwise
- $\lambda_{ij} = 1$ if the Leader establishes a direct connection between the nodes $i,j \in N$, where $(i,j) \in A$, and 0 otherwise
- $\zeta_{ij} = 1$ if the Follower establishes a direct connection between the nodes $i,j \in N$, where $(i,j) \in A$, and 0 otherwise

Denote $x = (x_i)_{i \in N}$, $y = (y_i)_{i \in N}$, $\lambda = (\lambda_{ij})_{i,j \in N}$, $\zeta = (\zeta_{ij})_{i,j \in N}$, for short. We propose the following model for the Leader;

$$\max \quad \sum_{(i,j,k,l) \in M} \mu_{ij/kl} w_{ij} u_{ij/kl} - \sum_{i \in N} f_i x_i - \sum_{(i,j) \in A} g_{ij} \lambda_{ij} \tag{1}$$

$$u_{ij/kl} = \frac{x_k x_l \lambda_{ik} \lambda_{kl} \lambda_{lj} e^{-\Theta(c_{ij/kl} + \mu_{ij/kl})}}{\displaystyle\sum_{(s,t) \in M_{ij}} x_s x_t \lambda_{is} \lambda_{st} \lambda_{tj} e^{-\Theta(c_{ij/st} + \mu_{ij/st})} + \gamma_{ij}^*}, \quad \forall (i,j,k,l) \in M \tag{2}$$

$$\gamma_{ij}^* = \sum_{(k,l) \in M_{ij}} y_k^* y_l^* \zeta_{ik}^* \zeta_{kl}^* \zeta_{lj}^* e^{-\Theta(c_{ij/kl} + \mu_{ij/kl})}, \quad \forall i,j \in N \tag{3}$$

$$(y^*, \zeta^*) \in F^*(x, \lambda) \tag{4}$$

$$x_i \in \{0,1\}, \quad \forall i \in N \tag{5}$$

$$\lambda_{ij} \in \{0,1\}, \quad \forall (i,j) \in A. \tag{6}$$

Here, y_i^*, ζ_{ij}^* $(i, j \in N)$ are composing the optimal solution for the Follower's problem, for which we propose the subsequent model;

$$\max \sum_{(i,j,k,l) \in M} \mu_{ij/kl} w_{ij/kl} v_{ij/kl} - \sum_{i \in N} f_i y_i - \sum_{(i,j) \in A} g_{ij} \zeta_{ij} \tag{7}$$

$$v_{ij/kl} = \frac{y_k y_l \zeta_{ik} \zeta_{kl} \zeta_{lj} e^{-\Theta(c_{ij/kl} + \mu_{ij/kl})}}{\sum_{(s,t) \in M_{ij}} y_s y_t \zeta_{is} \zeta_{st} \zeta_{tj} e^{-\Theta(c_{ij/st} + \mu_{ij/st})} + \eta_{ij}}, \quad \forall (i, j, k, l) \in M \tag{8}$$

$$\eta_{ij} = \sum_{(k,l) \in M_{ij}} x_k x_l \lambda_{ik} \lambda_{kl} \lambda_{lj} e^{-\Theta(c_{ij/kl} + \mu_{ij/kl})}, \quad \forall i, j \in N \tag{9}$$

$$y_i \in \{0, 1\}, \quad \forall i \in N \tag{10}$$

$$\zeta_{ij} \in \{0, 1\}, \quad \forall (i, j) \in A. \tag{11}$$

The objective functions (1) and (7) are representing the profits, which are calculated as a difference between the net income and the network installation costs. Feasible solutions are the tuples $(x, \lambda, y^*, \zeta^*)$ satisfying the constraints (2)–(6). Constraints (2) and (8) are representing the probabilities of choosing the respective routes, according to the logit model. The Eq. (3) represent the impact of the Follower on the Leader's market share. The Leader's impact on the Follower's market share is represented by the Eq. (9). Next, (4) indicates that the Follower chooses the optimal solution for any of the Leader's choice of hubs, where $F^*(x, \lambda)$ represents the set of the Follower's optimal solutions. The rest of the constraint sets are defining the variables' domains.

We note that the Follower's problem may have several optimal solutions, all feasible for a given (x, λ). As a result, the Leader's problem could be ill-posed. Thus, we distinguish two extreme cases:

- cooperative Follower's behaviour (altruistic Follower). In case of multiple optimal solutions, the Follower always selects the one providing the best objective function value for the Leader. We call it the *cooperative optimal solution* to the Follower's problem.
- non-cooperative Follower's behaviour (selfish Follower). In this case, the Follower always selects the solution that provides the worst objective function value for the Leader. We call it the *non-cooperative optimal solution* to the Follower's problem.

One can easily observe that the sum of objective functions in our bi-level program is not a constant. Therefore, the Follower's behaviour should be defined properly, i.e. an auxiliary optimization problem should be defined, as described in [22–24]. The corresponding optimal cooperative solution can be found using a two-stage algorithm.

At Stage 1, for a fixed solution (x, λ), we solve the Follower's problem and calculate the optimal value of its objective function $F(x, \lambda)$.

At Stage 2, for a fixed solution (x, λ), we solve the following *auxiliary problem*

$$\max \sum_{(i,j,k,l) \in M} \mu_{ij/kl} w_{ij} u_{ij/kl} \tag{12}$$

$$u_{ij/kl} = \frac{x_k x_l \lambda_{ik} \lambda_{kl} \lambda_{lj} e^{-\Theta(c_{ij/kl} + \mu_{ij/kl})}}{\sum\limits_{(s,t) \in M_{ij}} y_s y_t \zeta_{is} \zeta_{st} \zeta_{tj} e^{-\Theta(c_{ij/st} + \mu_{ij/st})} + \eta_{ij}}, \quad \forall (i,j,k,l) \in M \tag{13}$$

$$\sum_{(i,j,k,l) \in M} \mu_{ij/kl} w_{ij} v_{ij/kl} - \sum_{i \in N} f_i y_i - \sum_{(i,j) \in A} g_{ij} \zeta_{ij} \geq F(x, \lambda) \tag{14}$$

$$v_{ij/kl} = \frac{y_k y_l \zeta_{ik} \zeta_{kl} \zeta_{lj} e^{-\Theta(c_{ij/kl} + \mu_{ij/kl})}}{\sum\limits_{(s,t) \in M_{ij}} y_s y_t \zeta_{is} \zeta_{st} \zeta_{tj} e^{-\Theta(c_{ij/st} + \mu_{ij/st})} + \eta_{ij}}, \quad \forall (i,j,k,l) \in M \tag{15}$$

$$\eta_{ij} = \sum_{(k,l) \in M_{ij}} x_s x_t \lambda_{ik} \lambda_{kl} \lambda_{lj} e^{-\Theta(c_{ij/kl} + \mu_{ij/kl})}, \quad \forall i,j \in N \tag{16}$$

$$y_i \in \{0,1\}, \quad \forall i \in N \tag{17}$$

$$\zeta_{ij} \in \{0,1\}, \quad \forall (i,j) \in A \tag{18}$$

The corresponding optimal non-cooperative solution can be found using the same two-stage process, except we should solve the minimization problem, instead of the maximization. For a thorough understanding of this topic and the used terminology, we suggest the reader to examine the classic textbook of Dempe [25].

3 Mixed-Integer Linear Reformulation of the Follower's Problem

Suppose that the Leader has made his decision. To estimate his profit (and a market share) we need the Follower's optimal solution. Fortunately, the Follower's problem can be linearised to find the respective optimal solution by a solver.

Introducing a new variable R_{ijkl} (for $(i,j,k,l) \in M$), we can substitute the product $y_k y_l \zeta_{ik} \zeta_{kl} \zeta_{lj}$ in the constraint set (8). This substitution requires the additional sets of constraints

$$R_{ij/kl} - \frac{1}{5}(y_k + y_l + \zeta_{ik} + \zeta_{kl} + \zeta_{lj}) \leq 0, \quad \forall (i,j,k,l) \in M \tag{19}$$

$$R_{ij/kl} - y_k - y_l - \zeta_{ik} - \zeta_{kl} - \zeta_{lj} + 4 \geq 0, \quad \forall (i,j,k,l) \in M \tag{20}$$

$$R_{ij/kl} \in \{0,1\}, \quad \forall (i,j,k,l) \in M \tag{21}$$

where y_i, ζ_{ij} have the same meaning as in (7)–(11).

Now, only the constraints from (8) have non-linear terms. The literature knows several techniques for reformulating the logit-term, which are presented in [26–29]. Recently, Haase and Müller compared those approaches in [20] and their computational study, based on synthetic data, showed that the approach from [27] seems to be promising for solving larger problems. From (8), we directly obtain that the following holds:

$$v_{ij/kl} - \frac{e^{-\Theta(c_{ij/kl}+\mu_{ij/kl})}}{\eta_{ij} + e^{-\Theta(c_{ij/kl}+\mu_{ij/kl})}} R_{ij/kl} \leq 0, \quad \forall(i,j,k,l) \in M, \tag{22}$$

$$v_{ij/kl} \geq 0, \quad \forall(i,j,k,l) \in M. \tag{23}$$

The inequalities (22) are just tighter bounds on v_{ijkl}, than the obvious $v_{ijkl} \leq R_{ijkl}$. Basically, they state that a customer can only choose an established route. The domain of variables v_{ijkl} is specified in (23).

The ratio of the choice probabilities of the two alternatives is independent from other alternatives, i.e. we have that for some $v_{ij/kl}$ and $v_{ij/st}$, for which we know that $R_{ij/st} = 1$, the following identity holds

$$\frac{v_{ij/kl}}{v_{ij/st}} = \frac{e^{-\Theta(c_{ij/kl}+\mu_{ij/kl})}}{e^{-\Theta(c_{ij/st}+\mu_{ij/st})}}. \tag{24}$$

This property is called an Independence of Irrelevant Alternatives (IIA). From the previous equations, we conclude that the following constraints are valid

$$v_{ij/kl} \leq \frac{e^{-\Theta(c_{ij/kl}+\mu_{ij/kl})}}{e^{-\Theta(c_{ij/st}+\mu_{ij/st})}} v_{ij/st} + 1 - R_{ij/st}, \ \forall(k,l),(s,t) \in M_{ij}, \ \forall i,j \in N. \tag{25}$$

These inequalities are valid even if the impractical routes are included because their corresponding values for the choice probabilities v_{ijkl} and establishing the route R_{ijkl} could be both set to zero. It is easy to see that (24) is valid even if we use $u_{ij/kl}$ instead of $v_{ij/kl}$. Thus, we obtain an additional two sets of inequalities that are connecting the choice probabilities of the Follower's routes with the choice probabilities of the Leader's routes and describe the relation between the Leader's routes alone (as in (25) for the Follower). In other words, we have the following constraint sets to be valid for all OD pairs

$$u_{ij/kl} \leq \frac{e^{-\Theta(c_{ij/kl}+\mu_{ij/kl})}}{e^{-\Theta(c_{ij/st}+\mu_{ij/st})}} v_{ij/st} + 1 - R_{ij/st}, \quad \forall(k,l),(s,t) \in M_{ij}, \ \forall i,j \in N \tag{26}$$

$$T_{ij/st} u_{ij/kl} \leq T_{ij/kl} \frac{e^{-\Theta(c_{ij/kl}+\mu_{ij/kl})}}{e^{-\Theta(c_{ij/st}+\mu_{ij/st})}} u_{ij/st}, \quad \forall(k,l),(s,t) \in M_{ij}, \ \forall i,j \in N \tag{27}$$

$$u_{ij/kl} \geq 0, \quad \forall i,j,k,l \in N \tag{28}$$

where $T_{ij/st} = x_s x_t \lambda_{is} \lambda_{st} \lambda_{tj}$, for all $i,j,s,t \in N$. Note that the value of $u_{ij/kl}$ is not known until the Follower makes his move, which is not the case with the product $T_{ij/st}$. Now, for all OD pairs we have that sum of choice probabilities for both competitors is equal to one, that is

$$\sum_{(k,l)\in M_{ij}} u_{ij/kl} + \sum_{(k,l)\in M_{ij}} v_{ij/kl} = 1, \quad \forall i,j \in N. \tag{29}$$

We can introduce a new variable q_{ij} $(i,j \in N)$ to denote the cumulative choice probabilities of the Leader. Furthermore, we can derive new sets of constraints with fewer variables

$$\sum_{(k,l)\in M_{ij}} v_{ij/kl} + q_{ij} \le 1, \quad \forall i,j \in N \tag{30}$$

$$q_{ij} \ge 0, \quad \forall i,j \in N \tag{31}$$

where (31) defines the new variables' domains. As a matter of fact, (8) can be expressed solely in terms of v_{ijkl} and q_{ij} as a linear constraint

$$v_{ij/kl} - \frac{e^{-\Theta(c_{ij/kl}+\mu_{ij/kl})}}{\eta_{ij}} q_{ij} \le 0, \quad \forall(i,j,k,l) \in M. \tag{32}$$

One could notice that R_{ijkl} is omitted in the second term from the left-hand side. We do not need that binary variable because we already have that (22) must hold.

Now, taking all this into the account, we proved the following proposition.

Proposition 1. *The Follower's Problem (7)–(11) can be reformulated as a mix-integer linear program with the objective function*

$$\max \sum_{(i,j,k,l)\in M} w_{ij}\mu_{ij/kl}v_{ij/kl} - \sum_{i\in N} f_i y_i - \sum_{(i,j)\in A} g_{ij}\zeta_{ij} \tag{33}$$

subject to (9)–(11), (19)–(23), and (30)–(32).

Although this technique could be used to reformulate the problem of the Leader, there is a question of its usefulness, because the Leader is anticipating the move of the Follower. Nevertheless, it can be easily seen that the same approach will give us the reformulation of the auxiliary problem.

Proposition 2. *The auxiliary problem (12)–(18) can be reformulated as a mix-integer linear program with the objective function*

$$\max \sum_{(i,j,k,l)\in M} \mu_{ij/kl}w_{ij}u_{ij/kl} \tag{34}$$

subject to (14), (16)–(18), (19)–(23), (25)–(29).

The drawback of these reformulations is that they produce a large number of new constraints. The fact that we do not have a constraint on the number of hubs suggest that our models could still be difficult to solve by a solver even for smaller instances.

4   Computational Experiments

The central idea of the matheuristic we used is given in [30,31]. This is an alternating method, where for the solution of the Leader, we compute the best-possible solution for the Follower. Once this has been done, the Leader assumes the role of the Follower and re-optimizes his decision by solving the corresponding problem for the given solution. This process is then repeated until one of the Nash equilibria is discovered or the previously visited solution has been detected. The best what we have found for the Leader is returned as the result of the method. In the beginning, the Leader ignores the Follower.

We conducted the computational experiments to test the method using an artificially generated data. The Cartesian coordinates for the nodes $i \in N$ are randomly generated by a uniform distribution in the interval $[0, 100]$. The demand is also randomly generated using a (truncated) log-normal distribution on an interval $[1, 100]$, where numbers represent the flow in thousands. The log-normal distribution better corresponds to the real-world data then the uniform distribution when it comes to the passenger flows [32], air traffic demands [33], or airline business [34]. Next, following the work presented in [11,35,36], we took the hub location cost to be the same for all nodes. We could say that the cost of the hub location is proportional to the number of the passengers that will go through the hub. On the other hand, the hub location cost is inversely proportional to the number of hubs (because of competition and load shedding). Therefore, we have a range of cases, where only one hub exists in the market to the case with $|N|$ hubs (a point-to-point network as a trivial hub and spoke topology). Taking that into account, we took the following expression for hub location costs in our experiments $f_i = f = \beta \frac{H_{|N|}}{|N|} \sum_{i,j,k,l \in N} w_{ij}$. The sum represents the total amount of the passengers. In the average, that amount is distributed to the $H_{|N|}/|N|$ hubs, where H_n is the n-th harmonic number. As for $\beta > 0$, it is a coefficient that represents an operating cost per passenger. Considering the running time, we observed in our preliminary investigation that $\beta = 0.06$ happened to be a good choice. Also, we note that this is just a temporary solution for the hub location cost model. The cost of spoke allocation between the pair of nodes i and j was calculated using the expression $\zeta_{ij} = f \frac{c_{ij}/w_{ij}}{\max\limits_{(k,l)\in A} c_{kl}/w_{kl}}$, as in [11,35].

The travel cost is taken to be $c_{ij} = d_{ij}/\max_{i,j\in N} d_{ij}$, where d_{ij} is the Euclidean distance between pair of nodes i and j. This way, normalizing the travel cost, the interval from which we "harvested" the node coordinates becomes irrelevant. In our testing the discount cost values on consolidation and distribution links were $\aleph = \psi = 1$. The experiments were conducted on randomly generated graphs of 5, 6, 7, 8, 9 and 10 nodes. Three values $\alpha \in \{0.1, 0.5, 1.0\}$ are considered for the inter-hub discount factor, and three values, too, for the sensitivity parameter $\Theta \in \{0.25, 1.0, 4.0\}$. For a particular OD pair $(i, j) \in N^2$ all markups were taken to be equal, i.e. $\mu_{ij/kl} = \mu_{ij/st}$ for all $k, l, s, t \in N$. This approach is justified by the results presented in [11]. The markup for a particular OD pair is calculated as percentage of the travel cost of the corresponding non-stop flight. The percentages took values from the set $\{10, 25, 50\}$. For graphs of size 9 and 10 nodes, we used only the smallest and the biggest values of the parameters. In total, we tested 124 different instances. The alternating heuristic was implemented in Python 2.7 using Gurobi 6.5 as the solver, on a 64-bit Windows 8.1 Pro with two 2.00 GHz Six-Core AMD Opteron(tm) processors and 32 GiB of RAM.

It is worth noting that preliminary computational experiments on a model that included the impractical routes showed that they can be a cause for numerical instability, which could lead to wrong solutions and unreasonably long running times.

The Leader's Network Structure. In almost half of the instances tested, the best-reported solution for the Leader was the so-called Entry Deterrence. Slightly less than one-third of the cases had the Nash equilibrium as a solution. We observed from our testing sample that for stronger Leader's positions (lower markups or better developed networks), the harder it was for the Follower to obtain any profit at all. Something quite similar we observed for the running time of the algorithm. For stronger Leader's positions, Gurobi needed more time to find the exact solution and in some cases it even lasted the entire day.

The Role of Inter-hub Economies of Scale. Our computational investigation suggests that the inter-hub economies of scale have a minor impact on the Leader's profit. Unfortunately, we could not observe any solid pattern. It seems that greater values of the price sensitivity parameter combined with bigger markups can boost a little bit the role of the inter-hub economies of scale. For smaller values of markups and sensitivity parameter, the profit becomes more and more locked to one specific value. Also, the results of the computational tests suggest that economies of scale could have an effect to some extent on the Leader's networks solution. In our testing, we did not observe a significant difference in location of hubs for different values of the discount factor, but the resulting hub and spoke topologies were usually less developed for smaller values of the discount factor. We note that sometimes there was no difference at all, or it was the opposite. We could not observe that different values of α had any influence on the cycle length of the alternating heuristics.

The Effect of the Sensitivity Factor. It turns out that the sensitivity to the price differences could have a more significant role when it comes to the profit of the competitors. In most of the cases when the Nash equilibrium occurred as a solution the bigger sensitivity led to the greater profit for the Leader, although not always. The same could be said for the Entry Deterrence type of solution. When it comes to the resulting hub and spoke topologies, in most of the cases hub and spoke networks were the same for different values of Θ, especially for smaller markups. When they were not, the small values of the sensitivity factor usually corresponded to more developed (less translucent) networks. Also, it seems that the chance for the entry deterrence to occur as a solution is greater when Θ takes the smaller values. In contrast to that, the bigger values of Θ corresponded more often to the Nash equilibrium solutions. As in the previous analysis, we can not say that the cycle length is under the influence of the sensitivity factor.

The Role of the Fixed Markups. It is more likely that the Nash equilibrium as a solution will occur when the markups are bigger. For smaller markups, the Entry Deterrence appeared more often as a solution. It was not always the case that the bigger markup led to the greater profit, especially if the solution types were different, but for the same types the bigger markup corresponded to the higher obtained profit. We observed that for the smallest markups considered the alternating heuristic had the minimal number of steps, but for the considerably larger markups we could not observe a regular pattern.

5   Conclusion and Future Research

We present a novel approach to analyse a situation in which two companies compete in a transportation market in a sequential fashion. Here, the goal for the both companies is to maximize their profit by creating the optimal hub and spoke networks. It is assumed that the market is regulated. Because this research is quite fresh, the form of regulation is chosen to be the direct price setting. To be more precise, we assumed that all routes have fixed markups. We have to say that up to our knowledge no one has investigated the effects of regulations to the optimal hub location in a competitive environment. Next, we took that the customers' choice of provider and route depends solely on price and therefore it is possible to predict it by a simple logit model (although including other factors would be very easy). Upon that, we formulated a non-linear mixed integer bi-level program to model this Stackelberg competition. The choice of the solution approach was the alternating heuristic based on the Follower's best response.

The computational investigation showed that discount factor by itself has a relative impact to the solution and a not so sharp-cut role, as it was difficult to observe any regularity. It looks like the sensitivity factor plays a more significant role, but again it is hard to draw any specific conclusions. The markup has a significant effect, as expected. For smaller markups, there is a tendency towards the entry deterrence. Likewise, the bigger markups provided "more space" for the Nash equilibrium to occur. Loosely speaking, if the passengers in the market are less sensitive to the price differences and markups are quite small, than it could be expected that the first-to-enter company will be the only one providing the services. On the other hand, the price sensitive markets with bigger markups set are more "prone" to allow multiple competitors to operate. Colloquially speaking, both of these outcomes could be used as arguments in cooperative games.

From the purely computational point of view, we observed that it might be beneficial to derive a new model that would serve only to find the entry deterrence solution. Long running times confirmed our worries that this problem will not be easily solved for bigger instances by a commercial solver. Therefore, for a more thorough investigation, we have to find a better way to compute the exact solution for the Follower's problem. We intend to put our efforts in finding tighter reformulations of the Follower's model and designing a branch and bound based method that would utilize the structure of the program itself.

Recently, some new interesting results have been obtained by relating the bi-level programs with polynomial and approximation hierarchies [37–42]. The investigation of these relationships is an important area of research. Therefore, we plan to determine the position of our problem in each of them, too. Another direction of the research is oriented towards the other forms of regulation.

Acknowledgements. This research was partially supported by the RFBR grant 16-07-00319.

References

1. Marianov, V., Serra, D., ReVelle, C.: Location of hubs in a competitive environment. Eur. J. Oper. Res. **114**(2), 363–371 (1999). doi:10.1016/S0377-2217(98)00195-7
2. Eiselt, H., Marianov, V.: A conditional p-hub location problem with attraction functions. Comput. Oper. Res. **36**(12), 3128–3135 (2009). doi:10.1016/j.cor.2008.11.014
3. Gelareh, S., Nickel, S., Pisinger, D.: Liner shipping hub network design in a competitive environment. Trans. Res. Part E: Logistics Transp. Rev. **46**(6), 991–1004 (2010). doi:10.1016/j.tre.2010.05.005
4. Sasaki, M., Fukushima, M.: Stackelberg hub location problem. J. Oper. Res. Soc. Japan **44**(4), 390–405 (2001). doi:10.1016/S0453-4514(01)80019-3
5. Adler, N., Smilowitz, K.: Hub-and-spoke network alliances and mergers: price-location competition in the airline industry. Transp. Res. Part B **41**(4), 394–409 (2007). doi:10.1016/j.trb.2006.06.005
6. Sasaki, M., Campbell, J.F., Ernst, A.T., Krishnamoorthy, M.: Hub arc location with competition. Technical report, Nanzan Academic Society (2009)
7. de Menzes, A., Vieira, J.: Willingness to pay for airline services attributes: evidence from a stated preferences choice game. Eur. Transport **39**, 1–13 (2008)
8. Campbell, B., Vigar-Ellis, D.: The importance of choice attributes and the positions of the airlines within the South Africa domestic passenger airline industry as perceived by passengers at Durban International Airport. South. Afr. Bus. Rev. **16**(2), 97–119 (2012)
9. Cullinane, K., Notteboom, T., Sanchez, R., Wilmsmeier, G.: Costs, revenue, service attributes and competition in shipping. Marit. Econ. Logistics **14**, 265–273 (2012). doi:10.1057/mel.2012.7
10. Lambrecht, A., Seim, K., Vilcassim, N., Cheema, A., Chen, Y., Crawford, G., Hosanagar, K., Raghuram, I., Koenigsberg, O., Lee, R., Miravete, E., Sahin, O.: Price discrimination in service industries. Mark Lett. **23**, 423–438 (2012). doi:10.1007/s11002-012-9187-0
11. Lüer-Villagra, A., Marianov, V.: A competitive hub location and pricing problem. Eur. J. Oper. Res. **231**(3), 734–744 (2013). doi:10.1016/j.ejor.2013.06.006
12. Orbach, B.: What is Regulation? 30 Yale J. Regul. Online 1 (2012). Available at SSRN: http://ssrn.com/abstract=2143385
13. Hertog, J.: Review of Economic Theories of Regulation. Discussion Paper, 10–18, Tjalling C. Koopmans Research Institute, Utrecht University School of Economics, Utrecht University, Utrecht (2010)

14. Braithwaite, J., Drahos, P.: Global Business Regulation. Cambridge University Press, Cambridge (2000). ISBN -13: 978-0521784993

15. Khemani, R.S., Shapiro, D.M.: Glossary of Industrial Organisation Economics and Competition Law. OECD Publishing, Paris (1993). ISBN: 9789264137936

16. Treasury, A.: Price regulation of utilities. Econ. Roundup 1, 57–69 (1999)

17. Grammig, J., Hujer, R., Scheidler, M.: Discrete choice modelling in airline network management. J. Appl. Econ. 20, 467–486 (2005). doi:10.1002/jae.799

18. Ortúzar, J., Willumsen, L.: Modelling Trasnport, 4th edn. Wiley-Blackwell, West Sussex (2011)

19. Campbell, J., O'Kelly, M.: Twenty-five years of hub location research. Transp. Sci. 24(2), 153–169 (2012). doi:10.1287/trsc.1120.0410

20. Haase, K., Müller, S.: A comparison of linear reformulations for multinomial logit choice probabilities in facility location models. Eur. J. Oper. Res. 232, 689–691 (2014). doi:10.1016/j.ejor.2013.08.009

21. O'Kelly, M.E., Bryan, D., Skorin-Kapov, D., Skorin-Kapov, J.: Hub network design with single and multiple allocation: a computational study. Location Sci. 4(3), 125–138 (1996). doi:10.1016/S0966-8349(96)00015-0

22. Alekseeva, E., Kochetov, Y.: Matheuristics and exact methods for the discrete $(r|p)$-centroid problem. In: Talbi, E.-G., Brotcorne, L. (eds.) Metaheuristics for bilevel Optimization. SCI, vol. 482, pp. 189–220. Springer, Heidelberg (2013). doi:10.1007/978-3-642-37838-6

23. Beresnev, V., Melnikov, A.: Approximate algorithms for the competitive facility location problem. J. Appl. Ind. Math. 5(2), 180–190 (2012). doi:10.1134/S1990478911020049

24. Ben-Ayed, O.: Bilevel linear programming. Comput. Oper. Res. 20(5), 485–501 (1993)

25. Dempe, S.: Foundations of Bilevel Programming. Kluwer Academic, Dordrecht (2002). doi:10.1007/b101970

26. Benati, S., Hansen, P.: The maximum capture problem with random utilities: problem formulation and algorithms. Eur. J. Oper. Res. 143, 518–530 (2002). doi:10.1016/S0377-2217(01)00340-X

27. Haase, K.: Discrete location planning. Technical report WP-09-09. Institute for Transport and Logistics Studies, University of Sydney (2002)

28. Aros-Vera, F., Marianov, V., Mitchel, J.: p-Hub approach for the optimal park-and-ride facility location problem. Eur. J. Oper. Res. 226(2), 277–285 (2013). doi:10.1016/j.ejor.2012.11.006

29. Zhang, Y., Berman, O., Verter, V.: The impact of client choice on preventive healthcare facility network design. OR Spectr. 34, 349–370 (2012). doi:10.1007/s00291-011-0280-1

30. Carrizosa, E., Davydov, I., Kochetov, Y.: A new alternating heuristic for the $(r|p)$-centroid problem on the plane. In: Klatte, D., Lüthi, H.J., Schmedders, K. (eds.) Operations Research Proceedings 2011. Operations Research Proceedings, pp. 275–280 (2012). doi:10.1007/978-3-642-29210-1_44

31. Alekseeva, E., Kochetov, Y., Plyasunov, A.: An exact method for the discrete $(r|p)$-centroid problem. J. Global Optim. 63(3), 445–460 (2015). doi:10.1007/s10898-013-0130-6

32. Huang, Z., Wu, X., Garcia, A.J., Fik, T.J., Tatern, A.J.: An open-access modeled passenger flow matrix for the global air network in 2010. PLoS ONE 8(5), e64317 (2013). doi:10.1371/journal.pone.0064317

33. Knudsen, T.: Uncertainities in airport cost analysis. In: Visser, E. (ed.) Transport Decisions in an Age of Uncertainty. Proceedings of the Third World Conference on Transport Research, pp. 362–376. Kluwer Boston, The Hague (1977). doi:10.1007/978-94-009-9707-3

34. Benitez, C.: The Design of a Large Scale Airline Network, Dissertation. Delft University of Technology (2012). doi:10.4233/uuid:c28fef01-c47f-4f92-82ba-6d34d5c4daa4

35. Calik, H., Alumur, S., Kara, B., Karasan, O.: A tabu-search based heuristic for the hub covering problem over incomplete hub networks. Comput. Oper. Res. **36**(12), 3088–3096 (2009). doi:10.1016/j.cor.2008.11.023

36. O'Kelly, M.E.: Hub facility location with fixed costs. Pap. Reg. Sci. **71**(3), 293–306 (1992). doi:10.1007/BF01434269

37. Davydov, I., Kochetov, Y., Plyasunov, A.: On the complexity of the $(r|p)$-centroid problem in the plane. TOP **22**(2), 614–623 (2014). doi:10.1007/s11750-013-0275-y

38. Melnikov, A.: Computational complexity of the discrete competitive facility location problem. J. Appl. Ind. Math. **8**(4), 557–567 (2014). doi:10.1134/S1990478914040139

39. Panin, A.A., Pashchenko, M.G., Plyasunov, A.V.: Bilevel competitive facility location and pricing problems. Autom. Remote Control **75**(4), 715–727 (2014). doi:10.1134/S0005117914040110

40. Caprara, A., Carvalho, M., Lodi, A., Woeginger, G.J.: A study a computational complexity of the bi-level knapsack problem. Siam. J. Optim. **2**, 823–838 (2014). doi:10.1137/130906593

41. Iellamo, S., Alekseeva, E., Chen, L., Coupechoux, M., Kochetov, Y.: Competitive location in cognitive radio networks. 4OR **13**(1), 81–110 (2015). doi:10.1007/s10288-014-0268-1

42. Lavlinskii, S.M., Panin, A.A., Plyasunov, A.V.: A bilevel planning model for public-private partnership. Autom. Remote Control **76**(11), 1976–1987 (2015). doi:10.1134/S0005117915110077

Tabu Search Approach for the Bi-Level Competitive Base Station Location Problem

Ivan Davydov[1,2]([✉]), Marceau Coupechoux[3], and Stefano Iellamo[4]

[1] Sobolev Institute of Mathematics, Av. Koptyuga 4, Novosibirsk, Russia
vann.davydov@gmail.com
[2] Novosibirsk State University, ul. Pirogova 2, Novosibirsk, Russia
[3] Télécom ParisTech and CNRS LTCI, 6, Rue Barrault, Paris, France
marceau.coupechoux@telecom-paristech.fr
[4] Foundation for Research and Technology Hellas (FORTH), Leof. Plastira 100,
Iraklio, Greece
siellamo@ics.forth.gr

Abstract. This paper addresses the competitive base stations location problem with sharing. Two mobile operators, the leader and the follower, compete to attract customers from a given market of high-speed internet connection. The leader acts first by placing a number of base stations, anticipating that the follower will react to the decision by creating his own network. The Leader can share BS cites with the follower operator, receiving a rent payment from him. We propose new model of realistic clients behavior, when the choice of the operator is made upon the average quality of service. We provide a formulation of this problem as a nonlinear integer programming problem. We propose a fast tabu search heuristic for this problem and provide some computational results.

Keywords: Tabu search · Bilevel programming · Competitive location

1 Introduction

Mobile internet access is a highly required service nowadays. A lot of people use LTE/3G access to browse the world web an a lot more to join. Rapid development of new technologies force the mobile operators to update their equipment in order to provide better service to the clients. In this paper we consider new model of competitive base stations location. We assume, there are two competitive providers already operating a 4G network. One of them, the leader is the first to get an access to a new 5G technology, and construct a new high-speed network anticipating the reaction of the follower. The latter one is able not to use his own cites to set up an equipment but also is able to rent a cite at leader. For this cases leader sets the rent price, which has to be paid regularly. We suggest a new model of realistic clients behavior. Clients are mobile, so they experience and evaluate the quality of connection over the whole network. So the choice of the operator is made upon the average quality of service. Similar model for one-level problem

© Springer International Publishing Switzerland 2016
Y. Kochetov et al. (Eds.): DOOR 2016, LNCS 9869, pp. 364–372, 2016.
DOI: 10.1007/978-3-319-44914-2_29

was discussed in [3]. In the second section we provide the detailed description of the model. Third section provides a formulation of this problem as a nonlinear integer programming problem. We propose a fast tabu search heuristic for this problem in section four. Section five provides some computational results.

2   System Model

2.1   Network Model

We consider two competitive operators (called resp. O_1 and O_2), which we refer to as a leader and a follower due to their sequential entering a market. They compete to serve clients by installing and configuring 5G networks. The leader makes a decision first, taking into account the arrival of the follower. This problem may be considered as a competitive facility location problem [1].

Let $\mathcal{S}$ be the set of all sites, where base stations can be installed. This set is made of three subsets: $\mathcal{S} = \mathcal{S}_f \cup \mathcal{S}_1^o \cup \mathcal{S}_2^o$, where $\mathcal{S}_i^o$ is the set of sites having a 4G base station installed by O_i, $i \in \{1, 2\}$, and $\mathcal{S}_f$ is a set of free sites for potential new installations.

In a first phase, O_1 chooses a subset of $\mathcal{S}_1^n$ for the installation of its new 5G base stations subject to a budget constraint. This set is chosen among the free sites and the sites having old BSs of O_1: $\mathcal{S}_1^n \subset \mathcal{S}_f \cup \mathcal{S}_1^o$. Its goal is to capture a maximum number of users and to anticipate the arrival of O_2. At the end of the first phase, O_1 sets a rent price for every site having a base station installed, i.e. in the set $\mathcal{S}_1^n \cup \mathcal{S}_1^o$. In a second phase, O_2 is deploying its 5G network by choosing a set $\mathcal{S}_2^n$ for its 5G BSs. He has the choice of all the sites in $\mathcal{S}$, i.e. $\mathcal{S}_2^n \subset \mathcal{S}$. If a 5G BSs of O_2 is placed on a site in $\mathcal{S}_1^n$, he will have to pay for the installation and the rent price fixed by O_1. Otherwise, he will have to pay only the installation costs. At the end of the second phase, some users are leaving O_1 and take a subscription with O_2.

2.2   Propagation Model

In order to describe the propagation model let us consider a user located at x. Let us define the channel gain between x and BS b as $g_b(x)$ and let us assume that the transmit power of the antenna, located at BS b is P_b. The Signal to Interference plus Noise Ratio (SINR) of the considered user in x with respect to b is then given by:

$$\gamma_b(x) = \frac{P_b g_b(x)}{\sum_{i \neq b} P_i g_i(x) + N} , \tag{1}$$

where N is the thermal noise power in the band. This value define the real quality of the signal, obtained by the user due to interference between different base stations. Too many BS concentrated in a relatively small area make it hard to distinguish between signals thus lowering the quality of the connection.

User in x is said to be *covered* by b if $\gamma_b(x) \geq \gamma_{min}$ for some threshold γ_{min}. User in x is said to *served* by b if it is covered and $P_b g_b(x) \geq P_i g_i(x)$ for all $i \neq b$.

Note that at every location, users can be served by at most one BS from every operator.

For a user located in x and served by station b, the physical data rate achievable by this user is denoted $c_b(x)$, which is an increasing nonlinear function of $\gamma_b(x)$ with $c_b(x) = 0$ if $\gamma_b(x) < \gamma_{min}$.

2.3   Traffic Model

We assume there is a constant traffic demand in the network that operators will potentially serve. In every location x, there is a demand $\lambda(x)/\mu(x)$, where $\lambda(x)$ is the arrival rate and $1/\mu(x)$ is the average file size. Note that this demand in x is statistical and can be shared by O_1 and O_2 or not served at all. Let assume that x is covered by O_1 and a proportion $p_1(x)$ of the demand is served by BS b from O_1. A proportion $1 - p_1(x)$ of the demand is served by O_2. We focus in this paper on a specific case for p_1: If location x is not covered by O_1, $p_1(x) = 0$. Otherwise, p_1 does not depend on the location and depends only on the overall relative quality of service in the network O_1 compared to O_2. The idea behind this assumption is that users are mobile and they choose their operator not only with respect to the quality of service at a particular location but rather to the average experienced quality.

Then, the load created by x on b is $p_1(x)\varrho_{1b}(x)$, where $\varrho_{1b}(x) = \frac{\lambda(x)}{\mu(x)c_{1b}(x)}$, where $c_{1b}(x)$ is the physical data rate in x and is an increasing function of $\gamma_{1b}(x)$. The index 1 is here to recall that the SINR, so the physical data rate, and the load are computed in the network of O_1. This is important to specify, because in the rest of the paper station b is likely to be shared with O_2. We can now define the *load* of station b as: $p_1\rho_{1b}$, where $\rho_{1b} = \sum_{\mathcal{A}_{1b}} \varrho_{1b}(x)$, where $\mathcal{A}_{1b}$ is the serving area of b, i.e., the set of locations served by b in network O_1. BS b is stable if $p_1\rho_{1b} < 1$ and we will consider only scenarios where this condition is fulfilled.

2.4   Market Sharing

We assume that users are the players of an evolutionary game. In this framework, the choice of a single user does not influence the average throughput of an operator. An equilibrium is reached when average throughputs in both networks are the same. Let $\mathcal{A}_{ib}$ be the area *served* by $b \in O_i$ and $\mathcal{A}_i = \cup_{b \in \mathcal{S}_i^n} \mathcal{A}_{ib}$ be the overall area served by O_i. Let $\mathcal{A}$ be the whole network area. Let $\Lambda_i = \sum_{b \in \mathcal{S}_i^n} \sum_{\mathcal{A}_{ib}} p_i\lambda(x)$ be the global arrival rate in O_i, where p_i is the probability for a user to be with operator O_i. Let $\Lambda = \sum_{\mathcal{A}} \lambda(x)$ be the global demand over the entire network. A user of O_i has a null throughput with probability $\frac{\Lambda - \Lambda_i}{\Lambda}$ and with probability $\frac{\Lambda_i}{\Lambda}$ an average throughput of:

$$\frac{1}{\Lambda_i} \sum_{b \in \mathcal{S}_i^n} \sum_{\mathcal{A}_{ib}} p_i\lambda(x)(1 - p_i\rho_{ib})c_{ib}(x). \tag{2}$$

As a consequence, the average throughput of a user with O_i is:

$$t_i = \frac{1}{\Lambda} \sum_{b \in \mathcal{S}_i^n} \sum_{\mathcal{A}_{ib}} p_i \lambda(x)(1 - p_i \rho_{ib}) c_{ib}(x) \tag{3}$$

Let us denote for $i = 1, 2$:

$$I_{i1} = \sum_{b \in \mathcal{S}_i^n} \sum_{\mathcal{A}_{ib}} \lambda(x) \rho_{ib} c_{ib}(x) \tag{4}$$

$$I_{i2} = \sum_{b \in \mathcal{S}_i^n} \sum_{\mathcal{A}_{ib}} \lambda(x) c_{ib}(x) \tag{5}$$

We have $I_{i1} < I_{i2}$ because $\rho_{ib} < 1$. Now: $\Lambda t_i = -p_i^2 I_{i1} + p_i I_{i2}$, with $p_2 = 1 - p_1$. We have $t_1 \geq t_2$ iff $f(p_1) \geq 0$ with:

$$f(p_1) = p_1^2 (I_{21} - I_{11}) + p_1 (I_{12} - 2I_{21} + I_{22}) + I_{21} - I_{22}. \tag{6}$$

Let p_1^* be the operator O_1 market share at equilibrium. Several cases arise:

- If $f(p_1) > 0$ for all $p_1 \in [0; 1]$, then operator 1 is always preferred to operator 2, and $p_1^* = 1$.
- If $f(p_1) < 0$ for all $p_1 \in [0; 1]$, then $p_1^* = 0$.
- if $f(p_1) = 0$ for some $p_1 \in [0; 1]$, then there are one or several equilibrium points. In this case, we set $p_1^* = \max\{p_1 \in [0; 1] : f(p_1) = 0\}$. The assumption behind this choice is that operator 1 has come first on the market. The dynamics of p_1 thus starts from 1 and decreases to the first encountered equilibrium point.

2.5   Pricing Model

We distinguish between two types of costs: there is a fixed installation cost for BSs that have to be paid upfront at the beginning of the deployment and there are operational costs that has to be paid regularly. These operational costs include traditional costs like electricity, maintenance, site renting, and possibly a sharing price. The sharing price is paid by O_2 to O_1 for every site where BSs are shared. We assume that in the first phase O_1 does not have the possibility to choose sites owned by O_2 and thus has no sharing price to pay.

Let λ be the traditional operational cost per unit of time for a single operator BS. Let $(1 + \alpha)\lambda$ with $0 < \alpha < 1$ be the traditional operation cost for a shared BS. Let s_b be the sharing price set by O_1 for its BS b. Let κ be the installation cost of a new BS. Let $\beta\kappa$ with $0 < \beta < 1$ be the installation cost when the site is already occupied by a BS. We assume that O_1 has a total budget of K. Operational and installation costs are the same for O_2.

We assume that the revenues of both operators are proportional to their market share, i.e. $P_1 = p_1 * C$, and $P_2 = p_2 * C = (1 - p_1) * C$ where C is the total capacity of the market. The objective function is the revenue minus the operational costs.

3 Problem Formulation

As mentioned in the introduction, we are interested in the problem of strategic base station placement where two providers enter the market at different times (a leader and a follower), deploy their stations on possible candidate sites so as to maximize their profits. In this section, we model this problem as a bi-level optimization problem. Let us introduce two groups of the decision variables. The first group is the leader's variables:

$$x_j = \begin{cases} 1 & \text{if the leader installs antenna on a site } j \in \mathcal{S}_f \cup \mathcal{S}_1^o \\ 0, & \text{otherwise,} \end{cases}$$

$$x_{ij} = \begin{cases} 1 & \text{if the location } i \text{ is served from leader's station } j \\ 0, & \text{otherwise.} \end{cases}$$

After the deployment of antennas leader have to fix the rent price: $s_j \geq 0, j \in \mathcal{S}_1^n \cup \mathcal{S}_1^o$. The second group is the follower's variables:

$$y_j = \begin{cases} 1 & \text{if the follower installs antenna on a site } j \in \mathcal{S} \\ 0, & \text{otherwise,} \end{cases}$$

$$y_{ij} = \begin{cases} 1 & \text{if the location } i \text{ is served from follower's station } j \\ 0, & \text{otherwise.} \end{cases}$$

Now the competitive location problem can be written as a following bi-level mixed integer linear programming model:

$$\max_{x,y^*,s} \left(p_1 C + \sum_{j \in \mathcal{S}_f} s_j x_j y_j^* + \sum_{j \in \mathcal{S}_1^o} s_j y_j^* - \sum_{j \in \mathcal{S}_f \cup \mathcal{S}_1^o} [\lambda x_j(1 - y_j^*) + (1 + \alpha)\lambda x_j y_j^*] \right) \quad (7)$$

subject to

$$\sum_{j \in \mathcal{S}_f} \kappa x_j + \sum_{j \in \mathcal{S}_1^o} \beta \kappa x_j \leq K \quad (8)$$

$$P_b g_{ib} x_b \geq \gamma_{min} \sum_{j \in \mathcal{S}_f \cup \mathcal{S}_1^o, j \neq b} P_j g_{ij} x_j + \gamma_{min} N - \Gamma(1 - x_{ib}) \quad \forall b \in \mathcal{S}_f \cup \mathcal{S}_1^o, i \in I \quad (9)$$

$$P_b g_{ib} x_b \geq P_j g_{ij} x_j - \Gamma(1 - x_{ib}) \quad \forall b, j \in \mathcal{S}_f \cup \mathcal{S}_1^o, j \neq b, i \in I \quad (10)$$

$$x_{ij} \leq x_j \quad \forall i \in I, j \in \mathcal{S}_f \cup \mathcal{S}_1^o \quad (11)$$

where y^* is the optimal solution of the follower problem

$$\max_y \left((1 - p_1)C - \sum_{j \in \mathcal{S}_1^o} s_j y_j - \sum_{j \in \mathcal{S}_f} s_j x_j y_j - \sum_{j \in \mathcal{S}_f \cup \mathcal{S}_2^o} \lambda y_j(1 - x_j) \right) \quad (12)$$

subject to

$$P_b g_{ib} y_b \geq \gamma_{min} \sum_{j \in \mathcal{S}, j \neq b} P_j g_{ij} y_j + \gamma_{min} N - \Gamma(1 - y_{ib}) \quad \forall b \in \mathcal{S}, i \in I \quad (13)$$

$$P_b g_{ib} y_b \geq P_j g_{ij} y_j - \Gamma(1 - y_{ib}) \quad \forall b, j \in \mathcal{S}, i \in I \tag{14}$$

$$y_{ij} \leq y_j \quad \forall i \in I, j \in \mathcal{S} \tag{15}$$

Upper level (7)–(11) of the problem corresponds to the leader's problem and controls variables $\{x_j\}$, $\{x_{ij}\}$, and $\{s_j\}$. Lower level (12)–(15) formalizes the follower's problem and controls variables $\{y_j\}$, $\{y_{ij}\}$. The follower, at the lower level, maximizes its profit after the leader's decision, at the upper level. The leader maximizes its profit independently affected by the follower's reaction. We call the entire problem (7)–(15) as the leader's problem (LP) as well because our goal is to find the location of stations which provides the maximal leader's profit. Thereby a feasible solution to the LP is defined by the optimal solution to the follower's problem. The objective functions (7) and (12) can be understood as the total profit obtained respectively by the leader and the follower, computed as the difference between the expected revenue from clients served and the operational costs for the stations installed. The sharing payment gives additional profit to the leader, and reduces the gain of the follower. Constraint (8) limits the maximum number of stations that the leader can afford to install due to it's budget. Constraints (9) are the SINR conditions for a location to be covered. When $x_{ib} = 1$, the expression boils down to the SINR condition with respect to the SINR threshold γ_{min}. Whenever $x_{ib} = 0$ then the condition is always fulfilled because of the large value of constant Γ. Constraints (10) combined with (9) state that the location satisfying the minimal SINR constraint is served by a BS providing the most powerful signal. Constraints (13) and (14) have the same meaning in the follower problem. Constraints (11) and (15) state that a service is possible only if a station is installed. The value of p_1 is derived from the quadratic equation (6), as mentioned above.

4   Tabu Search Approach

Although, the constraints of the problem are linear, due to realistic model of clients behavior it is not the case for the goal function. Latter fact makes it hard to apply a broad variety of approaches, which works well with linear integer programming problems. In this study in order to tackle the problem we propose a double level metaheuristic based on the tabu search framework, which performs well on similar problems [4,6]. The tabu search method has been proposed by Fred Glover. It is a so called trajectory metaheuristic and has been widely used to solve hard combinatorial optimization problems [5]. The method is based on the original local search scheme that lets one "travel" from one local optimum to another looking for a global one, avoiding local optimum traps. The main mechanism that allows it to get out of local optima is a tabu list, which contains a list of solutions from previous iterations which are prohibited to be visited on the subsequent steps. We use well-known Flip and Swap neighborhoods to explore the search space. Together with the tabu list, we exploit the idea of randomized neighborhoods. This feature allows to avoid looping, significantly reduces the time per iteration, and improves search efficiency [2,8]. We denote by $Swap_q$ the part q of the $Swap$ neighborhood chosen at random. $Flip_q$ neighborhood is defined in the same way, but with the different value of parameter q. We apply tabu framework both to the lower and the upper level of the problem yet with different settings. Follower problem is the one to be solved a lot of times during the search, as it is needed to provide an estimation on the quality of leader's location. On the one hand better solutions provide better estimation of the goal function, but requires a lot of time to obtain. On the other hand, during the run-through of the neighborhood it is usually enough to make an approximate evaluation which would be refined at the *record update* step. The schematic view of the algorithm may be presented as follows:

Algorithm 1. Mathheuristic

1: Initialize; Read input data; Generate initial solution of the leader x_0, p_0 and the follower y_0, empty tabu lists.
2: **repeat**
3:     Generate the flip&swap neighborhood $N(x)$
4:     **for** each $x' \in N(x)Tabu$ **do**
5:         Apply TabuSearch heuristic to solve the follower problem for x'; Denote best found solution y'
6:         Calculate the goal function value for the leader and the follower
7:         **if** $\omega(x', y') > \omega(x^*, y^*)$ **then**
8:             Refine the solution to find the exact/approx solution of the follower problem
9:         **end if**
10:        **if** $\omega(x', y'_e) > \omega(x^*, y^*)$ **then**
11:            Rewrite the record $(x^*, y^*) := (x'_e, y'_e)$; $x := x'$
12:        **else**
13:            $x := x'$
14:        **end if**
15:    **end for**
16: **until** Given time limit is exceeded

The initial solution is chosen at random. The randomization parameters q for the neighborhoods are set to be sufficiently small. As the tabu list, we use an ordered list of units or pairs of the follower's facilities that have been closed and opened over the last few iterations. The length of the tabu list changes in a given interval during local search. If the best found solution value repeats too frequently, we increase the tabu list length by one; otherwise, we reduce it by one. The method stops after a given number of iterations or after a certain amount of computation time. Due to enormous amount or running time needed to obtain an optimal solution of the follower problem with commercial solvers, in this study at Line 8 we apply tabu search once again yet with another set of parameters in order try to refine the followers solution. This approach does not guarantee us a feasible solution for the leader's problem as the followers solution and thus the goal function value was obtained with an approximation.

5   Experimental Studies

The proposed approach has been implemented in C++ environment and tested on the randomly generated and real data instances. We generated 10 sets of instances with different number of client locations (20, 40, ..,200). All locations are chosen with the uniform distribution over the square area. The number of sites was 1/4 of number of clients locations. The demand data was generated at random, then normalized in order to satisfy the constraints on overload. In the beginning Leader occupies exactly half of the sites at random. In all the tests we assumed that the sharing prices are unique for all the leaders locations. All the test were performed with the following tabu search parameters. The randomization parameter was set to $q = 0.05$ for Swap neighborhood and $q = 0.2$ for Flip Neighborhood. The length of the tabu list was initially set to $l = m/4$ where $m = |\mathcal{S}|$. After each m iterations without new record the length l is increased by one until it reaches $m/2$. When a new record solution is found, tabu list length is set to it's initial value. While exploring the neighborhood on the upper level of the problem we use the followers solution obtained on the previous iteration as a starting point.

The aim of the first experiment was to study the behavior and convergence of the approach on the lower level problem. We run the algorithm on all the 100 instances, 10 runs per instance. Time limit was set to 5 sec. for each run. The algorithm has demonstrated strong convergence. Each time a random starting points was generated. Among all the instances there were only 3 examples, with different results on different runs. All of the examples were rather big (with 180, 200 and 200) locations. And the relative difference between outputs was less then 1 %. These results allow us to hope that the estimation of the leaders's goal function provided by this heuristic approach is not so far from real value.

The second experimental study concerns the real data. We use the client locations and base station cites of the part of 13-th district of Paris [7]. The geometric centers of the blocks are assumed to be client locations. We also use the coordinates of existing base stations in this area. The number of client locations, $n = 95$, the number of cites for possible BS placement, $m = 19$. Total budget of the market 3500. Table 1 contains the results of the dependance of followers behavior from the sharing price, proposed by the leader. It can be seen from the table that high sharing price is not always the optimal one for the leader. As the sharing price increase, the follower is forced to change the location of his antennas in order to avoid sharing.

In Table 2 we provide the resulting income of the leader depending on his initial budget. We consider the real data once again. There are 19 cites for BS installation. Seven of them are already occupied with leaders 4G equipment. The price of antenna installation in these locations is set to 50, while the price of installing new BS is 100.

We can see from the table that interference effect together with non-zero maintenance price force the leader not to use the whole budget as it will lower the overall income.

Table 1. Profit and market share influenced by rent price

Sharing price	Leader share (p_1)	Follower profit	Leader profit	N shared sites	N opened sites
200	0.327	1755	1305	2	7
220	0.327	1715	1345	2	7
250	0.333	1633	1427	2	7
280	0.408	1589	1471	1	6
310	0.481	1574	1446	0	6

Table 2. Leader profit from size

Budget	Leader share (p_1)	Leader profit	Follower profit	N old sites	N opened sites
300	0.445	1457.5	1822.5	4	5
400	0.451	1458.5	1801.5	4	6
500	0.473	1495.5	1724.5	6	8
600	0.492	1562	1678	5	8
700	0.492	1562	1678	5	8

6   Conclusions

We have considered new competitive base stations location problem with sharing. We have proposed a mathematical model for this problem and a tabu search based heuristic for obtaining approximate solutions. Computational results show the believability of the model. In the future research we are planning to implement the exact method for the follower problem in order to obtain feasible solutions of the leader's one.

Acknowledgements. The first author work was supported by Russian Science Foundation (project 15-11-10009).

References

1. Beresnev, V.L., Mel'nikov, A.A.: Approximate algorithms for the competitive facility location problem. Autom. Remote Control **5**(2), 180–190 (2011)
2. Davydov, I.A., Kochetov, Y., Mladenovic, N., Urosevic, D.: Fast metaheuristics for the discrete ($r|p$)-centroid problem. Autom. Remote Control **75**(4), 677–687 (2014)
3. Davydov, I., Coupechoux, M., Iellamo, S.: Tabu search heuristic for competitive base station location problem. In: OR2015 Proceedings. (accepted, in press) (2016)
4. Diakova, Z., Kochetov, Y.: A double VNS heuristic for the facility location and pricing problem. Electron. Notes Discrete Math. **39**, 29–34 (2012)
5. Glover, F., Laguna, M.: Tabu Search. Kluwer Academic Publishers, Dordrecht (1997)
6. Lellamo, S., Alekseeva, E., Chen, L., Coupechoux, M., Kochetov, Y.: Competitive location in cognitive radio networks. 4OR **13**, 81–110 (2015)
7. ANFR Cartoradio. http://www.cartoradio.fr/cartoradio/web/
8. Mel'nikov, A.A.: Randomized local search for the discrete competitive facility location problem. Autom. Remote Control **75**(4), 700–714 (2014)

Upper Bound for the Competitive Facility Location Problem with Quantile Criterion

Andrey Melnikov[1,2]($\boxtimes$) and Vladimir Beresnev[1,2]

[1] Sobolev Institute of Mathematics, Novosibirsk, Russia
{melnikov,beresnev}@math.nsc.ru
[2] Novosibirsk State University, Novosibirsk, Russia

Abstract. In this paper, we consider a competitive location problem in a form of Stackelberg game. Two parties open facilities with the goal to capture customers and maximize own profits. One of the parties, called Leader, opens facilities first. The set of customers is specified after Leader's turn with random realization of one of possible scenarios. Leader's goal is to maximize the profit guaranteed with given probability or reliability level provided that the second party, called Follower, acts rationally in each of the scenarios. We suggest an estimating problem to obtain an upper bound for Leader's objective function and compare the performance of estimating problem reformulations experimentally.

Keywords: Stackelberg game · Reformulation · Competitive location · Upper bound

1 Introduction

We deal with a bilevel location model firstly suggested in [7] as a deterministic reformulation of a stochastic competitive location problem. In this problem, two competing parties open their facilities in a finite discrete space with the goal to maximize their profits, i.e. the value of income from customers service minus the fixed costs of facilities opening. A decision making process is organized as a Stackelberg game [10]. One of the parties, called Leader, opens its facilities first.

It is assumed that the set of customers is unknown for Leader. Instead of this Leader is provided with a finite set of possible scenarios. Each scenario has known probability of realization and fully characterizes the set of customers.

After Leader opens facilities one of possible scenarios is realized and the set of customers becomes specified. This information is available for the second party called Follower, who opens own facilities with the goal to maximize profit as well.

In the model under consideration, each customer chooses the party to be served by according to his preferences. The facility assigned to serve the customer must be more preferable for him than any competitor's facility. This model of customers serving in competitive environment is referred to in [3] as a *free choice of supplier rule*. In the present paper we study a multi–scenario generalization of the problem in [3].

© Springer International Publishing Switzerland 2016
Y. Kochetov et al. (Eds.): DOOR 2016, LNCS 9869, pp. 373–387, 2016.
DOI: 10.1007/978-3-319-44914-2_30

Leader's goal in this situation is to maximize profit that can be guaranteed with given probability or *reliability level*. In other words, Leader selects a set of facilities to be opened such that there exists a subset of scenarios with total probability not less than the reliability level. In these scenarios, Leader gets a profit, which is not less than a certain value, and the goal is to make this value as big as possible. The scenarios participating in the calculation of guaranteed income are further referred to as *active scenarios*.

In [7] the value of income the customer brings to the facility is the same for all facilities. This assumption allows the authors to suggest upper and lower bounds for an optimum of the Leader's problem. In the case when the income depends on serving facility, the suggested upper bound is not valid. By using the technique from [3] we formulate an estimating problem in the form of MIP providing an upper bound in the case of facility–dependent income values. We suggest two reformulations of the estimating problem and perform numerical experiments to compare their efficiencies.

The rest of the paper is organized as follows. In Sect. 2 we present a mathematical model of the competitive facility location problem with quantile criterion (QCompFLP) in the form of a pessimistic bilevel mixed–integer programming problem [5] and discuss a concept of its pessimistic feasible solution. Also, we suggest a procedure to compute a pessimistic feasible solution for given values of Leader's location variables. Section 3 provides a formulation of the estimating problem for upper bound calculation. Two reformulations of it are presented as well. In Sect. 4 we compare the effectiveness of suggested formulations of the estimating problem and examine their qualities as providers of an upper bound for QCompFLP.

2   Mathematical Model

Let us introduce the necessary notations:

Sets:

$I = \{1, \ldots, m\}$ is a set of facilities or candidate sites for opening a facility;

$S = \{1, \ldots, l\}$ is a set of possible scenarios;

J_s is a finite set of customers in case when scenario $s \in S$ is realized. We assume that $J_{s_1} \cap J_{s_2} = \emptyset$ for each $s_1, s_2 \in S$, $s_1 \neq s_2$. The set of all possible customers is denoted with $J = \bigcup_{s \in S} J_s$. Without loss of generality we assume that $J = \{1, \ldots, n\}$.

Parameters:

f_i is a fixed cost of opening Leader's facility $i \in I$;

g_i is a fixed cost of opening Follower's facility $i \in I$;

c_{ij} is an income of Leader's facility $i \in I$ from customer $j \in J$;

d_{ij} is an income of Follower's facility $i \in I$ from customer $j \in J$;

p_s is a probability of realization of scenario $s \in S$;

p_0 is a reliability level.

Variables:

$$x_i = \begin{cases} 1, & \text{if Leader opens facility } i \\ 0, & \text{otherwise,} \end{cases}$$

$$x_{ij} = \begin{cases} 1, & \text{if Leader's facility } i \text{ is assigned to serve customer } j \\ 0, & \text{otherwise,} \end{cases}$$

$$z_i^s = \begin{cases} 1, & \text{if Follower opens facility } i \text{ in the case of scenario } s \text{ realization} \\ 0, & \text{otherwise,} \end{cases}$$

$$z_{ij} = \begin{cases} 1, & \text{if Follower's facility } i \text{ is assigned to serve customer } j \\ 0, & \text{otherwise,} \end{cases}$$

$$\delta_s = \begin{cases} 1, & \text{if scenario } s \text{ is active} \\ 0, & \text{otherwise,} \end{cases}$$

C stands for the value of guaranteed income.

In the QCompFLP model we use a binary customer patronizing rule [9]. It means that each customer $j \in J$ brings income to a single facility opened by either Leader or Follower. We assume that this facility is chosen according to the preferences of the customer. The preferences are represented with linear order $\succeq_j$ on the set I. Given $i_1, i_2 \in I$, the relation $i_1 \succeq_j i_2$ means that i_1 is not less preferable for j than i_2. If $i_1 \neq i_2$ then i_1 is strictly more preferable than i_2, and we denote it with $i_1 \succ_j i_2$. For a nonempty set $I_0 \subseteq I$ we denote with $i_j(I_0)$ such an element of I_0, for which $i_j(I_0) \succeq_j k$ for all $k \in I_0$. For a nonzero $(0,1)$–vector $v = (v_i), i \in I$ we use notation $i_j(v)$ for an element $i_j(\{k \in I | v_k = 1\})$.

If we are given with boolean vectors x and z of Leader's and Follower's location variables values respectively, then Leader's facility $i \in I$ can serve a customer $j \in J$ iff $i \succ_j i_j(z)$. Similarly, Follower's facility $i \in I$ can serve a customer $j \in J$ iff $i \succ_j i_j(x)$.

Using introduced notations the mathematical model of the QCompFLP is written as the following pessimistic bilevel program:

$$\max_{(x_i),(x_{ij}),(\delta_s),C} \quad \min_{(\tilde{z}_i^s),(\tilde{z}_{ij})} \left(-\sum_{i \in I} f_i x_i + C \right), \tag{1}$$

$$x_i \geq x_{ij}, \quad i \in I, \quad j \in J; \tag{2}$$

$$\sum_{i \in I} x_{ij} \leq \delta_s, \quad s \in S, \quad j \in J_s; \tag{3}$$

$$C \leq \sum_{i \in I} \sum_{j \in J_s} c_{ij} x_{ij} + M(1 - \delta_s), \quad s \in S; \tag{4}$$

$$\tilde{z}_i^s + \sum_{k \in I | i \succ_j k} x_{kj} \leq 1, \quad s \in S, j \in J_s; \tag{5}$$

$$\sum_{s \in S} p_s \delta_s \geq p_0; \tag{6}$$

$$x_i, \delta_s \in \{0,1\}; 0 \leq x_{ij} \leq 1, \quad i \in I, j \in J, s \in S; \tag{7}$$

$$\text{where } (\tilde{z}_i^s), (\tilde{z}_{ij}) \text{ solves} \tag{8}$$

$$\max_{(z_i^s),(z_{ij})} \sum_{s \in S} \left(-\sum_{i \in I} g_i z_i^s + \sum_{i \in I} \sum_{j \in J_s} d_{ij} z_{ij} \right), \tag{9}$$

$$z_i^s \geq z_{ij}, \quad i \in I, s \in S, j \in J_s; \tag{10}$$

$$x_i + \sum_{k \in I | i \succeq_j k} z_{kj} \leq 1, \quad j \in J; \tag{11}$$

$$x_i + z_i^s \leq 1, \quad i \in I, s \in S; \tag{12}$$

$$z_i^s, \in \{0, 1\}; 0 \leq z_{ij} \leq 1, \quad i \in I, j \in J, s \in S. \tag{13}$$

The objective function (1) of the upper–level problem represents the value of income, which is guaranteed with a probability p_0 reduced by the cost of opened facilities. Inequalities (2) forbid to serve customers with close facilities, (3) guarantee that customers from active scenarios cannot be served with more than one facility, (5) ensure that the customer is served with a facility which is more preferable than any of competitor's ones. Constraints (4) provide that the value of guaranteed income is not greater than the income realized in any of active scenarios. The term with a sufficiently large constant M excludes inactive scenarios from the consideration. Constraints (6) impose that the income value is guaranteed with probability p_0.

The lower–level objective function (9) is a sum of profits Follower obtains in all possible scenarios. Its maximization is equivalent to maximization of the profit for each scenario separately. The constraints (10) and (11) have the same meaning as the upper–level constraints (2) and (5), respectively. Finally, the inequalities (12) provide that Follower does not open facility in the place occupied by Leader.

For brevity let us denote the vector of values of x_i, $i \in I$ and z_i^s, $i \in I$, $s \in S$ with x and z correspondingly. Given x we denote the problem $\mathcal{F}$ with $\mathcal{F}(x)$. Analogously, the problem $\mathcal{L}$ with the value of z in the constraints (5) is denoted with $\mathcal{L}(z)$. A whole model (1)–(13) is referred to as $(\mathcal{L}, \mathcal{F})$.

2.1 Pessimistic Feasible Solutions of the Problem $(\mathcal{L}, \mathcal{F})$

Consider some (0,1)–vector $x = (x_i)$, $i \in I$ and a quadruple $\chi(x) = (X, \Delta, C, Z)$, where $X = ((x_i), (x_{ij}))$, $\Delta = (\delta_s)$, $Z = ((z_i^s), (z_{ij}))$, $i \in I$, $j \in J$, $s \in S$. We call quadruple $\chi(x)$ a feasible solution of the problem $(\mathcal{L}, \mathcal{F})$ if Z is an optimal solution of the problem $\mathcal{F}(x)$ and (X, Δ, C) is a feasible solution of the problem $\mathcal{L}(z)$, where $z = (z_i^s)$, $i \in I$, $s \in S$.

Let $Opt(x)$ be a set of optimal solutions of the problem $\mathcal{F}(x)$. Given $Z \in Opt(x)$, let $\chi(x, Z)$ denote a quadruple $(X(Z), \Delta(Z), C(Z), Z)$, where $(X(Z), \Delta(Z), C(Z))$ is an optimal solution of the problem $\mathcal{L}(z)$. We denote the value of objective function (1) on this solution with $L(\chi(x, Z))$. The solution $\chi(x, \check{Z})$ is called a *pessimistic feasible solution* of the problem $(\mathcal{L}, \mathcal{F})$ if $L(\chi(x, \check{Z})) \leq L(\chi(x, Z))$ for all $Z \in Opt(x)$. The problem $(\mathcal{L}, \mathcal{F})$ is equivalent to the problem of maximization of some implicitly given function $\check{f} : \{0,1\}^m \to \mathbb{R}$ such that $\check{f}(x) = L(\chi(x, \check{Z}))$ for each $(0,1)$–vector x.

The value $\check{f}(x)$ for a given x can be computed in two steps. At the first step the problem $\mathcal{F}(x)$ is solved, and let F^* be its optimum. At the second step an auxiliary MIP provides a pessimistic feasible solution and a value of $\check{f}(x)$. To formulate it we introduce new nonnegative variables u_j, $j \in J$. Variable u_j takes the value equals to Leader's income from the customer j.

$$\min_{(z_i^s),(z_{ij}),(u_j)} \sum_{j \in J} u_j \tag{14}$$

$$x_i + \sum_{k|i \succ_j k} z_{kj} \leq 1, \quad i \in I, \quad j \in J; \tag{15}$$

$$z_i^s \geq z_{ij}, \quad i \in I, s \in S, j \in J_s; \tag{16}$$

$$u_j \geq c_{ij}\Big(x_i - \sum_{k|k \succ_j i} z_k^s\Big), \quad i \in I, s \in S, j \in J_s; \tag{17}$$

$$\sum_{s \in S}\Big(-\sum_{i \in I} g_i z_i^s + \sum_{i \in I}\sum_{j \in J_s} d_{ij} z_{ij}\Big) \geq F^*; \tag{18}$$

$$x_i + z_i^s \leq 1, \quad i \in I, s \in S; \tag{19}$$

$$z_i^s \in \{0,1\}; 0 \leq z_{ij} \leq 1, \quad i \in I, s \in S, j \in J; \tag{20}$$

$$u_i \geq 0, \quad j \in J. \tag{21}$$

Let (Z, U), $Z = ((z_i^s), (z_{ij}))$, $U = (u_j)$ be an optimal solution of the problem (14)–(21), and let $z^s = (z_i^s)$. Notice that for solution (Z, U) the following equality holds for each $s \in S$, $j \in J_s$:

$$u_j = \max_{i|i \succ_j i_j(z^s)} \{c_{ij} x_i\}.$$

The value of income the Leader gets in the scenario $s \in S$ is calculated as follows: $C_s = \sum_{j \in J_s} u_j$. To choose the set of active scenarios one should sort values $\{C_s\}$ in descending order. Without loss of generality we can assume that $C_1 \geq C_2 \geq \cdots \geq C_l$. Let r be a such an index that $\sum_{s < r} p_s < p_0 \leq \sum_{s \leq r} p_s$. Then it is easy to see that $\check{f}(x) = -\sum_{i \in I} f_i x_i + C_r$.

Now we are able to construct a pessimistic feasible solution $\chi(x, Z)$. For every $s \leq r$ and all $j \in J_s$ such that $u_j > 0$ let us denote with i_j the index $i \in I$ for which the constraint (17) is active. Then we set $x_{i_j j} = 1$ and $\delta_s = 1$. For all other indexes $i \in I$, $j \in J$, $s \in S$ we set $x_{ij} = 0$ and $\delta_s = 0$. The quadruple $(((x_i), (x_{ij})), (\delta_s), C_r, (z_i^s), (z_{ij}))$ is a desired pessimistic feasible solution of the problem $(\mathcal{L}, \mathcal{F})$.

From the above we conclude that pessimistic QCompFLP is equivalent to the problem of maximization of implicitly given pseudo–boolean function. The function depends on m boolean variables. Its value on boolean vector $x = (x_i)$, $i \in I$ can be calculated by solving two mixed–integer linear programming problems.

3 Upper Bound

Consider the problem of finding the global maximum of pseudo–boolean function $\check{f} : \{0,1\}^m \rightarrow \mathbb{R}$ associating an arbitrary (0,1)–vector x with the value of objective function (1) on the corresponding pessimistic feasible solution $\chi(x, \check{Z})$. The method to calculate an upper bound for values of the function $\check{f}$ consists in constructing and solving of an auxiliary optimization problem, referred to as an *estimating problem*.

3.1 Estimating Problem

The basis of the estimating problem is a relaxation of the problem $(\mathcal{L}, \mathcal{F})$ obtained by removing the lower–level problem $\mathcal{F}$ and its variables. The resulting single–level mixed–integer problem models the situation there Leader is a monopolist. Obviously, an optimal value of the model is a valid upper bound for the function $\check{f}$, but its accuracy is insufficient for practical application.

Similarly to the earlier considered models of competitive location [2–4], the relaxation of $(\mathcal{L}, \mathcal{F})$ can be strengthened by using the system of estimating subsets $\{I_j\}$, $j \in J$. The construction of estimating subsets for the case of single scenario is presented in [3] and can be easily extended to the case of several scenarios. An algorithm of subsets construction allows to claim that if the most preferable for the customer j Leader's facility $i_j(x)$ is not in I_j, then Follower will open a facility which is more preferable for j than $i_j(x)$.

Following the method from [3] we transform an income matrix (c_{ij}) into a new matrix (c'_{ij}), which majorizes (c_{ij}) and is correlated with the preferences of customers. It means that $c'_{ij} \geq c_{ij}$ for all $i \in I$, $j \in J$ and for given $j \in J$ values (c'_{ij}) are monotone according to the order $\succeq_j$: given $i_1, i_2 \in I$ the relation $i_1 \succeq_j i_2$ implies that $c'_{i_1 j} \geq c'_{i_2 j}$. Such a matrix can be constructed by assuming that $c'_{ij} = \max_{k | i \succeq_j k} c_{kj}$ for all $i \in I$ and $j \in J$.

The algorithm of subsets $\{I_j\}$ construction for a single scenario case is presented in [3]. By considering an arbitrary scenario $s \in S$ separately we obtain the system of subsets $\{I_j\}$, where $j \in J_s$. By switching s one by one we obtain a subset I_j for every $j \in J$. An algorithm of construction ensures that the following inequality holds for every $j \in J$ and every pessimistic feasible solution $\chi(x, \check{Z})$ of the problem $(\mathcal{L}, \mathcal{F})$:

$$\sum_{i \in I} c_{ij} x_{ij} \leq \sum_{i \in I_j} c'_{ij} x_{ij}. \tag{22}$$

By assuming that for every $i \in I$ and $j \in J$

$$c''_{ij} = \begin{cases} c'_{ij}, & \text{if } i \in I_j \\ 0, & \text{otherwise} \end{cases},$$

we get the following estimating problem:

$$\max_{(x_i),(x_{ij}),(\delta_s),C} \left(-\sum_{i \in I} f_i x_i + C \right), \tag{23}$$

$$x_i \geq x_{ij}, \quad i \in I, \quad j \in J; \tag{24}$$

$$\sum_{i \in I} x_{ij} \leq \delta_s, \quad s \in S, \quad j \in J_s; \tag{25}$$

$$C \leq \sum_{i \in I} \sum_{j \in J_s} c''_{ij} x_{ij} + M(1 - \delta_s), \quad s \in S; \tag{26}$$

$$\sum_{s \in S} p_s \delta_s \geq p_0; \tag{27}$$

$$x_i, \delta_s \in \{0,1\}; 0 \leq x_{ij} \leq 1, \quad i \in I, j \in J, s \in S. \tag{28}$$

The model (23)–(28) is further referred to as $\mathcal{B}$. It is a relaxation of the bilevel problem $(\mathcal{L}, \mathcal{F})$, obtained by removing the lower–level problem $\mathcal{F}$ and its variables. Inequalities (26) are the corollary of estimating subsets properties (22).

Thus, the optimum of the constructed estimating problem is an upper bound for $\max_{x \in \{0,1\}^m} \check{f}(x)$. Its calculation is a time consuming procedure since the model $\mathcal{B}$ has a big integrality gap provided by inequalities (26), where the right–hand side can significantly change the value after relaxation of variables (δ_s). To find a compromise between accuracy of the upper bound and its calculation time we suggest two reformulations of the model $\mathcal{B}$.

3.2 Relaxation of a Large MIP

As it was mentioned, solving the problem $\mathcal{B}$ can be a time–consuming procedure. Let us introduce its reformulation involving exponential number of variables.

Let $\mathcal{R}$ be a set of all subsets of S. For each $R \in \mathcal{R}$ we introduce a new boolean variable u_R, which takes the value 1 if the set of active scenarios equals to R and 0 otherwise. Additionally we need a $(0,1)$–matrix (a_{sR}), $s \in S$, $R \in \mathcal{R}$ such that

$$a_{sR} = \begin{cases} 1, \text{ if } s \in R \\ 0 \text{ otherwise} \end{cases}.$$

Using the above definitions the reformulation of $\mathcal{B}$ is written as follows:

$$\max_{(x_i),(x_{ij}),(u_R),C} \left(-\sum_{i \in I} f_i x_i + C \right), \tag{29}$$

$$C \leq \sum_{i \in I} \sum_{j \in J_s} c''_{ij} x_{ij} + M\left(1 - \sum_{R \in \mathcal{R}} a_{sR} u_R\right), \quad s \in S; \tag{30}$$

$$\sum_{i \in I} x_{ij} \leq \sum_{R \in \mathcal{R}} a_{sR} u_R, \quad s \in S, j \in J_s; \tag{31}$$

$$x_i \geq x_{ij}, \quad i \in I, \quad j \in J; \tag{32}$$

$$\sum_{R \in \mathcal{R}} \sum_{s \in R} a_{sR} p_s u_R \geq p_0; \tag{33}$$

$$\sum_{R \in \mathcal{R}} u_R = 1; \tag{34}$$

$$x_i, u_R \in \{0,1\}; 0 \leq x_{ij} \leq 1, \quad i \in I, j \in J, R \in \mathcal{R}. \tag{35}$$

To deal with the linear relaxation of this large problem consider its dual:

$$\min_{(\alpha_s),(\beta_j),(\gamma_{ij}),\eta,\lambda} (-p_0\eta + \lambda + M) \tag{36}$$

$$\sum_{j \in J} \gamma_{ij} \leq f_i, \quad i \in I; \tag{37}$$

$$\beta_j + \gamma_{ij} \geq c''_{ij}\alpha_s, \quad i \in I, j \in J; \tag{38}$$

$$\sum_{s \in S} \alpha_s = 1, \tag{39}$$

$$\lambda \geq \sum_{s \in S} a_{sR} \left(p_s\eta + \sum_{j \in J_s} \beta_j - M\alpha_s \right), \quad R \in \mathcal{R}; \tag{40}$$

$$\alpha_s, \beta_j, \gamma_{ij}, \eta \geq 0. \tag{41}$$

Let $D(\mathcal{R}') = ((\alpha_s), (\beta_j), (\gamma_{ij}), \eta, \lambda)$ be an optimal solution of the problem (36)–(41), where exponentially large index set $\mathcal{R}$ in (40) is replaced with its relatively small subset $\mathcal{R}' \subseteq \mathcal{R}$. In the case when $D(\mathcal{R}')$ satisfies (40) for all $R \in \mathcal{R}$, it is an optimal solution of the dual problem and provides a required upper bound. Otherwise there exists a $(0,1)$–vector (δ_s) such that

$$\sum_{s \in S} \delta_s (p_s\eta + \sum_{j \in J_s} \beta_j - M\alpha_s) > \lambda. \tag{42}$$

The existence of such a vector can be checked by solving the following problem:

$$\max_{(\delta_s)} \sum_{s \in S} w_s\delta_s \tag{43}$$

$$\sum_{s \in S} \delta_s p_s \geq p_0 \tag{44}$$

$$\delta_s \in \{0,1\}, s \in S, \tag{45}$$

where $w_s = p_s\eta + \sum_{j \in J_s} \beta_j - M\alpha_s$.

Given an optimal solution (δ_s^*) of the problem (43)–(45), if the inequality $\sum_{s \in S} w_s\delta_s^* \leq \lambda$ holds, then the solution $D(\mathcal{R}')$ satisfies (40) for any $R \in \mathcal{R}$. Otherwise, one of constraints that the $D(\mathcal{R}')$ violates corresponds to the set of scenarios $\{s \in S | \delta_s^* = 1\}$. We include it into $\mathcal{R}'$ and get back to solving the dual problem with a new constraint of type (40).

Thus the cutting–plane (CP) scheme to calculate an upper bound for the value $\max_{x \in \{0,1\}} \check{f}(x)$ is an iterative process [6]. On each iteration, a restricted dual problem is being solved. We check its optimal solution for feasibility in the initial dual problem. In the case of positive answer, a valid upper bound is obtained and the procedure terminates. Otherwise, a cut is generated by solving a knapsack–type problem, and a new iteration begins.

At the first iteration, we must guarantee a feasibility of restricted dual problem by appreciate choice of $\mathcal{R}'$. In our experiments we initialize $\mathcal{R}'$ with a random subset $S' \subseteq S$ such that $\sum_{s \in S'} p_s \geq p_0$.

3.3   Reformulation of Bilevel Estimating Problem

Let us get back to the problem $\mathcal{B}$. Notice that if the location variables values are chosen, one can easily obtain an optimal facility assignment for each customer and calculate the income value for each possible scenario. Assume that $p_s = \frac{1}{l}$ for all $s \in S$. Then an optimal set of active scenarios contains exactly $\lceil lp_0 \rceil$ elements with the greatest total income. It leads us to the following bilevel formulation of the problem $\mathcal{B}$, which uses additional variables (C_s), $s \in S$ for the value of income in the corresponding scenario.

$$\max_{(x_i),(x_{ij}),(C_s),C} \left(-\sum_{i \in I} f_i x_i + C \right), \tag{46}$$

$$x_i \geq x_{ij}, \quad i \in I, \quad j \in J; \tag{47}$$

$$\sum_{i \in I} x_{ij} \leq 1, \quad s \in S, \quad j \in J_s; \tag{48}$$

$$C_s \leq \sum_{i \in I} \sum_{j \in J_s} c''_{ij} x_{ij}, \quad s \in S; \tag{49}$$

$$C \leq C_s + M(1 - \delta_s^*), \quad s \in S; \tag{50}$$

$$x_i \in \{0,1\}; x_{ij}, C, C_s \geq 0, \quad i \in I, j \in J, s \in S; \tag{51}$$

$$\text{where } (\delta_s^*) \text{ solves} \tag{52}$$

$$\max_{(\delta_s)} \sum_{s \in S} C_s \delta_s, \tag{53}$$

$$\sum_{s \in S} \delta_s = \lceil lp_0 \rceil; \tag{54}$$

$$0 \leq \delta_s \leq 1, s \in S. \tag{55}$$

In the lower–level problem (53)–(55) $\lceil lp_0 \rceil$ scenarios with the greatest income are chosen. Notice that here we can let δ_s, $s \in S$ take fractional values without loss of integrality of optimal solution.

Due to simplicity of the lower–level problem, we substitute it with complementary slackness conditions [1] to obtain a single–level reformulation of the problem $\mathcal{B}$. After linearization the resulting problem $\mathcal{B}'$ is written as follows:

$$\max_{(x_i),(x_{ij}),(\delta_s),(C_s),C} \left(-\sum_{i \in I} f_i x_i + C \right), \tag{56}$$

$$x_i \geq x_{ij}, \quad i \in I, \quad j \in J; \tag{57}$$

$$\sum_{i \in I} x_{ij} \leq 1, \quad s \in S, \quad j \in J_s; \tag{58}$$

$$C_s \leq \sum_{i \in I} \sum_{j \in J_s} c''_{ij} x_{ij}, \quad s \in S; \tag{59}$$

$$C \leq C_s + M(1 - \delta_s), \quad s \in S; \tag{60}$$

$$\sum_{s \in S} \delta_s = \lceil lp_0 \rceil; \tag{61}$$

$$u_s \leq M\delta_s, \quad s \in S; \tag{62}$$

$$u_s + w \geq C_s, \quad s \in S; \tag{63}$$

$$u_s + w \leq C_s + M(1 - \delta_s), \quad s \in S; \tag{64}$$

$$x_i, \delta_s \in \{0,1\}; x_{ij}, C_s, C \geq 0, \quad i \in I, j \in J, s \in S. \tag{65}$$

Here w and $(u_s), s \in S$ are dual variables for constraints (54) and (55), respectively. Variables $(\delta_s), s \in S$ are boolean in this model since they play role of indicator variables in complementary slackness conditions linearization. Model (56)–(65) by itself is further addressed as $\mathcal{B}'$.

4   Numerical Experiments

In this section, we present results of comparison of proposed estimating problem formulations. The section consists of two parts. In the first subsection, we compare models and their relaxations on randomly generated inputs.

In the second subsection, we consider a number of randomly generated instances of QCompFLP. We construct a system of estimating subsets and compare values of the upper bound provided by problems $\mathcal{B}$, $\mathcal{B}'$, their relaxations and cutting–plane procedure with the value of function $\check{f}$ on locally optimal solution.

Calculations are performed in a single thread by workstation with processors Intel Xeon X5675 3.07 GHz and 96 GB RAM. To solve mixed–integer programs we use Microsoft Solver Foundation 3.1 library with built–in Gurobi MIP solver.

4.1   Instances Not Induced by QCompFLP

To examine the efficiency of upper bound calculation procedures we generate a set of inputs for the estimating problem. For different values of m, n, and l a series of three tests was performed. In each test income matrix (c_{ij}), $i \in I$, $j \in J$ is filled with random integers uniformly distributed on the integer range $\{5, 6, \ldots, 15\}$. Fixed cost value f_i equals to 50 for all $i \in I$. The probability of scenario is generated in two stages. On the first stage an integer ρ_s, $s \in S$ is chosen from the range $\{1, 2, 3, 4\}$ with equal probabilities. The probability of realization of scenario s is set to be equal to $\rho_s / \sum_{l \in S} \rho_l$. In all instances of this bundle of tests $p_0 = 0.6$.

In the Table 1 we provide the following values:

Gap($\mathcal{B}$), relative integrality gap of the model $\mathcal{B}$, i.e. a difference between optimum of linear relaxation of $\mathcal{B}$ and its integer optimum divided by integer optimum, in percents;

Gap(CP), relative integrality gap of the model (29)–(35), i.e. a difference between the value provided by cutting–plane procedure and integer optimum divided by integer optimum, in percents;

OPT, value of integer optimum;

T($\mathcal{B}_{LR}$), calculation time for the linear relaxation of the model $\mathcal{B}$;

T(CP), calculation time for the cutting–plane procedure;

T($\mathcal{B}$), calculation time for model $\mathcal{B}$. It is limited by 10 min. The mark "TL" appears in the cases of reaching the time limit. Also in this cases the optimal value of objective function instead of relative integrality gap is presented in the columns Gap($\mathcal{B}$) and Gap(CP).

As Table 1 shows, the integrality gaps of both the model $\mathcal{B}$ and the model (29)–(35) are dramatically large. However on instances with $m = 10$ and $m = 15$ the relaxation of the second model outperforms the relaxation of $\mathcal{B}$: it can be solved by cutting–plane method in a comparable time and provide more accurate estimation of the optimum of $\mathcal{B}$. The calculation time for optimum of $\mathcal{B}$ is relatively big and grows rapidly while the dimensionality increases, thus

Table 1. Instances with different scenario probabilities

Gap($\mathcal{B}$)	Gap(CP)	OPT	T($\mathcal{B}_{LR}$)	T(CP)	T($\mathcal{B}$)
$m = 10, n = 150, l = 4$					
32 %	21 %	391	< 1"	< 1"	26"
22 %	21 %	426	< 1"	< 1"	33"
33 %	33 %	415	< 1"	< 1"	7"
$m = 15, n = 200, l = 6$					
41 %	41 %	323	1"	10"	41"
51 %	47 %	313	1"	9"	1'15"
73 %	66 %	288	< 1"	2"	1'55"
$m = 20, n = 250, l = 8$					
48 %	48 %	305	2"	31"	7'4"
86 %	82 %	276	2"	37"	6'3"
57 %	53 %	290	2"	23"	7'1"
$m = 25, n = 300, l = 10$					
441,00	435,00	–	8"	1'51"	TL
422,00	422,00	–	7"	1'42"	TL
438,00	438,00	–	6"	2'35"	TL

Table 2. Instances with equiprobable scenarios

Gap($\mathcal{B}$)	Gap($\mathcal{B}'$)	Gap(CP)	OPT	T($\mathcal{B}_{LR}$)	T($\mathcal{B}'_{LR}$)	T(CP)	T($\mathcal{B}$)	T($\mathcal{B}'$)
$m = 10$, $n = 150$, $l = 4$								
41 %	28 %	28 %	356	< 1"	< 1"	< 1"	11"	7"
69 %	53 %	53 %	300	< 1"	< 1"	< 1"	7"	22"
61 %	43 %	43 %	330	< 1"	< 1"	< 1"	13"	5"
$m = 15$, $n = 200$, $l = 6$								
31 %	26 %	26 %	321	< 1"	1"	4"	2'24"	1'36"
45 %	39 %	39 %	303	< 1"	2"	9"	1'40"	54"
53 %	44 %	44 %	310	1"	2"	3"	1'49"	43"
$m = 20$, $n = 250$, $l = 8$								
40 %	38 %	38 %	293	2"	6"	16"	7'55"	5'19"
472	459	459	–	2"	8"	20"	TL	TL
55 %	52 %	52 %	273	2"	7"	40"	TL	8'43"
$m = 25$, $n = 300$, $l = 10$								
388	388	388	–	5"	21"	1'34"	TL	TL
46 %	46 %	46 %	277	6"	15"	1'27"	TL	7'43"
404	404	404	–	6"	17"	4'49"	TL	TL

the model $\mathcal{B}$ can be utilized in a posteriori accuracy estimation of non-exact algorithms and in implicit enumeration schemes to solve small instances. The relaxations are more perspective for exact methods dealing with bigger instances.

Table 2 presents results of dealing with instances where scenarios are equiprobable. Additionally to columns from the Table 1 it includes:

Gap($\mathcal{B}'$), relative integrality gap of the model $\mathcal{B}'$;

T($\mathcal{B}'_{LR}$), calculation time for the linear relaxation of the model $\mathcal{B}'$;

T($\mathcal{B}'$), calculation time for the model $\mathcal{B}'$ (is bounded by 10 min as like as for $\mathcal{B}$).

According to the results from Table 2, the model $\mathcal{B}'$ outperforms $\mathcal{B}$ in relative integrality gap and calculation time. Its linear relaxation and linear relaxation of the model (29)–(35) provide the same results, but the first one consumes smaller amount of time. An important observation is that in the cases when $p_0 l$ is an integer number linear relaxations of the three proposed models provide the same results as it is in the case of the fourth series of tests, where $l = 10$ and $p_0 l = 6$.

4.2   Instances Induced by QCompFLP

In this subsection, we study the proposed models as an upper bound providers for the problem QCompFLP. Numerical data preparation consists of generating the QCompFLP instance, constructing an estimating subset system, and forming an input data for the estimating problem. For each generated instance of

QCompFLP we start a local search procedure using scheme from [8] to obtain a lower bound for the optimum. It allows us to estimate the quality of the upper bound.

QCompFLP instances are generated as follows. All operations of random choice are performed with uniform distribution on the corresponding domain. Fixed costs values f_i and g_i, $i \in I$ are calculated with formula $50+5\xi$, where value of ξ is randomly chosen from the integer range $\{-3, -2, \ldots, 3\}$ for each f_i and each g_i, $i \in I$. For each $i \in I$, $j \in J$ the income value p_{ij} is randomly chosen from the set $\{6, 7, \ldots, 14\}$. The value of q_{ij} is calculated with formula $p_{ij} + \zeta$, where ζ is randomly chosen from range $\{-3, -2, \ldots, 3\}$ each time. Scenario probabilities are equal. For each customer $j \in J$ we select at random the number of scenario the customer appears in and the order $\succeq_j$.

Table 3. Instances induced by QCompFLP

LB	$\mathcal{B}$	$\mathcal{B}'$	$\mathcal{B}_{LR}$	$\mathcal{B}'_{LR}$	CP	$T(\mathcal{B})$	$T(\mathcal{B}')$	$T(\mathcal{B}_{LR})$	$T(\mathcal{B}'_{LR})$	$T(CP)$
$m = 10, n = 160, l = 4$										
191	361	361	596	532	532	8"	2"	< 1"	< 1"	< 1"
178	358	358	592	524	524	33"	5"	2"	< 1"	< 1"
235	434	434	567	519	519	15"	9"	< 1"	2"	12"
$m = 10, n = 320, l = 8$										
130	379	379	579	568	568	1'17"	2'7"	1"	5"	25"
120	409	409	571	561	561	55"	40"	18"	7"	1'10"
150	387	387	571	561	561	3'50"	1'57"	< 1"	2"	10"
$m = 10, n = 640, l = 16$										
90	378	378	578	568	568	8'53"	6'23"	4"	10"	1'55"
121	391	391	590	579	579	9'0"	7'20"	4"	7"	1'32"
148	392	392	614	603	603	9'14"	6'19"	3"	6"	2'40"
$m = 15, n = 160, l = 4$										
166	379	379	619	555	555	55"	22"	< 1"	1"	< 1"
136	389	389	615	552	552	28"	24"	< 1"	1"	< 1"
97	430	430	565	521	521	38"	42"	< 1"	1"	1"
$m = 15, n = 320, l = 8$										
73	406	406	593	583	583	5'38"	2'25"	1"	4"	21"
129	432	432	605	597	597	2'47"	1'32"	2"	4"	18"
112	443	443	601	592	592	3'34"	2'34"	2"	4"	10"
$m = 15, n = 640, l = 16$										
89	–	434	621	611	611	TL	7'48"	6"	21"	3'8"
65	–	–	594	585	585	TL	TL	9"	19"	3'22"
58	404	404	625	615	615	9'16"	3'23"	6"	20"	3'43"

Table 4. Influence of reliability level p_0

p_0	LB	$\mathcal{B}$	$\mathcal{B}'$	$\mathcal{B}_{LR}$	$\mathcal{B}'_{LR}$	CP	T($\mathcal{B}$)	T($\mathcal{B}'$)	T($\mathcal{B}_{LR}$)	T($\mathcal{B}'_{LR}$)	T(CP)
0.5	116	412	412	618	618	618	6'12"	6'32"	2"	5"	1'15"
0.6	90	378	378	578	568	568	6'50"	5'16"	2"	8"	1'26"
0.7	66	364	364	539	519	519	7'34"	5'57"	3"	6"	46"
0.8	57	335	335	497	492	492	6'21"	3'9"	3"	7"	1'1"
0.9	34	301	301	440	407	407	8'49"	4'13"	3"	7"	13"
1.0	24	270	CR	293	293	293	8'11"	< 1"	3"	7"	1"

From Table 3 we see that revealed relations between performances of formulations of the estimating problem are retained for induced instances.

Column LB contains values of lower bound. In our case it is the best value of function $\tilde{f}$ the local search has found in 1 min for instances with $m = 10$ and in 10 min for instances with $m = 15$. The accuracy of the upper bound provided by estimating problem is comparable with results obtained for procedures using similar technique for previously studied bilevel location problems relative to QCompFLP [2,4].

Table 4 illustrates the influence of reliability level on the lower and upper bounds. We consider a single instance of QCompFLP, which is the first one in a bundle of tests with $m = 10$, $n = 640$, $l = 16$ from the Table 3. As we can see the values of upper and lower bounds decrease with the grow of reliability level. The formulation $\mathcal{B}'$ performs better in common. However, the last test with $p_0 = 1$ leads the solver to crush without apparent reasons from our side.

5   Conclusion and Discussion

In this paper, we investigate ways of calculation of an upper bound for the QCompFLP using reformulations of the estimating problem. Numerical experiments show that the suggested reformulations have smaller integrality gap. We highlighted a special case of QCompFLP where scenarios are equiprobable. This leads to a model, which outperforms the initial one in integrality gap and calculation time.

The quantile criteria is an interesting view on the robustness of solution. To operate with uncertainty in income from the customers we are to duplicate each customer for each possible value of the income. It leads us to the necessity of investigating techniques to deal with instances with large number of customers. Exact and approximate methods for the QCompFLP are subjects of the future research as well.

Acknowledgments. This research is supported by the Russian Science Foundation (grant 15-11-10009). We deeply grateful to Alexander Ageev for his assistance in preparing the English text.

References

1. Audet, C., Hansen, P., Jaumard, B., Savard, G.: Links between linear bilevel and mixed 0–1 programming problems. J. Optim. Theory Appl. **93**(2), 273–300 (1997)
2. Beresnev, V.: Branch-and-bound algorithm for competitive facility location problem. Comput. Oper. Res. **40**, 2062–2070 (2013)
3. Beresnev, V.: On the competitive facility location problem with a free choice of suppliers. Autom. Remote Control **75**, 668–676 (2014)
4. Beresnev, V., Mel'nikov, A.: The branch-and-bound algorithm for a competitive facility location problem with the prescribed choice of suppliers. J. Appl. Ind. Math. **8**, 177–189 (2014)
5. Dempe, S.: Foundations of Bilevel Programming. Kluwer Acad. Publ, Dortrecht (2002)
6. Feillet, D.: A tutorial on column generation and branch-and-price for vehicle routing problems. J. Oper. Res. **8**, 407–424 (2010)
7. Ivanov, S.V., Morozova, M.V.: Stochastic problem of competitive location of facilities with quantile criterion. Automation and Remote Control **77**, 451–461 (2016)
8. Mel'nikov, A.: Randomized local search for the discrete competitive facility location problem. Autom. Remote Control **75**, 700–714 (2014)
9. Santos-Penate, D.R., Suarez-Vega, R., Dorta-Gonzalez, P.: The leader follower location model. Netw. Spat. Econ. **7**, 45–61 (2007)
10. Stackelberg, H.: The Theory of the Market Economy. Oxford University Press, Oxford (1952)

Mathematical Programming

Fast Primal-Dual Gradient Method for Strongly Convex Minimization Problems with Linear Constraints

Alexey Chernov[1], Pavel Dvurechensky[2,3(✉)], and Alexander Gasnikov[4,5]

[1] Moscow Institute of Physics and Technology, Dolgoprudnyi 141700,
Moscow Oblast, Russia
alexmipt@mail.ru
[2] Weierstrass Institute for Applied Analysis and Stochastics, 10117 Berlin, Germany
pavel.dvurechensky@wias-berlin.de
[3] Institute for Information Transmission Problems, Moscow 127051, Russia
[4] Moscow Institute of Physics and Technology, Dolgoprudnyi 141700,
Moscow Oblast, Russia
[5] Institute for Information Transmission Problems, Moscow 127051, Russia
gasnikov@yandex.ru

Abstract. In this paper, we consider a class of optimization problems with a strongly convex objective function and the feasible set given by an intersection of a simple convex set with a set given by a number of linear equality and inequality constraints. Quite a number of optimization problems in applications can be stated in this form, examples being entropy-linear programming, ridge regression, elastic net, regularized optimal transport, etc. We extend the Fast Gradient Method applied to the dual problem in order to make it primal-dual, so that it allows not only to solve the dual problem, but also to construct nearly optimal and nearly feasible solution of the primal problem. We also prove a theorem about the convergence rate for the proposed algorithm in terms of the objective function residual and the linear constraints infeasibility.

Keywords: Convex optimization · Algorithm complexity · Entropy-linear programming · Dual problem · Primal-dual method

1 Introduction

In this paper, we consider a constrained convex optimization problem of the following form

$$(P_1) \qquad \min_{x \in Q \subseteq E} \{f(x) : A_1 x = b_1, A_2 x \leq b_2\},$$

where E is a finite-dimensional real vector space, Q is a simple closed and convex set, A_1, A_2 are given linear operators from E to some finite-dimensional real vector spaces H_1 and H_2 respectively, $b_1 \in H_1$, $b_2 \in H_2$ are given, $f(x)$ is a

© Springer International Publishing Switzerland 2016
Y. Kochetov et al. (Eds.): DOOR 2016, LNCS 9869, pp. 391–403, 2016.
DOI: 10.1007/978-3-319-44914-2_31

ν-strongly convex function on Q with respect to some chosen norm $\|\cdot\|_E$ on E. The last means that, for any $x, y \in Q$, $f(y) \geq f(x) + \langle \nabla f(x), y - x \rangle + \frac{\nu}{2}\|x - y\|_E^2$, where $\nabla f(x)$ is any subgradient of $f(x)$ at x and, hence, is an element of the dual space E^*. Also we denote the value of a linear function $g \in E^*$ at $x \in E$ by $\langle g, x \rangle$.

Problem (P_1) captures a broad set of optimization problems arising in applications. The first example is the classical entropy-linear programming (ELP) problem [1] which arises in different fields, such as econometrics [2], modeling in science and engineering [3], especially in modeling of traffic flows [4] and IP traffic matrix estimation [5,6]. Other examples are the ridge regression problem [7] and the elastic net approach [8], which are used in machine learning. Finally, the problem class (P_1) covers problems of regularized optimal transport (ROT) [9] and regularized optimal partial transport (ROPT) [10], which recently have become popular in application to image analysis.

Classical balancing algorithms such as [9,11,12] are very efficient for solving ROT problems or special types of ELP problem, but they can deal only with linear equality constraints of a special type and their rate of convergence estimates are rather impractical [13]. In [10], the authors provide a generalization, but only for ROPT problems which are a particular case of Problem (P_1) with linear inequalities constraints of a special type, and no convergence rate estimates are provided. Unfortunately, the existing balancing-type algorithms for ROT and ROPT problems become very unstable when the regularization parameter is chosen very small, which is the case when one needs to calculate a good approximation to the solution of an optimal transport (OT) or an optimal partial transport (OPT) problem.

In practice, typical dimensions of the spaces E, H_1, H_2 range from thousands to millions, which makes it natural to use a first-order method to solve Problem (P_1). A common approach to solve such large-scale Problem (P_1) is to make the transition to the Lagrange dual problem and solve it by some first-order method. Unfortunately, the existing methods, which elaborate this idea, have at least two drawbacks. Firstly, the convergence analysis of the Fast Gradient Method (FGM) [14] can not be directly applied since it is based on the assumption of boundedness of the feasible set in both the primal and the dual problem, which does not hold for the Lagrange dual problem. A possible way to overcome this obstacle is to assume that the solution of the dual problem is bounded and add some additional constraints to the Lagrange dual problem in order to make the dual feasible set bounded. But, in practice, the bound for the solution of the dual problem is usually unknown. In [15], the authors use this approach with additional constraints and propose a restart technique to define the unknown bound for the optimal dual variable value. The authors consider classical ELP problems only with equality constraints and do not discuss any possibility of application of their technique to Problem (P_1) with inequality constraints. Secondly, it is important to estimate the rate of convergence not only in terms of the error in the solution of the Lagrange dual problem, as it is done in [16,17], but also in terms of the objective residual in the primal problem[1] $|f(x_k) - Opt[P_1]|$ and the linear constraints infeasibility $\|A_1 x_k - b_1\|_{H_1}$, $\|(A_2 x_k - b_2)_+\|_{H_2}$, where vector v_+ denotes the vector with components $[v_+]_i = (v_i)_+ = \max\{v_i, 0\}$, x_k is the output of the algorithm on the k-th iteration, $Opt[P_1]$ denotes the optimal function value for Problem (P_1). Alternative approaches [18, 19], based on the idea of the method of multipliers, and the quasi-Newton methods such as L-BFGS also do not allow to obtain the convergence rate for the primal problem residual and the linear constraints infeasibility.

Our contributions in this work are the following. We extend the Fast Gradient Method [14, 20], applied to the dual problem, in order to make it primal-dual, so that it allows not only to solve the dual problem, but also to construct nearly optimal and nearly feasible solution to the primal problem (P_1). We also equip our method with a stopping criterion, which allows an online control of the quality of the approximate primal-dual solution. Unlike [9, 10, 15–19], we provide the estimates for the rate of convergence in terms of the primal objective residual $|f(x_k) - Opt[P_1]|$ and the linear constraints infeasibility $\|A_1 x_k - b_1\|_{H_1}$, $\|(A_2 x_k - b_2)_+\|_{H_2}$. In the contrast to the estimates in [14], our estimates do not rely on the assumption that the feasible set of the dual problem is bounded. At the same time, our approach is applicable for the wider class of problems defined by (P_1) than the approaches in [9, 15]. In the computational experiments, we show that our approach allows to solve ROT problems more efficiently than the algorithms of [9, 10, 15] when the regularization parameter is small.

2 Preliminaries

2.1 Notation

For any finite-dimensional real vector space E, we denote by E^* its dual. We denote the value of a linear function $g \in E^*$ at $x \in E$ by $\langle g, x \rangle$. Let $\|\cdot\|_E$ denote some norm on E and $\|\cdot\|_{E,*}$ denote the norm on E^*, which is dual to $\|\cdot\|_E$

$$\|g\|_{E,*} = \max_{\|x\|_E \le 1} \langle g, x \rangle.$$

In the special case of a Euclidean space E, we denote the standard Euclidean norm by $\|\cdot\|_2$. Note that, in this case, the dual norm is also Euclidean. By $\partial f(x)$ we denote the subdifferential of a function $f(x)$ at a point x. Let E_1, E_2 be two finite-dimensional real vector spaces. For a linear operator $A : E_1 \to E_2$, we define its norm as follows

$$\|A\|_{E_1 \to E_2} = \max_{x \in E_1, u \in E_2^*} \{\langle u, Ax \rangle : \|x\|_{E_1} = 1, \|u\|_{E_2,*} = 1\}.$$

For a linear operator $A : E_1 \to E_2$, we define the adjoint operator $A^T : E_2^* \to E_1^*$ in the following way

$$\langle u, Ax \rangle = \langle A^T u, x \rangle, \quad \forall u \in E_2^*, \quad x \in E_1.$$

---

[1] The absolute value here is crucial since x_k may not satisfy linear constraints and, hence, $f(x_k) - Opt[P_1]$ could be negative.

We say that a function $f : E \to \mathbb{R}$ has a L-Lipschitz-continuous gradient if it is differentiable and its gradient satisfies Lipschitz condition

$$\|\nabla f(x) - \nabla f(y)\|_{E,*} \leq L\|x - y\|_E.$$

We characterize the quality of an approximate solution to Problem (P_1) by three quantities $\varepsilon_f, \varepsilon_{eq}, \varepsilon_{in} > 0$ and say that a point $\hat{x}$ is an $(\varepsilon_f, \varepsilon_{eq}, \varepsilon_{in})$-solution to Problem (P_1) if the following inequalities hold

$$|f(\hat{x}) - Opt[P_1]| \leq \varepsilon_f, \quad \|A_1\hat{x} - b_1\|_2 \leq \varepsilon_{eq}, \quad \|(A_2\hat{x} - b_2)_+\|_2 \leq \varepsilon_{in}, \quad (1)$$

where $Opt[P_1]$ denotes the optimal function value for Problem (P_1) and, for any vector v, the vector v_+ denotes the vector with components $[v_+]_i = (v_i)_+ = \max\{v_i, 0\}$. Also, for any $t \in R$, we denote by $\lceil t \rceil$ the smallest integer greater than or equal to t.

2.2   Dual Problem

Let us denote $\Lambda = \{\lambda = (\lambda^{(1)}, \lambda^{(2)})^T \in H_1^* \times H_2^* : \lambda^{(2)} \geq 0\}$. The Lagrange dual problem to Problem (P_1) is

$$(D_1) \quad \max_{\lambda \in \Lambda} \left\{ -\langle \lambda^{(1)}, b_1 \rangle - \langle \lambda^{(2)}, b_2 \rangle + \min_{x \in Q} \left(f(x) + \langle A_1^T \lambda^{(1)} + A_2^T \lambda^{(2)}, x \rangle \right) \right\}.$$

We rewrite Problem (D_1) in the equivalent form of a minimization problem.

$$(P_2) \quad \min_{\lambda \in \Lambda} \left\{ \langle \lambda^{(1)}, b_1 \rangle + \langle \lambda^{(2)}, b_2 \rangle + \max_{x \in Q} \left(-f(x) - \langle A_1^T \lambda^{(1)} + A_2^T \lambda^{(2)}, x \rangle \right) \right\}.$$

We denote

$$\varphi(\lambda) = \varphi(\lambda^{(1)}, \lambda^{(2)}) = \langle \lambda^{(1)}, b_1 \rangle + \langle \lambda^{(2)}, b_2 \rangle + \max_{x \in Q} \left(-f(x) - \langle A_1^T \lambda^{(1)} + A_2^T \lambda^{(2)}, x \rangle \right). \quad (2)$$

Note that the gradient of the function $\varphi(\lambda)$ is equal to (see e.g. [14])

$$\nabla \varphi(\lambda) = \begin{pmatrix} b_1 - A_1 x(\lambda) \\ b_2 - A_2 x(\lambda) \end{pmatrix}, \quad (3)$$

where $x(\lambda)$ is the unique solution of the problem

$$\max_{x \in Q} \left(-f(x) - \langle A_1^T \lambda^{(1)} + A_2^T \lambda^{(2)}, x \rangle \right). \quad (4)$$

It is important that $\nabla \varphi(\lambda)$ is Lipschitz-continuous (see e.g. [14]) with the constant

$$L = \frac{1}{\nu} \left(\|A_1\|_{E \to H_1}^2 + \|A_2\|_{E \to H_2}^2 \right). \quad (5)$$

Obviously, we have

$$Opt[D_1] = -Opt[P_2], \quad (6)$$

where by $Opt[D_1]$, $Opt[P_2]$ we denote the optimal function value in Problem (D_1) and Problem (P_2) respectively. Finally, the following inequality follows from the weak duality

$$Opt[P_1] \geq Opt[D_1]. \quad (7)$$

2.3   Main Assumptions

We make the following two main assumptions

1. The problem (4) is simple in the sense that for any $x \in Q$ it has a closed form solution or can be solved very fast up to a machine precision.
2. The dual problem (D_1) has a solution $\lambda^* = (\lambda^{*(1)}, \lambda^{*(2)})^T$ and there exist some $R_1, R_2 > 0$ such that

$$\|\lambda^{*(1)}\|_2 \leq R_1 < +\infty, \quad \|\lambda^{*(2)}\|_2 \leq R_2 < +\infty. \tag{8}$$

2.4   Examples of Problem (P_1)

In this subsection, we describe several particular problems, which can be written in the form of Problem (P_1).

Entropy-linear programming problem [1].

$$\min_{x \in S_n(1)} \left\{ \sum_{i=1}^{n} x_i \ln (x_i / \xi_i) : Ax = b \right\}$$

for some given $\xi \in \mathbb{R}^n_{++} = \{x \in \mathbb{R}^n : x_i > 0, i = 1, ..., n\}$. Here $S_n(1) = \{x \in \mathbb{R}^n : \sum_{i=1}^{n} x_i = 1, x_i \geq 0, i = 1, ..., n\}$.

Regularized optimal transport problem [9].

$$\min_{X \in \mathbb{R}^{p \times p}_+} \left\{ \gamma \sum_{i,j=1}^{p} x_{ij} \ln x_{ij} + \sum_{i,j=1}^{p} c_{ij} x_{ij} : Xe = a_1, X^T e = a_2 \right\}, \tag{9}$$

where $e \in \mathbb{R}^p$ is the vector of all ones, $a_1, a_2 \in S_p(1)$, $c_{ij} \geq 0, i, j = 1, ..., p$ are given, $\gamma > 0$ is the regularization parameter, X^T is the transpose matrix of X, x_{ij} is the element of the matrix X in the i-th row and the j-th column.

Regularized optimal partial transport problem [10].

$$\min_{X \in \mathbb{R}^{p \times p}_+} \left\{ \gamma \sum_{i,j=1}^{p} x_{ij} \ln x_{ij} + \sum_{i,j=1}^{p} c_{ij} x_{ij} : Xe \leq a_1, X^T e \leq a_2, e^T Xe = m \right\},$$

where $a_1, a_2 \in \mathbb{R}^p_+$, $c_{ij} \geq 0, i, j = 1, ..., p$, $m > 0$ are given, $\gamma > 0$ is the regularization parameter and the inequalities should be understood component-wise.

3   Algorithm and Theoretical Analysis

We extend the Fast Gradient Method [14,20] in order to make it primal-dual, so that it allows not only to solve the dual problem (P_2), but also to construct a nearly optimal and nearly feasible solution to the primal problem (P_1). We also equip it with a stopping criterion, which allows an online control of the quality of the approximate primal-dual solution. Let $\{\alpha_i\}_{i\geq 0}$ be a sequence of coefficients satisfying

$$\alpha_0 \in (0,1], \quad \alpha_k^2 \leq \sum_{i=0}^{k} \alpha_i, \quad \forall k \geq 1.$$

We define also $C_k = \sum_{i=0}^{k} \alpha_i$ and $\tau_i = \frac{\alpha_{i+1}}{C_{i+1}}$. Usual choice is $\alpha_i = \frac{i+1}{2}$, $i \geq 0$. In this case $C_k = \frac{(k+1)(k+2)}{4}$. Next, let us define Euclidean norm on $H_1^* \times H_2^*$ in a natural way

$$\|\lambda\|_2^2 = \|\lambda^{(1)}\|_2^2 + \|\lambda^{(2)}\|_2^2,$$

for any $\lambda = (\lambda^{(1)}, \lambda^{(2)})^T \in H_1^* \times H_2^*$. Unfortunately, we can not directly use the convergence results of [14, 20] for the reason that the feasible set Λ in the dual problem (D_1) is unbounded and the constructed sequence $\hat{x}_k$ may possibly not satisfy the equality and inequality constraints.

ALGORITHM 1. Fast Primal-Dual Gradient Method

Input: The sequence $\{\alpha_i\}_{i\geq 0}$, Lipschitz constant L (5), accuracy $\tilde{\varepsilon}_f, \tilde{\varepsilon}_{eq}, \tilde{\varepsilon}_{in} > 0$.
Output: The point $\hat{x}_k$.
Choose $\lambda_0 = (\lambda_0^{(1)}, \lambda_0^{(2)})^T = 0$.
Set $k = 0$.
repeat
    Find

$$\eta_k = (\eta_k^{(1)}, \eta_k^{(2)})^T = \arg\min_{\lambda \in \Lambda} \left\{ \varphi(\lambda_k) + \langle \nabla\varphi(\lambda_k), \lambda - \lambda_k \rangle + \frac{L}{2}\|\lambda - \lambda_k\|_2^2 \right\}.$$

$$\zeta_k = (\zeta_k^{(1)}, \zeta_k^{(2)})^T = \arg\min_{\lambda \in \Lambda} \left\{ \sum_{i=0}^{k} \alpha_i \left(\varphi(\lambda_i) + \langle \nabla\varphi(\lambda_i), \lambda - \lambda_i \rangle \right) + \frac{L}{2}\|\lambda\|_2^2 \right\}.$$

Set
$$\lambda_{k+1} = (\lambda_{k+1}^{(1)}, \lambda_{k+1}^{(2)})^T = \tau_k \zeta_k + (1 - \tau_k)\eta_k,$$

where $\tau_k = \frac{\alpha_{k+1}}{\sum_{i=0}^{k+1} \alpha_i}$.
Set
$$\hat{x}_k = \frac{1}{\sum_{i=0}^{k} \alpha_i} \sum_{i=0}^{k} \alpha_i x(\lambda_i) = (1 - \tau_{k-1})\hat{x}_{k-1} + \tau_{k-1}x(\lambda_k).$$

Set $k = k + 1$.
until $|f(\hat{x}_k) + \varphi(\eta_k)| \leq \tilde{\varepsilon}_f$, $\|A_1\hat{x}_k - b_1\|_2 \leq \tilde{\varepsilon}_{eq}$, $\|(A_2\hat{x}_k - b_2)_+\|_2 \leq \tilde{\varepsilon}_{in}$;

Theorem 1. Let the assumptions listed in the Subsect. 2.3 hold and $\alpha_i = \frac{i+1}{2}$, $i \geq 0$ in Algorithm 1. Then Algorithm 1 stops after not more than

$$N_{stop} = \max \left\{ \left\lceil \sqrt{\frac{8L(R_1^2 + R_2^2)}{\tilde{\varepsilon}_f}} \right\rceil, \left\lceil \sqrt{\frac{8L(R_1^2 + R_2^2)}{R_1\tilde{\varepsilon}_{eq}}} \right\rceil, \left\lceil \sqrt{\frac{8L(R_1^2 + R_2^2)}{R_2\tilde{\varepsilon}_{in}}} \right\rceil \right\} - 1$$

iterations. Moreover, after not more than

$$N = \max \left\{ \left\lceil \sqrt{\frac{16L(R_1^2 + R_2^2)}{\varepsilon_f}} \right\rceil, \left\lceil \sqrt{\frac{8L(R_1^2 + R_2^2)}{R_1 \varepsilon_{eq}}} \right\rceil, \left\lceil \sqrt{\frac{8L(R_1^2 + R_2^2)}{R_2 \varepsilon_{in}}} \right\rceil \right\} - 1$$

iterations of Algorithm 1, the point $\hat{x}_N$ will be an approximate solution to Problem (P_1) in the sense of (1).

Proof. From the complexity analysis of the FGM [14,20], one has

$$C_k \varphi(\eta_k) \leq \min_{\lambda \in \Lambda} \left\{ \sum_{i=0}^{k} \alpha_i \left(\varphi(\lambda_i) + \langle \nabla \varphi(\lambda_i), \lambda - \lambda_i \rangle \right) + \frac{L}{2} \|\lambda\|_2^2 \right\}. \quad (10)$$

Let us introduce a set

$$\Lambda_R = \{ \lambda = (\lambda^{(1)}, \lambda^{(2)})^T : \lambda^{(2)} \geq 0, \|\lambda^{(1)}\|_2 \leq 2R_1, \|\lambda^{(2)}\|_2 \leq 2R_2 \},$$

where R_1, R_2 are given in (8). Then, from (10), we obtain

$$C_k \varphi(\eta_k) \leq \min_{\lambda \in \Lambda} \left\{ \sum_{i=0}^{k} \alpha_i \left(\varphi(\lambda_i) + \langle \nabla \varphi(\lambda_i), \lambda - \lambda_i \rangle \right) + \frac{L}{2} \|\lambda\|_2^2 \right\}$$

$$\leq \min_{\lambda \in \Lambda_R} \left\{ \sum_{i=0}^{k} \alpha_i \left(\varphi(\lambda_i) + \langle \nabla \varphi(\lambda_i), \lambda - \lambda_i \rangle \right) + \frac{L}{2} \|\lambda\|_2^2 \right\}$$

$$\leq \min_{\lambda \in \Lambda_R} \left\{ \sum_{i=0}^{k} \alpha_i \left(\varphi(\lambda_i) + \langle \nabla \varphi(\lambda_i), \lambda - \lambda_i \rangle \right) \right\} + 2L(R_1^2 + R_2^2). \quad (11)$$

On the other hand, from the definition (2) of $\varphi(\lambda)$, we have

$$\varphi(\lambda_i) = \varphi(\lambda_i^{(1)}, \lambda_i^{(2)}) = \langle \lambda_i^{(1)}, b_1 \rangle + \langle \lambda_i^{(2)}, b_2 \rangle$$

$$+ \max_{x \in Q} \left(-f(x) - \langle A_1^T \lambda_i^{(1)} + A_2^T \lambda_i^{(2)}, x \rangle \right)$$

$$= \langle \lambda_i^{(1)}, b_1 \rangle + \langle \lambda_i^{(2)}, b_2 \rangle - f(x(\lambda_i)) - \langle A_1^T \lambda_i^{(1)} + A_2^T \lambda_i^{(2)}, x(\lambda_i) \rangle.$$

Combining this equality with (3), we obtain

$$\varphi(\lambda_i) - \langle \nabla \varphi(\lambda_i), \lambda_i \rangle = \varphi(\lambda_i^{(1)}, \lambda_i^{(2)}) - \langle \nabla \varphi(\lambda_i^{(1)}, \lambda_i^{(2)}), (\lambda_i^{(1)}, \lambda_i^{(2)})^T \rangle$$

$$= \langle \lambda_i^{(1)}, b_1 \rangle + \langle \lambda_i^{(2)}, b_2 \rangle - f(x(\lambda_i)) - \langle A_1^T \lambda_i^{(1)} + A_2^T \lambda_i^{(2)}, x(\lambda_i) \rangle$$

$$- \langle b_1 - A_1 x(\lambda_i), \lambda_i^{(1)} \rangle - \langle b_2 - A_2 x(\lambda_i), \lambda_i^{(2)} \rangle = -f(x(\lambda_i)).$$

Summing these inequalities from $i = 0$ to $i = k$ with the weights $\{\alpha_i\}_{i=1,\ldots k}$, we get, using the convexity of $f(\cdot)$,

$$\sum_{i=0}^{k} \alpha_i \left(\varphi(\lambda_i) + \langle \nabla\varphi(\lambda_i), \lambda - \lambda_i \rangle \right)$$

$$= -\sum_{i=0}^{k} \alpha_i f(x(\lambda_i)) + \sum_{i=0}^{k} \alpha_i \langle (b_1 - A_1 x(\lambda_i), b_2 - A_2 x(\lambda_i))^T, (\lambda^{(1)}, \lambda^{(2)})^T \rangle$$

$$\leq -C_k f(\hat{x}_k) + C_k \langle (b_1 - A_1 \hat{x}_k, b_2 - A_2 \hat{x}_k)^T, (\lambda^{(1)}, \lambda^{(2)})^T \rangle.$$

Substituting this inequality to (11), we obtain

$$C_k \varphi(\eta_k) \leq -C_k f(\hat{x}_k)$$
$$+ C_k \min_{\lambda \in \Lambda_R} \left\{ \langle (b_1 - A_1 \hat{x}_k, b_2 - A_2 \hat{x}_k)^T, (\lambda^{(1)}, \lambda^{(2)})^T \rangle \right\} + 2L(R_1^2 + R_2^2).$$

Finally, since

$$\max_{\lambda \in \Lambda_R} \left\{ \langle (-b_1 + A_1 \hat{x}_k, -b_2 + A_2 \hat{x}_k)^T, (\lambda^{(1)}, \lambda^{(2)})^T \rangle \right\}$$
$$= 2R_1 \|A_1 \hat{x}_k - b_1\|_2 + 2R_2 \|(A_2 \hat{x}_k - b_2)_+\|_2,$$

we obtain

$$\varphi(\eta_k) + f(\hat{x}_k) + 2R_1 \|A_1 \hat{x}_k - b_1\|_2 + 2R_2 \|(A_2 \hat{x}_k - b_2)_+\|_2 \leq \frac{2L(R_1^2 + R_2^2)}{C_k}. \quad (12)$$

Since $\lambda^* = (\lambda^{*(1)}, \lambda^{*(2)})^T$ is an optimal solution of Problem (D_1), we have, for any $x \in Q$,

$$Opt[P_1] \leq f(x) + \langle \lambda^{*(1)}, A_1 x - b_1 \rangle + \langle \lambda^{*(2)}, A_2 x - b_2 \rangle.$$

Using the assumption (8) and that $\lambda^{*(2)} \geq 0$, we get

$$f(\hat{x}_k) \geq Opt[P_1] - R_1 \|A_1 \hat{x}_k - b_1\|_2 - R_2 \|(A_2 \hat{x}_k - b_2)_+\|_2. \quad (13)$$

Hence,

$$\varphi(\eta_k) + f(\hat{x}_k) = \varphi(\eta_k) - Opt[P_2] + Opt[P_2] + Opt[P_1] - Opt[P_1] + f(\hat{x}_k) \overset{(6)}{=}$$

$$= \varphi(\eta_k) - Opt[P_2] - Opt[D_1] + Opt[P_1] - Opt[P_1] + f(\hat{x}_k) \overset{(7)}{\geq}$$

$$\geq -Opt[P_1] + f(\hat{x}_k) \overset{(13)}{\geq} -R_1 \|A_1 \hat{x}_k - b_1\|_2 - R_2 \|(A_2 \hat{x}_k - b_2)_+\|_2. \quad (14)$$

This and (12) give

$$R_1 \|A_1 \hat{x}_k - b_1\|_2 + R_2 \|(A_2 \hat{x}_k - b_2)_+\|_2 \leq \frac{2L(R_1^2 + R_2^2)}{C_k}. \quad (15)$$

Hence, we obtain

$$\varphi(\eta_k) + f(\hat{x}_k) \overset{(14),(15)}{\geq} -\frac{2L(R_1^2 + R_2^2)}{C_k}. \tag{16}$$

On the other hand, we have

$$\varphi(\eta_k) + f(\hat{x}_k) \overset{(12)}{\leq} \frac{2L(R_1^2 + R_2^2)}{C_k}. \tag{17}$$

Combining (15), (16), (17), we conclude

$$\|A_1\hat{x}_k - b_1\|_2 \leq \frac{2L(R_1^2 + R_2^2)}{C_k R_1},$$

$$\|(A_2\hat{x}_k - b_2)_+\|_2 \leq \frac{2L(R_1^2 + R_2^2)}{C_k R_2},$$

$$|\varphi(\eta_k) + f(\hat{x}_k)| \leq \frac{2L(R_1^2 + R_2^2)}{C_k}. \tag{18}$$

As we know, for the chosen sequence $\alpha_i = \frac{i+1}{2}, i \geq 0$, it holds that $C_k = \frac{(k+1)(k+2)}{4} \geq \frac{(k+1)^2}{4}$. Then, in accordance to (18), after given in the theorem statement number N_{stop} of the iterations of Algorithm 1, the stopping criterion is fulfilled and Algorithm 1 stops.

Now let us prove the second statement of the theorem. We have

$$\varphi(\eta_k) + Opt[P_1] = \varphi(\eta_k) - Opt[P_2] + Opt[P_2] + Opt[P_1] \overset{(6)}{=}$$

$$= \varphi(\eta_k) - Opt[P_2] - Opt[D_1] + Opt[P_1] \overset{(7)}{\geq} 0.$$

Hence,

$$f(\hat{x}_k) - Opt[P_1] \leq f(\hat{x}_k) + \varphi(\eta_k). \tag{19}$$

On the other hand,

$$f(\hat{x}_k) - Opt[P_1] \overset{(13)}{\geq} -R_1\|A_1\hat{x}_k - b_1\|_2 - R_2\|(A_2\hat{x}_k - b_2)_+\|_2. \tag{20}$$

Note that, since the point $\hat{x}_k$ may not satisfy the equality and inequality constraints, one can not guarantee that $f(\hat{x}_k) - Opt[P_1] \geq 0$. From Equation (19), (20), we can see that if we set $\tilde{\varepsilon}_f = \varepsilon_f, \tilde{\varepsilon}_{eq} = \min\{\frac{\varepsilon_f}{2R_1}, \varepsilon_{eq}\}, \tilde{\varepsilon}_{in} = \min\{\frac{\varepsilon_f}{2R_2}, \varepsilon_{in}\}$, and run Algorithm 1 for N iterations, where N is given in the theorem statement, we obtain that (1) fulfills and $\hat{x}_N$ is an approximate solution to Problem (P_1) in the sense of (1). □

We point that other authors [9,10,15–19] do not provide the complexity analysis for their algorithms when the accuracy of the solution is defined by (1).

4   Preliminary Numerical Experiments

To compare our algorithm with the existing algorithms, we choose the problem (9) of regularized optimal transport [9], which is a special case of Problem (P_1). The first reason for this choice is that, despite insufficient theoretical analysis, the existing balancing-type methods for solving this class of problems are known to be very efficient in practice [9] and provide a kind of benchmark for any new method. The second reason is that ROT problem have recently become very popular in application to image analysis based on Wasserstein spaces geometry [9,10].

Our numerical experiments were carried out on a PC with CPU Intel Core i5 (2.5 Hgz), 2 Gb of RAM using Matlab 2012 (8.0). We compare proposed in this article Algorithm 1 (below we refer to it as FGM) with the following algorithms

- Applied to the dual problem (D_1), Conjugate Gradient Method in the Fletcher–Reeves form [21] with the stepsize chosen by one-dimensional minimization. We refer to this algorithm as CGM.
- The algorithm proposed in [15] and based on the idea of Tikhonov's regularization of the dual problem (D_1). In this approach the regularized dual problem is solved by the Fast Gradient Method [14]. We will refer to this algorithm as REG;

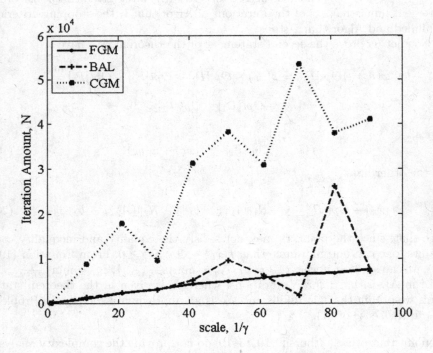

Fig. 1. Complexity of FGM, BAL and CGM as γ varies

- Balancing method [9,12] which is a special type of a fixed-point-iteration method for the system of the optimality conditions for ROT problem. It is referred below as BAL.

The key parameters of the ROT problem in the experiments are as follows

- $n := dim(E) = p^2$ – problem dimension, varies from 2^4 to 9^4;
- $m_1 := dim(H_1) = 2\sqrt{n}$ and $m_2 = dim(H_2) = 0$ – dimensions of the vectors b_1 and b_2 respectively;
- $c_{ij}, i,j = 1,...,p$ are chosen as squared Euclidean pairwise distance between the points in a $\sqrt{p} \times \sqrt{p}$ grid originated by a 2D image [9,10];
- a_1 and a_2 are random vectors in $S_{m_1}(1)$ and $b_1 = (a_1, a_2)^T$;
- the regularization parameter γ varies from 0.001 to 1;
- the desired accuracy of the approximate solution in (1) is defined by its relative counterpart ε_f^{rel} and ε_g^{rel} as follows

$$\varepsilon_f = \varepsilon_f^{rel} \cdot f(x(\lambda_0)) \quad \varepsilon_{eq} = \varepsilon_g^{rel} \cdot \|A_1 x(\lambda_0) - b_1\|_2,$$

where λ_0 is the starting point of the algorithm. Note that $\varepsilon_{in} = 0$ since no inequality constraints are present in ROT problems.

Figure 1 shows the number of iterations for the FGM, BAL and CGM methods depending on the inverse of the regularization parameter γ. The results for

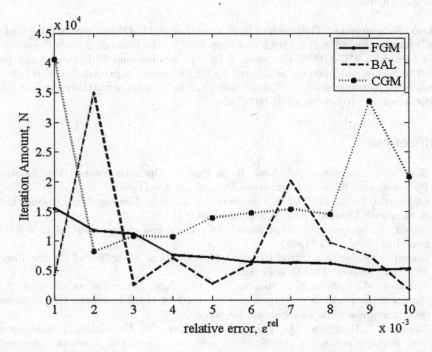

Fig. 2. Complexity of FGM, BAL and CGM as the desired relative accuracy varies

the REG are not plotted since this algorithm required one order of magnitude more iterations than the other methods. In these experiment we chose $n = 2401$ and $\varepsilon_f^{rel} = \varepsilon_g^{rel} = 0.01$. One can see that the complexity of the FGM (i.e. proposed Algorithm 1) depends nearly linearly on the value of $1/\gamma$, and that this complexity is smaller than that of the other methods when γ is small.

Figure 2 shows the number of iterations for the FGM, BAL and CGM methods depending on the relative error ε^{rel}. The results for the REG are not plotted since this algorithm required one order of magnitude more iterations than the other methods. In these experiment we chose $n = 2401$, $\gamma = 0.1$ and $\varepsilon_f^{rel} = \varepsilon_g^{rel} = \varepsilon^{rel}$. One can see that in half of the cases the FGM (i.e. proposed Algorithm 1) performs better or equally to the other methods.

5   Conclusion

This paper proposes a new primal-dual approach to solve a general class of problems stated as Problem (P_1). Unlike the existing methods, we managed to provide the convergence rate for the proposed algorithm in terms of the primal objective residual $|f(\hat{x}_k - Opt[P_1]|$ and the linear constraints infeasibility $\|A_1\hat{x}_k - b_1\|_2$, $\|(A_2\hat{x}_k - b_2)_+\|_2$. Our numerical experiments show that our algorithm performs better than existing methods for problems of regularized optimal transport, which are a special instance of Problem (P_1) for which there exist efficient algorithms.

Acknowledgements. The research by A. Gasnikov and P. Dvurechensky presented in Sect. 3 was conducted in IITP RAS and supported by the Russian Science Foundation grant (project 14-50-00150), the research by A. Gasnikov and P. Dvurechensky presented in Sect. 4 was partially supported by RFBR, research project No. 15-31-20571 mol_a_ved. The research by A. Chernov presented in Sect. 4 was partially supported by RFBR, research project No.14-01-00722-a.

References

1. Fang, S.-C., Rajasekera, J., Tsao, H.-S.: Entropy Optimization and Mathematical Programming. Kluwers International Series, Boston (1997)
2. Golan, A., Judge, G., Miller, D.: Maximum Entropy Econometrics: Robust Estimation with Limited Data. Wiley, Chichester (1996)
3. Kapur, J.: Maximum entropy models in science and engineering. John Wiley & Sons Inc., New York (1989)
4. Gasnikov, A., et al.: Introduction to Mathematical Modelling of Traffic Flows. MCCME, Moscow (2013). (in russian)
5. Rahman, M.M., Saha, S., Chengan, U., Alfa, A.S.: IP traffic matrix estimation methods: comparisons and improvements. In: 2006 IEEE International Conference on Communications, Istanbul, pp. 90–96 (2006)
6. Zhang, Y., Roughan, M., Lund, C., Donoho, D.: Estimating point-to-point and point-to-multipoint traffic matrices: an information-theoretic approach. IEEE/ACM Trans. Networking **13**(5), 947–960 (2005)

7. Hastie, T., Tibshirani, R., Friedman, R.: The Elements of Statistical Learning: Data Mining, Inference and Prediction. Springer, Heidelberg (2009)
8. Zou, H., Hastie, T.: Regularization and variable selection via the elastic net. J. Roy. Stat. Soc.: Seri. B (Stat. Methodol.) **67**(2), 301–320 (2005)
9. Cuturi, M.: Sinkhorn distances: lightspeed computation of optimal transport. In: Advances in Neural Information Processing Systems, pp. 2292–2300 (2013)
10. Benamou, J.-D., Carlier, G., Cuturi, M., Nenna, L., Peyre, G.: Iterative bregman projections for regularized transportation problems. SIAM J. Sci. Comput. **37**(2), A1111–A1138 (2015)
11. Bregman, L.: The relaxation method of finding the common point of convex sets and its application to the solution of problems in convex programming. USSR Comput. Math. Math. Phys. **7**(3), 200–217 (1967)
12. Bregman, L.: Proof of the convergence of Sheleikhovskii's method for a problem with transportation constraints. USSR Comput. Math. Math. Phys. **7**(1), 191–204 (1967)
13. Franklin, J., Lorenz, J.: On the scaling of multidimensional matrices. Linear Algebra Appl. **114**, 717–735 (1989)
14. Nesterov, Y.: Smooth minimization of non-smooth functions. Math. Program. **103**(1), 127–152 (2005)
15. Gasnikov, A., Gasnikova, E., Nesterov, Y., Chernov, A.: About effective numerical methods to solve entropy linear programming problem. Comput. Math. Math. Phys. **56**(4), 514–524 (2016). http://arxiv.org/abs/1410.7719
16. Polyak, R.A., Costa, J., Neyshabouri, J.: Dual fast projected gradient method for quadratic programming. Optim. Lett. **7**(4), 631–645 (2013)
17. Necoara, I., Suykens, J.A.K.: Applications of a smoothing technique to decomposition in convex optimization. IEEE Trans. Autom. Control **53**(11), 2674–2679 (2008)
18. Goldstein, T., O'Donoghue, B., Setzer, S.: Fast Alternating Direction Optimization Methods. Technical Report, Department of Mathematics, University of California, Los Angeles, USA, May (2012)
19. Shefi, R., Teboulle, M.: Rate of convergence analysis of decomposition methods based on the proximal method of multipliers for convex minimization. SIAM J. Optim. **24**(1), 269–297 (2014)
20. Devolder, O., Glineur, F., Nesterov, Y.: First-order methods of smooth convex optimization with inexact oracle. Math. Program. **146**(1–2), 37–75 (2014)
21. Fletcher, R., Reeves, C.M.: Function minimization by conjugate gradients. Comput. J. **7**, 149–154 (1964)

An Approach to Fractional Programming via D.C. Constraints Problem: Local Search

Tatiana Gruzdeva[✉] and Alexander Strekalovsky[✉]

Matrosov Institute for System Dynamics and Control Theory of SB RAS,
Lermontov Str., 134, 664033 Irkutsk, Russia
gruzdeva@icc.ru, strekal@icc.ru
http://nonconvex.isc.irk.ru

Abstract. We consider the problem of optimizing the sum of several rational functions via reduction to a problem with d.c. constraints. We propose a method of finding a local solution to the fractional program which can be subsequently used in the global search method based on the global optimality conditions for a problem with nonconvex (d.c.) constraints [21–23]. According to the theory, we construct explicit representations of the constraints in the form of differences of two convex functions and perform a local search method that takes into account the structure of the problem in question. This algorithm was verified on a set of low-dimensional test problems taken from literature as well as on randomly generated problems with up to 200 variables and 200 terms in the sum.

Keywords: Nonconvex optimization · Rational optimization · d.c. representation · Local search · Linearized problems

1 Introduction

The fractional optimization is quite challenging and arises in various on a both economic and non-economic applications, whenever one or several ratios are required to be optimized. Let us mention a few examples mainly following the surveys by Schaible [8,19,20], where numerous other applications can be found. Numerators and denominators in ratios may represent cost, capital, profit, risk or time, etc. Fractional programs is closely related to the associated multiple-objective optimization problem, where a number of ratios are to be maximized simultaneously. Thus, the objective function in a fractional program can be considered as a utility function expressing a compromise between the different objective functions of the multiple-objective problem. Other applications include a multistage stochastic shipping problem [1,7], profit maximization under fixed cost [4], various models in cluster analysis [17], multi-objective bond portfolio [13], and queuing location problems [5].

As known, without assumptions, the sum-of-ratios program is NP-complete [9]. Surveys on methods for solving this problem can be found in [8,20,28,29]. According to the surveys, the majority of the methods make restrictive assumptions either on the concavity or linearity of the ratios. When the

© Springer International Publishing Switzerland 2016
Y. Kochetov et al. (Eds.): DOOR 2016, LNCS 9869, pp. 404–417, 2016.
DOI: 10.1007/978-3-319-44914-2_32

ratios are nonlinear, the most popular techniques are based on the Branch and Bound approach, see e.g. [2,6].

We propose reducing a fractional problem to the optimization problem with nonconvex constraints [10,23], as it was mentioned in [6], with a subsequent application of the Global search theory for solving this class of nonconvex problems [21–23,25,26].

The outline of the paper is as follows. In Sect. 2 we substantiate the reduction of the sum-of-ratios fractional problem to the optimization problem with nonconvex constraints. Then in Sect. 3 we recall the local search method from [21], which implies linearization of the functions defining the basic non-convexity of the problem with d.c. constraints in the current point. In Sect. 4 we show how to explicitly represent the nonconvex functions, describing the constraints of the problem in question, as differences of two convex functions (the d.c. representation). The final section offers computational testing of the local search method on fractional program instances with a small number of variables and terms in the sum. We use the examples found in the literature as well as randomly generated problems of higher dimension.

2   Reduction to the Problem with Nonconvex Constraints

Now consider the following problem of the fractional optimization [3,20]

$$f(x) := \sum_{i=1}^{m} \frac{\psi_i(x)}{\varphi_i(x)} \downarrow \min_x, \ x \in S, \qquad (\mathcal{P}_0)$$

where $\psi_i : \mathbb{R}^n \to \mathbb{R}$, $\varphi_i : \mathbb{R}^n \to \mathbb{R}$, $\varphi_i(x) > 0$, $\forall x \in S$, $i = 1, \ldots, m$, and $S \subset \mathbb{R}^n$ is a convex set.

Proposition 1. *(i) Let the pair $(x_*, \alpha_*) \in \mathbb{R}^n \times \mathbb{R}^m$ be a solution to the following problem:*

$$\sum_{i=1}^{m} \alpha_i \downarrow \min_{(x,\alpha)}, \ x \in S, \ \frac{\psi_i(x)}{\varphi_i(x)} = \alpha_i, \ i = 1, \ldots, m. \qquad (\mathcal{P}_1)$$

Then x_* is the solution to Problem $(\mathcal{P}_0)$ and $f(x_*) = \sum_{i=1}^{m} \alpha_{*i}$.

(ii) Conversely, if x_* is the solution to Problem $(\mathcal{P}_0)$ $(x_* \in Sol(\mathcal{P}_0))$, then the vector $\alpha_* = (\alpha_{*1}, \ldots, \alpha_{*m})^T \in \mathbb{R}^m$ defined as $\alpha_{*i} = \dfrac{\psi_i(x_*)}{\varphi_i(x_*)}$, $i = 1, \ldots, m$, is part of the solution (x_*, α_*) to Problem $(\mathcal{P}_1)$.

Proof. (i) Let $(x_*, \alpha_*) \in Sol(\mathcal{P}_1)$, i.e. $\alpha_{*i} = \dfrac{\psi_i(x_*)}{\varphi_i(x_*)}$, $i = 1, \ldots, m$, $x_* \in S$, and

$$\sum_{i=1}^{m} \alpha_{*i} \leq \sum_{i=1}^{m} \alpha_i \text{ for all } \alpha_i : \ \exists x \in S, \ \frac{\psi_i(x)}{\varphi_i(x)} = \alpha_i, i = 1, \ldots, m,$$

$$f(x_*) = \sum_{i=1}^{m} \frac{\psi_i(x_*)}{\varphi_i(x_*)} = \sum_{i=1}^{m} \alpha_{*i} \leq \sum_{i=1}^{m} \alpha_i = \sum_{i=1}^{m} \frac{\psi_i(x)}{\varphi_i(x)} = f(x) \quad \forall x \in S.$$

Therefore, x_* is the solution to Problem $(\mathcal{P}_0)$.

(ii) Now let $x_* \in Sol(\mathcal{P}_0)$. Then

$$f(x_*) = \sum_{i=1}^{m} \frac{\psi_i(x_*)}{\varphi_i(x_*)} \le \sum_{i=1}^{m} \frac{\psi_i(x)}{\varphi_i(x)} = f(x) \quad \forall x \in S. \tag{1}$$

Define $\alpha_{*i} := \dfrac{\psi_i(x_*)}{\varphi_i(x_*)}$, $i = 1, \ldots, m$, and consider the set

$$D_\alpha = \left\{ \alpha \in I\!R^m \; : \; \exists x \in S, \; \alpha_i = \frac{\psi_i(x)}{\varphi_i(x)} \right\}.$$

Then (1) implies $\sum_{i=1}^{m} \alpha_{*i} = f(x_*) \le f(x) = \sum_{i=1}^{m} \alpha_i \; \forall \alpha \in D_\alpha, \; \forall x \in S$. Therefore $x_* \in Sol(\mathcal{P}_0)$. □

Proposition 2. *Let the pair* $(x_*, \alpha_*) \in I\!R^n \times I\!R^m$ *be a solution to the following problem:*

$$\sum_{i=1}^{m} \alpha_i \downarrow \min_{(x,\alpha)}, \quad x \in S, \quad \frac{\psi_i(x)}{\varphi_i(x)} \le \alpha_i, \; i = 1, \ldots, m. \tag{$\mathcal{P}_2$}$$

Then

$$\frac{\psi_i(x_*)}{\varphi_i(x_*)} = \alpha_{*i}, \; i = 1, \ldots, m. \tag{2}$$

Proof. Let $(x_*, \alpha_*) \in Sol(\mathcal{P}_2)$. Suppose that

$$\exists j \in \{1, 2, \ldots, m\} : \; \frac{\psi_j(x_*)}{\varphi_j(x_*)} < \alpha_{*j}, \tag{3}$$

and construct $\hat{\alpha}$: $\hat{\alpha}_i = \alpha_{*i} \; \forall i \ne j$, $\hat{\alpha}_j = \dfrac{\psi_j(x_*)}{\varphi_j(x_*)}$. It can be readily seen that $(x_*, \hat{\alpha})$ is a feasible pair to Problem $(\mathcal{P}_2)$. The assumption (3) implies

$$\sum_{i=1}^{m} \alpha_{*i} = \sum_{i \ne j} \alpha_{*i} + \alpha_{*j} > \sum_{i=1}^{m} \hat{\alpha}_i.$$

It means that in Problem $(\mathcal{P}_2)$ the pair $(x_*, \hat{\alpha})$ is better than the pair (x_*, α_*), so, $(x_*, \alpha_*) \notin Sol(\mathcal{P}_2)$. This contradiction proves the equality (2). □

Corollary 1. *Any solution* $(x_*, \alpha_*) \in I\!R^n \times I\!R^m$ *to Problem* $(\mathcal{P}_2)$ *will be a solution to Problem* $(\mathcal{P}_1)$, *and, therefore, will be a solution to Problem* $(\mathcal{P}_0)$.

Remark 1. The inequality constraints in Problem $(\mathcal{P}_2)$ can be replaced by equivalent constraints $\psi_i(x) - \alpha_i \varphi_i(x) \le 0$, $i = 1, \ldots, m$, since $\varphi_i(x) > 0 \; \forall x \in S$. It leads us to the following problem with m nonconvex constraints:

$$\sum_{i=1}^{m} \alpha_i \downarrow \min_{(x,\alpha)}, \quad x \in S, \quad \psi_i(x) - \alpha_i \varphi_i(x) \le 0, \; i = 1, \ldots, m. \tag{$\mathcal{P}$}$$

It is easy to see that Problem $(\mathcal{P})$ is a global optimization problem with the nonconvex feasible set (see, e.g., [10,23]), and we can apply the Global Search Theory for solving this class of nonconvex problems [21–23,25,26]. The Theory allows one to construct an algorithm for solving problems with nonconvex constraints. The algorithm contains two principal stages: (a) local search, which provides an approximately critical point; (b) procedures of escaping from critical points.

In the next section, we shall consider a local search method.

3   Local Search for Problem with D.C. Constraints

The local search methods (LSMs) play an important role in searching for the global solution to nonconvex problems, since it provides the so-called critical (stationary) points that might be considerably be better than a simple feasible point. Moreover, if a starting point occurs rather close to the global solution, then the LSMs can provide the global solution.

In order to find a local solution to Problem $(\mathcal{P})$, we apply a special LSM [21].

Let us consider the following problem with d.c. constraints:

$$\left.\begin{array}{l} f_0(x) \downarrow \min_x, \quad x \in S, \\ f_i(x) := g_i(x) - h_i(x) \leq 0, \ i \in I \triangleq \{1, \ldots, m\}, \end{array}\right\} \quad (4)$$

where the functions f_0 and $g_i, h_i, \ i \in I$, as well as the set $S \subset \mathbb{R}^n$, are convex. Further, suppose that the feasible set $D := \{\, x \in S \mid f_i(x) \leq 0, \ i \in I \,\}$ of the problem (4) is not empty and the optimal value $\mathcal{V}\,(4) := \inf_x\{f_0(x) \mid x \in D\}$ of the problem (4) is finite: $\mathcal{V}\,(4) > -\infty$.

Furthermore, assume that a feasible starting point $x^0 \in D$ is given and, in addition, after several iterations it has derived the current iterate $x^s \in D$, $s \in \mathbb{Z}_+ = \{0, 1, 2, \ldots\}$.

In order to propose a local search method for the problem (4), apply a classical idea of linearization with respect to the basic nonconvexity of the problem (i.e. with respect to $h_i(\cdot), \ i \in I$) at the point x^s [21]. Thus, we obtain the following linearized problem:

$$\left.\begin{array}{l} f_0(x) \downarrow \min_x, \quad x \in S, \\ \varphi_{is}(x) := g_i(x) - \langle \nabla h_i(x^s), x - x^s \rangle - h_i(x^s) \leq 0, \quad i \in I. \end{array}\right\} \quad (\mathcal{PL}_s)$$

Suppose the point x^{s+1} is provided by solving Problem $(\mathcal{PL}_s)$, so that

$$x^{s+1} \in D_s = \{x \in S \mid g_i(x) - \langle \nabla h_i(x^s), x - x^s \rangle - h_i(x^s) \leq 0, \quad i \in I\}$$

and inequality $f_0(x^{s+1}) \leq \mathcal{V}(\mathcal{PL}_s) + \delta_s$ holds. Here $\mathcal{V}(\mathcal{PL}_s)$ is the optimal value to Problem $(\mathcal{PL}_s)$:

$$\mathcal{V}_s := \mathcal{V}(\mathcal{PL}_s) \triangleq \inf_x\{f_0(x) \mid x \in S, \ \varphi_{is}(x) \leq 0, \ i \in I\},$$

and the sequence $\{\delta_s\}$ satisfies the following condition functions $\sum_{s=0}^{\infty} \delta_s < +\infty$.
It can be easily seen that $D_s \subset D$, so x^{s+1} turns out to be feasible in the
problem (4). Actually, since the functions $h_i(\cdot)$, $i \in I$ are convex, the following
inequalities hold

$$0 \geq g_i(x^{s+1}) - \langle h_i'(x^s), x^{s+1} - x^s \rangle - h_i(x^s) = \varphi_{is}(x^{s+1}) \geq$$
$$\geq g_i(x^{s+1}) - h_i(x^{s+1}) = f_i(x^{s+1}), \quad i \in I.$$

Therefore, the LSM generates the sequence $\{x^s\}$, $x^s \in D_s$, $s = 0, 1, 2, \ldots$, of
solutions to Problems $(\mathcal{PL}_s)$. As it was proven in [21], the cluster point $x_* \in D_*$
of the sequence $\{x^s\}$ is a solution to the linearized Problem $(\mathcal{PL}_*)$ (which is
Problem $(\mathcal{PL}_s)$ with x^s instead of x_*), and x_* can be called the critical point
with respect to the LSM. Thus, the algorithm constructed in this way provides
critical points by employing suitable convex optimization methods [15] for any
given accuracy τ. The following inequality:

$$f_0(x^s) - f_0(x^{s+1}) \leq \frac{\tau}{2}, \quad \delta_s \leq \frac{\tau}{2},$$

can be chosen as a stopping criterion for the LSM [21].

In order to implement the LSM, we need an explicit d.c. representation of
$f_i(\cdot)$, i.e. $f_i(\cdot) = g_i(\cdot) - h_i(\cdot)$, $i \in I$.

4   D.C. Representation of the Constraints

The first stage of any algorithm developed according the Global search theory
is the decomposition of a nonconvex function as a difference of two convex functions. Such decomposition is constructing in several different ways depending on
the functions $\psi_i(\cdot)$ and $\varphi_i(\cdot)$.

4.1   Affine Functions

Let the functions $\psi_i(\cdot)$ and $\varphi_i(\cdot)$ be constructed by means of the vectors
$a^i, c^i \in \mathbb{R}^n$, and numbers $b_i, d_i \in \mathbb{R}$,

$$\psi_i(x) = \langle a^i, x \rangle + b_i, \quad \varphi_i(x) = \langle c^i, x \rangle + d_i > 0, \quad i \in I.$$

In this case, the basic nonconvexity of Problem $(\mathcal{P})$ is the bilinear term
$\alpha_i \varphi_i(x) = \langle \alpha_i c^i, x \rangle + \alpha_i d_i$ in each constraint $(i \in I)$. Then, the bilinear function can be represented as a difference of two convex functions in the following
way [27]:

$$\langle \alpha_i c^i, x \rangle = \frac{1}{4} \parallel \alpha_i c^i + x \parallel^2 - \frac{1}{4} \parallel \alpha_i c^i - x \parallel^2, \quad i \in I. \tag{5}$$

Hence, the functions $f_i(\cdot)$ have the form $f_i(x, \alpha_i) = g_i(x, \alpha_i) - h_i(x, \alpha_i)$, where

$$g_i(x, \alpha_i) = \frac{1}{4} \parallel \alpha_i c^i - x \parallel^2 - \alpha_i d_i + \langle a^i, x \rangle + b_i,$$
$$h_i(x, \alpha_i) = \frac{1}{4} \parallel \alpha_i c^i + x \parallel^2 . \tag{6}$$

Taking into account the d.c. representation (6), the linearized Problem $(\mathcal{PL}_s)$ has the following form

$$\left. \begin{array}{l} \sum_{i=1}^{m} \alpha_i \downarrow \min_{(x,\alpha)}, \quad x \in S, \\[2mm] \frac{1}{4} \parallel \alpha_i c^i - x \parallel^2 + \langle a^i, x \rangle - \alpha_i d_i \langle \nabla h_i(x^s, \alpha_i^s), (x, \alpha_i) \rangle + \mathcal{C}_{is} \leq 0, \ i \in I, \end{array} \right\} \quad (7)$$

where $\mathcal{C}_{is} = b_i + \langle \nabla h_i(x^s, \alpha_i^s), (x^s, \alpha_i^s) \rangle - \frac{1}{4} \parallel \alpha_i^s c^i + x^s \parallel^2$,

$$\nabla h_i(x^s, \alpha_i^s) = (\nabla h_{ix}, \nabla h_{i\alpha})^T,$$
$$\nabla h_{ix} = \frac{1}{2}(\alpha_i^s c^i + x^s), \quad \nabla h_{i\alpha} = \frac{1}{2}(\alpha_i^s \parallel c^i \parallel^2 + \langle c^i, x^s \rangle). \quad (8)$$

The problem (7) is a convex optimization problem and it can be solved by an appropriate convex optimization method [15] at a given accuracy: $\delta_s > 0$, $s = 0, 1, \ldots$.

Further, we will consider Problem $(\mathcal{P})$ where $\psi_i(\cdot)$ are convex quadratic functions and $\varphi_i(\cdot)$ are affine functions, $i \in I$.

4.2   Quadratic/Affine Functions

Suppose we are given symmetric positive definite matrices A^i $(n \times n)$, vectors $p^i, c^i \in \mathbb{R}^n$, and scalars $q_i, d_i \in \mathbb{R}$,

$$\psi_i(x) = \langle x, A^i x \rangle + \langle p^i, x \rangle + q_i, \quad \varphi_i(x) = \langle c^i, x \rangle + d_i > 0, \ i \in I.$$

As has been done in Subsect. 4.1, we represent the bilinear term $\alpha_i \varphi_i(x)$ as the difference of two convex functions, which yields us the d.c. representations $f_i(x, \alpha_i) = g_i(x, \alpha_i) - h_i(x, \alpha_i)$, $i \in I$, where

$$\begin{array}{l} g_i(x, \alpha_i) = \langle x, A^i x \rangle + \langle p^i, x \rangle + q_i + \frac{1}{4} \parallel \alpha_i c^i - x \parallel^2 - \alpha_i d_i, \\[2mm] h_i(x, \alpha_i) = \frac{1}{4} \parallel \alpha_i c^i + x \parallel^2 . \end{array} \quad (9)$$

Taking into account the d.c. representation (9), the linearized Problem $(\mathcal{PL}_s)$ takes the following form

$$\left. \begin{array}{l} \sum_{i=1}^{m} \alpha_i \downarrow \min_{(x,\alpha)}, \quad x \in S, \\[2mm] \langle x, A^i x \rangle + \langle p^i, x \rangle + \frac{1}{4} \parallel \alpha_i c^i - x \parallel^2 - \alpha_i d_i \\[2mm] - \langle \nabla h_i(x^s, \alpha_i^s), (x, \alpha_i) \rangle + \mathcal{C}_{is} \leq 0, \ i \in I, \end{array} \right\} \quad (10)$$

where the gradient $\nabla h_i(x^s, \alpha_i^s)$ is calculated by the formula (8), and $\mathcal{C}_{is} = q_i + \langle \nabla h_i(x^s, \alpha_i^s), (x^s, \alpha_i^s) \rangle - \frac{1}{4} \parallel \alpha_i^s c^i + x^s \parallel^2$.

The problem (10), as well as (7), can be solved by a suitable convex optimization method [15].

Remark 2. If the symmetric matrices A^i in the quadratic functions $\psi_i(\cdot)$ are indefinite, then one can represent A^i as the difference of two symmetric positive definite matrices $A^i = A_1^i - A_2^i$, $A_1^i, A_2^i > 0$, using, for example, a simple method from [24]. Afterwards, it is possible to construct functions $g_i(\cdot)$ and $h_i(\cdot)$ as follows: for all $i \in I$ add the convex part with the matrix A_1^i to the function $g_i(\cdot)$ and the nonconvex part with the matrix A_2^i to $h_i(\cdot)$.

In what follows, we will examine the case where $\psi_i(\cdot)$ and $\varphi_i(\cdot)$ are convex quadratic functions, $i \in I$.

4.3 Quadratic Functions

Now let us consider the following functions:

$$\psi_i(x) = \langle x, A^i x \rangle + \langle p^i, x \rangle + q_i, \quad \varphi_i(x) = \langle x, B^i x \rangle + \langle c^i, x \rangle + d_i > 0,$$

A^i and B^i are positive definite $(n \times n)$ matrices, $p^i, c^i \in \mathbb{R}^n$, $q_i, d_i \in \mathbb{R}$, $i \in I$. Therefore, Problem $(\mathcal{P})$ has the following term

$$\alpha_i \varphi_i(x) = \alpha_i \langle x, B^i x \rangle + \alpha_i \langle c^i, x \rangle + \alpha_i d_i, \qquad (11)$$

which generate nonconvexity in every constraint $(i \in I)$.

The term $\alpha_i \langle c^i, x \rangle$ in (11) can be presented in the d.c. form by the formula (5).

Further, let us denote $r_i := \langle x, B^i x \rangle$. Then, the product $\alpha_i u_i$ can be expressed by formula (5) as follows

$$\alpha_i r_i = \frac{1}{4}(\alpha_i + r_i)^2 - \frac{1}{4}(\alpha_i - r_i)^2$$
$$= \frac{1}{4}\left(\alpha_i + \langle x, B^i x \rangle\right)^2 - \frac{1}{4}\left(\alpha_i - \langle x, B^i x \rangle\right)^2, \quad i \in I.$$

If B^i, $i \in I$, are positive definite matrices and the following conditions hold

$$\alpha_i + \langle x, B^i x \rangle \geq 0, \quad \alpha_i - \langle x, B^i x \rangle \geq 0 \ \forall x \in S, \ i \in I, \qquad (12)$$

then

$$g_i(x, \alpha_i) = \frac{1}{4}\left(\alpha_i - \langle x, B^i x \rangle\right)^2 + \frac{1}{4}\parallel \alpha_i c^i - x \parallel^2 - \alpha_i d_i + \psi_i(x),$$
$$h_i(x, \alpha_i) = \frac{1}{4}\left(\alpha_i + \langle x, B^i x \rangle\right)^2 + \frac{1}{4}\parallel \alpha_i c^i + x \parallel^2$$

are convex functions. Hence, we obtain the following d.c. representation:

$$f_i(x, \alpha_i) = g_i(x, \alpha_i) - h_i(x, \alpha_i), \quad i \in I, \qquad (13)$$

and the following linearized Problem $(\mathcal{PL}_s)$

$$\left.\begin{array}{c} \sum\limits_{i=1}^{m} \alpha_i \downarrow \min\limits_{(x,\alpha)}, \quad x \in S, \\ \langle x, A^i x \rangle + \langle p^i, x \rangle + \frac{1}{4}\left(\alpha_i - \langle x, B^i x \rangle\right)^2 + \frac{1}{4}\parallel \alpha_i c^i - x \parallel^2 - \alpha_i d_i \\ -\langle \nabla h_i(x^s, \alpha_i^s), (x, \alpha_i) \rangle + C_{is} \leq 0, \quad i \in I, \end{array}\right\} \qquad (14)$$

where $\mathcal{C}_{is} = q_i + \langle \nabla h_i(x^s, \alpha_i^s), (x^s, \alpha_i^s) \rangle - \frac{1}{4}\left(\alpha_i^s + \langle x^s, B^i x^s \rangle\right)^2 - \frac{1}{4}\parallel \alpha_i^s c^i + x^s \parallel^2$,
$\nabla h_i(x^s, \alpha_i^s) = (\nabla h_{ix}, \nabla h_{i\alpha})^T$ is the gradient of the function $h(\cdot)$:

$$\nabla h_{ix} = \left(\alpha_i^s + \langle x^s, B^i x^s \rangle\right) B^i x^s + \frac{1}{2}(\alpha_i^s c^i + x^s),$$
$$\nabla h_{i\alpha} = \frac{1}{2}\left(\alpha_i^s + \langle x^s, B^i x^s \rangle\right) + \frac{1}{2}\left(\alpha_i^s \parallel c^i \parallel^2 + \langle c^i, x^s \rangle\right).$$

If the conditions (12) are not satisfied, one can construct the d.c. representation (13) by decomposition of the trilinear term $\alpha_i \langle x, B^i x \rangle = \sum_{l=1}^{n}\sum_{j=1}^{n} b_{lj}^i x_l x_j \alpha_i$
using the following equality holding for the product of three variables (for example, u, v, w):

$$uvw = \frac{1}{8}\hat{g}(u, v, w) - \frac{1}{8}\hat{h}(u, v, w),$$

$$\hat{g}(u, v, w) = ((u + v)^2 + (1 + w)^2)^2 + (1 + w)^4 + (u^2 + w^2)^2$$
$$+ (v^2 + w^2)^2 + 2(u^2 + v^2),$$

$$\hat{h}(u, v, w) = w^4 + ((u + v)^2 + w^2)^2 + (u^2 + (1 + w)^2)^2$$
$$+ (v^2 + (1 + w)^2)^2 + 2(u + v)^2.$$

Therefore, we get (13), where

$$g_i(x, \alpha_i) = \psi_i(x) + \frac{1}{4}\parallel \alpha_i c^i - x \parallel^2 - \alpha_i d_i + \frac{1}{8}\sum_{l=1}^{n}\sum_{j=1}^{n} b_{lj}^i \hat{g}(x_l, x_j, \alpha_i),$$

$$h_i(x, \alpha_i) = \frac{1}{4}\parallel \alpha_i c^i + x \parallel^2 + \frac{1}{8}\sum_{l=1}^{n}\sum_{j=1}^{n} b_{lj}^i \hat{h}(x_l, x_j, \alpha_i).$$

Obviously, in this case the linearized Problem $(\mathcal{PL}_s)$ is the problem of minimization of the linear function over the convex feasible set given by more complicated nonlinear functions $\varphi_{ik}(x, \alpha_i)$ in comparison with the problems (7), (10) or even (14). At the same time, the linearized problems are convex, and therefore can be solved by a suitable convex optimization method [15].

Remark 3. If the symmetric matrices A^i and B^i in the functions $\psi_i(\cdot)$ and $\varphi_i(\cdot)$, respectively, are indefinite, then this case is already described above in Remark 2.

5 Computational Simulations

The algorithm of the local search method (LSM) from Sect. 3 was coded in C++ language and was tested with various starting points. All computational experiments were performed on the Intel Core i7-4790K CPU 4.0 GHz.

At each iteration of the LSM, the convex Problem $(\mathcal{PL}_s)$ was solved by the software package IBM ILOG CPLEX 12.6.2 [11]. The accuracy of the LSM was $\tau = 10^{-6}$. The accuracy of the solution to the linearized problems $(\mathcal{PL}_s)$ increased during the LSM. Thus, we solved $(\mathcal{PL}_s)$ at a low accuracy at the first steps; further, the accuracy δ_s was gradually improved $(\delta_s \downarrow 0)$, i.e., $\delta_0 = 0.1$, $\delta_{s+1} = 0.5\delta_s$, until the condition $\delta_s \leq \frac{\tau}{2}$ was fulfilled with a given accuracy $\tau > 0$.

At the first stage, we numerically solved several instances of fractional programming problems from [2,3,7,14,16,18] with a small number of variables.

5.1 Low-Dimensional Fractional Program with Affine Functions

Tables 1 and 2 represent the results of the computational testing of the LSM and employ the following designations:

name is the name of the test example;
n is the number of variables (problem's dimension);
m is the number of terms in the sum;

Table 1. Low-dimensional fractional program. Minimization.

name	n	m	$f_0(x_0)$	$f_0(z)$	it	Time	x_0	z
Prob3 [18]	2	1	0.400	0.333	6	0.01	(1.000; 0.000)	(0.000; 0.000)
			0.750	0.333	7	0.01	(1.000; 1.000)	(0.000; 0.000)
			1.000	0.333	6	0.01	(0.000; 1.000)	(0.000; 0.000)
			0.333	0.333	1	0.00	(0.000; 0.000)	(0.000; 0.000)
			4.500	4.500	1	0.00	(2.000; 1.000)	(2.000; 1.000)
Prob5 [18]	2	2	4.156	4.500	5	0.01	(2.250; 1.250)	(2.000; 1.000)
			6.500	4.500	7	0.01	(1.000; 4.000)	(2.000; 1.000)
			5.000	4.500	5	0.01	(1.000; 1.000)	(2.000; 1.000)
			1.733	1.623	12	0.02	(0.000; 0.000)	(0.000; 0.284)
Prob3 [7]	2	2	2.758	1.623	11	0.02	(0.750; 0.750)	(0.000; 0.284)
			2.400	1.623	14	0.02	(0.500; 1.000)	(0.000; 0.284)
			4.250	1.623	18	0.02	(0.000; 1.000)	(0.000; 0.284)
			2.830	2.830	1	0.00	(1.500; 1.500)	(1.500; 1.500)
Prob3 [14]	2	2	3.524	2.830	6	0.01	(3.000; 4.000)	(1.500; 1.500)
			3.129	2.830	5	0.01	(2.000; 2.333)	(1.500; 1.500)
			3.070	3.000	4	0.01	(0.314; 0.842; 0.427)	(0.437; 0.000; 0.000)
Prob6 [14]	3	3	3.035	3.000	3	0.01	(0.900; 0.000; 0.633)	(0.952; 0.000; 0.000)
			3.000	3.000	1	0.00	(1.100; 0.000; 0.000)	(1.098; 0.000; 0.000)
			2.895	2.889	4	0.01	(0.000; 0.000; 2.000)	(0.513; 0.000; 1.795)
Prob7 [14]	3	3	2.890	2.889	3	0.01	(0.431; 0.000; 1.828)	(0.513; 0.000; 1.795)
			3.000	2.889	4	0.01	(0.000; 1.111; 0.000)	(0.513; 0.000; 1.795)

Table 2. Low-dimensional fractional program. Maximization.

name	n	m	$f_0(x_0)$	$f_0(z)$	it	Time	x_0	z
Prob1 [18]	2	1	1.500	3.714	31	0.04	(0.000; 2.500)	(30.000; 0.000)
			2.000	3.714	31	0.04	(0.000; 0.000)	(30.000; 0.000)
			2.143	3.714	25	0.03	(9.000; 7.000)	(30.000; 0.000)
Prob6 [2]	2	2	4.500	**5.000**	6	0.01	(2.000; 1.000)	**(1.000; 1.000)**
			4.156	**5.000**	7	0.01	(2.250; 1.250)	**(1.000; 1.000)**
			6.500	6.500	1	0.00	(1.000; 4.000)	(1.000; 4.000)
			5.000	**5.000**	1	0.00	(1.000; 1.000)	**(1.000; 1.000)**
Prob1 [7]	2	2	4.913	5.000	30	0.04	(1.500; 1.500)	(3.000; 4.000)
			5.000	5.000	1	0.00	(3.000; 4.000)	(3.000; 4.000)
			4.946	5.000	11	0.02	(2.000; 2.333)	(3.000; 4.000)
Prob2 [7]	3	2	1.348	2.471	6	0.01	(1.003; 0.731; 1.184)	(1.000; 0.000; 0.000)
			1.879	2.471	6	0.01	(1.500; 0.000; 0.500)	(1.000; 0.000; 0.000)
			2.471	2.471	1	0.00	(1.000; 0.000; 0.000)	(1.000; 0.000; 0.000)
			2.107	2.471	5	0.01	(0.750; 0.000; 0.250)	(1.000; 0.000; 0.000)
Prob1 [16]	3	3	2.988	3.003	30	0.05	(0.414; 1.954; 0.000)	(0.000; 3.333; 0.000)
			3.003	3.003	1	0.00	(0.000; 3.333; 0.000)	(0.000; 3.333; 0.000)
			2.947	**3.000**	4	0.01	(0.512; 0.000; 0.610)	**(0.000; 0.082; 0.000)**
			2.963	**3.000**	4	0.01	(1.000; 0.000; 0.000)	**(0.000; 0.114; 0.000)**
Prob2 [16]	3	4	3.967	4.091	5	0.01	(0.000; 0.000; 2.000)	(1.111; 0.000; 0.000)
			4.000	4.091	4	0.01	(0.000; 0.000; 0.000)	(1.111; 0.000; 0.000)
			4.091	4.091	1	0.00	(1.111; 0.000; 0.000)	(1.111; 0.000; 0.000)
			3.868	4.091	5	0.01	(0.000; 0.625; 1.875)	(1.111; 0.000; 0.000)

$f_0(x^0)$ is the value of the goal function to Problem $(\mathcal{P})$ at the starting point;
$f_0(z)$ is the value of the function at the critical point provided by the LSM;
it is the number of linearized problems solved (iterations of the LSM);
$Time$ stands for the CPU time of computing (seconds);
x^0 stands for the starting point chosen in the test problem;
z is the critical point provided by the LSM.

Note that in the problems "Prob6 [2]" and "Prob1 [16]" in Table 2, local solutions derived by the LSM are not global (shown in bold).

Known global solutions to all problem instances were found just by the local search that confirms the computational effectiveness of the LSM. All test problems were successfully solved.

Further, we study if the LSM performance is affected by the increase in dimension of the variable x and the number of terms in the sum.

5.2   Randomly Generated Problems with Affine and Quadratic Functions

In this subsection, we will report computational results of testing the LSM on randomly generated problems of the form

$$f_0(x) := \sum_{i=1}^{m} \frac{\langle a^i, x \rangle + b_i}{\langle c^i, x \rangle + d_i} \uparrow \max_x, \quad \langle \bar{A}, x \rangle \leq \bar{b}, \quad x \geq 0. \tag{15}$$

Data a^i_j, c^i_j, $\bar{A}_{lj} \in [0, 10]$ were uniformly random numbers, $b_i = d_i = 10$, $\bar{b}_l = 10$, $i = 1, \ldots, m$, $j = 1, \ldots, n$, $l = 1, \ldots, L$.

Results of the computational testing of the LSM on fractional problems (15) up to 100 variables and 100 terms in the sum are listed in Table 3. The denotations in Table 3 are the same as in Tables 1 and 2.

Table 3. Randomly generated problems (15) with affine functions

n	m	$f_0(x_0)$	$f_0(z)$	it	Time
5	5	5.000000	5.659817	14	0.02
5	10	10.000000	11.399243	15	0.03
5	50	50.000000	56.107594	16	0.08
5	100	100.000000	106.644654	19	0.16
10	5	5.000000	5.560987	27	0.06
10	10	10.000000	12.368279	18	0.05
10	50	50.000000	57.873668	27	0.20
10	100	100.000000	106.665004	89	1.27
50	5	5.000000	7.286323	85	1.83
50	10	10.000000	12.572450	282	13.63
50	50	50.000000	58.460209	158	22.65
50	100	100.000000	109.059418	224	24.12
100	5	5.000000	6.809288	265	17.05
100	10	10.000000	13.774653	251	42.11
100	50	50.000000	56.692829	340	64.94
100	100	100.000000	109.858345	589	209.59

Moreover, we have carried out testing of the LSM on fractional problems with quadratic functions in the numerators of ratios. We generated the problems from [12] up to 200 variables and 200 terms in the sum:

$$f_0(x) := \sum_{i=1}^{m} \frac{\frac{1}{2}\langle x, A^i x \rangle + \langle p^i, x \rangle}{\langle c^i, x \rangle} \downarrow \min_x, \quad \langle \bar{A}, x \rangle \leq \bar{b}, \quad x \in [1, 5]^n, \tag{16}$$

Table 4. Randomly generated problems (16) with quadratic functions

n	m	$f_0(x_0)$	$f_0(z)$	it	Time
10	5	17.747630	15.607047	14	0.04
10	10	37.690672	35.242307	16	0.08
10	50	165.797927	155.938010	44	0.62
10	100	312.867828	296.561267	130	3.35
10	200	616.334973	601.908457	202	10.80
50	5	17.940091	15.460443	15	0.46
50	10	35.338013	30.663274	15	0.91
50	50	158.226948	151.224157	35	5.28
50	100	306.427872	297.535557	50	15.13
50	200	607.096322	589.882128	141	92.93
100	5	18.943321	15.771841	16	1.57
100	10	33.902949	29.500002	17	4.46
100	50	156.222645	148.923647	25	15.40
100	100	305.948925	296.692149	49	61.85
100	200	608.664712	591.959402	104	263.58
200	5	19.018280	15.512586	17	8.04
200	10	34.139152	29.172948	16	22.70
200	50	155.645907	146.144178	19	44.37
200	100	306.462441	295.558108	39	188.26
200	200	603.988798	587.865367	69	701.85

where $A^i = U_i D^i U_i^T$, $U_i = Q_1 Q_2 Q_3$, $i = 1, \ldots, m$, $Q_j = I - 2\frac{w_j w_j^T}{\|w_j\|^2}, j = 1, 2, 3$ and $w_1 = -i + rand(n,1)$, $w_2 = -2i + rand(n,1)$, $w_3 = -3i + rand(n,1)$, $D^i = rand(n,n)$, $c^i = i - i \cdot rand(n,1)$, $p^i = i + i \cdot rand(n,1)$, $i = 1, \ldots, m$, $\bar{A} = -1 + 2 \cdot rand(5,n)$, $\bar{b} = 2 + 3 \cdot rand(5,1)$ [12]. (We denote by $rand(k_1, k_2)$ the random matrix with k_1 rows, k_2 columns and elements generated randomly on $[0,1]$.)

As it is shown in Table 4, the number of iteration (it) of the LSM is almost independent of the number of variables (n) but approximately proportional to the number of terms in the sum (m). The run-time increased proportionally to n and m.

Computational simulations confirm the efficiency of the LSM developed, the performance of which naturally depends on the choice of the method or the software package (IBM ILOG CPLEX) employed to solve auxiliary problems.

Thus, LSM can be applied in future implementations of the global search algorithm for solving the sum of ratios fractional problems via problems with d.c. constraints.

6   Conclusions

In this paper, we considered the fractional programming problem as an optimization problem with d.c. constraints. To this end, we carried out the explicit representation of nonconvex functions as differences of two convex functions and applied the local search algorithm based on linearization of the functions defining the basic non-convexity of the problem under study.

We investigated the effectiveness of the local search method for solving problems with d.c. constraints that generate the nonconvexity of the feasible set.

The numerical experiments demonstrated that the local search algorithm can globally solve low-dimensional sum-of-ratios test problems. Moreover, the local search algorithm developed in this paper turned out to be rather efficient at finding critical points in randomly generated fractional programming problems of high dimension.

Therefore, the method developed can be applied within the global search procedures for fractional programming problems.

Acknowledgments. This work has been supported by the Russian Science Foundation, Project N 15-11-20015.

References

1. Almogy, Y., Levin, O.: Parametric analysis of a multistage stochastic shipping problem. In: Lawrence, J. (ed.) Operational Research, vol. 69, pp. 359–370. Tavistock Publications, London (1970)
2. Benson, H.P.: Global optimization algorithm for the nonlinear sum of ratios problem. J. Optim. Theory Appl. **112**(1), 1–29 (2002)
3. Bugarin, F., Henrion, D., Lasserre, J.-B.: Minimizing the sum of many rational functions. Math. Prog. Comp. **8**, 83–111 (2016)
4. Colantoni, C.S., Manes, R.P., Whinston, A.: Programming, profit rates and pricing decisions. Account. Rev. **13**, 467–481 (1969)
5. Drezner, Z., Schaible, S., Simchi-Levi, D.: Queuing-location problems on the plane. Naval Res. Log. **37**, 929–935 (1990)
6. Dur, M., Horst, R., Thoai, N.V.: Solving sum-of-ratios fractional programs using efficient points. Optimization **49**, 447–466 (2001)
7. Falk, J.E., Palocsay, S.W.: Optimizing the sum of linear fractional functions. In: Floudas, C.A., Pardalos, P.M. (eds.) Recent Advances in Global Optimization, Princeton Series in Computer Science, pp. 221–257. Princeton University Press, Stanford (1992)
8. Frenk, J.B.G., Schaible, S.: Fractional programming. In: Hadjisavvas, S.S.N., Komlosi, S. (eds.) Handbook of Generalized Convexity and Generalized Monotonicity, Series Nonconvex Optimization and Its Applications, vol. 76, pp. 335–386. Springer, Heidelberg (2002)
9. Freund, R.W., Jarre, F.: Solving the sum-of-ratios problem by an interior-point method. J. Global Optim. **19**(1), 83–102 (2001)
10. Horst, R., Tuy, H.: Global Optimization. Deterministic approaches. Springer-Verlag, Berlin (1993)

11. Ibm, I.: High-performance mathematical programming solver for linear programming, mixed integer programming, and quadratic programming. http://www-03.ibm.com/software/products/en/ibmilogcpleoptistud

12. Jong, Y.-Ch.: An efficient global optimization algorithm for nonlinear sum-of-ratios problems. Optim. Online. http://www.optimization-online.org/DB_HTML/2012/08/3586.html

13. Konno, H.: Watanabe: Bond portfolio optimization problems and their applications to index tracking. J. Oper. Res. Soc. Japan **39**, 295–306 (1996)

14. Ma, B., Geng, L., Yin, J., Fan, L.: An effective algorithm for globally solving a class of linear fractional programming problem. J. Softw. **8**(1), 118–125 (2013)

15. Nocedal, J., Wright, S.J.: Numerical Optimization, 2nd edn. Springer, New York (2006)

16. Kuno, T.: A branch-and-bound algorithm for maximizing the sum of several linear ratios. J. Global Optim. **22**, 155–174 (2002)

17. Rao, M.R.: Cluster analysis and mathematical programming. J. Amer. Statist. Assoc. **66**, 622–626 (1971)

18. Raouf, O.A., Hezam, I.M.: Solving fractional programming problems based on swarm intelligence. J. Ind. Eng. Int. **10**, 56–66 (2014)

19. Schaible, S.: Fractional programming. In: Horst, R., Pardalos, P.M. (eds.) Handbook of Global Optimization, pp. 495–608. Kluwer Academic Publishers, Dordrecht (1995)

20. Schaible, S., Shi, J.: Fractional programming: the sum-of-ratios case. Optim. Methods Softw. **18**, 219–229 (2003)

21. Strekalovsky, A.S.: On local search in d.c. optimization problems. Appl. Math. Comput. **255**, 73–83 (2015)

22. Strekalovsky, A.S.: On solving optimization problems with hidden nonconvex structures. In: Rassias, T.M., Floudas, C.A., Butenko, S. (eds.) Optimization in Science and Engineering, pp. 465—502. Springer, New York (2014)

23. Strekalovsky, A.S.: Minimizing sequences in problems with D.C. constraints. Comput. Math. Math. Phys. **45**(3), 418–429 (2005)

24. Strekalovsky, A.S.: Elements of nonconvex optimization [in Russian]. Nauka, Novosibirsk (2003)

25. Strekalovsky, A.S., Gruzdeva, T.V.: Local search in problems with nonconvex constraints. Comput. Math. Math. Phys. **47**(3), 381–396 (2007)

26. Strekalovsky, A.S., Gruzdeva, T.V., Ulianova, N.Y.: Optimization Problems with Nonconvex Constraints [in Russian]. Irk. State University, Irkutsk (2013)

27. Strekalovsky, A.S., Orlov, A.V.: Bimatrix games and bilinear programming [in Russian]. FizMatLit, Moscow (2007)

28. Strekalovsky, A.S., Yakovleva, T.V.: On a local and global search involved in nonconvex optimization problems. Autom. Remote Control **65**, 375–387 (2004)

29. Wu, W.-Y., Sheu, R.-L., Birbil, I.S.: Solving the sum-of-ratios problem by a stochastic search algorithm. J. Global Optim. **42**(1), 91–109 (2008)

Partial Linearization Method for Network Equilibrium Problems with Elastic Demands

Igor Konnov and Olga Pinyagina[✉]

Institute of Computational Mathematics and Information Technologies,
Kazan Federal University, Kremlevskaya st. 18, 420008 Kazan, Russia
konn-igor@yandex.ru, Olga.Piniaguina@kpfu.ru
http://kpfu.ru

Abstract. We suggest a partial linearization method for network equilibrium problems with elastic demands, which can be set-valued in general. The main element of this method is a partially linearized auxiliary problem. We propose a simple solution method for the auxiliary problem, which is based on optimality conditions. This method can be viewed as alternative to the conditional gradient method for the single-valued case. Some results of preliminary calculations which confirm efficiency of the new method are also presented.

Keywords: Network equilibrium problem · Elastic demand · Set-valued demand · Partial linearization

1 Introduction

The partial linearization approach for optimization problems was proposed in [1,2] and developed in [3] for variational inequalities. This approach has advantages when the feasible set of the considered problem has a relatively simple structure and the objective function can be decomposed into two parts, one of them is suitable for linearization, and the other is sufficiently simple. Then the use of specific properties of a problem makes it possible to suggest more efficient solution methods (see [1]–[3]).

In the present paper, we apply the partial linearization approach to the network equilibrium problem with set-valued elastic demand. We formulate a partially linearized auxiliary problem and propose a simple solution method based on the optimality conditions of the initial problem. This method can be considered as alternative to the conditional gradient method for the single-valued case. Results of preliminary calculations confirm usefulness of the partial linearization method.

Let us first remind partial linearization and conditional gradient methods in the general case.

In this work, the authors were supported by Russian Foundation for Basic Research, project No 16-01-00109. The first author was also supported by grant No 297689 from Academy of Finland.

2 Preliminaries

We consider the following optimization problem

$$\min_{x \in D} \longrightarrow \mu(x), \tag{1}$$

where the objective function $\mu : R^n \to R$ is the sum of two functions $\mu(x) = f(x) + h(x)$, $f : R^n \to R$, $h : R^n \to R$, the first of them f is smooth and the second one h is convex, and the feasible domain D is a convex closed set in R^n.

Let us describe the partial linearization method for problem (1). Let at the kth iteration, $k = 0, 1, \ldots$, we have a point $x^k \in D$. Define $z^k \in D$ as a solution to the auxiliary problem

$$\min_{x \in D} \longrightarrow \left\{ \langle f'(x^k), z \rangle + h(z) \right\},$$

set $d^k = z^k - x^k$ and define the next iterate $x^{k+1} = x^k + \lambda_k d^k$, where the step $\lambda_k \in [0, 1]$ can be found as a solution of the one-dimensional minimization problem

$$\min_{\lambda \in [0,1]} \longrightarrow \mu(x^k + \lambda d^k).$$

If h is smooth, we can also use the inexact linesearch approach instead of the exact search. Find the smallest nonnegative number n that it holds

$$\mu(x^k + \delta^n d^k) - \mu(x^k) \le \beta \delta^n \langle \mu'(x^k), d^k \rangle,$$

where $\delta \in (0, 1), \beta \in (0, 1)$ are given parameters. Define the next iterate $x^{k+1} = x^k + \lambda d^k$, where $\lambda = \delta^n$.

The above-stated methods converge to a stationary point of the problem provided that the feasible set D is bounded [2].

Another variant of partial linearization is possible under the assumption of strong convexity of the function h. Then the point z^k is uniquely defined, the lower level sets of the function μ are compact, problem (1) has a unique solution, and the descent method converges to a stationary point.

We also recall the general scheme of the conditional gradient method, which was originally proposed by M.Frank and Ph.Wolfe in [4] for the quadratic programming problems and developed in [5]. Let $f : R^n \to R$ be a smooth function, D be a convex closed bounded set in R^n. We consider the following constrained optimization problem

$$\min_{x \in D} \longrightarrow f(x) \tag{2}$$

and the corresponding linearized problem

$$\min_{y \in D} \longrightarrow \langle f'(x), y \rangle. \tag{3}$$

Under the given assumptions, both problems (2), (3) have solutions, they are nonunique in general. We denote by $Z(x)$ the set of solutions to problem (3).

At the kth iteration of method, $k = 0, 1, \ldots$, we have a point $x^k \in D$. If $x^k \in Z(x^k)$, problem (2) is solved and x^k is its exact solution. Otherwise we find a point $z^k \in Z(x^k)$ as a solution of problem (3) with $x = x^k$. Then we construct the descent direction $d^k = z^k - x^k$ and solve the one-dimensional optimization problem

$$\min_{\lambda \in [0,1]} \longrightarrow f(x^k + \lambda d^k). \tag{4}$$

using a suitable linesearch. We set $x^{k+1} = x^k + \lambda_k d^k$, where λ_k is a solution to problem (4). The inexact linesearch approach can also be used.

We will apply the above-stated methods to the network equilibrium problem with elastic (variable) demand. In the next section, we give a description of this problem.

3 Network Equilibrium Problem with Elastic Demand

Let us formulate the network equilibrium problem with elastic demand (see, for example, [6–8]).

Let $\mathcal{V}$ be a set of network nodes. Some (or all) nodes are connected by directed arcs. We denote the set of arcs by $\mathcal{A}$. Let a set of origin-destination (O/D) pairs $\mathcal{M} \subseteq \mathcal{V} \times \mathcal{V}$ be given. We denote by y_m a nonnegative variable demand (bid) for each pair $m \in \mathcal{M}$. We assume this demand is bounded from above, $y_m \leq \gamma_m$, $\gamma_m > 0$ for all $m \in \mathcal{M}$. For each O/D pair $m \in \mathcal{M}$ we have a set $\mathcal{P}_m$ of simple directed paths joining m, the correspondence of paths and arcs are given by the incidence matrix A with elements

$$\alpha_{pa} = \begin{cases} 1 & \text{if arc } a \text{ belongs to path } p, \\ 0 & \text{otherwise.} \end{cases}$$

Let x_p denotes a variable flow value on the path p, for all $p \in \mathcal{P}_m, m \in \mathcal{M}$. The feasible set of problem has the form:

$$\mathcal{W} = \left\{ (x, y) \ \middle| \ \sum_{p \in \mathcal{P}_m} x_p = y_m, \ x_p \geq 0, \ 0 \leq y_m \leq \gamma_m, \ p \in \mathcal{P}_m, m \in \mathcal{M} \right\}. \tag{5}$$

Then the values of arc flows, $a \in \mathcal{A}$, are defined as follows:

$$f_a = \sum_{m \in \mathcal{M}} \sum_{p \in \mathcal{P}_m} \alpha_{pa} x_p. \tag{6}$$

Let for each arc a a continuous cost function C_a be known, depending on the flow f_a. Then the summary cost for the path p has the form:

$$G_p(x) = \sum_{a \in \mathcal{A}} \alpha_{pa} C_a(f_a).$$

We note that in this case the mapping G is potential. In the general case, when the functions C_a may depend on all arcs flows, this assertion is not true.

For each O/D pair $m \in \mathcal{M}$ we have a so-called disutility function τ_m, depending on the demand value y_m, which is supposed to be continuous.

For finding an equilibrium state of this network, one can solve the following variational inequality (VI for short): Find a vector $(x^*, y^*) \in \mathcal{W}$ such that

$$\langle G(x^*), x - x^* \rangle - \langle \tau(y^*), y - y^* \rangle \geq 0 \quad \forall (x, y) \in \mathcal{W}, \tag{7}$$

where the feasible set $\mathcal{W}$ is defined in (5), the vectors G and τ are composed of components G_p, τ_m, respectively, $p \in \mathcal{P}_m, m \in \mathcal{M}$.

Under the above assumptions, problem (7) has a solution.

In paper [9], it was shown that the above network equilibrium problem can be treated as a two-side multicommodity auction equilibrium problem. The optimality conditions for the network equilibrium problem with elastic demand have the form stated below.

A point $(x^*, y^*) \in \mathcal{W}$ constitutes network equilibrium, if for each $m \in \mathcal{M}$ there exists a number λ_m such that

$$G_p(x^*) \begin{cases} \geq \lambda_m & \text{if } x_p^* = 0, \\ = \lambda_m & \text{if } x_p^* > 0, \end{cases} \quad \forall p \in \mathcal{P}_m, \tag{8}$$

$$\tau_m(y^*) \begin{cases} \leq \lambda_m & \text{if } y_m^* = 0, \\ = \lambda_m & \text{if } y_m^* \in (0, \gamma_m), \\ \geq \lambda_m & \text{if } y_m^* = \gamma_m. \end{cases} \tag{9}$$

For each point $(x^*, y^*) \in \mathcal{W}$ conditions (8)–(9) are equivalent to (7). We note that due to the separability and continuity of functions C_a and τ_m they are integrable, i.e., there exist functions

$$\mu_a(f_a) = \int_0^{f_a} C_a(t) dt \quad \forall a \in \mathcal{A},$$

$$\sigma_m(y_m) = \int_0^{y_m} \tau_m(t) dt \quad \forall m \in \mathcal{M}.$$

Therefore we can say that VI (7) also gives an optimality condition of the following optimization problem:

$$\min_{(x,y) \in \mathcal{W}} \longrightarrow \left\{ \sum_{a \in \mathcal{A}} \mu_a(f_a) - \sum_{m \in \mathcal{M}} \sigma_m(y_m) \right\}, \tag{10}$$

where $f_a, \forall a \in \mathcal{A}$ are defined in (6).

Hence, each solution to optimization problem (10) is a solution to VI (7), the reverse assertion is true if for instance the vectors G and $-\tau$ in (7) are monotone.

Now we assume that all the functions $-\sigma_m$ in (10) are convex, but can be nonsmooth. Hence for each $m \in \mathcal{M}$ there exists the subdifferential $T_m = -\partial(-\sigma_m)$

and instead of functions τ_m we have mappings $T_m : R \to \Pi(R)$, $\forall m \in \mathcal{M}$. Therefore VI (7) takes the form: Find a vector $(x^*, y^*) \in \mathcal{W}$ such that $\exists \tau_m^* \in T_m(y_m^*)$, $\forall m \in \mathcal{M}$,

$$\langle G(x^*), x - x^* \rangle - \langle \tau^*, y - y^* \rangle \geq 0 \quad \forall(x, y) \in \mathcal{W}, \tag{11}$$

where the feasible set $\mathcal{W}$ is defined in (5), G and τ^* are composed of components G_p, τ_m^*, respectively, $p \in \mathcal{P}_m, m \in \mathcal{M}$.

At the same time, equilibrium conditions (9) are reduced to the form

$$\exists \tau_m^* \in T_m(y_m^*), \ \tau_m^* \begin{cases} \leq \lambda_m & \text{if } y_m^* = 0, \\ = \lambda_m & \text{if } y_m^* \in (0, \gamma_m), \\ \geq \lambda_m & \text{if } y_m^* = \gamma_m. \end{cases} \tag{12}$$

In the following section, we remind the solution method for the inner auxiliary problem in the conditional gradient method, proposed by T. Magnanti in [6].

4   Conditional Gradient Method for the Network Equilibrium Problem

In this section, we consider the smooth case of problem (10).

The auxiliary problem in the conditional gradient method for the network equilibrium problem has the following form. At the kth iteration ($k = 0, 1, \dots$) of the main process, we have the vector of path flows x^k and demands y^k. We calculate the values of cost functions $G_p(x^k)$, $\tau_m(y^k)$ for all $p \in \mathcal{P}_m, m \in \mathcal{M}$. It is required to find a vector $(\bar{x}^k, \bar{y}^k) \in \mathcal{W}$, which is a solution to the auxiliary linearized VI:

$$\sum_{m \in \mathcal{M}} \left[\sum_{p \in \mathcal{P}_m} G_p(x^k)(x_p - \bar{x}_p^k) - \tau_m(y^k)(y_m - \bar{y}_m^k) \right] \geq 0 \quad \forall(x, y) \in \mathcal{W}, \tag{13}$$

or the equivalent optimization problem

$$\min_{(x,y) \in \mathcal{W}} \longrightarrow \sum_{m \in \mathcal{M}} \left[\sum_{p \in \mathcal{P}_m} G_p(x^k)x_p - \tau_m(y^k)y_m \right]. \tag{14}$$

For problems (13) or (14) we can use equilibrium conditions (8) and (9). Then we obtain an independent problem for each O/D pair and consider the following simple algorithm for its solution; see [6].

Algorithm A
For each O/D pair $m \in \mathcal{M}$ we calculate a set of shortest paths $\bar{\mathcal{P}}_m^k$ with cost values $G_p(x^k)$. Let $\tilde{\lambda}_m = G_p(x^k)$, $\forall p \in \bar{\mathcal{P}}_m^k$. Then for all $m \in \mathcal{M}$ the following three cases are possible.

(1) If $\tau_m(y^k) < \tilde{\lambda}_m$, then we set $\bar{y}_m^k = 0$, $\bar{x}_p^k = 0$, $\forall p \in \bar{\mathcal{P}}_m^k$, $\lambda_m \in [\tau_m(y^k), \tilde{\lambda}_m]$.

(2) If $\tau_m(y^k) > \tilde{\lambda}_m$, then we set $\bar{y}_m^k = \gamma_m$, distribute the demand value $\bar{y}^{km}$ among paths $p \in \bar{\mathcal{P}}_m^k$ (it is possible to associate the whole demand with one path), and set $\lambda_m = \tilde{\lambda}_m$.

(3) Otherwise we have $\tau_m(y^k) = \tilde{\lambda}_m$, then we choose a feasible demand $\bar{y}^{km} \in [0, \gamma_m]$, distribute this value $\bar{y}_m^k$ among paths $p \in \bar{\mathcal{P}}_m^k$, as above, and set $\lambda_m = \tilde{\lambda}_m$.

This method is very simple, for example, in comparison with the simplex method, the inner problem of the conditional gradient method is reduced to the problem of the shortest path finding.

But equilibrium conditions (8)–(9) can be also useful, if we partially linearize the objective function in the network equilibrium problem.

5 Partial Linearization Method for the Network Equilibrium Problem

In this section, we first also consider the smooth case of problem (10). We suppose that $\tau_m(y) = \tau_m(y_m)$, τ_m are monotonically decreasing functions, $\forall m \in \mathcal{M}$.

We apply the partial linearization method and obtain the following auxiliary problem. At the kth iteration ($k = 0, 1, \dots$) of the main process we have the vector of path flows x^k. We calculate the values of cost functions $G_p(x^k)$, for all $p \in \mathcal{P}_m$, $m \in \mathcal{M}$. It is required to find a vector $(\bar{x}^k, \bar{y}^k) \in \mathcal{W}$ which is a solution to the auxiliary linearized VI:

$$\sum_{m \in \mathcal{M}} \left[\sum_{p \in \mathcal{P}_m} G_p(x^k)(x_p - \bar{x}_p^k) - \tau_m(\bar{y}^k)(y_m - \bar{y}_m^k) \right] \geq 0 \quad \forall(x, y) \in \mathcal{W}, \quad (15)$$

or the equivalent optimization problem

$$\min_{(x,y) \in \mathcal{W}} \longrightarrow \sum_{m \in \mathcal{M}} \left[\sum_{p \in \mathcal{P}_m} G_p(x^k)x_p - \sigma_m(y_m) \right], \quad (16)$$

where $\tau_m(y_m) = \sigma'_m(y_m)$. These problems are also decomposed into a family of independent problems for each O/D pair. Hence the algorithm has the following simple form.

Algorithm B1

We calculate for each O/D pair $m \in \mathcal{M}$ the set of shortest paths $\bar{\mathcal{P}}_m^k$ with costs values $G_p(x^k)$. Let $\tilde{\lambda}_m = G_p(x^k)$, $\forall p \in \bar{\mathcal{P}}_m^k$. Hence for all $m \in \mathcal{M}$ the following three cases are possible.

(1) If $\tau_m(0) \leq \tilde{\lambda}_m$, then we set $\bar{y}_m^k = 0$, $\bar{x}_p^k = 0$, $\forall p \in \bar{\mathcal{P}}_m^k$, $\lambda_m \in [\tau_m(0), \tilde{\lambda}_m]$.

(2) If $\tau_m(\gamma_m) \geq \tilde{\lambda}_m$, then we set $\bar{y}_m^k = \gamma_m$, distribute the demand value γ_m among paths $p \in \bar{\mathcal{P}}_m^k$ (it is possible to associate the whole demand with one path), and set $\lambda_m = \tilde{\lambda}_m$.

(3) Otherwise we have $\tau_m(\gamma_m) < \tilde{\lambda}_m < \tau_m(0)$, then we find the value of demand $\bar{y}_m^k \in [0, \gamma_m]$ such that $\tau_m(\bar{y}_m^k) = \tilde{\lambda}_m$, distribute the demand $\bar{y}_m^k$ among paths $p \in \bar{\mathcal{P}}_m^k$, as above, and set $\lambda_m = \tilde{\lambda}_m$.

Now we consider the other case of problem (10) where functions σ_m are nonsmooth. Then auxiliary linearized VI (15) takes the form: Find a vector $(\bar{x}^k, \bar{y}^k) \in \mathcal{W}$ such that $\exists \tau_m^k \in T_m(\bar{y}_m^k), \forall m \in \mathcal{M}$,

$$\sum_{m \in \mathcal{M}} \left[\sum_{p \in \mathcal{P}_m} G_p(x^k)(x_p - \bar{x}_p^k) - \tau_m^k(y_m - \bar{y}_m^k) \right] \geq 0 \quad \forall (x, y) \in \mathcal{W}. \tag{17}$$

The following algorithm presents a modified variant of Algorithm B1 for the nonsmooth case.

Algorithm B2

We calculate for each O/D pair $m \in \mathcal{M}$ the set of shortest paths $\bar{\mathcal{P}}_m^k$ with costs values $G_p(x^k)$. Let $\tilde{\lambda}_m = G_p(x^k)$, $\forall p \in \bar{\mathcal{P}}_m^k$. Hence for all $m \in \mathcal{M}$ the following three cases are possible.

(1) If $\exists \tau' \in T_m(0)$ such that $\tau' \leq \tilde{\lambda}_m$, then we set $\bar{y}_m^k = 0$, $\bar{x}_p^k = 0$, $\forall p \in \bar{\mathcal{P}}_m^k$, $\lambda_m \in [\tau', \tilde{\lambda}_m]$.

(2) If $\exists \tau'' \in T_m(\gamma_m)$ such that $\tau'' \geq \tilde{\lambda}_m$, then we set $\bar{y}_m^k = \gamma_m$, distribute the demand value γ_m among paths $p \in \bar{\mathcal{P}}_m^k$ (it is possible to associate the whole demand with one path), and set $\lambda_m = \tilde{\lambda}_m$.

(3) Otherwise we have $\tau'' < \tilde{\lambda}_m < \tau'$, $\forall \tau' \in T_m(0)$, $\forall \tau'' \in T_m(\gamma_m)$, then we find the value of demand $\bar{y}_m^k \in [0, \gamma_m]$ such that $\exists \tau \in T_m(\bar{y}_m^k)$, $\tau = \tilde{\lambda}_m$, distribute the demand $\bar{y}_m^k$ among paths $p \in \bar{\mathcal{P}}_m^k$, as above, and set $\lambda_m = \tilde{\lambda}_m$.

6   Implementation of the Methods

In this section, we describe some implementations of the above stated methods for the network equilibrium problem with elastic demand. We first consider the smooth case. We denote

$$\varphi(x, y) = \sum_{a \in \mathcal{A}} \mu_a(f_a) - \sum_{m \in \mathcal{M}} \sigma_m(y_m),$$

where $f_a, \forall a \in \mathcal{A}$ are defined in (6). In the main algorithm, we use a variant of Armijo inexact linesearch.

Partial linearization algorithm 1 (PLA1)

Step 0. Choose a stop criterion $\varepsilon > 0$, numbers $\delta \in (0, 1), \beta \in (0, 1)$. Choose nonempty $\mathcal{P}_m^0 \subset \mathcal{P}_m$ for all $m \in \mathcal{M}$. Set $x_p^0 = 0$, $y_m^0 = 0$ for all $p \in \mathcal{P}_m^0$, $m \in \mathcal{M}$. Set $k = 0$.

Step 1. Using Algorithm B1, find the sets of shortest paths $\bar{\mathcal{P}}_m^k$, for all $m \in \mathcal{M}$ and the vector $(\bar{x}^k, \bar{y}^k) \in \mathcal{W}$, which is a solution to problem (15).

Step 2. Set $\mathcal{P}_m^{k+1} = \mathcal{P}_m^k \cup \bar{\mathcal{P}}_m^k$, for all $m \in \mathcal{M}$. Reduce the points x^k and $\bar{x}^k$ to the equivalent dimensions, if necessary, initializing missing components by zeros. If $(x^k, y^k) \equiv (\bar{x}^k, \bar{y}^k)$, then we have found the exact solution to the initial problem, the iterative process stops.

Step 3. Set the descent direction $(u^k, v^k) = (\bar{x}^k, \bar{y}^k) - (x^k, y^k)$. Find the smallest nonnegative number n that it holds

$$\varphi((x^k, y^k) + \delta^n(u^k, v^k)) - \varphi(x^k, y^k) \leq \beta \delta^n \langle \varphi'(x^k, y^k), (u^k, v^k) \rangle .$$

Set $\lambda = \delta^n$, $(x^{k+1}, y^{k+1}) = (x^k, y^k) + \lambda(u^k, v^k)$.

Step 4. If $|\langle \varphi'(x^k, y^k), (u^k, v^k) \rangle| < \varepsilon$, then the iterative process stops, we have achieved the desired accuracy ε.

Step 5. Set $k = k + 1$ and go to Step 1.

We note also that the dimension of the network equilibrium problem (the number of feasible paths for all O/D pairs) is usually great, but the solution often contains many zero values. Therefore in practice we use the following approach. On the initial stage, we choose some nonempty subset $\mathcal{P}_m^0 \subset \mathcal{P}_m$ for all $m \in \mathcal{M}$ and on each iteration they can increase including new shortest paths. At some moment, the subsets $\mathcal{P}_m^K$ stop to increase. Moreover, if some path has zero flow during long time, we exclude it from the current set of paths.

We also consider the implementation of the conditional gradient method, using Algorithm A. It differs from the PLA1 in Steps 1 and 5 only:

Conditional gradient algorithm (CGA)

Step 1'. Using Algorithm A, find the sets of shortest paths $\bar{\mathcal{P}}_m^k$, for all $m \in \mathcal{M}$ and the vector $(\bar{x}^k, \bar{y}^k) \in \mathcal{W}$, which is a solution to problem (13).

Step 5'. Set $k = k + 1$ and go to Step 1'.

At last, we formulate the implementation of the partial linearization method for nonsmooth problems. In the main iteration process of algorithm, we use the exact linesearch.

Partial linearization algorithm 2 (PLA2)

Step 0. Choose a stop criterion $\varepsilon > 0$. Choose nonempty $\mathcal{P}_m^0 \subset \mathcal{P}_m$ for all $m \in \mathcal{M}$. Set $x_p^0 = 0$, $y_m^0 = 0$ for all $p \in \mathcal{P}_m^0$, $m \in \mathcal{M}$. Set $k = 0$.

Step 1. Using Algorithm B2, find the sets of shortest paths $\bar{\mathcal{P}}_m^k$, for all $m \in \mathcal{M}$ and the vector $(\bar{x}^k, \bar{y}^k) \in \mathcal{W}$, which is a solution to problem (17).

Step 2. Set $\mathcal{P}_m^{k+1} = \mathcal{P}_m^k \cup \bar{\mathcal{P}}_m^k$, for all $m \in \mathcal{M}$. Reduce the points x^k and $\bar{x}^k$ to the equivalent dimensions, if necessary, initializing missing components by zeros. If $(x^k, y^k) \equiv (\bar{x}^k, \bar{y}^k)$, then we have found the exact solution to the initial problem, the iterative process stops.

Step 3. Set the descent direction $(u^k, v^k) = (\bar{x}^k, \bar{y}^k) - (x^k, y^k)$. Find λ_k as a solution to the one-dimensional problem

$$\min_{\lambda \in [0,1]} \longrightarrow \varphi((x^k, y^k) + \lambda(u^k, v^k))$$

Set $(x^{k+1}, y^{k+1}) = (x^k, y^k) + \lambda_k(u^k, v^k)$.

Step 4. If $|\varphi(x^{k+1}, y^{k+1}) - \varphi(x^k, y^k)| < \varepsilon$, then the iterative process stops, we have achieved the desired accuracy ε.

Step 5. Set $k = k + 1$ and go to Step 1.

In the following section, we present some results of preliminary computational experiments on test examples.

7   Computational Experiments

The program has been written in Visual C++, tested on an AMD Athlon at 2.33 GHz, 1.93 Gb, running under Windows XP.

The first three examples are smooth. For smooth cases we use the stop criterion $|\langle \varphi'(x^k, y^k), (u^k, v^k) \rangle| < \varepsilon$, $k = 0, 1, \ldots$ Let us consider the network composed of 18 nodes (Fig. 1). All arcs are bi-directional. The set of OD-pairs contains 5 elements, $\mathcal{M} = \{(1, 17), (2, 13), (3, 8), (16, 13), (12, 18)\}$. The cost functions are $C_a(f_a) = 1 + f_a$ for all a. The disutility functions are $\tau_m(y_m) = 15 - 0.5 y_m$ for all $m \in \mathcal{M}$. $\gamma_m = 10$, for all $m \in \mathcal{M}$. Beside the

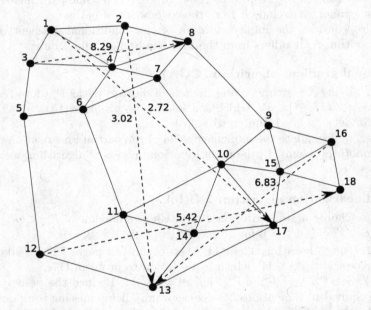

Fig. 1. Network example 1, 18 nodes, 5 O/D pairs

Table 1. Numbers of iterations and time for Example 1

Accuracy	*CGA*	*PLA*1
0.1	6973 it, 1750 ms	3065 it, 750 ms
0.01	23730 it, 6343 ms	15615 it, 4203 ms
0.001	31137 it, 8406 ms	26706 it, 7313 ms

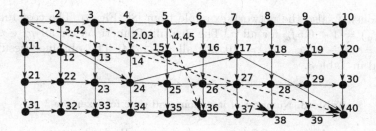

Fig. 2. Network example 2, 40 nodes, 3 O/D pairs

Table 2. Numbers of iterations and time for Example 2

Accuracy	CGA	PLA1
0.1	2815 it, 1172 ms	516 it, 187 ms
0.01	15233 it, 6875 ms	4924 it, 2032 ms
0.001	21659 it, 9953 ms	13836 it, 6031 ms

network structure, on the figure we show O/D pairs as dash directed links and the obtained elastic demand values.

We have applied the described implementation of both methods to this problem for different accuracy values and obtained the results presented in Table 1.

Let us consider another example, which has a more structured form (Fig. 2). The set of OD-pairs contains 3 element, $\mathcal{M} = \{(1, 40), (3, 38), (5, 36)\}$. The cost functions are $C_a(f_a) = 1 + f_a$ for all a. The disutility functions are $\tau_m(y_m) = 15 - 0.5 y_m$ for all $m \in \mathcal{M}$. $\gamma_m = 10$, for all $m \in \mathcal{M}$. The arc directions are shown on the figure by arrows.

The obtained results are presented in Table 2.

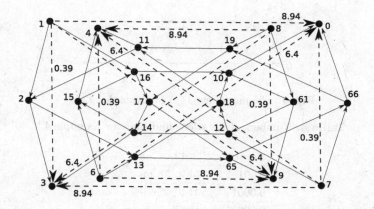

Fig. 3. Network example 3, 22 nodes, 12 O/D pairs

We consider also the network example from [8] (Fig. 3). The cost functions are $C_a(f_a) = 1 + 0.5f_a$ for all a. The disutility functions for all $m \in \mathcal{M}$ are $\tau_m(y_m) = 15 - 0.5y_m$. $\gamma_m = 10$, for all $m \in \mathcal{M}$. The obtained results are presented in Table 3.

Table 3. Numbers of iterations and time for Example 3

Accuracy	CGA	$PLA1$
0.1	5018 it, 1625 ms	17 it, <16 ms
0.01	48320 it, 16750 ms	18 it, <16 ms
0.001	>200000 it, >100000 ms	18 it, 16 ms

The partial linearization method with Algorithm B1 often shows better results than the conditional gradient method with Algorithm A: we usually obtain lower numbers of iterations, lower time, and lower values of the objective function φ.

Now we consider the nonsmooth form of problem (10). For nonsmooth cases we apply the stop criterion $|\varphi(x^{k+1}, y^{k+1}) - \varphi(x^k, y^k)| < \varepsilon$, $k = 0, 1, \ldots$ In Example 4 we use the network structure from Example 1. The cost functions are $C_a(f_a) = 1 + 0.5f_a$ for all a. $\gamma_m = 20$, for all $m \in \mathcal{M}$. We set $\sigma_m(y_m) = \min\{10y_m, 5y_m + 25\}$, for all $m \in M$. Figure 4 shows the form of mappings $T_m(y_m)$.

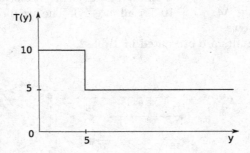

Fig. 4. The mappings T(y) in Examples 4, 5

Table 4. Numbers of iterations and time for Example 4

Accuracy	$PLA2$
0.001	144 it, 78 ms
0.0001	504 it, 282 ms
0.00001	1418 it, 781 ms
0.000001	3811 it, 2219 ms

Table 5. Numbers of iterations and time for Example 5

Accuracy	PLA2
0.001	69 it, 32 ms
0.0001	318 it, 188 ms
0.00001	893 it, 547 ms
0.000001	1000 it, 641 ms

We apply partial linearization algorithm 2 to solve this problem. The results for different accuracy values are presented in Table 4. We obtain the demand vector $(4.09, 5, 5, 5, 5)$ and the cost functions values $(10, 9.83, 5.72, 7, 7.78)$.

In Example 5 we use the network structure from Example 3. The set of OD-pairs is $\mathcal{M} = \{(1, 6), (1, 7), (1, 8), (2, 5), (2, 7), (2, 8), (3, 5), (3, 6), (3, 8), (4, 5), (4, 6), (4, 7)\}$. The cost functions are $C_a(f_a) = 1 + 0.5 f_a$ for all a. $\gamma_m = 20$, $\sigma_m(y_m) = \min\{10 y_m, 5 y_m + 25\}$, for all $m \in M$.

The results for different accuracy values are presented in Table 5. We obtain the demand vector $(5, 1, 4, 5, 4, 1, 1, 4, 5, 4, 1, 5)$ and the cost functions values $(7, 10, 10, 7, 10, 10, 10, 10, 7, 10, 10, 7)$.

The conducted calculations on test examples have shown the applicability of the proposed approaches to the network equilibrium problems with elastic demands.

References

1. Fukushima, M., Mine, H.: A generalized proximal point algorithm for certain non-convex minimization problems. Int. J. Syst. Sci. **12**, 989–1000 (1981)
2. Mine, H., Fukushima, M.: A minimization method for the sum of a convex function and a continuously differentiable function. J. Optim. Theor. Appl. **33**, 9–23 (1981)
3. Patriksson, M.: Nonlinear Programming and Variational Inequality Problems: a Unified Approach. Kluwer, Dordrecht (1999)
4. Frank, M., Wolfe, P.: An algorithm for quadratic programming. Nav. Res. Logistics Q. **3**, 95–110 (1956)
5. Levitin, E.S., Polyak, B.T.: Constrained minimization methods. USSR Comput. Math. Math. Phys. **6**(5), 1–50 (1966)
6. Magnanti, T.L.: Models and Algorithms for Predicting Urban Traffic Equilibria. In: Florian, M. (ed.) Transportation Planning Models, pp. 153–185. North-Holland, Amsterdam (1984)
7. Patriksson, M.: The Traffic Assignment Problem. Models and Methods. VSP, Utrecht (1994)
8. Nagurney, A.: Network Economics: A Variational Inequality Approach. Kluwer, Dordrecht (1999)
9. Konnov, I.V.: On Auction Equilibrium Models with Network Applications. Netnomics **16**, 107–125 (2015)

Multiple Cuts in Separating Plane Algorithms

Evgeni Nurminski(✉)

School of Natural Sciences, Far Eastern Federal University, Ajax Street, Vladivostok, Russky Island, Russia
nurminskiy.ea@dvfu.ru

Abstract. This paper presents an extended version of the separation plane algorithm for subgradient-based finite-dimensional nondifferentiable convex optimization. The extension introduces additional cuts for epigraph of the conjugate of objective function which improve the convergence of the algorithm. The case of affine cuts is considered in more details and it is shown that it requires solution of an auxiliary convex subproblem the dimensionality of which depends on the number of additional cuts and can be kept arbitrary low. Therefore algorithm can make use of the efficient algorithms of low-dimensional nondifferentiable convex optimization which overcome known computational complexity bounds for the general case.

Keywords: Convex optimization · Conjugate function · Cutting plane · Separating plane · Center of gravity algorithm

1 Introduction and Notations

We consider a finite-dimensional nondifferentiable convex optimization (NCO) problem

$$\min_{x \in E} f(x) = f_\star = f(x^\star), x^\star \in X_\star, \tag{1}$$

where E denotes a finite-dimensional space of primal variables and $f : E \to \mathbb{R}$ is a finite convex function, not necessarily differentiable. As we are interested in computational issues related to solving (1) mainly we assume that this problem is solvable and has nonempty set of solutions $X_\star$.

This problem enjoys a considerable popularity due to its important theoretical properties and numerous applications in large-scale structured optimization, Lagrange relaxation in discrete optimization, exact penalization in constrained optimization, and others. This led to the development of different algorithmic ideas, starting with the subgradient algorithm due to Shor [1] and Polyak [2] and followed by cutting plane [3], conjugate subgradient [4], bundle methods [13], ellipsoid and space dilatation [5–7], ϵ-subgradient methods [8,9], VU-methods [10] and many others. This paper describes an extended version of the separation plane algorithm (SPA) [14] which differs from the original idea in that

E. Nurminski—This work is supported by RFBR grant 13-07-1210.

it introduces several additional cuts for epigraph of the conjugate of objective function. The simplest form of SPA with just one additional cut was considered in all details including computational experiments in [15–17]. The positive experience with this algorithm raised some hopes that introduction of more cuts will improve the computational efficiency further on.

Throughout the paper we use the following notations: $\dim(E)$ is the dimensionality of E, $|I|$ is the cardinality of a finite set I, xy is the inner product of x, y from E, $\|x\| = \sqrt{xx}$. The set of nonnegative vectors of E is denoted as E_+ or E_+^n if the dimensionality n of E has to be specified.

We use also the distance function $\mathrm{dist}(X, Y) = \inf_{x \in X, y \in Y} \|x - y\| = \mathrm{dist}(Y, X)$ between $X \subset E, Y \subset E$. If X is a singleton $\{x\}$ we will write just $\mathrm{dist}(x, Y)$.

A vector of ones of a suitable dimensionality is denoted by $e = (1, 1, \ldots, 1)$. A standard simplex $\{x : x \geq 0, xe = 1\}$ with $x \in E, \dim(E) = n$ is denoted by Δ_n.

2   Separating Plane Algorithms

One of the ways to represent the popular bundle [13] and the other methods of NCO is to view them as a projection algorithms for computing

$$f^\star(0) = -\min_x f(x) = -f_\star = - \inf_{(0,\mu) \in \mathrm{epi}\, f^\star} \mu,$$

where $f^\star(g) = \sup_x \{xg - f(x)\}$ is a Fenchel-Moreau conjugate of f, $\mathrm{epi}\, f^\star = \{(g, \mu') : \mu \geq f^\star(g)\} \subset E^\star \times \mathbb{R}$ is the epigraph of $f^\star(g)$, and $g \in E^\star$, the space of conjugate variables (gradients). This idea, presented originally in [14], unifies a number of known NCO techniques and suggests some new computational ideas.

The general idea of SPA is to bound the epigraph $\mathrm{epi}\, f^\star$ of the conjugate function $f^\star$ from below and above (in terms of set-theoretical inclusion) by the approximations L_f and U_f:

$$L_f \subset \mathrm{epi}\, f^\star \subset U_f.$$

These approximations provide lower and upper estimates for $f^\star(0)$:

$$\inf_{(0,\mu) \in U_f} \mu = v_U \leq -f^\star(0) \leq \inf_{(0,\mu) \in L_f} \mu = v_L \qquad (2)$$

and are gradually refined in the vicinity of the vertical axis $\{0\} \times \mathbb{R} \subset E^\star \times \mathbb{R}$ to make at least one of v_U or v_L converge to $f^\star(0)$.

The iterations of SPA consist in recursive application of the update procedure to L_f and U_f which is given in more details further on. This procedure is based on computed values of conjugate function $f^\star$ at certain points of the conjugate space, determined by the procedure itself. As a result at k-th iteration of SPA we have the bundle of accumulated information on $\mathrm{epi}\, f^\star$ which consists of pairs of conjugate variables and values of conjugate function at these points. This bundle will be denoted as $\mathcal{B}_I^\star = \{(g^i, f^\star(g^i)), i \in I\}$ where $I = \{1, 2, \ldots, k\}$ and $g^i, f^\star(g^i)$ are conjugate variables and the value of conjugate function, computed at i-th iteration. In other words $\mathcal{B}_I^\star$ contains all information available up to the current iteration k, however some selection can be performed to save memory.

For technical reasons we assume also that $\mathcal{B}_I^\star$ contains a special pair $(0, \alpha)$ with $\alpha > f^\star(0)$. In terms of the original problem (1) it means that we assume a certain lower bound $-\alpha$ for $f_\star$ to be known. It may be a very crude estimate and introduced mainly for formal reasons, but it is necessary to avoid in a simplest way certain ill-defined subproblems in the algorithm. Notice that by construction $(0, \alpha) \in \text{epi } f^\star$.

The points in the bundle $\mathcal{B}_I^\star$ have their natural counterparts $\{(x^i, f(x^i)), i \in I\}$ in the extended space of primal variables $E \times \mathbb{R}$ with $g^i \in \partial f(x^i)$, $f^\star(g^i) = x^i g^i - f(x^i)$. In fact the algorithms based on the bundle $\mathcal{B}_I^\star$ can be considered as based on the primal bundle $\mathcal{B}_I = \{(x^i, f(x^i)), i \in I\}$ and operating on the primal variables and the original objective function. Notice that the bundle $\mathcal{B}_I$ provides information on the support function of epi $f^\star$, that is the hyperplane

$$P_i = \{(g, \mu) : g\hat{x}^i - \mu = f(\hat{x}^i) = \sup_{(g,\mu) \in \text{epi } f^\star} \{g\hat{x}^i - \mu\}\} \tag{3}$$

is a supporting plane of epi $f^\star$ at the point $(g^i, f^\star(g^i))$.

Due to convexity the natural way to construct L_f and U_f is to use the inner and outer approximations:

$$L_f = \text{co}\{(g^i, f^\star(g^i)), i \in I\} + \{0\} \times \mathbb{R}_+ \subset \text{epi } f^\star \tag{4}$$

and

$$U_f = \cap H_i, \, i \in I \supset \text{epi } f^\star \tag{5}$$

where

$$H_i = \{(g, \mu) : \mu \geq f^\star(g^i) + x^i(g - g^i), x^i \in \partial(g^i)\} \supset \text{epi } f^\star$$

are the half-spaces, generated by supporting planes P_i (3) to epi $f^\star$ at the points $(g^i, f^\star(g^i))$.

The general scheme to update L_f and U_f at k-th iteration with $I = \{1, 2, \ldots, k\}$ is described in the Algorithm 1.

For better understanding the sequence of major steps in the update process is illustrated on Figs. 1, 2, 3 and 4.

From computational point of view the separating plane $H_{\hat{x}}$ in the **Step 2** **(Separate)** can be obtained for the finite value of v_U by solving the projection problem

$$\min_{(z,\mu) \in L_{f^\star}} \|z\|^2 + (v_U - \mu)^2 = \|\hat{z}\|^2 + (v_U - \hat{\mu})^2 \tag{7}$$

and appropriate normalization: $\hat{x} = -\hat{z}/(v_U - \hat{\mu})$.

The **Support** step of the algorithm is just the computation of the objective function and its subgradient at the point $\hat{x}$ as demonstrated by (3).

Data: The bundle $\mathcal{B}_I^\star$, the upper and low approximations U_f, L_f of epi $f^\star$.
Result: The updated: set I, approximations L_f, U_f and the bundle $B_f^\star$.
Step 1. Estimate: estimate the lower bound for $f^\star(0)$. Compute

$$v_U = \inf_{(0,\mu)\in U_f} \mu \leq= \inf_{(0,\mu)\in\text{epi } f^\star} \mu = f^\star(0).$$

It can be set to $-\infty$ if U_f is taken to be the trivial upper approximation $E \times \mathbb{R}$ at the start of SPA.
Step 2. Separate: strictly separate $(0, v_U)$ from L_f with a separating plane $H_{\hat{x}} = \{(g,\mu) : g\hat{x} - \mu = -\hat{v}_U\}$, parameterized by the support vector $(\hat{x}, -1)$ and $\hat{v}_U$ to be found. If $v_U = -\infty$ just take an arbitrary $\hat{x}$. If strict separability is impossible, that is $\hat{v}_U = f^\star(0) = -f_\star$, then we are done, otherwise continue.
Step 3. Support: for a given $\hat{x}$, found at the previous step, find the supporting hyperplane $P_{\hat{x}}^\star$ for epi $f^\star$:

$$P_{\hat{x}}^\star = \{(g,\mu) : g\hat{x} - \mu = \sup_{(g,\epsilon)\in\text{epi } f^\star}\{\hat{x}g - \epsilon\} = \\ \sup_g\{\hat{x}g - f^\star(g)\} = \hat{x}\hat{g} - f^\star(\hat{g}) = f(\hat{x})\} \tag{6}$$

with $\hat{g} \in \partial f(\hat{x})$. Notice, that this is just the calculation of $f(\hat{x})$ and $\hat{g} \in \partial f(\hat{x})$. The hyperplane $P_{\hat{x}}^\star$ defines the "upper" half-space $H_{\hat{x}}^\star$ which contains epi $f^\star$:

$$H_{\hat{x}}^\star = \{(g,\mu) : \mu \geq g\hat{x} - f(\hat{x})\} \supset \{(g,\mu) : \mu \geq \sup_x\{gx - f(x)\}\} = \\ \{(g,\mu) : \mu \geq f^\star(g)\} = \text{epi } f^\star$$

and hence $H_{\hat{x}}^\star$ can be safely added to the cuts of the upper approximation U_f.
Step 4. Update: perform the update of the basic data structures of SPA:
    the bundle: $\mathcal{B}_I^\star \to \mathcal{B}_I^\star \cap \{(\hat{g}, f^\star(\hat{g}))\}$,
    the approximations: redefine L_f and U_f according to (4) and (5)

$$L_f \to \text{co}(L_f, (\hat{g}, \hat{\epsilon})), \quad U_f \to U_f \cap S_{\hat{x}}^\star$$

    the index set: $I \to I \cup \{k+1\}$.

Algorithm 1. The generic structure of update step for the upper and low approximations of epi $f^\star$

Notice that after the update of U_f in any way we obtain a new upper estimate for $f_\star$ which is not worse that the previous:

$$v_U' = \inf_{(0,\mu)\in U_f\cap S_{\hat{x}}^\star} \mu \geq \max\{\inf_{(0,\mu)\in U_f} \mu, \inf_{(0,\mu)\in S_{\hat{x}}^\star} \mu\} = \max\{v_U, -f(\hat{x})\} \geq v_U$$

and may be better if $f(\hat{x})$ sets a new record. Unfortunately we can not guarantee that this will be just the case and so the algorithm is not monotone in terms of the objective function. This may be one of the factors which slows down the practical convergence of SPA, and it seems to be possible to improve it by adding an additional cut or cuts on epi $f^\star$.

That was the original idea, tested with positive results in [15, 16] when just the single extra cut generated by the auxiliary subproblem of cutting plane method was added. Here we consider some aspects of adding several extra cuts.

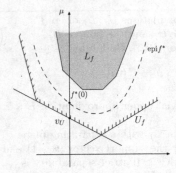

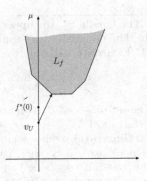

Fig. 1. Basic algorithm objects: L_f, U_f are lower and upper approximations, v_U approximates $f^\star(0)$ from below.

Fig. 2. Projection: determines the (normalized) vector $(\hat{x}, -1)$ such that $g\hat{x} - \mu \leq -v_U$ for any $(g, \mu) \in \mathrm{epi}\, f^\star$.

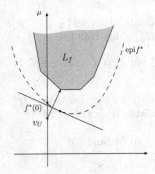

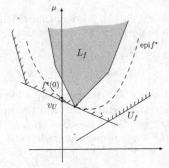

Fig. 3.   Support: compute $\sup_{(g,\mu) \in \mathrm{epi}\, f^\star} \{\hat{x}g - \mu\} = f(\hat{x})$ and the corresponding subgradient $\hat{g} \in \partial f(\hat{x})$.

Fig. 4. Update: the lower L_f and the upper U_f approximations are updated with the help of a new $(g, f^\star(g))$ and cutting support plane at $(g, f^\star(g))$.

3   Multiple Additional Cuts

From the formal point of view the additional cuts for $\mathrm{epi}\, f^\star$ can be considered as a a certain subset Q of $E \times \mathbb{R}$ which is superimposed on $\mathrm{epi}\, f^\star$. It means that now instead of $\mathrm{epi}\, f^\star$ in the **Support** step of the Algorithm 1. we are going to use $\mathrm{epi}\, f^\star \cap Q$

In this case a new supporting hyperplane $P_{\hat{x}}^\star = \{(g, \mu) : g\hat{x} - \mu = \bar{\mu}\}$ will have $\bar{\mu} \geq \hat{\mu}$:

$$-\bar{\mu} = \sup_{(g,\mu) \in \mathrm{epi}\, f^\star \cap Q} \{g\hat{x} - \mu\} \leq \sup_{(g,\mu) \in \mathrm{epi}\, f^\star} \{g\hat{x} - \mu\} = -\hat{\mu} = f(\hat{x})$$

and therefore we have a better chance to improve v_U':

$$\bar{v}_U' = \max\{v_U, \bar{\mu}\} \geq \max\{v_U, \hat{\mu}\} = v_U'$$

There is a great flexibility in the choice of Q, the only essential requirement is to ensure that the solution $(0, -f_\star)$ still belongs to epi $f^\star \cap Q$.

The updated iteration of the separating plane algorithms with cuts is represented in Algorithm 2.

Data: The bundle $\mathcal{B}_f^\star$, the upper and low approximations U_f, L_f of epi $f^\star$, and the cut $Q \subset E \times \mathbb{R}$.

Result: The updated index set I, approximations L_f, U_f and the bundle $B_f^\star$.

Step 1. Estimate: Unchanged.

Step 2. Separate: Unchanged.

Step 3. Support: Modified to include the cut Q. For a given $\hat{x}$, found at the previous step, find the supporting hyperplane $H_{\hat{x}}^\star$ for epi $f^\star \cap Q$:

$$P_{\hat{x}}^\star = \{(g,\mu) : g\hat{x} - \mu = \sup_{\substack{(g,\epsilon)\, \in\, \text{epi}\, f^\star \\ (g,\epsilon)\, \in\, Q}} \{\hat{x}g - \epsilon\}\}. \tag{8}$$

The details of these calculations depend upon the definition of the cut set Q and are discussed further on.

The hyperplane $H_{\hat{x}}^\star$ defines the "upper" half-space $H_{\hat{x}}^\star$ which contains epi $f^\star$:

$$S_{\hat{x}}^\star = \{(g,\mu) : \mu \geq g\hat{x} - f(\hat{x})\} \supset \{(g,\mu) : \mu \geq \sup_x\{gx - f(x)\}\} = \{(g,\mu) : \mu \geq f^\star(g)\} = \text{epi}\, f^\star$$

and hence $S_{\hat{x}}^\star$ can be safely added to the cuts of the upper approximation U_f.

Step 4. Update: Unchanged.

Algorithm 2. The generic structure of update step for the upper and low approximations of epi $f^\star$ in SPA with multiple cuts.

From practical point of view it is convenient to have Q described by a system of convex inequalities $Q = \{(g,\mu) : h_i(g,\mu) \leq 0, i = 1, 2, \ldots, m\}$, each of which can be considered as a separate cut, applied to epi $f^\star$. Therefore we call this type of algorithms as separating plane algorithm with multiple cuts (SPA-MC).

In the simplest case all $h_i(g,\mu)$ are affine functions:

$$h_i(g,\mu) = \hat{x}^i g + \mu - \bar{\mu}_i, \tag{9}$$

where $\hat{x}^i$ represent some trial points in the space of the original primal variables.

The support problem of the **Step 3** in SPA-MC for the case of affine cuts can be written as

$$w_U = \sup\{xg - \mu] \tag{10}$$
$$\mu \geq f^\star(g)$$
$$\hat{x}^i g + \mu \leq \bar{\mu}_i, i = 1, 2, ; m$$

which can be transformed into the dual form

$$w_U = \sup_{\mu \geq f^\star(g)} \inf_{\lambda \geq 0} \{xg - \mu - \sum_{i=1}^m \lambda_i(\hat{x}^i g + \mu - \bar{\mu}_i)\}, \tag{11}$$

where $\lambda = (\lambda_1, \lambda_2, \ldots, \lambda_m)$ is a nonnegative vector of Lagrange multipliers.

By convexity

$$w_U = \inf_{\lambda \geq 0} \sup_{\mu \geq f^\star(g)} \{xg - \mu - \sum_{i=1}^m \lambda_i(\hat{x}^i g + \mu - \bar{\mu}_i)\} =$$

$$\inf_{\lambda \geq 0} \{\sum_{i=1}^m \lambda_i \bar{\mu}_i + \sup_{\mu \geq f^\star(g)} \{(x - \sum_{i=1}^m \lambda_i \hat{x}^i)g - (1 + \sum_{i=1}^m \lambda_i)\mu)\} =$$

$$\inf_{\lambda \geq 0} \{\sum_{i=1}^m \lambda_i \bar{\mu}_i + (1 + \sum_{i=1}^m \lambda_i) \sup_{\mu \geq f^\star(g)} \{\frac{x - \sum_{i=1}^m \lambda_i \hat{x}^i}{1 + \sum_{i=1}^m \lambda_i} g - \mu)\}\} =$$

$$\inf_{\lambda \geq 0} \{\sum_{i=1}^m \lambda_i \bar{\mu}_i + (1 + \sum_{i=1}^m \lambda_i) \sup_g \{\frac{x - \sum_{i=1}^m \lambda_i \hat{x}^i}{1 + \sum_{i=1}^m \lambda_i} g - f^\star(g)\} =$$

$$\inf_{\lambda \geq 0} \{\sum_{i=1}^m \lambda_i \bar{\mu}_i + (1 + \sum_{i=1}^m \lambda_i) f(\frac{x - \sum_{i=1}^m \lambda_i \hat{x}^i}{1 + \sum_{i=1}^m \lambda_i})\} = \inf_{\lambda \geq 0} \Xi(\lambda),$$

where

$$\Xi(\lambda) = \sum_{i=1}^m \lambda_i \bar{\mu}_i + (1 + \sum_{i=1}^m \lambda_i) f(\frac{x - \sum_{i=1}^m \lambda_i \hat{x}^i}{1 + \sum_{i=1}^m \lambda_i})$$

has a controllable dimensionality m which is determined by the number of additional cuts and can be set to any value.

Therefore $\Xi(\lambda)$ can be minimized by specific algorithms, tailored to this particular dimensionality. An appropriate example of such algorithms is the center of gravity method (CGM) by Levin [11] and Newmann [12] which is easily implemented at least in 2-dimensional case and provides a geometric rate of convergence independent of properties of objective function and feasibility set. Hopefully the efficient and practical methods may appear or already exist, unknown to the author, in higher dimensions.

The essential part of $\Xi(\lambda)$ which may create different problems with the following minimization is the nonlinear term $(1 + \sum_{i=1}^m \lambda_i) f((x - \sum_{i=1}^m \lambda_i \hat{x}^i)/(1 + \sum_{i=1}^m \lambda_i))$. Fortunately it inherits a convexity of the original problem which follows from its definition as a supremum of linear forms in λ. Nevertheless it is useful for the further maximization to consider the nonlinear part of $\Xi(\lambda)$ as a generic function

$$\phi(\theta) = \left(\sum_{i=1}^m \theta_i\right) f\left(\frac{\sum_{i=1}^m \theta_i \hat{x}^i}{\sum_{i=1}^m \theta_i}\right) \tag{12}$$

for $\theta = (\theta_1, \theta_2, \ldots, \theta_m) \in E_+^m$ and $\theta \neq 0$. It makes sense to complement the definition of $\phi(\cdot)$ at 0 as $\phi(0) = 0$ without loosing the continuity. Then ϕ becomes defined on the whole E_+^m and its convexity properties are covered by the following lemma which might be of a separate interest.

Lemma 1. *Let $f : E \to \mathbb{R}$ is a convex finite function, $\hat{x}^i, i = 1, 2, \ldots m$ — a collection of m points in E, and $\theta = (\theta_1, \theta_2, \ldots, \theta_m) \in E_+^m$ — a vector of nonnegative variables. Then $\phi(\theta)$ defined by (12) is a convex function of θ on E_+^m.*

Proof. Denote $\sum_{i=1}^{m} \theta_i = \sigma(\theta)$. Then

$$\phi(\theta) = \sigma(\theta) f\left((\sum_{i=1}^{m} \theta_i \hat{x}^i)/\sigma(\theta)\right)$$

for $\sigma(\theta) > 0$ and $\phi(0) = 0$ by definition. Let $\alpha \in [0, 1]$ and $\theta', \theta'' \in E_+^m$. Next we show that $\phi(\cdot)$ satisfies the Jensen inequality $\phi(\alpha\theta' + (1 - \alpha)\theta'') \leq \alpha\phi(\theta') + (1 - \alpha)\phi(\theta'')$.

Notice first, that ϕ is positive homogeneous of degree 1: $\phi(\nu\theta) = \nu\phi(\theta)$ for $\nu \geq 0$ hence the case when either $\theta' = 0$ or $\theta'' = 0$ is trivial.

Assume further on that $\sigma(\theta')\sigma(\theta'') > 0$. Let us fix α and denote $\kappa = \alpha\sigma(\theta') + (1 - \alpha)\sigma(\theta'') > 0$. Then

$$\phi(\alpha\theta' + (1 - \alpha)\theta'') = \kappa f\left((\alpha \sum_{i=1}^{m} \theta'_i x^i + (1 - \alpha) \sum_{i=1}^{m} \theta''_i x^i)/\kappa\right) =$$
$$\kappa f\left(\alpha(\sum_{i=1}^{m} \theta'_i x^i)/\kappa + (1 - \alpha)(\sum_{i=1}^{m} \theta''_i x^i)/\kappa\right) =$$
$$\kappa f\left(\alpha \frac{\sum_{i=1}^{m} \theta'_i x^i}{\sigma(\theta')} \frac{\sigma(\theta')}{\kappa} + (1 - \alpha) \frac{\sum_{i=1}^{m} \theta''_i x^i}{\sigma(\theta'')} \frac{\sigma(\theta'')}{\kappa}\right) = \kappa f(\gamma' \bar{x}' + \gamma'' \bar{x}''),$$

where

$$\gamma' = \alpha\sigma(\theta')/\kappa, \quad \gamma'' = \alpha\sigma(\theta'')/\kappa,$$

and

$$\bar{x}' = \sum_{i=1}^{m} \theta'_i x^i/\sigma(\theta'), \quad \bar{x}'' = \sum_{i=1}^{m} \theta''_i x^i/\sigma(\theta''),$$

As $\gamma' + \gamma'' = \alpha\sigma(\theta')/\kappa + (1 - \alpha)\sigma(\theta'')/\kappa = 1$ and $\gamma', \gamma'' \geq 0$. then

$$\phi(\alpha\theta' + (1 - \alpha)\theta'') \leq \kappa f(\gamma' \bar{x}' + \gamma'' \bar{x}'') \leq \kappa(\gamma' f(\bar{x}') + \gamma'' f(\bar{x}'')) =$$
$$\alpha\sigma(\theta')\kappa f(\bar{x}')/\kappa + (1 - \alpha)\sigma(\theta'')\kappa f(\bar{x}'')/\sigma(\theta'')/\kappa = \alpha\phi(\theta') + (1 - \alpha)\phi(\theta'').$$

which completes the proof.

By setting $z^1 = x$, $z^{i+1} = -\hat{x}^i, i = 1, 2, \ldots, m$ and applying Lemma 1 to $\phi(\theta) = \sigma(\theta) f\left((\sum_{i=1}^{m+1} \theta_i z^i)/\sigma(\theta)\right)$ with $\theta_1 = 1$ we obtain convexity of $\Xi(\theta)$.

4 Convergence

The following theorem establishes the convergence of SPA-MC.

Theorem 1. *Let $\{v_U^k\}$ be the sequence of lower estimates*

$$v_U^k = \inf_{(0,\mu) \in U_f^k} \mu \leq f^\star(0),$$

of the optimal value in the problem (1) which are generated by SPA-MC as prescribed by Algorithms 1–2. Then

$$\lim_{k \to \infty} v_U^k = f^\star(0) = -\min_x f(x).$$

Proof. Let $k = 1, 2, \ldots$ number the sequence of the update iterations of SPA-MC which are prescribed by Algorithms 1–2, and let U_f^k, L_f^k are the corresponding upper and lower approximations of epi $f^\star$ at the beginning of k-th iteration. Naturally, updated U_f^k, L_f^k become U_f^{k+1}, L_f^{k+1}.

By construction $U_f^k \supset$ epi $f^\star \supset L_f^k$ and

$$\text{epi } f^\star \subset U_f^{k+1} \subset U_f^k, \quad L_f^k \subset L_f^{k+1} \subset \text{epi } f^\star$$

so both these sequences have Kuratovski limits, which we denote as $U_f^\bullet, L_f^\bullet$ respectively.

Observe that $v_U^k \le v_U^{k+1} \le f^\star(0)$ hence the sequence $\{v_U^k\}$ has a limit which we denote as $v_U^\bullet$.

Convergence of SPA-MC means that $v_U^k \to f^\star(0)$ or, equivalently, $\text{dist}(\bar{v}_U^k, \text{epi } f^\star) = \text{dist}((0, v_U^k), \text{epi } f^\star) \to 0$ when $k \to \infty$. As $\text{dist}(\bar{v}_U^k, \text{epi } f^\star) \le \text{dist}(\bar{v}_U^k, L_f^k)$ it is sufficient to show that $\text{dist}(\bar{v}_U^k, L_f^k) \to 0$.

Denote $\bar{V}_U^k = v_U^k - \{0\} \times \mathbb{R}_+$ and notice that

$$\text{dist}(\bar{v}_U^k, L_f^k) = \text{dist}(\bar{v}_U^k - 0 \times R_+, L_f^k) = \text{dist}(\bar{V}_U^k, L_f^k).$$

As $\bar{V}_U^{k+1} \supset \bar{V}_U^k$ and $L_f^{k+1} \supset L_f^k$ the distance $\text{dist}(\bar{V}_U^k, L_f^k)$ is non-increasing:

$$\text{dist}(\bar{V}_U^{k+1}, L_f^{k+1}) \le \text{dist}(\bar{V}_U^k, L_f^k) \le \text{dist}(\bar{V}_U^0, (0, \kappa) + 0 \times \mathbb{R}_+) = \|v_U^0 - \kappa\|$$

hence the norms of all vectors $z^k = \Pi_{L_f^k}(\bar{V}_U^k) - \bar{V}_U^k$ are uniformly bounded and have the same limit $\rho_z = \lim_{k \to \infty} \|z^k\|$. The key question is however what is the value of ρ_z. If $\rho_z = 0$ then $\text{dist}(\bar{V}_U^k, \text{epi } f^\star) \to 0$ and

$$\lim_{k \to \infty} v_U^k = v_U^\bullet = f^\star(0)$$

which establishes convergence of SPA-MC.

To show that this is just the case assume contrary: $\rho_z > 0$. Then the sequence $\{z^k\}$ due to its boundness has at least one limit point, which we denote as $z^\bullet$ with a certain subsequence $\{z^{k_t}, t = 1, 2, \ldots\} \to z^\bullet$.

The **Support** and **Update** steps of the Algorithms 1–2 redefines v_U^k in a following way:

1. Solve

$$\inf_{\bar{g}=(g,\mu)\in\text{epi } f^\star \cap Q_k} z^k \bar{g} = z^k \bar{g}^k = \gamma_k,$$

where $\bar{g}^k = (g^k, \mu_k)$.
2. If $\gamma_k > v_U^k$ redefine $v_U^{k+1} = \gamma_k$. Otherwise $v_U^{k+1} = v_U^k$.

In any case $\bar{g}^k z^k \le \bar{v}_U^{k+1} z^k$ and passing in this inequality to the limit along the subsequence where all limits exist obtain

$$\bar{g}^\bullet z^\bullet \le v_U^\bullet z^\bullet. \tag{13}$$

On the other hand as z^k is obtained by projection of $(0, v_U^k)$ on L_f^k and taking into account that $\bar{g}^k \in L_f^{k+1}$ we have $(\bar{g}^k - \bar{v}^{k+1})z^{k+1} \geq \|z^{k+1}\|^2$. Passing to the limit gives $(\bar{g}^\bullet - \bar{v}^\bullet)z^\bullet \geq \|z^\bullet\|^2 \geq \gamma > 0$ or

$$\bar{g}^\bullet z^\bullet \geq \bar{v}_U^\bullet z^\bullet + \gamma > \bar{v}_U^\bullet z^\bullet. \tag{14}$$

Obviously (13) and (14) contradict each other and it proves the theorem.

5 Conclusion

We present in this work the general scheme for modification of separating plane algorithms which provides additional possibilities for improving relaxational properties of algorithms of nonsmooth optimization. It is based on imposing additional cuts in the dual space of conjugate variables which restrict the test area and may additionally localize the extremum. The scheme allows also more sophisticated low-dimensional local search procedures to be applied on each iteration to speed up convergency.

References

1. Shor, N.Z.: Primenenije metoda gradientnogo spuska dlya reshenija setevoj transportnoj zadachi. Materialy nauchn. seminara po teor. i prikl. vopr. kibernetiki i issledovanija operacij, pp. 9–17 (1962)
2. Polyak, B.T.: Minimization of unsmooth functionals. USSR Comput. Math. Math. Phys. **9**, 14–29 (1969)
3. Kelley, J.E.: The cutting-plane method for solving convex programs. J. Soc. Ind. Appl. Math. **8**(4), 703–712 (1960)
4. Wolfe, P.: A method of conjugate subgradients for minimizing nondifferentiable functions. Math. Programm. Study **3**, 145–173 (1975)
5. Shor, N.Z.: Utilization of the operation of space dilatation in the minimization of convex functions. Cybernetics **6**, 7–15 (1970)
6. Shor, N.Z., Zhurbenko, N.G.: The minimization method using space dilatation in direction of difference of two sequential gradients. Kibernetika **7**(3), 51–59 (1971)
7. Shor, N.Z., Zhurbenko, N.G., Likhovid, A.P., Stetsyuk, P.I.: Algorithms of nondifferentiable optimization: development and application. Cybern. Syst. Anal. **39**(4), 537–548 (2003)
8. Kiwiel, K.C.: An algorithm for nonsmooth convex minimization with errors. Math. Comput. **45**(171), 173–180 (1985)
9. Rzhevskiy, S.V.: ϵ-subgradient method for the solution of a convex programming problem. USSR Comput. Math. Math. Phys. **21**(5), 51–57 (1981)
10. Mifflin, R., Sagastizábal, C.A.: A VU-algorithm for convex minimization. Math. Program. **104**, 583–606 (2005)
11. Levin, A.Yu.: Ob odnom algoritme miniminzacii vypuklykh funkcij. DAN SSSR **160**(6), 1244–1247 (1965)
12. Newman, D.J.: Location of the maximum on unimodal surfaces. J. Assoc. Comput. Mach. **12**(3), 395–398 (1965)

13. Lemarechal, C.: An extension of Davidon methods to non-differentiable problems. Math. Program. Study **3**, 95–109 (1975)
14. Nurminski, E.A.: Separating plane algorithms for convex optimization. Math. Program. **76**, 375–391 (1997)
15. Vorontsova, E.A.: A projective separating plane method with additional clipping for non-smooth optimization. WSEAS Trans. Math. **13**, 115–121 (2014)
16. Vorontsova, E.A., Nurminski, E.A.: Synthesis of cutting and separating planes in a nonsmooth optimization method. Cybern. Syst. Anal. **51**(4), 619–631 (2015)
17. Vorontsova, E.A.: Extended separating plane algorithm and NSO-solutions of PageRank problem. In: Kochetov, Y., Khachay, M., Beresnev, V., Nurminski, E., Pardalos, P. (eds.) DOOR 2016. LNCS vol. 9869, pp. 547–560. Springer, Heidelberg (2016)

On the Parameter Control of the Residual Method for the Correction of Improper Problems of Convex Programming

Vladimir D. Skarin[1,2(✉)]

[1] Krasovskii Institute of Mathematics and Mechanics UB RAS, Ekaterinburg, Russia
skavd@imm.uran.ru
[2] Ural Federal University, Ekaterinburg, Russia

Abstract. The residual method which is one of the standard regularization procedures for ill-posed optimization problems is applied to an improper convex programming problem. A typical problem for the residual method is reduced to the minimization problem for the quadratic penalty function. For this approach, we establish convergence conditions and estimates for the approximation accuracy. Further, here we present an algorithm for the practical realization of the proposed method.

Keywords: Convex programming · Improper problems · Optimal correction · Residual method · Penalty function

1 Introduction

In this paper, we proceed the research from [1], which is connected with an application of the residual method [2] which is one of the standard regularization procedures for ill-posed optimization problems for the correction of improper problems [3] of convex programming.

Models with inconsistent constraints form a very important class of improper problems of linear and convex programming, they often arise in mathematical modelling of complex real-life operation research problems.

Reasons for the appearance of inconsistencies could be the approximate character of input data, the lack of necessary resources or excessive requirements imposed on quality of the solution.

Due to the frequency of occurrence of improper problems it becomes important to develop a theory and methods for their numerical approximation (correction), i.e. objective procedures for the "resolution" of conflicting constraints, transformation of an improper model into a set of solvable problems and choice of an optimal correction among them. The correction procedure should be objective in the sense of requiring no preliminary information on the consistency of the constraints system in the original problem and producing the solution of the original problem in the case of its feasibility.

© Springer International Publishing Switzerland 2016
Y. Kochetov et al. (Eds.): DOOR 2016, LNCS 9869, pp. 441–451, 2016.
DOI: 10.1007/978-3-319-44914-2_35

Research on the theory of improper problems and the construction of effective procedures for their correction retain their importance in modern mathematical programming. Along with [3], it could be important to notice the following papers [4–9]. Further, we note that there exist the close connection between the improper problem of mathematical programming and the problem for the treatment of experimental data with the aid of so-called Total Least Squares (TLS) [10, 11]. The method TLS is studied active in last time. It can be considered as a way of matrix correction with respect to the Euclidean norm for the infeasible systems of linear equations and inequalities [12, 13]. We note also the widespread regularization algorithms based on TLS [14, 15].

Since problems with inconsistent constraints can arise due to the approximate character of input data, which is connected with issues of stability of the solution, such problems are of interest for the theory and methods of ill-posed optimization problems [16–18]. This is the reason why it makes sense to consider the standard means of regularization of ill-posed models such as the residual method for the analysis of improper problems.

In this paper, we study the possibility of applying the residual method to correct improper convex programming problems. In the case of constraints inconsistence, for the typical scheme of residual method, we register the restrictions of this problem by the quadratic penalty function. The estimates characterising the convergence of the penalty function minimizer to appropriate approximate solutions of the improper problem are given. Besides, we propose an iterative algorithm for realization of this approach.

2   Problem Statement and the Residual Method

Consider the convex programming problem

$$\min\{f_0(x) \mid x \in X\}, \tag{1}$$

where $X = \{x \mid f(x) \leq 0\}$, $f(x) = [f_1(x), \ldots, f_m(x)]$, the functions $f_i(x)$ are convex in $\mathbb{R}^n$ for $i = 0, 1, \ldots, m$.

The residual method applied to the regularization of a feasible convex programming problem of form (1) consists [2] in solving a sequence of problems that depend on some numerical parameter δ:

$$\min\{\|x\|^2 \mid x \in X \cap M_\delta\}, \tag{2}$$

here $M_\delta = \{x \mid f_0(x) \leq \delta\}$, $\delta \geq \bar{f}$, where $\bar{f}$ is the optimal value of problem (1). In this case, problem (2) has a unique solution $\bar{x}_\delta$ for any δ. Since $M_{\delta_1} \supset M_{\delta_2} \supset \ldots \supset M_{\bar{f}}$ for $\delta_1 \geq \delta_2$, we have $\|\bar{x}_{\delta_1}\| \leq \|\bar{x}_{\delta_2}\| \leq \ldots \leq \|\bar{x}_0\|$, where $\bar{x}_0$ is the solution of (1) with minimal norm (so-called the normal solution), $\|\cdot\|$ denotes the Euclidean norm of a vector.

Thus, all the points $\bar{x}_\delta$ lie in the compact set $\{x \mid \|x\| \leq \|\bar{x}_0\|\}$. There exists a limit point $\tilde{x}$ of the sequence $\{\bar{x}_\delta\}$ as $\delta \to \bar{f}$, $\tilde{x} \in X$, $f_0(\tilde{x}) = \bar{f}$ and $\|\tilde{x}\| \leq \|\bar{x}_0\|$. It follows from the uniqueness of $\bar{x}_0$ that $\tilde{x} = \bar{x}_0$ and $\lim_{\delta \to \bar{f}} \bar{x}_\delta = \bar{x}_0$.

Define $\Lambda = \{\lambda \in \mathbb{R}_+^m \mid \inf_x L(x,\lambda) > -\infty\}$, where $L(x,\lambda) = f_0(x) + (\lambda, f(x))$ is the Lagrange function corresponding to problem (1), $x \in \mathbb{R}^n$, $\lambda \in \mathbb{R}_+^m$.

In a problem with inconsistent constraints, $X = \emptyset$. If, in addition, $\Lambda \neq \emptyset$, then (1) is called [3] *an improper convex programming problem of the first kind.*

Problems of this type are the most frequently encountered in practice. They are characterized by the property that if the set X is replaced by a set $X_\xi = \{x \mid f(x) \leq \xi\}$, $\xi \in \mathbb{R}_+^m$, such that $X_\xi \neq \emptyset$, then $\inf\{f_0(x) \mid x \in X_\xi\} > -\infty$.

Let $E = \{\xi \in \mathbb{R}_+^m : X_\xi \neq \emptyset\}$ and $\bar{\xi} = \arg\min\{\|\xi\| \mid \xi \in E\}$. Along with (1) we consider the problem

$$\min\{f_0(x) \mid x \in X_{\bar{\xi}}\}. \tag{3}$$

If $X \neq \emptyset$ in problem (1), then we have $\bar{\xi} = 0$ and problems (1) and (3) coincide. Otherwise, (3) is an example of possible correction for improper problem (1), and we may accept the solution of (3) as a generalized (approximative) solution of improper problem (1).

Assume that the set X_ξ is nonempty and bounded for some $\xi = \xi_0$. Then, E is convex and closed, and that guarantees the existence and uniqueness of the vector $\bar{\xi}$. In this case, it is easy to see that $\bar{\xi} = f^+(\bar{x})$, where $\bar{x} \in \bar{X} = \operatorname{Arg\,min} g(x)$, $g(x) = \|f^+(x)\|^2$, and $\bar{X} = X_{\bar{\xi}}$. Note that $X_{\bar{\xi}}$ is bounded and problem (3) is feasible.

3 Treatment of the Constraints by Means of a Quadratic Penalty Function

In the study of regularization methods as applied to ill-posed problems of constrained optimization, the constraints of a model are typically taken into account by means of a certain penalty function. One of the widespread modifications of the penalty function method is the quadratic penalty method (see, e.g. [19–21]). On the one hand, this method provides only an asymptotic equivalence between the original problem and the problem with penalty as the penalty parameter tends to infinity. On the other hand, the quadratic penalty function is smooth enough to make it attractive from the point of view of numerical minimization.

To problem (2), we assign the problem of finding

$$\min_x F_\delta(x,r), \tag{4}$$

where $F_\delta(x,r) = \|x\|^2 + \rho\|f^+(x)\|^2 + \rho_0(f_0(x) - \delta)^{+^2}$, $r = [\rho, \rho_0] > 0$, $\delta \in \mathbb{R}^1$.

The function $F_\delta(x,r)$ is strongly convex with respect to $x \in \mathbb{R}^n$. Hence problem (4) has a unique solution $\bar{x}_{r,\delta}$ for every $r \in \mathbb{R}^2$, $r > 0$ and $\delta \in \mathbb{R}^1$, including the case $X = \emptyset$, in contrast with problem (2). Therefore, the function $F_\delta(x,r)$ can be used for the analysis of improper convex programming problems.

Suppose that continuous functions $f_i^\varepsilon(x)$, defined on $\mathbb{R}^n$, such that

$$|f_i(x) - f_i^\varepsilon(x)| < \varepsilon \qquad (\forall\, x \in \mathbb{R}^n,\ i = 0, 1, \ldots, m), \qquad \varepsilon > 0,$$

are known in problems (1) and (2) instead of the functions $f_i(x)$. Then, we have the following problem instead of (4)

$$\min_x F_\delta^\varepsilon(x, r),$$ (5)

where $F_\delta^\varepsilon(x, r)$ is obtained from (4) by change of the functions $f_i(x)$ for $f_i^\varepsilon(x)$, i.e. $F_\delta^\varepsilon(x, r) = \|x\|^2 + \rho\|f^{\varepsilon^+}(x)\|^2 + \rho_0(f_0^\varepsilon(x) - \delta)^{+^2}$, $r = [\rho, \rho_0] > 0$, $\varepsilon, \delta \in \mathbb{R}^1$, $\varepsilon > 0$.

Problem (5) is solvable for any r, ε, δ (see Lemma 1 in [9]). Let $\bar{x}_{r,\delta}^\varepsilon$ be a solution of problem (5).

Let us investigate the connection between problems (3) and (5).

Theorem 1. *Suppose that (1) is an improper convex programming problem of the first kind, $\bar{x}$ is an optimal point of problem (3), $\bar{f} = f_0(\bar{x})$ and $\Delta = \bar{f} - \delta$. Then, for any $r = [\rho, \rho_0] > 0$, $\delta \in \mathbb{R}^1$, $0 < \varepsilon \leq 1$ the following estimates are valid:*

$$\|(f^\varepsilon(\bar{x}_{r,\delta}^\varepsilon) - \bar{\xi})^+\| \leq \frac{1}{\sqrt{\rho}} B(r, \delta, \varepsilon),$$ (6)

$$f_0^\varepsilon(\bar{x}_{r,\delta}^\varepsilon) \leq \bar{f} + \frac{1}{\sqrt{\rho_0}} B(r, \delta, \varepsilon) - \Delta,$$ (7)

$$\|\bar{x}_{r,\delta}^\varepsilon\| \leq B(r, \delta, \varepsilon),$$ (8)

where $B(r, \delta, \varepsilon) = \left[\|\bar{x}\|^2 + \varepsilon\rho(4\|\bar{\xi}\|_1 + m\varepsilon) + \rho_0(\Delta^+ + \varepsilon)^2\right]^{1/2}$.

Proof. By the definition of the point $\bar{x}_{r,\delta}^\varepsilon$, we have

$$\|\bar{x}_{r,\delta}^\varepsilon\|^2 + \rho\|f^{\varepsilon^+}(\bar{x}_{r,\delta}^\varepsilon)\|^2 + \rho_0(f_0^\varepsilon(\bar{x}_{r,\delta}^\varepsilon) - \delta)^{+^2} \leq \|\bar{x}\|^2 + \rho\|f^{\varepsilon^+}(\bar{x})\|^2 + \rho_0(f_0^\varepsilon(\bar{x}) - \delta)^{+^2}.$$

Since

$$|f_i^{\varepsilon^+}(\bar{x}) - f_i^+(\bar{x})| \leq |f_i^\varepsilon(\bar{x}) - f_i(\bar{x})| < \varepsilon, \qquad i = 1, \ldots, m,$$

$$|(f_0^\varepsilon(\bar{x}) - \delta)^+ - (f_0(\bar{x}) - \delta)^+| \leq |f_0^\varepsilon(\bar{x}) - f_0(\bar{x})| < \varepsilon,$$

we have

$$\|\bar{x}_{r,\delta}^\varepsilon\|^2 + \rho\|f^{\varepsilon^+}(\bar{x}_{r,\delta}^\varepsilon)\|^2 + \rho_0(f_0^\varepsilon(\bar{x}_{r,\delta}^\varepsilon) - \delta)^{+^2}$$

$$\leq \|\bar{x}\|^2 + \rho\|\bar{\xi}\|^2 + \rho\varepsilon(2\|\bar{\xi}\|_1 + m\varepsilon) + \rho_0(\Delta^+ + \varepsilon)^2.$$ (9)

Let us estimate the difference $\|\bar{\xi}\|^2 - \|f^{\varepsilon^+}(\bar{x}_{r,\delta}^\varepsilon)\|^2$. Since $\bar{x} \in \bar{X}$, we have $0 \in \partial g(\bar{x})$, where $\partial g(\bar{x})$ denotes the sub-differential of $g(x)$ at the point $\bar{x}$. Choose $\bar{e}_i(\bar{x}) \in \partial f_i^+(\bar{x})$ for $i = 1, \ldots, m$ such that

$$0 = 2\sum_{i=1}^m f_i^+(\bar{x})\bar{e}_i(\bar{x}) = 2\sum_{i=1}^m \bar{\xi}_i\bar{e}_i(\bar{x}),$$

here $\bar{\xi}_i$ is the i-th component of the vector $\bar{\xi}$, $i = 1, \ldots, m$. From the convexity of the functions $f_i(x)$, $i = 1, \ldots, m$, we obtain

$$\sum_{i=1}^{m} \bar{\xi}_i f_i(\bar{x}) \leq \sum_{i=1}^{m} \bar{\xi}_i f_i(x) + \sum_{i=1}^{m} \bar{\xi}_i (\bar{e}_i(\bar{x}), \bar{x} - x),$$

$$\|\bar{\xi}\|^2 = (\bar{\xi}, f(\bar{x})) \leq (\bar{\xi}, f(x)) \qquad (\forall x \in \mathbb{R}^n). \tag{10}$$

Further, we apply the immediately verified inequality $(a - b)^{+^2} \leq (a^+ - b)^2$, which holds for all real numbers a and b. Applying it and inequality (10), we find for any $x' \in \mathbb{R}^n$ that

$$\|(f^\varepsilon(x') - \bar{\xi})^+\|^2 \leq \|(f^{\varepsilon^+}(x') - \bar{\xi}\|^2 = \|(f^{\varepsilon^+}(x')\|^2 - 2(\bar{\xi}, f^{\varepsilon^+}(x')) + \|\bar{\xi}\|^2$$

$$< \|f^{\varepsilon^+}(x')\|^2 - 2(\bar{\xi}, f^+(x')) + 2\varepsilon\|\bar{\xi}\|_1 + \|\bar{\xi}\|^2$$

$$\leq \|f^{\varepsilon^+}(x')\|^2 - \|\bar{\xi}\|^2 + 2\varepsilon\|\bar{\xi}\|_1. \tag{11}$$

Hence, for $x' = \bar{x}$ we have

$$\|(f^\varepsilon(\bar{x}) - \bar{\xi})^+\|^2 - 2\varepsilon\|\bar{\xi}\|_1 \leq \|f^{\varepsilon^+}(\bar{x})\|^2 - \|\bar{\xi}\|^2, \tag{12}$$

$$\|\bar{\xi}\|^2 - \|f^{\varepsilon^+}(\bar{x})\|^2 \leq 2\varepsilon\|\bar{\xi}\|_1. \tag{13}$$

Combining estimates (12), (13), with (9) we obtain (6) and (8). Inequality (9) implies that $\rho_0(f_0^\varepsilon(\bar{x}_{r,\delta}^\varepsilon) - \delta)^{+^2} \leq B(r, \delta, \varepsilon)$. Since $f_0^\varepsilon(\bar{x}_{r,\delta}^\varepsilon) - \bar{f} \leq (f_0^\varepsilon(\bar{x}_{r,\delta}^\varepsilon) - \delta)^+ - \Delta$, we get (7).

The theorem is proved.

Corollary 1. *Suppose that, in problem (5) $\rho \to \infty$, $\rho_0 \to \infty$, $\varepsilon \to 0$, $\Delta \to 0$, $\varepsilon\rho \to 0$, $\rho_0(\Delta^+ + \varepsilon)^2 \to 0$. Then $\bar{x}_{r,\delta}^\varepsilon \to \bar{x}_0$, where $\bar{x}_0$ is the normal solution of problem (3).*

Indeed, suppose that $\bar{x} = \bar{x}_0$. In view of inequality (8) and under conditions on the parameters r, δ, ε formulated above, we have $\lim B(r, \delta, \varepsilon) = \|\bar{x}_0\|$. This implies the boundedness of the sequence $\{\bar{x}_{r,\delta}^\varepsilon\}$. Denote by $\tilde{x}$ the limit point of $\{\bar{x}_{r,\delta}^\varepsilon\}$. By (6)–(8), we obtain $\tilde{x} \in X_{\bar{\xi}} \cap M_{\bar{f}}$, $\|\tilde{x}\| = \|\bar{x}_0\|$. Since problem (3) has the unique normal solution $\bar{x}_0$, we have $\tilde{x} = \bar{x}_0$ and $\lim \bar{x}_{r,\delta}^\varepsilon = \bar{x}_0$.

Corollary 2. *Suppose that there exists a saddle point $[\bar{x}_0, \bar{\lambda}]$ of the function $L_{\bar{\xi}}(x, \lambda) = f_0(x) + (\lambda, f(x) - \bar{\xi})$ in the domain $\mathbb{R}^n \times \mathbb{R}_+^m$. Then, together with (7), the following inequality holds*

$$\bar{f} - f_0^\varepsilon(\bar{x}_{r,\delta}^\varepsilon) \leq \frac{\|\bar{\lambda}\|}{\sqrt{\rho}} B(r, \delta, \varepsilon) + \varepsilon(1 + \|\bar{\lambda}\|_1). \tag{14}$$

From the definition of the saddle point $[\bar{x}_0, \bar{\lambda}]$, we have for any $x \in \mathbb{R}^n$ the relation

$$\bar{f} = f_0(\bar{x}_0) \leq f_0(x) + (\bar{\lambda}, f(x) - \bar{\xi}) < f_0^\varepsilon(x) + (\bar{\lambda}, (f^\varepsilon(x) - \bar{\xi})^+) + \varepsilon(1 + \|\bar{\lambda}\|_1). \tag{15}$$

Hence, in view of (6), we obtain (14).

4   An Iterative Algorithm for the Correction of the Improper Convex Programming Problem

The main difficulty by the practical application of the residual method is connected with the necessity to implement the condition $\delta \to \bar{f}$. Estimates (6)–(8) show that this problem arise usually for the case when $\Delta > 0$, i.e. when $\delta < \bar{f}$. Note that problem (2) is improper, although problem (1) may be solvable. In this case there is the connection between method (5) and the parameter-free penalty function methods [20] (methods of objective function parametrization [2,22]). In this methods, we assign to (1) the problem of finding

$$\inf_{x \in \mathbb{R}^n} \varphi(x, \delta), \tag{16}$$

where $\varphi(x, \delta) = \rho_0[(f_0(x) - \delta)^+]^{p_0} + \rho \sum_{i=1}^{m} [f_i^+]^{p_i}(x)$, $\delta \in \mathbb{R}^1$, $\delta < \bar{f}$. Above, p_i, ρ_0 and ρ are parameters of the method, $p_i \geq 1$, $i = 0, 1, \ldots, m$; $\rho_0 > 0$, $\rho > 0$.

   The advantage of method (16) is that the parameters ρ_0 and ρ must not be tended to $+\infty$. It suffices to generate the bounded sequence of values δ, which converges to the optimal value of problem (1). In addition, the points $x_\delta = \arg\min_x \varphi(x, \delta)$ converge to a solution of (1).

   Below, we propose an algorithm for the control of the parameters in problem (5) (first at all of the parameter δ), that warrants the necessary convergence. On the one hand, this algorithm presents an iterative scheme of the residual method for the improper convex programming problem. On the other hand, this method is a regularized variant of the parameter-free penalty function method (see also [23]).

   Consider the sequence of positive numbers τ_k that $\tau_k \searrow 0$ $(k \to \infty)$. Select the initial value δ_0 of the parameter δ such that $\delta_0 < \bar{f} - \tau_0$. Assume that $\delta_k < \bar{f} - \tau_k$. Define

$$\delta_{k+1} = \delta_k + (f_0^{\varepsilon_k}(\bar{x}_k) - \delta_k - \tau_k)^+, \qquad k = 0, 1, \ldots, \tag{17}$$

where $\bar{x}_k = \arg\min F_k(x)$, $F_k(x) = \widetilde{F}_{\delta_k}^{\varepsilon_k}(x, r_k, \tau_k) = \|x\|^2 + \rho_k\|f^{\varepsilon_k^+}(x)\|^2 + \rho_k^0(f_0^{\varepsilon_k}(x) - \delta_k - \tau_k)^+$, $r_k = [\rho_k, \rho_k^0] > 0$, $\varepsilon_k \geq 0$.

Theorem 2. *Suppose that the conditions of Theorem 1 are satisfied and there exists a saddle point $[\bar{x}_0, \bar{\lambda}]$ of the function $L_{\bar{\xi}}(x, \lambda)$ in the domain $\mathbb{R}^n \times \mathbb{R}_+^m$. Assume that parameters τ_k, ρ_k, ρ_k^0, ε_k are chosen such that*

$$\tau_k \searrow 0, \qquad \rho_k \nearrow +\infty, \qquad \rho_k^0 \nearrow +\infty, \qquad \frac{\rho_k^0}{\rho_k} \searrow 0, \qquad \varepsilon_k \searrow 0,$$

$$\rho_k \varepsilon_k \searrow 0, \qquad \rho_k^0(\tau_k - \tau_{k+1} - \varepsilon_k) \nearrow +\infty \qquad (k \to \infty). \tag{18}$$

Then $\lim_{k \to \infty} \delta_k = \bar{f}$ in method (17) and any limit point of the sequence $\{\bar{x}_k\}$ solves problem (3).

   Moreover, if additional condition $\lim_{k \to \infty} \dfrac{\rho_k^{0^2}}{\rho_k} = 0$ holds, then $\lim_{k \to \infty} \bar{x}_k = \bar{x}_0$, where $\bar{x}_0$ is the normal solution of problem (3).

Proof. By definition of the point $\bar{x}_k$, it holds that

$$\|\bar{x}_k\|^2 + \rho_k\|f^{\varepsilon_k^+}(\bar{x}_k)\|^2 + \rho_k^0(f_0^{\varepsilon_k}(\bar{x}_k) - \delta_k - \tau_k)^+$$
$$\leq \|\bar{x}_0\|^2 + \rho_k(\|\bar{\xi}\|^2 + 2\varepsilon_k\|\bar{\xi}\|_1 + m\varepsilon_k^2) + \rho_k^0(\bar{f} - \delta_k - \tau_k) + \rho_k^0\varepsilon_k. \tag{19}$$

From (11) for $x' = \bar{x}_k$, we obtain

$$\|\bar{\xi}\|^2 - \|f^{\varepsilon_k^+}(\bar{x}_k)\|^2 \leq 2\varepsilon_k\|\bar{\xi}\|_1.$$

Hence, it follows from (19) that

$$(f_0^{\varepsilon_k}(\bar{x}_k) - \delta_k - \tau_k)^+ \leq \frac{\|\bar{x}_0\|^2}{\rho_k^0} + (\bar{f} - \delta_k - \tau_k) + \varepsilon_k\left[1 + (m\varepsilon_k + 4\|\bar{\xi}\|_1)\frac{\rho_k}{\rho_k^0}\right].$$

Since $\rho_k\varepsilon_k \to 0$ as $k \to \infty$, we find such a number k_0, that $(m\varepsilon_k + 4\|\bar{\xi}\|_1)\rho_k\varepsilon_k \leq 1$ for $k \geq k_0$. Thus,

$$\delta_k \leq \delta_{k+1} = \delta_k + (f_0^{\varepsilon_k}(\bar{x}_k) - \delta_k - \tau_k)^+ \leq \frac{\|\bar{x}_0\|^2 + 1}{\rho_k^0} + \bar{f} - \tau_k + \varepsilon_k. \tag{20}$$

It follows from conditions (18) that the number $k_1 \geq k_0$ exists such that $\rho_k^0(\tau_k - \tau_{k+1} - \varepsilon_k) \geq \|\bar{x}_0\|^2 + 1$ for all $k \geq k_1$.

Hence, $\delta_k \leq \delta_{k+1} < \bar{f} - \tau_{k+1}$ for all sufficiently large k. This relation implies the existence of limit $\lim_{k \to \infty} \delta_k = \bar{\delta}$, where $\bar{\delta} \leq \bar{f}$. Simultaneously, the limit relation

$$\overline{\lim_{k \to \infty}}(f_0^{\varepsilon_k}(\bar{x}_k) - \delta_k - \tau_k)^+ = 0 \tag{21}$$

is fulfilled. From (19), we deduce the inequality

$$\rho_k(\|f^{\varepsilon_k^+}(\bar{x}_k)\|^2 - \|\bar{\xi}\|^2)$$
$$\leq \rho_k^0(\bar{f} - f_0^{\varepsilon_k}(\bar{x}_k)) + \|\bar{x}_0\|^2 + \rho_k\varepsilon_k(2\|\bar{\xi}\|_1 + m\varepsilon_k) + \rho_k^0\varepsilon_k. \tag{22}$$

Since $[\bar{x}_0, \bar{\lambda}]$ is a saddle point of the function $L_{\bar{\xi}}(x, \lambda)$, we have in view of analogy to (15)

$$\bar{f} - f_0^{\varepsilon_k}(\bar{x}_k) \leq \|\bar{\lambda}\|\|(f^{\varepsilon_k}(\bar{x}_k) - \bar{\xi})^+\| + \varepsilon_k(1 + \|\bar{\lambda}\|_1). \tag{23}$$

Applying this inequality and estimate (11) for $x' = \bar{x}_k$, we obtain from (22) that

$$\rho_k\|(f^{\varepsilon_k}(\bar{x}_k) - \bar{\xi})^+\|^2 - \rho_k^0\|\bar{\lambda}\|\|(f^{\varepsilon_k}(\bar{x}_k) - \bar{\xi})^+\|$$
$$\leq \|\bar{x}_0\|^2 + \rho_k\varepsilon_k(4\|\bar{\xi}\|_1 + m\varepsilon_k) + \rho_k^0\varepsilon_k(2 + \|\bar{\lambda}\|_1).$$

Reforming the left part of this relation, we find that

$$\|(f^{\varepsilon_k}(\bar{x}_k) - \bar{\xi})^+\| \leq \frac{\rho_k^0}{2\rho_k}\|\bar{\lambda}\| + \sqrt{B_1(r_k, \varepsilon_k)}, \tag{24}$$

where $B_1(r_k, \varepsilon_k) = \dfrac{\rho_k^{0\,2}}{4\rho_k^2}\|\bar\lambda\|^2 + \dfrac{\|\bar x_0\|^2}{\rho_k} + \varepsilon_k(4\|\bar\xi\|_1 + m\varepsilon_k) + \dfrac{\rho_k^0}{\rho_k}\varepsilon_k(2 + \|\bar\lambda\|_1)$. It follows from (24) that

$$f_i(\bar x_k) \le \bar\xi_i + \frac{\rho_k^0}{2\rho_k}\|\bar\lambda\| + \sqrt{B_1(r_k, \varepsilon_k)} + \varepsilon_k, \qquad i = 1, \ldots, m. \qquad (25)$$

If conditions (18) are fulfilled, then $\lim\limits_{k\to\infty} B_1(r_k, \varepsilon_k) = 0$ is valid. Therefore, there is the number $k_2 \ge k_1$ such that all points $\bar x_k$ lie in a set $X_{\xi'}$, where $\xi' > \bar\xi$, $k \ge k_2$. From the assumption for X_{ξ_0} the set $X_{\xi'}$ is bounded. This implies the boundedness of the sequence $\{\bar x_k\}$. Let us denote by x^* any limit point of $\{\bar x_k\}$. Then, because of (25) we obtain $x^* \in X_{\bar\xi}$ and $f_0(x^*) \ge \bar f$. On the other hand, it follows from (21) that $f_0(x^*) \le \bar\delta \le \bar f$. Thus, $\lim\limits_{k\to\infty} \delta_k = \bar\delta = \bar f = f_0(x^*)$.

Let us supplement conditions (18) such that $\dfrac{\rho_k^{0\,2}}{\rho_k} \to 0$ as $k \to \infty$. Then the sequence $\{\bar x_k\}$ will be converge to the normal solution of problem (3). Indeed, it follows from (19) that

$$\|\bar x_k\|^2 \le \|\bar x_0\|^2 + \rho_k(\|\bar\xi\|^2 - \|f^{\varepsilon_k^+}(\bar x_k)\|^2) + \rho_k^0(\bar f - f_0^{\varepsilon_k}(\bar x_k))$$
$$+ 2\rho_k\varepsilon_k\|\bar\xi\|_1 + \rho_k\varepsilon_k^2 m + \rho_k^0\varepsilon_k.$$

Applying here estimates (11) (for $x' = \bar x_k$) and (23), we obtain that

$$\|\bar x_k\|^2 \le \|\bar x_0\|^2 - \rho_k\|(f^{\varepsilon_k}(\bar x_k) - \bar\xi)^+\|^2 + \rho_k^0\|\bar\lambda\|\|(f^{\varepsilon_k}(\bar x_k) - \bar\xi)^+\| + B_2(r_k, \varepsilon_k),$$

where $B_2(r_k, \varepsilon_k) = \rho_k\varepsilon_k(4\|\bar\xi\|_1 + m\varepsilon_k) + \rho_k^0\varepsilon_k(\|\bar\lambda\|_1 + 2)$. Thus, $\|\bar x_k\|^2 \le \|\bar x_0\|^2 + \dfrac{\rho_k^{0\,2}}{4\rho_k}\|\bar\lambda\|^2 + B_2(r_k, \varepsilon_k)$. Since from (18) $\lim\limits_{k\to\infty}\dfrac{\rho_k^0}{\rho_k} = \lim\limits_{k\to\infty}\rho_k\varepsilon_k = 0$, we have $\lim\limits_{k\to\infty}\rho_k^0\varepsilon_k = 0$. Therefore, $\lim\limits_{k\to\infty}B_2(r_k, \varepsilon_k) = 0$. The last inequality implies that $\|x^*\| = \|\bar x_0\|$. Consequently, by the uniqueness of the normal solution of problem (3) we obtain $\lim\limits_{k\to\infty}\bar x_k = \bar x_0$.

The theorem is proved.

Let us estimate the values $|\delta_k - \bar f|$ and $|f_0^{\varepsilon_k}(\bar x_k) - \bar f|$ for algorithm (17).

Theorem 3. *Suppose that the conditions of Theorem 2 are satisfied. There exists the number K such that for $k \ge K$ the sequences δ_k and $\bar x_k$ from algorithm (17) satisfy the inequalities*

$$|\delta_{k+1} - \bar f| \le B_3(r_k, \varepsilon_k) + \tau_k, \qquad (26)$$

$$|f_0^{\varepsilon_k}(\bar x_k) - \bar f| \le \max\{B_3(r_k, \varepsilon_k), \tau_k - \tau_{k+1}\}, \qquad (27)$$

where $B_3(r_k, \varepsilon_k) = \|\bar\lambda\|\left(\dfrac{\rho_k^0}{2\rho_k}\|\bar\lambda\| + \sqrt{B_1(r_k, \varepsilon_k)}\right) + \varepsilon_k(1 + \|\bar\lambda\|_1)$, $B_1(r_k, \varepsilon_k)$ *is from (24), $B_3(r_k, \varepsilon_k) \to 0$ $(k \to \infty)$.*

Proof. From (17) we have

$$\delta_{k+1} - \bar{f} = \delta_k + (f_0^{\varepsilon_k}(\bar{x}_k) - \delta_k - \tau_k)^+ - \bar{f} \geq f_0^{\varepsilon_k}(\bar{x}_k) - \bar{f} - \tau_k.$$

Hence, in view of (23) and (24) we find $\bar{f} - \delta_{k+1} \leq \bar{f} - f_0^{\varepsilon_k}(\bar{x}_k) + \tau_k \leq B_3(r_k, \varepsilon_k) + \tau_k$. Since, from (17) $\delta_{k+1} < \bar{f}$, we can write estimate (26).

Further, let us estimate the difference $f_0^{\varepsilon_k}(\bar{x}_k) - \bar{f}$. We apply inequality (20) for $k \geq k_1 = K$. Then $(\|\bar{x}_0\|^2 + 1)(\rho_k^0)^{-1} \leq \tau_k - \tau_{k+1} - \varepsilon_k$ and thus the relation

$$f_0^{\varepsilon_k}(\bar{x}_k) - \tau_k \leq \delta_k + (f_0^{\varepsilon_k}(\bar{x}_k) - \delta_k - \tau_k)^+ \leq \frac{\|\bar{x}_0\|^2 + 1}{\rho_k^0} + \bar{f} - \tau_k + \varepsilon_k \leq \bar{f} - \tau_{k+1}$$

holds. From here and (23), (24) we find estimate (27).

The theorem is proved.

In conclusion, we give an example of the sequences τ_k, ρ_k, ρ_k^0, ε_k, such that satisfy conditions (18).

Let us choose the positive numbers α, β, γ such that $\frac{\alpha}{3} < \gamma < \beta < \alpha \leq \frac{1}{2}$. Define

$$\tau_k = \alpha^{k+1}, \quad \rho_k = \frac{1}{\gamma^{k+1}}, \quad \rho_k^0 = \frac{1}{\beta^{k+1}}, \quad \varepsilon_k = \left(\frac{\alpha}{3}\right)^{k+1}, \quad k = 0, 1, 2, \ldots.$$

Obviously, $\lim_{k \to \infty} \tau_k = \lim_{k \to \infty} \varepsilon_k = 0$, $\lim_{k \to \infty} \rho_k = \lim_{k \to \infty} \rho_k^0 = +\infty$, $\frac{\rho_k^0}{\rho_k} = \left(\frac{\gamma}{\beta}\right)^{k+1} \to 0$, $\rho_k \varepsilon_k = \left(\frac{\alpha}{3\gamma}\right)^{k+1} \to 0$, $\rho_k^0(\tau_k - \tau_{k+1} - \varepsilon_k) = \left(\frac{\alpha}{\beta}\right)^{k+1}\left(1 - \alpha - \frac{1}{3^{k+1}}\right) > \frac{1}{6}\left(\frac{\alpha}{\beta}\right)^{k+1} \to +\infty$ ($k \to \infty$). The numbers α, β and γ may be chosen as $\alpha = 1/2$, $\beta = 1/3$, $\gamma = 1/4$.

The convergence $\{\tilde{x}_k\}$ to $\bar{x}_0$ requires the condition $\lim_{k \to \infty} \frac{\rho_k^{0^2}}{\rho_k} = 0$. For this, we may assume that $\rho_k = \frac{1}{\gamma^{2(k+1)}}$. Then $\frac{\rho_k^{0^2}}{\rho_k} = \left(\frac{\gamma}{\beta}\right)^{2(k+1)} \to 0$ ($k \to \infty$). For the satisfaction of conditions (18), we need also take $\varepsilon_k = \left(\frac{\alpha}{3}\right)^{2(k+1)}$.

5   Conclusion

In the present paper, we offer an approach to the optimal correction of improper programming problem based on the classical residual method for the regularization of ill-posed optimization problems. The inconsistent constraints for the typical scheme of the residual method are eliminated by the regularized quadratic penalty function. The estimates characterising the connection between solutions of the problem with penalty and appropriate problem for approximation of initial improper model are obtained. Finally, we propose an iterative algorithm for the realization of this approach.

Acknowledgements. This work was supported by Russian Science Foundation. Grant N 14–11–00109.

References

1. Skarin, V.D.: On the application of the residual method for the correction of inconsistent problems of convex programming. Proc. Steklov Inst. Math. **289**(Suppl. 1), 182–191 (2015)
2. Vasil'ev, F.P.: Optimization Methods. Faktorial Press, Moscow (2002). (in Russian)
3. Eremin, I.I., Mazurov, Vl.D., Astaf'ev, N.N.: Improper Problems of Linear and Convex Programming. Nauka, Moscow (1983). (in Russian)
4. Eremin, I.I.: Systems of Linear Inequalities and the Linear Optimization. UrO RAN, Ekaterinburg (2007). (in Russian)
5. Mazurov, Vl.D.: Committee Method for Problems of Optimiation and Classification. Nauka, Moscow (1990). (in Russian)
6. Khachay, M.Yu.: On approximate algorithm of a minimal committee of a linear inequalities system. Pattern Recogn. Image Anal. **13**(3), 459–464 (2003)
7. Popov, L.D.: Use of barrier functions for optimal correction of improper problems of linear programming of the 1st kind. Autom. Remote Control **73**(3), 417–424 (2012)
8. Erokhin, V.I., Krasnikov, A.S., Khvostov, M.N.: Matrix corrections minimal with respect to the Euclidean norm for linear programming problems. Autom. Remote Control **73**(3), 219–231 (2012)
9. Skarin, V.D.: On the application of the regularization method for the correction of improper problems of convex programming. Proc. Steklov Inst. Math. **283**(Suppl. 1), 126–138 (2013)
10. De Groen, P.: An introduction to total least squares. Niew Archief voor Wiskunde **14**(2), 237–254 (1996)
11. Rosen, J.B., Park, H., Glick, J.: Total least norm formulation and solution for structured problems. SIAM J. Matrix Anal. Appl. **17**(1), 110–128 (1996)
12. Amaral, P., Barahona, P.: Connections between the total least squares and the correction of an infeasible system of linear inequalities. Linear Algebra Appl. **395**, 191–210 (2005)
13. Dax, A.: The smallest correction of an inconsistent system of linear inequalities. Optim. Eng. **2**, 349–359 (2001)
14. Golub, G.H., Hansen, P.C., O'Leary, D.P.: Tikhonov regularization and total least squares. SIAM J. Matrix Anal. Appl. **21**(1), 185–194 (1999)
15. Renaut, R.A., Guo, H.: Efficient algorithms for solution of regularized total least squares. SIAM J. Matrix Anal. Appl. **26**(2), 457–476 (2005)
16. Tikhonov, A.N., Vasil'ev, F.P.: Methods for solving ill-posed extremal problems. In: Mathematics Models and Numerical Methods, vol. 3, pp. 291–348. Banach Center Publ., Warszava (1978)
17. Bakushinskii, A.B., Goncharskii, A.V.: Ill-posed Problems: Theory and Application. Kluwer Acad. Publ., Dordrecht (1994)
18. Kaltenbacher, B., Neubauer, A., Scherzer, O.: Iterative Regularization Methods in Nonlinear Ill-Posed Problems. W. de Gruyter, New York (2008)
19. Eremin, I.I.: The penalty method in convex programming. Soviet Math. Dokl. **8**, 459–462 (1967)
20. Bourkary, D., Fiacco, A.V.: Survey of penalty, exact-penalty and multiplier methods from 1968 to 1993. Optimization **32**, 301–334 (1995)

21. Auslender, A., Cominetti, R., Haddou, M.: Asymptotic analysis of penalty and barrier methods in convex and linear programming. Math. Oper. Res. **22**(1), 1–18 (1997)
22. Evtushenko, J.G.: Methods for Solving Extremal Problems and its Application in Optimization Systems. Nauka, Moscow (1982). (in Russian)
23. Popov, L.D.: On the adoptation of the least squares method to improper problems of mathematical programming. Trudy Inst. Math. Mech. **19**(2), 247–255 (2013). UrO RAN. (in Russian)

On the Merit and Penalty Functions
for the D.C. Optimization

Alexander S. Strekalovsky[✉]

Institute for System Dynamics and Control Theory SB RAS,
Lermontov street, 134, Irkutsk 664033, Russia
strekal@icc.ru

Abstract. This paper addresses a rather general problem of nonlinear optimization with the inequality constraints and the goal function defined by the (d.c.) functions represented by the difference of two convex functions. In order to reduce the constrained optimization problem to an unconstrained one, we investigate three auxiliary problems with the max-merit, Lagrange and penalty goal functions. Further, their relations to the original problem are estimated by means of the new Global Optimality Conditions and classical Optimization Theory as well as by examples.

Keywords: Nonlinear programming · d.c. functions · Global optimality conditions · The lagrange function · Penalty functions

1 Introduction

It is well-known that the contemporary Optimality Conditions (OC) theory [1–13], a considerable part of which is presented by modern generalizations of the KKT-theorem [1–13], turns out to be ineffective when it comes to a characterization of a global solution to nonconvex problems. Meanwhile, real-life applied problems might have a lot (often a huge number!) of local extrema [15–22].

On the other hand, new attractive and promising areas for investigations in optimization in the 21st century arise from various applications. Among others, let us mention the following problems: the search for equilibrium in competitions; hierarchical optimization problems; dynamical control problems. However, as it has been shown in [17], it turns out that all these new problems are related to nonconvexity. Hence, it becomes obvious that we need new mathematical tools (optimality conditions, numerical procedures etc.) that would allow us to escape stationary or local solutions and construct numerical procedures able to jump out of local pits simultaneously improving the goal functions. The first attempts to propose such an apparatus have been undertaken in [17–22] for special d.c. optimization problems such as d.c. minimization, convex maximization, reverse-convex problems etc [15–22]. To deal with this class of problems, the Global Search Theory [17,21] has been developed, which comprises local search methods [17–19,21] and global search procedures [17,21,22] that employ classical methods.

© Springer International Publishing Switzerland 2016
Y. Kochetov et al. (Eds.): DOOR 2016, LNCS 9869, pp. 452–466, 2016.
DOI: 10.1007/978-3-319-44914-2_36

In this paper, we address more general problems with d.c. functions in inequality constraints and goal functions and investigate the reduction of the constrained problem to three unconstrained ones formed by the max-merit, Lagrange and penalty goal functions. The study is performed, in particular, by means of the new Global Optimality Conditions for three different auxiliary unconstrained problems. After the statement of the problem in Sects. 2 and 3 considers the max-merit function and an auxiliary problem with Example 1 to illustrate inadequacy of the max-merit function to the original Problem.

Further, we study the well-known Lagrange function and show by means of the Global Optimality Conditions (GOC) and examples that $\mathcal{L}(x, \lambda)$ seeks a saddle point (which does not always exist) but not a solution to the original problem.

In Sect. 5, we consider a penalization approach and provide some necessary information. We study the GOC for the penalized problem and use examples to demonstrate that the new tools are effective.

Section 6 provides the conclusion to summarize the content of the paper.

2   Statement of Problem

Let us consider the following problem:

$$(\mathcal{P}): \quad \left. \begin{array}{l} f_0(x) := g_0(x) - h_0(x) \downarrow \min_{x}, \quad x \in S, \\ f_i(x) := g_i(x) - h_i(x) \leq 0, \quad i \in I = \{1, \ldots, m\}; \end{array} \right\} \quad (1)$$

where the functions $g_i(\cdot)$, $h_i(\cdot)$, $i = 0, 1, \ldots, m$, are convex on $I\!R^n$, so that the functions $f_i(\cdot)$, $i = 0, 1, \ldots, m$, are the (d.c.) functions of A.D. Alexandrov represented as a difference of two convex functions [1, 2, 4, 5].

In order to avoid some singularities [1, 4, 5], we assume that

$$S \subset \left[\bigcap_{i=0}^{m} int(dom\ g_i(\cdot)) \right] \cap \left[\bigcap_{i=0}^{m} int(dom\ h_i(\cdot)) \right] \neq \emptyset,$$

where the set $S \subset I\!R^n$ is convex.

Further, let the following assumptions hold:

$$D := \{x \in S \mid f_i(x) \leq 0, \quad i \in I\} \neq \emptyset,$$

$$\mathcal{V}(\mathcal{P}) := \inf(f_0, D) \triangleq \inf_{x}\{f_0(x) \mid x \in S, \ f_i(x) \leq 0, \ i \in I\} > -\infty,$$

$$Sol(\mathcal{P}) := \{z \in D \mid f_0(z) = \mathcal{V}(\mathcal{P})\} \neq \emptyset.$$

3   The Max-Merit Function

Let us consider the following function [5–9]

$$F(x, \eta) := \max\{f_0(x) - \eta; f_1(x), \ldots, f_m(x)\}, \quad (2)$$

where $\eta \in I\!R$. Below, for a feasible (in $(\mathcal{P})$) point $z \in D$, we denote $\zeta := f_0(z)$.

Proposition 1 ([5–9]). *Suppose that a point z is a solution to Problem $(\mathcal{P})$: $z \in Sol(\mathcal{P})$. Then, the point z is a solution to the following auxiliary problem*

$$(\mathcal{P}_\zeta): \qquad F(x, \zeta) \downarrow \min_x, \quad x \in S. \tag{3}$$

Proposition 2 ([5]). *Suppose the parameter η is equal to the optimal value of Problem $(\mathcal{P})$–(1), $\eta = \mathcal{V}(\mathcal{P})$. Then, the point $z \in D$ is a solution to Problem $(\mathcal{P})$ if and only if z is a solution to the auxiliary problem $(\mathcal{P}_\zeta)$ with $\zeta := f_0(z) = \mathcal{V}(\mathcal{P})$. Under latter conditions, the equality $Sol(\mathcal{P}) = Sol(\mathcal{P}_\zeta)$ holds.*

Lemma 1. *Suppose that the point $z \in D$ is not a solution to problem $(\mathcal{P}_\eta)$ with $\eta = \zeta := f_0(z)$, so that there exists a point $u \in S$, such that*

$$F(u, \zeta) < 0 = F(z, \zeta).$$

Then, the point $z \in D$ cannot be a solution to Problem $(\mathcal{P})$: $z \notin Sol(\mathcal{P})$.

Proof. From the inequality $F(u, \zeta) < 0$ it follows that $u \in S$, $f_i(u) < 0$, and $f_0(u) < f_0(z) = \zeta$, so that u is feasible for Problem $(\mathcal{P})$. Hence, $z \notin Sol(\mathcal{P})$. □

Note that it is not difficult to show that the objective function $F(x, \eta)$ of Problem $(\mathcal{P}_\eta)$–(3), given in (2), is a d.c. function.

Remark 1. Let us now pay attention to the fact that Proposition 2 provides the sufficient conditions for $z \in S$ to be a solution to Problem $(\mathcal{P}_\zeta)$, $\zeta = f_0(z)$, but not to the original Problem $(\mathcal{P})$. Only if we added the equality $\zeta = \mathcal{V}(\mathcal{P})$ (see Proposition 2), which is, in particular, rather difficult to verify in the majority of the applied problems, then we would be able to make a conclusion about the global solution property in Problem $(\mathcal{P})$ of the feasible point $z \in D$ under investigation. On the other hand, the equality $f_0(z) = \zeta = \mathcal{V}(\mathcal{P})$ for a feasible point $z \in D$ immediately provides that $z \in Sol(\mathcal{P})$ without any supplementary conditions. So, this condition appears to be incorrect. In order to see the adequateness of $F(x, \eta)$ with respect to Problem $(\mathcal{P})$ let us consider an example. □

Example 1. Consider the problem

$$\left. \begin{array}{l} f_0(x) = \frac{1}{2}(x_1 - 4)^2 + (x_2 + 2)^2 \downarrow \min_x, \\ f_1(x) = (x_1 - 1)^2 - (x_2 + 1)^2 \le 0, \\ f_2(x) = (x_2 - 2)^2 - (x_1 + 2)^2 \le 0. \end{array} \right\} \tag{4}$$

It is easy to see that the point $z_* = (4, -2)^\top$ is the global minimum of the strongly convex function $f_0(\cdot)$ on $\mathbb{R}^2$, and $f_0(z_*) = 0$ provides a lower bound for $\mathcal{V}(4) = \inf(f_0, D) \ge 0$. Note that z_* is unfeasible in (4), since $f_1(z_*) = 8 > 0$. Let us consider another point $z = (\frac{4}{3}, -\frac{2}{3})^\top$ which is feasible for (4), since $f_1(z) = 0$ and $f_2(z) = -4 < 0$. In addition, it can be readily seen that z satisfies the KKT-conditions with $\lambda_0 = 1$, $\lambda_1 = 4 > 0$, $\lambda_2 = 0$, $\zeta := f_0(z) = 5\frac{1}{3}$.

Further, since

$$f_1(x) = (x_1 - x_2 - 2)(x_1 + x_2) \le 0, \quad f_2(x) = (x_2 - x_1 - 4)(x_2 + x_1) \le 0,$$

it can be readily shown that the feasible set $D := \{x \in I\!\!R^2 \mid f_i(x) \leq 0, \ i = 1, 2\}$ is represented by the union of the two convex parts: $D = D_1 \cup D_2$, $D_1 = \{x \mid x_1 + x_2 = 0\}$, $D_2 = \{x \mid x_1 + x_2 \geq 0, \ x_2 - x_1 - 4 \leq 0, \ x_1 - x_2 - 2 \leq 0\}$.

Hence, from the geometric view-point, it is easy to see that the point $z_0 = (\frac{8}{3}, -\frac{8}{3})^\top$ is the global solution to (4) with the optimal value $\mathcal{V}(4) = f_0(z_0) =: \zeta_0 = \frac{4}{3}$. So, the point $z = (\frac{4}{3}, -\frac{2}{3})^\top$ is not to be a global solution to (4), since $f_0(z) = 5\frac{1}{3} = \zeta$. However, the goal function $F(x, \zeta)$ of Problem $(\mathcal{P}_\zeta)$ does not distinguish between these two points. Actually,

$$F(z_0, \zeta) = \max\{f_0(z_0) - \zeta; f_1(z_0); f_2(z_0)\} = 0 =$$
$$\max\{f_0(z) - \zeta; f_1(z); f_2(z)\} = F(z, \zeta),$$

because $z \in D$, $z_0 \in D = \{x \in I\!\!R^2 \mid f_i(x) \leq 0, \ i = 1, 2\}$.

Moreover, for all feasible (in Problem (4)) points, which are better (in the sense of the problem (4)) than the point z, i.e. $u \in \{x \in I\!\!R^2 \mid x_1 + x_2 = 0, f_0(x) < \zeta = 5\frac{1}{3}\}$, we have the same results: $F(u, \zeta) = 0$, because $f_1(u) = 0 = f_2(u)$. For instance, for any point $x(\alpha)$ of the form

$$x(\alpha) = (v(\alpha), -v(\alpha))^\top, \quad v(\alpha) = 1.1\alpha + 4.2(1 - \alpha), \quad \alpha \in [0, 1],$$

we have $F(x(\alpha), \zeta) = 0$. Meanwhile, $f_0(x(\alpha)) < f_0(z) = 5\frac{1}{3} \ \forall \alpha \in [0, 1]$.

Therefore, one can see that there exist a great number of points better than z in the sense of Problem (4). □

So, Example 1 demonstrates that Problem $(\mathcal{P}_\eta)$ is not sufficiently adequate to Problem $(\mathcal{P})$. More precisely, taking into account Propositions 1 and 2, it is easy to see that the set $Sol(\mathcal{P}_\zeta)$ might contain a lot of points not belonging to $Sol(\mathcal{P})$, so that the inclusion $Sol(\mathcal{P}) \subset Sol(\mathcal{P}_\zeta)$ may be really proper. Moreover, the inequality $\zeta > \zeta_* = \mathcal{V}(\mathcal{P})$ holds together with the inclusion $Sol(\mathcal{P}) \subset Sol(\mathcal{P}_\zeta)$, which is inadmissible. Therefore, we move on to another type of the merit (or penalty, in a rough sense) function.

4   The Lagrange Function

Consider the standard (normal) Lagrange function for Problem $(\mathcal{P})$

$$\mathcal{L}(x, \lambda) = f_0(x) + \sum_{i=1}^{m} \lambda_i f_i(x). \tag{5}$$

It is common to call a pair (z, λ) a saddle point of the Lagrange function $\mathcal{L}(x, \lambda)$: $(z, \lambda) \in Sdl(\mathcal{L})$, if the following two inequalities are satisfied [1–9]:

$$\forall \mu \in I\!\!R_+^m: \quad \mathcal{L}(z, \mu) \leq \mathcal{L}(z, \lambda) \leq \mathcal{L}(x, \lambda) \quad \forall x \in S. \tag{6}$$

Lemma 2 ([1,2,4–9]). *For a pair $(z, \lambda) \in S^* \times I\!\!R_+^m$ the following two assertions are equivalent:*

$$(i) \qquad \max_{\mu}\{\mathcal{L}(z, \mu) \mid \mu \in I\!\!R_+^m\} = \mathcal{L}(z, \lambda); \tag{7}$$

$$(ii) \quad z \in D, \quad \lambda \in I\!\!R_+^m, \quad \lambda_i f_i(z) = 0, \quad i = 1, \ldots, m. \tag{8}$$

Recall that a vector $\lambda \in I\!\!R_+^m$ satisfying the KKT-conditions, including (8), is usually called a Lagrange multiplier [1,2,4–9] at a point $z \in D$. The set of all Lagrange multipliers at z will be denoted below by $\mathcal{M}(z)$.

Remember, in addition, that for a convex optimization problem $(\mathcal{P})$–(1), when $h_i(x) \equiv 0 \ \forall i \in \{0\} \cup I$, we have $\mathcal{M}(z_1) = \mathcal{M}(z_2) = \mathcal{M}$, if $z_i \in Sol(\mathcal{P})$, $i = 1, 2$ [5, Chapter VII].

Proposition 3 ([1,2,4,5,7–9]). *If the pair $(z, \lambda) \in S \times I\!\!R_+^m$ is a saddle point of the Lagrange function $\mathcal{L}(x, \mu)$ on the set $S \times I\!\!R_+^m$, then the point z is a global solution to Problem $(\mathcal{P})$.*

In what follows, we will employ this assertion in a different form.

Proposition 4. *Suppose $z \in D$, z is a KKT-point but not a global solution to Problem $(\mathcal{P})$. Then, there exists no Lagrange multiplier $\lambda \in \mathcal{M}(z)$ such that $(z, \lambda) \in Sdl(\mathcal{L})$.*

Further, since $f_i(x) = g_i(x) - h_i(x)$, $i = 0, 1, \ldots, m$, $\mathcal{L}(x, \lambda)$ has a very simple, clear and suitable d.c. representation

$$\left. \begin{aligned} &(a) \qquad\qquad \mathcal{L}(x, \lambda) = G_\lambda(x) - H_\lambda(x), \\ &(b)\ G_\lambda(x) = g_0(x) + \sum_{i=1}^m \lambda_i g_i(x), \quad H_\lambda(x) = h_0(x) + \sum_{i=1}^m \lambda_i h_i(x). \end{aligned} \right\} \tag{9}$$

Taking into account (9), let us look at the normal Lagrange function (5) from the view-point of the Global Optimality Conditions (GOC) [17,18,20,21].

Theorem 1. *Suppose $(z, \lambda) \in Sdl(\mathcal{L})$, $\lambda_0 = 1$, $\zeta := f_0(z)$. Then, $\forall (y, \beta) \in I\!\!R^n \times I\!\!R$ such that*

$$H_\lambda(y) := \sum_{i=0}^m \lambda_i h_i(y) = \beta - \zeta, \tag{10}$$

the following inequality holds

$$G_\lambda(x) - \beta \geq \sum_{i=0}^m \lambda_i \langle h_i'(y), x - y \rangle \quad \forall x \in S, \tag{11}$$

for any subgradients $h_i'(y) \in \partial h_i(y)$ of the functions $h_i(\cdot)$ at the point y, $i \in I \cup \{0\}$.

Proof. According to the assumption, we have the chain

$$\zeta := f_0(z) = \sum_{i=0}^{m} \lambda_i f_i(z) = \mathcal{L}(z, \lambda) \le \sum_{i=0}^{m} \lambda_i f_i(x) = \mathcal{L}(x, \lambda) \quad \forall x \in S,$$

from which, due to (9) and (10), it follows

$$\beta - H_\lambda(y) = \zeta \le \mathcal{L}(x, \lambda) = G_\lambda(x) - H_\lambda(x) \quad \forall x \in S.$$

Further, by the convexity of $H_\lambda(\cdot) = \sum_{i=0}^{m} \lambda_i h_i(\cdot)$, $\lambda_i \ge 0$, $i \in I$, we obtain

$$G_\lambda(x) - \beta \ge H_\lambda(x) - H_\lambda(y) \ge \sum_{i=0}^{m} \lambda_i \langle h_i'(y), x - y \rangle \quad \forall x \in S,$$

which coincides with (11). $\square$

Remark 2. Due to Proposition 3, it can be readily seen that for a global solution $z \in Sol(\mathcal{P})$, for which one can find a multiplier $\lambda \in \mathcal{M}(z)$ such that $(z, \lambda) \in Sdl(\mathcal{L})$, the conditions (10)–(11) turn out to be necessary global optimality conditions.

Remark 3. It is clear that Theorem 1 reduces the nonconvex problem

$$(\mathcal{L}): \qquad \mathcal{L}(x, \lambda) \downarrow \min_x, \quad x \in S,$$

to the verification of the principal inequality (PI) (11) for the family of parameters (y, β): $H_\lambda(y) = \beta - \zeta$, or, more precisely, to solving the family of the convex linearized problems

$$(\mathcal{LL}(y)): \qquad \Phi_\lambda(x) = G_\lambda(x) - \langle H_\lambda'(y), x \rangle \downarrow \min_x, \quad x \in S, \qquad (12)$$

with the subsequent verification of PI (11) with $x = u = u(y, \beta) \in Sol(\mathcal{LL}(y))$.

Remark 4. Furthermore, suppose that there exists a tuple (y, β, u), such that (y, β) satisfies the equality (10) and violates the PI (11), i.e. $G_\lambda(u) - \beta < \langle H_\lambda'(y), u - y \rangle$. Whence, due to convexity of $H_\lambda(\cdot) = \sum_{i=0}^{m} \lambda_i h_i(\cdot)$, it follows

$$G_\lambda(u) - \beta < H_\lambda(u) - H_\lambda(y).$$

Next, on account of (9) and (10), we have

$$\mathcal{L}(u, \lambda) = G_\lambda(u) - H_\lambda(u) < \beta - H_\lambda(y) = \zeta = f_0(z) = \mathcal{L}(z, \lambda),$$

where $\lambda \in \mathcal{M}(z)$. Hence, the right-hand-side inequality in (6) is violated with $u \in S$. It means that the pair (z, λ) is not a saddle point: $(z, \lambda) \notin Sdl(\mathcal{L})$. $\square$

Therefore, from the point-of-view of optimization theory [1, 2, 4–12, 17, 18, 20, 21] the conditions (10)–(11) of Theorem 1 possess the constructive property. Nevertheless, it is not clear whether there exists a tuple (y, β, u) that violates (11). The answer is given by the following result.

Theorem 2. *Let there be given a KKT-point $z \in D$ with the corresponding multipliers $\lambda \in \mathcal{M}(z)$, $\lambda_0 = 1$. In addition, let the following assumption take place*

$$(\mathcal{H}): \qquad \exists v \in \mathbb{R}^n: \quad \mathcal{L}(v, \lambda) > \mathcal{L}(z, \lambda) = f_0(z) =: \zeta. \tag{13}$$

Besides, suppose that the pair (z, λ) is not a saddle point of $\mathcal{L}(z, \lambda)$ on $S \times \mathbb{R}^m_+$.

Then, one can find a tuple (y, β, u), where $(y, \beta) \in \mathbb{R}^{n+1}$, $u \in S$, and a fixed ensemble of subgradients $\{h'_{00}(y), h'_{10}(y), \ldots, h'_{m0}(y)\}$, $h'_{io}(y) \in \partial h_i(y)$, $i \in \{0\} \cup I$, such that

$$H_\lambda(y) \triangleq \sum_{i=0}^m \lambda_i h_i(y) = \beta - \zeta, \tag{14}$$

$$G_\lambda(y) \leq \beta, \tag{15}$$

$$G_\lambda(u) - \beta < \sum_{i=0}^m \lambda_i \langle h'_{io}(y), u - y \rangle. \tag{16}$$

$\square$

Let us verify the effectiveness of the constructive property of the GOC of Theorems 1 and 2 by the example.

Example 1 (Revisited). As it has been shown above, the point $z = (\frac{4}{3}, -\frac{2}{3})^\top$ is the KKT-point with $\lambda_0 = 1$, $\lambda_1 = 4$, $\lambda_2 = 0$. Recall that $\zeta := f_0(z) = 5\frac{1}{3}$. Meanwhile, there exist points feasible in the problem (4) and better than z in the sense of the problem (4).

Now, let us apply the GOC of Theorems 1 and 2 in order to improve the point z. For this purpose, employ the Lagrange function $\mathcal{L}(x, \lambda)$ with $\lambda = (1, 4, 0) \in \mathcal{M}(z)$:

$$\mathcal{L}(x, \lambda) = f_0(x) + \lambda f_1(x) = \frac{1}{2}(x_1 - 4)^2 + (x_2 + 2)^2 + 4[(x_1 - 1)^2 - (x_2 + 1)^2].$$

Then we have $\mathcal{L}(x, \lambda) = G_\lambda(x) - H_\lambda(x)$, where

$$G_\lambda(x) = \frac{1}{2}(x_1 - 4)^2 + (x_2 + 2)^2 + 4(x_1 - 1)^2, \quad H_\lambda(x) = 4(x_2 + 1)^2. \tag{17}$$

Let us choose $y = (0, -\frac{1}{2})^\top$, $u = (\frac{4}{3}, 0)^\top$, $f_1(u) = -\frac{8}{9} < 0$, $f_2(u) = -7\frac{1}{9} < 0$.

$$\nabla H_\lambda(x) = \begin{bmatrix} 0 \\ 8(x_2 + 1) \end{bmatrix}, \quad \nabla H_\lambda(y) = \begin{bmatrix} 0 \\ 8(y_2 + 1) \end{bmatrix} = \begin{pmatrix} 0 \\ 4 \end{pmatrix}.$$

Further, we obtain that $\beta = H_\lambda(y) + \zeta = 4(-\frac{1}{2} + 1)^2 + 5\frac{1}{3} = 6\frac{1}{3}$,

$$\langle \nabla H_\lambda(y), u - y \rangle = \langle \begin{pmatrix} 0 \\ 4 \end{pmatrix}, \begin{pmatrix} \frac{4}{3} \\ 0.5 \end{pmatrix} \rangle = 2, \quad \psi(y, \beta) \triangleq \beta + \langle \nabla H_\lambda(y), u - y \rangle = 8\frac{1}{3}.$$

$$G_\lambda(u) = \frac{1}{2}(\frac{4}{3} - 4)^2 + 2^2 + 4(\frac{4}{3} - 1)^2 = \frac{32}{9} + 4 + \frac{4}{9} = 8.$$

Hence, we see that

$$G_\lambda(u) = 8 < 8\frac{1}{3} = \beta + \langle \nabla H_\lambda(y), u - y \rangle,$$

and the PI (11) is violated. Due to Theorems 1 and 2, it means that (z, λ) is not a saddle point of the Lagrange function, which can be proved as follows:

$$\mathcal{L}(u, \lambda) \triangleq f_0(u) + 4f_1(u) = 7\frac{5}{9} + 4 \cdot \left(-\frac{8}{9}\right) = 4 < 5\frac{1}{3} = f_0(z) = \mathcal{L}(z, \lambda).$$

On the other hand, it is easy to compute that

$$f_0(u) = \frac{1}{2}\left(\frac{4}{3} - 4\right)^2 + 2^2 = \frac{32}{9} + 4 = 7\frac{5}{9} > f_0(z) = 5\frac{1}{3},$$

so that there is no improvement at all in the original problem (4). Consequently, we see that the GOC of Theorems 1 and 2 allow us to improve the point z in the sense of the Lagrange function, since they are striving to minimize the Lagrange function with respect to the variable x. However, they do not aim at minimizing the function $f_0(x)$ over the feasible set D, i.e. at solving Problem $(\mathcal{P})$.   □

The next result will be useful below.

Lemma 3 ([9, 10, 12]). *A point z is a solution to the problem*

$$(Q): \quad \begin{array}{l} S(x) := \varphi(x) + f(x) \downarrow \min_x, \quad x \in S, \\ \varphi(x) := \max_j \{\varphi_j(x) \mid j \in J = \{1, \dots, N\}\}. \end{array} \right\} \quad (18)$$

if and only if the pair (z, t_*) is a solution to the problem

$$(QA): \quad \Phi(x, t) := t + f(x) \downarrow \min_{(x,t)}, \quad x \in S, \ t \in \mathbb{R}, \quad \varphi_j(x) \leqslant t, \ j \in J, \quad (19)$$

where

$$t_* = \varphi(z) = \max_j \{\varphi_j(z) \mid j \in J\}. \quad (20)$$

Lemma 4 *Let the quadratic function $q(x) := \frac{1}{2}\langle x, Ax \rangle - \langle b, x \rangle$ with the positive definite matrix $A = A^T > 0$, $b \in \mathbb{R}^n$, be given. Consider the optimization problem (with a parameter $u \in \mathbb{R}^n$)*

$$Q(y, \beta) := \beta + \langle \nabla q(y), u - y \rangle \uparrow \max_{(y,\beta)}, \quad (y, \beta) \in \mathbb{R}^{n+1}: \quad q(y) = \beta - \gamma. \quad (21)$$

Then, the solution (y_*, β_*) to (21) is provided by the equalities

$$y_* = u, \quad \beta_* = q(y_*) + \gamma. \quad (22)$$

Example 2 (of G.R. Walsh, [23], p. 67). Consider the problem

$$\begin{array}{l} f_0(x) = x_1 x_2 - 2x_1^2 - 3x_2^2 \downarrow \min_x, \\ f_1(x) = 3x_1 + 4x_2 - 12 \leq 0, \quad f_2(x) = x_2^2 - x_1^2 + 1 \leq 0, \\ f_3(x) = -x_1 \leq 0, \quad f_4(x) = x_1 - 4 \leq 0, \\ f_5(x) = -x_2 \leq 0, \quad f_6(x) = x_2 - 3 \leq 0. \end{array} \right\} \quad (23)$$

Let us study the unique solution $z = (4,0)^\top$ to (23), $f_0(z) = \zeta = -32$. Clearly, the Lagrange function at z takes the form $\mathcal{L}(x,\lambda) = x_1 x_2 - 2x_1^2 - 3x_2^2 + \lambda_1(3x_1 + 4x_2 - 12) + \lambda_4(x_1 - 4) - \lambda_5 x_2$, $S = \mathbb{R}^2$, because $\lambda_2 = \lambda_3 = \lambda_6 = 0$, due to the complementarity conditions: $\lambda_i f_i(x) = 0$, $i = \overline{1,6}$. In addition, with the help of the KKT-conditions at $z = (4,0)^\top$, we derive that $\lambda_1 = 2$, $\lambda_4 = 10$, $\lambda_5 = 12$, so that

$$\mathcal{L}(x,\lambda) = x_1 x_2 - 2x_1^2 - 3x_2^2 + 16x_1 - 4x_2 - 64. \qquad (24)$$

Besides, $\mathcal{L}(z,\lambda) = -32 = \zeta = f_0(z)$, as it should be. Further, it can be readily seen that the function $x \mapsto \mathcal{L}(x,\lambda)$ is a d.c. one. We will use the d.c. representation as follows: $\mathcal{L}(x,\lambda) = G_\lambda(x) - H_\lambda(x)$, where

$$G_\lambda(x) = 2(x_1^2 + x_2^2) + 16x_1 - 4x_2 - 64, \quad H_\lambda(x) = 4x_1^2 + 5x_2^2 - x_1 x_2. \quad (25)$$

Therefore, one can see that $\beta = H_\lambda(y) + \zeta = H_\lambda(y) - 32$, $\nabla H_\lambda(x) = (8x_1 - x_2, 10x_2 - x_1)^\top$, $\langle \nabla H(y), u - y \rangle = 8y_1 u_1 - y_2 u_1 + 2y_1 y_2 - 8y_1^2 - 10y_2^2 + 10y_2 u_2 - y_1 u_2$, from which it follows that

$$\begin{aligned} \theta(y,\beta) &:= \beta + \langle \nabla H(y), u - y \rangle = \\ \psi(y) &:= (8u_1 - u_2)y_1 + (10u_2 - u_1)y_2 + y_1 y_2 - 4y_1^2 - 5y_2^2 - 32. \end{aligned} \qquad (26)$$

Furthermore, Lemma 4 leads us to the equality: $y_* = u$. Now, let us choose the vector u as $u = (-\frac{1}{5}, -\frac{4}{5})^\top$, since $S = \mathbb{R}^2$. Then one can compute that $G_\lambda(u) = -62\frac{16}{25} < -42 = \psi(y_*) = \theta(y_*, \beta_*)$, so that the inequality (11) is violated. Therefore, due to Theorems 1 and 2, we see that the pair (z,λ) is not a saddle point of $\mathcal{L}(u,\lambda)$. The latter assertion can be easily verified by the direct calculations $\mathcal{L}(u,\lambda) = -65\frac{21}{25} < -32 = \mathcal{L}(z,\lambda)$.

To sum up, we see that there does not exist a Lagrange multiplier λ such that $(z,\lambda) \in Sdl(\mathcal{L})$ even for the unique global solution $z \in Sol(\mathcal{P})$. $\qquad \square$

So, the max-merit function $F(x,\eta)$ defined in (2), as well as the Lagrange function, possesses some shortcomings, and both functions do not reflect completely the properties of Problem $(\mathcal{P})$. Hence, we have to undertake further investigations, perhaps, with different penalty or merit functions [10–13].

5   Penalty Functions

Now, let us introduce now the l_∞-penalty function [10–13] for Problem $(\mathcal{P})$–(1)

$$W(x) := \max\{0, f_1(x), \ldots, f_m(x)\} =: \max\{0, f_i(x), i \in I\}. \qquad (27)$$

Further, consider the penalized problem as follows

$$(\mathcal{P}_\sigma): \qquad \theta_\sigma(x) = f_0(x) + \sigma W(x) \downarrow \min, \quad x \in S. \qquad (28)$$

As well-known [5,6,11–13], if $z \in Sol(\mathcal{P}_\sigma)$, and $z \in D := \{x \in S \mid f_i(x) \le 0, \quad i \in I\}$, then $z \in Sol(\mathcal{P})$. On the other hand, if $z \in Sol(\mathcal{P})$, then, under supplementary conditions [5,6,10–13], for some $\sigma_* \ge \|\lambda_z\|_1$ (where

$\lambda_z \in \mathbb{R}^m$ is the KKT-multipliers corresponding to z), the inclusion $z \in Sol(\mathcal{P}_\sigma)$ holds $\forall \sigma > \sigma_*$.

Furthermore [5, Lemma 1.2.1, Chapter VII], $Sol(\mathcal{P}) = Sol(\mathcal{P}_\sigma)$, so that Problems $(\mathcal{P})$ and $(\mathcal{P}_\sigma)$ happen to be equivalent $\forall \sigma \geq \sigma_*$.

Before we move on any further, a few words should be said about "supplementary conditions". First of all, let us mention the notion of calmness introduced by R.T. Rockafellar [1] (see F. Clarke [6] and J.V. Burke [13]). To begin with, instead of $(\mathcal{P})$, consider the perturbed d.c. optimization problem $(v \in \mathbb{R}^m)$

$$(\mathcal{P}(v)): \qquad f_0(x) \downarrow \min_x, \quad x \in S, \quad f_i(x) \leq v_i, \ i \in I. \qquad (29)$$

Let $x_* \in S$, $v_* \in \mathbb{R}^m$ be such that $f_i(x_*) \leq v_{*i}$, $i \in I$. Then, Problem $(\mathcal{P}(v_*))$ is said to be calm at x_*, if there exist constants $\varkappa \geq 0$ and $\rho > 0$ such that $\forall (x,v) \in S \times \mathbb{R}^m$ with $\|x - x_*\| \leq \rho$ and $f_i(x) \leq v_i$, $i \in I$. We have

$$f_0(x_*) \leq f_0(x) + \varkappa \|v - v_*\|. \qquad (30)$$

The constants $\varkappa$ and ρ are called the modulus and the radius of calmness for $(\mathcal{P}(v_*))$ at x_*, respectively. Observe that, if $(\mathcal{P}(v_*))$ is calm at x_*, then x_* is a ρ-local solution to $(\mathcal{P}(v_*))$, i.e. $(v = v_*)$

$$f_0(x_*) \leq f_0(x) \quad \forall x \in S: \ \|x - x_*\| \leq \rho, \quad \text{and } f_i(x) \leq v_{*i}, i \in I,$$

i.e. $x \in D(v_*) \cap \mathbb{B}_x(x_*, \rho)$. For instance, when $v_* = 0$ and $(\mathcal{P}(0))$ is calm (with $\varkappa \geq 0$ and $\rho > 0$), x_* is the ρ-local solution to $(\mathcal{P}(0)) := (\mathcal{P})$.

The most fundamental finding consists in the equivalence of the calmness of $(\mathcal{P}(v_*))$ at x_* (with $\varkappa \geq 0$ and $\rho > 0$) and the fact that x_* is a ρ-local minimum of the following penalized function with $\sigma \geq \varkappa$

$$\theta_\sigma(x; v_*) := f_0(x) + \sigma \operatorname{dist}(F(x) \mid \mathbb{R}^m_- + v_*), \qquad (31)$$

where $F(x) = (f_1(x), \ldots, f_m(x))^\top$, $\mathbb{R}^m_- = \{y \in \mathbb{R}^m \mid y_i \leq 0, \ i \in I\}$, $\operatorname{dist}(y_0 \mid C) = \inf\{\|y - y_0\| \ : \ y \in C\}$. If one takes, for instance, $\|y\|_\infty \triangleq \max_i \{|y_i| \ : \ i \in I\}$, then it is obvious that

$$W(x) := \max\{0, f_1(x), \ldots, f_m(x)\} = \operatorname{dist}_\infty(F(x) \mid \mathbb{R}^m_-).$$

Various conditions can be found in the literature (see [13] and the references therein) that ensure that the calmness hypothesis is satisfied. All of these conditions are related to the regularity of the constraint system of Problem $(\mathcal{P})$:

$$x \in S, \quad f_i(x) \leq 0, \ i \in I. \qquad (32)$$

Recall that the system (32) is said to be regular at the solution x_0 (i.e. $x_0 \in S$, $F(x_0) \leq 0_m$) if there exist some constants $M > 0$ and $\varepsilon > 0$ such that $\operatorname{dist}_x(x \mid D(v)) \leq M \operatorname{dist}_y(F(x) \mid \mathbb{R}^m_- + v) \ \forall x \in S \cap \mathbb{B}_x(x_0, \varepsilon)$ and $\forall v \in \mathbb{B}_y(0, \varepsilon)$ where

$$D(v) := \{x \in S \mid F(x) \in \mathbb{R}^m_- + v\}, \quad \mathbb{B}_x(x_0, \varepsilon) := \{x \in \mathbb{R}^n \mid \|x - x_0\|_x \leq \varepsilon\}.$$

The conditions yielding the regularity of system (32) and more general systems have been studied in [14]. In the optimization literature such conditions are often called the constraint qualifications conditions, e.g. the Slater and Mangasarian-Fromovitz conditions (etc. see [10–14]).

To sum up, one can say that, under well-known regularity conditions at the global solution $z \in Sol(\mathcal{P})$, Problem $(\mathcal{P})$ $(v_* = 0_m)$ is calm at $z \in Sol(\mathcal{P})$ (with the corresponding $\varkappa \geq 0$ and $\rho > 0$), and, therefore, the goal function of the penalized problem $(\forall \sigma \geq \varkappa \geq 0)$

$$(\mathcal{P}_\sigma): \quad \theta_\sigma(x) = f_0(x) + \sigma \, \mathrm{dist}_\infty(F(x) \mid I\!\!R^m_-) = f_0(x) + \sigma W(x) \downarrow \min_x, \quad x \in S$$

attains at z its global minimum over S.

Furthermore, it can be readily seen that the penalized function $\theta_\sigma(\cdot)$ is a d.c. function, because the functions $f_i(\cdot)$, $i \in I \cup \{0\}$, are the same. Actually, since $\sigma > 0$,

$$\theta_\sigma(x) = G_\sigma(x) - H_\sigma(x), \tag{33}$$

$$H_\sigma(x) := h_0(x) + \sigma \sum_{i \in I} h_i(x), \tag{34}$$

$$G_\sigma(x) := \theta_\sigma(x) + H_\sigma(x)$$
$$= g_0(x) + \sigma \max \left\{ \sum_{i=1}^m h_i(x); \max_{i \in I}[g_i(x) + \sum_{\substack{j \in I \\ j \neq i}}^{j \neq i} h_j(x)] \right\}, \tag{35}$$

it is clear that $G_\sigma(\cdot)$ and $H_\sigma(\cdot)$ are convex functions.

For $z \in S$, denote $\zeta := \theta_\sigma(z)$.

Now we can formulate the major result of the paper.

Theorem 3 *If $z \in Sol(\mathcal{P}_\sigma)$, then*

$$\forall(y, \beta): \ H_\sigma(y) = \beta - \zeta \tag{36}$$

the following inequality holds

$$G_\sigma(x) - \beta \geq \langle \nabla h_0(y) + \sigma \sum_{i \in I} \nabla h_i(y), x - y \rangle \quad \forall x \in S. \tag{37}$$

$\square$

It is easy to see that Theorem 3 reduces the nonconvex (d.c.) Problem $(\mathcal{P}_\sigma)$ to solving the family of convex linearized problems of the form

$$(\mathcal{P}_\sigma\mathcal{L}(y)): \quad G_\sigma(x) - \langle \nabla H_\sigma(y), x \rangle \downarrow \min_x, \quad x \in S, \tag{38}$$

depending on the parameters (y, β) fulfilling (36).

If for such a pair (y, β) and some $u \in S$ (u may be a solution to $(\mathcal{P}_\sigma L(y))$) the inequality (37) is violated, i.e.

$$G_\sigma(u) < \beta + \langle \nabla H_\sigma(y), u - y \rangle, \tag{39}$$

then, due to convexity of $H_\sigma(\cdot)$ and with the help of (36), we obtain that

$$G_\sigma(u) < \beta + H_\sigma(u) - H_\sigma(y) = H_\sigma(u) + \zeta.$$

The latter implies that $\theta_\sigma(u) = G_\sigma(u) - H_\sigma(u) < \zeta := \theta_\sigma(z)$, so that $u \in S$ is better than z, i.e. $z \notin Sol(\mathcal{P}_\sigma)$.

It means that the Global Optimality Conditions (36), (37) of Theorem 3 possess the constructive (algorithmic) property allowing to design local and global search methods for Problem $(\mathcal{P}_\sigma)$ [17, 18, 20–22].

In particular, they enable us to escape a local pit of $(\mathcal{P}_\sigma)$ to reach a global solution. The question arises whether such a tuple (y, β, u) exists. The answer is given by the following result.

Theorem 4 *Let for a point $z \in S$ there exists $w \in \mathbb{R}^n$ such that*

$$(\mathcal{H}): \qquad \theta_\sigma(w) > \theta_\sigma(z).$$

If z is not a solution to Problem $(\mathcal{P}_\sigma)$, then one can find a pair $(y, \beta) \in \mathbb{R}^{n+1}$, satisfying (36), and a point $u \in S$ such that the inequality (39) holds. $\qquad\square$

Now let us set $y = z$ in (38). Then from (36) it follows that $\beta = \theta_\sigma(z) + H_\sigma(z) = G_\sigma(z)$. Furthermore, from (37) we derive

$$G_\sigma(x) - G_\sigma(z) \geq \langle \nabla H_\sigma(z), x - z \rangle \quad x \in S,$$

that yields that z is a solution to the convex linearized problem

$$(\mathcal{P}_\sigma \mathcal{L}(z)): \quad G_\sigma(x) - \langle \nabla H_\sigma(z), x \rangle \downarrow \min_x, \quad x \in S,$$

With the help of Lemma 3 and due to (33)–(35), we see that the latter problem amounts to the next one

$$\left.\begin{array}{c} g_0(x) - \langle \nabla H_\sigma(z), x \rangle + \sigma t \downarrow \min\limits_{(x,t)}, \quad x \in S, \ t \in \mathbb{R}, \\ \sum\limits_{i \in I} h_i(x) \leq t, \quad g_i(x) + \sum\limits_{j \neq i} h_i(x) \leq t, \quad i \in I. \end{array}\right\} \tag{40}$$

Moreover, the KKT-conditions to Problem (40) provide for the KKT-conditions at z for the original Problem $(\mathcal{P})$.

So, the Global Optimality Conditions (36), (37) of Theorems 3 and 4 are connected with classical optimization theory [1–13, 15].

Example 1 *(Revisited).* Let us return to problem (4), where the point $z = (\frac{4}{3}, -\frac{2}{3})^\top$, with $f_1(z) = 0$ and $f_2(z) = -4 < 0$, $\zeta := f_0(z) = 5\frac{1}{3}$, satisfies the KKT-conditions with $\lambda_0 = 1$, $\lambda_1 = 4 > 0$, $\lambda_2 = 0$.

On the other hand, the point $z_0 = (\frac{8}{3}, -\frac{8}{3})^\top$ is the global solution to (4) with the optimal value $\mathcal{V}(4) = f_0(z_0) =: \zeta_0 = \frac{4}{3}$.

In this example, $h_0(x) \equiv 0$, $g_0(x) = f_0(x) = \frac{1}{2}(x_1 - 4)^2 + (x_2 + 2)^2$, $g_1(x) = (x_1 - 1)^2$, $h_1(x) = (x_2 + 1)^2$, $g_2(x) = (x_2 - 2)^2$, $h_2(x) = (x_1 + 2)^2$. Therefore, taking into account (34) and (35), one can see that

$$H_\sigma(x) := h_0(x) + \sigma \sum_{i \in I} h_i(x) = \sigma\left[(x_1 + 2)^2 + (x_2 + 1)^2\right],$$

$$G_\sigma(x) := g_0(x) + \sigma \max\left\{\sum_{i=1}^{2} h_i(x); g_1(x) + h_2(x); g_2(x) + h_1(x)\right\} =$$

$$\tfrac{1}{2}(x_1 - 4)^2 + (x_2 + 2)^2 +$$

$$\sigma \max\left\{(x_1 + 2)^2 + (x_2 + 1)^2; (x_1 - 1)^2 + (x_1 + 2)^2; (x_2 - 2)^2 + (x_2 + 1)^2\right\}.$$

Set $\sigma := 5 = \|\lambda\|_1$, $y = (1.5, -2)^\top \notin D$. Then, we have $\beta = H_\sigma(y) + \zeta = 5\left[(1.5 + 2)^2 + (-2 + 1)^2\right] + 5\tfrac{1}{3} = 71\tfrac{7}{12}$. Now set $u = (2, -2)^\top \in D$.

Then, according to Theorem 3, it is not difficult to compute that

$$\beta + \langle \nabla H_\sigma(y), u - y \rangle = 89\frac{1}{12}.$$

On the other hand, we see that

$$G_\sigma(u) = 87 < 89\frac{1}{12} = \beta + \langle \nabla H_\sigma(y), u - y \rangle.$$

It means that the principal inequality (37) of the GOC is violated, so that $z \notin Sol(\mathcal{P}_\sigma)$. Consequently, $z \notin Sol(\mathcal{P})$, since $2 = f_0(u) = \theta_\sigma(u) < \theta_\sigma(z) = f_0(z) = 5\tfrac{1}{3}$, because u and z are feasible.

So, the Global Optimality Conditions (GOC) of Theorems 3 and 4 allow us not only to show that the KKT-point $z = (\tfrac{4}{3}, -\tfrac{2}{3})^T$ is not a global solution to the problem (7), but, in addition, to construst a feasible point $u = (2, 2)^T \in D$ which is better than z and closer to the global solution $z_0 = (\tfrac{8}{3}, -\tfrac{8}{3})^T$. Remember, the max-merit function $F(x, \zeta)$ does not differ between these two points:

$$F(z, \zeta) = 0 = F(z_0, \zeta).$$

Besides, in order to find a saddle point, the Lagrange function aims at improving exactly itself but not at solving Problem $(\mathcal{P})$.   □

Hence, the exact penalization approach demonstrated some advantages in comparison with the max-merit and Lagrange functions.

In addition, employing the constructive property of the GOC of Theorems 3 and 4, we can design the Special Local Search and Global Search Methods. The latter one can escape local pits and attain global solutions in general d.c. optimization problems.

6   Conclusion

We considered the reduction of the constrained optimization problem with the d.c. goal function and d.c. inequality constraints to three auxiliary unconstrained problems with different objective functions: the max-merit function, the Lagrange function and an exact penalty function.

The comparison was based on the level of adequacy to the original problem and has been carried out by means of the classical tools, the new Global Optimality Conditions (GOC) and a few examples.

The results showed certain advantages of the exact penalization approach that facilitates development of the new local and global search methods for solving the original problem [17–22].

Acknowledgments. This work has been supported by the Russian Science Foundation, project No. 15-11-20015.

References

1. Rockafellar, R.T.: Convex Analysis. Princeton University Press, Princeton (1970)
2. Rockafellar, R.T., Wets, R.J.-B.: Variational Analysis. Springer, New York (1998)
3. Demyanov, V.F., Rubinov, A.M.: Constructive Nonsmooth Analysis. Peter Lang, Frankfurt (1995)
4. Hiriart-Urruty, J.-B.: Generalized differentiability, duality and optimization for problems dealing with difference of convex functions. In: Ponstein, J. (ed.) Convexity and Duality in Optimization. Lecture Notes in Economics and Mathematical Systems, vol. 256, pp. 37–70. Springer, Heidelberg (1985)
5. Hiriart-Urruty, J.-B., Lemaréchal, C.: Convex Analysis and Minimization Algorithms. Springer, Heidelberg (1993)
6. Clarke, F.H.: Optimization and Nonsmooth Analysis. Wiley, New York (1983)
7. Ioffe, A.D., Tihomirov, V.M.: Theory of Extremal Problems. Elsevier Science Ltd., Amsterdam (1979)
8. Alekseev, V.M., Tikhomirov, V.M., Fomin, S.V.: Optimal Control. Springer, New York (1987)
9. Hiriart-Urruty, J.-B.: Optimisation et Analyse Convex. Presses Universitaires de France, Paris (1998)
10. Nocedal, J., Wright, S.J.: Numerical Optimization. Springer, New York (2006)
11. Bonnans, J.-F., Gilbert, J.C., Lemaréchal, C., Sagastizábal, C.A.: Numerical Optimization: Theoretical and Practical Aspects, 2nd edn. Springer, Heidelberg (2006)
12. Izmailov, A.F., Solodov, M.V.: Newton-Type Methods for Optimization and Variational Problems. Springer, New York (2014)
13. Burke, J.V.: An exact penalization viewpoint of constrained optimization. SIAM J. Control Optim. **29**(4), 968–998 (1991)
14. Borwein, J.M.: Stability and regular points of inequality systems. J. Optim. Theory Appl. **48**, 9–52 (1986)
15. Horst, R., Tuy, H.: Global Optimization Deterministic Approaches. Springer, Heidelberg (1993)
16. Tuy, H.: D.c. optimization: theory, methods and algorithms. In: Horst, R., Pardalos, P.M. (eds.) Handbook of Global Optimization, pp. 149–216. Kluwer Academic Publisher, Dordrecht (1995)
17. Strekalovsky, A.S.: On solving optimization problems with hidden nonconvex structures. In: Rassias, T.M., Floudas, C.A., Butenko, S. (eds.) Optimization in Science and Engineering, pp. 465–502. Springer, New York (2014)
18. Strekalovsky, A.S.: On local search in D.C. optimization problems. Appl. Math. Comput. **255**, 73–83 (2015)

19. Strekalovsky, A.S., Gruzdeva, T.V.: Local search in problems with nonconvex constraints. Comput. Math. Math. Phys. **47**(3), 381–396 (2007)
20. Strekalovsky, A.S.: On problem of global extremum. Proc. USSR Acad. Sci. **292**(5), 1062–1066 (1987)
21. Strekalovsky, A.S.: Elements of Nonconvex Optimization. Nauka, Novosibirsk (2003). [In Russian]
22. Strekalovsky, A.S.: Minimizing sequences in problems with D.C. constraints. Comput. Math. Math. Phys. **45**(3), 418–429 (2005)
23. Walsh, G.R.: Methods of Optimization. Wiley, New York (1975)

Mathematical Economics and Games

Application of Supply Function Equilibrium Model to Describe the Interaction of Generation Companies in the Electricity Market

Natalia Aizenberg[✉]

Melentiev Energy Systems Institute SB RAS, Lermontov St., 130, Irkutsk, Russia
ayzenebrg.nata@gmail.com
http://www.isem.irk.ru

Abstract. The paper studies the trade in the spot electricity market based on submitting bids of energy consumers and producers to the market operator. We investigate supply function equilibrium (SFE) model, in which generation capacities are integrated into large generation companies that have a common purpose of maximizing their profits. For this case we prove the existence and uniqueness of equilibrium for a linear function of aggregate demand and quadratic costs. The mechanism is tested on the basis of the Siberian electric power system, Russia.

Keywords: Electricity market · Models of imperfect markets · Oligopoly · Model of supply function equilibrium · Liberalization

1 Introduction

The restructuring of Russia's electric power industry was carried out without a preliminary testing of the mechanisms to be implemented [1]. Taking into account the successful experience of other countries we chose a double auction market for the wholesale trading [2]. The market was supposed to ensure the right incentives for the behavior of participants. For testing, we need models to analyze the specific features of the market architecture and rules of organizing the interaction among the market participants. This mathematical model allows to describe, on the one hand, the actions of the generating companies, and on the other hand - the actions of consumers, under the conditions of a liberalized market.

Interaction in the electricity market can be described by a model of mixed oligopoly [3,4] or a conjectured supply function model [5–10]. In this approach, each agent models its own strategy on the basis of the assumption about the actions of its competitors. It is important that the information is not perfect, which makes the model more real. One of the special cases of the expected supply function model is the supply function equilibrium (SFE, or rather its linear variant). For inelastic demand we can apply the Cournot competition (which is characteristic of the spot electricity market). All these approaches can take into account the capacity constraints.

© Springer International Publishing Switzerland 2016
Y. Kochetov et al. (Eds.): DOOR 2016, LNCS 9869, pp. 469–479, 2016.
DOI: 10.1007/978-3-319-44914-2_37

We develop a model of supply function equilibrium by introducing into consideration the generation companies. Generating capacities are integrated into large generation companies to jointly maximize their profit. At the same time, each capacity submits bids to the commercial operator of the market. Modeling of these conditions requires additional tools, taking into account the common goals of several capacities. It is not enough to simply combine them into a common profit function. In this case, we deal with the underdetermined system of nonlinear equations. It is necessary to impose additional conditions to make it possible to identify missing variables. In the present paper, we propose a solution to this problem.

The goal of the suggested model is to solve the problem of modeling the electricity market functioning, analyze the situation, determine the most powerful players in the market, and find out the stability of strategies chosen by the producers.

2   The Model

The pricing in the electricity spot market is organized as a double auction. Namely, there is a strategic interaction between the firms that generate power (power plants), and the price is derived from the equality of demand and supply. Power plants are different in their technological capabilities and type of costs. Therefore, it is reasonable to describe the market behavior by applying the models that divide firms into strategic producers (that can affect the market price) and competitive firms (that do not take part in the auction, but take the price as fixed). This makes sense because the RF Law regulating the electric power industry allows the participants of these two types to take part in the auction [11].

We study the model that considers influence of the amount of power generation by competing firms to establish the market price, taking into account the fact that this price will affect the output of each generator. Supply functions are called conjectured, since the firms can only guess the reaction of their competitors. These functions have the so-called coefficients of influence of each participant on the situation in general.

All consumers are aggregated by the total nonincreasing demand function $D\left(p\right)$. The supply functions of individual firms are $q_i\left(p\right)$, $i = \overline{1,n}$; $n \geq 2$ is the number of firms in the market. We assume that supply function $q_i(p)$ of firm i is non-negative and non-decreasing. The price is determined by a market-clearing condition. The total output supplied at the market-clearing price must be equal to the demand

$$\sum_{i=1}^{n} q_i\left(p\right) = D\left(p\right).$$

The inverse demand function is $p = D^{-1}\left(\sum_{i=1}^{n} q_i\right)$. We will define the outputs of competitors of firm i as $q_{-i}\left(p\right) = \sum_{j \neq i} q_j\left(p\right)$, this is the total output excluding

$q_i(p)$. Thus, the residual demand of the generation company i is $q_i(p) = D(p) - q_{-i}(p)$. Here $p \in R_+^1$ is the price formed through the interaction of agents in the market. Cost functions $C_i(q_i)$ are convex and increasing, $q_i \geq 0$, $i = \overline{1, n}$. Generation companies seek to maximize the profit on the residual demand such that, in equilibrium, the demand will equal to the total output of the companies:

$$\pi(P, q_i) = p \cdot (D(p) - q_{-i}(p)) - C_i(D(p) - q_{-i}(p)). \tag{1}$$

Each firm i determines the supply functions of all other firms and uses this information when maximizing its profit on residual demand. It is important that the scale of these reactions is assumed by firm i itself. Hence, the answers can differ from the real reactions of competitors. Here, many researchers see interconnection between this model and the problem the Stackelberg Competition. A simplified form of this model is the SFE model [5,6] where all interested all stakeholders have information about the competitors. This assumption significantly simplifies the model. Changing market firms influence leads to different equilibrium models: the maximum market power will correspond to the Cournot model, and minimum influence – to the model of perfect competition.

Consider the example of our models by making the following assumptions: demand function is linear, generation companies are heterogeneous in convex quadratic operating cost functions and constraints on power generation.

Let

$$D(p) = N - \gamma \cdot p \tag{2}$$

be the linear function of total demand, where $\gamma > 0$. Let

$$C_i(q_i) = \frac{1}{2}c_i q_i^2 + a_i q_i, \tag{3}$$

$c_i > 0$, $a_i \neq 0$, $i = \overline{1, n}$, be the production costs of firm i, quadratic and strictly convex. Let $a_1 \leq a_2 \leq ... \leq a_n$.

The models presented below have a common idea. In these models, each producer i defines the strategy of behavior by the described supply function $q_i(p)$, which includes possible reactions of all competitors on the change in the production of the firm i. In order to rationally choose an optimal volume of production each firm solves the profit maximization problem (1) on residual demand $D(p) - q_{-i}(p)$ for firm i under conditions (2), (3):

$$\pi_i(p) = (p - a_i) \cdot (D(p) - q_{-i}(p)) - \frac{1}{2}c_i(D(p) - q_{-i}(p))^2 \to \max_p, \quad i = \overline{1, n}.$$

Each firms profit function $\pi_i(p)$, $i = \overline{1, n}$, is concave with respect to p under (2)-(3), and hence, has a unique maximum.

First-order optimality condition:

$$\frac{\partial \pi_i}{\partial p} = 0, \implies q_i(p) = (p - C_i'(q_i(p)))\left(-\frac{dD}{dp} + \frac{dq_{-i}}{dp}\right), \quad i = \overline{1, n}.$$

2.1 Models of Linear Supply Function Equilibrium

Here we consider the generation companies strategies that can be applied to the electricity market. These strategies correspond to the model of linear supply function equilibrium [9,10]. The firms are interacting with the competitive environment on the auction. It should be noted that the model described below is a special case of the SFE model (namely, we consider the special supply functions). The SFE model deals with the problem of multiple equilibria and the complexity of their location. The only way to solve it is to use linear supply functions of competitive firms. To use this model is reasonable for the electric power industry, since the type of submitted bids is usually determined by the rules of auctions. These bids can be either in the form of step functions or in the form of linear functions [7].

It is assumed that the market rules determine the supply function for each firm in the linear form

$$q_i(p) = \beta_i(p - \alpha_i), \tag{4}$$

$\beta_i \geq 0$, $i = \overline{1, n}$, where parameters α_i and β_i are chosen by firm i. At the given price level the supply functions cannot be negative.

The profit function for firm i is concave on the residual demand:

$$\pi_i(p) = (p - a_i)\left(N - \gamma p - \sum_{j \neq i} \beta_j(p - \alpha_j)\right) - \frac{c_i}{2}\left(N - \gamma p - \sum_{j \neq i} \beta_j(p - \alpha_j)\right)^2. \tag{5}$$

The function is concave because the coefficient of p^2 is negative:

$$-\left(\gamma + \sum_{j \neq i} \beta_j\right) - \frac{c_i}{2}\left(\gamma + \sum_{j \neq i} \beta_j\right)^2 < 0.$$

From the first-order optimality condition we have

$$\beta_i(p - \alpha_i) = \left(\gamma + \sum_{j \neq i} \beta_j\right)(p - a_i - c_i \beta_i(p - \alpha_i)). \tag{6}$$

The solution has a different form depending on the interval where the equilibrium is located [9] or on demand. We consider the simple case when (i) the market does not have much excess of capacity, (ii) all generators is involved in the selection, (iii) they take into account each other when forming its supply function. The solving of the problems arising in the situation of small demand is difficult and is described in detail in [7]. Our version of the model corresponds to proposition 1 from paper [7]. In [7,10] the solution of problem (5) is suggested for the case of linear supply functions when $\max_{1 \leq i \leq n} a_i < p < p_{max}$, where p_{max} is determined from the maximum possible demand. In [6] this solution was refined to the marginal cost of type (3):

$$\beta_i = \left(\gamma + \sum_{j \neq i} \beta_j\right)(1 - c_i\beta_i), \quad \alpha_i = a_i, \quad \beta_i > 0, \quad i = \overline{1, n}. \tag{7}$$

The proof of the uniqueness is similar to the proposed by Rudkevich [12]. We can transform (7) into quadratic equation with respect to β_i:

$$\beta_i^2 - \left((B + \gamma) + \frac{2}{c_i}\right)\beta_i + \frac{(B + \gamma)}{c_i} = 0.$$

where $B = \sum_{i=1}^{n} \beta_i$. We select the equation root that satisfies the condition of positivity $B > 0$. Thus, the solution is unique and positive:

$$\beta_i = \frac{B + \gamma}{2} + \frac{1}{c_i} - \sqrt{\frac{(B + \gamma)^2}{4} + \left(\frac{1}{c_i}\right)^2},$$

where B is the unique solution of equation

$$B = \sum_{i=1}^{n} \frac{1}{c_i} + n \cdot \frac{B + \gamma}{2} - \sum_{i=1}^{n} \sqrt{\frac{(B + \gamma)^2}{4} + \left(\frac{1}{c_i}\right)^2}. \tag{8}$$

In right-hand side the function is concave w.r.t. B, in the left-hand side the function is linear. The solution will be unique for positive value because the value of the right-hand side of Eq. (8) is higher than the left-hand side one at $B = 0$.

Taking into account the constraints on power generation, the supply function for producers and equilibrium prices take the following form for $i = \overline{1, n}$:

$$q_i(p) = \begin{cases} 0, & p \leq a_i, \\ \beta_i(p - a_i), & a_i < p \leq p_{max}. \end{cases}$$

The market price is $p^* = (N + \gamma \sum_i \beta_i a_i) / (\gamma \sum_i \beta_i + 1)$. The condition of equilibrium $\alpha_i = a_i$ for each firm i means that the auction participants have an incentive to make the values of their coefficients a_i, (in the functions of marginal costs) generally known.

Therefore, this strategy will mean that the producer is guided by the price p^*, demand elasticity at this price, and some reaction of the competitors to the change in its price and supply.

It should be mentioned that the considered model of linear supply function equilibrium is true only when the prices are higher than all coefficients a_i (non-negativity of production volumes is provided). Otherwise, it is necessary to use piecewise-linear approximations.

2.2 Models of Linear Supply Function Equilibrium with Generation Company

Here we study the models in terms of the integration of several capacities into generation companies. Accordingly, the profit function of such integration is the sum of profits of individual capacities. The generation company acts as a multi-capacity monopoly, i.e., on the basis of the principle of equality of the marginal costs of individual productions. Generating capacities are integrated into large generation companies that have a common purpose to maximize their profit. At the same time, each capacity submits bids to the commercial operator of the market, despite the fact that major players interact in the market. Let capacities $k = \overline{1, m}$ belong to one generation company G. This problem has the form:

$$\pi_G\left(q_k, q_{-k}, p\right) = p \cdot \sum_{k \in G} q_k - \sum_{k \in G} C_k\left(q_k\right) \to \max,$$

$$q_k \geq 0, \quad \beta_k > 0, \quad \alpha_k \geq 0 \ \forall k \in G,$$

where q_k has the form (4). The described problem, as well as the model of supply function equilibrium, has a unique solution under positive β_k, $k \in G$. Additionally, it is necessary to take into consideration the profit maximization condition, i.e., the equality of marginal production costs within the company:

$$a_k + c_k \beta_k\left(p - a_k\right) = a_l + c_l \beta_l\left(p - a_l\right) \quad \forall l \in G.$$

In this case the following statement holds.

Proposition. *In the model of supply function equilibrium with n capacities in the case of participation of the generation company G the equilibrium exists and is unique under the constructed supply functions $q_i = \beta_i\left(P - \alpha_i\right)$, $i = \overline{1, n}$, and demand function $D\left(p\right) = N - \gamma \cdot p$ is linear. The equilibrium is defined by the coefficients for $a_i < p$*

$$\alpha_i = a_i, \quad \beta_i = \left(1 + c_i\left(\gamma + \sum_{j \neq i} \beta_j\right)\right)^{-1}, \quad i \notin G, \tag{9}$$

$$\alpha_l = \sum_{k \in G} a_k - 1 + \frac{\displaystyle\sum_{k \in G, k \neq l} \frac{(a_l - a_k)}{c_k}}{c_l \cdot \beta_l \cdot \displaystyle\sum_{k \in G} \frac{1}{c_k}}, \quad l \in G, \tag{10}$$

$$\beta_l = \left(\left(1 + \sum_{k \in G} \frac{c_l}{c_k} + c_l\left(\gamma + \sum_{j \notin G} \beta_j\right)\right)\right)^{-1}, \quad l \in G. \tag{11}$$

The proof is presented in the Appendix.

Theoretical analysis of the models has showed that

(a) competitive environment increases production and decreases the equilibrium price in comparison with the one-level interaction of strategic firms;
(b) from consumers' viewpoint the competition among the linear supply functions of electricity producers is more preferable than Cournot competition. In this case the market achieves equilibrium at lower prices and with larger production volumes. Despite the fact that each firm makes less profit than in the case of Cournot competition, the social welfare generally increases due to the consumer surplus, and the equilibrium price tends to Walrasian price.

Theoretical conclusions have confirmed by test examples, this allows us to compare the market mechanisms according to the criterion of social welfare maximization.

3 Modeling of Interaction in the Siberian Electricity Market

Here we consider only modeling of the spot market and analyze the strategies of economic agents and their market power. We discus the modeling of the interaction without network constraints, based on the models described in Sect. 2.

We use a scheme with 15 nodes and model formation of the price taking into account the strategic interaction of producers in the market. We use the main characteristics of generation and consumption at the nodes of system, average hourly consumption, and average annual costs of producers for 2011 in the system Sibir [13].

The demand function in our model is linear. As usual, we assume that the demand of electricity has low elasticity close to zero. The demand elasticity in [14] is -0.165 for urban population and -0.28 for rural population. These data have calculated for Siberia (Novosibirskenergo). In this case the parameter of the slope of the demand function corresponding to the elasticity of -0.3 is $\gamma = 40$.

Producers (power generators) are divided into strategic producers (that significantly influence on the price, the price makers) and the price takers. The second group consists of hydropower plants (HPPs) that are assumed to have zero marginal costs; they are present in the market by providing information only on the volumes of generated power (Krasnoyarsk, Sayano–Shushensk, Bratsk and Ust-Ilimsk HPPs). All plants have constraints on generation. Transmission losses are considered across the whole system. Prices were calculated on the basis of Cournot model and LSFE model with competitive environment.

The resulting strategy are presented in the Table 1, where columns show an excess supply function coefficients β of generation companies over marginal costs.

In the experiment, Gusinoozersk and Kharanorsk CPPs were merged into one company, taking into account generation companies. The columns represent the excess (in percentage) of the values of supply function over the marginal costs for the LSFE (7), LSFE(GenCo) (9)–(11) and Cournot models that characterize market power of individual companies.

Table 1. The coefficients of supply function of generation companies corresponding to different market models.

$(\frac{1}{\beta_i c_i} - 1)(\%)$	LSFE model	GenCo model	Cournot model
Irkutsk HPP	17	16	28
Gusinoozer CPP	6	4	11
Kharanorsk CPP	7	8	4
Krasnoyarsk HPP	26	26	38
Novosibirsk HPP	21	20.5	24
Kuzbass HPP	3	3	22

4   Conclusion

We propose a modification of the model of supply functions equilibrium with the introduction the generation companies. The generation companies are major players interacting in the electricity spot market. Each capacity has the supply function in the market and submits it to the commercial operator. We found a unique equilibrium for the case of the linear supply functions, the quadratic costs and the linear aggregate demand.

As a result of modeling the strategic interaction in the wholesale market of the electric power system "Sibir" using the linear supply functions models with generation companies, we found out that:

(1) the lowest equilibrium price is achieved in the models that use equilibrium supply functions with the competitive environment (linear supply functions in our case);
(2) the firms that work under the conditions of inelastic demand can significantly overprice regarding the marginal costs (Walrasian prices);
(3) the presence of generation companies changes market power, the integration of small capacitites into a generation company reduces the market power of other companies;
(4) the merge of individual capacities into big companies leads to an increase the market power and prices.

Acknowledgments. This work was partially supported by the Russian Foundation for Basic Research, grant 16-06-00071.

Appendix: Proof of the Proposition

Let a generation company have two capacities; $G = \{l, k\}$. Then the profit function is:

$$\pi_{l+k}\left(q_k, q_l, q_{-(k+l)}, p\right) = p \cdot q_k + p \cdot q_l - C_k\left(q_k\right) - C_l\left(q_l\right).$$

Since the company redistributes the output (a residual demand) inside according to the condition of cost optimization it is necessary to equate the marginal revenue (which in this case is the same for any sold unit of commodity) to the marginal costs. For two capacities this is:

$$\begin{cases} \frac{\partial \pi_{k+l}}{\partial q_k} = \frac{\partial p}{\partial q_k} (q_k + q_l) - MC_k (q_k) = 0, \\ \frac{\partial \pi_{k+l}}{\partial q_l} = \frac{\partial p}{\partial q_l} (q_k + q_l) - MC_l (q_l) = 0. \end{cases}$$

Hence, the main condition is: $MC_k (q_k) = MC_l (q_l)$. The impact of each unit of output of companies on the market price is equivalent to $\frac{\partial p}{\partial q_l} = \frac{\partial p}{\partial q_k}$. For the linear supply functions this condition is:

$$a_k + c_k \beta_k (p - \alpha_k) = a_l + c_l \cdot \beta_l (p - \alpha_l).$$

Then

$$\beta_k = \frac{\beta_l \cdot c_l}{c_k}, \qquad \alpha_k = \alpha_l + \frac{a_k - a_l}{\beta_l \cdot c_l}. \tag{12}$$

Using (12), the supply function can be written through the costs of the capacities

$$q_k + q_l = \beta_l \left(1 + \frac{c_l}{c_k}\right) \cdot p - \beta_l \left(1 + \frac{c_l}{c_k}\right) \alpha_l - \frac{a_k - a_l}{c_k}.$$

Hence the linear supply function, submitted by the generation company to operator, is:

$$q_{k+l} = \beta_g (p - \alpha_g),$$

where

$$\alpha_g = \alpha_l + \frac{a_k - a_l}{\beta_l \cdot (c_k + c_l)}, \quad \beta_g = \beta_l \cdot \left(1 + \frac{c_l}{c_k}\right).$$

The total costs of the generation company is the sum of costs of separate capacities:

$$TC (q_l + q_k) = a_k q_k + 0.5 \cdot c_k q_k^2 + a_l q_l + 0.5 \cdot c_l q_l^2.$$

Substitute $q_i = \beta_i (p - \alpha_i)$, $i \in \{l, k\}$, and (12):

$$TC (q_l + q_k) =$$

$$\left(1 + \frac{c_l}{c_k}\right) \cdot \beta_l (p - \alpha_l) - 0.5 \cdot c_l \left(1 + \frac{c_l}{c_k}\right) (\beta_l (p - \alpha_l))^2 \alpha_l - \frac{(a_k - a_l)(a_k + a_l)}{2c_k}.$$

Using the general form of the cost function

$$TC (q_g) = a_g \beta_g (p - \alpha_g) + 0.5 \cdot c_g (\beta_g (p - \alpha_g))^2,$$

we obtain a system of equations from which it is possible to determine the required parameters for the aggregate cost function of the generation company:

$$\begin{cases} a_g - c_g \frac{a_k - a_l}{c_k} = 1, \\ c_g = \frac{c_k \cdot c_l}{c_k + c_l}, \\ 2a_g - c_g \frac{a_k - a_l}{c_k - c_l} = a_k + a_l. \end{cases}$$

Thus, we can reduce our problem to the problem with separate capacities. In this case, the competitors consider generating company with the supply function of the form $q_{k+l} = \beta_g (p - \alpha_g)$. The company problem is:

$$\begin{cases} \pi (q_g, q_{-g}, p) = p \cdot \beta_g (p - \alpha_g) - a_g \beta_g (p - \alpha_G) - 0.5 \cdot c_g (\beta_g (p - \alpha_g))^2 \to \max_P; \\ a_g = a_k + a_l - 1, \; c_g = \frac{c_k \cdot c_l}{c_k + c_l}, \\ \alpha_g = \alpha_l + \frac{a_k - a_l}{\beta_l \cdot (c_k + c_l)} = \alpha_k + \frac{a_l - a_k}{\beta_k \cdot (c_k + c_l)}, \\ \beta_g = \beta_l \cdot \left(1 + \frac{c_l}{c_k}\right) = \beta_k \cdot \left(1 + \frac{c_k}{c_l}\right). \end{cases}$$

The profit function is concave, therefore the maximum is attained. As in [7], we prove the uniqueness of the positive solution for the coefficients $\beta_1, ..., \beta_g, ..., \beta_n$. From the uniqueness of the solution and (7) it follows the uniqueness of a solution for the case with generation companies. The coefficients of the supply function is:

$$q_i = \beta_i (P - \alpha_i), \; \alpha_i = a_i, \; \beta_i = \frac{1}{1 + c_i \left(\gamma + \sum_{j \neq i} \beta_j\right)}, \; i \notin G$$

$$\alpha_g = a_k + a_l - 1, \; \beta_g = \left(\left(1 + \frac{c_l \cdot c_k}{c_l + c_k} \left(\gamma + \sum_{j \notin G} \beta_j\right)\right)\right)^{-1}.$$

Hence $q_k = \beta_k (p - \alpha_k), \; q_l = \beta_l (p - \alpha_l)$, where

$$\beta_l = \left(\left(1 + \frac{c_l}{c_k} + c_l \left(\gamma + \sum_{j \notin G} \beta_j\right)\right)\right)^{-1}, \; l, k \in G,$$

$$\beta_k = \left(\left(1 + \frac{c_k}{c_l} + c_k \left(\gamma + \sum_{j \notin G} \beta_j\right)\right)\right)^{-1}, \; l, k \in G,$$

$$\alpha_l = a_k + a_l - 1 - \frac{a_k - a_l}{\beta_l (c_l + c_k)}, \quad \alpha_k = a_k + a_l - 1 - \frac{a_l - a_k}{\beta_k (c_l + c_k)}.$$

The special case considered above can be extended to the general one if the generation company has set G of individual capacities. Then the coefficients for the general supply function of the generation company are:

$$\alpha_g = \sum_{k \in g} a_k - 1, \; \beta_g = \left(\left(1 + \frac{1}{\sum_{k \in G} \frac{1}{c_k}} \left(\gamma + \sum_{j \notin G} \beta_j\right)\right)\right)^{-1}.$$

From this we can obtain all coefficients for the supply functions of individual productions.

The Proposition is proved.

References

1. Aizenberg, N.I.: Analysis of the mechanisms of functioning of wholesale electricity markets. ECO **6**, 97–112 (2014). (in Russian)
2. Joskow, P.L.: Lessons learned from electricity market liberalization. Energy J. **29**(2), 9–42 (2008). The Future of Electricity: Papers in Honor of David Newbery
3. Kalashnikov, V.V., Bulavsky, V.A., Kalashnykova, N.I., Castillo, F.J.: Mixed oligopoly with consistent conjectures. Eur. J. Oper. Res. **210**, 729–735 (2011)
4. Kalashnikov, V.V., Bulavsky, V.A., Kalashnykova, N.I.: Structure of demand and consistent conjectural variations equilibrium (CCVE) in a mixed oligopoly model. Ann. Oper. Res. **217**, 281–297 (2014)
5. Klemperer, P., Meyer, M.: Supply function equilibria in oligopoly under uncertainty. Econometrica **57**, 1243–1277 (1989)
6. Baldick, R., Grant, R., Kahn, E.: Theory and application of linear supply function equilibrium in electricity markets. J. Regul. Econ. **25**, 143–167 (2004)
7. Vasin, A., Dolmatova, M., Gao, H.: Supply function auction for linear asymmetric oligopoly: equilibrium and convergence. Procedia Comput. Sci. **55**, 112–118 (2015)
8. Ayzenberg, N., Kiseleva, M., Zorkaltsev, V.: Models of imperfect competition in analysis of siberian electricity market. J. New Econ. Assoc. **18**(2), 62–88 (2013). (in Russian)
9. Anderson, E., Hu, X.: Finding supply function equilibria with asymmetric firms. Oper. Res. **56**, 697–711 (2008)
10. Green, R.: Increasing competition in the British electricity spot market. J. Ind. Econ. **44**(2), 205–216 (1996)
11. Davidson M.R., Dogadushkina Yu.V., Kreines E.M., Novikova N.M., Udaltsov Yu.A., Shiryaeva L.V.: Mathematical model of competitive wholesale electricity market in Russia. In: Proceedings of the Russian Academy of Sciences. Theory and Systems of Control, vol. 3, pp. 72–83 (in Russian) (2004)
12. Rudkevich, A., Supply function equilibrium in power markets: learning all the way. TCA Technical paper, pp. 1299–1702 (1999)
13. Annual Report: The operator of power systems Siberia 2011. Kemerovo (2011)
14. Nahata, B., Izyumov, A., Busygin, V., Mishura, A.: Application of Ramsey model in transition economy: a Russian case study. Energy Econ. **29**, 105–125 (2007)

Chain Store Against Manufacturers: Regulation Can Mitigate Market Distortion

Igor Bykadorov[1,2,3]($\boxtimes$), Andrea Ellero[4], Stefania Funari[4], Sergey Kokovin[2,5], and Marina Pudova[3]

[1] Sobolev Institute of Mathematics SB RAS,
4 Acad. Koptyug avenue, 630090 Novosibirsk, Russia
bykadorov.igor@mail.ru
[2] Novosibirsk State University, 2 Pirogova Street, 630090 Novosibirsk, Russia
[3] Novosibirsk State University of Economics and Management,
56 Kamenskaja Street, 630099 Novosibirsk, Russia
pudova@ngs.ru
[4] Department of Management, Ca' Foscari University of Venice,
Cannaregio 873, 30121 Venice, Italy
{ellero,funari}@unive.it
[5] National Research University Higher School of Economics,
16 Soyuza Pechatnikov Street, 190068 Saint Petersburg, Russia
skokov7@gmail.com

Abstract. Contemporary domination of chain-stores in retailing is modeled, perceiving a monopolistic retailer as a *market leader*. A myriad of her suppliers compete in a monopolistic competitive sector, displaying quadratic consumers' preferences for a differentiated good. The leader announces her markup before the suppliers choose their prices/quantities. She may restrict the range of suppliers or allow for free entry. Then, a market distortion, stemming from double marginalization and excessive variety would be *softened* whenever the government allows the retailer to apply an entrance fee to the suppliers, or/and per-quantity sales subsidies (doing the opposite to usual Russian regulation).

Keywords: Industrial organization theory · Vertical relations · Chain stores · Dominant retailer · Entrance fee · Welfare

1 Introduction

Emerging domination of chain-stores like Wal-Mart in retailing has raised some controversial questions for consumers and regulating authorities. A political struggle against chain-stores resulted in some restrictions on their construction or/and operations in US and Europe. Similarly, the Russian Retailing law (2010) restricts each store's share in a city district. It also forbids entrance fees, previously required by retailers from manufacturers. Can such measures be really welfare-enhancing, as claimed by politicians?

Y. Kochetov et al. (Eds.): DOOR 2016, LNCS 9869, pp. 480–493, 2016.
DOI: 10.1007/978-3-319-44914-2_38

To sharpen this question, we model the dominant market position of a big retailer by a limiting case where only *one* monopolistic retailer faces *numerous* manufacturers (in contrast with some papers which consider a single manufacturer facing many retailers). In the present paper, alike our previous works [1,2] (see also [3]) and [4], the retailer plays as a leader against her followers – a continuum of suppliers. They compete in a monopolistically competitive sector with free entry, that involves endogenous diversity of goods, as in [5], but the representative consumer has quasi-linear quadratic preferences for a differentiated good, as in [6]. The retailer announces her markup before the suppliers choose their prices. They observe the markup and the current level of competition (price-index), correctly anticipating the demand, but ignoring the (negligible) influence on each other. The retailer correctly anticipates the equilibrium and can also restrict the mass of manufacturers entering the industry (if she finds such restriction profitable). It is a sort of two-tier monopoly, because each supplier practices monopolistic pricing on her variety of the good and the retailer's markup is added to the supplier's markup. Related market distortion can be a reason for some governmental regulation.

The first research question concerns *sales tax*: does it soften or enforce the market imperfections arising in such two-tier monopoly? We start with finding a closed-form characterization of equilibria and welfare in our market and comparing these to the socially-optimal firm size (output) and diversity (mass of firms). Rather naturally, in the absence of regulation, the direction of market distortion turns out to be two-fold: the firms tend to be too *small* and the diversity is also *insufficient* (see the explanation after Lemma 1 and Proposition 1). Thus, monopolism suppresses the market in both dimensions, and there is room for market regulation. Implementing this idea, we first find a welfare-maximizing tax level under any (possibly zero) entrance fee. As one could expect, the socially-optimal sales tax rate is negative, that means (politically problematic) *sales subsidization* (Proposition 2). The explanation suggested after Proposition 1 exploits the typical arguments of a regulated monopoly but also involves new considerations of optimal diversity.

The second research question concerns the entrance fee imposed by a retailer onto her suppliers (by tax or subsidy). Such practice was very common in Russia until the 2010 law on retailing had forbidden an explicit fee. However, the retailers responded by imposing an indirect entrance fee in the form of obligatory advertising by a manufacturer. This practice still continues, showing its importance for the market. Is an entrance fee welfare-enhancing, as the legislators supposed? Lemma 2 and Proposition 3 support the entrance fee as a socially desirable practice. The subsequent explanation emphasizes arguments in favour of two-part tariffs, usual for any monopoly. This tool "integrates the industry" to eliminate the loss due to non-cooperative vertical relations. This logic remains true in our case, in spite of complications with product diversity and monopolistic competition, unusual for Industrial Organization (IO), see [5].

Comparing our findings to those obtained in the literature on manufacturer-retailer interactions (see detailed comparison in the Online Appendix), we should cite Spengler [7] on the general idea of eliminating "double marginalization". More recent studies involve product differentiation; Perry and Groff [8] use a model where a monopolistic manufacturer signs a contract with each retailer; the number of goods is less than the socially optimal one. According to reviews on the theory of vertical integration, as in [9, 10], the manufacturer is typically the leader of the supply chain and moves first, whereas the retailer is the follower; our timing and distribution of power is exactly the opposite. Although in a different context than here, Choi [11] assumed retailer's leadership. The effects on retail prices and social welfare strongly depend on the characteristics of the demand function: social welfare is enhanced by leadership in case of nonlinear demand. More closely to the point of view of the present paper, Bykadorov et al. [3] and [4] compare the equilibrium obtained with a leading retailer with those obtained either when the retailer is the follower or when a "myopic" behavior like Nash equilibrium prevails; retailer's leadership has a positive effect on welfare. The present paper supplements this literature by studying both the optimal fiscal policy and the entrance fee, which have been not addressed so far in this context.

In the sequel, Sect. 2 formulates the model, Sect. 3 characterizes the equilibrium as a function of tax and entrance fee, and compares consumption and diversity with the socially-optimal ones. Sections 4 and 5 find the socially-optimal taxation and the impact of the entrance fee on welfare, respectively. A conclusion section summarizes and briefly comments the results. Most part of the proofs are contained in an Online Appendix [12].

2   Model

As we have pointed out, real-life vertical relations in the contemporary retailing industry typically include several huge chain-stores like Wal-Mart, who display essential bargaining power against hundreds of manufacturers-suppliers. Indeed, when rejected by a dominant chain-store, a small producer could be forced to sharply squeeze her output and sell to small shops only. Our stylized model exaggerates such retailer's bargaining power by assuming that there is only *one* retailer in a "leader's" position. She faces a mass of manufacturers, the "followers", competing in a market of differentiated goods, sold for money (the numeraire). Such monopolistic-competition framework includes several traditional assumptions: (i) a differentiated good in the sector studied, i.e., incomplete substitution among "varieties" of a good, each produced by a single firm; (ii) big number (continuum) of producers with free entry and exit in the industry; (iii) sequential moves in the game, where the bigger players move first, being able to foresee the consequences of their moves, whereas each small player ignores her (negligible) impact on the market.

In other words, the timing is the following:

(1) the retailer chooses her markup and the number of firms to deal with (i.e. the diversity in the industry);

(2) each manufacturer sets her price, considering the competition level as a parameter; whenever the current competition among firms yields negative profits, some manufacturers exit until zero-profit situation establishes;

(3) a representative consumer chooses her consumption vector treating prices, markups and variety of goods as given, paying with a numeraire.

We construct the related subgame-perfect Nash equilibrium by backward induction, starting from the last stage of the game (consumption) and describing the behavior of agents.

2.1   The Consumer-Manufacturer-Retailer Triad

Consumers. As rather usual in the modern theory of differentiated goods (see [6] and also [13]), we model a homogeneous consumers' population by a *quadratic* utility function

$$U(q, N, m) = \alpha \int_0^N q(i)di - \frac{\beta - \gamma}{2} \int_0^N [q(i)]^2 di - \frac{\gamma}{2} \left[\int_0^N q(i)di \right]^2 + m,$$

where $N > 0$ is the length of the product line, i.e., the mass of varieties in the market, so that the firms belong to the continuum $[0, N]$, which approximates a "very high" number of firms. α, β, γ are some positive parameters with the following meaning: $\alpha > 0$ is a choke-price (maximal price that consumers tolerate), $\beta - \gamma > 0$ reflects satiability of demand and, remarkable, $\gamma > 0$ reflects the degree of substitution among the N varieties. Moreover, $q(i) \geq 0$ is the consumption of i-th variety; $m \geq 0$ denotes the money remaining in the pocket after shopping.

To formulate the budget constraint, we denote the wholesale price by $p(i)$, while $r(i)$ is the retailer's markup, levied on i-th variety, and τ is the uniform sales tax (positive, zero or negative). As a result, $\breve{p}(i) = p(i) + r(i) + \tau$ is the final sale price of the i-th variety. Normalizing the wage ($w \equiv 1$) and the numeraire price ($P_A \equiv 1$), we formulate the budget constraint of the representative consumer as

$$\int_0^N \breve{p}(i)q(i)di + m \leq Y, \quad \breve{p}(i) = p(i) + r(i) + \tau,$$

where the income $Y = L + \int_0^N \pi_{\mathcal{M}}(i)di + \pi_{\mathcal{R}}$ is made up of wage L, the total profit of the manufacturers and the profit $\pi_{\mathcal{R}}$ of the retailer. Income composition plays no role in our analysis, whereas it is important to underline that income is assumed to be "sufficient" and is treated parametrically by the consumer. The right-hand side of the budget constraint can be considered as the Gross Domestic Product (GDP) of the economy measured by income, while the left-hand side represents expenditures.

Maximizing the concave utility subject to the budget constraint, we obtain the FOC (first-order condition) $\partial U / \partial q(i) = \alpha - (\beta - \gamma) q(i) - \gamma \int_0^N q(j)dj = \breve{p}(i)$. Using it and defining the composite parameters

$$a \equiv \frac{\alpha}{\beta + \gamma \cdot (N - 1)}, \quad b \equiv \frac{\gamma}{(\beta - \gamma) \cdot [\beta + \gamma \cdot (N - 1)]}, \tag{1}$$

we express the demand function $q(i) \equiv q(i, \check{p}, P, N)$ for each variety $i \in [0, N]$ as

$$\hat{q}(i, p, r, P, N) = a - \frac{1}{\beta - \gamma} \cdot [p(i) + r(i) + \tau] + bP. \qquad (2)$$

where P is the market aggregate, i.e., the utility-specific price index $P \equiv \int_0^N [p(j) + r(j) + \tau] dj$. The higher is P, the lower is the total demand for all varieties, which positively influences each individual demand for a variety, due to substitution among them ($\gamma > 0$).

Manufacturers. To formulate the problem faced by the i-th manufacturer, c will denote the marginal cost, i.e., the labor required to produce one unit of the differential good. The fixed cost $f_{\mathcal{M}}$ of manufacturing (measured in labor) is assumed to be the same for each producer. Similarly, $f_E \geq 0$ is the entrance fee for any manufacturer, possibly imposed by the retailer or by the government which can impose a licensing cost in the industry (we shall study both these market structures). Each manufacturer knows the pre-determined retailer's markup $r(i)$ and the demand function (treating the market aggregate P parametrically). So, her profit maximization can be formulated as

$$\pi_{\mathcal{M}}(i, p, r(i), P) = (p(i) - c)\,\hat{q}(i, p(i), r(i), P, N) - f_{\mathcal{M}} - f_E \to \max_{p(i)}.$$

Since function $\pi_{\mathcal{M}}$ is quadratic and strictly concave in $p(i)$, we can easily maximize it by differentiation. From FOC, we find the function of producer's response to any retailer's markup choice $r(i)$ and market index P

$$p^*(i, r(i), P, N) = \arg\max_p \pi_{\mathcal{M}}(i, p, r(i), P, N) =$$

$$= 0.5 \cdot [(\beta - \gamma) \cdot (a + bP) - r(i) - \tau + c], \qquad (3)$$

where a and b depend on N as in (1). Since this price does not depend explicitly on i, from now on, taking into account profit concavity and symmetric (homogeneous) producers, it is reasonable to consider only *symmetric* markups ($r(i) = r$, outputs ($q(i) = q$) and prices ($p(i) = p$) across all firms. So, we drop index i in the sequel. Then, plugging solution p^* into output q^* yields

$$q^*(p, r, P, N) = \frac{0.5}{\beta - \gamma} \cdot [(\beta - \gamma) \cdot (a + bP) - r - \tau - c], \qquad (4)$$

which implies, as one can derive, the equilibrium price index P as a function of τ, r and N: $P = \frac{N}{2 - (\beta - \gamma)bN} \cdot [(\beta - \gamma) \cdot a + r + \tau + c]$.

Further, plugging a, b, P into the firm's output q yields the equilibrium production which represents the response of the market to the chosen markup and to variety (r, N):

$$q^{\#}(r, N) = \frac{\alpha - (r + \tau + c)}{2\beta + \gamma \cdot (N - 2)}. \qquad (5)$$

Similarly, the equilibrium price can be expressed, using (3), as a function of r and N:

$$p^{\#}(r, N) = (\beta - \gamma) \cdot \frac{\alpha - (r + \tau + c)}{2\beta + \gamma \cdot (N - 2)} + c. \tag{6}$$

Retailer. To formulate the retailer's optimization problem, we recall that the retailer plays first, correctly anticipating the subsequent behavior of the producers and equilibrium values of $p^{\#}(r, N), q^{\#}(r, N), P$, i.e., she knows the response function (5). In the following $c_{\mathcal{R}}$ will denote the retailer's marginal cost necessary to sell one unit of the differential good of each type (it corresponds to the labor of the salesmen), and $f_{\mathcal{R}}$ will be the fixed cost, necessary to sell each type (the labor necessary to maintain the i-th shelf in the shop). The retailer simultaneously chooses the markup r and the range N of varieties, correctly anticipating the subsequent manufacturers' response and the consumers' demand. Namely, her profit can be maximized with respect to N and to the individualized markups $\mathbf{r} = \{r_i\}_{i \in [0,N]}$:

$$\pi_{\mathcal{R}} = \int_0^N [r(i) - c_{\mathcal{R}}]q^*(i, r, P(N), N)di - \int_0^N f_{\mathcal{R}}di \to \max_{\mathbf{r}, N}, \tag{7}$$

$$\text{subject to} \quad \pi_{\mathcal{M}}(r(i), N) \geq 0 \; \forall i. \tag{8}$$

However, due to the concavity property of the profit function and reasonably assuming symmetric retailer's policy with $r_1 = r_2 = \ldots = r_N = r$, we can simplify the maximization problem as:

$$\pi_{\mathcal{R}} = N \cdot \left[(r - c_{\mathcal{R}}) \cdot q^{\#}(r, N) - f_{\mathcal{R}} \right] = \tag{9}$$

$$= N \cdot \left[(r - c_{\mathcal{R}}) \cdot \frac{\alpha - (r + \tau + c)}{2\beta + \gamma \cdot (N - 2)} - f_{\mathcal{R}} \right] \to \max_{r, N}, \tag{10}$$

$$\text{subject to} \quad \pi_{\mathcal{M}}(r, N) \geq 0. \tag{11}$$

According to the constraint (11), the producers cannot make losses; this way, in view of possible free exits, the retailer will not rise her markup too much.

2.2 Equilibrium and Welfare

(Symmetric) Equilibrium. $(\bar{r}, \bar{N}, \bar{p}, \bar{q})$ is a bundle comprising the solution $(\bar{r}, \bar{N})$ to the retailer's problem (7)–(8), price $\bar{p} = p^{\#}(r, N)$ defined by (6) and each variety consumption $\bar{q} = q^{\#}(r, N)$, described in (5).

One can show (see Lemma 1 below) that non negativity constraint on the manufacturer's profit becomes active or inactive at the equilibrium, depending on the value of the "investment ratio" parameter $F \equiv \frac{f_{\mathcal{R}} - f_E}{2(f_{\mathcal{M}} + f_E)}$. We will in fact distinguish two cases: (1) "Positive-Profit equilibria" which occurs when $F \equiv \frac{f_{\mathcal{R}} - f_E}{2(f_{\mathcal{M}} + f_E)} > 1$, which means *very high retailer's fixed cost* $f_{\mathcal{R}}$, that is more than one half of the total industry fixed cost (since $f_E \geq 0$). In this case the manufacturers' no-loss condition (11) can be ignored during retailer's maximization ($\pi_{\mathcal{M}} > 0$: PP case); (2) "Zero-Profit equilibria" which occurs when $F \leq 1$ (small retailer's fixed cost); here the no-loss condition becomes active at the maximimum ($\pi_{\mathcal{M}} = 0$: ZP case).

In other words, we can say that, it is the size of the retailer's fixed cost $f_{\mathcal{R}}$, in comparison to other costs, that determines the type of market equilibrium. The retailer's fixed cost can be interpreted as the shelf-maintenance cost. Typically, in the real life the markup does not exceed the double of the manufacturers cost, therefore, case ZP looks closer to reality.

Welfare and Social Optimum. In order to investigate the social benefits of taxation/subsidization or entrance fee, we consider the social welfare function W, that measures society prosperity in a symmetric equilibrium, encompassing the consumer surplus and profits. The social welfare function for an industry (as function of consumption and variety) is built subtracting all costs from the total utility of the representative consumer:

$$W(q, N) = (\alpha - c - c_{\mathcal{R}})Nq - \frac{\beta - \gamma}{2N}N^2q^2 - \frac{\gamma}{2}N^2q^2 - (f_{\mathcal{M}} + f_{\mathcal{R}})N.$$

We may note that welfare is quadratic in N, thus ensuring a unique maximum in q under any given N. By substituting in W the equilibrium values of $\bar{q}(\tau)$ and $\bar{N}(\tau)$, which depend on the tax rate, we maximize the welfare.

To compare various market outcomes with the first-best social optimum of this industry, we find now the optimum. We maximize the quadratic function $W(q, N)$ w.r.t. both, the size of each firm q and the diversity of products N, obtaining the first-order conditions

$$(\alpha - c - c_{\mathcal{R}})N - \frac{\beta - \gamma}{N}N^2q - \gamma N^2q = (f_{\mathcal{M}} + f_{\mathcal{R}})N, \tag{12}$$

$$(\alpha - c - c_{\mathcal{R}})q - \frac{\beta - \gamma}{2}q^2 - Nq^2 = (f_{\mathcal{M}} + f_{\mathcal{R}}), \tag{13}$$

which will be used further on.

3 Equilibria Compared with Social Optimum

For both investment ratio regions, i.e., $F \equiv \frac{f_{\mathcal{R}} - f_E}{2(f_{\mathcal{M}} + f_E)} > 1$ and $0 \leq F \leq 1$, it is possible to obtain the equilibria in closed form.[1] From (10) we deduve that the retailer's objective function $\pi_{\mathcal{R}}(r, N)$ is quadratic in r, and the maximum with respect to r can be found independently from N. In the positive-profit case ($F > 1$), we can find the unconstrained maximum of $\pi_{\mathcal{R}}$. It turns out that maximization with respect to r, which does not depend from N, is equivalent to

---

[1] We find the equilibrium in three steps. First we derive the consumer's demand (2) to find $q(p(\cdot), r, N)$, then we solve the manufacturer's problem to find the pricing rule $p(r, N)$ as in (3) and finally, plugging $q(p(r, N), r, N)$ and $p(r, N)$ into the retailer's problem (7)–(8), we solve it determining the best choice (r, N).

finding a maximum of a concave parabolic function that yields a unique solution $\bar{r} = (\alpha - c - \tau + c_{\mathcal{R}})/2$. Substituting into the profit function, we obtain the following function

$$\pi_{\mathcal{R}}(N) = (\bar{r} - c_{\mathcal{R}}) \cdot \frac{N \cdot (\alpha - \bar{r} - \tau - c)}{2(\beta - \gamma) + \gamma N} - N \cdot f_{\mathcal{R}},$$

that is concave in N (here all the coefficients are positive). From the first-order conditions:

$$r + \frac{\tau - \alpha + c - c_{\mathcal{R}}}{2} + B\sqrt{(\beta - \gamma)(f_{\mathcal{M}} + f_E)} \cdot \left(1 - \frac{f_{\mathcal{R}} - f_E}{2(f_{\mathcal{M}} + f_E)}\right) = 0.$$

we find the unique maximum

$$\bar{N} = N_{pp} = \frac{\beta - \gamma}{\gamma} \cdot \left(\frac{\alpha - c - c_{\mathcal{R}} - \tau}{\sqrt{2(\beta - \gamma)(f_{\mathcal{R}} - f_E)}} - 2\right).$$

In the zero-profit case ($0 \leq F \leq 1$), from (4), exploiting constraint $\pi_{\mathcal{M}}(r, N) = 0$ we can express the variety as a function of markup r:

$$\bar{N} = N_{zp}(r) = \frac{\beta - \gamma}{\gamma} \cdot \left(\frac{\alpha - c - \tau - r}{\sqrt{(\beta - \gamma)(f_{\mathcal{R}} - f_E)}} - 2\right)$$

The retailer maximizes her profit

$$\pi_{\mathcal{R}} = N_{zp}(r) \cdot [r - c_{\mathcal{R}}]q(r, N_{zp}(r)) - N^*(r)f_{\mathcal{R}}di$$

choosing the markup r.

The equilibrium values of quantities, prices, markups and variety are summarized in the following lemma.

Lemma 1. *Under taxation $\tau \geq 0$, the equilibrium magnitudes $\bar{q}, \bar{p}, \bar{r}, \bar{N}$ and under the retailer's profit $\pi_{\mathcal{R}}$ are parametrically determined, as shown in the following table, for both cases PP ($F \equiv \frac{f_{\mathcal{R}} - f_E}{2(f_{\mathcal{M}} + f_E)} > 1$) and ZP ($F \leq 1$)*

	$\bar{q}$	$\bar{p} - c$	$\bar{r} - c_{\mathcal{R}}$	N	$\pi_{\mathcal{R}}$
PP	$\dfrac{B\sqrt{F}}{\beta - \gamma}$	$B\sqrt{F}$	D	$\dfrac{(\beta - \gamma)(D - 2B\sqrt{F})}{\gamma \cdot B\sqrt{F}}$	$\dfrac{\left(D - 2B\sqrt{F}\right)^2}{\gamma}$
ZP	$\dfrac{B}{\beta - \gamma}$	B	$D - (1 - F)B$	$\dfrac{(\beta - \gamma)(D - (1 + F)B)}{\gamma \cdot B}$	$\dfrac{(D - (1 + F)B)^2}{\gamma}$

where

$$A \equiv \alpha - c - c_{\mathcal{R}}, \quad B \equiv \sqrt{(\beta - \gamma) \cdot (f_{\mathcal{M}} + f_E)}, \quad D = \frac{A - \tau}{2}.$$

The welfare in the PP case is $\bar{W}_{pp} = \dfrac{(D - 2B\sqrt{F}) \cdot \left(\tau + 3A - 2B \cdot \frac{(2 + 3F)}{\sqrt{F}}\right)}{4\gamma}$ while in the ZP it becomes $\bar{W}_{zp} = \dfrac{1}{4\gamma} \cdot (D - B \cdot (1 + F)) \cdot (\tau + 3A - 2B \cdot (2 + 3F))$. The manufacturer's profit in the PP case is $\pi_{\mathcal{M}} = (f_{\mathcal{M}} + f_E) \cdot (F - 1)$.

Impact of parameters on equilibria and optimum. We observe that the equilibria in the two cases are rather different. When the manufacturer's profit is positive, her fixed cost f_M has no direct impact on retailer's behavior and equilibria, being just a constant in her objective function. Similarly the retailer's fixed cost f_R, when it is small enough (ZP case), has no impact on $\bar{q}, \bar{N}$.

Moreover, from Lemma 1 we have that the entrance fee f_E affects the size of the firm q in an opposite way in PP and ZP cases, lowering and increasing it, respectively. By contrast, in both cases, taxes *do not* influence the size of the equilibrium firm $\bar{q}$ and the mill price $\bar{p}$. However, markup r moves in the opposite direction than the tax: it *decreases* under growing tax $\tau > 0$ but *increases under subsidy*, as if the retailer expropriates half of the government's subsidy from the consumer, softening thereby the change in the final sales price $(\bar{p}+r+\tau)$ imposed by τ. The impact of subsidy, i.e. $\tau < 0$, on quantity $\bar{q}$ works through *increasing* diversity $\bar{N}(\tau)$. This shift, in turn, pulls *up* the price index P in such a way, that the demand for a variety in formula (2) remains unaffected. Similarly, a positive tax $\tau > 0$ should affect both the retailer's markup r and the variety N *negatively*. Which direction is socially beneficial then?

Solving the system (12)–(13) w.r.t. q, N, and comparing the market equilibrium values $\bar{q}, \bar{N}$ (regulated by tax τ and entrance fee f_E) with the social optimum, we immediately obtain the following proposition

Proposition 1. *(i)The firm's socially-optimal size and diversity are:*

$$q^{sopt} = \sqrt{2(f_M + f_R)/(\beta - \gamma)}, \tag{14}$$

$$N^{sopt} = \frac{\beta - \gamma}{\gamma} \cdot \left(\frac{\alpha - c - c_R}{\sqrt{2 \cdot (f_M + f_R)(\beta - \gamma)}} - 1 \right). \tag{15}$$

(ii) The firm's socially-optimal size q^{sopt} can be bigger than unregulated equilibrium outputs; in particular, $q^{sopt} > 2\bar{q}_{PP}$ and $q^{sopt} > \sqrt{2}\bar{q}_{ZP}$ when $f_E = 0$. However, a moderate positive entrance fee (i.e. $f_E \in (0, f_M + 2f_R]$) softens this kind of distortion in ZP case.

The distortion of output mentioned in Proposition 1 (ii) can be explained by the usual reasoning on monopoly. To cover fixed costs, each producer sets a big market price *exceeding* the marginal cost. Then, the potential mutual benefit between the firm and consumer is forgone, sales are (inefficiently) lowered by a too high price. Therefore, the firm becomes (inefficiently) small. This kind of distortion can be partially cured by an entrance fee. For instance, to reach the optimal firm size (14) in the (plausible) ZP case, the government should allow for a positive entrance fee. When it reaches magnitude $f_E = f_M + 2f_R$, the firm size could become optimal, but a positive entrance fee simultaneously changes the mass of firms, and its overall impact on welfare seems to be unclear. We will come back to this issue in another section.

Comparing the socially optimal product diversity (15) with the market outcomes ($\bar{N}$) of Lemma 1, we obtain the following claim.

Claim. *Under zero regulation* $(\tau = 0, f_E = 0)$ *and the restriction on retailer's fixed cost* $f_{\mathcal{R}} < f_{\mathcal{M}}$, *the market equilibrium diversity (mass of firms) in* ZP *case appears smaller than the socially-optimal diversity, i.e.* N^{sopt}.

Thus, in a rather reasonable case, the *market diversity appears insufficient* provided our hypothesis of non-additive quasilinear utility; with an additive non-quasilinear utility, instead, the opposite distortion appears when the demand is sufficiently flat (Dixit-Stiglitz [5]). One can also see that *a subsidy pulls the diversity up, as well as the output.*

Somewhat surprisingly, the parameter γ reflecting substitution among the varieties, does not influence the sign of the diversity distortion, but only its magnitude. The direction of distortion can be explained by (inefficiently) high market price and low quantity, as already discussed. Indeed, according to (13), the social planner, when choosing N^{sopt}, compares the left-hand side (the marginal benefit from launching one additional firm), with the right-hand side, the industry fixed cost $f_{\mathcal{M}} + f_{\mathcal{R}}$. In other words, the consumer surplus per variety is compared with a sort of marginal cost of increasing variety N.[2] According to Lemma 1, the market chooses N differently: a firm enters the market only when the gross operating profit per variety $(p + r - c - c_R)q$ weakly exceeds the gross fixed cost $f_{\mathcal{M}} + f_{\mathcal{R}}$. Clearly, a smaller q and the inability to completely absorb the potential consumer surplus, both affect negatively the market choice of diversity N, in comparison with social decision N^{sopt}. Until now we have explored only the ZP case; similar considerations could be made also in the PP case.

In general, we see that the benevolent government should choose some *different* variables (q, N) than the market does. Since a direct dictate over the industry is politically infeasible, we discuss from now on the indirect tools. These are taxation, allowing entrance fee, and licensing, used to reach a "second-best optimum".

4   Welfare Consequences of Taxation

Sales taxes are the rule in many countries, but are they welfare-enhancing under a leadership of the retailer? In this section, we focus only on taxation in order to separately investigate its effects on welfare. Therefore we assume here that entrance fee f_E is fixed (even equal to zero). Being τ the tax that a retailer pays for each unit sold, the amount $N\tau q$ is transferred to the government (or to consumers in the lump-sum form).

The tax can influence the market in the Pigouvian way, by reducing sales, while using the same logic, sales can be stimulated by sales subsidization (financed from the budget as a lump-sum taxation on consumers in the amount $N\tau q$, or by licensing).

To find a fiscal policy that maximizes social welfare, we use the formulae in Lemma 1 that depend on tax $s\tau$, and maximize the equilibrium welfare $\bar{W}$

---

[2] A negative impact on the consumer's satisfaction from other varieties (through parameter γ) could be taken into account as "a cost of launching a new firm," but with a continuum of firms this impact is negligible.

w.r.t. τ. In both cases, PP and ZP, we can obtain the explicit formulae for optimal taxation τ_W^* or *subsidization* of the retailer.

Proposition 2. *In any case it is optimal to subsidize, in the PP case the socially-optimal value of τ is $\tau_{pp}^* = \left(\sqrt{F} + \frac{2}{\sqrt{F}}\right) \cdot B - A < 0$, while in the ZP case, the optimal choice is $\tau_{zp}^* = (1 + 2 \cdot F) \cdot B - 2 \cdot \delta < 0$. Corresponding diversity and welfare magnitudes at the regulated equilibrium are:*

$$\bar{W}_{pp}\left(\tau_{pp}^*\right) = \frac{1}{8 \cdot \gamma} \cdot \left(\frac{2B}{\sqrt{F}} - 2A + 5B \cdot \sqrt{F}\right)^2,$$

$$\bar{W}_{zp}\left(\tau_{zp}^*\right) = \frac{1}{8\gamma} \cdot \left((3 + 4 \cdot F) \cdot B - 2A\right)^2,$$

$$\bar{N}_{zp}\left(\tau_{zp}^*\right) = \frac{\beta - \gamma}{\gamma} \cdot \left(\frac{A}{B} - \frac{3 + 4 \cdot F}{2}\right). \tag{16}$$

What is the motivation behind the optimality of welfare-enhancing subsidies? One side of the story is the well-known effect in monopolistic regulation. A monopolist tends to reduce the output and rise her price above the socially-efficient level (namely, under linear demand she reduces the quantity twice). As a response, the government can soften this tendency by artificially increasing the marginal revenue from each unit sold. It means a sales subsidy. This politically tough measure can be compensated by expensive license, i.e., a lump-sum payment by the monopolist to finance the market stimulation activity.

This usual monopolistic scenario is complemented in the specific context studied in this paper by a two-tier monopoly and by asking to stimulate a "socially-efficient diversity" (see Dixit and Stiglitz [5]). In this paper, under additive non-quasi-linear utility, the diversity is excessive whenever the demand function is more flat ("less log-convex"), than a power function. Under our non-additive preferences with direct substitution, we have observed the opposite tendency: *insufficient* product diversity, too few producers. As we have seen, a sales subsidy of reasonable size mitigates both kinds of distortion in outputs and in diversity. However, the first-best optimum is not reached by such a measure, as one can see comparing (15) to (16), and recalling that output is not affected by a tax.

From the above considerations, we may deduce that small outputs and insufficient diversity leads to distortion also in presence of regulation by subsidies. There is room for a heavier governmental intervention compensating both these market imperfections. Next section shows that allowing the retailer to impose an entrance fee further enhances welfare.

5   Entrance Fee and Welfare

In real life, the relationship between manufacturers and retailers has several dimensions. In particular, a manufacturer in Russia typically pays to the retailer some entrance fee monthly or annually, independently from sales volume. The fee was formally banned by the federal law, but now it takes the form of obligatory advertising by the manufacturer or/and low-price campaigns. The shops argue that the fee is necessary, it is the way to cover the fixed cost of retailing, and we shall see that such "compensation" is optimal at the equilibrium.

The entrance fee value f_E and taxes have been considered as fixed in this section. Now, unlike (9)–(11), we enrich the retailer's profit maximization problem with f_E as an additional variable:

$$\pi_{\mathcal{R}} = N \cdot (r - c_{\mathcal{R}}) \cdot q(r, N) - N \cdot (f_{\mathcal{R}} - f_E) \rightarrow \max_{r, N, f_E} \qquad (17)$$

$$\pi_{\mathcal{M}} = (p(r, N) - c)q(r, N) - (f_{\mathcal{M}} + f_E) \geq 0. \qquad (18)$$

In the Online Appendix [12], we show that asymmetric optimization (treating similar producers in a different way) cannot bring a higher profit than with the symmetric optimization above considered. Moreover, only the ZP kind of equilibria exists since $\frac{\partial \pi_{\mathcal{R}}}{\partial f_E} = N > 0$. Hence, the maximum of the objective function is achieved on the boundary, where $\pi_{\mathcal{M}} = 0$ and case PP becomes impossible. So (17)–(18) has a closed form solution obtained expressing the optimal size N via markup r and entrance fee f_E as $N = N(r, f_E) = \frac{\beta - \gamma}{\gamma} \cdot \left(\frac{A + c_{\mathcal{R}} - (r + \tau)}{B} - 2 \right)$. Substituting $f_E = f_E(N)$ into $\pi_{\mathcal{R}}$, we can simplify the profit maximization problem as

$$\pi_{\mathcal{R}}(r, f_E) = -\frac{(r - c_{\mathcal{R}} + \tau - A + 2B(f_E))(r - c_{\mathcal{R}} - 2F(f_E) \cdot B(f_E))}{\gamma} \rightarrow \max_{r, f_E}.$$

Solving the above problem, we obtain a closed-form expression for the equilibrium price and corresponding quantity and variety.

Lemma 2. *When an entrance fee $f_E > 0$ is chosen and kept by the retailer, it exactly compensates the retailer's fixed cost ($f_E = f_{\mathcal{R}}$), whereas the equilibrium quantity, price, markup, variety and profit are expressed as*

q^E	p^E	r^E	N^E	f_E	$\pi_{\mathcal{R}}^E$
$\frac{B}{\beta - \gamma}$	$B + c$	$D + c_{\mathcal{R}} - B$	$\frac{\beta - \gamma}{\gamma \cdot B} \cdot (D - B)$	$f_{\mathcal{R}}$	$\frac{1}{\gamma} \cdot (D - B)^2$

while the welfare is $W^E = -\dfrac{1}{8\gamma} \cdot (\tau - A + 2 \cdot B) \cdot (\tau + 3 \cdot A - 4 \cdot B)$.

By comparing the values p^E, q^E, W^E in Lemma 2 with the ZP-equilibrium values $\bar{p}, \bar{q}, \bar{W}$ in Lemma 1, we obtain the following conclusion about the benefits of a two-part tariff practice.

Proposition 3. *When (under any given tax τ) the retailer exploits an entrance fee $f_E > 0$ levied on the manufacturer price, the equilibrium values p^E, q^E, r^E differ from the ZP-equilibrium values p^{ZP}, q^{ZP} obtained without a fee ($f_E = 0$); more precisely: (a) $p^E > p^{ZP}$, i.e., entrance fee pushes up the mill price; (b)*

$r^E < r^{ZP}$, i.e., it pulls down the markup; (c) $p^E + r^E < p^{ZP} + r^{ZP}$, i.e., it pulls down the retail price; (d) $q^E > q^{ZP}$, i.e., it pushes up output; (e) $q^E N^E > q^{ZP} N^{ZP}$, i.e., it pushes up total consumption; (f) $W^E > W^{ZP}, \pi_\mathcal{R}^E > \pi_\mathcal{R}^{ZP}$, it pushes up social welfare and the retailer's profit.

The main finding here is that *entrance fees improve social welfare*, in contrast with the ideas of Russian legislators, prohibiting them in 2010. The explanation is the following. The relations between the retailer and the manufacturers are a sort of "two-tier monopoly" and therefore generate the well-known "double marginalization" effect. It means that each producer sets a high monopolistic price (being a monopolist for her specific variety), and the retailer adds her markup above the supplier's markup. As a result, under linear demand, the quantity consumed can become four times less than the socially-efficient one. Instead, the entrance fee or two-part-tariff is a pricing tool bringing the retailer to keep part of this social loss and thereby increase total welfare.

In essence, the entrance fee works as if the retailer should indirectly collect ownership of the industry in her hands. It is a well-known effect in IO theory, but usually it is analysed with the opposite kind of leadership: a single producer collecting the profits of her multiple retailers. Now, after several huge chain-stores became the leaders, such relations turned upside-down. This paper shows that neither the version of leadership, nor the monopolistic competition among the producers (the issue of diversity and free entry) can overthrow the benefits of the two-part tariff, with or without governmental taxation or subsidies. Primarily, to better exploit the two-part tariff benefits, the government may wish to directly choose an entrance fee (license fee), instead of allowing the retailer to choose it. Should such indirect measure be sufficient or not to attain a true optimum is a topic of further study. Other possible extensions are to study a retailing context under general additive preferences as in [14] and trade as in [15,16].

6   Conclusions

The age of dominant chain-stores requires to consider the case of a monopolistic retailer as a *leader* in the vertical retailing relations, under monopolistic competition among the producers. The retailer "plans the market," chooses the diversity and attempts to collect the benefits. Such monopolism yields socially *insufficient* mass of firms (diversity) and *insufficient* output of each firm. In this situation, negative sales tax (i.e., subsidized sales) improves welfare in both respects, output and diversity. Moreover, if the retailer imposes an entrance fee for producers (two-part tariff) to cover her fixed costs, this turns out to be beneficial for the society. This pricing tool partially overcomes "double marginalization" distortion, because it indirectly merges the industry. Thus, under our assumptions, welfare considerations *cannot justify common legal restrictions* on the chain-stores and positive sales-tax, suggesting instead the opposite public measures.

Acknowledgments. The study has been funded by the Russian Academic Excellence Project '5-100'. We gratefully acknowledge partial financing this project by grant SSD SECS-S/06, 571/2014 from Department of Management of University Ca' Foscari Venezia, by grants 15-06-05666, 16-01-00108 and 16-06-00101 from RFBR. Also we acknowledge to many our colleagues for useful comments and discussions.

References

1. Bykadorov, I.A.: Product Diversity in a Vertical Distribution Channel under Monopolistic Competition. The Economic Education and Research Consortium (EERC), Working Paper 10/03E (2010)
2. Bykadorov, I.A., Kokovin, S.G., Zhelobodko, E.V.: Product diversity in a vertical distribution channel under monopolistic competition. Math. Game Theory Appl. **2**(2), 3–41 (2010). (In Russian)
3. Bykadorov, I.A., Kokovin, S.G., Zhelobodko, E.V.: Product diversity in a vertical distribution channel under monopolistic competition. Autom. Remote Control. **75**, 1503–1524 (2014)
4. Bykadorov, I.A., Kokovin, S.G.: Effectiveness of retailer's market power: monopolistic competition of producers. Vestnik NSUEM **1**, 326–337 (2014). (In Russian)
5. Dixit, A.K., Stiglitz, J.E.: Monopolistic competition and optimum product diversity. Am. Econ. Rev. **67**, 297–308 (1977)
6. Ottaviano, G.I.P., Tabuchi, T., Thisse, J.-F.: Agglomeration and trade revised. Int. Econ. Rev. **43**, 409–436 (2002)
7. Spengler, J.J.: Vertical integration and antitrust policy. J. Polit. Economy. **53**, 347–352 (1950)
8. Perry, M.K., Groff, R.H.: Resale price maintenance and forward integration into a monopolistically competitive industry. Q. J. Econ. **100**, 1293–1311 (1985)
9. Tirole, J.: The Theory of Industrial Organization. MIT Press, Cambridge (1988)
10. Ingene, C.A., Parry, M.E.: Mathematical Models of Distribution Channels. Kluwer Academic Publishers, New York (2004)
11. Choi, S.C.: Price competition in a channel structure with a common retailer. Mark. Sci. **10**, 271–296 (1991)
12. Bykadorov, I., Ellero, A., Funari, S., Kokovin, S., Pudova, M.: Chain store against manufacturers: regulation can mitigate market distortion. Online appendix (2016). https://www.hse.ru/en/org/persons/38617864-Other
13. Combes, P.-P., Mayer, T., Thisse, J.-F.: Economic Geography. The Integration of Regions and Nations. Princeton University Press, Princeton (2008)
14. Zhelobodko, E., Kokovin, S., Parenti, M., Thisse, J.-F.: Monopolistic competition in general equilibrium: Beyond the Constant Elasticity of Substitution. Econometrica **80**(6), 2765–2784 (2012)
15. Antoshchenkova, I.V., Bykadorov, I.A.: Monopolistic competition model: the influence of technological progress on equilibrium and social optimality. Math. Game Theory Appl. **6**(2), 3–31 (2014). (In Russian)
16. Bykadorov, I., Gorn, A., Kokovin, S., Zhelobodko, E.: Why are losses from trade unlikely? Econ. Lett. **129**, 35–38 (2015)

On the Existence of Immigration Proof Partition into Countries in Multidimensional Space

Valeriy M. Marakulin[1,2(✉)]

[1] Sobolev Institute of Mathematics RAS,
4 Acad. Koptyug Avenue, Novosibirsk, Russia
marakulv@gmail.com
[2] Novosibirsk State University, 2 Pirogova Street, Novosibirsk, Russia

Abstract. The existence of immigration proof partition for communities (countries) in a multidimensional space is studied. This is a Tiebout type equilibrium its existence previously was stated only in one-dimensional setting. The migration stability means that the inhabitants of a frontier have no incentives to change jurisdiction (an inhabitant at every frontier point has equal costs for all possible adjoining jurisdictions). It means that inter-country boundary is represented by a continuous curve (surface).

Provided that the population density is measurable two approaches are suggested: the first one applies an one-dimensional approximation, for which a fixed point (via Kakutani theorem) can be found after that passing to limits gives the result; the second one employs a new generalization of Krasnosel'skii fixed point theorem for polytopes. This approach develops [8] and extends the result to an arbitrary number of countries, arbitrary dimension, possibly continuous dependence on additional parameters and so on.

Keywords: Country formation · Alesina and Spolaore's world · Migration · Stable partitions · Multidimensional space · Krasnosel'skii fixed point theorem

1 Introduction

In the seminal paper [1] a basic model of country formation was offered. In this model, the cost of the population individuum is described as the sum of the two values—the ratio of total costs and the total weight of the population plus transportation costs to the center of the state. This model has been investigated in a number of subsequent studies, but in each of them deals with the case of one-dimensional region and the interval-form countries (country formation on the interval $[0, 1]$).

A progress in the resolution of the problem of existence was made in [2], where the well known Gale–Nikaido–Debreu lemma was applied to state the existence of *nontrivial* immigration proof partition for interval countries, i.e. such that no one has incentive to change their country of residence. In [2] rather strong

Y. Kochetov et al. (Eds.): DOOR 2016, LNCS 9869, pp. 494–508, 2016.
DOI: 10.1007/978-3-319-44914-2_39

assumptions were made on the distribution of the population—continuous density, separated from zero. Next, in [5] the mathematical part of the approach was significantly strengthened and extended to the case of distribution of the population, described as a Radon measure (probability measure defined on the Borel σ-algebra). In [8] a new significant advancement was suggested; it disseminates the result (existence theorem) to the case of 2 or higher dimensional region. The proof in [8] is very elegant and is based on the application of KKM-lemma (Knaster–Kuratowski–Mazurkiewicz), but the result is essentially limited by the presence of capitals with fixed positions in the space. In this paper, I intend to take the next step and let capitals (or other relevant parameters) be changed continuously in space, which is important for example in the context of party formation. The proof is based on a new original generalization of Krasnosel'skii fixed point theorem, which is extended to the case of a convex polytope (bounded polyhedron) that is interesting in its own right.

In the second section, we consider a particular case of division of a rectangular area into two countries at a given measurable random distribution of the population. Here a basic one-dimensional approximation is described, for which a fixed point (via Kakutani theorem) can be found, and then the limit process gives the result.

The third section provides further generalization of the existence result which is extended to an arbitrary number of countries, arbitrary dimension, and possibly continuous dependence on a finite number of significant parameters for country formation (capitals and so on).

2   The Partition into Two Countries on the Plane via One-Dimensional Approximation

The division of the one-dimensional world on countries surely cannot be considered as a satisfactory solution of the problem. However, 2-dimensional formulation seems to be a fundamentally more difficult problem. Now, for a particular example of division of a rectangular area in two countries, we consider an approximating design allowing to find a solution by passing to the limit.

First, we define the principle of stability which is applied to countries located on the plane. As in the case of one-dimensional world, it must be such division that boundary residents have no incentive to change their jurisdiction. Thus, the costs for any boundary resident should be the same with respect to any of the possible for her/him adjoining jurisdictions. It is assumed that the boundaries between two countries allow continuous parametrization, i.e. they are an image of the interval from $\mathbb{R}$ for some *continuous one-to-one* mapping. As a result, as in the one-dimensional case, the function of individual costs of inhabitants should be continuous on *the whole field* of country division, that is, country partition must implement continuous "gluing" of country-depended individual costs.

For the sake of simplicity, we consider now a particular case of a rectangular area of possible settlement represented by rectangle $\square ABCD$ in the Fig. 1. We assume that $c_i(\cdot)$, $i = 1, \ldots, n$, are functions of individual costs, depending on the place of individual location, given by coordinates $(x, y) \in \square ABCD$, the weight of the resident jurisdiction $\mu_i(S_i)$, the location of its center $r_c(S_i)$, metrics $\rho(\cdot, \cdot)$ (to specify the distance to the center) and so on. The basic model representation of these cost functions is

$$c_i(x, y, \delta_i, r_c(S_i)) = \frac{g_i}{\delta_i} + \rho((x, y), r_c(S_i)), \quad g_i > 0, \quad i \in N = \{1, 2, \ldots, n\}. \quad (1)$$

Here scalar variables $\delta_i > 0$ are associated with the i-th country mass of population, i.e. $\delta_i = \mu_i(S_i)$; $g_i > 0$ is an expenditure (costs) on the maintenance of government and they are uniformly distributed among the country citizens. The second summand $\rho((x, y), r_c(S_i))$ presents an individual expenditure specified by inhabitant location at the point $(x, y) \in \square ABCD$. In general, cost functions may have sufficiently general form but they always continuously depend on certain country parameters and obey some other specific assumptions (see Sect. 3). Everywhere below we assume

(**P**) *The distribution of population is described by an absolutely continuous probability measure μ such that* $\mathrm{supp}(\mu) = \square ABCD$.[1]

The idea of approach is that given coordinate system (potentially curved), a stable partition, specified for one-dimensional world, must be implemented along every coordinate line. At the same time, the function of individual costs must be calculated relative to the position of "center" of the country and the general population distributed in *two-dimensional space*. It is not easy to find such a partition. To solve the problem we apply a special "one-dimensional approximation", relatively which a country partition can be found by a fixed point theorem (Brouwer or Kakutani).

The construction is as follows: specify $m - 2$ straight lines parallel to the base of the rectangle, $m \geq 3$. Let the lower base have the number m, the top one—the number 1, and all others be numbered from the top to the bottom. Each i-th segment is divided into two parts by the point x_i, which can be considered the point from interval $[0, 1]$ (length of the base $\square ABCD$), $i = 1, \ldots, m$. Straight line segments connecting consecutive points $x_1, \ldots, x_m$, form a polygon line, which we accept as the boundary between the left and right countries. Now, if density $f(x, y)$ is presented then it is possible to integrate it over each of the country area, finding the weights (size) $\mu(S)$ of their populations.

Within each country its "center" (the capital) $r_c(S) \in S$ is specified. We assume these positions *depend continuously* from a given country settings $\mathbf{x} = (x_1, \ldots, x_m) \in [0, 1]^m$. Thus we have:

$$\mu(\mathbf{S_{left}}) = \int_{\mathbf{S_{left}}} f(x, y) dx dy \geq 0, \quad r_c(\mathbf{S_{left}}) = r_{\mathbf{left}}(x_1, \ldots, x_m) \in \mathbf{S_{left}}$$

$$\mu(\mathbf{S_{right}}) = \int_{\mathbf{S_{right}}} f(x, y) dx dy \geq 0, \quad r_c(\mathbf{S_{right}}) = r_{\mathbf{right}}(x_1, \ldots, x_m) \in \mathbf{S_{right}}.$$

---

[1] This being combined means that $\mu(A) > 0 \iff \int_A dx dy > 0$ for every measurable $A \subseteq \square ABCD$.

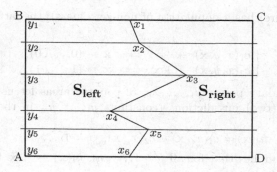

Fig. 1. Possible division into two countries of the rectangular area $ABCD$, $m = 6$.

Moreover, without loss of generality

$$\mu(\mathbf{S_{left}}) + \mu(\mathbf{S_{right}}) = 1.$$

The fact that we talk about the "mass of the population" and the "distance to the center" (transport availability of capital) as the main parameters determining the costs of individuals in a country is only an interpretation of the cost function in the context of the main model variant. The same can be said about the property of the center of the country be located on its territory—it is just a natural variant of content, from a mathematical point of view, the center could be anywhere. The really important fact is (described below) certain specific properties of individual costs.

Next we consider a point-to-set mapping, whose fixed point gives the desired country partition. The construction of mapping applies the ideas borrowed from the one-dimensional case, see [6]. Define

$$X = [0, 1]^m.$$

Now we specify a point-to-set mapping of X into itself.

Let $c_1(\cdot)$, $c_2(\cdot)$ be the functions of individual costs depending on the weight of the jurisdiction population $\mu_1(\mathbf{x})$, $\mu_2(\mathbf{x})$, location of its center $r_c(S_1)$, $r_c(S_2)$, metrics $\rho(\cdot, \cdot)$ (to determine the distance to the center) and a place of the individual location specified by coordinates $(x, y) \in \square ABCD$. The basic model representation of these functions is (1). Now we shall consider that they are functions of a general form continuously depending on $\mathbf{x} = (x_1, \ldots, x_m) \in [0, 1]^m$ for $\mu(S_k(\mathbf{x})) > 0$, $k = 1, 2$. Additionally, we assume that

(i) $c_k(x, y, \mathbf{x}) > 0$ for $\mu(S_k) \neq 0$ and
(ii) $c_k(x, y, \mathbf{x}) \to +\infty$ if $\mu(S_k) \to 0$, $k = 1, 2$.

For the functions of (1) this condition is always satisfied. At the same time, if the density $f(\cdot)$ of the population is so that $\int_A dxdy > 0$ implies $\int_A f(x, y)dxdy > 0$ for every measurable subset $A \subset \square ABCD$ (i.e. each subset of nonzero area

(Lebesgue measure) has a population of non-zero mass), the latter requirement is equivalent to

$$\begin{cases} c_1(x, y, \mathbf{x}) \to +\infty \iff \mathbf{x} \to (0, \ldots, 0), \\ c_2(x, y, \mathbf{x}) \to +\infty \iff \mathbf{x} \to (1, \ldots, 1). \end{cases} \tag{2}$$

For the boundary points $x_1, \ldots, x_m$ of country areas let us find an excess cost of possible (two) jurisdictions (constants $y_1, \ldots, y_m$ in the argument are excluded)

$$h_i(\mathbf{x}) = c_1(x_i, \mathbf{x}) - c_2(x_i, \mathbf{x}), \quad i = 1, \ldots, m.$$

Notice that (2) implies that for all $i = 1, \ldots, m$, $h_i(\mathbf{x}) \to +\infty$ for $\mathbf{x} \to 0$, and $\mathbf{x} \to 1$ when $h_i(\mathbf{x}) \to -\infty$.

Next we define the (single-valued) map $\varphi : X \to X = [0, 1]^m$ putting

$$\varphi_i(\mathbf{x}) = \begin{cases} x_i - \frac{x_i}{2} \cdot \frac{h_i(\mathbf{x})}{1 + h_i(\mathbf{x})}, & \text{for } h_i(\mathbf{x}) \geq 0, \\ x_i + \frac{1 - x_i}{2} \cdot \frac{h_i(\mathbf{x})}{h_i(\mathbf{x}) - 1}, & \text{for } h_i(\mathbf{x}) \leq 0. \end{cases} \tag{3}$$

By construction, this mapping is well defined everywhere on X with the exception of two points $\mathbf{x} = \mathbf{0} = (0, \ldots, 0)$ and $\mathbf{x} = \mathbf{1} = (1, \ldots, 1)$, values of which can be defined by continuity:

$$\varphi(\mathbf{0}) = (0, \ldots, 0), \quad \varphi(\mathbf{1}) = (1, \ldots, 1).$$

It is obvious that according to the construction these points are *trivial* fixed points of $\varphi(\cdot)$, that does not comply with the requirements of the division of rectangular area. Further construction and analysis will focus on finding of the *nontrivial* fixed point corresponding to the division of the area into two countries with non-zero masses of the population.

Now we define a point-to-set mapping Φ from $\mathfrak{X} = X \times \Delta$ to X, where $\Delta = \{(\mu_1, \mu_2) \mid \mu_1 + \mu_2 = 1, \ \mu_1 \geq 0, \ \mu_2 \geq 0\}$, by formula: for $(\mu_1, \mu_2) = (\mu(\mathbf{S}_{\mathbf{left}}(\mathbf{x})), \mu(\mathbf{S}_{\mathbf{right}}(\mathbf{x})))$ specify

$$\Phi(\mathbf{x}, \nu) = \begin{cases} \{\frac{\nu_1}{\mu_1} \varphi(\mathbf{x})\}, & \text{for } \nu_1 \leq \mu_1, \ \mu_1 \neq 0, \\ \{\frac{\nu_2}{\mu_2} \varphi(\mathbf{x}) + \frac{\mu_2 - \nu_2}{\mu_2} (1, \ldots, 1)\}, & \text{for } \nu_2 \leq \mu_2, \ \mu_2 \neq 0, \\ X, & \text{for } \nu_1 = \mu_1 = 0 \text{ or } \nu_1 = \mu_1 = 1. \end{cases} \tag{4}$$

The second mapping $\Psi : X \Rightarrow \Delta$ is specified as follows

$$\Psi(\mathbf{x}) = \operatorname*{argmax}_{\nu \in \Delta} \langle H(\mathbf{x}), \nu \rangle. \tag{5}$$

where $H(\mathbf{x}) = (H_1(\mathbf{x}), H_2(\mathbf{x}))$ and

$$I_+ = \{i \mid h_i(\mathbf{x}) \geq 0, \ i = 1, \ldots, n\}, \quad I_- = \{i \mid h_i(\mathbf{x}) \leq 0, \ i = 1, \ldots, n\}$$

are defined by formulas[2]

$$H_1(\mathbf{x}) = [\inf_{i=1,\ldots,m} h_i(\mathbf{x})]^+ + \sum_{i \in I_+} x_i \frac{h_i(\mathbf{x})}{h_i(\mathbf{x}) + 1}, \qquad I_+ \neq \emptyset$$

$$H_2(\mathbf{x}) = [\sup_{i=1,\ldots,m} h_i(\mathbf{x})]^- + \sum_{i \in I_-} (1 - x_i) \frac{h_i(\mathbf{x})}{h_i(\mathbf{x}) - 1}, \quad I_- \neq \emptyset.$$

---

[2] We use standard notations $z^+ = \sup\{z, 0\}$ and $z^- = \sup\{(-z), 0\}$ for any real z.

If $I_+ = \emptyset$ or $I_- = \emptyset$, then by definition $H_1(\mathbf{x}) = 0$ and $H_2(\mathbf{x}) = 0$, respectively. Constructed map is well defined everywhere excepting $\mathbf{0}$ and $\mathbf{1}$ for which we postulate

$$\Psi(\mathbf{0}) = (1,0), \quad \Psi(\mathbf{1}) = (0,1).$$

Finally, we define the resulting mapping

$$\Upsilon : \mathfrak{X} \Rightarrow \mathfrak{X}, \quad \Upsilon(\mathbf{x}, \nu) = \Phi(\mathbf{x}, \nu) \times \Psi(\mathbf{x}, \nu);$$

its fixed points give us the desired result. The following lemma describes the important properties of the mapping $\Upsilon(\cdot)$.

Lemma 1. *The mapping* $\Upsilon : \mathfrak{X} \Rightarrow \mathfrak{X}$ *is a Kakutani map, i.e. it has closed graph and for every* $\kappa \in \mathfrak{X}$ *takes non-empty convex values.*

Proof of Lemma 1. We check the properties of $\Psi(\cdot)$. We need to show that it has a closed graph. First, we establish the continuity of $H = (H_1, H_2)$. To this end, we consider the functions

$$g^-(t) = \begin{cases} \frac{t}{t-1}, & \text{for } t \leq 0, \\ 0, & \text{for } t \geq 0, \end{cases} \qquad g^+(t) = \begin{cases} \frac{t}{t+1}, & \text{for } t \geq 0, \\ 0, & \text{for } t \leq 0, \end{cases}$$

which obviously are continuous on $[-\infty, +\infty]$. From the construction one can now derive

$$H_1(\mathbf{x}) = [\inf_{i=1,\dots,m} h_i(\mathbf{x})]^+ + \sum_{i=1}^m x_i \cdot g^+(h_i(\mathbf{x})),$$

$$H_2(\mathbf{x}) = [\sup_{i=1,\dots,m} h_i(\mathbf{x})]^- + \sum_{i=1}^m (1 - x_i) g^-(h_i(\mathbf{x})).$$

This form of representation clearly implies the continuity of $H(\cdot)$ at all points except for $\mathbf{0}$ and $\mathbf{1}$. So, everywhere on X, excepting these points, $\Psi(\cdot)$ is closed. It is also closed at zero, since by construction (due to the first term) $H_1(\mathbf{x}) > 0$ and $H_2(\mathbf{x}) = 0$ for all $\mathbf{x}$ sufficiently close to zero. Consequently, in some neighborhood of zero $\Psi(\mathbf{x}) \equiv (1,0)$, which means that the closure of $\Psi(\cdot)$ at $\mathbf{0}$. Closeness at $\mathbf{1}$ is stated in a similar way.

All other required properties of the mapping $\Upsilon(\cdot)$ are established by a routine checking of definitions. Lemma is proved.

Lemma 2. *Under the above assumptions, the map* $\varphi(\cdot)$ *has **nontrivial** fixed point in X such that the mass of the population of each country is **nonzero**.*

Proof of Lemma 2. Consider any fixed point

$$(\bar{\mathbf{x}}, \bar{\nu}) \in \Upsilon(\bar{\mathbf{x}}, \bar{\nu}),$$

which does exist due to Lemma 1 and Kakutani fixed point theorem. Let us show that this point satisfies

$$0 < \bar{\nu}_1 < 1 \quad \& \quad \bar{\mathbf{x}} \neq \mathbf{0}, \quad \bar{\mathbf{x}} \neq \mathbf{1}. \tag{6}$$

Suppose that the first country has zero mass of the population, that is $\mu(\mathbf{S}_{\mathbf{left}}(\bar{\mathbf{x}})) = \mu_1 = 0$. This is possible only if $\bar{\mathbf{x}} = 0$ that implies $h_i(\bar{\mathbf{x}}) = +\infty$ $\forall i = 1, \ldots, m \Rightarrow H_1(\bar{\mathbf{x}}) > 0$ and $H_2(\bar{\mathbf{x}}) = 0$. Now by formula (5) and properties of the fixed point we conclude $\bar{\nu} = (\bar{\nu}_1, \bar{\nu}_2) = (1, 0)$ that due to (4) in the case $\nu_1 = 1 \geq 0 = \mu_1$ and $\mu_2 = 1$, $\nu_2 = 0$ implies

$$\frac{\bar{\nu}_2}{\mu_2}\varphi(\mathbf{x}) + \frac{\mu_2 - \bar{\nu}_2}{\mu_2}(1, \ldots, 1) = (1, \ldots, 1) \neq 0 = \bar{\mathbf{x}}.$$

This contradiction proves $\bar{\mathbf{x}} \neq 0$.

The case of the second country with zero population mass is considered in a similar way:

$$\mu(\mathbf{S}_{\mathbf{right}}(\bar{\mathbf{x}})) = \mu_2 = 0 \iff \bar{\mathbf{x}} = (1, \ldots, 1) \Rightarrow h_i(\bar{\mathbf{x}}) = -\infty \; \forall i = 1, \ldots, m.$$

Therefore, $H_2(\bar{\mathbf{x}}) > 0$ and $H_1(\bar{\mathbf{x}}) = 0$, that due to (5) implies $\bar{\nu} = (\bar{\nu}_1, \bar{\nu}_2) = (0, 1)$. By construction (4) in the case $\nu_1 \leq \mu_1$ and $\mu_1 = 1$, $\nu_1 = 0$ one has

$$\frac{\bar{\nu}_1}{\mu_1}\varphi(\bar{\mathbf{x}}) = (0, \ldots, 0) \neq (1, \ldots, 1) = \bar{\mathbf{x}},$$

that proves (6). This due to (5) allows to conclude $H_1(\bar{\mathbf{x}}) = H_2(\bar{\mathbf{x}})$. Let us show now that $H_1(\bar{\mathbf{x}}) = H_2(\bar{\mathbf{x}}) = 0$ is the only possibility.

Suppose $H_1(\bar{\mathbf{x}}) = H_2(\bar{\mathbf{x}}) \neq 0$. Firstly notice that $[\inf_{i=1,\ldots,n} h_i(\bar{\mathbf{x}})]^+ > 0$ is now impossible since otherwise $H_1(\bar{\mathbf{x}}) > 0$ and $H_2(\bar{\mathbf{x}}) = 0$ that is invalid. Likewise, it is impossible $[\sup_{i=1,\ldots,n} h_i(\bar{\mathbf{x}})]^- > 0$. Therefore, both of these terms in the definition of H vanish. Now, from the definition of H one can conclude that there are i, j such that

$$h_i(\bar{\mathbf{x}}) > 0 \; \& \; \bar{x}_i \cdot \frac{h_i(\bar{\mathbf{x}})}{h_i(\bar{\mathbf{x}}) + 1} \neq 0 \Rightarrow \bar{x}_i > 0,$$

$$h_j(\bar{\mathbf{x}}) < 0 \; \& \; (1 - \bar{x}_j) \cdot \frac{h_j(\bar{\mathbf{x}})}{h_j(\bar{\mathbf{x}}) - 1} \neq 0 \Rightarrow \bar{x}_j < 1.$$

Next, we turn again to the properties of the fixed point and the formula (4). In the first case, for $0 < \nu_1 \leq \mu_1 < 1 \Rightarrow 0 < \lambda = \frac{\nu_1}{\mu_1} \leq 1$, via $\bar{x}_i > 0$ we have

$$\bar{x}_i > \Phi_i(\bar{\mathbf{x}}, \bar{\nu}) = \lambda\varphi_i(\bar{\mathbf{x}}) = \lambda\left[\bar{x}_i - \frac{\bar{x}_i}{2} \cdot \frac{h_i(\bar{\mathbf{x}})}{h_i(\bar{\mathbf{x}}) + 1}\right].$$

In the second case, for $0 < \lambda = \frac{\nu_2}{\mu_2} \leq 1$, via $\bar{x}_j < 1$ we have

$$\bar{x}_j < \Phi_i(\bar{\mathbf{x}}, \bar{\nu}) = \lambda\varphi_j(\bar{\mathbf{x}}) + 1 - \lambda = \lambda\left[x_j + \frac{1 - \bar{x}_j}{2} \cdot \frac{h_j(\bar{\mathbf{x}})}{h_j(\bar{\mathbf{x}}) - 1}\right] + 1 - \lambda.$$

Both cases are impossible. Consequently, it is proved $H_1(\bar{\mathbf{x}}) = H_2(\bar{\mathbf{x}}) = 0$. By construction, this is equivalent to

$$\bar{x}_i\frac{h_i(\bar{\mathbf{x}})}{h_i(\bar{\mathbf{x}}) + 1} = 0, \; h_i(\bar{\mathbf{x}}) \geq 0 \; \forall i = 1, \ldots, m,$$

$$(1 - \bar{x}_j)\frac{h_j(\bar{\mathbf{x}})}{h_j(\bar{\mathbf{x}}) + 1} = 0 \quad h_j(\bar{\mathbf{x}}) \leq 0 \quad \forall j = 1, \ldots, m.$$

Now due to (3) this means that $\bar{\mathbf{x}} \in X$ is a nontrivial fixed point of $\varphi(\cdot)$.

Theorem 1. *Let the individual costs be given by (1) and centers be situated on a line parallel to the axis of abscissa. Then for each positive integer $m \in \mathbb{N}$ there exists the partition of $\square ABCD$ into two countries $\mathbf{S}_{\mathbf{left}}(\mathbf{x})$ and $\mathbf{S}_{\mathbf{right}}(\mathbf{x})$, with piecewise linear boundary formed by the points $x_k, \ldots, x_l$, $1 < k+1 \leq l-1 < m$ where all $x_{k+1}, \ldots, x_{l-1}$ are immigration proof.*

Corollary 1. *Let the costs in formula (1) be calculated relative to the Euclidean distance. Then in the conditions of Theorem 1 boundary points $x_{k+1}, \ldots, x_{l-1}$ are suited on classical hyperbola. In the case of a more general form of the metric (for example, for p-norm), these points belong to a generalized hyperbola.*

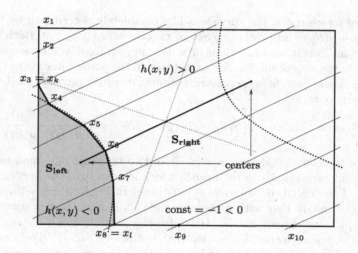

Fig. 2. Partition according to (i)–(ii) for *const* $< 0 \iff g_2\mu(\mathbf{S}_{\mathbf{left}}) < g_1\mu(\mathbf{S}_{\mathbf{right}})$.

Proof of Theorem 1. Consider a fixed point

$$\mathbf{x} = (x_1, \ldots, x_n) = \varphi(\mathbf{x}),$$

which satisfies the conclusion of Lemma 2. Note that from the construction of $\varphi(\cdot)$ for each $i = 1, \ldots, m$ (formula (3)) only one of three possibilities is realized:

(i) $h_i(\mathbf{x}) = 0$,
(ii) $h_i(\mathbf{x}) > 0 \Rightarrow x_i = 0$,
(iii) $h_i(\mathbf{x}) < 0 \Rightarrow x_i = 1$.

Indeed, for example consider alternative (ii). Assuming the contrary, one concludes $x_i \frac{h_i(\mathbf{x})}{h_i(\mathbf{x})+1} > 0$, that implies $x_i > \varphi_i(\mathbf{x})$—contradiction with fixed point. Alternative (iii) is checked out in a similar manner.

Next we consider alternative (i) and the corresponding set of points x_i onto coordinate segments. All these points can be described as the intersection of the coordinate segments with the curve described by the equation

$$h(x,y) = c_1(x,y,\mathbf{x}) - c_2(x,y,\mathbf{x}) = 0,$$

where $\mathbf{x}$ can be treated as a constant. Specifically, $\mathbf{x} = (x_1, \ldots, x_m)$ plays the role of parameters defining curve in the most general terms. To illustrate the idea and to formulate concrete result we turn to the analysis of a particular case, given in formula (1), recall:

$$c_k(x,y,\mu(S_k),r_c(S_k)) = \frac{g_k}{\mu(S_k)} + \rho((x,y),r_c(S_k)), \quad g_k > 0, \quad k = 1,2.$$

Here we are interested in the curve which is completely determined by the population mass $\mu(S_k)$ and centers $r_c(S_k)$ of two countries $k = 1,2$. Both of these parameters are continuous functions of $\mathbf{x}$. For a given fixed point they are fixed. Therefore, in the case of the Euclidean distance in the plane equation of the curve defines the classic *hyperbola* (geometric definition), which presents the boundary between two countries:

$$h(x,y) = 0 \iff ||(x,y)-r_c(S_1)||_2 - ||(x,y)-r_c(S_2)||_2 = \frac{g_2}{\mu(S_2)} - \frac{g_1}{\mu(S_1)} = const.$$

The sign of the constant determines which of two branches one must take: negative constant corresponds to the branch which is nearest to the first center and vice versa. The described situation is illustrated in Fig. 2 which represents the hyperbolic boundary case with a negative right hand side. The alternatives (i)–(ii) are implemented now and (i), (iii)—for the polar case. Of course, case (i) is possible in a pure form. Notice also that options (ii) and (iii) do not occur simultaneously: this follows from the convexity of the rectangular area and the convexity of one of the areas bounded by the hyperbola.

Finally, as soon as the centers of the country are located on a common straight line parallel to the base, this line is parallel to the coordinate segments and therefore each of these segments has the *only* point of intersection with the hyperbola or do not intersect it at all (the cases (ii) and (iii)). It establishes the existence of numbers k and l from the theorem statement. Theorem 1 is proved.

Theorem 2. *Let for the rectangle $\square ABCD$ individual costs be defined by (1) and centers of the country be located on a line parallel to the axis of abscissa. Then there is immigration proof division into two countries $\mathbf{S_{left}}$ and $\mathbf{S_{right}}$ with a continuous boundary.*

Remark 1. It is an immaterial fact that the considered area is a rectangular. This result holds for any convex closed bounded domain. So, this result can be generalized, and the continuous dependence on parameters defining the country center is the only requirement, but it will require substantial transformation of presented proof. The simplest method is to consider moving coordinate lines parallel to the line passing through the centers of countries.   □

Proof of Theorem 2. Let us consider an increasing family

$$Y_\xi \subset Y_{\xi+1} \subset [0,1], \quad \xi = 1, 2, \ldots$$

of points on the y-axis defining intercountry piecewise–linear boundary. We choose a family so that

$$\mathrm{cl}\left(\bigcup_{\xi \in \mathbb{N}} Y_\xi\right) = [0,1].$$

For every $\xi \in \mathbb{N}$, Lemma 1 takes place, that implies: for every ξ hyperbola is specified by the parameters of the country centers (focuses) $r_c(\mathbf{S}^\xi_{\mathbf{left}})$, $r_c(\mathbf{S}^\xi_{\mathbf{right}})$ and "population masses" $\mu(\mathbf{S}^\xi_{\mathbf{left}})$, $\mu(\mathbf{S}^\xi_{\mathbf{right}})$. These parameters vary under limits and therefore they contain convergent subsequences. Without loss of generality we can assume that already presented sequences are converged. Limit values

$$\bar{\mathbf{r}}_\mathbf{k} = \lim_\xi r_c(\mathbf{S}^\xi_\mathbf{k}), \quad \bar{\mu}_\mathbf{k} = \lim_\xi \mu_k(\mathbf{S}^\xi_\mathbf{k}), \quad k = 1, 2$$

define a limit hyperbola. For this hyperbola one can easily prove two key facts that give the desired result:

(i) $\bar{\mu}_\mathbf{k} \neq 0$, $k = 1, 2$, proof by contradiction with the fixed point property $\mathbf{x}_\xi \in \varphi(\mathbf{x}_\xi)$ for all $\xi \in \mathbb{N}$.

(ii) Let $\bar{\xi} \in \mathbb{N}$ and $y_{\bar{\xi}} \in Y_{\bar{\xi}}$ be fixed. As soon as $Y_{\bar{\xi}} \subset Y_\xi \; \forall \xi \geq \bar{\xi}$, then a sequence $(x_\xi, y_{\bar{\xi}})$, $\xi \geq \bar{\xi}$ of points is defined; they satisfy all fixed point relations. In the rectangle they are, starting with some number, either points located on the left or on the right hand side, or couple $(x_\xi, y_{\bar{\xi}})$ is placed on ξ-th hyperbola (the intersection of $\bar{\xi}$-th segment with the hyperbola). Since hyperbola converge to the limit option, then their (the only!) points of intersection with a fixed line will be convergent, i.e. $(x_\xi, y_{\bar{\xi}}) \to (\bar{x}_{\bar{\xi}}, y_{\bar{\xi}})$, $\xi \to \infty$. Consequently, the limit values of the population $\bar{\mu}_\mathbf{k} = \lim_\xi \mu_k(\mathbf{S}^\xi_\mathbf{k})$, $k = 1, 2$ for countries with piecewise-linear boundaries *coincide* with the value (mass) of the population of marginal hyperbola areas.

Thus, we have found a nontrivial fixed point this is the map (in the space of continuous functions) whose graph consists of a (non-empty) intersection of the hyperbola with the area, and possibly two vertical segments. This fragment of the hyperbola is the desired boundary between two countries. Theorem 2 is proved.

3 General Partition into Three or More Countries

Now we consider a general method that allows us to establish the existence of immigration proof division into n countries not only on the plane, but in any finite-dimensional space. It is not a possible generalization only, but an opportunity in its context to consider more general problems, e.g. partition according to party affiliation.

The initial construction is similar to the one proposed in [8]. We need to divide the area $\mathcal{A} \subset \mathbb{R}^l$ into n counties, $N = \{1, \ldots, n\}$. The difference is that the cost function $c_i(\cdot)$ may depend not only on the mass $\delta_i \in [0, 1]$ of country, individual location $x \in \mathcal{A}$, but also additional parameters $y \in Y$, which can be changed according to a partition configuration. In particular, y can be used as a center of the country as well as other important for country formation parameters. It is assumed that the cost functions depend continuously on $\delta \in \Delta^{(n-1)}$ and $y \in Y$; moreover Y (the range of y) is convex and compact. More specifically, in addition to assumption (**P**) (page 3) we impose

(**C**) *For each $i \in N$ costs $c_i(\cdot)$ are defined and continuous on*

$$\mathcal{A} \times Y \times (\Delta^{(n-1)} \setminus F_i), \text{ where } F_i = \{\delta \in \Delta^{(n-1)} \mid \delta_i = 0\},$$

and obey

(i) $c_i(x, y, \delta_1, \ldots, \delta_n) \to +\infty$ *when* $(x, y, \delta_i, \delta_{-i}) \to (\bar{x}, \bar{y}, 0, \bar{\delta}_{-i})$, *i.e.* $\bar{\delta}_i = 0$;
(ii) *the set of indifferent agents*

$$A_{ij}(y, \delta) = \{x \in \mathcal{A} \mid c_i(x, y, \delta) = c_j(x, y, \delta)\}$$

has zero Lebesgue measure $\forall j \neq i$, *and for all fixed* $(y, \delta) \in Y \times \Delta^{(n-1)}$.

Note the difference between our assumption and the one in [8]: the continuity relative to all variables and for item (ii)—the set $A_{ij}(y, \delta)$ may depend on $y \in Y$ and masses of other jurisdictions δ_k, $k \neq i, j$.

The idea of the proof is that for a collection $(\delta_1, \ldots, \delta_n, y)$ of *nominal* parameters one can put into correspondence a similar collection of *real* parameters, calculated for an immigration stable partition defined by nominal ones. While doing so, we define a mapping with a nontrivial fixed point which obeys all requirements of country partition we seek for. Now we consider this construction in more details.

Let us consider a standard simplex $\Delta^{(n-1)} = \{\delta \in \mathbb{R}^n \mid \sum \delta_i = 1, \ \delta_i \geq 0 \ \forall i\}$, the mappings $S_i : (\delta, y) \to S_i(\delta, y) \subset \mathcal{A}$, $(\delta, y) \in \Delta^{(n-1)} \times Y$, $i \in N$, and $\mathcal{M} : (S_i)_{i \in N} \to (\mu_i)_{i \in N}$ defined by formulas[3]:

$$S_i(\delta, y) = \{x \in \mathcal{A} \mid c_i(x, \delta, y) = \min_{j \in N} c_j(x, \delta, y)\}, \quad \mu_i(\delta, y) = \mu(S_i(\delta, y)), \quad i \in N.$$

---

[3] Here as above $\mu(\cdot)$ is absolutely continuous measure on $\mathcal{A}$, specifying the resettlement of the population.

Assuming also that there is a continuous $\mathcal{F} : \Delta^{(n-1)} \times Y \to Y$, we obtain the resulting map

$$[\mathcal{M} \times \mathcal{F}](\delta, y) = \mathcal{M}(\delta, y) \times \mathcal{F}(\delta, y), \quad (\delta, y) \in \Delta^{(n-1)} \times Y.$$

Clearly, it suffices to find a *nontrivial* fixed point $\bar{\delta} = (\bar{\delta}_1, \dots, \bar{\delta}_n) \in \Delta^{(n-1)}$, $\bar{y} \in Y$ of this map, i.e.

$$\bar{y} = \mathcal{F}(\bar{\delta}, \bar{y}), \quad \mu_i(\bar{\delta}, \bar{y}) = \bar{\delta}_i, \ \forall i \in N, \ \text{ such that } \ \bar{\delta} = (\bar{\delta}_1, \dots, \bar{\delta}_n) \gg 0.$$

It is proven in [8] that[4] that: for some $0 < \varepsilon < 1$

(i) the map $\mathcal{M}(\cdot)$ is continuous on $\Delta_\varepsilon^{(n-1)}$ and
(ii) $\mathcal{M}(\cdot)$ maps the ε-sub-simplex

$$\Delta_\varepsilon^{(n-1)} = \{\delta \in \mathbb{R}^n \mid \sum \delta_i = 1, \ \delta_i \geq \varepsilon \ \forall i \in N\}$$

so that the faces of $\Delta_\varepsilon^{(n-1)}$ pass into the corresponding faces of initial simplex, i.e.

$$\delta = (\delta_1, \dots, \delta_n) \in \Delta_\varepsilon^{(n-1)} \ \& \ \delta_i = \varepsilon \Rightarrow \mu_i(\delta) = 0, \ \mathcal{M}(\delta) = (\mu_1(\delta), \dots, \mu_n(\delta)).$$

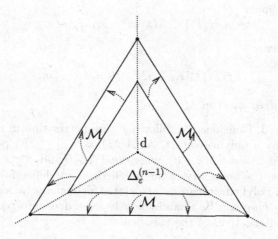

Fig. 3. Initial and embedded sub-simplexes $\Delta_\varepsilon^{(n-1)}$ and the mapping $\mathcal{M}(\cdot)$.

The properties (i), (ii) can be easily extended to our case although now we need to find a fixed point of the map $\mathcal{M} \times \mathcal{F}$. So, the existence of the required fixed point can be proved via (i), (ii). Note that the Brouwer fixed point theorem (and similar theorems) cannot be applied in the case, because $\mathcal{M}(\cdot)$ being defined on

---

[4] This is Lemma 1 from [7], where its comprehensive proof is also presented.

$\Delta_\varepsilon^{(n-1)} \times Y$ *is not* a mapping *into itself,* i.e. requirement $\mathcal{M}(\Delta_\varepsilon^{(n-1)} \times Y) \subseteq \Delta_\varepsilon^{(n-1)}$ *is not true.* Moreover, I do not know any other workable theorem for our case. In [8] further argumentation is based on the application of KKM-lemma (Knaster–Kuratowski–Mazurkiewicz), which is rather elegant solution to the issue, but it is limited to a particular case of fixed (unchanged) capitals. In our case this means the lack of postulated dependence of $\mathcal{M}$ on $y \in Y$.

The foregoing reasons yield the need in additional analysis and proving of the following theorem, which can be viewed as a (new) generalization of Krasnosel'skii's theorem in case of a bounded polyhedron (simplex) see [3] and its generalizations in [4].

Let $M \subset \mathbb{R}^n$ be a *convex bounded* polyhedron and $A(M)$ be its *affine hull.* Let $d \in \mathrm{ri}M$ be a point in the relative interior of a polyhedron M, and F_t, $t = 1, \ldots, m$ its non-trivial faces of a maximum dimension (one less than M). With every facet associate cone $K_t \subset A(M)$ with a vertex at d:

$$K_t = \{d + \lambda(\kappa - d) \mid \kappa \in F_t, \ \lambda \geq 0\} \quad \Rightarrow \quad A(M) = \bigcup_{t=1,\ldots,m} K_t.$$

Theorem 3. *Let* $f : M \to A(M)$ *be a continuous mapping defined on a polyhedron* M *and* $d \in \mathrm{ri}M$, $A(M)$, F_t, K_t *be defined as described above. Let one of the conditions hold:*

(i) *Compressive form*

$$f(F_t) \subset M, \quad \forall t = 1, \ldots, m. \tag{7}$$

(ii) *Expansive form*

$$f(F_t) \subset K_t \setminus \mathrm{ri}M, \quad \forall t = 1, \ldots, m. \tag{8}$$

Then $f(\cdot)$ *has a fixed point in* M.

Proof of Theorem 3. Consider the following parametrization in the affine space $A(M)$, spanned by a polyhedron M. As $A(M) = \cup_{t=1,\ldots,m} K_t$ point $x \in A(M)$ can be specified as $x = d + \lambda(\kappa - d)$, where real $\lambda > 0$ and, for $x \neq d$, the vector $\kappa \in \cup_{t=1,\ldots,m} F_t$ on the boundary of the polyhedron are defined one-to-one. Now the points of the polyhedron can be associated with pairs (λ, κ) for $0 \leq \lambda \leq 1$ and a continuous map can be unambiguously extended onto pairs (λ, κ). Next we consider the alternatives of the theorem.

(i) *Compressive form.* Let $f(\lambda, \kappa) = (\lambda', \kappa')$. We now define a new mapping[5] $g(\lambda, \kappa) = (1 \wedge \lambda', \kappa')$. Obviously, $g : M \to M$ is continuous and due to Brouwer theorem it has a fixed point $\bar{x} = (\bar{\lambda}, \bar{\kappa}) = g(\bar{\lambda}, \bar{\kappa})$. Let us show that this point is also a fixed point of f. Indeed, the difference in the values of f and g can be revealed only if $\bar{\lambda} < 1$ and $\bar{\lambda}' > 1$. But then $\bar{\lambda} = 1 \wedge \bar{\lambda}' = 1$, that cannot be true for the fixed point.

---

[5] Here $a \wedge b = \min\{a, b\}$.

(*ii*) *Expansive form.* Without loss of generality we can assume that $f(\lambda, \kappa) = (\lambda', \kappa')$ and $\lambda' \leq 2$. Otherwise, consider the new mapping $f'(\lambda, \kappa) = (2 \wedge \lambda', \kappa')$, which has the same fixed points on M as the original one. Next, we define $g(\lambda, \kappa) = (2\lambda - \lambda', \kappa')$. For $(\lambda, \kappa) \in F_t$ we have $\lambda = 1$, $1 \leq \lambda' \leq 2$ and therefore $0 \leq 2\lambda - \lambda' \leq 1$, which implies $g(F_t) \subset M$ $\forall t$. By the above item (*i*), $g(\cdot)$ has in M a fixed point $(\bar{\lambda}, \bar{\kappa})$ i.e. there is $(\bar{\lambda}, \bar{\kappa}) = (2\bar{\lambda} - \lambda', \kappa')$. Writing this componentwise we have $\bar{\kappa} = \kappa'$ and $\bar{\lambda} = 2\bar{\lambda} - \lambda' \Rightarrow \bar{\lambda} = \lambda'$, but this means that $(\bar{\lambda}, \bar{\kappa})$ is a fixed point of f. Theorem 3 is proved.

Remark 2. Note that we apply the parametrization $A(M)$ via (λ, κ) only to specify a transformation of the initial function f which does not change fixed points. A new function defined in this way is continuous and maps M into itself.

Notice that the assumption $d \in \text{ri}M$ is essential—without it the theorem statement becomes wrong, appropriate examples can be easily constructed. The analysis of the proof shows that Theorem 3 can be generalized to the case of the Cartesian product of maps provided that the first satisfies the condition of Theorem 3 and the second map obeys the conditions of Brouwer theorem or it can be reducible to it.   □

So now we can formulate the main result. In the case of our interest we have

$$\mathcal{M} : \Delta_\varepsilon^{(n-1)} \times Y \to \Delta^{(n-1)}.$$

If as a central point $d \in M = \Delta_\varepsilon^{(n-1)}$ one considers the center of simplex $(\frac{1}{n}, \ldots, \frac{1}{n}) = d$ then, by expansive property (*ii*) of the map $\mathcal{M}$, condition (*ii*) of Theorem 3 is fulfilled. Now if

$$\mathcal{F} : \Delta_\varepsilon^{(n-1)} \times Y \to Y$$

is any continuous map, then the map $\mathcal{M} \times \mathcal{F}$ has a fixed point in $X = \Delta_\varepsilon^{(n-1)} \times Y$. As a result we proved the following

Theorem 4. *Let $\mathcal{A}$ be a compact subset of a finite dimensional linear space and μ be a measure on $\mathcal{A}$. If assumptions* (**P**), (**C**) *are satisfied, then the area $\mathcal{A}$ can be nontrivially partitioned into any number of immigration proof communities. This partition can also obey any consistent continuous requirements.*

Notice that this result *does not imply* Theorem 2, in which we did not require restrictive assumption **C**(*ii*). Thus Theorems 2 and 4 complement each other.

The mapping $\mathcal{F}$, introduced into the design of the search of a fixed point, expresses some additional requirements for cross-country division. For example, one can impose requirements on the centers (the capital) of countries. In particular, one can require the capital be located in the center of gravity of the countries and so on.

References

1. Alesina, A., Spolaore, E.: On the number and size of nations. Q. J. Econ. **113**, 1027–1056 (1997)
2. Le Breton, M., Musatov, M., Savvateev, A., Weber, S.: Rethinking Alesina and Spolaore's "Uni-Dimensional World": existence of migration proof country structures for arbitrary distributed populations. In: Proceedings of XI International Academic Conference on Economic and Social Development. University - Higher School of Economics, Moscow, 6–8 April 2010 (2010)
3. Krasnoselskii, M.A.: Fixed points of cone-compressing or cone-extending operators. Proc. USSR Acad. Sci. **135**(3), 527–530 (1960)
4. Kwong, M.K.: On Krasnoselskiis cone fixed point theorem. J. Fixed Point Theor. Appl. **2008**, Article ID 164537, 18 p. (2008)
5. Marakulin, V., M.: On the existence of migration proof country structures. Novosibirsk. Preprint No 292, Sobolev Institute of Mathematics SB RAS, p. 12 (2014). (in Russian)
6. Marakulin, V., M.: Spatial equilibrium: the existence of immigration proof partition into countries for one-dimensional space. Siberian J. Pure Appl. Math., 15 p. (submitted on 04 April 2016). (in Russian)
7. Marakulin, V.M.: Generalized Krasnosel'skii fixed point theorem for polytopes and spatial equilibrium. Siberian Math. J., 7 p. (submitted on 24 April 2016). (in Russian)
8. Savvateev, A., Sorokin, C., Weber, S.: Multidimensional Free-Mobility Equilibrium: Tiebout Revisited. Mimeo, 23 pages (2016)

Search of Nash Equilibrium in Quadratic n-person Game

Ilya Minarchenko[(⊠)]

Melentiev Energy Systems Institute of Siberian Branch of the Russian Academy
of Sciences, Lermontov Street 130, 664033 Irkutsk, Russia
eq.progr@gmail.com

Abstract. This paper is devoted to Nash equilibrium search in quadratic n-person game, where payoff function of each player is quadratic with respect to its strategic variable. Interactions between players are defined by corresponding bilinear terms in the payoffs. First, the statement is considered without any assumptions on payoffs' concavity. We use Nikaido-Isoda approach in order to reduce Nash equilibrium problem to optimization problem with nonconvex implicitly defined objective function. We propose global search algorithm based on the linearization of implicit part of the objective by linear support minorants. This technique allows to determine numerically whether the game has no equilibria. Then payoffs are assumed to be concave with respect to its strategic variables, and we suggest d.c. decomposition of the objective, thus corresponding local search method is applicable. Computational results are provided in the paper. Local search method is compared with extragradient equilibrium search algorithm.

Keywords: Nash equilibrium · Nikaido-Isoda function · Support function · d.c. decomposition · Extragradient method

1 Introduction

A lot of investigations are devoted to methods for computing Nash equilibrium, and between them the so-called Nikaido-Isoda approach is widely used, especially for generalized Nash equilibrium problems (GNEP). GNEP differs from standard Nash equilibrium problem (NEP) by coupled strategy sets. It means that in GNEP for every player his or her strategy set depends on rival players choices. Nikaido-Isoda approach leads to an optimization problem which turns out to be equivalent to NEP (in particular sense).

One of the first investigations devoted to GNEP and Nash equilibrium computing can be found in [1]. In [2] one can find algorithms for computing generalized Nash equilibrium using convex programming techniques. In both papers, convergence of algorithms depends on identical conditions. More recent investigations are concerned with algorithms based on Nikaido-Isoda approach, ideas

This work was supported by the RFBR grant no. 15-07-08986.

Y. Kochetov et al. (Eds.): DOOR 2016, LNCS 9869, pp. 509–521, 2016.
DOI: 10.1007/978-3-319-44914-2_40

of regularization of Nikaido-Isoda function, and relaxation methods (see, for example, [3–8]). Gradient-type approach to equilibrium programming problems was proposed in [9]. It is worth to note that all of mentioned papers examine problems with imposed concavity assumption on payoffs.

Games with quadratic payoffs and coupled constraints have been considered in [10,11]. In these papers, the concavity assumption on payoffs is used as well. 2-person game with quadratic payoffs is described in [12] as a particular case of bilinear equilibrium programming problem.

The present investigation is devoted to Nash equilibrium search in quadratic n-person game with independent strategy sets. In contrast to papers on Nash equilibrium search known for the author, we in the first part of our paper do not make any assumptions on concavity of players' payoff functions. It is well-known that payoff concavity with respect to player's variable is a standard proposition widely used for algorithms' convergence. It also ensures existence of Nash equilibrium due to Kakutani's fixed point theorem [13] (if some additional presumptions on strategy sets are made). We abandon the concavity assumption and propose numerical procedure based on Nikaido-Isoda approach for computing Nash equilibrium in quadratic game of certain form. Moreover, such a procedure is able to determine whether the game has no equilibria if this is the case. Preliminary research on equilibrium computing in quadratic 2-person game with non-concave payoffs were made by the author in [14]. Here we present more detailed discussion provided with some illustrative examples and extended computational results. Also we consider the case with (player-) concave payoffs and propose d.c. decomposition of the objective function. It allows us to use local search method for d.c. functions. D.c. decomposition is based on the Lagrange dual problem for certain optimization problem gained by Nikaido-Isoda approach.

Following [10–12], in the framework of this paper, quadratic game is a game, where every player's payoff function is quadratic with respect to its strategic variable. In addition, payoff function of every player is assumed to be linear with respect to another players' variables, and it is defined by corresponding bilinear terms. In other words, quadratic game under consideration generalizes bilinear n-person game by adding quadratic term to every payoff function.

At the same time even the class of bilinear games is rather wide. It includes in particular the mixed extension of bimatrix games as well as of polymatrix games, and differs from these classes by an arbitrary strategy set for every player [15]. In polymatrix game players choose their strategies from symplexes. As we show in the paper, it makes significant impact on the difficulty of Nash equilibrium search.

1.1   Nikaido-Isoda Approach

Now we discuss an approach associating a NEP with an optimization problem. One of the main feature of the approach is as follows: if the set of Nash equilibria is nonempty then the corresponding optimization problem is turn out to be equivalent to the NEP.

Let us consider a noncooperative n-person game in strategic form. The set of players is $N = \{1, 2, \ldots, n\}$, the set of strategies of ith player is $X_i \subset \mathbb{R}^{n_i}$, and the payoff function of ith player is $f_i \colon X \to \mathbb{R}$, where $X = X_1 \times \cdots \times X_n$ is the set of strategy profiles, and $\mathbb{R}$ denotes the set of real numbers. We suppose that X_i is a compact set for every $i \in N$, the function $\sum_{i=1}^{n} f_i$ is continuous over the set X, and $f_i(x_i, \cdot)$ is continuous function over the set $X_1 \times \cdots \times X_{i-1} \times X_{i+1} \times \cdots \times X_n$ for each fixed $x_i \in X_i$, and for every $i \in N$. The problem is to find Nash equilibrium for the given game, i.e. to find a point $x^* \in X$ meeting the following conditions:

$$f_i(x_i, x_{-i}^*) \leqslant f_i(x^*) \quad \forall x_i \in X_i, \quad \forall i \in N, \tag{1}$$

where $(x_i, x_{-i}^*) = (x_1^*, \ldots, x_{i-1}^*, x_i, x_{i+1}^*, \ldots, x_n^*)$.

A function $\varphi \colon X \times X \to \mathbb{R}$ such as

$$\varphi(x, y) = \sum_{i \in N} [f_i(y_i, x_{-i}) - f_i(x)] \tag{2}$$

is referred to as *Nikaido-Isoda function*. It is known, that a point $x^* \in X$ is Nash equilibrium in the game stated above if and only if [16]

$$\max_{y \in X} \varphi(x^*, y) = 0. \tag{3}$$

Note that

$$F(x) = \max_{y \in X} \varphi(x, y) \geqslant 0 \quad \forall x \in X. \tag{4}$$

Then the next result immediately follows. If the set of Nash equilibria in the game stated above is nonempty, then it coincides with the set of solutions of the following optimization problem:

$$F(x) \to \min, \quad x \in X. \tag{5}$$

On the other hand, the set of Nash equilibria in the game is empty if and only if

$$F(x) > 0 \quad \forall x \in X. \tag{6}$$

Consequently, every solution of (5), where objective function attains zero value, is Nash equilibrium in the game. If objective does not attain zero over the feasible set X, then the game has no equilibrium points. Hence, Nikaido-Isoda theorem gives an opportunity to apply optimization techniques to NEP.

One of the main difficulty on this way is implicitly defined objective function F. In some particular cases, the "inner" maximization problem that defines objective F can be solved analytically. The examples of such cases are the mixed extensions of bimatrix games and, more general, of polymatrix games. In these classes of games payoff functions have bilinear form and the strategy set of every player is a simplex. These are the key features that allow an explicit form of F. Problem (5) in explicit form for bimatrix games first was given in [17,18]. This result then was used for comprehensive research on computation of Nash equilibrium for the mixed extension of bimatrix games in [5]. Detailed investigation of equivalent problem (5) for polymatrix games one can find in [19]. However, quadratic games with arbitrary polyhedra as the strategy sets (even if they are box-constrained) do not admit explicit form of F.

The second issue to deal with is nonconvexity of objective F. As we show in the paper, there can be local minima of F that are not Nash equilibria.

It should be mentioned that equivalence between NEP and optimization problem of form (5) in general is only correct for the games with independent strategy sets. In the games with coupled players' strategies Nash equilibria may exist that do not solve (5) (see [20] for the example).

2   Global Search

Here we consider quadratic game without the concavity assumption and propose a method for Nash equilibrium search based on Nikaido-Isoda approach. First we customize n-person game from Sect. 1.1 as the following quadratic game. All notations stay the same. Let the strategy set of ith player be non-empty compact set of form

$$X_i = \{x_i \in \mathbb{R}^{m_i} \mid A_i x_i \leqslant b_i\}, \tag{7}$$

and let ith player's payoff function be defined as

$$f_i(x) = x_i^\top \left(\frac{1}{2} B_i x_i + d_i\right) + \sum_{j \neq i} x_i^\top C_{ij} x_j, \tag{8}$$

where all matrices and vectors have proper sizes. $(a^\top)$ is the transpose of a. Consider the following block-structured matrices:

$$C = \begin{pmatrix} 0 & C_{12} & \dots & C_{1n} \\ C_{21} & 0 & \dots & C_{2n} \\ \dots & \dots & \dots & \dots \\ C_{n1} & C_{n2} & \dots & 0 \end{pmatrix}, \; B = \begin{pmatrix} B_1 & 0 & \dots & 0 \\ 0 & B_2 & \dots & 0 \\ \dots & \dots & \dots & \dots \\ 0 & 0 & \dots & B_n \end{pmatrix}, \; A = \begin{pmatrix} A_1 & 0 & \dots & 0 \\ 0 & A_2 & \dots & 0 \\ \dots & \dots & \dots & \dots \\ 0 & 0 & \dots & A_n \end{pmatrix}.$$

Then Nikaido-Isoda function for the game (7)–(8) can be written as

$$\varphi(x, y) = \frac{1}{2} y^\top B y + y^\top (Cx + d) - x^\top \left(\frac{1}{2} B + C\right) x - x^\top d,$$

and the equivalent problem (5) takes the form

$$F(x) = \max_{y \in X} \left[\frac{1}{2} y^\top B y + y^\top (Cx + d)\right] - x^\top \left(\frac{1}{2} B + C\right) x - x^\top d \rightarrow \min_{x \in X}, \tag{9}$$

where $d = (d_1, d_2, \dots, d_n)$, $X = \{x \in \mathbb{R}^m \mid Ax \leqslant b\}$, and $m = m_1 + \dots + m_n$.

In order to handle with implicitly defined objective we propose to make approximation of implicit term by linear support minorant functions. Recall that support minorant for some function ρ in some feasible point $\widetilde{x}$ refers to a function l such that for every feasible x

$$l(x) \leqslant \rho(x), \quad l(\widetilde{x}) = \rho(\widetilde{x}).$$

Let us make the following notations:

$$\psi(x, y) = \frac{1}{2} y^\top B y + y^\top (Cx + d),$$

$$\rho(x) = \max_{y \in X} \psi(x, y).$$

It is clear, that ρ is convex function over the set X. Hence linearization of ρ in arbitrary point from X gives us linear support minorant for ρ. Let x^k be the kth iteration point. Then support minorant for ρ in x^k is the function $\psi(x, y^k)$, where

$$y^k \in \operatorname{Arg}\max_{y \in X} \psi(x^k, y).$$

Optimization problem with approximated objective function F on the kth iteration takes the form

$$\widetilde{F}_k(x) = \max_{0 \leqslant i \leqslant k} \{\psi(x, y^i)\} - x^\top \left(\frac{1}{2} B + C\right) x - x^\top d \to \min_{x \in X}. \tag{10}$$

One can easily reformulate problem (10) as

$$\theta_k(\alpha, x) = \alpha - x^\top \left(\frac{1}{2} B + C\right) x - x^\top d \to \min_{(\alpha, x)}, \tag{11}$$

$$\alpha \geqslant \psi(x, y^i), \quad 0 \leqslant i \leqslant k, \tag{12}$$

$$x \in X. \tag{13}$$

Let vector (α^*, x^*) be a solution of (11)–(13). Then, obviously, x^* solves (10), moreover $\widetilde{F}_k(x^*) = \theta_k(\alpha^*, x^*)$. The next point x^{k+1} is chosen as a vector x^*. The numerical procedure should be stopped if

$$F^* - \widetilde{F}_k(x^{k+1}) \leqslant \varepsilon,$$

where F^* is the best known objective value (record) and ε is a small positive real number. The following algorithm represents our discussion.

Global Search Algorithm.

1. Choose an initial point $x^0 \in X$, and small positive numbers ε_1 and ε_2. Set $F^* = +\infty$, and $k = 0$.
2. Solve the problem
$$\psi(x^k, y) \to \max, \quad y \in X,$$
and assign the solution to y^k.
3. Compute the record: if $\varphi(x^k, y^k) \leqslant F^*$ then assign $\varphi(x^k, y^k)$ to F^*.
4. Solve the problem (11)–(13). Let (α^*, x^*) be a solution. Assign x^* to x^{k+1}.
5. If $F^* - \widetilde{F}_k(x^{k+1}) \leqslant \varepsilon_1$ then STOP, else set $k = k + 1$ and go to the Step 2.
6. If $F^* \leqslant \varepsilon_2$ then x^{k+1} is Nash equilibrium for the game (7)–(8), else the game has no equilibrium points.

Therefore, it needs to solve two nonconvex quadratic programming problems on each iteration of the algorithm. These problems can be solved, for example, by excess search between all stationary points for small dimension and box-constrained strategy sets. Another way is to use some of existing global solvers. We program the algorithm in GAMS with Couenne as a solver for nonconvex problems [21]. Convergence of described algorithm to global solution is established in [22,23].

Let us consider two examples of the game (7)–(8) with two players and indefinite matrix B.

Example 1. $N = \{1, 2\}$, $X_1 = X_2 = [-1, 1]$,
$f_1(x_1, x_2) = x_1^2 + x_1 x_2$, $f_2(x_1, x_2) = -x_2^2 + \frac{1}{2}x_1 x_2$.
This game has two Nash equilibria: $(1, \frac{1}{4})$ and $(-1, -\frac{1}{4})$. The plot of function F is presented on the Fig. 1 (on the left). Let $(1, 1)$ be initial point, and $\varepsilon_1 = \varepsilon_2 = 10^{-5}$, then global search algorithm needs only two iterations for converging to equilibrium point. Results of computing one can find in the Table 1.

Example 2. $N = \{1, 2\}$, $X_1 = X_2 = [-1, 1]$,
$f_1(x_1, x_2) = x_1^2 + x_1 x_2$, $f_2(x_1, x_2) = -x_2^2 - x_1 x_2$.
This game does not have any equilibria since $\min_{x \in X} F(x) = 0.25 > 0$. The plot of function F is presented on the Fig. 1 (on the right). Algorithm parameters have the same values as in the Example 1. The Table 2 presents iterations of global search.

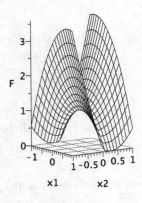
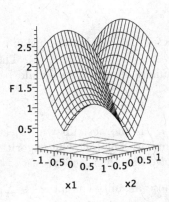

Fig. 1. Plot of the function $F(x_1, x_2)$ for the game with two Nash equilibria (left) and for the game without Nash equilibria (right).

In the Table 3, we gather computational results of global search for randomly generated problems with box-constrained strategy sets. Tolerance parameters ε_1 and ε_2 were set to 10^{-4} and 10^{-5} respectively. Notation in the table is as follows: Dimension is a total dimension of the game, Problems is a number of games generated for a given dimension, With NE is a number of problems with equilibrium (as the algorithm shows), Without NE is a number of problems without

Table 1. Iterations of global search algorithm for the Example 1

k	x^k	F^*	$\widetilde{F}_k(x^{k+1})$
0	$(1,1)$	0.5625	−1.6875
1	$(-1,-1)$	0.5625	0
2	$(-1,-0.25)$	0	0

Table 2. Iterations of global search algorithm for the Example 2

k	x^k	F^*	$\widetilde{F}_k(x^{k+1})$
0	$(1,1)$	2.25	−1.00
1	$(-1,-0.5)$	1.00	0.25
2	$(-1,0)$	0.25	0.25

Table 3. Global search for randomly generated problems

Dimension	Problems	With NE	Without NE	It	It (with NE)	It (without NE)
4	50	28	22	5	4	5
6	50	25	25	7	6	8
8	30	16	14	9	6	12
10	30	26	4	10	9	16
12	10	7	3	7	5	11
14	10	10	0	11	11	—

equilibria, It is an average number of iterations, It (with NE) and It (without NE) are the average numbers of iterations for problems with and without equilibrium respectively.

Systematic solving global optimization problems is computationally expensive procedure, what may be unacceptable for large scale problems. Hence it is worth to take into account one particular case of the game that considerably lowers computational time spending on equilibrium search. If for every pair of players mixed partial derivatives of their payoffs are equal to each other, i.e.

$$C_{ij} = C_{ji}^\top, \quad i \in N, \; j \in N, \; i \neq j,$$

then the game (7)–(8) is a potential game [24]. In such a case every (global) solution of the problem

$$P(x) \to \max, \quad x \in X$$

is Nash equilibrium. Here P is explicit (generally nonconvex) function reffered to as a *potential*. Potential function for the game (7)–(8) is defined by the formula

$$P(x) = \sum_{i \in N} \left[\frac{1}{2} x_i^\top B_i x_i + x_i^\top d_i + \frac{1}{2} \sum_{j \neq i} x_i^\top C_{ij} x_j \right].$$

Note that the set of local non-global maxima of a potential may contain equilibrium points as well.

3 Local Search

In this section we consider the game (7)–(8) with strictly concave players' payoffs with respect to its strategic variables. The following discussion is the further investigation for that made in [25]. In the present paper, it is extended by improved computational results and a comparison of proposed method with extragradient algorithm. For convenience we assume that f_i from (8) is the loss function for ith player. Hence, the concavity assumption should be replaced by the convexity one. It is equivalent to the condition that for every $i \in N$ matrix B_i is positively definite. It implies that matrix B is positively definite too. Then the equivalent optimization problem (9) should be rewritten as

$$F(x) = x^\top (Cx + d) + \frac{1}{2} x^\top B x + \max_{y \in X} \left[-y^\top (Cx + d) - \frac{1}{2} y^\top B y \right] \to \min_{x \in X}, \quad (14)$$

The convexity assumption provides strict concavity of inner maximization problem in (14):

$$\psi(x, y) = -y^\top (Cx + d) - \frac{1}{2} y^\top B y \to \max, \quad y \in X, \quad \text{for some fixed } x \in X.$$

Moreover, taking into account non-emptiness, compactness, and convexity of strategy space X_i for every $i \in N$ one can immediately conclude that equilibrium point always exists due to Kakutani's fixed point theorem. It implies that $\min_{x \in X} F(x) = 0$. Hence we propose to use local search method joined with multistart from random initial point, since we can easily verify whether a point given by the procedure is Nash equilibrium.

As $\psi(x, \cdot)$ is strictly concave function and the set X is defined by linear constraints, we have

$$\max_{y \in X} \psi(x, y) = \min_{\lambda \geqslant 0} \max_{y \in \mathbb{R}^m} L(y, \lambda; x) \quad \text{for every} \quad x \in X,$$

where λ is a vector of Lagrange multipliers, and

$$L(y, \lambda; x) = -y^\top (Cx + d) - \frac{1}{2} y^\top B y - \lambda^\top (Ay - b)$$

is Lagrange function. Obviously, $L(\cdot, \lambda; x)$ attains its maximum over R^m in the point, where the first derivative with respect to variable y equals to zero. Then in view of non-singularity of matrix B which is ensured by its positive definiteness, we have $y^* = -B^{-1}(Cx + d + A^\top \lambda)$ such as

$$\max_{y \in X} \psi(x, y) = \min_{\lambda \geqslant 0} L(y^*, \lambda; x) \quad \text{for every} \quad x \in X. \quad (15)$$

Executing substitution, we get:

$$L(y^*, \lambda; x) = \frac{1}{2} (Cx + d)^\top B^{-1} (Cx + d)$$

$$+ (Cx + d)^\top B^{-1} A^\top \lambda + \frac{1}{2} \lambda^\top AB^{-1} A^\top \lambda + \lambda^\top b. \qquad (16)$$

Then the equalities (15) and (16) imply

$$\max_{y \in X} \psi(x, y) = \frac{1}{2} (Cx + d)^\top B^{-1} (Cx + d) +$$

$$\min_{\lambda \geqslant 0} \left[(Cx + d)^\top B^{-1} A^\top \lambda + \frac{1}{2} \lambda^\top AB^{-1} A^\top \lambda + \lambda^\top b \right] \quad \forall x \in X.$$

$$(17)$$

With respect to (17) the problem (14) may be rewritten as

$$F(x) = g(x) - h(x) \to \min_{x \in X}, \qquad (18)$$

where

$$g(x) = x^\top \left(C + \frac{1}{2} B + \frac{1}{2} C^\top B^{-1} C \right) x + x^\top \left(C^\top B^{-1} d + d \right) + \frac{1}{2} d^\top B^{-1} d,$$

$$h(x) = - \min_{\lambda \geqslant 0} v(x, \lambda),$$

$$v(x, \lambda) = \lambda^\top AB^{-1} Cx + \frac{1}{2} \lambda^\top AB^{-1} A^\top \lambda + \lambda^\top \left(b + AB^{-1} d \right).$$

Next, we formulate the statement that allows us to proceed to numerical method for solving (18).

Proposition 1. *Functions g and h are convex.*

Proof. Using the symmetry of matrix B and denoting $z = (B + C)x$, quadratic part of g can be easily represented as

$$x^\top \left(C + \frac{1}{2} B + \frac{1}{2} C^\top B^{-1} C \right) x = \frac{1}{2} x^\top (B + C)^\top B^{-1} (B + C) x = \frac{1}{2} z^\top B^{-1} z.$$

Since B is positive definite, then $z^\top B^{-1} z > 0$ for any non-zero z, at the same time $z(x)^\top B^{-1} z(x) \geqslant 0$ for any x. Hence, g is convex. Function v is linear with respect to x then h is convex too.

Thus, we represent F as a difference of two convex functions (d.c. decomposition). Our further suggestion is to use for solving problem (18) the well-known iterative local search d.c. algorithm [26]. Its main idea is a linearization of concave term of objective in current iteration point and solving derived convex optimization problem. In such a way, original nonconvex problem reduces to series of convex problems. Next, we describe the steps of this algorithm as it applies to (18). We propose to use linearization at the same manner as in the previous section. Thus we omit detailed discussion and proceed with an algorithm.

Local Search Algorithm.

1. Choose initial point $x^0 \in X$, and small numbers $\varepsilon_1 > 0$, $\varepsilon_2 > 0$. Set $k = 0$.
2. Get λ^{k+1} as a solution of convex minimization problem:

$$\lambda^{k+1} = \arg\min_{\lambda \geqslant 0} v\left(x^k, \lambda\right).$$

3. Get x^{k+1} as a solution of convex linearized problem:

$$x^{k+1} = \arg\min_{x \in X} \left[g\left(x\right) + v\left(x, \lambda^{k+1}\right)\right].$$

4. If $F(x^{k+1}) \leqslant \varepsilon_1$ then STOP: x^{k+1} is a global solution of (18) and is Nash equilibrium in the game. Else if $\left\|x^{k+1} - x^k\right\| \leqslant \varepsilon_2$ then STOP: x^{k+1} is a local solution and is not an equilibrium for the game. Otherwise, set $k = k+1$ and go to the Step 2.

3.1   Computational Experiment

Proposed local search for d.c. functions was compared with extragradient method [12]. Every iteration of extragradient algorithm consists of two half-steps, and for the game (7)–(8) (where f_i is a loss function) has the form

$$\overline{x}^k = \pi_X\left(x^k - t_k\left((B+C)x^k + d\right)\right),$$

$$x^{k+1} = \pi_X\left(x^k - t_k\left((B+C)\overline{x}^k + d\right)\right).$$

Here $\pi_X(\cdot)$ denotes projection operator onto the set X, and step length t_k is chosen meeting the following condition:

$$2\,t_k^2\,\|(B+C)(\overline{x}^k - x^k)\|^2 \leqslant (1-\delta)\|\overline{x}^k - x^k\|^2, \quad 0 < \delta < 1.$$

The stop criterion for extragradient method we used:

$$\|x^{k+1} - x^k\| \leqslant \varepsilon_3.$$

It should be noted that the convergence of extragradient algorithm to equilibrium point is ensured if $(B+C)$ is positive semi-definite. As matrix C is arbitrary, it is not the case in general. However, since extragradient algorithm does not need to solve any optimization problems during computation, it consumes significantly less time.

Computational experiment was made for randomly generated problems with two players. For all problems we set

$$X = \{x \in \mathbb{R}^m \mid -10 \leqslant x_i \leqslant 10, i = 1, \ldots, m\},$$

and also set $\varepsilon_1 = 10^{-4}$, $\varepsilon_2 = \varepsilon_3 = 10^{-6}$. Both methods started from identical randomly generated initial points (multistart). Program was composed in MatLab. Convex mathematical programming problems on the Step 2 and the

Table 4. Computational results of local search for d.c. functions and extragradient method

Dimension	MS	Method	It(av)	NE	Loc	NE(uniq)	Loc(uniq)	Time(av)	Max(it)
2	20	dcls	71	20	0	3	—	0.47	0
		exgr	77	20	0	2	—	0.02	0
4	20	dcls	123	15	5	3	1	0.81	0
		exgr	74	20	0	1	—	0.02	0
6	20	dcls	94	17	3	5	2	0.65	0
		exgr	126	20	0	2	—	0.03	0
8	20	dcls	26	20	0	3	—	0.19	0
		exgr	85	20	0	2	—	0.02	0
10	20	dcls	96	12	8	1	3	0.71	0
		exgr	—	0	0	—	—	—	20
10	20	dcls	79	6	14	3	7	0.59	0
		exgr	123	20	0	2	—	0.03	0
12	20	dcls	1051	3	17	3	8	8.24	0
		exgr	—	0	0	—	—	—	20
14	20	dcls	273	14	6	2	4	2.14	0
		exgr	142	20	0	2	—	0.04	0
16	20	dcls	230	6	14	1	6	1.88	0
		exgr	—	0	0	—	—	—	20
18	20	dcls	811	12	8	3	6	6.74	0
		exgr	—	0	0	—	—	—	20
20	20	dcls	1281	15	5	6	3	11.33	0
		exgr	1343	20	0	2	—	0.37	0
30	20	dcls	3308	8	12	5	11	39,23	0
		exgr	325	20	0	2	—	0,11	0
40	20	dcls	3244	0	20	—	19	54,20	0
		exgr	—	0	0	—	—	—	20
50	10	dcls	3062	0	10	—	8	57.06	0
		exgr	440	2	0	1	—	0.15	8
60	10	dcls	2656	0	10	—	10	61.30	0
		exgr	2072	2	0	1	—	0.74	8
70	10	dcls	5504	0	10	—	10	168.14	0
		exgr	2356	10	—	2	—	0.88	0
80	10	dcls	2499	0	10	—	10	88.25	0
		exgr	9908	10	—	1	—	3.93	0
90	10	dcls	3536	0	10	—	10	160.99	0
		exgr	—	0	0	—	—	—	10
100	10	dcls	3928	0	10	—	10	216.26	0
		exgr	—	0	0	—	—	—	10

Step 3 were solved by standard solver Quadprog. Computation was performed on the PC with AMD FX-8350 4.00 GHz CPU. Maximal number of iteration was set to 40000. Results are gathered in the Table 4.

Notation is as follows: Dimension denotes a total dimension of the game, MS is a number of initial points, dcls denotes local search for d.c. functions, exgr denotes extragradient method, It(av) is an average number of iteration, NE is a number of starts that gain equilibrium point, Loc is a number of starts that gain non-equilibrium point, NE(uniq) is a number of unique equilibrium points gained by multistart, Loc(uniq) is a number of unique non-equilibrium points gained by multistart, Time(av) is an average time spending for a single start (in seconds), Max(it) is a number of starts where stop criterion does not hold. For computing the number of iterations and time, we took into account starts with fulfilled stop criterion only.

Computational experiment shows that both algorithms joined with multistart are able to find more than one unique Nash equilibrium. However, there are the problems without any equilibria found by both methods. Also experiment shows that extragradient algorithm which is suitable for equilibrium search converges to equilibrium points only, while local search for d.c. functions converges to stationary points of the objective function F, and there may exist non-equilibrium points among them.

4   Conclusion

In the paper we investigate two statements of quadratic n-person game. The first one is the game without the concavity assumption on every player's payoff with respect to its strategic variable. Whereas in the second statement this assumption takes place. In both cases Nikaido-Isoda approach leads to optimization problem with nonconvex implicitly defined objective function to be minimized over the set of game strategy profiles. For the first statement, we propose global search method which either computes Nash equilibrium or finds out that the game has no equilibria if this is the case. Global search algorithm represents the series of nonconvex quadratic programming problems. For the second statement, we construct d.c. decomposition of implicit objective function in corresponding optimization problem. It allows us to use local search which is based on linearization of concave term in d.c. decomposition. It is worth to implement local search with multistart as it can converge to non-equilibrium local minima.

References

1. Rosen, J.B.: Existence and uniqueness of equilibrium points for concave n-person games. Econometrica **33**(3), 520–534 (1965)
2. Zukhovitskiy, S.I., Polyak, R.A., Primak, M.E.: Many-person convex games. Economica i Mat. Metody **7**(6), 888–900 (1971). (in Russian)
3. Krawczyk, J., Uryasev, S.: Relaxation algorithms to find nash equilibria with economic applications. Environ. Model. Assess. **5**, 63–73 (2000)

4. Krawczyk, J.: Numerical solutions to coupled-constraint (or generalized nash) equilibrium problems. CMS **4**, 183–204 (2007)
5. Strekalovskiy, A.S., Orlov, A.V.: Bimatrix Games and Bilinear Programming. FIZMATLIT, Moscow (2007) (in Russian)
6. Flam, S.D., Ruszczynski, A.: Finding normalized equilibrium in convex-concave games. Int. Game Theory Rev. **10**(1), 37–51 (2008)
7. von Heusinger, A., Kanzow, C.: Relaxation methods for generalized nash equilibrium problems with inexact line search. J. Optim. Theory Appl. **143**, 159–183 (2009)
8. Dreves, A., von Heusinger, A., Kanzow, C., Fukushima, M.: A globalized newton method for the computation of normalized nash equilibria. J. Glob. Optim. **56**, 327–340 (2013)
9. Antipin, A.S.: Equilibrium programming: gradient methods. Autom. Remote Control **58**(8), 1337–1347 (1997)
10. Schiro, D.A., Pang, J.-S., Shanbhag, U.V.: On the solution of affine generalized nash equilibrium problems with shared constraints by Lemke's Method. Math. Program., Ser. A. **142**, 1–46 (2013)
11. Dreves, A.: Finding all solutions of affine generalized nash equilibrium problems with one-dimensional strategy sets. Math. Meth. Oper. Res. **80**, 139–159 (2014)
12. Antipin, A.S.: Gradient and Extragradient Approaches in Bilinear Equilibrium Programming. Dorodnitsyn Computing Center RAS, Moscow (2002). (in Russian)
13. Kakutani, S.: A generalization of Brouwer's fixed point theorem. Duke Math. J. **8**, 457–458 (1941)
14. Minarchenko, I.M.: Support function method in bilinear two-person game. Bull. Tambov State Univ. **20**(5), 1312–1316 (2015). (in Russian)
15. Garg, J., Jiang, A.X., Mehta, R.: Bilinear games: polynomial time algorithms for rank based subclasses. In: Chen, N., Elkind, E., Koutsoupias, E. (eds.) Internet and Network Economics. LNCS, vol. 7090, pp. 399–407. Springer, Heidelberg (2011)
16. Nikaido, H., Isoda, K.: Note on noncooperative convex games. Pac. J. Math. **5**(5), 807–815 (1955)
17. Mills, H.: Equilibrium points in finite games. J. Soc. Ind. Appl. Math. **8**(2), 397–402 (1960)
18. Mangasarian, O.L.: Equilibrium points of bimatrix games. J. Soc. Ind. Appl. Math. **12**, 778–780 (1964)
19. Strekalovskiy, A.S., Enkhbat, R.: Polymatrix games and optimization problems. Autom. Remote Control **75**(4), 632–645 (2014)
20. Cavazutti, E., Flam, S.D.: Evolution to selected Nash equilibria. In: Giannessi, F. (ed.) Nonsmooth Optimization Methods and Applications, pp. 30–41. Gordon and Breach, London (1992)
21. The General Algebraic Modeling System. http://www.gams.com
22. Bulatov, V.P.: Numerical methods for solving the multiextremal problems connected with the inverse mathematical programming problems. J. Glob. Optim. **12**, 405–413 (1998)
23. Khamisov, O.V.: A Global Optimization Approach to Solving Equilibrium Programming Problems. Series on Computers and Operations Research 1: Optimization and Optimal Control, pp. 155–164 (2003)
24. Monderer, D., Shapley, L.S.: Potential games. Games Econ. Behav. **14**, 124–143 (1996)
25. Khamisov, O.V., Minarchenko, I.M.: Local search in bilinear two-person game. Bull. Siberian State Aerosp. Univ. **17**(1), 91–96 (2016)
26. Strekalovskiy, A.S.: On local search in d.c. optimization problems. Appl. Math. Comput. **255**, 73–83 (2015)

Applications of Operational Research

Convergence of Discrete Approximations of Stochastic Programming Problems with Probabilistic Criteria

Andrey I. Kibzun[1] and Sergey V. Ivanov[1,2(✉)]

[1] Moscow Aviation Institute, Volokolamskoe Shosse 4, A-80, GSP-3,
Moscow, Russia 125993
kibzun@mail.ru, sergeyivanov89@mail.ru
[2] Sobolev Institute of Mathematics, Acad. Koptyug Avenue, 4,
Novosibirsk, Russia 630090

Abstract. We consider stochastic programming problems with probabilistic and quantile objective functions. The original distribution of the random variable is replaced by a discrete one. We thus consider a sequence of problems with discrete distributions. We suggest conditions, which guarantee that the sequence of optimal strategies converges to an optimal strategy of the original problem. We consider the case of a symmetrical distribution, the case of the loss function increasing in the random variable, and the case of the loss function increasing in the optimization strategy.

Keywords: Stochastic programming · Probabilistic criterion · Value-at-risk · Quantile function · Convergent discretization

1 Introduction

The probabilistic and quantile (value-at-risk) objective functions [1] are used in stochastic programming to take into account risks. As it is described in [1], they can be applied in different economic and engineering systems, i.e., for planning a budget, for correction of a satellite orbit, etc.

Let us consider the stochastic programming problem in general formulation

$$\mathbf{G}[\Phi(u, Y)] \to \min_{u \in U}, \tag{1}$$

where u is a strategy, U is a set of feasible strategies, $\Phi(u, x)$ is a loss function, $\mathbf{G}$ is a probabilistic functional (e.g., probability, quantile, or expectation), Y is a random variable defined on a probability space $(\Omega_0, \mathcal{F}_0, \mathbf{P}_0)$. We suppose that Y has realizations $x \in \mathcal{X} \subset \mathbb{R}^1$. The distribution function of the random variable Y is denoted by $F(x) \triangleq \mathbf{P}_0\{Y \leq x\}$. In this paper, we consider stochastic programming problems with probabilistic and quantile objective functions.

S.V. Ivanov—Supported by Russian Science Foundation (project 15-11-10009).

Y. Kochetov et al. (Eds.): DOOR 2016, LNCS 9869, pp. 525–537, 2016.
DOI: 10.1007/978-3-319-44914-2_41

It is known that these problems are difficult to solve if the random parameters are continuous. However, stochastic programming problems with probabilistic criteria can be reduced to mixed integer programming problems if the random parameters are discrete [2]. Mixed integer programming problems can be solved using available software. So we can try to replace the original continuous random variable Y by a discrete one. We thus get a discrete approximation of original problem (1).

Discrete approximations of stochastic programming problems with expectation criterion have been considered in many works (see, e.g., [4–7]). Methods described in these works are based on approximate computation of integrals. Approximations of stochastic programming problems with probabilistic criteria are less researched. We can emphasize works [8,9]. The discretization of the quantile objective function in [8] is also made by approximate computation of integrals. Also, we would like to notice that in this work the strategy depends on random variables. In the present paper, we assume that the strategy is chosen before a realization of the random variable appears. In [9] convergence of confidence strategies, i.e., guaranteeing an upper bound of the optimal value of the quantile objective function, is proven.

Let us describe our technique of the discretization. We suppose that a sequence of independent random variables $\{Y_k\}$, $k \in \mathbb{N}$, with the distribution function $F(x)$ is given. Let this sequence be defined on the probability space $(\Omega_0, \mathcal{F}_0, \mathbf{P}_0)$.

It is known that the distribution function $F(x)$ can be estimated with the sample distribution function

$$\hat{F}^{(n)}(x) \triangleq \frac{M(x)}{n}, \tag{2}$$

where $M(x) \triangleq \sum_{k=1}^n \theta(x - Y_k)$,

$$\theta(x) \triangleq \begin{cases} 0, & x < 0; \\ 1, & x \geq 0. \end{cases} \tag{3}$$

It is easily seen that $M(x)$ is the random number of elements of the sample $Y_1, \ldots, Y_n$ such that $Y_k \leq x$, $k = \overline{1, n}$. By the Glivenko-Cantelli theorem,

$$\sup_{x \in \mathbb{R}^1} |\hat{F}^{(n)}(x) - F(x)| \xrightarrow{a.s.} 0 \text{ as } n \to \infty. \tag{4}$$

Notice that $\hat{F}^{(n)}(x)$ is random. Let us denote by $\hat{F}^{(n)}(x, \omega)$ the realization of the sample distribution function for the elementary event $\omega \in \Omega_0$. From convergence (4) it follows that the probability $\mathbf{P}_0$ of the event

$$\lim_{n \to \infty} \hat{F}^{(n)}(x, \omega) = F(x) \tag{5}$$

is equal to one.

Let us fix an elementary event $\omega_0 \in \Omega_0$. For this elementary event we introduce a new probability space $(\Omega, \mathcal{F}, \mathbf{P})$. Let us introduce random variables X_n,

$n \in \mathbb{N}$, defined on the probability space $(\Omega, \mathcal{F}, \mathbf{P})$. We suppose that the random variable X_n has the distribution function

$$F_n^{\omega_0}(x) \triangleq \frac{M(x, \omega_0)}{n}, \tag{6}$$

where $M(x, \omega_0) \triangleq \sum_{k=1}^{n} \theta(x - Y_k(\omega_0))$. Notice that for every fixed $\omega_0 \in \Omega_0$ we have

$$F_n^{\omega_0}(x) = \mathbf{P}\{X_n \leq x\} = \hat{F}^{(n)}(x, \omega_0),\ n \in \mathbb{N}. \tag{7}$$

Let us define the following sequence of discrete approximations of original problem (1)

$$u_n(\omega_0) \in \text{Arg} \min_{u \in U} \mathbf{G}[\Phi(u, X_n)],\ n = 1, 2, 3, \ldots \tag{8}$$

From (5) it follows that X_n tends in distribution to a random variable X having the same distribution function $F(x)$ as the random variable Y. The random variable X is defined on the probability space $(\Omega, \mathcal{F}, \mathbf{P})$. We can suppose that the distribution functions of $\{X_n\}$ are known. In this paper, we do not require that random variables X_n are discrete, they can have arbitrary distributions. Our purpose is to obtain conditions guaranteeing that optimal strategies $u_n(\omega_0)$ of these problems converge to an optimal strategy of problem (1). In this paper, we are going to develop this technique for problems that can be solved using deterministic equivalents [3]. This is the first step to solve the problem for an arbitrary statement.

2 Statement of the Problem

Let a probability space $(\Omega, \mathcal{F}, \mathbf{P})$ be given. Let X be a random variable defined on the probability space $(\Omega, \mathcal{F}, \mathbf{P})$. Let X have realizations $x \in \mathcal{X} \subset \mathbb{R}^1$ and a distribution function $F(x)$. Let $u \in \mathbb{R}^m$ be an optimization strategy. Let $U \subset \mathbb{R}^m$ be a set of feasible optimization strategies. We assume that the set U is compact. Let $\Phi(u, x) \colon U \times \mathcal{X} \to \mathbb{R}^1$ be a loss function. Let us consider the probability function

$$P_\varphi(u) \triangleq \mathbf{P}\{\Phi(u, X) \leq \varphi\}, \tag{9}$$

where φ is a fixed parameter. Let us define the quantile function

$$\varphi_\alpha(u) \triangleq \min\{\varphi \colon P_\varphi(u) \geq \alpha\}, \tag{10}$$

where α is a fixed probability level.

Let us consider the probability maximization problem

$$U_\varphi^* \triangleq \text{Arg} \max_{u \in U} P_\varphi(u) \tag{11}$$

and the quantile minimization problem

$$U_\alpha^* \triangleq \text{Arg} \min_{u \in U} \varphi_\alpha(u). \tag{12}$$

Let $\{X_n\}$ be a random sequence defined on the probability space $(\Omega, \mathcal{F}, \mathbf{P})$ such that $X_n \xrightarrow{d} X$ as $n \to \infty$, i.e., X_n converges in distribution to X. The distribution functions of X_n are assumed to be known and are denoted by $F_n(x)$. For all $n \in \mathbb{N}$ let us define the probability function

$$P_\varphi^{(n)}(u) \triangleq \mathbf{P}\{\Phi(u, X_n) \leq \varphi\} \tag{13}$$

and the quantile function

$$\varphi_\alpha^{(n)}(u) \triangleq \min\{\varphi \colon P_\varphi^{(n)}(u) \geq \alpha\}. \tag{14}$$

Consider the sequence of probability maximization problems

$$u_\varphi^{(n)} \in \operatorname*{Arg\,max}_{u \in U} P_\varphi^{(n)}(u), \ n = 1, 2, 3, \ldots, \tag{15}$$

and the sequence of quantile minimization problems

$$u_\alpha^{(n)} \in \operatorname*{Arg\,min}_{u \in U} \varphi_\alpha(u), \ n = 1, 2, 3, \ldots \tag{16}$$

Since the set U is compact, the sequence $\{u_\varphi^{(n)}\}$ has a convergent subsequence $\{u_\varphi^{(n_k)}\}$. Also, the sequence $\{u_\alpha^{(n)}\}$ has a convergent subsequence $\{u_\alpha^{(n_k)}\}$. Let us denote the limits of the sequences $\{u_\varphi^{(n_k)}\}$ and $\{u_\alpha^{(n_k)}\}$ by $\bar{u}_\varphi$ and $\bar{u}_\alpha$ respectively. Our purpose is to suggest conditions, which guarantee that the solutions $\bar{u}_\varphi$ and $\bar{u}_\alpha$ are optimal to original problems (9) and (10), i.e., $\bar{u}_\varphi \in U_\varphi^*$, $\bar{u}_\alpha \in U_\alpha^*$.

Remark 1. Let the random sequence $\{X_n\}$ be constructed according to (6). Then the sequences $\{u_\varphi^{(n)}\}$, $\{u_\alpha^{(n)}\}$ and the limits $\bar{u}_\varphi$, $\bar{u}_\alpha$ depend on $\omega_0 \in \Omega_0$. To show it, we can denote them by $\{u_\varphi^{(n)}(\omega_0)\}$, $\{u_\alpha^{(n)}(\omega_0)\}$, $\bar{u}_\varphi(\omega_0)$, $\bar{u}_\alpha(\omega_0)$ respectively. Hence we can consider these sequences to be random. If we guarantee that $\bar{u}_\varphi(\omega_0) \in U_\varphi^*$, $\bar{u}_\alpha(\omega_0) \in U_\alpha^*$ for almost all $\omega_0 \in \Omega_0$, then we will be able to guarantee that

$$\mathbf{P}_0\{\omega_0 \colon \bar{u}_\varphi(\omega_0) \in U_\varphi^*\} = 1, \tag{17}$$

$$\mathbf{P}_0\{\omega_0 \colon \bar{u}_\alpha(\omega_0) \in U_\alpha^*\} = 1, \tag{18}$$

i.e., $\{u_\varphi^{(n)}\}$, $\{u_\alpha^{(n)}\}$ have subsequences converging almost surely to optimal solutions of the original problems. To simplify notation, we omit dependence strategies on $\omega_0 \in \Omega_0$. All the results below are obtained for a fixed elementary event $\omega_0 \in \Omega_0$.

3   Auxiliary Propositions

3.1   Uniform Convergence of Distribution Functions

Let us consider a random sequence $\{X_n\}$ converging in distribution to a random variable X. Recall that $F_n(\cdot)$ and $F(\cdot)$ are the distribution functions of X_n and X respectively. The following theorem provides the uniform convergence of the distribution functions.

Theorem 1 ([10, p. 152]). *Let $X_n \xrightarrow{d} X$ as $n \to \infty$. Assume that the random variable X is continuous. Then*

$$\lim_{n \to \infty} \sup_{x \in \mathbb{R}^1} |F_n(x) - F(x)| = 0. \tag{19}$$

3.2 Probability Function Maximization

Let us consider the following auxiliary problem

$$U_F^* \triangleq \operatorname{Arg\,max}_{u \in U} F(r(u)), \tag{20}$$

where $r(\cdot) \colon \mathbb{R}^m \to \mathbb{R}^1$ is a continuous function, $F(\cdot)$ is the distribution function of the random variable X. Also, let a sequence of optimization problems

$$U_F^{(n)} \triangleq \operatorname{Arg\,max}_{u \in U} F_n(r(u)) \tag{21}$$

be given. Here, $F_n(\cdot)$ is the distribution function of the random variable X_n. Let us assume that there exists a convergent sequence $\{u_n\}$ such that $u_n \in U_F^{(n)}$. Its limit is denoted by $\bar{u}$.

Lemma 1. *Let the following conditions hold:*

  (i) *The function $F(\cdot)$ is continuous;*
 (ii) *The function $r(\cdot) \colon \mathbb{R}^m \to \mathbb{R}^1$ is continuous;*
(iii) $X_n \xrightarrow{d} X$;
 (iv) $u_n \in U_F^{(n)}$;
  (v) $u_n \to \bar{u}$ as $n \to \infty$.

Then $\bar{u} \in U_F^*$

Proof. Let us first prove that

$$\lim_{n \to \infty} F_n(r(u_n)) = F(r(\bar{u})). \tag{22}$$

To simplify notation, let us denote $g_n(u) \triangleq F_n(r(u))$, $g(u) \triangleq F(r(u))$. From (i) and (ii) it follows that $g(\cdot)$ is continuous. Then

$$|F_n(r(u_n)) - F_n(r(\bar{u}))| = |g_n(u_n) - g(\bar{u})| \le |g_n(u_n) - g(u_n) + g(u_n) - g(\bar{u})| \le$$
$$|g_n(u_n) - g(u_n)| + |g(u_n) - g(\bar{u})|. \tag{23}$$

Since $g(\cdot)$ is continuous, $|g(u_n) - g(\bar{u})| \to 0$ as $n \to \infty$. Notice that

$$|g_n(u_n) - g(u_n)| \le \sup_{x \in \mathbb{R}^1} |F_n(x) - F(x)|. \tag{24}$$

According to Theorem 1, $\sup_{x \in \mathbb{R}^1} |F_n(x) - F(x)| \to 0$ as $n \to \infty$. Therefore equality (22) holds.

To prove the lemma, assume the converse, i.e., $\bar{u} \notin U_F^*$. Then there exists a strategy $u^* \in U$ such that $g(u^*) = g(\bar{u}) + c$, where $c > 0$. According to (i) and (v), $g_n(u^*) \to g(u^*)$ as $n \to \infty$. Hence there exists a number N such that for any $n > N$ we have

$$|g_n(u^*) - g(u^*)| = |g_n(u^*) - (g(\bar{u}) + c)| \leq \frac{c}{3}, \tag{25}$$

$$|g_n(u_n) - g(\bar{u})| \leq \frac{c}{3}. \tag{26}$$

Then

$$g(\bar{u}) + \frac{2}{3}c \leq g_n(u^*) \leq g(\bar{u}) + \frac{4}{3}c, \tag{27}$$

$$g(\bar{u}) - \frac{1}{3}c \leq g_n(u_n) \leq g(\bar{u}) + \frac{1}{3}c. \tag{28}$$

Hence $g_n(u^*) > g_n(u_n)$, i.e., the strategy u_n is not optimal to problem (21). This contradicts to assumption (iv).

3.3 Quantile Function Minimization

Let us denote by $[X]_\alpha$ the α-quantile of a random variable X. Consider the following problem

$$U_r^* \triangleq \operatorname*{Arg\,min}_{u \in U} r(u, [X]_\alpha), \tag{29}$$

where $r(\cdot) \colon U \times \mathbb{R}^1 \to \mathbb{R}^1$ is a continuous function. Let a random sequence $\{X_n\}$ be given. Let us consider the sequence of problems

$$U_r^{(n)} \triangleq \operatorname*{Arg\,min}_{u \in U} r(u, [X_n]_\alpha). \tag{30}$$

Let $u_n \in U_r^{(n)}$. We assume that the sequence $\{u_n\}$ converges to a value $\bar{u}$ as $n \to \infty$.

Lemma 2. *Let the following conditions hold:*

(i) $[X_n]_\alpha \to [X]_\alpha$ *as* $n \to \infty$;
(ii) The function $r(\cdot) \colon U \times \mathbb{R}^1 \to \mathbb{R}^1$ *is continuous;*
(iii) $u_n \in U_r^{(n)}$;
(iv) $u_n \to \bar{u}$ *as* $n \to \infty$.

Then $\bar{u} \in U_r^*$.

Proof. Let us assume the converse, i.e., $\bar{u} \notin U_r^*$. Hence there exist $c > 0$ and $u^* \in U_r^*$ such that $r(u^*, [X]_\alpha) = r(\bar{u}, [X]_\alpha) - c$. Since $r(\cdot)$ is continuous,

$$\lim_{n \to \infty} r(u_n, [X_n]_\alpha) = r(\bar{u}, x_\alpha). \tag{31}$$

It follows that there exists a number N such that for any $n > N$ we have

$$|r(u^*, [X_n]_\alpha) - r(u^*, [X]_\alpha)| = |r(u^*, [X_n]_\alpha) - (r(\bar{u}, [X]_\alpha) - c)| \le \frac{c}{3}, \qquad (32)$$

$$|r(u_n, [X_n]_\alpha) - r(\bar{u}, [X_n]_\alpha)| \le \frac{c}{3}. \qquad (33)$$

From this, we conclude that

$$r(u^*, [X_n]_\alpha) < r(u_n, [X_n]_\alpha). \qquad (34)$$

This contradicts the optimality of u_n.

4   Deterministic Equivalents

In this section, we give deterministic equivalents to original problem (9) and (10). We will use these equivalents to prove theorems in the next section.

4.1   Case of Symmetrical Distribution

Let us consider the case when

$$\Phi(u, x) = r(s(u)(x + c)), \qquad (35)$$

where $s(\cdot): U \to \mathbb{R}^1$ is a function, $r(\cdot): \mathbb{R}^1 \to \mathbb{R}^1$ is a strictly increasing, continuous function, c is a real number. We suppose that $s(u) \ne 0$ for all $u \in U$. Let us assume that the distribution of X is symmetrical, i.e., for all $x \in \mathbb{R}^1$:

$$F(x) \triangleq \mathbf{P}\{X \le x\} = \mathbf{P}\{-X \le x\}. \qquad (36)$$

A similar case has been considered in [3] where X is a continuous random vector with spherically symmetrical distribution, the function $s(\cdot)$ is linear.

Theorem 2. *If the loss function is given by (35), $s(u) \ne 0$ for all $u \in U$, the distribution of X is symmetrical, then*

$$P_\varphi(u) = F\left(\frac{r^{-1}(\varphi) - cs(u)}{|s(u)|}\right) \qquad (37)$$

is the probability function, where the function $r^{-1}(\cdot)$ is the inverse of $r(\cdot)$;

$$\varphi_\alpha(u) = r\left(|s(u)|[X]_\alpha + cu\right) \qquad (38)$$

is the quantile function.

Proof. Let us first prove (37). We have

$$P_\varphi(u) = \mathbf{P}\left\{\Phi(u, X) \le \varphi\right\} = \mathbf{P}\left\{r(s(u)(X + c)) \le \varphi\right\} =$$
$$\mathbf{P}\left\{s(u)(X + c) \le r^{-1}(\varphi)\right\}. \qquad (39)$$

If $s(u) > 0$, then

$$\mathbf{P}\left\{s(u)(X + c) \leq r^{-1}(\varphi)\right\} = \mathbf{P}\left\{X \leq \frac{r^{-1}(\varphi) - cs(u)}{s(u)}\right\} =$$
$$F\left(\frac{r^{-1}(\varphi) - cs(u)}{|s(u)|}\right). \quad (40)$$

If $s(u) < 0$, then

$$\mathbf{P}\left\{s(u)(X + c) \leq r^{-1}(\varphi)\right\} = \mathbf{P}\left\{X \geq \frac{r^{-1}(\varphi) - cs(u)}{s(u)}\right\} =$$
$$\mathbf{P}\left\{-X \leq \frac{r^{-1}(\varphi) - cs(u)}{|s(u)|}\right\} = F\left(\frac{r^{-1}(\varphi) - cs(u)}{|s(u)|}\right). \quad (41)$$

Let us prove (38):

$$\varphi_\alpha(u) = \min\left\{\varphi\colon F\left(\frac{r^{-1}(\varphi) - cs(u)}{|s(u)|}\right) \geq \alpha\right\} =$$
$$\min\left\{\varphi\colon \frac{r^{-1}(\varphi) - cs(u)}{|s(u)|} \geq [X]_\alpha\right\} = \min\left\{\varphi\colon r\left(|s(u)|[X]_\alpha + cu\right) \leq \varphi\right\} =$$
$$r\left(|s(u)|[X]_\alpha + cu\right). \quad (42)$$

Therefore, in this case, original problem (9) is equivalent to the problem

$$U_\varphi^* = \operatorname*{Arg\,max}_{u \in U} F\left(\frac{r^{-1}(\varphi) - cs(u)}{|s(u)|}\right), \quad (43)$$

problem (10) is equivalent to

$$U_\alpha^* = \operatorname*{Arg\,min}_{u \in U} r\left(|s(u)|[X]_\alpha + cu\right). \quad (44)$$

4.2   Case of Loss Function Increasing in Random Variable

Consider the case when the function $\Phi(u, x)$ is strictly increasing in x and continuous in x. Let us denote by $\Phi_x^{-1}(u, \varphi)$ the inverse of $\Phi(u, x)$ with respect to x, i.e.,

$$\Phi(u, \Phi_x^{-1}(u, \varphi)) = \varphi. \quad (45)$$

Theorem 3 ([3]). *Let the function $\Phi(u, x)$ be strictly increasing in x and continuous in x. Then the probability function is*

$$P_\varphi(u) = F(\Phi_x^{-1}(u, \varphi)) \quad (46)$$

and the quantile function is

$$\varphi_\alpha(u) = \Phi(u, [X]_\alpha). \quad (47)$$

Remark 2. It has been proven [11] that (47) is valid if $\Phi(u, x)$ is nondecreasing in x.

In this case, we can write deterministic equivalents

$$U_\varphi^* = \text{Arg} \max_{u \in U} F(\Phi_x^{-1}(u, \varphi)), \tag{48}$$

$$U_\alpha^* = \text{Arg} \min_{u \in U} \Phi(u, [X]_\alpha). \tag{49}$$

4.3   Case of Loss Function Increasing in Optimization Strategy

Let

$$\Phi(u, x) = r(s(u), x), \tag{50}$$

where $s(\cdot)\colon \mathbb{R}^m \to \mathbb{R}^1$, the function $r(s, x)\colon \mathbb{R}^1 \times \mathbb{R}^1 \to \mathbb{R}^1$ is strictly increasing in s and continuous in s.

Theorem 4 ([3]). *Let $\Phi(u, x)$ be given by (50), then*

$$P_\varphi(u) = F_\xi(-s(u)), \tag{51}$$

where

$$F_\xi(x) = \mathbf{P}\{\xi \le x\}, \tag{52}$$

$\xi \triangleq -r_s^{-1}(\varphi, X)$, $r_s^{-1}(\varphi, x)$ *is the inverse of $r(s, x)$ with respect to s.*

Under assumptions of Theorem 4, original problem (10) is equivalent to

$$U_\varphi^* = \text{Arg} \max_{u \in U} F_\xi(-s(u)), \tag{53}$$

5   Main Results

Let us consider the sequences $\{u_\varphi^{(n)}\}$ and $\{u_\alpha^{(n)}\}$ defined in Sect. 2. As we noticed above, these sequences have convergent subsequences $\{u_\varphi^{(n_k)}\}$ and $\{u_\alpha^{(n_k)}\}$. We will suggest conditions, which guarantee that these sequences converge, i.e., every partial limit of the sequence $\{u_\varphi^{(n)}\}$ is optimal to problem (9) and every partial limit of $\{u_\alpha^{(n)}\}$ is optimal to (10).

5.1   Case of Symmetrical Distribution

Firstly, consider the case introduced in Sect. 4.1 when

$$\Phi(u, x) = r(s(u)(x + c)) \tag{54}$$

where $s(\cdot)\colon U \to \mathbb{R}^1$ such that $s(u) \ne 0$ for all $u \in U$, $r(\cdot)\colon \mathbb{R}^1 \to \mathbb{R}^1$ is a strictly increasing, continuous function, the distribution of X is symmetrical.

Theorem 5. *Let the following conditions hold:*

(i) $\Phi(u, x)$ is defined by (54);
(ii) The function $s(\cdot) \colon U \to \mathbb{R}^1$ is continuous such that $s(u) \neq 0$ for all $u \in U$;
(iii) The distributions of X and X_n are symmetrical;
(iv) $X_n \xrightarrow{d} X$ as $n \to \infty$;
(v) The random variable X is continuous;
(vi) $u_\varphi^{(n)} \in \text{Arg}\max_{u \in U} P_\varphi^{(n)}(u)$.

Then every partial limit $\bar{u}_\varphi$ of the sequence $\{u_\varphi^{(n)}\}$ belongs to the set U_φ^*.

Proof. Let $\{u_\varphi^{(n_k)}\}$ be a convergent subsequence of the sequence $\{u_\varphi^{(n)}\}$. From Theorem 2 it follows that

$$U_\varphi^* = \text{Arg}\max_{u \in U} F\left(\frac{r^{-1}(\varphi) - cs(u)}{|s(u)|}\right), \tag{55}$$

$$u_\varphi^{(n)} \in \text{Arg}\max_{u \in U} F_n\left(\frac{r^{-1}(\varphi) - cs(u)}{|s(u)|}\right), \ n \in \mathbb{N}. \tag{56}$$

Since $s(\cdot)$ is continuous and $s(u) \neq 0$, the function $u \mapsto \frac{r^{-1}(\varphi) - cs(u)}{|s(u)|}$ is also continuous. Then, according to Lemma 1, $\bar{u}_\varphi = \lim_{k \to \infty} u_\varphi^{(n_k)} \in U_\varphi^*$.

Theorem 6. Let the following conditions hold:

(i) $\Phi(u, x)$ is defined by (54);
(ii) The function $s(\cdot) \colon U \to \mathbb{R}^1$ is continuous such that $s(u) \neq 0$ for all $u \in U$;
(iii) The distributions of X and X_n are symmetrical;
(iv) $[X_n]_\alpha \to [X]_\alpha$ as $n \to \infty$;
(v) $u_\alpha^{(n)} \in \text{Arg}\min_{u \in U} \varphi_\alpha^{(n)}(u)$.

Then every partial limit $\bar{u}_\alpha$ of $\{u_\alpha^{(n)}\}$ belongs to U_α^*.

Proof. Let $\{u_\alpha^{(n_k)}\}$ be a convergent subsequence of the sequence $\{u_\alpha^{(n)}\}$. From Theorem 2 we have

$$U_\alpha^* = \text{Arg}\min_{u \in U} r\left(|s(u)|[X]_\alpha + cu\right), \tag{57}$$

$$u_\alpha^{(n)} \in \text{Arg}\min_{u \in U} r\left(|s(u)|[X_n]_\alpha + cu\right). \tag{58}$$

Since the function $r(\cdot)$ is strictly increasing and continuous, according to Lemma 2, we conclude that $\bar{u}_\alpha = \lim_{k \to \infty} u_\alpha^{(n_k)} \in U_\alpha^*$.

5.2 Case of Loss Function Increasing in Random Variable

Consider the case introduced in Sect. 4.2.

Theorem 7. *Let the following conditions hold:*

(i) *The function $\Phi(u,x)$ is strictly increasing in x and continuous in x;*
(ii) *The function $\Phi_x^{-1}(u,\varphi)$ is continuous in $u \in U$;*
(iii) *$X_n \xrightarrow{d} X$ as $n \to \infty$;*
(iv) *The random variable X is continuous;*
(v) *$u_\varphi^{(n)} \in \operatorname{Arg\,max}_{u \in U} P_\varphi^{(n)}(u)$.*

Then every partial limit $\bar{u}_\varphi$ of $\{u_\varphi^{(n)}\}$ belongs to U_φ^*.

Proof. From Theorem 3 it follows that

$$U_\varphi^* = \operatorname{Arg\,max}_{u \in U} F(\Phi_x^{-1}(u,\varphi)), \tag{59}$$

$$u_\varphi^{(n)} \in \operatorname{Arg\,max}_{u \in U} F_n(\Phi_x^{-1}(u,\varphi)), \ n \in \mathbb{N}. \tag{60}$$

Since the function $\Phi_x^{-1}(u,\varphi)$ is continuous in x, we conclude from Lemma 1 that $\bar{u}_\varphi \in U_\varphi^*$.

Theorem 8. *Let the following conditions hold:*

(i) *The function $\Phi(u,x)$ is nondecreasing in x and continuous in (u,x);*
(ii) *$[X_n]_\alpha \to [X]_\alpha$ as $n \to \infty$;*
(iii) *$u_\varphi^{(n)} \in \operatorname{Arg\,max}_{u \in U} P_\varphi^{(n)}(u)$.*

Then every partial limit $\bar{u}_\alpha$ of $\{u_\alpha^{(n)}\}$ belongs to U_α^*.

Proof. From Theorem 3 and Remark 2 it follows that

$$U_\alpha^* = \operatorname{Arg\,min}_{u \in U} \Phi(u,[X]_\alpha), \tag{61}$$

$$u_\alpha^{(n)} \in \operatorname{Arg\,min}_{u \in U} \Phi(u,[X_n]_\alpha), \ n \in \mathbb{N}. \tag{62}$$

Since the function $\Phi(u,x)$ is continuous, according to Lemma 2, we have $\bar{u}_\alpha \in U_\alpha^*$

5.3 Case of Loss Function Increasing in Optimization Strategy

Consider the case introduced in Sect. 4.3. Let us denote by $F_{\xi_n}(x)$ the probability function of the random variable $\xi_n \triangleq -r_s^{-1}(\varphi, X_n)$. We recall that $\xi \triangleq -r_s^{-1}(\varphi, X)$, $F_\xi(\cdot)$ is the distribution function of the random variable ξ.

Theorem 9. *Let the following conditions hold:*

(i) *$\Phi(u,x) = r(s(u),x)$, where the function $s(\cdot) \colon \mathbb{R}^m \to \mathbb{R}^1$ is continuous, the function $r(s,x) \colon \mathbb{R}^1 \times \mathbb{R}^1 \to \mathbb{R}^1$ is strictly increasing in s and continuous in s;*
(ii) *$\xi_n \xrightarrow{d} \xi$ as $n \to \infty$;*

(iii) The random variable ξ is continuous;

(iv) $u_\varphi^{(n)} \in \operatorname{Arg\,max}_{u \in U} P_\varphi^{(n)}(u)$.

Then every partial limit $\bar{u}_\varphi$ of the sequence $\{u_\varphi^{(n)}\}$ belongs to U_φ^*.

Proof. From Theorem 4 it follows that

$$U_\varphi^* = \operatorname{Arg\,max}_{u \in U} F_\xi(-s(u)), \tag{63}$$

$$u_\varphi^{(n)} \in \operatorname{Arg\,max}_{u \in U} F_{\xi_n}(-s(u)), \; n \in \mathbb{N}. \tag{64}$$

Since the functions $\Phi(\cdot)$ and $s(\cdot)$ are continuous, according to Lemma reflemma1, we conclude that $\bar{u}_\varphi \in U_\varphi^*$.

6   Conclusion

In this paper, we have suggested conditions guaranteeing the convergence of discrete approximation of stochastic programming problems with probabilistic and quantile objective function. We should notice that these conditions describe very particular cases of these problems. Therefore, more general conditions of convergence should be the topic of future works. We hope that obtained results can be expanded to the case of continuous probability and quantile functions. However, from the proofs of the suggested theorems it follows that the conditions $X_n \xrightarrow{d} X$ and $[X_n]_\alpha \to X_\alpha$ cannot be restricted.

References

1. Kibzun, A.I., Kan, Yu.S: Stochastic Programming Problems with Probability and Quantile Functions. Wiley, New York (1996)
2. Kibzun, A.I., Naumov, A.V., Norkin, V.I.: On reducing a quantile optimization problem with discrete distribution to a mixed integer programming problem. Autom. Remote Control. **74**, 951–967 (2013)
3. Vishnaykov, B.V., Kibzun, A.I.: Deterministic equivalents for the problems of stochastic programming with probabilistic criteria. Autom. Remote Control. **67**, 945–961 (2006)
4. Lepp, R.: Projection and discretization methods in stochastic programming. J. Comput. Appl. Math. **56**, 55–64 (1994)
5. Pennanen, T.: Epi-convergent discretizations of multistage stochastic programs via integration quadratures. Math. Program. Ser. B. **116**, 461–479 (2009)
6. Choirat, C., Hess, C., Seri, R.: Approximation of stochastic programming problems. In: Niederreiter, H., Talay, D. (eds.) Monte Carlo and Quasi-Monte Carlo Methods 2004, pp. 45–59. Springer, Heidelberg (2006)
7. Pflug, G.Ch.: Scenario tree generation for multiperiod financial optimization by optimal discretization. Math. Program. **89**, 251–271 (2001)
8. Lepp, R.: Approximation of value-at-risk problems with decision rules. In: Uryasev, S.P. (ed.) Probabilistic Constrained Optimization. pp. 186–197. Kluwer Academic Publishers, Norwell (2000)

9. Kibzun, A., Lepp, R.: Discrete approximation in quantile problem of portfolio selection. In: Uryasev, S., Pardalos, P.M. (eds.) Stochastic Optimization: Algorithms and Applications, pp. 119–133. Kluwer Academic Publishers, Norwell (2000)
10. Chistyakov, V.P.: Probability Theory Course. Nauka, Moscow (1978). (in Russian)
11. Kibzun, A.I., Naumov, A.V., Ivanov, S.V.: Bilevel optimization problem for railway transport hub planning. Upravlenie bol'simi sistemami **38**, 140–160 (2012). (in Russian)

A Robust Leaky-LMS Algorithm for Sparse System Identification

Cemil Turan[1(⊠)] and Yedilkhan Amirgaliev[1,2]

[1] Department of Computer Engineering,
Suleyman Demirel University, Almaty, Kazakhstan
{cemil.turan,amirgaliev.yedilkhan}@sdu.edu.kz
[2] Institute of Information and Computational Technologies, Almaty, Kazakhstan

Abstract. In this paper, a new Leaky-LMS (LLMS) algorithm that modifies and improves the Zero-Attracting Leaky-LMS (ZA-LLMS) algorithm for sparse system identification has been proposed. The proposed algorithm uses the sparsity of the system with the advantages of the variable step-size and l_0-norm penalty. We compared the performance of our proposed algorithm with the conventional LLMS and ZA-LLMS in terms of the convergence rate and mean-square-deviation (MSD). Additionally, the computational complexity of the proposed algorithm has been derived. Simulations performed in MATLAB showed that the proposed algorithm has superiority over the other algorithms for both types of input signals of additive white Gaussian noise (AWGN) and additive correlated Gaussian noise (ACGN).

Keywords: Adaptive filters · Sparse system identification · Leaky LMS · L_0-norm penalty

1 Introduction

In adaptive filtering technology, the least-mean-square (LMS) algorithm is a commonly used algorithm for system identification (see Fig. 1), noise cancellation or channel equalization models [1]. Although it is a very simple and robust algorithm, its performance deteriorates for some applications which have high correlation, long filter length, sparse signals etc. In the literature, many different LMS-type algorithms were proposed to improve the performance of the standard LMS algorithm.

Leaky-LMS algorithm was proposed [2, 3] to overcome the issues when the input signal is highly correlated, by using shrinkage in its update equation. Another LMS based algorithm VSSLMS uses a variable step-size in update equation of the standard LMS to increase the convergence speed at the beginning stages of the iterations and decrease MSD at later iterations [4, 5]. In order to improve the performance of the LMS algorithm when the system is sparse (most of the system coefficients are zeroes), ZA-LMS algorithm was proposed in [6].

In [7], the author proposed the ZA-LLMS algorithm which combines the LLMS algorithm and the ZA-LMS algorithm for sparse system identification. A better performance was obtained for AWGN and ACGN input signals. In [8], a high performance algorithm called zero-attracting function-controlled variable step-size LMS

© Springer International Publishing Switzerland 2016
Y. Kochetov et al. (Eds.): DOOR 2016, LNCS 9869, pp. 538–546, 2016.
DOI: 10.1007/978-3-319-44914-2_42

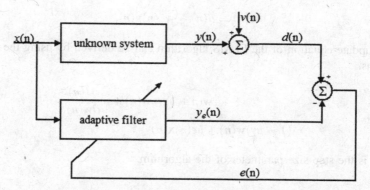

Fig. 1. Block diagram of the system identification process.

(ZAFC-VSSLMS) was proposed by using the advantages of variable step-size and l_0-norm penalty. We were motivated by the inspiration of the combination of these two algorithms. So in this paper we proposed a new algorithm that combines the ZA-LLMS and ZAFC-VSSLMS algorithms. In the next section, a brief review of the LLMS and ZA-LLMS algorithms is provided. We describe the proposed algorithm in Sect. 3 with computational complexity and convergence analysis. In Sect. 4, the simulations are presented and the performance of the algorithm is compared. Conclusions are drawn in the last section.

2 Review of the Related Algorithms

2.1 Leaky-LMS (LLMS) Algorithm

In a system identification process, the desired signal is defined as,

$$d(n) = \mathbf{w}_0^T \mathbf{x}(n) + v(n) \tag{1}$$

where $\mathbf{w}_0 = [w_{00}, \ldots, w_{0N-1}]^T$ are the unknown system coefficients with length N, $\mathbf{x}(n) = [x_0, \ldots, x_{N-1}]^T$ is the input-tap vector and $v(n)$ is the additive noise. In addition to being independent of the noise sample $v(n)$ with zero mean and variance of σ_v^2, the input data sequence $\mathbf{x}(n)$ and the additive noise sample $v(n)$ are also assumed to be independent.

The cost function of the LLMS algorithm is given by,

$$J_1(n) = \frac{1}{2}e^2(n) + \gamma \mathbf{w}^T(n)\mathbf{w}(n) \tag{2}$$

where $\mathbf{w}(n)$ is the filter-tap vector at time n, γ is a positive constant called 'leakage factor' and $e(n)$ is the instantaneous error and given by,

$$e(n) = d(n) - \mathbf{w}^T(n)\mathbf{x}(n) \tag{3}$$

The update equation of the LLMS algorithm can be derived by using the gradient method as,

$$\mathbf{w}(n+1) = \mathbf{w}(n) + \mu \frac{\partial J(n)}{\partial \mathbf{w}(n)} \tag{4}$$

$$= (1 - \mu\gamma)\mathbf{w}(n) + \mu e(n)\mathbf{x}(n)$$

where μ is the step-size parameter of the algorithm.

2.2 Zero-Attracting Leaky-LMS (ZA-LLMS) Algorithm

The cost function of the LLMS algorithm was modified by adding the log-sum penalty of the filter-tap vector as given below:

$$J_2(n)) = \frac{1}{2}e^2(n) + \gamma \mathbf{w}^T(n)\mathbf{w}(n) + \gamma' \sum_{i=1}^{N}\left(1 + \frac{|w_i|}{\xi'}\right) \tag{5}$$

where γ' and ξ' are positive parameters. Taking the gradient of the cost function and subtracting from the previous filter-tap vector iteratively, then the update equation was derived as follows [7]:

$$\mathbf{w}(n+1) = (1 - \mu\gamma)\mathbf{w}(n) + \mu e(n)\mathbf{x}(n) - \rho \frac{\text{sgn}[\mathbf{w}(n)]}{1 + \xi|\mathbf{w}(n)|} \tag{6}$$

where $\rho = \frac{\mu\gamma'}{\xi'}$ is the zero-attracting parameter, $\xi = \frac{1}{\xi'}$ and sgn(.) operation is defined as,

$$\text{sgn}(x) = \begin{cases} \frac{x}{|x|} & \text{if } x \neq 0 \\ 0 & \text{if } x = 0 \end{cases} \tag{7}$$

3 The Proposed Algorithm

3.1 Derivation of the Proposed Algorithm

An improved sparse LMS-type algorithm was proposed in [8] by exploiting the advantages of variable step-size and recently proposed [9] l_0-norm which gives an approximate value of $\|\cdot\|_0$. We modify the cost function of that algorithm by adding the weight vector norm penalty as,

$$J_3(n)) = \frac{1}{2}e^2(n) + \gamma \mathbf{w}^T(n)\mathbf{w}(n) + \varepsilon \|\mathbf{w}(n)\|_0 \tag{8}$$

where ε is a small positive constant and $\|\mathbf{w}(n)\|_0$ denotes the l_0-norm of the weight vector given as,

$$\|\mathbf{w}(n)\|_0 \simeq \sum_{k=0}^{N-1}(1 - e^{-\lambda|\mathbf{w}(n)|}) \tag{9}$$

where λ is a positive parameter. Deriving (8) with respect to $\mathbf{w}(n)$ and substituting in the update equation we get,

$$\mathbf{w}(n+1) = (1 - \mu(n)\gamma)\mathbf{w}(n) + \mu(n)e(n)\mathbf{x}(n) - \rho(n)\mathrm{sgn}[\mathbf{w}(n)]e^{-\lambda|\mathbf{w}(n)|} \tag{10}$$

where $\rho(n) = \mu(n)\varepsilon\lambda$ is the sparsity aware parameter and depends on the positive constant λ and $\mu(n)$ which is the variable step-size and given in [6] as

$$\mu(n+1) = \alpha\mu(n) + \gamma_s f(n)\frac{e(n)^2}{\hat{e}_{ms}^2(n)} \tag{11}$$

where $0 < \alpha < 1$, $\gamma_s > 0$ are some positive constants and $e(n)$ is a mean value of the error vector. $\hat{e}_{ms}^2(n)$ is the estimated mean-square-error (MSE) and is defined as

$$\hat{e}_{ms}^2(n) = \beta\hat{e}_{ms}^2(n-1) + (1 - \beta)e(n)^2 \tag{12}$$

where β is a weighting factor given as $0 < \beta < 1$ and $f(n)$ is a control function given below

$$f(n) = \begin{cases} 1/n & n < L \\ 1/L & n \geq L \end{cases} \tag{13}$$

A summary of the algorithm is given in Table 1.

It is seen that the update equation of the ZA-LLMS algorithm has been modified by changing the constant step-size μ with $\mu(n)$ given in [8] and the zero-attractor $\rho\frac{\mathrm{sgn}[\mathbf{w}(n)]}{1+\xi|\mathbf{w}(n)|}$ with $\rho(n)\mathrm{sgn}[\mathbf{w}(n)]e^{-\lambda|\mathbf{w}(n)|}$.

3.2 Computational Complexity

The update equation of the conventional LMS algorithm has $O(N)$ complexity and has been calculated as $2N + 1$ multiplications and $2N$ additions at each iteration [10]. For $(1 - \mu\gamma)\mathbf{w}(n)$ in LLMS $N + 1$ extra multiplications and one addition are required. In the update equation of the estimated MSE used in the algorithm proposed in [8], 3 multiplications and 2 additions are required additionally to compute the $\hat{e}_{ms}^2(n)$. For update equation of $\mu(n)$, we need 5 multiplications and one addition. The

Table 1. Summary of the SBFC-VSSLMS algorithm.

---

define $N, L, \varepsilon, \alpha, \gamma_s, \gamma, \lambda$ *and* β

initialize $\mathbf{w}(0) = 0,\ \mu(0) = 0,\ \hat{e}_{ms}(0) = 0$

for $n = 1, 2, \ldots$

$\mathbf{w}(n+1) = (1 - \mu(n)\gamma)\mathbf{w}(n) + \mu(n)e(n)\mathbf{x}(n) - \rho(n)\,\mathrm{sgn}[\mathbf{w}(n)]e^{-\lambda|\mathbf{w}(n)|}$

where

$e(n) = d(n) - \mathbf{w}^T(n)\mathbf{x}(n)$

$\mu(n+1) = \alpha\mu(n) + \gamma_s f(n)\dfrac{e(n)^2}{\hat{e}_{ms}^2(n)}$

$\hat{e}_{ms}^2(n) = \beta_B\,\hat{e}_{ms}^2(n-1) + (1 - \beta_B)e(n)^2$

---

computational complexity of the zero attractor, $\rho(n)\mathrm{sgn}[\mathbf{w}(n)]e^{-\lambda|\mathbf{w}(n)|}$, requires N multiplications for $\lambda|\mathbf{w}(n)|$, N additions for $e^{-\lambda|\mathbf{w}(n)|}$ (taking the first two terms of Taylor series), N multiplications for $\rho(n)\mathrm{sgn}[\mathbf{w}(n)]$, N multiplications for one by one element product of $\lambda|\mathbf{w}(n)|$ by $\rho(n)\mathrm{sgn}[\mathbf{w}(n)]$ and N comparisons for $\mathrm{sgn}[\mathbf{w}(n)]$. So, overall complexity of the zero attractor is $3N$ multiplications, N additions and N comparisons. The overall computational complexity of the proposed algorithm requires $6N + 10$ multiplications, $3N + 4$ additions and N comparisons, that is, (O(N) complexity.

4   Simulation Results

In this section, we compare the performance of the proposed algorithm with LLMS and ZA-LLMS algorithms in high-sparse and low-sparse system identification settings. Two different experiments are performed for each of AWGN and ACGN input signals. To increase the reliability of the expected ensemble average, experiments were repeated by 200 independent Monte-Carlo runs. The constant parameters are found by extensive tests of simulations to obtain the optimal performance as follows: For LLMS: $\mu = 0.002$ and $\gamma = 0.001$. For ZA-LLMS: $\mu = 0.002$, $\gamma = 0.001$, $\rho = 0.0005$ and $\xi = 30$. For the proposed algorithm: $\rho = 0.0005$ and $\lambda = 8$.

In the first experiment, all algorithms are compared for 90 % high-sparsity and 50 % low-sparsity of the system with 20 coefficients having in the first part two '1' and 18 '0'; in the second part ten '1' and ten '0' for 5000 iterations. Signal-to-noise ratio (SNR) is kept at 10 dB by regulating the variances of the input signal and the additive noise. The performance of the algorithm is compared in terms of convergence speed and $MSD = E\left\{\|\mathbf{w}_0 - \mathbf{w}(n)\|^2\right\}$. Figures 2 and 3 give the MSD vs. iteration number of the three algorithms for 90 % sparsity and 50 % sparsity levels respectively. In Fig. 2,

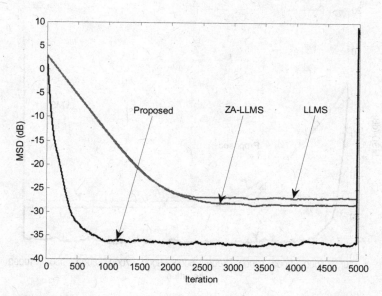

Fig. 2. Steady state behavior of the LLMS, ZA-LLMS and the proposed algorithm for 90 % sparsity with AWGN.

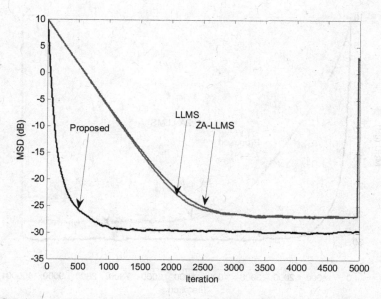

Fig. 3. Steady state behavior of the LLMS, ZA-LLMS and the proposed algorithm for 50 % sparsity with AWGN.

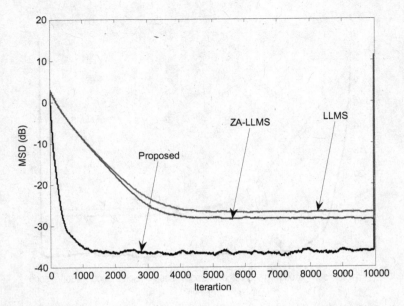

Fig. 4. Steady state behavior of the LLMS, ZA-LLMS and the proposed algorithm for 90 % sparsity with ACGN.

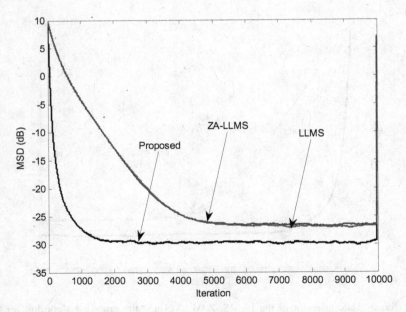

Fig. 5. Steady state behavior of the LLMS, ZA-LLMS and the proposed algorithm for 50 % sparsity with ACGN.

the proposed algorithm has a convergence speed of around 1000 iterations and MSD about -37 dB, while the others are close to each other at a convergence speed of 2500 iterations and MSD of $-26 \sim 28$ dB. Figure 3 shows that the proposed algorithm has a convergence speed of 1100 iterations and MSD of -29.9 dB while the other algorithms have convergence speed and MSD about 3000 iterations and -26 dB, respectively. The figures show that, the proposed algorithm has a fairly fast convergence with lower MSD than that of the other algorithms.

In the second experiment, all conditions are kept as same as in the previous experiment except the input signal type. A correlated signal is created by the AR(1) process as $x(n) = 0.4x(n-1) + v_0(n)$ and the normalized. Figures 4 and 5 shows that the proposed algorithm has a faster convergence and lower MSD than the other algorithms for 90 % sparsity and 50 % sparsity levels respectively.

5 Conclusions

In this work, we proposed a modified leaky-LMS algorithm for sparse system identification. It was derived by combining the ZA-LLMS and ZAFC-LMS algorithms. The performance of the proposed algorithm was compared with LLMS and ZA-LLMS algorithms for 90 % and 50 % sparsity levels of the system with AWGN and ACGN input signals in two different experiments performed in MATLAB. Additionally, the computational complexity of the proposed algorithm has been derived. It was shown that the computational complexity of the proposed algorithm is O(N) as same as in other LMS-type algorithms. Besides, the simulations showed that the proposed algorithm has a very high performance with a quite faster convergence and lower MSD than that of the other algorithms. As a future work, it is recommended that the proposed algorithm can be modified for transform domain or be tested for non-stationary systems.

References

1. Zaknich, A.: Principles of Adaptive Filters and Self-learning Systems. Springer, London (2005)
2. Mayyas, K.A., Aboulnasr, T.: Leaky-LMS: a detailed analysis. In: Proceedings of IEEE International Symposium on Circuits and Systems, vol. 2, pp. 1255–1258 (1995)
3. Sowjanya, M., Sahoo, A. K., Kumar, S.: Distributed incremental leaky LMS. In: International Conference on Communications and Signal Processing (ICCSP), pp. 1753–1757 (2015)
4. Chen, W.Y., Haddad, R.: A variable step size LMS algorithm. In: IEEE Proceedings of 33rd Midwest Symposium on Circuits and Systems, Calgary, vol. 1, pp. 423–426 (1990)
5. Won, Y.K., Park, R.H., Park, J.H., Lee, B.U.: Variable LMS algorithms using the time constant concept. IEEE Trans. Consum. Electron. 40(4), 1083–1087 (1994)
6. Chen, Y., Gu, Y., Hero, A. O.: Sparse LMS for system identification. In: IEEE International Conference Acoustic, Speech and Signal Processing, pp. 3125–3128 (2009)

7. Salman, M.S.: Sparse leaky-LMS algorithm for system identification and its convergence analysis. Int. J. Adapt. Control Sig. Process. **28**(10), 1065–1072 (2013)

8. Turan, C., Salman, M. S.: Zero-attracting function controlled VSSLMS algorithm with analysis. In: Circuits, Systems, and Signal Processing, vol. 34, no. 9, pp. 3071–3080. Springer (2015)

9. Sing-Long, C.A., Tejos, C.A., Irarrazaval, P.: Evaluation of continuous approximation functions for the l_0-norm for compressed sensing. In: Proc. Int. Soc. Mag. Reson. Med. 17: 4585 (2009)

10. Dogancay, K.: Partial-Update Adaptive Filters and Adaptive Signal Processing: Design, Analysis and Implementation. Elsevier, Hungary (2008)

Extended Separating Plane Algorithm and NSO-Solutions of PageRank Problem

Evgeniya Vorontsova$^{(\boxtimes)}$

Far Eastern Federal University, Suhanova Street 8, 690950 Vladivostok, Russia
vorontsovaea@gmail.com

Abstract. The separating plane method with additional clippings and stockpiling (SPACLIP-S) for nonsmooth optimization is proposed in this paper. Both the theoretical and experimental investigations of the method showed that this method is efficient and widely applicable to nonsmooth optimization problems with convex objective functions. Computational experiments demonstrated a rather high performance of SPACLIP-S when applied to the Web page ranking problem. Web page ranking approach is widely used by search engines such as Google and Yandex to order Web pages. Page ranking problem is one of the most important problems in information retrivial due to wide range of applications. In this paper an iterative regularization method with a new penalty function for solving PageRank problem is also presented. Finally, experimental results of comparison of SPACLIP-S and other algorithms for solving test PageRank problems are provided.

Keywords: Nonsmooth convex optimization · Subgradient methods · Black-box minimization · Pagerank problem

1 Introduction

We consider the following problem of unconstrained convex nondifferentiable optimization:

$$\min_{x \in \mathbb{R}^n} f(x), \tag{1}$$

where $f(x)$ is a convex nonsmooth objective function, $x = (x^1, x^2, \ldots, x^n) \in \mathbb{R}^n$. It is assumed that this problem is solvable.

This kind of problems arises in many scientific and engineering areas, for example, in interval analysis [1], economics [2], image denoising [3], control theory [4], neural network training, data mining, computational chemistry and physics, etc. Certain techniques (for instance, decompositions, dual formulations, penalty functions methods) for solving smooth problems leads directly to the necessity to solve nonsmooth problems. Nonsmooth optimization (NSO) addresses such kind of problems.

There are several approaches to solve NSO problems. First, derivative-free methods like [5] can be used. However, these methods become inefficient for large-scale minimization problems. There are different kinds of smoothing techniques

© Springer International Publishing Switzerland 2016
Y. Kochetov et al. (Eds.): DOOR 2016, LNCS 9869, pp. 547–560, 2016.
DOI: 10.1007/978-3-319-44914-2_43

(see, for example, [6]) but they are not, in general, as efficient as the nonsmooth approach [7]. Special methods for solving NSO problems can be divided at two main groups: subgradient methods (see e.g. [2]) and bundle methods (see e.g. [8–12]). All these methods are based on the assumption that the entire accessible information on the objective function $f(x)$ of the problem (1) is provided by a subgradient oracle, and at any arbitrary point $\bar{x}$ only the objective function value $f(\bar{x})$ and a single subgradient $g \in \partial f(\bar{x})$ (generalized gradient [13]) arbitrarily chosen from the subdifferential $\partial f(\bar{x})$ (that is, the set of subgradients) of the function $f(x)$ can be found.

The subgradient method proposed for the first time by Shor [14] has the simplest computational scheme

$$x_{k+1} = x_k - h_k g_k, \quad g_k \in \partial f(x_k), \quad k = 0, 1, \ldots,$$

where the conditions $\sum_{k=1}^{\infty} h_k = \infty$, $h_k \to +0$, $\sum_{k=1}^{\infty} h_k^2 < \infty$ for step multipliers h_k are sufficient for convergence to the solution of (1). However, the rate of convergence with such step size rule is very low. The best choice rule for the step size h_k is known as the B.T. Polyak rule [15] but under the condition that the optimum value $f^* = \min_{x \in \mathbb{R}^n} f(x)$ is known in advance. Among other methods of step size control, the technique from [16] can also be mentioned. However, various numerical experiments and theoretical analysis demonstrated that subgradient methods may have low convergence and many of them suffer from serious drawbacks. That is why search for new, more efficient algorithms continues.

The next stage of the development of NSO methods is connected with the emergence of the bundle methods, for example, the level method [17] developed in 1995. The basic idea of bundle methods is to approximate the subdifferential of the objective function by gathering subgradients from previous iterations into a bundle. Among recent publications in the field of bundle methods, the methods with inexact oracle (if most of the time the available information from oracle is inaccurate) [18–20] should be mentioned. For more details about bundle methods see, for example, [11,12] and the references therein.

This article is devoted to the further investigation and improvement of separating plane (SP) methods [21–25] for solving unconstrained problems of multidimensional convex NSO that does not require additional information on the internal structure of the function being optimized, and are the representatives of the so-called black-box optimization. And at the same time SP methods can be classified as bundle methods, but with an important theoretical feature. Methods work in the extended conjugate space of subgradients and the Legendre-Fenchel conjugate of $f(x)$ [26] $f^*(g) = \sup_x \{gx - f(x)\}$. More precisely, SP methods replace the problem (1) with computing the corresponding Legendre-Fenchel conjugate function at the origin of the conjugate space. As the investigations of SP methods showed, this idea of replacement increased the rate of convergence.

The rest of this paper is organized as follows. The SP algorithm with additional clippings is described in Sect. 2. Some improvements of the original SP algorithm is proposed in Sect. 3 where new, modified SP algorithm with clippings and stockpiling is presented. The algorithm is called SPACLIP-S. In Sect. 4.1, we give a description of the so-called PageRank problem, which can be reformulated as NSO problem and can be solved by SP method with additional clippings and by SPACLIP-S. In this section an iterative regularization method with a new penalty function for solving PageRank problem is also presented. In Sect. 4.2, we test the performance of the proposed algorithm and compare it with other numerical methods. Section 5 concludes the paper.

2   Separating Plane Method with Clippings

Separating plane methods are based on the idea of replacement of the initial minimization problem (1) by the problem of computing the Legendre-Fenchel conjugate at zero:

$$f(x^*) = \min_{x \in \mathbb{R}^n} f(x) = -\sup_x \{0 \cdot x - f(x)\} = -f^*(0) \tag{2}$$

where the function $f^*(g) = \sup_x \{gx - f(x)\}$ is the Legendre-Fenchel conjugate of the function $f(x)$. Problem (2) can be interpreted as the problem of searching the intersection point of the conjugate function's graph with the vertical line $\{0\} \times \mathbb{R}_+$. The optimal point x^* can be obtained as a subgradient of f^*: $x^* \in \partial f^*(0)$.

The SP methods construct sequences of outer and inner approximations of the epigraph of f^* (epi $f^* = \{(\mu, g) : \mu \geq f^*(g)\}$) by means of convex polyhedral sets U and D. At each iteration of the algorithm the approximations are gradually refined. Eventually we obtain converging lower and upper estimates for $f^*(0)$. The set D is modified by the addition of new points $(g_k, f^*(g_k))$ located on the graph of f^*.

In [23,24] the following changes for SP methods were suggested. At each iteration we execute an additional step that removes the upper part of epi f^*:

$$\sup_{\substack{(g, \varepsilon) \in \text{epi } f^* \\ \varepsilon \leq v}} \{gx - \varepsilon\}, \tag{3}$$

where the upper-bound estimate v can be obtained as the solution of a linear-programming problem and is also an estimate found by the Kelley's cutting plane method [27]:

$$v = \min_{(0,\,\varepsilon)\,\in\,D} \varepsilon = \min_{\substack{(0,\,\tau)\,\in\,\text{conv}\{(g_i,\,f^*(g_i)),\,i=1,\,2,\,...,\,k\}\,+\,\{0\}\,\times\,\mathbb{R}_+}} \tau =$$

$$= \min_{\substack{\tau = \sum_{i=1}^{k} \lambda_i f^*(g_i), \\ 0 = \sum_{i=1}^{k} \lambda_i g_i, \\ \lambda_i \geq 0,\, \sum_{i=1}^{k} \lambda_i = 1}} \tau = \min_{\substack{0 = \sum_{i=1}^{k} \lambda_i g_i, \\ \lambda_i \geq 0,\, \sum_{i=1}^{k} \lambda_i = 1}} \sum_{i=1}^{k} \lambda_i f^*(g_i),$$

$$\tag{4}$$

where conv A denotes the convex hull of a set A (the intersection of all convex sets that contain the set A).

Such clippings localizes possible points of the epigraph of f^* and the inner approximation D more precisely. The SP algorithm with additional clippings is illustrated in Fig. 1.

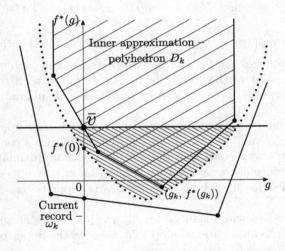

Fig. 1. Typical iteration of separating plane algorithm with additional clippings (SPACLIP)

3   Separating Plane Method with Clippings and Stockpiling

This section is devoted to a new modification of SP method with additional clippings, namely, SP method with clippings and stockpiling. The main idea of *stockpiling* is to use the additional data returned by the subgradient oracle during the operations of the line-search algorithm.

As in the standard SP method, problem (3) may be easily transferred to the space of primal variables $x \in \mathbb{R}^n$, but, in this case, an auxiliary one-dimensional minimization problem arises with a rather unexpected objective function. In fact, denoting a dual variable u for an additional constraint, we obtain

$$\sup_{(g,\varepsilon)\,\in\,\text{epi}\,f^*;\;\varepsilon\,\leq\,\bar{v}} \{gx - \varepsilon\} = \sup_g \inf_{u\,\geq\,0} \{gx - f^*(g) + u(\bar{v} - f^*(g))\}$$

$$= \inf_{u\,\geq\,0} \left\{u\bar{v} + \sup_g \{gx - (u+1)f^*(g)\}\right\} = \inf_{u\,\geq\,0} \left\{u\bar{v} + (1+u)\,f\left(\frac{x}{1+u}\right)\right\}$$

$$= -\bar{v} + \inf_{\check{u}\,\geq\,1} \{\check{u}(\bar{v} + f(\check{u}^{-1}x))\} = -\bar{v} + \inf_{\check{u}\,\geq\,1} \varphi(\check{u}, x), \qquad (5)$$

where $\check{u} = (1 + u)$ and $\varphi(\check{u}, x) = \check{u}(\bar{v} + f(\check{u}^{-1}x))$.

It is easy to show that the function $\varphi(\check{u}, x)$ is convex with respect to the set of variables $(\check{u}, x)$ if $f(x)$ is a convex function. The proof of this particular case see, for example, in [24].

It was suggested to solve one-dimensional NSO problem (5) by a fast line-search algorithm [28] since it can achieve superlinear or even quadratic convergence rate under favorable conditions. A special implementation of the fast line-search algorithm was created for the SPACLIP algorithm. The results of numerous computational experiments showed that this algorithm is rather efficient as a computational block for solving (5) within SPACLIP.

Another opportunity to improve convergence rate of SPACLIP is to use trial points of one-dimensional search as additional source of information for construction of the inner approximation D. During every iteration of SPACLIP the line-search procedure calculates a series of values of the objective function $f(x)$ and subgradients of $f(x)$ for its own purpose. These values can be added to the inner approximation D of the epigraph of the conjugate f^* of $f(x)$. This modification of the SPACLIP can be called *stockpiling*.

Heuristic justification of the effect of such *stockpiling* is as follows. Calculated in the iteration process of the line-search algorithm new points are added to D. Next, it is needed to make a projection onto D from the point of the current record of the objective function f, taken with the opposite sign (see Fig. 2). The projection is performed using the suitable affine subspace method (SimPro [29]), which has a globally higher-than-linear convergence rate. The numerical experiments showed that the presence of additional points in the convex set D, onto which the projection is performed do increase the convergence rate of SimPro method. One of the reasons for such effect is that the algorithm performs a very few removals of vertices from the basis. So, the added points are usually good perspective points, they refine inner approximation of epi f^* and improve the convergence rate of SimPro method for finding the projection.

3.1   Algorithm of SP Method with Clippings and Stockpiling

Finally, the algorithm of SP Method with clippings and stockpiling will look like the following.

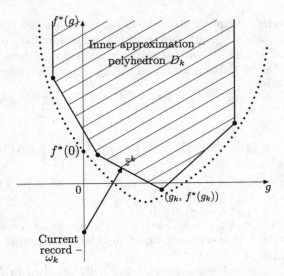

Fig. 2. Projection onto D_k from the point of the current record of the objective function, taken with the opposite sign

Step 0. **Initialization.** Set the iterations' counter $k = 0$ and determine an initial point $x_0 \in \text{dom} f$ of the minimizing sequence $\{x_0, x_1, x_2, \ldots\}$.

Step 1. D_0 **creation.** In order to avoid future problems with projection onto the polyhedron D_k, it is suggested to add an auxiliary point $(0, \alpha)$, where $\alpha \in \mathbb{R}$, $\alpha > 0$, such that $\alpha > f^*(0) = -\min f(x)$, to D_0. By construction, $(0, \alpha)$ belongs to epi f^*. If this addition is made, the problem (4) will always have a solution.

Step 2. **Current record.** Compute $\inf\limits_{0 \in U_k(\omega)} \omega = \omega_k$, where U_k is the kth outer approximation of the epi f^*. The latter problem can be solved recurrently as follows:

$$\omega_k = \max\{\omega_{k-1}, -f(x_{k-1})\}, \quad k \geq 1. \tag{6}$$

And $\omega_0 = -\infty$ for $k = 0$. In fact, $-\omega_k$ is the current record of the objective function f (see Fig. 1).

Step 3. **Projection.** Find the vector $\bar{z}^k = (z^k, \xi_k)$ – the projection of a point $(0, \omega_k)$ onto the polyhedron D_k. As mentioned above, the polyhedron D_k is the inner approximation of the epigraph of f^*. To solve this problem, the suitable affine subspace method [29] is used.

Step 4. **Update.** Compute the next element of the minimizing sequence

$$x_k = -z^k / \xi_k.$$

Step 5. **Clipping.** Determine a cutting level of the upper part of epi f^*, i.e., $\bar{v}$. The value $\bar{v}$ is found by solving the linear programming problem (4).

Step 6. **One-dimensional minimization.** Solve the one-dimensional non-smooth minimization problem (5). Let $\check{u}_k$ be a computed at the kth iteration solution of (5).

Step 7. Stockpiling. Add new points $(g_i \in \partial f(x_i), f^*(g_i))$, $i = i_1, \ldots, i_m$ to the inner approximation, i.e., the polyhedron D_k. These new points were calculated in the iteration process of the fast line-search algorithm during *Step 6.*

Step 8. Update 2. Compute $x_k = \breve{u}_k^{-1} x_k$.

Step 9. Inner approximation renewal. Add a pair $(g_k \in \partial f(x_k), f^*(g_k))$ to the inner approximation, i.e., the polyhedron D_k.

Step 10. Stop. If any of completion conditions is satisfied, then quit. Otherwise, increase the iterations' counter k by one and go to the *Step 1.*

The next section contains computational results for this new algorithm and experimental comparison with the old one.

4   Numerical Experiments: PageRank Computation

We conclude this paper with the results of numerical experiments. Numerical experiments demonstrated quite satisfactory computational performance of SP Algorithm with CLIPpings and Stockpiling (SPACLIP-S). Moreover, the algorithm described above is compared with the SPACLIP method.

The codes were written by the author in Octave programming language [30] under a Linux operating system. The syntax of Octave is very close to MATLAB, and this system is a convenient tool for developing first versions of computational algorithms.

4.1   PageRank

A set of pages in the Web may be modeled as nodes in a directed graph $G = (V, E)$. The edges, E, between nodes (vertices, V) represent references from one web page to another. A graph of a simple 3-page Web is shown in Fig. 3. The directed edge from Node 1 to Node 2 signifies that Page 1 has a link to Page 2, and so on. However, Page 1 has no link to Page 3, so there is no edge from Node 1 to Node 3. All the nodes in this graph have loops, i.e. the pages have links to themselves. Clearly, Page 3 is the least important in this graph. But how to measure the importance of each page (or nodes in graphs) in general?

The most acknowledged methods of measuring importance of nodes in graphs are based on random surfer models. First of all, PageRank [31] and HITS [32] should be mentioned. The other important approaches of ordering the pages according to relevance which are based on machine learning technics can be found, for example, in [33,34]. In this section we consider the classical PageRank model only and how the problem of measuring importance of nodes in graphs can be solved by a number of methods including SPACLIP-S and SPACLIP.

The PageRank algorithm constructs web pages' importance hierarchies based upon the link structure of the Web. The score of a node equals to its probability in the stationary distribution of a Markov process. Calculating the probability vector corresponds to finding a maximal eigenvalue of web-graph transition matrix. This can be done by means of iterative numerical methods.

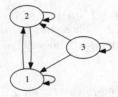

$$A = \begin{pmatrix} 1 & 1 & 0 \\ 1 & 1 & 0 \\ 1 & 1 & 1 \end{pmatrix} \qquad (7)$$

$$P = \begin{pmatrix} 1/2 & 1/2 & 0 \\ 1/2 & 1/2 & 0 \\ 1/3 & 1/3 & 1/3 \end{pmatrix} \qquad (8)$$

Fig. 3. Directed graph representing Web of three pages

A transition matrix P is constructed as follows:

$$p_{ij} = \begin{cases} \dfrac{1}{\deg\,(v_i)}, & \text{if the edge } (v_i, v_j) \text{ exists,} \\ 0, & \text{otherwise,} \end{cases}$$

where deg (v) is the degree of vertex v, i.e. the number of outbound links from page v. For the 3-node graph in Fig. 3, the adjacency matrix A is shown in equation (7), and the transition matrix P is shown in equation (8).

Brin and Page [35] add an adjustment matrix $\frac{1}{n}ee^T$ to P, where n is the order of P and e is a all-ones vector. Sometimes the matrix ee^T can be replaced with a matrix ve^T, where v is a *personalization vector* (for more details see [36]). To ensure the random surfer does not get stuck in a *dangling node* (node with no outgoing edges) the following suggestions were made. At each node,

1. With probability $1 - \alpha$ the surfer jumps to a random place on the Web,
2. With probability α the surfer decides to choose, uniformly at random, an outlink of the current node.

Google reportedly uses $\alpha = 0.85$. Thus, we construct a new stochastic matrix M_α as

$$M_\alpha = \alpha\, P^T + (1 - \alpha)\, \frac{1}{n}ee^T.$$

The PageRank vector for a graph with transition matrix P and damping factor α is the unique probabilistic eigenvector x_α of the matrix M_α, corresponding to the eigenvalue 1 (for a proof and additional details, see, for example, a survey [37], the book [36] or [38]). However, the eigenvector x_α can differ strongly from the eigenvector of P [37].

Because of huge size of transition matrix it is impossible to solve systems $x = x\,P$ or $x = M_\alpha\,x$ for the real Web graph. The alternative way of computing the probabilistic eigenvector corresponding to the eigenvalue 1 is given by well-known *Power method*

$$x_{k+1} = x_k\,P, \quad k = 0, 1, 2, \ldots$$

The method works for the regular stochastic matrices, but its convergence rate can not be fast enough. In this section it is suggested to use an idea of ℓ_1-regularization by Polyak B.T., Timonina A.V. and Tremba A.A. [39,40] (PTT method) for solving PageRank problem with some modifications. The ℓ_1-regularization approach and its modifications proposed in this article can be used to solve PageRank problem in the case when the spectral gap condition on transition matrix is not satisfied. Spectral gap of a matrix is the difference between the two largest modulus eigenvalues. Practical experiments in [41] showed that the PTT method is the best choice when there is no guarantee that the spectral gap of P is big enough. In other cases, when the spectral gap condition is satisfied, other methods can be chosen (for example, the Reduced Power Method [42], Markov Chain Monte Carlo method [41], see also the references in [41]). It should also be mentioned that an algorithm for robust eigenvector approximation is proposed in [43]. But robust alternatives to the standard PageRank technique is a different subject, which is not covered in this article.

We consider the task of computing the probabilistic eigenvector as optimization problem

$$\min \quad \|M_\alpha x - x\|_1 + \beta \left(\left| \sum_{i=1}^{n} x^i - 1 \right| + \sum_{i=1}^{n} [x^i]^- \right), \tag{9}$$

where $\|y\|_1$ is ℓ_1-norm of vector $y = (y^1, y^2, \ldots, y^n) \in \mathbb{R}^n$, i.e. $\sum_{i=1}^{n} |y^i|$; $\beta > 0$ is a penalty parameter and $[a]^- = \max\{0, -a\}$.

Numerical experiments showed that optimization algorithms for finding optimal value of objective function in (9) can be used for solving web page ranking problem. We used SPACLIP and SPACLIP-S methods for solving (9). Indirect benefits from the applying of nonsmooth optimization methods for solving the problem (9) is that it is possible to use nonsmooth penalty functions to guarantee that the solution vector x is a standard simplex in $\mathbb{R}^n$. Nonsmooth penalty functions are *exact*, which means that, for certain choices of their penalty parameters, a minimization with respect to x can yield the exact solution of the nonlinear optimization problem [44].

4.2   Experimental Results

In first test problem we found PageRank vector for the simple 3-page Web shown in Fig. 3 by SPACLIP, SPACLIP-S methods and Power method. The code of Power method was taken from [36]. Running the code of Power method in Octave 3.6.4 with a convergence tolerance of 10^{-10} and $\alpha = 0.85$ produces the following results:

Iter.	PageRank vector			$\|x\,P - x\|_1$	$\|M_\alpha\,x - x\|_1$
0	[1/3,	1/3,	1/3]	8.377580	6.2832
1	[0.427777778,	0.427777778,	0.144444444]	0.192593	1.0704e-01
2	[0.454537037,	0.454537037,	0.090925926]	0.121235	3.0327e-02
3	[0.462118827,	0.462118827,	0.075762346]	0.101016	8.5927e-03
4	[0.464267001,	0.464267001,	0.071465998]	0.095288	2.4346e-03
5	[0.464875650,	0.464875650,	0.070248699]	0.093665	6.8980e-04
6	[0.465048100,	0.465048100,	0.069903798]	0.093205	1.9544e-04
7	[0.465096962,	0.465096962,	0.069806076]	0.093075	5.5376e-05
8	[0.465110806,	0.465110806,	0.069778388]	0.093038	1.5690e-05
9	[0.465114728,	0.465114728,	0.069770543]	0.093027	4.4454e-06
10	[0.465115840,	0.465115840,	0.069768221]	0.093024	1.2595e-06
		. . .			
19	[0.465116279,	0.465116279,	0.069767442]	0.093023	1.4821e-11

Then the same test problem was solved by SPACLIP and SPACLIP-S methods. The initial vector x_0 was the same, and the penalty parameter β was equal to 0.3. The results are as follows:

Iter.	PageRank vector			$\|M_\alpha\,x - x\|_1$
	SPACLIP			
1	[0.310571,	0.310571,	0.046318]	–
2	[0.021836,	0.021836,	0.178326]	–
3	[0.465116,	0.465116,	0.069767]	1.9429e-16
4	[0.465126,	0.465126,	0.069747]	–
	SPACLIP-S			
1	[−0.29508,	−0.29508,	−2.40984]	–
2	[0.310571,	0.310571,	0.046318]	–
3	[0.021836,	0.021836,	0.178326]	–
4	[0.465116,	0.465116,	0.069767]	3.3723e-15

The last iteration from SPACLIP method was unsuccessful, so the answer was taken from the previous iteration. Because the methods are not monotone, such situations sometimes arise. In that case SPACLIP and SPACLIP-S methods return a point with the objective function's minimal value. It happened in this test solved by SPACLIP method only. The overall results of all three methods are shown in Table 1. The Power method was the fastest, but SPACLIP-S took second place. The answer (PageRank vector) was the same for all three methods. This simple test showed that SP methods can be used for the solution of PageRank problem. But the dimension of the first test problem was really small. In second test problem we used SP methods for for finding large-size PageRank vector.

Table 2 and Fig. 5 present results of our experiments on a test problem 2 – Hollins University Web graph. The dataset of the hollins.edu Web site crawled on January, 2004 and consists of 6012 nodes. It is now available on

Table 1. Comparison of the methods on the test problem 1 (P matrix 3×3)

Method	Time, s	Number of iterations	$\|M_\alpha x - x\|_1$
Power method	9.25e-4	19	1.4821e-11
SPACLIP	0.122	4	1.9429e-16
SPACLIP-S	0.059	4	3.3723e-15

http://www.limfinity.com/ir/data/hollins.dat.gz. The structure of the adjacency matrix of this graph is shown in Fig. 4. The matrix is sparse, so each dot represents a nonzero in the matrix.

Table 2. Comparison of the methods and β choice on the test problem 2 (P matrix 6012×6012)

Method	Time, s	Number of iterations	$\|M_\alpha x - x\|_1$
Power method	0.0765	58	21.948
SPACLIP-S, $\beta = 0.3$	4499.8	276	1.2256
SPACLIP-S, $\beta = 0.2$	4354.4	260	1.0646
SPACLIP-S, $\beta = 0.1$	3950.5	252	1.3047
SPACLIP-S, $\beta = 6.6e-4$	4445.6	285	0.51836

Power method failed the test problem 2. Its results, which are shown in Table 2, were obtained only after normalizing the vector x on each iteration by dividing all the components of vector x by vector's sum, otherwise the method diverged. The other experiments were devoted to the selection of the penalty parameter β. This parameter should be chosen as follows: $\beta \leq 1$ (the maximum

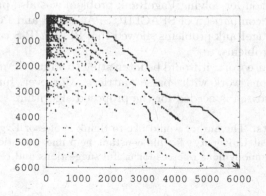

Fig. 4. Adjacency matrix of the graph of hollins.edu Web site

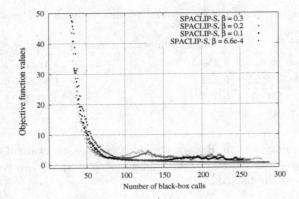

Fig. 5. Convergence results for SP methods on the test problem 2 (P matrix 6012 × 6012)

element of the matrix P) and $\beta \geq 6.6\mathrm{e}{-4}$ (the average number of incoming or outgoing links 3.97 divided by 6012).

5   Conclusions

The separating plane method with additional clippings and stockpiling (SPACLIP-S) for nonsmooth optimization is proposed in this paper. The method belongs to the family of separating plane methods. Both the theoretical and experimental investigations of the method showed that this method is efficient and widely applicable to nonsmooth optimization problems with convex objective functions.

Proposed optimization algorithm was applied to the Web page ranking problem. The second part of the article relating to PageRank problem, was made under the impression from [33, 45]. The iterative regularization method with a new penalty function for solving PageRank problem was also presented. Experimental results of comparison of SPACLIP, SPACLIP-S and Power method for solving two test PageRank problems showed that SPACLIP-S can be applied for solving ranking problems.

In terms of future work, it would be interesting to adjust SP methods for more effective calculations work with sparse matrices. Moreover, further theoretical and numerical research is needed for the proposed algorithm.

Acknowledgments. The author would like to thank Professor Evgeni A. Nurminski (Far Eastern Federal University) for his constant help and attention. The author is also greatful to the anonymous referees for useful suggestions and references.

References

1. Shary, S.P.: Maximum consistency method for data fitting under interval uncertainty. J. Glob. Optim. **63**, 1–16 (2015)
2. Shor, N.Z.: Minimization Methods for Non-differentiable Functions. Springer Science & Business Media, Berlin (2012)
3. Jung, M., Kang, M.: Efficient nonsmooth nonconvex optimization for image restoration and segmentation. J. Sci. Comput. **62**(2), 336–370 (2015)
4. Clarke, F.N., Ledyaev, Y.S., Stern, R.J., Wolenski, P.R.: Nonsmooth Analysis and Control Theory. Springer, Heidelberg (1998)
5. Powell, M.J.D.: A View of Algorithms for Optimization without Derivatives. DAMTp. 2007/NA03. Technical Report, Cambridge University (2007)
6. Nesterov, Y.: Smooth minimization of non-smooth functions. Math. Program. **103**, 127–152 (2005)
7. Mäkelä, M.M., Neittaanmäki, P.: Nonsmooth Optimization: Analysis and Algorithms with Applications to Optimal Control. World Scientific Publishing Co., Singapore (1992)
8. Frangioni, A.: Generalized bundle methods. SIAM J. Optim. **13**, 117–156 (2002)
9. Haarala, N., Miettinen, K., Mäkelä, M.M.: Globally convergent limited memory bundle method for large-scale nonsmooth optimization. Math. Program. **109**, 181–205 (2007)
10. Helmberg, C., Rendl, F.: A spectral bundle method for semidefinite programming. SIAM J. Optim. **10**, 673–696 (1999)
11. Bagirov, A., Karmitsa, N., Mäkelä, M.M.: Introduction to Nonsmooth Optimization: Theory, Practice and Software. Springer, Heidelberg (2014)
12. Hiriart-Urruty, J.B., Lemaréchal, C.: Convex Analysis and Minimization Algorithms II: Advanced Theory and Bundle Methods. Fundamental Principles of Mathematical Sciences. Springer, Berlin (1993)
13. Clarke, F.H.: Optimization and Nonsmooth Analysis. Wiley-Interscience, New York (1983)
14. Shor, N.Z.: Minimization Methods for Non-differentiable Functions. Springer, New York (1985)
15. Polyak, B.T.: Minimization of nonsmooth functionals. Comp. Math. Math. Phys. **9**, 509–521 (1969)
16. Nurminski, E.A.: Envelope stepsize control for iterative algorithms based on Fejer processes with attractants. Optimiz. Meth. Softw. **25**, 97–108 (2010)
17. Lemaréchal, C., Nemirovskii, A., Nesterov, Y.: New variants of bundle methods. Math. Program. **69**(1), 111–147 (1995)
18. de Oliveira, W., Sagastizábal, C.: Level bundle methods for oracles with on-demand accuracy. Optim. methods Softw. **29**(6), 1180–1209 (2014)
19. de Oliveira, W., Sagastizábal, C.: Bundle methods in the XXIst century: a Birds'-eye view. Pesquisa Operacional **34**(3), 647–670 (2014)
20. de Oliveira, W., Sagastizábal, C., Lemaréchal, C.: Convex proximal bundle methods in depth: a unified analysis for inexact oracles. Math. Program. **148**, 241–277 (2014)
21. Nurminski, E.A.: Separating plane algorithms for convex optimization. Math. Program. **76**, 373–391 (1997)
22. Nurminski, E.A.: Separating plane method with bounded memory for the solution of convex nonsmooth optimization problems. Comput. Meth. Program. **7**, 133–137 (2006)

23. Vorontsova, E.A.: A projective separating plane method with additional clipping for non-smooth optimization. WSEAS Trans. Math. **13**, 115–121 (2014)
24. Vorontsova, E.A., Nurminski, E.A.: Synthesis of cutting and separating planes in a nonsmooth optimization method. Cybern. Syst. Anal. **51**(4), 619–631 (2015)
25. Nurminski, E.: Multiple cuts in separating plane algorithms. In: Kochetov, Y., et al (eds.) DOOR 2016. LNCS, vol. 9869, pp. 430–440. Springer, Switzerland (2016)
26. Rockafellar, R.T.: Convex Analysis. Princeton University Press, Princeton (1970)
27. Kelley, J.E.: The cutting plane method for solving convex programs. J. SIAM **8**(4), 703–712 (1960)
28. Vorontsova, E.A.: Modified fast line-search algorithm for non-smooth optimization. Inf. Sci. Control Syst. **2**, 39–48 (2012). (in Russian)
29. Nurminskii, E.A.: Convergence of the suitable affine subspace method for finding the least distance to a simplex. Comp. Math. Math. Physics. **45**(11), 1915–1922 (2005)
30. Octave Page. http://www.gnu.org/software/octave/
31. Page, L., Brin, S., Motwani, R., Winograd, T.: The Pagerank Citation Ranking: Bringing Order to the Web. Technical Report, Computer Science Department, Stanford University (1998)
32. Kleinberg, J.M.: Authoritative sources in a hyperlinked environment. J. ACM **46**(5), 604–632 (1999)
33. Bogolubsky, L., Dvurechensky, P., Gasnikov, A., Gusev, G., Nesterov, Y., Raigorodskii, A., Tikhonov, A., Zhukovskii, M.: Learning Supervised PageRank with Gradient-Based and Gradient-Free Optimization Methods (2016). arXiv:1603.00717
34. Teo, C.H., Vishwanathan, S.V.N., Smola, A.J., Le, Q.V.: Bundle methods for regularized risk minimization. J. Mach. Learn. Res. **11**, 311–365 (2010)
35. Brin, S., Page, L.: The anatomy of a large-scale hypertextual web search engine. Comput. Netw. ISDN Syst. **30**, 107–117 (1998)
36. Langville, A.N., Meyer, C.D.: Googles PageRank and Beyond: The Science of Search Engine Rankings. Princeton University Press, Princeton (2011)
37. Langville, A.N., Meyer, C.D.: Deeper inside pagerank. Internet Math. **1**(3), 335–380 (2005)
38. Thorson, K.: Modeling the Web and the Computation of PageRank. Undergraduate thesis. Hollins University (2004). www.limfinity.com/ir/kristen_thesis.pdf
39. Polyak, B.T., Timonina, A.V.: PageRank: new regularizations and simulation models. In: Proceedings of 18th IFAC World Congress, pp. 11202–11207. Milano, Italy (2011)
40. Polyak, B.T., Tremba, A.A.: Regularization-based solution of the PageRank problem for large matrices. Autom. Remote Control **73**(11), 1877–1894 (2012)
41. Gasnikov, A.V., Dmitriev, D.Y.: On efficient randomized algorithms for finding the PageRank vector. Comp. Math. Math. Phys. **55**(3), 349–365 (2015)
42. Nesterov, Y., Nemirovski, A.: Finding the stationary states of Markov chains by iterative methods. Appl. Math. Comput. **255**, 58–65 (2015)
43. Juditsky, A., Polyak, B.: Robust eigenvector of a stochastic matrix with application to PageRank (2012). arXiv:1206.4897
44. Nocedal, J., Wright, S.J.: Numerical Optimization. Springer Series in Operations Research and Financial Engineering. Springer, New York (2006)
45. Anikin, A., Gasnikov, A., Gornov, A., Kamzolov, D., Maximov, Y., Nesterov, Y.: Effective Numerical Methods for Huge-Scale Linear Systems with Double-Sparsity and Applications to PageRank (2016). arXiv:1508.07607v3

Short Communications

Location, Pricing and the Problem of Apollonius

André Berger[1], Alexander Grigoriev[1(✉)],
Artem Panin[2,3], and Andrej Winokurow[1]

[1] Department of Quantitative Economics,
Maastricht University School of Business and Economics,
P.O. Box 616, 6200MD Maastricht, The Netherlands
{a.berger,a.grigoriev,a.winokurow}@maastrichtuniversity.nl
[2] Sobolev Institute of Mathematics, Acad. Koptyug Ave. 4,
630090 Novosibirsk, Russia
aapanin1988@gmail.com
[3] Novosibirsk State University, 2 Pirogov Str., 630090 Novosibirsk, Russia

Abstract. In Euclidean plane geometry, Apollonius' problem is to construct a circle in a plane that is tangent to three given circles. We will use a solution to this ancient problem to solve several versions of the following geometric optimization problem. Given is a set of customers located in the plane, each having a demand for a product and a budget. A customer is satisfied if her total, travel and purchase, costs do not exceed her budget. The task is to determine location of production facilities in the plane and one price for the product such that the revenue generated from the satisfied customers is maximized.

Keywords: Pricing problem · Facility location · Apollonius' problem · Complexity · Exact algorithm

1 Introduction

Consider the following geometric Stackelberg game. A *leader* in the game is a company producing a single product in large quantities in uncapacitated production facilities. The company has to determine location of m facilities in a Euclidean plane and a selling price p per unit of the product, one for all facilities. *Followers* in this game are n customers of the company. Let J denotes the set of these customers. Each customer $j \in J$ is situated in the plane and her coordinates are given by a point $x_j \in \mathbb{Q}^2$. Each customer $j \in J$ is *single-minded*, i.e., she is willing to purchase either her full demand $d_j \in \mathbb{Z}_+$, known to the company, or nothing. Moreover, each customer $j \in J$ announces to the company her budget $b_j \in \mathbb{Z}_+$ indicating that the product will be purchased only if the sum of travel and purchase costs does not exceed her budget, i.e.,

$$d_j \times p + c_j \times ||x_j - y|| \leq b_j, \tag{1}$$

A. Panin—This work was partially supported by RFBR grants 15-37-51018 mol_nr and 16-07-00319.

Y. Kochetov et al. (Eds.): DOOR 2016, LNCS 9869, pp. 563–569, 2016.
DOI: 10.1007/978-3-319-44914-2_44

where c_j is the customer travel cost per distance unit, $|| \cdot ||$ is the Euclidean norm and y is the closest to the customer facility. Here, we also assume that the travel costs are known to the company. If the *budget constraint* expressed by the Eq. (1) for customer $j \in J$ is satisfied, we call the customer a *winner*. The winner purchases the product and the company generates revenue of $d_j \times p$ from that winner. The problem is to find a revenue maximizing strategy for the leader, i.e., to determine location of facilities and the price such that the total revenue generated from the winners is maximized.

Further we denote the set of facilities by I, the location of facility $i \in I$ by $y_i \in \mathbb{R}^2$, a feasible strategy of the leader by (y, p), and the set of winners by $J(y, p) \subseteq J$. We refer to this problem as the *location-pricing* problem. Furthermore, without loss of generality we assume $c_j = 1$ for all customers $j \in J$, otherwise we normalize the instance dividing the demands and the budgets by the customer travel costs.

In this work, we first address the location-pricing problem with one facility and three customers having unit demands. This problem is solved by using a solution to the Apollonius' problem. Then, we extend the algorithm to the problem with n customers having arbitrary demands. We present an algorithm solving the location-pricing problem with m facilities. We conclude with discussion and some open problems.

2   A Single Facility Case with Three Customers

Consider a simple special case of the location-pricing problem with one facility and three customers. Notice, in any optimal solution, the budget constraint of at least one of the customers must be tight, otherwise the company can increase the revenue by increasing the price. Thus, for three customers we have three distinct cases: one-, two-, or all three budget constraints are tight. We present derivations for the latter case with all three budget constraints being tight. The other two cases are even simpler though treated similarly.

Theorem 1 (Generalized Apollonius' Problem). *Given positive integers* $b_1, b_2, b_3, d_1, d_2, d_3$ *and three points* $x_1, x_2, x_3 \in \mathbb{Q}^2$ *in general positions, the following system of equalities has at most eight solutions, all expressible in a closed analytic form.*

$$||x_j - y||^2 = (b_j - d_j \times p)^2, \quad j \in \{1, 2, 3\}.$$

Proof. The geometric intuition of the proof can be illustrated on a special case with $d_1 = d_2 = d_3 = 1$. We defer the formal proof of the general case with arbitrary values for d_1, d_2, d_3 and derivation of the closed form formulas to the complete journal version of this paper.

For $d_1 = d_2 = d_3 = 1$, a solution to the system is a solution to the famous *Apollonius' Problem* [1]: Given three circles, B_1, B_2, B_3, in a plane with centers x_1, x_2, x_3 and of radii b_1, b_2, b_3, respectively, find a circle B_0 tangent to all three

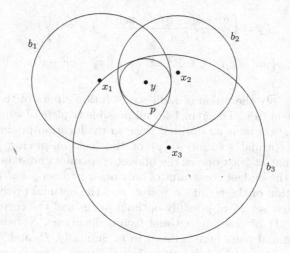

Fig. 1. Apollonius problem

given circles; for illustration see Fig. 1. Here, the solution to the system and to the Apollonius problem is a circle B_0 of radius p centered at y.

Intuitively, any three distinct circles generically have eight different circles that are tangent to them. The number eight comes from the fact that solution circles can enclose or exclude the three given circles in eight different ways: (1) B_0 does not enclose the whole interior of neither of the given circles; (2–4) B_0 encloses the entire interior of exactly one of the circles, B_1 or B_2 or B_3; (5–7) B_0 encloses the entire interior of exactly two out of three circles; (8) B_0 encloses all three given circles.   □

3   Discretization of the Location-Pricing Problem

In fact, Theorem 1 allows us to restrict the search for optimal solutions to the location-pricing problem among finitely many location-pricing pairs, making the location-pricing problem a combinatorial optimization problem. Even stronger, the number of possibly optimal location-pricing pairs is cubic in the number of customers and independent on the number of facilities.

Theorem 2 (Location-Price Discretization). *For any instance of the location-pricing problem with n customers, there is a set S of pairs $(y, p) \in \mathbb{R}^2 \times \mathbb{R}^+$ of size $O(n^3)$ such that some pair $(y, p) \in S$ is an optimal solution to the location-pricing problem.*

Proof. Let $P = P_1 \cup P_2 \cup P_3$, where

$$P_1 = \left\{ \frac{b_j}{d_j} : j \in J \right\},$$

$$P_2 = \left\{ \frac{b_j + b_k - ||x_j - x_k||}{d_j + d_k} : j, k \in J \right\},$$

$$P_3 = \bigcup_{J' \subset J: \; |J'|=3} \left\{ p \in \mathbb{R}_+ : \; \exists y \in \mathbb{R}^2 \text{ s.t. } b_j - ||x_j - y|| - d_j p = 0, \; \forall j \in J' \right\}.$$

Here, set P_3 is simply the union of radii p of solution circles of the Generalized Apollonius Problem from Theorem 1 for all possible triplets of winners in J. We shall argue that P contains an optimal price to the location-pricing problem.

Consider an optimal solution (y, p) of the location-pricing problem. We already argued that at least one of the budget constraints must be tight.

Assume that the budget constraint of only one customer $j \in J$ is tight. Then the optimal location of the facility is $y = x_j$ and the optimal price is $p = b_j/d_j$. This forms the first set P_1 of possibly optimal prices and the corresponding set $Y_1 = \{x_j : \; j \in J\}$ of possibly optimal facility locations. Notice, the number of potentially optimal pairs here is linear in n. Similarly, P_2 and P_3 are formed when the number of tight budget constraints is 2 and 3, respectively. Since for P_2 and for P_3 we have to consider all possible pairs and triplets of tight budget constraints, respectively, the cardinality of P_2 is quadratic and the cardinality of P_3 is cubic in n.

In the case when more than three budget constraints are tight, the system of equations becomes overdetermined: we have only three variables, p and two coordinates of y, while the number of equations is at least four. In this case, if a solution to the system exists, it will be fully determined by a subset consisting of only three equations. This brings us back to the case with only triplets of budgetary constraints and the set of possibly optimal prices P_3. Potentially optimal facility locations are found by solving the Generalized Apollonius Problem for y. Since there are at most eight solutions to the Generalized Apollonius Problem for every triplet of constraints, we have a cubic number of potentially optimal facility locations. This proves the theorem. □

For a single facility case of the location-pricing problem we can straightforwardly enumerate all $O(n^3)$ possible facility locations together with prices from P. For every location-pricing pair, the revenue is evaluated in $O(n)$ time. Then, the maximum revenue solution is the optimal solution to the problem and we have proven the following theorem.

Theorem 3. *The location-pricing problem with one facility and n customers can be solved in $O(n^4)$ time.*

4 Multiple Facilities

Not surprisingly, the geometric location-pricing problem becomes intractable when the number of facilities becomes an input parameter.

Theorem 4. *The location-pricing problem with m facilities and n clients is strongly NP-hard.*

Proof. Consider the following geometric *minimum hitting set* problem on unit disks: given n unit disks in the plane and a positive integer K, does there exist a *hitting set* of size K, i.e., a set of points in the plane such that every given disk contains at least one point of the set? Here, the disks are given by rational coordinates of their centers.

We reduce this minimum hitting set problem, known to be strongly NP-complete [2–4], to the location-pricing problem with $m = K + 1$ facilities. Given n unit disks in the minimum hitting set problem, we construct the input of the location-pricing problem as follows. For every disk we create a *disk customer* located in the center of the disk. Every disk customer j has budget $b_j = 2$, and unit demand and transportation costs, $d_j = c_j = 1$. Pick a point z on distance at least 3 from all disk customers. Create a *controlling customer* at point z, having budget of 1, travel costs of 1 and demand of M, where M is a sufficiently large number to be defined later. We claim that, for a suitable choice of M, a hitting set of size K exists if and only if the location-pricing problem on $K + 1$ facilities has a solution with revenue $M + n$.

We first argue that, for sufficiently large M, the optimal price is exactly one per unit of the product. For contradiction, consider two cases: the optimal price is $p^0 > 1$ and the optimal price is $0 < p^0 < 1$.

Let the optimal price p^0 be strictly greater than 1. Then, the product is not affordable for the controlling customer, as her budget is 1. From the disk customers the company can get revenue at most $2n$, as the budget of each disk customer is 2. On the other hand, if the company sets the price to be 1, it can generate revenue of M from the controlling customer alone. For $M > 2n$, this implies $p^0 > 1$ is a suboptimal price.

Let $0 < p^0 < 1$, then at price p^0, the company generates revenue $p^0 M$ from the controlling customer and kp^0 from the disk customers, where k is the number of winners among the disk customers. By Theorem 2, we may assume that p^0 is at most $p^* = \max\{p \in P : \ p < 1\}$. Thus, the optimal revenue is at most $(n + M)p^0 \le (n + M)p^*$. On the other hand, the optimal revenue is at least M as the unit price is a feasible solution. Thus, $M \le (n + M)p^*$ or equivalently $M \le np^*/(1 - p^*)$. However, if we set $M > np^*/(1 - p^*)$, we have that the revenue of M generated by unit price is greater than the optimal revenue, a contradiction.

Now, if the price is fixed at $p = 1$, the company gets revenue M from the controlling customer and revenue 1 from each disk customer who is the winner. Any disk customer cannot afford traveling further than one distance unit as her budget is 2 and for the product she already must pay 1. This implies that a customer is a winner if and only if a facility is located within radius 1 from the customer location, or equivalently, if a facility hits the customer's disk. Notice, one facility must be reserved to cover the controlling customer. Thus, to maximize the revenue, the other K facilities must hit the maximum number of n unit disks around the disk customers. Therefore, for n unit disks in the plane a geometric hitting set of size K exists if and only if the corresponding instance of the location-pricing problem has a solution of revenue $M + n$. □

On the positive side, if the number m of facilities is fixed (a constant), we can enumerate all possible choices for m facilities among $O(n^3)$ options of the set S as in Theorem 2. Again, choosing the maximum revenue option, we obtain an optimal solution to the problem. This result is reported in the following theorem.

Theorem 5. *The location-pricing problem with m facilities and n customers can be solved in $n^{O(m)}$ time.*

5   Location and Pricing on the Line

Consider a special case of the location-pricing problem, where all input points of the customers are co-linear. In this case, the location-pricing problem can be solved in polynomial time.

The most important ingredient of the algorithm is again the discretization theorem.

Theorem 6. *For any instance of the location-pricing problem with n co-linear customers, there is a set S of pairs $(y, p) \in \mathbb{R}^2 \times \mathbb{R}^+$ of size $O(n^2)$ such that some pair $(y, p) \in S$ is an optimal solution to the location-pricing problem.*

Proof. The proof of this theorem repeats the lines of the proof of Theorem 2. Moreover, from the co-linearity of the customers we know that the locations of the facilities will also be co-linear with the customers. Thus, the locations of the customers as well as facility locations are specified by a single coordinate. Hence, a system with already three tight budgetary constraints becomes overdetermined as we have only two variables: one for price and one for a point coordinate. Therefore, we have to consider only the doubles of tight budgetary constraints that reduces the size of the set S to $O(n^2)$. □

Now, for an arbitrary number m of facilities, we can construct a dynamic program that finds the optimal solution, i.e., m location-pricing pairs, in polynomial time. Without loss of generality we assume that all customers are situated on a line. For simplicity, let set S be specified by y-coordinates of the points and the price set P. Let N denote the size of set S. Again, without loss of generality, we assume that coordinates of points in set S are arranged in non-decreasing order: $y_1 \leq y_2 \leq \ldots \leq y_N$.

For every price $p \in P$, we compute facility locations maximizing the total revenue. Let $g(p, k, j)$, $0 \leq k < j \leq N + 1$, denote the revenue generated by customers/winners from the interval $[y_k, y_j]$, i.e., the customers who can afford buying the product at price p at the nearest endpoint of the interval, either in y_k or in y_j. We assume here that $y_0 = -\infty$ and $y_{N+1} = +\infty$. Given $p \in P$, all values for g can be computed in $O(n^5)$ time. By Theorem 6, the number of possible prices is quadratic in n. Thus, computing all values for g for all possible prices p takes $O(n^7)$ time.

Let $f(p, i, j)$, $i = 1, 2, \ldots, m$, $j = 1, 2, \ldots, N$, denote the maximal revenue generated at price p by all customers from the interval $(-\infty, y_j]$ subject to allocation of the i-th facility at y_j. To initialize the lookup table, we define the values $f(p, 1, j) = g(p, 0, j)$ for all $j = 1, 2, \ldots, N$. The following recursive formula completes the dynamic programming table. For all $i = 2, 3, \ldots m$ and $j = 1, 2, \ldots, N + 1$ compute

$$f(p, i, j) = \max_{1 \leq k < j} \{f(p, i - 1, k) + g(p, k, j)\} \tag{2}$$

Then, $f(p, m, N+1)$ is the maximum revenue generated at price p. Notice, filling-in the lookup table requires $O(mn^5)$ time. Enumerating over all possible prices solves the location-pricing problem. Therefore, the dynamic programming takes $O(mn^7)$ time and we have the following theorem.

Theorem 7. *For any instance of the location-pricing problem with n co-linear customers, an optimal solution to the location-pricing problem can be constructed in $O(mn^7)$ time.*

6 Open Problems

Interestingly, if the company is allowed to differentiate the prices among the facilities, even the problem on the line becomes highly non-trivial. This is due to the fact that the customers with sufficiently large budgets can afford traveling to further facilities, i.e., not the nearest ones, to get there the product at a lower price. Thus, our first open question is whether or not the problem on the line with price differentiation is polynomially solvable.

The second open problem is to construct an FPT algorithm for the geometric location-pricing problem parametrized by the number m of facilities, i.e., running in time $f(m)poly(n)$ for some function $f(\cdot)$.

References

1. Dörrie, H.: The tangency problem of Apollonius. In: 100 Great Problems of Elementary Mathematics: Their History and Solutions, pp. 154–160. Dover, New York (1965)
2. Hochbaum, D.S., Maass, W.: Fast approximation algorithms for a nonconvex covering problem. J. Algorithms **8**(3), 305–323 (1987)
3. Maass, W.: On the complexity of nonconvex covering. SIAM J. Comput. **15**(2), 453–467 (1986)
4. Mustafa, N.H., Ray, S.: Improved results on geometric hitting set problems. Discrete Comput. Geom. **44**, 883–895 (2010)

Variable Neighborhood Search Approach for the Location and Design Problem

Tatyana Levanova[(✉)] and Alexander Gnusarev

Sobolev Institute of Mathematics, Omsk Branch Pevtsova str. 13,
644043 Omsk, Russia
levanova@ofim.oscsbras.ru, alexander.gnussarev@gmail.com

Abstract. In this paper the location and design problem is considered. The point of this is that a Company is going to open markets to attract the largest share of total customers demand. This share varies flexibly depending on the markets location and its design variant. The Company vies for consumers demand with some pre-existing competitors markets. The mathematical model is nonlinear, therefore, there are difficulties in the application of exact methods and commercial solvers for it. The ways of constructing upper bounds of the objective function are described. Two algorithms based on the Variable Neighborhood Search approach are proposed. To study the algorithms a series of test instances similar to the real data of the applied problem has been constructed, experimental analysis is carried out. The results of these studies are discussed.

Keywords: Discrete optimization · Integer programming · Upper bound · Location problems · Variable neighborhood search

1 Introduction

A lot of economic situations are described with a help of mathematical models of discrete location problems. Often they are quite hard both from theoretical and practical points of view. Obtaining optimal solutions to such problems with the help of exact algorithms, including software packages, may require significant time and computer resources. In recent years, methods for the approximate solution of various applied problems have been actively developed. They include the local search algorithms [1]. In this article we develop Variable Neighborhood Search Approach for the location and design problem. Two versions of the algorithm proposed for the problem are under consideration. Some numerical experiments have been carried out on the test instances. For the analyses of the quality of obtained solutions, the upper bounds are used.

2 Problem Formulation

The new Company plans to locate its facilities (supermarkets), which differ from one another in design: size, range, etc. Clients of each point choose to use the

A. Gnusarev—This research was supported by the Russian Foundation for Basic Research, grant 15-07-01141.

Y. Kochetov et al. (Eds.): DOOR 2016, LNCS 9869, pp. 570–577, 2016.
DOI: 10.1007/978-3-319-44914-2_45

facilities of Company or its competitors depending on their attractiveness and distance. The Company's goal is to attract a maximum number of customers, i.e. to serve the largest share of total demand. This share for the Company is not fixed. It depends on where and what design options open up new enterprises and, as a result, whose markets will be chosen by customers. Therefore, it can be classified as a multivariant location problem.

For the first time this problem is described by Kraas, O. Berman, R. Aboolian in [3]. Let us write out the mathematical model according to [2]. Let R be the set of facility designs, $r \in R$. There are w_i customers at the point i of discrete set $N = \{1, 2, \ldots, n\}$. All customers have the same demand, so each item can be considered as one client with weight w_i. The distance d_{ij} between the points i and j is measured, for example, in Euclidean metric or equals to the shortest distance in the corresponding graph. Let $P \subseteq N$ be the set of potential facility locations. It is assumed that $C \subset P$ is the set of pre-existing competitor facilities. The Company may open its markets in $S = P \setminus C$ taking into account the budget B and the cost of opening c_{jr} facility $j \in S$ with design $r \in R$.

Such flexible choice of customers is represented in the gravity-type spatial interaction models. These models are known as the brand share models in the marketing literature [4]. According to these models the utility u_{ij} for a customer at point $i \in N$ of a facility at location $j \in S$ can be written as an exponential function.

Let $x_{jr} = 1$, if facility j is opened with design variant r and $x_{jr} = 0$ otherwise, $j \in S, r \in R$. To determine the usefulness u_{ij} of the facility $j \in S$ for the customer $i \in N$ the supplementary coefficients k_{ijr} are introduced: $k_{ijr} = a_{jr}(d_{ij}+1)^{-\beta}$. They depend on the sensitivity β of customers to distance to facility and attractiveness a_{jr}. These two parameters are used in the spatial interaction models. In practice to determine the measures of attractiveness of each facility survey of the population is conducted. The survey data are processed using regression analysis.

Utility $u_{ij} = \sum_{r=1}^{R} k_{ijr}x_{jr}$. The total utility for the customers in point $i \in N$ from the facilities controlled by the competitors is $U_i(C) = \sum_{j \in C} u_{ij}$.

The demand function is

$$g(U_i) = 1 - \exp\left(-\lambda_i U_i\right),$$

where λ_i is the characteristic of flexible demand in point i, $\lambda_i > 0$; U_i is the total utility for a customer at $i \in N$ from all open facilities:

$$U_i = \sum_{j \in S} \sum_{r=1}^{R} k_{ijr}x_{jr} + U_i(C) = U_i(S) + U_i(C).$$

The company's total share of facility $i \in N$:

$$MS_i = \frac{U_i(S)}{U_i(S) + U_i(C)} = \frac{\sum_{j \in S} \sum_{r=1}^{R} k_{ijr}x_{jr}}{\sum_{j \in S} \sum_{r=1}^{R} k_{ijr}x_{jr} + \sum_{j \in C} u_{ij}}.$$

Then the mathematical model looks like:

$$\max \sum_{i \in N} w_i \cdot g(U_i) \cdot MS_i, \tag{1}$$

$$\sum_{j \in S} \sum_{r \in R} c_{jr} x_{jr} \leq B, \tag{2}$$

$$\sum_{r \in R} x_{jr} \leq 1, j \in S, \tag{3}$$

$$x_{jr} \in \{0, 1\}, \quad r \in R, j \in S. \tag{4}$$

Based on above notation, the objective function (1) looks as follows:

$$\max \sum_{i \in N} w_i \left(1 - \exp\left(-\lambda_i \left(\sum_{j \in S} \sum_{r=1}^{R} k_{ijr} x_{jr} + U_i(C) \right) \right) \right) \cdot \tag{5}$$

$$\cdot \left(\frac{\sum_{j \in S} \sum_{r=1}^{R} k_{ijr} x_{jr}}{\sum_{j \in S} \sum_{r=1}^{R} k_{ijr} x_{jr} + \sum_{j \in C} u_{ij}} \right).$$

The objective function (5) reflects the Company's goal to maximize the share of customers demand. Inequality (2) takes into account the available budget. Condition (3) shows that for each facility only one variant of the design can be selected.

3   Upper Bounds

It is known that the location problem considered in this paper is NP-hard [5]. Sience the objective function (5) is non-linear, it is impossible to use the linear programming methods to solve problem (2)–(5). In this case, the calculation of estimates of the objective function becomes relevant. Below there is an observation for constructing the upper bound for (5) proposed by Yu. Kochetov.

Let us consider the objective function (1) of a location and design problem. When λ_i is close to 1, the multiplier of the objective function behaves as a constant:

$$1 - \exp\left(-\lambda_i \left(\sum_{j \in S} \sum_{r \in R} k_{ijr} x_{jr} + U_i(C) \right) \right) \approx 1.$$

Then the initial problem is equivalent to the following one:

$$\max \sum_{i \in N} w_i \cdot \left(\frac{\sum_{j \in S} \sum_{r \in R} k_{ijr} x_{jr}}{\sum_{j \in S} \sum_{r \in R} k_{ijr} x_{jr} + U_i(C)} \right), \tag{6}$$

$$\sum_{j \in S} \sum_{r \in R} c_{jr} x_{jr} \leq B, \tag{7}$$

$$\sum_{r \in R} x_{jr} \leq 1, \quad j \in S, \tag{8}$$

$$x_{jr} \in \{0, 1\}, \quad r \in R, j \in S. \tag{9}$$

It can be reduced to a problem of mixed-integer linear programming. To do this, we represent the objective function (6) in the following way:

$$\max \sum_{i \in N} w_i \cdot \left(\frac{\sum_{j \in S} \sum_{r \in R} k_{ijr} x_{jr}}{\sum_{j \in S} \sum_{r \in R} k_{ijr} x_{jr} + U_i(C)} \right) =$$

$$= \sum_{i \in N} \sum_{j \in S} \sum_{r \in R} \frac{w_i k_{ijr} x_{jr}}{\sum_{j \in S} \sum_{r \in R} k_{ijr} x_{jr} + U_i(C)} = \sum_{i \in N} \sum_{j \in S} \sum_{r \in R} z_{ijr},$$

where

$$z_{ijr} = \frac{w_i k_{ijr} x_{jr}}{\sum_{j \in S} \sum_{r \in R} k_{ijr} x_{jr} + U_i(C)}, i \in N, j \in S, r \in R.$$

The nonlinear model (2)–(5) is reduced to the mixed-integer programming model by introducing service variables:

$$y_i = \frac{w_i}{\sum_{j \in S} \sum_{r \in R} k_{ijr} x_{jr} + U_i(C)}, i \in N.$$

Then z_{ijr} is given by the inequalities:

$$k_{ijr} y_i + m(x_{jr} - 1) \leq z_{ijr} \leq k_{ijr} y_i,$$

$$z_{ijr} \leq x_{jr} w_i,$$

$$\text{where} \quad m = \max \frac{w_i k_{ijr}}{U_i(C)}, i \in N, j \in S, r \in R. \tag{10}$$

The following linear model is obtained:

$$\max \sum_{i \in N} \sum_{j \in S} \sum_{r \in R} z_{ijr}, \tag{11}$$

$$k_{ijr} y_i + m(x_{ir} - 1) \leq z_{ijr} \leq k_{ijr} y_i, i \in N, j \in S, r \in R, \tag{12}$$

$$z_{ijr} \leq x_{jr} w_i, i \in N, j \in S, r \in R, \tag{13}$$

$$\sum_{r \in R} \sum_{j \in S} z_{ijr} + y_i U_i(C) = w_i, i \in N, \tag{14}$$

$$\sum_{j \in S} \sum_{r \in R} c_{jr} x_{jr} \leq B, \tag{15}$$

$$x_{jr} \in \{0, 1\}, \quad j \in S, r \in R. \tag{16}$$

Condition (14) says that every customer spends the budget proportionally to the utility either from the Company's facilities or from a competitor. Problem (11)–(16) can be solved exactly.

Note that the constant m in (10) may be selected in various ways. This will determine the accuracy of the upper bounds.

4   Variable Neighborhood Search Approach

The use of solver CoinBonmin [6] for location and design problem can calculate only a record but not the optimal solution even when CPU time $t \rightarrow \infty$ [7].

Solving such problems requires a significant investment of time and computing resources. In this regard, one of the approaches to its solution is the use of approximate methods. In this paper we develop Variable Neighborhood Search approach [8,9] for the considered problem. The frame of Variable Neighborhood Search algorithm (VNS) is the following [8].

Scheme of VNS algorithm

Initialization. Select a set of neighborhood structures $N_k, k = 1, \ldots, k_{max}$, that will be used in the search; find an initial solution x; choose a stopping condition.

Repeat the following until the stopping condition is met:
(1) Set $k := 1$.
(2) Until $k = k_{max}$, repeat the following steps:

(a) *Shaking.* Generate a point $x\prime$ at random from the k-th neighborhood of x ($x\prime \in N_k(x)$);

(b) *Local search.* Apply some local search method with $x\prime$ as initial solution; denote the obtained local optimum by $x\prime\prime$;

(c) *Move or not.* If this local optimum $x\prime\prime$ is better than the best incumbent, move to $x := x\prime\prime$, and continue the search with $N_1, k := 1$; otherwise, set $k := k + 1$.

We propose a variant of VNS approach which is called the Relaxed Neighborhood Search Algorithm (RVNS). Unlike the basic VNS there is no step "Local search" in RVNS. The basic idea of VNS approach is to explore a set of predefined neighborhoods successively to provide a better solution. Therefore, an important step is the choice of neighborhoods set. Here the new types of neighborhoods are used for the algorithm [10]. They will be described below.

Let the vector $z = (z_i)$ be such that z_i corresponds to facility i: $z_i = r$ iff $x_{ir} = 1$. The feasible initial solution z is obtained using special deterministic procedure.

Neighborhood 1 (N1). Feasible solution z' is called neighboring for z if it can be obtained with the following moves:

(a) choose one of the open facilities p with design variant z_p and close it;
(b) select the facility q which is closed and has highest attractiveness; then open facility q with the design variant z_p.

Neighborhood 2 (N2). Feasible solution z' is called neighboring for z if it can be obtained with the following operations:

(a) choose one of the open facilities p with design variant z_p and reduce the number of design variant;
(b) select the facility q and increase the number of design variant of it.

Neighborhood 3 (N3). Unlike Neighborhood 2 on the step b) select the facility q which is closed; then open the facility q with the design variant z_p.

Lin-Kernighan neighborhood (LK) was applied as *Neighborhood 4* [11].

5   Experimental Study

The validation of the VNS algorithm was conducted for the following data: the neighborhoods N1, N2, N3 and LK were used; the local descent was carried out with the help of the neighborhood Lin-Kernighan, it consists of 9 elements. RVNS algorithm uses neighborhoods N1, N2, LK; Lin-Kernighan neighborhood consists of 9 elements. The facilities p in the neighborhood N1, the facilities p, q in N2 and q in N3 were selected randomly. Number of restarts shaking trials without improvement is limited by 100 for both algorithms. Stopping criteria for VNS and RVNS was an exploration of neighborhoods without improvement of the solution.

To study the algorithms a series of test instances similar to the real data of the applied problem has been constructed [2]. The test instances consist of two sets with Euclidean and arbitrary distances. They contain 96 instances for location of 60, 80, 100, 150, 200 and 300 facilities; 3 types of design variants are used, the budget of 3, 5, 7, 9 is limited; the demand parameter is $\lambda_i = 1, i \in N$; the customer sensitivity to the distance is high ($\beta = 2$).

The parameter m was calculated by formula (10) and $m = \max w_i k_{ijr}, i \in N$, $j \in S, r \in R$. Therefore, the two values UB1 and UB2 for the upper bound were obtained respectively. It should be noted that UB1 coincides with UB2 for all tasks with arbitrary distances. For all tasks with Euclidean distances the UB2 is closer to optimal solution than UB1. It is interesting to note that deviation UB2 from UB1 increases with the dimensions of problems. Thus the maximal

Table 1. Best known solutions

Tests	Arbitrary distances				Euclidean distances				
	UB1 = UB2	GAMS	VNS	RVNS	UB1	UB2	GAMS	VNS	RVNS
300.3.1	36.151	36.150	36.143	36.143	43.560	43.180	—	35.183	35.183
300.3.2	57.166	54.12	57.158	57.158	65.977	65.256	—	54.446	54.446
300.3.3	75.524	74.41	75.513	74.615	86.951	86.228	—	73.053	73.053
300.3.4	96.111	94.47	96.097	96.097	107.227	106.504	—	91.081	91.081
300.5.1	30.332	30.02	30.334	28.463	36.502	36.193	—	30.899	30.899
300.5.2	51.054	51.04	51.051	51.051	58.776	58.371	—	50.514	50.514
300.5.3	67.507	66.99	67.503	67.503	80.024	79.418	—	69.360	69.360
300.5.4	86.437	81.93	86.312	83.178	100.468	99.785	—	87.756	87.848
300.7.1	36.427	36.43	36.427	36.427	43.957	43.730	—	36.154	36.154
300.7.2	55.631	53.69	55.627	55.627	69.487	69.179	—	57.568	57.568
300.7.3	74.619	74.62	74.610	74.177	93.207	92.748	—	77.670	77.787
300.7.4	95.276	92.83	95.266	95.266	115.810	115.256	—	96.992	97.119
300.9.1	31.827	31.83	31.823	31.823	36.107	36.041	—	32.093	32.093
300.9.2	51.279	48.65	51.274	51.274	58.020	57.934	—	51.503	51.503
300.9.3	69.004	67.09	69.000	67.420	79.212	79.086	—	70.294	70.293
300.9.4	88.340	85.87	85.603	81.537	99.909	99.773	—	88.947	88.947

and average deviations are equal to 27.4 % and to 11.9 % for $|N| = 60$, and they are equal to 1.09 % and to 0.56 % for $|N| = 300$ respectively.

Table 1 shows the values of the upper bounds (UB1, UB2), the best solutions are found by CoinBonmin solver built into GAMS [7] and by algorithms VNS and RVNS for tests with the largest dimension from sets with Euclidean and arbitrary distances.

Table 2 contains some information about minimal (min), average (av) and maximal (max) CPU time (in seconds) of the proposed algorithms for test problems using a PC Intel i5-2450M, 2.50 GHz, memory 4 GB.

The test instances with Euclidean distances proved to be difficult for the Coin-Bonmin solver. In particular the maximum CPU time for test instances with $|N| = 60$ was more than 63 h. In all instances with a dimension of 300 the CoinBonmin solver could not find a feasible solutions in 25 min. While maximal running time of the proposed algorithms does not exceed 25 min. In the remaining test instances algorithms have been compared to the upper bound and among themselves. In 5 test instances with Euclidean distances VNS algorithm improved the record values found by RVNS. The average deviation of the VNS from the upper bound UB2 was 12.1 % (RVNS 12.1 %). The average time of the VNS algorithm until the stopping criterion triggered was 181.35 s (RVNS 4.28 s).

In the test instances with arbitrary distances CoinBonmin found the records for all tasks. The average time until the stopping criterion triggered was 139.39 s (RVNS 4.69 sec.). During this time in 27 test instances with arbitrary distances VNS algorithm improved RVNS records. The average deviation from the upper bound obtained by VNS was 0.14 % (RVNS 0.7 %).

Analyzing the results we can say that in 15 test instances the records of VNS coincided with upper bounds in the test instances with arbitrary distances. Therefore we can conclude that the VNS algorithm found the optimal solutions for 15 test instances. In addition CoinBonmin has found 10 optimal solutions and RVNS has found 9 such solutions. Optimal solutions have not been obtained on the series with the Euclidean distance. The confidence interval for the probability of obtaining an optimum (the confidence level is 95 %) is between 0.56 and 0.94.

In general we can say that VNS obtains solutions closer to the optimum, while RVNS algorithm is faster in comparison with other considered algorithms.

Table 2. CPU Time (sec)

Tests	Arbitrary distances						Euclidean distances					
	VNS			RVNS			VNS			RVNS		
$|N|$	min	av	max	min	av	max	min	av	max	min	av	max
60	10.94	21.92	46.16	0.36	0.75	1.33	12.29	20.52	39.26	0.34	0.70	1.05
80	21.32	34.16	86.34	0.66	1.41	2.57	23.04	36.49	83.92	0.62	1.15	2.02
100	38.01	60.95	152.16	1.06	1.92	2.76	32.58	48.47	100.74	0.74	1.50	2.83
150	76.11	97.48	141.28	2.32	4.21	6.53	75.41	145.97	451.60	1.64	3.64	6.18
200	115.05	183.43	295.40	3.61	7.17	11.74	109.51	222.46	447.04	2.80	5.33	8.22
300	265.45	438.38	643.23	5.56	12.70	22.49	268.04	614.21	1408.92	7.65	13.35	21.27

6   Conclusion

In this paper we have developed two variants of algorithms based on the Variable Neighborhood Search approach for the location and design problem. New neighborhoods of a special type allowed us to find the optimal solutions. Computational experiment was carried out on a series of test examples based on real data. The ways of constructing upper bounds of the objective function have described. The proposed algorithms found new best known solutions or solutions with smaller relative error. Having in mind the complexity and size of the considered problem we can conclude that the computational times are rather good.

The obtained results indicate the usefulness of the Variable Neighborhood Search approach for solving the commercial-size problem.

Acknowledgments. We would like to thank Prof. N. Mladenovic and Prof. Yu. Kochetov for their attention to our paper and helpfull comments.

References

1. Aarts, E., Lenstra, J.K.: Local Search in Combinatorial Optimization. Wiley, Hoboken (1997)
2. Aboolian, R., Berman, O., Krass, D.: Competitive facility location and design problem. Eur. J. Oper. Res. **182**(1), 40–62 (2007)
3. Aboolian, R., Berman, O., Krass, D.: Competitive facility location model with concave demand. Eur. J. Oper. Res. **181**, 598–619 (2007)
4. Naret, P., Weverbergh, M.: On the predictive power of market share attraction models. J. Mark. Res. **18**, 146–153 (1981)
5. Aboolian, R., Berman, O., Krass, D.: Capturing market share: facility location and design problem. In: International Conference on Discrete Optimization and Operations Research, pp. 7–11. Sobolev Institute of Mathematics, Novosibirsk (2013)
6. Bonami, P., Biegler, L.T., Conn, A.R., Cornuéjols, G., Grossmann, I.E., Laird, C.D., Lee, J., Lodi, A., Margot, F., Sawaya, N., Wächter, A.: An algorithmic framework for convex mixed integer nonlinear programs. Discrete Optim. **5**(2), 186–204 (2008)
7. The General Algebraic Modeling System (GAMS). http://www.gams.com
8. Hansen, P., Mladenovic, N.: Variable neighborhood search: principles and applications (invited review). Eur. J. Oper. Res. **130**(3), 449–467 (2001)
9. Hansen, P., Mladenovic, N., Moreno-Perez, J.F.: Variable neighbourhood search: algorithms and applications. Ann. Oper. Res. **175**, 367–407 (2010)
10. Levanova, T., Gnusarev, A.: Heuristic algorithms for the location problem with flexible demand. In: 42th International Symposium on Operations Research, SYMOP-IS 2015, Belgrad, Serbia, pp. 245–247 (2015)
11. Kochetov, Y., Alekseeva, E., Levanova, T., Loresh, M.: Large neighborhood local search for the p-median problem. Yugosl. J. Oper. Res. **15**(2), 53–64 (2005)

On a Network Equilibrium Problem
with Mixed Demand

Olga Pinyagina[(✉)]

Institute of Computational Mathematics and Information Technologies,
Kazan Federal University, Kremlevskaya St. 18, 420008 Kazan, Russia
Olga.Piniaguina@kpfu.ru
http://kpfu.ru

Abstract. In the present paper, we formulate the network equilibrium problem with mixed demand containing the fixed and variable components. We present the equilibrium conditions and the conditions for existence of solution of this problem. In addition, we show that the network equilibrium problem with mixed demand generalizes the network equilibrium problems with fixed demand and elastic demand and establish the connection with the auction equilibrium problem. Preliminary computational experiments are also presented.

Keywords: Network equilibrium problem · Fixed demand · Elastic demand · Mixed demand

1   Introduction

The network equilibrium problems with fixed and elastic demand, which arise in different areas, including telecommunication and transport networks, have long been known and investigated in detail (see, for example, [1]–[3]).

In the present paper, we generalize the network equilibrium problems with fixed and elastic demand and propose the network equilibrium problem with mixed demand containing the fixed and variable components. For this problem, we present the equilibrium conditions and the conditions for existence of solution and establish the connection with the auction equilibrium problem.

In the next section, we remind the definitions of the network equilibrium problems with fixed and elastic demand.

2   Preliminaries

We remind that the network equilibrium problem with *fixed* demand is to find a point $x^* \in X$ such that

$$\langle G(x^*), x - x^* \rangle \geq 0 \qquad \forall x \in X, \tag{1}$$

This work is supported by Russian Foundation for Basic Research, project No 16-01-00109.

where $\quad X = \left\{ x \ \middle| \ \sum_{p \in P_w} x_p = d_w, x_p \geq 0, \ p \in P_w, w \in W \right\}$,

G is a cost mapping, which will be defined below. Here V is the set of network nodes, A is the set of directed links, W is the set of origin-destination nodes (O/D-pairs) (i,j), $i,j \in V$. For each $w \in W$ the set P_w of simple directed paths joining w and the fixed demand value $d_w > 0$ are given.

The problem is to distribute the demand flow d_w for each O/D-pair $w \in W$ among the given set of paths P_w, using the equilibrium criterion. x_p denotes the variable value of flow passing through the path p. Therefore, the set X is a Cartesian product of simplices, its dimension equals $\sum_{w \in W} |P_w|$.

Paths and links are connected by the incidence matrix with elements

$$\alpha_{pa} = \begin{cases} 1, & \text{if link } a \text{ belongs to path } p , \\ 0, & \text{otherwise.} \end{cases}$$

The values of link flows are defined as follows: $f_a = \sum_{w \in W} \sum_{p \in P_w} \alpha_{pa} x_p$, $a \in A$. For each link a a continuous cost function C_a is given, in the general case this function can depend on all link flows. The summary cost function for path p has the form $G_p(x) = \sum_{a \in A} \alpha_{pa} C_a(f)$. The equilibrium state for this network is such an element $x^* \in X$ that

$$\forall w \in W, q \in P_w, x_q^* > 0 \implies G_q(x^*) = \min_{p \in P_w} G_p(x^*). \tag{2}$$

The network equilibrium problem with *elastic* demand is to find such a vector $(x^*, d^*) \in K$ that

$$\langle G(x^*), x - x^* \rangle - \langle \lambda(d^*), d - d^* \rangle \geq 0 \quad \forall (x,d) \in K, \tag{3}$$

where $\quad K = \left\{ (x,d) \ \middle| \ \sum_{p \in P_w} x_p = d_w, x_p \geq 0, \ p \in P_w, w \in W \right\}$.

In this model the demand d_w is a variable value for all $w \in W$. For each O/D-pair $w \in W$ a so-called disutility function λ_w of d_w is given, which is supposed to be continuous.

For this problem the equilibrium conditions have the following form: a vector composed of path flow variables and demand variables $(x^*, d^*) \in K$ is a solution to problem (3), if the correlations hold:

$$G_p(x^*) \begin{cases} = \lambda_w(d^*) & \text{if} x_p^* > 0, \\ \geq \lambda_w(d^*) & \text{if} x_p^* = 0. \end{cases} \tag{4}$$

3 Main Results

In the present paper, we consider the network equilibrium problem with **mixed** demand. In this problem, the fixed and variable components of demand are simultaneously used. The feasible set takes the form

$$K_M = \left\{ (x,d) \;\middle|\; \sum_{p \in P_w} x_p = d_w^{const} + d_w, x_p \geq 0, \; p \in P_w, d_w \geq 0, w \in W \right\},$$

where $d_w^{const} \geq 0 \; \forall w \in W$; d_w^{const} and d_w are the constant and variable demands for $w \in W$, respectively. The formulation of variational inequality has the form similar to the problem with elastic demand: Find a vector $(x^*, d^*) \in K_M$ such that

$$\langle G(x^*), x - x^* \rangle - \langle \lambda(d^*), d - d^* \rangle \geq 0 \quad \forall (x,d) \in K_M. \tag{5}$$

The proof of solution existence for the considered problem is based on the results obtained by I.V.Konnov in paper [4] for the network equilibrium problem with elastic demand.

We will use the coercivity condition in the following form [4]:
(C1) *There exists a number $r > 0$ such that for any point $(x,d) \in K_M$ and each $w \in W$ the following implication holds true*

$$d_w > r \implies \exists p \in P_w \text{ such that } x_p > 0, G_p(x) \geq \lambda_w(d).$$

Theorem 1. *Let the feasible set K_M be nonempty, functions $C_a \; \forall a \in A$ and $\lambda_w \; \forall w \in W$ be continuous. If condition (C1) is fulfilled, then problem (5) has a solution.*

The proof of this theorem follows Theorem 2 from [4].

We remind also that in paper [4] the equivalence of auction equilibrium problems and network equilibrium problems with fixed and elastic demand has been proved. Concerning the network equilibrium problem with mixed demand, we note that it corresponds to a two-side multi-commodity auction, where each product is associated with multiple sellers and one buyer, the bid/offer volumes are bounded from below by zero and unbounded from above, and an outer demand for each product can also exist.

The following theorem presents the equilibrium conditions for problem (5).

Theorem 2. *A vector $(x^*, d^*) \in K_M$ is a solution to problem (5) if and only if it satisfies conditions*
   (a) *if $x_p^* > 0$, then $G_p(x^*) = \min\limits_{q \in P_w} G_q(x^*)$,*
i.e., only paths with minimal costs have nonzero flows;
   (b) *if $x_p^* > 0$ and $d_w^* > 0$, then $G_p(x^*) = \lambda_w(d^*)$,*
i.e., for nonzero variable demand the cost values for paths with nonzero flows are equal to the value of disutility function for this O/D-pair;
   (c) *if $x_p^* = 0$ or $d_w^* = 0$, then $G_p(x^*) \geq \lambda_w(d^*)$,*
i.e., the value of disutility function cannot exceed the cost values for paths of this O/D-pair.

The proof of this theorem follows from [4], Proposition 1.

At last, we show that the network equilibrium problem with mixed demand generalizes network equilibrium problems with fixed and elastic demand.

Proposition 1. *1. In network equilibrium problem with mixed demand (5) we set $d_w^{const} = 0$ for all $w \in W$. We obtain network equilibrium problem with elastic demand (3).*

2. Let in network equilibrium problem with mixed demand (5) the cost functions be given so that $G_p(x) > 0$ with $x_p > 0$, for all $x_p \in P_w$, $p \in P_w, w \in W$. We set disutility functions λ_w identically equal to zero for all $w \in W$. We obtain network equilibrium problem with fixed demand (1).

The first assertion of Proposition 1 is evident. The second assertion follows from Theorems 1 and 2.

When for some O/D pair the nonzero fixed and variable demands are simultaneously used, one can interpret the value of fixed demand as a guaranteed low bound of demand, which must be satisfied in any case. Therefore, the system can provide flows on the most important directions.

On the other hand, the network equilibrium problem with mixed demand can be interpreted as some compromise between interests of separate O/D-pairs and the whole system. Then the values of fixed demand can be given "from below", as requests of O/D-pairs, and the disutility functions can be obtained "from above", as a regulation mechanism for the whole system.

4 Numerical Experiments

For solution of the network equilibrium problem (NEP) with fixed or elastic demand, the projection method is commonly used, which was originally proposed for constrained minimization problems (see [5]). We have applied a modification of this method from [6].

For simplicity we denote by π the projection operator onto the set K_M.

Gradient projection algorithm

Step 0. Let an accuracy $\varepsilon > 0$, parameters $\alpha \in (0,1)$, $\beta \in (0,1)$, and an initial point $(x^0, d^0) \in K_M$ be given. We set $k = 0$.

Step 1. If $\|(x^k, d^k) - \pi[(x^k, d^k) - (G(x^k), -\lambda(d^k))]\| \leq \varepsilon$, then we obtain the given accuracy level, the calculation process stops.

Step 2. We find the minimal nonnegative integer m, which satisfies the inequality

$$\langle (G(x^{k,m}), -\lambda(d^{k,m})), (x^{k,m}, d^{k,m}) - (x^k, d^k) \rangle \leq$$

$$\alpha \langle (G(x^k), -\lambda(d^k)), (x^{k,m}, d^{k,m}) - (x^k, d^k) \rangle,$$

where $(x^{k,m}, d^{k,m}) = \pi[(x^k, d^k) - \beta^m(G(x^k), -\lambda(d^k))]$.

Step 3. We set $(x^{k+1}, d^{k+1}) = (x^{k,m}, d^{k,m})$, $k = k + 1$ and go to Step 1.

Let us consider the network example from [7] with the non-potential mapping G (Fig. 1). The network is composed of 25 nodes and 40 links. 5 nodes of the set $\tilde{V} = \{1, \ldots, 5\}$ are terminal nodes, O/D pairs can include these nodes only. An additional constraint requires that nodes from the set $\tilde{V}$ cannot be used as intermediate nodes in any path. Beside links directed from and to terminal nodes, there exist inner and outer loops, which are clockwise and anticlockwise oriented, respectively.

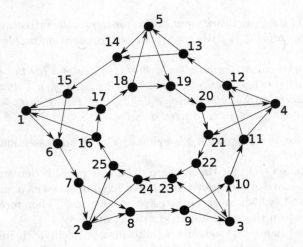

Fig. 1. Bertsekas–Gafni network, 25 nodes, 40 links, 5 O/D pairs

The set of O/D pairs $W = \{(1,4), (2,5), (3,1), (4,2), (5,3)\}$.

The set of links A contains 4 subsets:

a) highway links: $A_h = \{(6,7), (8,9), (10,11), (12,13), (14,15), (17,18),$ $(19,20), (21,22), (23,24), (25,16)\}$;

b) exit links: $A_x = \{(16,1), (15,1), (24,2), (7,2), (22,3), (9,3), (20,4), (11,4),$ $(18,5), (13,5)\}$;

c) entrance links: $A_e = \{(1,6), (1,17), (2,25), (2,8), (3,23), (3,10), (4,21),$ $(4,12), (5,19), (5,14)\}$;

d) bypass links: $A_b = \{(15,6), (7,8), (9,10), (11,12), (13,14), (16,17),$ $(18,19), (20,21), (22,23), (24,25)\}$.

Let some function $z : R \to R$ and scale coefficient $\mu > 0$ be given. Then the link cost function C_a of link $a = (i,j)$ depends on the link flows vector f and is defined as follows:

$$
C_a = \begin{cases}
z(f_a), & \text{if } a \in A_x \cup A_b, \\
10z(f_a) + 2\mu z(f_{\tilde{a}}), & \text{if } a \in A_h, \text{ where } \tilde{a} \in A_x, \tilde{a} = (j,s), \\
z(f_a) + \mu z(f_{\tilde{a}}), & \text{if } a \in A_e, \text{ where } \tilde{a} \in A_b, \tilde{a} = (s,j).
\end{cases}
$$

We present numerical results for several examples of problem with the accuracy value $\varepsilon = 0.0001$ and different initial data. In all examples, $z(f_a) = 1 + f_a$, $\alpha = 0.5$, $\beta = 0.5$, $\mu = 0.5$.

Example 1. The problem with fixed demand. The values of fixed demand are (1, 4, 1, 4, 1). We obtain the minimal values of cost functions (82.37, 101.13, 97.63, 101.49, 81.88).

Example 2. The problem with elastic demand (i.e., the fixed demands are null). The disutility functions $\lambda_w(d_w) = 100 - 0.5d_w$ for all $w \in W$.

We obtain the demand vector (2.39, 2.39, 2.39, 2.39, 2.39), the minimal values of cost functions are (98.8, 98.8, 98.8, 98.8, 98.8).

Example 3. The problem with mixed demand. The fixed demands are $(1, 4, 1, 4, 1)$. The disutility functions $\lambda_w(d_w) = 100 - 0.5(d_w + d_w^{const})$ for all $w \in W$.

We obtain the demand vector $(1.81, 4, 1, 4, 1.87)$, the values of cost functions $(99.09, 108.14, 99.57, 108.7, 99.07)$. In this example, for the second, third, and fourth O/D-pairs the fixed demand is satisfied only, and for other pairs there exists the additional nonzero variable demand value.

Example 4. The problem with mixed demand. The fixed demands are $(1, 2.5, 1, 2.5, 1)$. The disutility functions $\lambda_w(d_w) = 100 - 0.5d_w$ for all $w \in W$.

We obtain the demand vector $(2.36, 2.54, 2.3, 2.53, 2.38)$, the values of cost functions are $(99.31, 99.98, 99.35, 99.98, 99.31)$. In this example, for each OD-pair the fixed demand is satisfied and there exists the additional nonzero elastic demand value.

In Table 1 we compare calculation results for problems with fixed, elastic, and mixed demand and present the numbers of iterations for different accuracy values.

Table 1. Numbers of iterations for different accuracy values

Accuracy	0.001	0.0001	0.00001	0.000001
NEP with fixed demand (Example 1)	24	29	33	38
NEP with elastic demand (Example 2)	24	30	37	43
NEP with mixed demand (Example 3)	26	31	36	46

In conclusion we note that, in our opinion, the proposed model of network equilibrium with mixed demand is promising for the further investigation and can be used in practical applications.

References

1. Dafermos, S.: Traffic equilibrium and variational inequalities. Transp. Sci. **14**, 42–54 (1980)
2. Dafermos, S.: The general multimodal network equilibrium problem with elastic demand. Networks **12**, 57–72 (1982)
3. Nagurney, A.: Network Economics: A Variational Inequality Approach. Kluwer, Dordrecht (1999)
4. Konnov, I.V.: On auction equilibrium models with network applications. Netnomics **16**, 107–125 (2015)
5. Levitin, E.S., Polyak, B.T.: Constrained minimization methods. USSR Comput. Math. Math. Phys. **6**(5), 1–50 (1966)
6. Konnov, I.V.: On an approach to the solution of flow equilibrium problems. Issledovaniya po Informatike (in Russian). **2**, 125–132 (2000)
7. Bertsekas, D., Gafni, E.: Projection methods for variational inequalities with application to the traffic assignment problem. Math. Progr. Study. **17**, 139–159 (1982)

Author Index

Ageev, Alexander   93, 259
Aizenberg, Natalia   469
Amirgaliev, Yedilkhan   538

Bampis, Evripidis   3
Batsyn, Mikhail   244
Batsyna, Ekaterina   244
Beresnev, Vladimir   325, 373
Berger, André   563
Bredereck, Robert   105
Brimberg, Jack   336
Bulteau, Laurent   105
Bykadorov, Igor   480

Censor, Yair   15
Chentsov, Alexander G.   121
Chernov, Alexey   391
Chernykh, Ilya   272, 284
Coupechoux, Marceau   364
Čvokić, Dimitrije D.   350

Davydov, Ivan   364
Dubinin, Roman   193
Dvurechensky, Pavel   391

Ellero, Andrea   480
Eremeev, Anton V.   298
Erzin, Adil   220

Funari, Stefania   480

Gasnikov, Alexander   391
Gimadi, Edward Kh.   136, 148
Glebov, Aleksey N.   159
Gnusarev, Alexander   570
Gordeeva, Anastasiya V.   159
Grigoriev, Alexander   563
Grigoryev, Alexey M.   121
Gruzdeva, Tatiana   404

Iellamo, Stefano   364
Il'ev, Victor   25
Il'eva, Svetlana   25

Istomin, Alexey M.   136
Ivanov, Mikhail   259
Ivanov, Sergey V.   525

Kel'manov, Alexander   171, 182
Khachay, Michael   193
Khamidullin, Sergey   171
Khandeev, Vladimir   171
Kibzun, Andrey I.   525
Kochetov, Yury A.   350
Kokovin, Sergey   480
Komusiewicz, Christian   105
Konnov, Igor   418
Kononov, Alexander   25, 309
Kovalenko, Yulia V.   298, 309

Levanova, Tatyana   570
Lgotina, Ekaterina   284
Lien, Jaimie W.   37

Marakulin, Valeriy M.   494
Mazalov, Vladimir V.   37
Melnik, Anna V.   37
Melnikov, Andrey   325, 373
Mikhailova, Ludmila   171
Minarchenko, Ilya   509
Mladenović, Nenad   220, 336
Motkova, Anna   182

Nikolaev, Andrei   206
Nurminski, Evgeni   430

Panin, Artem   563
Pardalos, Panos M.   50
Pinyagina, Olga   418, 578
Plotnikov, Roman   220
Plyasunov, Aleksandr V.   350
Pudova, Marina   480

Ravetti, Martín   50
Rykov, Ivan   148

Schieber, Tiago   50
Shakhlevich, Natalia V.   74

Shioura, Akiyoshi   74
Shmyrev, Vadim I.   61
Simanchev, Ruslan   233
Skarin, Vladimir D.   441
Strekalovsky, Alexander S.   404, 452
Strusevich, Vitaly A.   74

Talmon, Nimrod   105
Todosijević, Raca   336
Tsidulko, Oxana Yu.   136
Turan, Cemil   538

Urazova, Inna   233
Urošević, Dragan   336
Utkina, Irina   244

van Bevern, René   105
Vorontsova, Evgeniya   547

Winokurow, Andrej   563
Woeginger, Gerhard J.   105

Zheng, Jie   37
Zur, Yehuda   15

Printed in the United States
By Bookmasters